PREDICTION AND PERFORMANCE IN GEOTECHNICAL ENGINEERING

PROCEEDINGS OF THE INTERNATIONAL SYMPOSIUM ON PREDICTION AND PERFORMANCE IN GEOTECHNICAL ENGINEERING / CALGARY / 17-19 JUNE 1987

Prediction and Performance in Geotechnical Engineering

Edited by
RAMESH C.JOSHI & FRED J.GRIFFITHS
University of Calgary, Alberta, Canada

Sponsored by the Department of Civil Engineering and the Faculty of Continuing Education at the University of Calgary, and the Canadian Geotechnical Society, Calgary Chapter

A.A.BALKEMA / ROTTERDAM / BOSTON / 1987

ORGANIZING COMMITTEE

NOTICE

The papers in this volume of proceedings express the opinions of the authors. The organizing or paper review committee in no way takes responsibility for any opinions or accuracy of the data presented in the papers of the different authors. Readers should use judgement in utilizing the data for design or other purposes.
The texts of the various papers in this volume were set individually by typists under the supervision of each of the authors concerned.

ISBN 90 6191 707 7

Distributed in USA & Canada by: A.A.Balkema Publishers, P.O.Box 230, Accord, MA 02018
Printed in the Netherlands

Table of contents

Soil improvement

General prediction and performance of soils

Prediction and performance of unique soils

Prediction and performance of retaining structure behaviour

Prediction and performance of tunnels

Environmental geotechnology

Triaxial testing of soils

Centrifuge model testing

Risk analysis in geotechnical engineering

Foreword

Over the last few years much research has been conducted in the area of prediction of properties and performance of soils. During this period significant advances have been made, resulting in improvements to predictive techniques. For the practising engineer, a reduction in field and laboratory investigation costs is now possible in conjunction with increased confidence in the results.

Results of all these research projects have not generally been available to the engineering community. Therefore, the theme of the present symposium is to emphasize the relation between theory and application through compilation of all this newly available information and its discussion in a joint forum.

This symposium consisted of 11 sessions in two concurrent programs, as well as keynote addresses by several internationally recognized academics, researchers, and practising engineers in 3 plenary sessions.

These proceedings contain most of the papers accepted for presentation. The reviewing procedure included review of both the submitted abstracts and papers by the reviewing committee. Their objective was to assure scientific quality, clear and reasonable presentation, and satisfactory use of the English language. The papers in these proceedings were not reviewed technically with the same standards as applicable to international journals of high quality, nonetheless, the papers meet the criteria set for the conference.

The members of the review committee were: K.Bean and W.Burwash (Golder Associates Ltd., Calgary), F.Griffiths (Department of Civil Engineering, University of Calgary), P.Hannak and R.Johnson (Alberta Environmental Centre, Vegreville), R.Joshi and B.Karney (Department of Civil Engineering, University of Calgary), B.Leach (Golder Associates Ltd., Calgary), S.Masoumzadeh (Department of Civil Engineering, University of Calgary), S.McKeown (Golder Associates Ltd., Calgary), J.Oswell and J.Sladen (EBA Engineering Consultants Ltd., Calgary), B.Smith (Thurber Consultants Ltd., Calgary), M.Wilson (Alberta Environmental Centre, Vegreville).

The Organizing Committee wishes to acknowledge the efforts of the authors, the reviewers, session chairmen, the staff of the Faculty of Continuing Education at the University of Calgary, the Calgary Chapter of the Canadian Geotechnical Society, and A.A.Balkema Publishers in putting the symposium and the proceedings together. Acknowledgement must also be made to Dr. N.R.Morgenstern for graciously delivering the banquet address.

The organization of such a symposium is made possible by the cooperation and tireless efforts of many individuals and organizations. Professor Allan Torvi, Ms. Marion Vavra and Mrs. Pat Braid of the Faculty of Continuing Education deserve special mention for their assistance. Financial assistance provided by the Research Grants Committee, Dean, Faculty of Engineering, and Head, Department of Civil Engineering at the University of Calgary is greatly appreciated.

R.C.Joshi
Chairman of the Symposium Organizing Committee

Keynote addresses

Prediction and Performance in Geotechnical Engineering / Calgary / 17-19 June 1987

Non-destructive characterization of particulate systems for soil classification and in situ prediction of soil properties and soil performance

K.Arulanandan
University of California, Davis, USA

ABSTRACT: A non-destructive method of characterizing particulate systems using the electrical properties of soil is presented. This method of soil characterization is used to develop a fundamental classification of soils. The capability of predicting λ, κ, M, K_0, O.C.R., e and anisotropy index A using the electrical method is demonstrated. The application of using input properties representative of in situ conditions in constitutive models is demonstrated by comparing the predicted steady state line and steady state strength line with those measured.

1 INTRODUCTION

In geotechnical engineering, soil characterization for soil classification and for the prediction of soil properties is necessary to permit reliable representation of the complex conditions and problems presented by nature.

In 1908, Atterberg had worked out his classification of size fractions based on decimal multiples of 2µ and 6µ with "clay fraction" defined as the percentage by weight of particles smaller than 2µ. However, he realized that particle size alone provided an insufficient basis for classification of cohesive soils, and decided that a measure of their "plasticity" in terms of liquid and plastic limits fulfilled to a large extent the required additional criterion. This system with some modifications in nomencleature has been generally used in soil mechanics for the past 50 years.

In a series of five papers, Lambe and Martin (1953-1957) reported compositional data and their relationships to engineering properties for a large number of soils. Their study showed that the plasticity characteristics of natural soils are less than that predicted using a knowledge of the percentage of clay minerals as shown, for example, by the data in Fig. 1. Such effects have been ascribed to the interstratification of clay minerals. In addition, the percentage minerals is likely to be considerably more or less than the percentage clay size (2 µm) as shown, for example, by the data in Fig. 2.

Inspite of the significant observations reported by Lambe and Martin (1953-1957), engineers have continued to use Atterberg Limits and particle size as parameters to classify soils and to predict soil behavior such as swell and erosion potentials of soils with serious consequences, Lambe and Martin (1953-1957), Arulanandan (1973), Basu and Arulanandan (1973), Muraleetharan and Arulanandan (1986), Arulanandan and Perry (1983). Hence a fundamental approach to the characterization of particulate system for soil classification is necessary. The electrical properties of soils, which are functions of mineralogy and pore fluid composition, provide a fundamental basis for soil classification and prediction of soil properties Arulanandan et al. (1973).

Mineralogy, along with the interaction with the fluid phase, determine swelling, erosion, compression, strength and permeability of soils. The initial state of a soil defined by the specific volume v = 1+e, the anisotropy index A, K_0, O.C.R. and mean normal stress, p′ and the soil properties, the compression index λ, swell index κ, and the slope of the critical state line, M are necessary parameters to establish the stress path and the stress-strain relationship of soils.

The objectives of this paper are 1) to characterize soil nondestructively by determining the electrical properties of soils in order to quantify the mineralogy

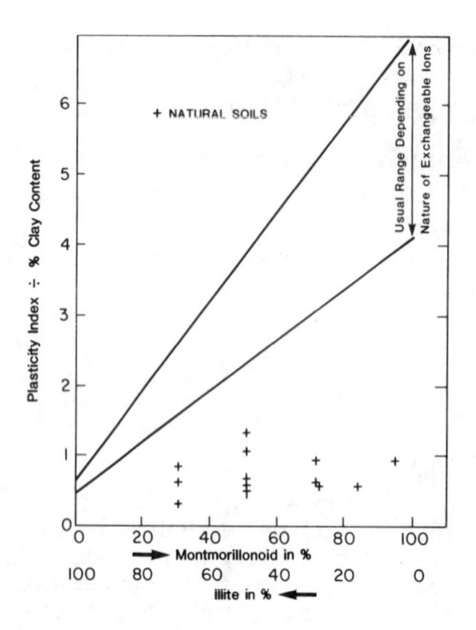

Fig. 1 Activity vs. clay composition (After Lambe and Martin, 1953-1957)

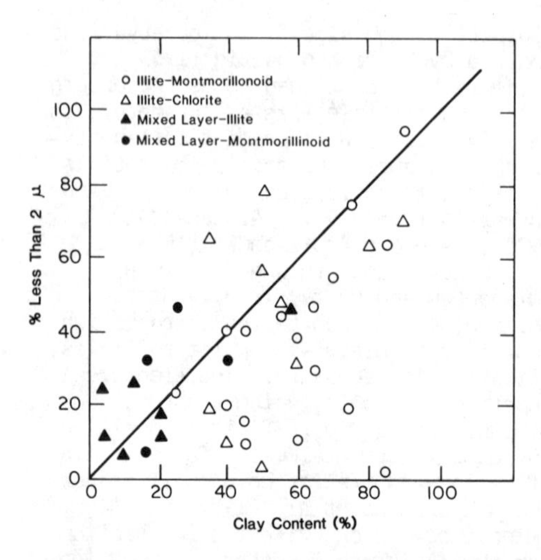

Fig. 2 Relation between clay content and clay size for several soils (After Lambe and Martin, 1953-1957 and Basu and Arulanandan, 1973)

of a soil for the establishment of a fundamental approach to soil classification and, 2) to predict properties such as porosity, K_0, O.C.R, λ, κ, M and anisotropy index A by in situ testing for the prediction of the steady state line and the steady state strength line.

2 ELECTRICAL CONDUCTION THROUGH HETEROGENEOUS MEDIA

Maxwell (1881) derived an expression for the conductivity of heterogeneous media consisting of spherical particles immersed in a solution. The suspension was assumed to be dilute such that the effects of electrical interference due to charges in a sphere does not affect the field in another sphere. If there are n spheres of radius 'a' in the solution at such distances from each other that their effects in disturbing the course of the current may be taken as independent of each other, and if we assume that all these spheres are inside a large sphere of radius a', and if we define the particle volume concentration of spheres as $\rho = na^3/a'^3$, and for the continuity of current and potential within the large sphere, it can be shown that

$$\frac{\sigma^*}{\sigma_1^*} = \frac{2\sigma_1^* + \sigma_2^* - 2\rho(\sigma_1^* - \sigma_2^*)}{2\sigma_1^* + \sigma_2^* + \rho(\sigma_1^* - \sigma_2^*)} \qquad (1)$$

where σ^* - complex conductivity of the suspension
σ_1^* - complex conductivity of the suspending medium
σ_2^* - complex conductivity of suspended particles

The complex conductivity σ^* could be written as follows:

$$\sigma^* = \sigma + j\omega\varepsilon$$

where σ is the conductivity
ε is the dielectric constant
ω is angular frequency of the applied field.

Therefore, one could write

$$\sigma_1^* = \sigma_1 + j\omega\varepsilon_1 \quad , \quad \sigma^* = \sigma_2 + j\omega\varepsilon_2 \quad \text{and}$$

$$\sigma^* = \sigma + j\omega\varepsilon$$

Substituting these in equation (1) and equating the real and imaginary parts on both L.H.S. and R.H.S., we can obtain,

$$\sigma = \sigma_1 [\{(2\sigma_1+\sigma_2) - 2\rho(\sigma_1-\sigma_2)\}\{(2\sigma_1+\sigma_2) + \rho(\sigma_1-\sigma_2)\} + \omega^2\{(2\varepsilon_1+\varepsilon_2) - 2\rho(\varepsilon_1-\varepsilon_2)\}\{(2\varepsilon_1+\varepsilon_2) + \rho(\varepsilon_1-\varepsilon_2)\}]$$

$$\frac{- 3\omega^2\varepsilon_1\rho[(2\varepsilon_1+\varepsilon_2)(\sigma_1-\sigma_2) -}{\{(2\sigma_1+\sigma_2) + \rho(\sigma_1-\sigma_2)\}^2 +}$$

$$\frac{(\varepsilon_1-\varepsilon_2)(2\sigma_1+\sigma_2)]}{\omega^2\{(2\varepsilon_1+\varepsilon_2) + \rho(\varepsilon_1-\varepsilon_2)\}^2} \qquad (2)$$

and

$$\varepsilon = \varepsilon_1 [\{(2\sigma_1+\sigma_2) - 2\rho(\sigma_1-\sigma_2)\}\{(2\sigma_1+\sigma_2) + \rho(\sigma_1-\sigma_2)\} + \omega^2\{(2\varepsilon_1+\varepsilon_2) - 2\rho(\varepsilon_1-\varepsilon_2)\}\{(2\varepsilon_1+\varepsilon_2) + \rho(\varepsilon_1-\varepsilon_2)\}]$$

$$\frac{+ 3\sigma_1\rho[(2\varepsilon_1+\varepsilon_2)(\sigma_1-\sigma_2)-(\varepsilon_1-\varepsilon_2)(2\sigma_1+\sigma_2)]}{\{(2\sigma_1+\sigma_2)+\rho(\sigma_1-\sigma_2)\}^2+\omega^2\{(2\varepsilon_1+\varepsilon_2)+\rho(\varepsilon_1-\varepsilon_2)\}^2} \qquad (3)$$

The above equations express the variation of conductivity and dielectric constant as a function of frequency and particle volume concentration ρ. The above expression is also shown to be valid for dense suspension, Fricke (1953).

For different values of particle conductivity, σ_2, depending on the mineral solution interface characteristics, the variations of dielectric constant and conductivity as a function of frequency are shown in Fig. 3. This result suggests that different electrical dispersion characteristics for different soils depend on the type of clay minerals present, Arulanandan et.al. (1986).

Fricke (1924) extended Maxwell's relationship for ellipsoidal particles with the electric field applied parallel to any of the axes. He showed that a similar dispersion behavior could be obtained for ellipsoidal particles.

3 SOIL CLASSIFICATION

Due to the interaction of clay mineral surface with the pore fluid, clay minerals will have a surface conductance, i.e., $\sigma_2 \neq 0$. Clay minerals, which have surface conductance, show electrical dispersion in the radio frequency range according to Maxwell's equations (2) and (3). This was

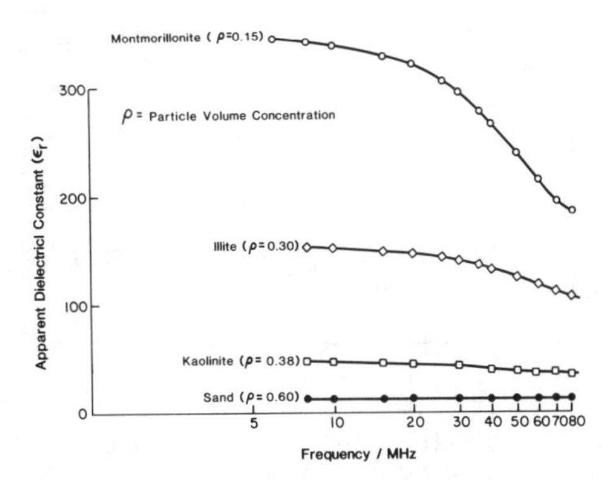

Fig. 3 Theoretical variation of dielectric constant as a function of frequency obtained from Maxwell's equation for different soil types having different values of particle conductivity (σ_2) (After Arulanandan et al., 1986)

verified experimentally for clays as shown in Fig. 4.

For non-clay minerals the interaction of solution with the surface is nonexistent, i.e. σ_2=0. Non-clay minerals do not show electrical dispersion in the radio frequency range according to Maxwell's equations (2) and (3). This was verified for sands and silts experimentlly as shown in Fig. 5. Hence soils could be classified broadly into two groups, based on the dispersion behavior in the radio frequency range. Clay minerals which exhibit electrical dispersion behavior and non-clay minerals which do not exhibit electrical dispersion behavior.

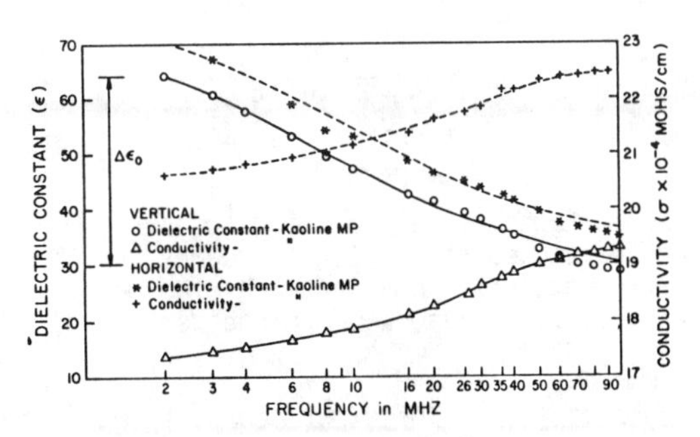

Fig. 4 Variation of dielectric constant and conductivity as a function of frequency for Kaoline MP.

From the dielectric dispersion curves (variation of the dielectric constant with frequency) a fundamental index, magnitude

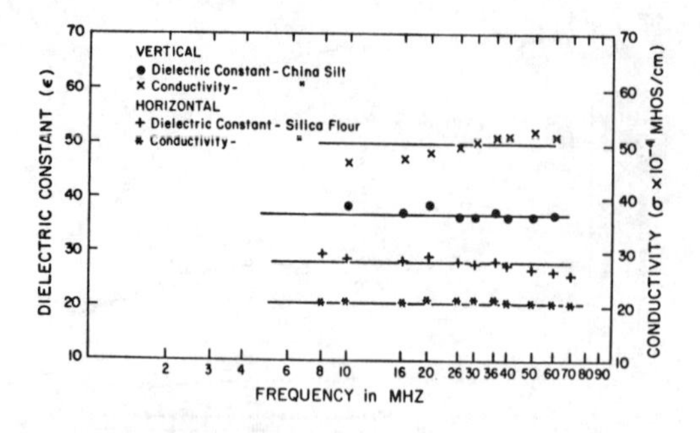

Fig. 5 Variation of dielectric constant and conductivity as a function of frequency for sands and silts.

of dielectric dispersion ($\Delta\varepsilon_0$), is defined as the difference in apparent dielectric constant at about 2 MHz and at about 32 MHz (Fig. 4). As shown theoretically in Fig. 3 (Arulanandan et al. 1986) and experimentally in Fig. 6 (Arulanandan et al. 1973, Basu and Arulanandan 1973), $\Delta\varepsilon_0$ is zero for silica flour and varies in the ascending order as Kaolinite < Illite < Montmorillonite. Fig. 7 shows variation of $\Delta\varepsilon_0$ with the amount of clay content and varying pore fluid composition for Montmorillonite, Illite, and Kaolinite. $\Delta\varepsilon_0$ is shown to increase with increasing clay content and decreasing pore fluid

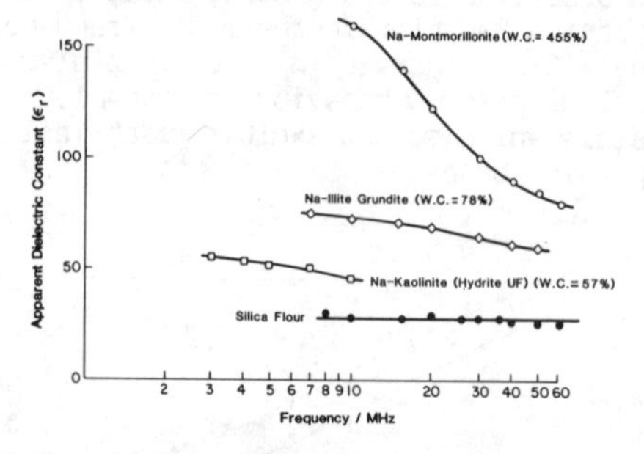

Fig. 6 Experimental variation of dielectric constant with frequency for different soil types (After Arulanandan et al., 1973, Basu and Arulanandan, 1973)

concentration. The experimental trends shown in Fig. 7 also have been verified by theoretical studies by extending Maxwell's (1881) pioneering work on electrical conduction through a heterogeneous particulate media (Arulanandan, et al., 1986). Thus, it can be concluded that the fundamental index $\Delta\varepsilon_0$ is a function of

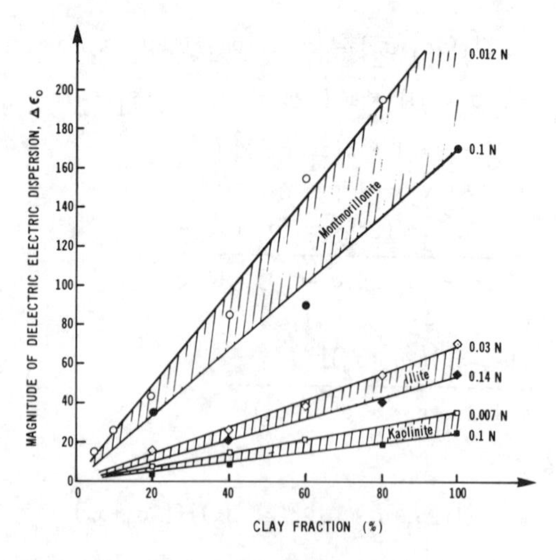

Fig. 7 Variation of magnitude of dielectric dispersion ($\Delta\varepsilon_0$) with % clay mineral and pore fluid concentration for different types of clay minerals (Gu and Arulanandan unpublished data)

amount and type of clay mineral and the pore fluid composition. Therefore Δεo can be used to obtain properties of a soil which depends on type and amount of clay minerals and the pore fluid composition such as specific surface area.

It can be shown theoretically by extending Maxwell's (1881) work on electrical conduction through a heterogeous particulate media, that the dielectric constants at frequencies greater than 30 MHz is a unique function of porosity for many geologic materials (Arulanandan 1986). Application of this method to the in situ prediction of porosity is shown in Fig. 14.

The value of $\Delta\varepsilon_0$ and dielectric constant at 32 MHz frequency can be measured in situ using the dielectric probe (Arulanandan, et al. 1985).

4 FUNDAMENTAL ELECTRICAL PROPERTIES

The formation factor, F, defined as the ratio of the conductivity of the electrolyte which saturates a particulate medium to the conductivity of the mixture at low frequencies (≃1kHZ), Archie (1942), Arulanandan, et al. (1978), has been shown theoretically and experimentally to depend on the porosity, particle shape and size distribution and the direction of measurement, Arulanandan, et al. (1979),

Arulmoli (1980), Dafalias, et al. (1979), and Kutter (1978).

The formation factor was shown to be a tensorial parameter with tensorial components related to the microstructural features in sands, Dafalias et al., (1979a). The average formation factor, $\bar{F}$, and the anisotropy index, A, are defined for a transversely isotropic soil as follows:

$$\bar{F} = (F_V + 2F_H)/3 \quad (1)$$

$$A^2 = F_V/F_H \quad (2)$$

where F_V = formation factor in the vertical direction and F_H = formation factor in the horizontal direction.

An integration technique proposed by Bruggeman (1935) was used by Dafalias and Arulanandan (1978) to derive an expression for average formation factor, $\bar{F}$, as a function of porosity, n, and average shape factor, $\bar{f}$, as

$$\bar{F} = n^{-\bar{f}} \quad (3)$$

The average shape factor $\bar{f}$ is the negative slope of the log $\bar{F}$-log n plot. It is the first invariant of the second order shape factor tensor f and it relates the electric fields inside and outside the sand particles. It has been shown both theoretically and experimentally that the shape factor is direction dependent and depends on gradation and particle shape and orientation, Arulanandan (1979), Arulmoli (1980), Dafalias, et al. (1979), and Kutter (1978). Since the average formation factor is independent of orientation of particles, the average shape factor for a given sand is expected to be a function of the shape of particles in the case of noncemented particles. In the case of cementation $\bar{F} \simeq an^{-\bar{f}}$ where a defines cementation (Report to ESSO Resources, Canada, 1986).

The electrical parameters $\bar{F}$, A, $\bar{f}$ of sand deposits are governed by the grain and aggregate characteristics of the particles.

The preceeding paragraphs have shown that $\bar{F}$ is a unique function of porosity, 'A' quantifies particle's orientation and $\bar{f}$ is a measure of the shape of the particles. Thus $\bar{F}$ and A may be used to quantify the aggregate property. The aggregate property is sensitive to sampling disturbance and needs to be measured in situ. Grain property (shape) is insensitive to sampling disturbance and can be determined on disturbed samples.

Mechanical properties of soils are strong functions of grain and aggregate characteristics. It is shown that fundamental electrical properties $\bar{F}$, A and $\bar{f}$ are also functions of grain and aggregate characteristics of a soil. Hence it is possible to correlate various combinations of $\bar{F}$, A and $\bar{f}$ with various mechanical properties of a soil. $\bar{F}$ and A can be predicted in situ non-destructively using a conductivity probe, Arulanandan (1977).

5 PROPERTY CORRELATIONS

5.1 In situ porosity

In situ porosity can be predicted using the equation $\bar{F} = n^{-\bar{f}}$ if average formation factor ($\bar{F}$) can be measured in situ non-destructively. For uncemented soils average shape factor ($\bar{f}$) is only a function of shape of the particles and can be measured on disturbed samples. In situ prediction of porosities made using the conductivity and dielectric probes at Jackson Dam site, Wyoming are compared with measurements made on undisturbed samples as shown in Fig. 14. The reasonable agreement observed provide support to the electrical methods of predicting porosity.

5.2 Slope of isotropic consolidation line in e-ℓnp' space (λ)

Bolt (1956) attempted to predict the compressibility characteristics of clays based on the concept of osmotic pressure using Gouy-Chapman diffuse double layer theory and Van't Hoff's theories of parallel platy particles. Experimental compression characteristics of Na-montmorillonite and Na-illite were found to be close to the predicted relationships. The theory, however, was found to be valid only for clays exhibiting very strong colloidal properties such as montmorillonite. Deviations from Bolt's findings have been reported by Mitchell (1960) and Olson and Mitronovas (1962) and are ascribed mainly to particle orientation. Quigley and Thompson (1966) have observed fabric changes in natural Leda clay during consolidation using X-ray diffraction methods. It has been shown by Rosenquist (1958) that the compressibility of clay is dependent on the type as well as the valence and concentration of ions adsorbed on the surface of the clay particles. Further, Olson and Mesri (1970) have concluded that both mechanical and physico chemical factors influence the compressibility of soils in general, although one or the other may dominate depending on the soil type.

It has been shown that $\Delta\varepsilon_0$ is significantly influenced by type and amount of clay mineral. The values of $\Delta\varepsilon_0$ are shown to increase in the sequence kaolinite < illite < montmorillonite. The compression indices of these soils also increase in this sequence. The magnitude of dielectric dispersion decreases with an increase in percentage of sand in sand-clay mixtures (Fig. 7) so does the compression index as it is widely known. Olson, et al. (1970) have shown that the compression index of kaolinite is decreased when the electrolyte concentration is increased from 0.0001 N Sodium to 1.0 N Sodium and Fig. 7 shows that $\Delta\varepsilon_0$ also decreased with increasing electrolyte concentration.

The preceding discussion suggests that the factors influencing the mechanisms controlling the compression of clays and magnitude of dielectric dispersion are the same. Based on this mutual dependency, $\Delta\varepsilon_0$ has been correlated with λ. Sharlin (1972) has shown that there is a linear relationship between $\Delta\varepsilon_0$ and λ for natural clays as shown in Fig. 8. Further results confirm the general validity of this relationship between $\Delta\varepsilon_0$ and λ, Anandarajah (1983).

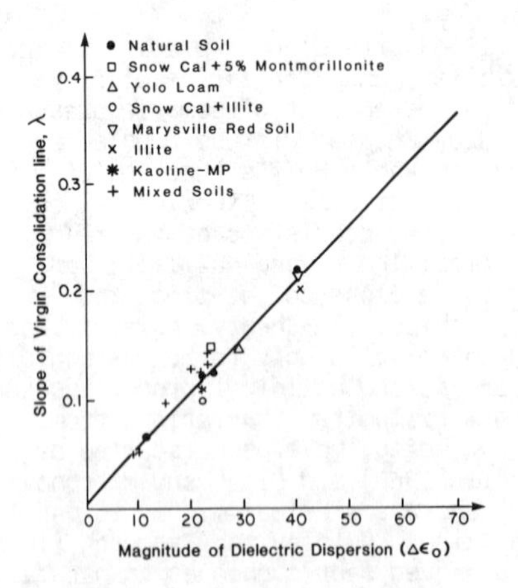

Fig. 8 Correlation between the slope of the isotropic consolidation line in e-ℓnp′ space (λ) and the magnitude of dielectric dispersion ($\Delta\varepsilon_0$) (After Arulanandan et al., 1983a)

5.3 Slope of isotropic swelling line in e-ℓnp space (κ)

The swelling characteristic of saturated clays due to the removal of external load has been investigated by many, either by mechanical models such as the one used by Terzaghi (1929) where swelling is assumed to result from elastic rebound of bent particles or by physico-chemical models such as the one used by Bolt (1956) where osmotic repulsive forces are assumed to be responsible for swelling. Although it has been possible to explain the mechanisms controlling swelling characteristics by the above concepts, it was not very successful owing to the complicated structural arrangements of particles in clays.

The concept of clusters in fine grained soils, Michael, et al. (1954) and Quirk (1959), has been utilized by Olsen (1961) in his study of hydraulic flow through saturated clays and he concluded that the discrepancy between the measured permeability and the one predicted by Kozeny-Carman equation in clays is mainly due to unequal pore sizes due to grouping of clay particles in clusters. The existence of primary particles aggregation has been observed by many using electron microscope, Quigley, et al. (1966).

Figure 9 shows variation of intra (e_I) and inter-cluster (e_p) void ratios, obtained using a three element model (Arulanandan et al. 1983) with total void ratio using electrical dispersion data for Snow Cal (95%) + Montmorillonite (5%). The results corresponding to measurements made in the vertical and horizontal directions are identical and is very similar to the one predicted by Olsen (1961), Meegoda (1983), and Abdullah (1983).

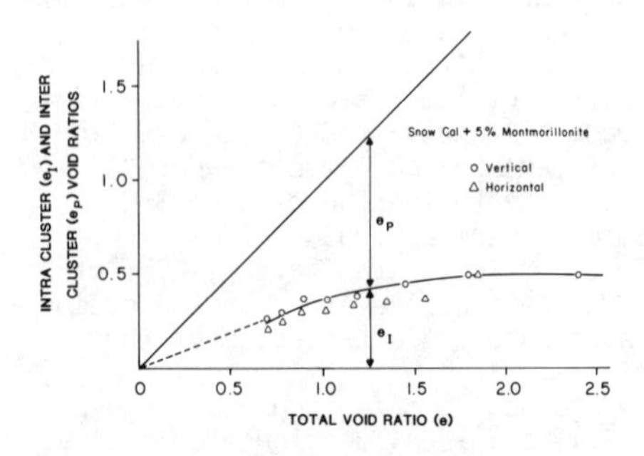

Fig. 9 Variation of intra-cluster (e_I) and inter-cluster (e_p) void ratio with total void ratio (e) for 95% Snow Cal + 5% Montmorillonite (after Meegoda, 1983)

It has been shown that the swelling of fine grained soils is caused by swelling of clusters, Smith and Arulanandan (1981), and the decrease in inter-cluster pores during compression is irreversible, Meegoda (1983).

If the ratio of intra-cluster to total void ratio is large for a given soil, the elastic compression due to an increase in the external load would be high and consequently swelling would also be high when the load is removed. Assuming this mechanism of swelling, the ratio e_I/e is correlated with κ as shown in Fig. 10.

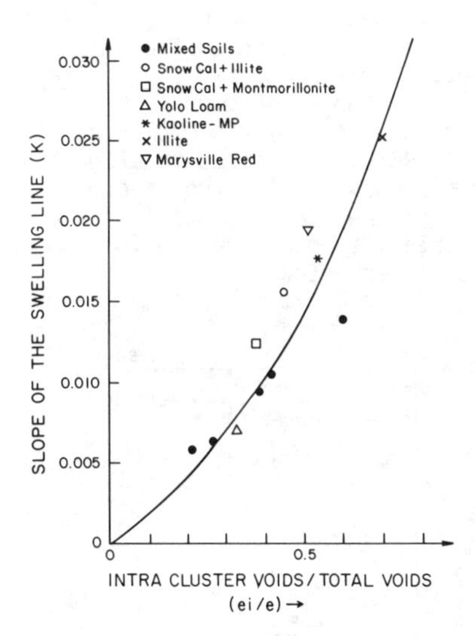

Fig. 10 Correlation between the slope of the isotropic swelling line in e-ℓnp′ space (κ) and the ratio of intra-cluster to total void ratio (e_I/e)

5.4 Slope of Critical State Line M

When a soil element is sheared under drained or undrained conditions, experimental results indicate that the soil element fails when the stress path reaches the critical state line independent of the initial stress state of the soil element, Schofield, et al. (1968). At failure, the void ratio, e, and the effective mean normal pressure, p′, lie on a unique line, referred to as a critical state line on the e-p′ space. This concept is widely known as the critical state concept, Roscoe, et al. (1968). The slope of the critical state line, on the p′-q space, therefore, represents the ultimate shear strength of soils.

Lambe (1960) has discussed the factors controlling ultimate shear resistance of fine grained soil, which is considered to be due to friction and interference between particles. These components of ultimate shear resistance and hence M would depend on many factors such as particle size, shape, surface texture and the structure of the soils resulting from the attractive and repulsive forces between the adjacent clay particles. In fine grained soils, the shape factor ($\bar{f}$) would reflect in addition to shape of the particles, the physico-chemical interaction between particles. In other words, the shape factor, $\bar{f}$, is a function of the shape, composition and arrangement of particles. Therefore the shape factor ($\bar{f}$), is correlated to M and the correlation is given in Fig. 11.

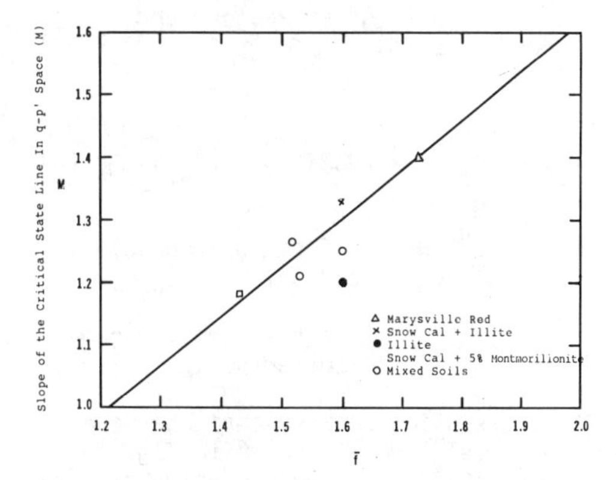

Fig. 11 Correlation between the slope of the critical state line in p′-q space (M) and the average shape factor ($\bar{f}$)

The values of M are obtained from the normally consoldiated undrained test results.

5.5 Coefficient of earth pressure at rest (K_o)

Meegoda and Arulanadan (1986) established a correlation between the coefficient of earth pressure at rest, K_o, and an electrical index, $A^4\bar{f}$ for normally consolidated clays. This correlation is given in Fig. 12.

5.6 Overconsolidation ratio (O.C.R.)

K_o for overconsolidated soils is given by the following expression (Schmidt, 1966):

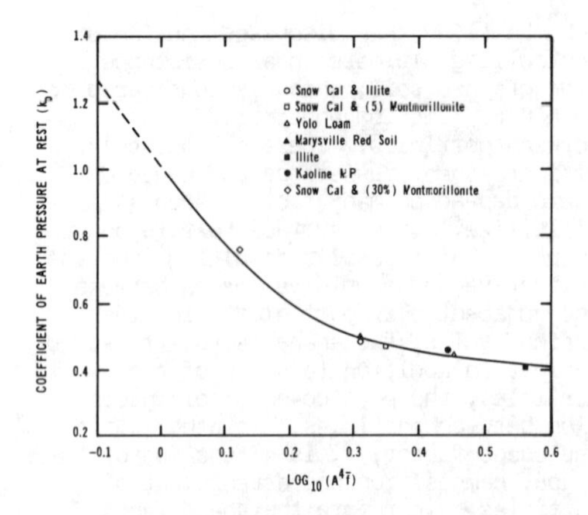

Fig. 12 Correlation between the coefficient of earth pressure at rest (K_0) for normally consolidated soils and the electrical index $A^4\bar{f}$ (After Meegoda and Arulanandan, 1986)

$$(K_o)_{OC} = (K_o)_{NC} \times (OCR)^{\alpha} \quad (4)$$

where $(K_o)_{OC}$ = K_o of the overconsolidated soil

$(K_o)_{NC}$ = K_o of the normally consolidated soil

The correlation between parameter α and an electrical index is shown in Fig. 13 (Meegoda, 1983). An average value of 0.45 for α can be used in (1) in order to obtain the value of OCR if $(K_0)_{OC}$ and $(K_0)_{NC}$ are known.

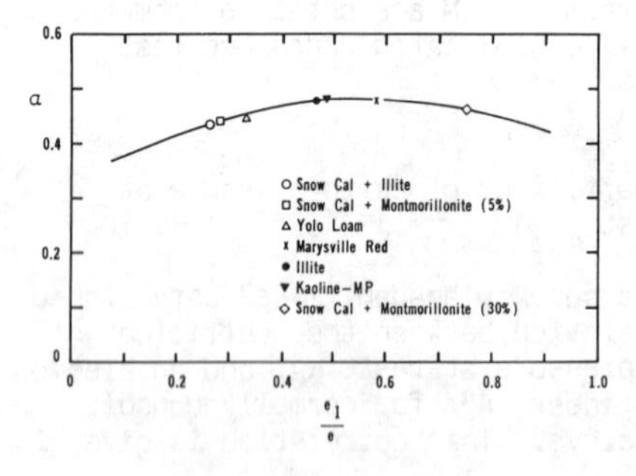

Fig. 13 Correlation between the parameter α (see equation 1) and the ratio of intracluster (e_I) to total void ratio (e) (After Meegoda, 1983)

6 IN SITU APPLICATIONS

Using the conductivity probe, average formation factors ($\bar{F}$) and electrical anisotropy indices (A) were measured at the Jackson dam site, Wyoming. Average shape factors ($\bar{f}$) were also obtained by laboratory testing of samples. Using these measurements and the given correlations, in situ void ratio (e_0), K_0 and O.C.R. were predicted and compared with the values predicted by other methods (Figs. 14,15,16). Steady state line (Fig. 17) and steady state strength line (Fig. 18) were also predicted using the predicted values of K_0, O.C.R., e_0, M, λ, and κ, and the bounding surface plasticity model. These predicted lines were compared with those measured using remoulded specimens by the Poulos et al. approach (1985). The close relationship between the measured and the predicted positions of the steady state line and the steady state strength line provide support to the proposed method.

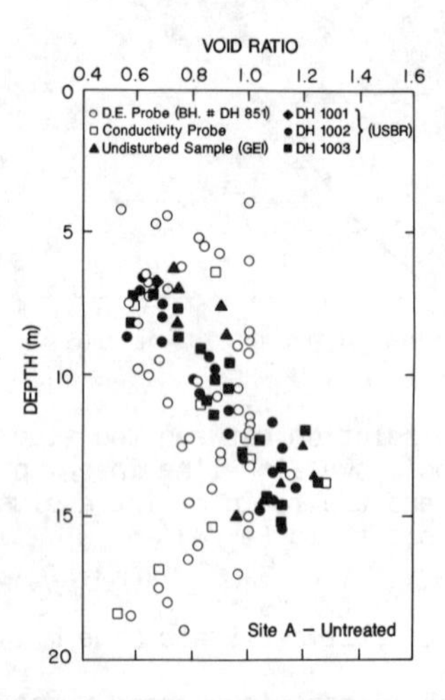

Fig. 14 Variation of void ratio as a function of depth predicted by electrical methods and measured from undisturbed samples at site A (untreated), Jackson Dam, Wyoming (from the report to USBR, 1987)

7 SUMMARY AND CONCLUSIONS

The variation of conductivity and dielectric constant with frequency in the radio frequency range (electrical dispersion) has been shown to classify soils into two broad categories, clay minerals and non clay minerals. It has been shown theoretically, by extending Maxwell's (1881) theory for electrical conduction

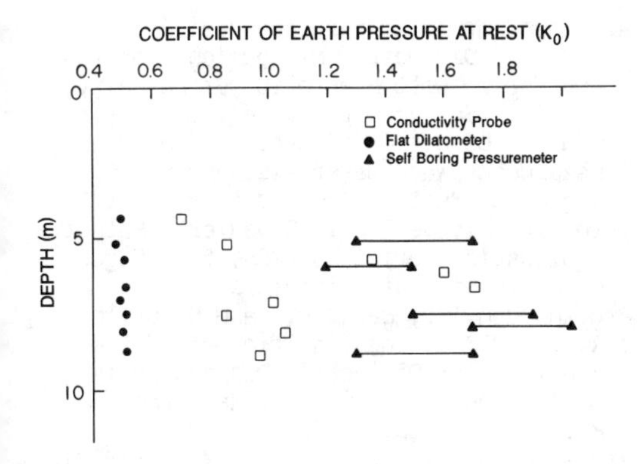

Fig. 15 Comparison of K_0 predicted by the conductivity probe with K_0 predicted by flat plate dilatometer and self boring pressuremeter at Sector C (Area 3), Jackson Dam, Wyoming (from the report to USBR, 1987)

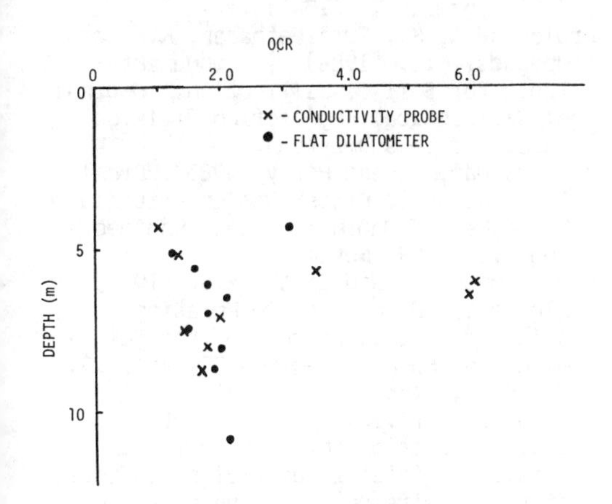

Fig. 16 Comparison of OCR predicted by the conductivity probe with OCR predicted by flat plate dilatometer at Sector C (Area 3), Jackson Dam, Wyoming (from the report to USBR, 1987)

through heterogeneous media, and experimentally that non-clay minerals exhibit no electrical dispersion behavior where as clay minerals exhibit electrical dispersion behavior. The magnitude of dielectric dispersion ($\Delta\varepsilon_0$) (difference in dielectric constants between 2 MHz and 32 MHz) has been shown to be a function of amount and type of clay minerals and the pore fluid composition. A unique correlation between $\Delta\varepsilon_0$ and the slope of

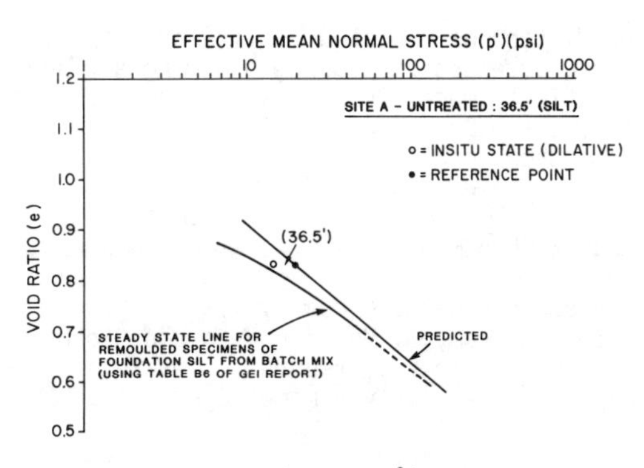

Fig. 17 Predicted (at 36.5') and measured steady state lines for silt at Site A - untreated, Jackson Dam, Wyoming (from the report to USBR, 1987)

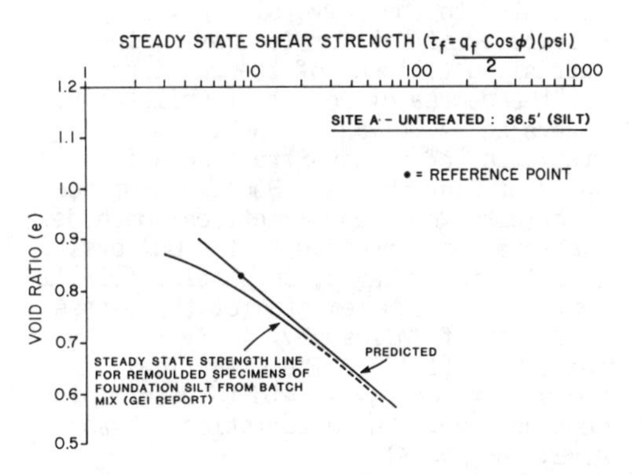

Fig. 18 Predicted (at 36.5') and measured steady state strength line for silt at Site A - untreated, Jackson Dam, Wyoming (from the report to USBR, 1987)

the isotropic consolidation line in e-lnp' space (λ) for clay minerals is also obtained.

The electrical parameters, the average formation factor $\bar{F}$, the electrical anisotropy index A, and the average shape factor $\bar{f}$ have been shown to predict the in situ porosity, K_0, O.C.R., and M, the slope of the critical state line. The dielectric constant value at frequencies greater than 32 MHz has also been used to predict the in situ porosity at Jackson Dam site, U.S.A. The prediction of κ, the swell index is postulated utilizing two levels of porosities, the inter cluster e_p, and the intra cluster e_I, porosities.

The in situ application of the non destructive method of soil characterization using electrical properties for the prediction of the steady state are steady state strength line using bounding surface plasticity model for soils encountered at Jackson Dam site, U.S.A. is demonstrated.

ACKNOWLEDGMENTS

The assistance provided by K.K. Muraleetharan, C. Yogachandran and Dinah Greenstreet in the preparation of this paper is gratefully acknowledged.

REFERENCES

Abdullah, A. (1983). A Fundamental Approach for the Characterization of Mixed Soils to Predict Stress-Strain Behavior In Situ. Dissertation presented to the University of University of California, Davis, in partial fulfillment of the requirements for the degree of Doctor of Philosophy.

Anandarajah, A. (1982). In Situ Prediction of Stress-Strain Relationships of Clays Using a Bounding Surface Plasticity Model and Electrical Methods. Dissertation presented to the University of California, Davis, in partial fulfillment of the requirements for the degree of Doctor of Philosophy.

Archie, G.E. (1942). The Electrical Resistivity Log as an Aid in Determining Some Reservoir Characteristics. Trans. AIME, 146, 54-61.

Arulanandan, K. (1973). Swell Potential Behavior of Marysville Soil - Department of Civil Engineering, UCD Report.

Arulanandan, K. (1977). Method and Apparatus for Measuring In situ Density and Fabric of Soils. Patent Applications, Regents of the University of California.

Arulanandan, K., Anandarajah, A., and Meegoda, N.J. (1983a). Soil Characterization for Non Destructive In Situ Testing. Symposium Proceedings Part 2, The Interaction of Non Nuclear Munitions with Structures, U.S. Air Force Academy, Colorado, pp. 69-75.

Arulanandan, K., Anandarajah, A., and Meegoda, N.J. (1983b). Quantification of Inter- and Intra-cluster Void Ratios Using Three Element Electrical Model. Report, Department of Civil Engineering, University of California, Davis.

Arulanandan, K. and Arulanandan, S. (1985). Dielectric Methods and Apparatus for In situ Prediction of Porosity, Specific Surface Area (i.e., soil type) and for Detection of Hydrocarbon, Hazardoous Wate Materials, and the Degree of Melting Ice and to Predict In situ Stress-Strain Behavior. Patent Application Serial No. 709, 592, Regents of University of California.

Arulanandan, K., Basu, R., and Scharlin, R.J. (1973). The Significance of the Magnitude of Dielectric Dispersion in Soil Technology. Highway Research Record, 426, pp. 23-32.

Arulanandan, K. and Dafalias, Y.F. (1979). Significance of Formation Factor in Sand Structure Characterization. Letters in App. and Eng. Sciences, Vol. 17, pp. 109-112.

Arulanandan, K. and Kutter, B.L. (1978). A Directional Structure Index Related to Sand Liquefaction. Proc. ASCE, Geotechnical Engineering Division Specialty Conference, Earthquake Eng. Soil Dynamics 1, 213-230.

Arulanandan, K., Muraleetharan, K.K. and Meegoda, N.J. (1986). A Fundamental Index for Soil Classification. Journal of Geotechnical Engineering Division, ASCE, (to be submitted).

Arulanandan, K. and Perry (1983)."Erosion in Relation to Filter Design Criteria in Earth Dams," Journal of Geo. Engineering, Vol. 109, November.

Arulanandan, K. and Smith, S.S. (1973). Electrical Dispersion in Relation to Soil Structure. Journal of the Soil Mech. and Found. Divs., ASCE, Vol. 99, No. SM12, Proc. Paper 10235.

Arulmoli, K. (1980). Sand Structure Characterization for In Situ Testing. Thesis submitted in partial satisfaction of the requirements for the degree of Master of Science in Engineering, University of California, Davis.

Basu, R. and Arulanandan, K. (1973). A New Approach for the Identification of Swell Potential of Soils. Proceedings of the Third International Conference on Expansive Soils, Haifa, Israel, Vol. 1, pp. 1-11.

Bolt, G.H. (1956). Physico Chemical Analysis of the Compressibility of Pure Clays. Geotechnique, Vol. 6, pp. 86-93.

Bruggeman, D.A.G. (1935). Berechung Verschiuedenez Physika Lischev Konstanten Von Heterogenen Substanzen. Ann. Phys. Lp z.5, Vol. 24, pp. 636.

Dafalias, Y.F. and Arulanandan, K. (1978). The Structure of Anisotropic Sands in Relation to Electrical Measurements. Mechanics Research Communications, Vol. 5, No. 6, pp. 325-330.

Dafalias, Y.F. and Arulanandan, K. ((1979a). The Formation Factor Tensor in Relation to Structural Characteristics of Anisotropic Granular Soils. Proc., Colloque International due C.N.R.S., Euromech Colloquium 115, Villard-de-Lans, France.

Dafalias, Y.F. and Arulanandan, K. (1979). Electrical Characterization of Transversely Isotropic Sands. Archives of Mechanics, 31, 5, pp. 723-739, Warsaw.

Dafalias, Y.F. and Herrmann, L.R. (1980). Bounding Surface Formulation of Soil Plasticity. Soils Under Cyclic and Transient Loading, pp. 335-345, Swansea, U.K.

Kutter, B.L. (1978). Electrical Properties in Relation to Structure Cohesionless Soils. Thesis submitted in partial satisfaction of the requirements for the degree of Master of Science in Engineering, Univ. of California, Davis.

Lambe, T.W. and Martin, R.T., (1953-1957). "Composition and Engineering Properties of Soils," Highway Research Board Proceedings, I-1953, II-1954, III-1955, IV-1956, and V-1957.

Lambe, T.W. (1960). A Mechanistic Picture of Shear Strength in Clay. Proceedings Research Conf. on Shear Strength of Cohesive Soils, Soil Mech. and Found. Div., ASCE, University of Colorado, Boulder, Colorado.

Maxwell, J.C. (1881). A Treatise on Electricity and Magnetism. 2nd Ed., pp. 398, Clarendon Press, Oxford.

Meegoda, N.J. (1983). Prediction of In Situ Stress State Using Electrical Method. Thesis submitted in partial satisfaction of the requirements for the degree of Master of Science in Engineering, University of California, Davis.

Meegoda, N.J., Anandarajah, A. and Arulanandan, K. (1985). Verification of In Situ Prediction of Stress-Strain Behavior by Laboratory and Centrifuge Tests. Proceedings of Second Symposium on the Interaction of Non-Nuclear Munitions with Structures, pp. 441-446, Panama City Beach, Florida.

Meegoda, N.J., and Arulanandan, K. (1986). "Electrical Method of Predicting In Situ Stress State of Normally Consolidated Clays," Proceedings of the In Situ 86', ASCE Specialty Conference, Blacksburg, VA, pp. 794-808.

Michaels, A.S. and Lin, C.S. (1954). The Permeability of Kaolinite. Ind. and Eng. Chem., Vol. 46, pp. 1239-1246.

Mitchell, J.K. (1960). The Application of Colloidal Theory to the Compressibility of Clays. Proc., Seminar on Interparticle Forces in Clay-Water-Electrolyte System, Commonwealth Scientific and Industrial Research Organization, pp. 292-297, Melbourne, Australia.

Muraleetharan, K.K. and Arulanandan, K. (1986). Site Characterization in Foundation Engineering with Reference to Compressibility and Swell. Asian Regional Symposium on Geotechnical Problems and Practices in Foundation Engineering, Colombo, Sri Lanka.

Olsen, H.W. (1961). Hydraulic Flow Through Saturated Clays. Sc.D. thesis, M.I.T., Cambridge, Mass.

Olson, R.E. and Mesri, G. (1970). Mechanisms Controlling Compressibility of Clays. ASCE, J. of Soil Mech. and Found. Div., SM6.

Olson, R.E. and Mitronovas, F. (1962). Shear Strength and Consolidation Characteristics of Calcium and Magnesium Illite. Proc. 9th Nat. Conf. Clays and Clay Minerals, pp. 185-209.

Quigley, R.M. and Thompson, C.D. (1966). The Fabric of Anisotropically Consolidated Sensitive Marine Clay. Canadian Geotechnical Journal 3, Vol. 2, pp. 61-73.

Quirk, J.P. (1959). Permeability of Porous Media. Nature, Vol. 183, pp. 387-388.

Roscoe, K.H. and Burland, J.B. (1968). On the Generalized Stress-Strain Behavior of 'Wet Clay'. Eng. Plasticity, Ed. J. Heyman and F.A. Leckie, Cambridge University Press, pp. 535-609.

Rosenqvist, I. Th. (1958). Physico-Chemical Properties of Soils: Soil Water Systems. Journal of Soil Mechanics and Foundation Division, ASCE, Vol. 85, No. SM1, Proc. Paper 2000, pp. 31-53.

Poulos, S.J., Castro, G., and France, J.W. (1985). "Liquefaction Evaluation Procedure," Journal of Geotechnical Engineering Division, ASCE, Vol. 111, No. 6, pp. 772-792.

Scharlin, R. (1972). "A New Approach to Soil Classification," M.S. Thesis, Univ. of California, Davis, June 1971.

Schofield, A.N. and Wroth, C.P. (1968). Critical State Soil Mechanics.

Smith, S.S. and Arulanandan, K. (1981). Relationship of Electrical Dispersion to Soil Properties, J. of Geotechnical Eng. Div., ASCE, Vol. 107, No. GT5.

Terzaghi, K. (1929). Technish-Geologische Beschreibung der Bodenbeschoffenheit fur Bautechnische Zwecke. Chap. IX, Part A, Ingeneiurgeologie, by K.A. Redlich, K. Terzaghi and R. Kempe, Julio Springer, Wein and Berlin.

Prediction and Performance in Geotechnical Engineering / Calgary / 17-19 June 1987

Geotechnical predictions in ice affected marine environments

J.I.Clark
Centre for Cold Ocean Resources Engineering (C-CORE), St.John's, Newfoundland, Canada

I.J.Jordaan
Memorial University of Newfoundland, St.John's, Canada

ABSTRACT: The Canadian experience with bottom founded offshore marine structures is limited to artificial islands for exploration in the Beaufort Sea and jack up rigs on the east coast. No structures are yet in place for production. Most of the Canadian east coast and all of the arctic waters are affected by either sea ice and/or glacial ice for at least part of each year. Interaction of ice with the structure is an important design consideration but interaction of ice with the seabed also has a major influence in the design of production facilities. The presence of ice introduces geotechnical considerations of ice properties and ice loading, that are not major considerations in other presently active offshore production areas elsewhere in the world. Many other countries are planning offshore production in polar regions in the future; hence, the Canadian experience will be of considerable interest. It is important that designers understand the reliability of the data that models the geotechnical environment, the material properties and design analysis to predict performance. This paper presents a brief review of the arctic and east coast environments, the factors that affect design and performance prediction and summarizes techniques that might be implemented to overcome some of the design constraints.

1 INTRODUCTION

Geotechnical predictions on land usually center on the properties and behavior of soils or rocks and the water and/or gas that fills their pore spaces. In permafrost regions ice in the soil becomes a consideration and even there it is the effects of thermal perturbations that are most often of concern. When designing for structures in water, in most of Canada, ice by itself becomes a major consideration. This is particularly true for the Arctic and East Coast waters.

The Canadian experience with bottom founded offshore marine structures is limited to artificial islands for exploration in the Beaufort Sea and some jack-up exploration platforms on the east coast. A gravity-based structure is planned for the Hibernia development on the Grand Banks of Newfoundland and artificial islands or gravity-based structures are being considered for production platforms in the Beaufort Sea. Within current planning, the first offshore oil production in Canada may occur in 1988 from the Amauligak Field where Gulf Canada Corporation is planning seasonal production from the Molikpaq structure. The Molikpaq is a sand-core steel-hull structure which has been used as an exploration platform and which has already yielded a great deal of performance data (Rogers et al. 1986).

Of all the offshore production fields in the world, the environment most similar to that of the East Coast of Canada is that of the North Sea. The major difference between the two areas is that most of the waters of Canada's East Coast are covered by pack ice or invaded by icebergs for at least part of every year. There is no production occurring elsewhere in the world in an environment similar to the Canadian Arctic waters. Hence there is not an extensive experience base with which to test predictions of performance for production structures in these waters. Many countries are at present planning production facilities in ice affected waters so that the Canadian experience in the coming years will be of considerable interest.

There are at least 74 bottom-founded production structures in the North Sea. Of these, structures in three fields are experiencing geotechnical difficulties: (1) Ekofisk, which has settled over 3.5m (Fact Sheet, 1986), (2) Valhalla, which is settling at the same rate as Ekofisk, i.e. 400mm per year (Offshore Engineer, 1986), and (3) Frigg, which has experienced problems related to erosion. The rapid settlement at Ekofisk and Valhalla is related to reservoir depletion, and was not predicted; obviously the problems at Frigg were also not predicted. Nevertheless, there have been no serious foundation failures in the North Sea where loss of life or loss of the structure occurred.

Within the Canadian Beaufort Sea, over 30 artificial islands have been built. It would probably be fair to say that, of these, about 20 to 25 percent have experienced unpredicted geotechnical problems, although very little has been published either describing or explaining the problems. Exceptions are the Nerlerk berm failure in 1984 which has been well-documented (Mitchell, 1984; Sladen et al., 1985), and the Molikpaq sand core partial liquefaction of 1986 which has been recognized and discussed at several conferences, including the IAHR Symposium on Ice 1986 in Iowa City and the NRC Workshop on Extreme Ice Features, also in 1986.

Difficulties with predicting performance may arise as a result of uncertainties in the geological setting and site investigation or from uncertainties in the models used to characterize the loads and resistance or from errors in the analytical methods.

Ice-structure interaction and soil-structure interaction are only two of the engineering aspects requiring detailed design consideration and performance prediction in the Canadian offshore. Ice-seabed interaction is a feature of the Canadian Arctic and East Coast that has a significant impact on the development of production systems in both regions. On the east coast, icebergs impact the seabed and create either linear scours or localized pits (eg. Fader and King, 1981; Lewis and Barrie, 1981). Two models have been developed to predict depth of scour and are in the public domain (the two are the work-energy model of Chari (1979) and the Fenco-APOA formulation of 1975 of both work-energy and time step dynamic models, with the latter being limited to 3 degrees of freedom). Both models consider the ice to be much stronger than the soil it scours, i.e. it is assumed to act as a rigid indentor. Recent analysis suggests that the depth of scours or pits is strongly influenced by the strength of the ice (Clark and Landva, 1986). The present models appear to substantially overpredict scour.

Geotechnical designs and predictions for the development of Canadian offshore petroleum resources require prediction of not only the interaction of the structure with the environment, but also interaction of components of the environment with each other. There are a number of constraints that make such predictions a formidable challenge. These include:

1. very little performance data for structures in the Canadian Beaufort and for structures on the East Coast.
2. very little geotechnical data, compared to what is normally available for similar scale structures on land.
3. the exceptionally high cost of acquiring geotechnical data.
4. very little information on environmental loadings, particularly global ice loads.
5. very little information on global ice strength

Although unexpected problems have been experienced with exploration platforms in the Beaufort Sea, the experience provides a data base for material property models and analytical models that should reduce uncertainties of design criteria for production platforms. Probabilistic methods can be applied to determine risk associated with a specific design.

An example of improved reliability with the experience built up over a period of about 15 years for 4 gravity based structures in the North Sea is given in a review by Wu (1986). He shows that for the cases considered, the uncertainty from mapping and spacial variability (which depend on site investigation techniques and geologic history) is about equal to the uncertainty due to errors in the model for the material properties (which depend upon sample quality and test procedures) and that due to model error for analysis. Research has progressively reduced the model error for material strength. Wu (1986) demonstrates that even though a structure may have a safety factor according to modern day models that is lower than that based on early models, the new models have a reliability index that is larger than that of the earlier models and the design is in fact safer than could be demonstrated for the earlier model with a higher factor of safety. The reliability is equal to (1-P) where P is the probability of failure. For one case that Wu analyzes (Brent B-G) he demonstrates that additional boreholes would not significantly reduce the uncertainty but that further investigation of material properties would. This type of analysis helps identify the most important aspects of site investigation and material characterization and would appear to be of even greater value when loading associated with ice is considered.

This paper presents a brief review of the East Coast and Beaufort Sea experience, the present constraints to making accurate predictions for design, and some of the techniques that might be implemented to overcome some of the above constraints.

BEAUFORT SEA

Ice Loads

Ice loading data on structures in the Beaufort Sea have been collected over a period of several years by several operators. Most of these have not been published. What data are available seem to support the concept that small scale strength tests substantially overestimate ice strength and tend to agree with the simplified universal ice pressure curve reproduced in Figure 1 (Croasdale 1985).

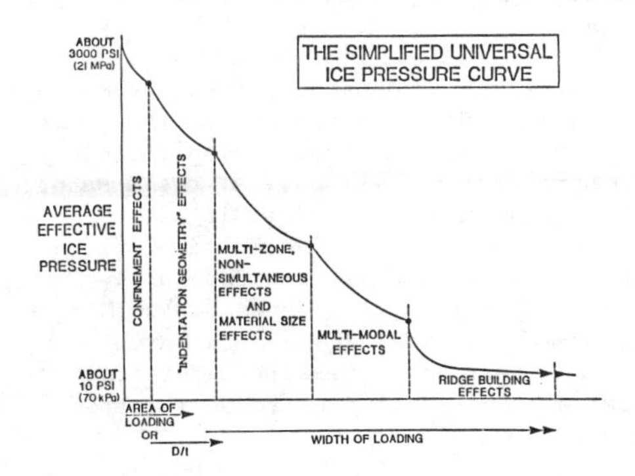

Figure 1. Simplified universal ice pressure curve (from Croasdale 1985).

All recent artificial island structures in the Beaufort Sea have been instrumented to measure ice loading. An example of instrumentation has been described by Rogers et al. (1986). Proprietary data are gradually being released but global ice loads on structures not protected by a rubblemound remains an important subject requiring more research.

Where the water depths and structure design are such that a rubble field will build up around an artificial island, the load transmitted through the rubblemound to the structure may be very low. Apparently part of the load is transmitted from the pack ice to the seabed by means of the grounded rubblemound. The properties of the rubblemound are not well known nor is the mechanism by which the load is transferred from the pack ice to the seabed. This is the subject of current research under a joint project by Esso, the National Research Council of Canada, the Faculty of Engineering and Applied Science at Memorial University of Newfoundland, and the Centre for Cold Ocean Resources Engineering (C-CORE).

As an alternative to a rubblemound, spray ice has been used. A large mass of spray ice is gradually built up around the island causing the ice surface to sink and eventually contact the seabed. Apparently the spray ice barrier acts in the same manner as a rubblemound in protecting the structure by transmitting load from the pack ice to the seabed. Spray ice has also been used to construct drilling platforms (Goff et al. 1986).

ICE SCOUR

Ice scour of the seabed in the Beaufort Sea occurs most frequently in water depths of 15m to 45m (Hnatiuk and Wright 1983). The scours may be caused by deep keels associated with ice pressure ridges, by ice islands or thick ice pieces, by rafted ice or by ice floes as they move into shallow water (Fenco 1975). Hnatiuk and Wright (1983) observe that 97% of the scours are between 0.6m and 2.1m deep. This is an interesting observation given the variable shapes and sizes of the ice features that impact the seabed.

Within the water depth range of 15m to 45m much of the seabed has been reworked by multiple contacts of ice with the seabed. Most of the scours appear to be modern, based on the fact that 2% to 6% of the seabed is reworked by ice features each year (Barnes et al. 1983). No predictive models for scour depths associated with deep keels at pressure ridges exist. Models developed for iceberg scour (described later in this paper) would be appropriate for scour prediction for discrete ice features such as ice islands or thick ice pieces but these models would obviously have the same limitations as for iceberg scour predictions. At present, scour depth predictions in the Beaufort Sea are based on phenomenology.

FOUNDATION CONDITIONS

The foundation conditions in the Canadian Beaufort Sea are extremely variable, reflecting a complex geologic history. Conditions may vary from very good to poor. The presence of permafrost at some locations as well as gas hydrates are complicating features.

The geologic setting and stress history of the Holocene Sediments in the Canadian Beaufort Sea have been recently presented by Christian and Morgenstern (1986). Although not published in the Proceedings, Blasco presented an excellent summary of the geotechnical, geophysical and marine geology setting and interrelationships at the Third Canadian Conference on Marine Geotechnical Engineering in St. John's, in June of 1986. At the same conference, Hill et al. (1986) describe physical and sedimentological properties of nearshore sediments in the southern Beaufort Sea and Crooks et al. deal with geotechnical properties of the clays found in the Southern Beaufort. In summary, the geologic setting is complex and material properties are extremely variable, in comparison with the soil conditions on the Grand Banks, for example. There is also a large amount of geotechnical data and geological analysis available in the public literature, produced by government researchers, oil company personnel and consultants.

EAST COAST

Ice Loading:

In contrast to the Beaufort Sea experience, little or no data are available for ice-structure loading on the East Coast. The main reason for this lack of experience is that only drillships and semi-submersibles have been used for exploratory drilling in ice affected areas to date, which by their design require removal whenever pack ice or icebergs encroach the rig. The proposed gravity base production structure for the Hibernia field will be the first of its kind and will be carefully monitored for both ice and wave loading.

Investigations of the properties of iceberg ice are ongoing, but few results have been published to date. A survey of iceberg ice properties by Nadreau (1986) shows that published data are practically non-existent for large-scale mechanical properties and for instantaneous velocities, both of which are very important for the design of "ice-resistant" exploration and production structures.

It is well known that the strength of ice is dependent on temperature and strain rate. Diemand (1984) shows that the temperature of the central core of an iceberg remains stable and may be as low as $-20^{o}C$ even though the outer layers ablate due to melting. Tests by Lachance (1985) on small iceberg ice samples indicate that the crushing strength increases from 5.5MPa at $-5^{o}C$ to 8MPa at $-20^{o}C$.

Unconfined and triaxial compressive strength tests carried out at different strain rates seem to indicate that ice strength increases approximately seven-fold as the rate of strain is increased from $10^{-6}\ s^{-1}$ to $10^{-3}\ s^{-1}$ but then decreases by 50% when the strain rate is increased to $10^{-1}\ s^{-1}$. Test results such as these must be considered when designing a structure to resist iceberg impact.

Very little work has been published on the strength or dynamics of the pack ice cover off the East Coast, even though it is usually present from December/January to April/May and can have a thickness of several metres. The first in situ strength testing of the pack ice in the marginal ice zone was carried out by C-CORE as a part of the Labrador Ice Margin Experiment (LIMEX) Project in March of 1987. These were pressure meter tests that are being analyzed to be reported later.

Iceberg Scours and Pits

A feature common to all ice-infested Canadian marine environments is ice scour, caused by the interaction of ice and the seabed. In the Beaufort Sea (as previously described), ice scours are caused mainly by the deep keels of ice pressure ridges and by thick ice cover moving into shallow water. In East Coast waters, the vast majority are caused by icebergs. When the southward-drifting bergs reach the continental shelf area (Figure 2), they often contact the seabed, creating features that range from long linear furrows to localized pits.

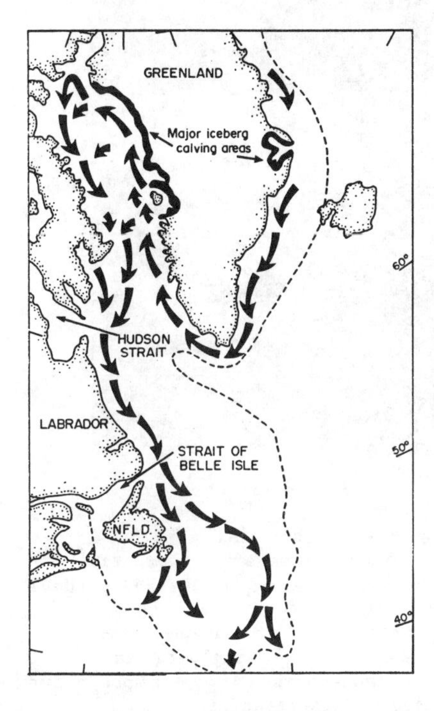

Figure 2. Iceberg drift patterns off the Canadian East Coast.

The characteristics of each feature depend on the type of interaction, the properties of the seabed at the contact point and the properties of the iceberg ice. Over 95% of the ice created seabed features in the Grand Banks are less than 2m deep. Given the large variety of shapes and icebergs that impact the seabed, the variation in configuration of iceberg scour tracks is remarkably small, suggesting that the process is controlled by factors more significant than iceberg mass and shape (Clark and Landva, 1986).

Clark and Landva (1986) describe three possible ice-seabed failure mechanisms which result in scours and pits. The most common is probably the trailing keel mechanism, in which a freely-floating iceberg becomes grounded (by either drifting to shallower waters or rolling), rotates about a vertical axis to an orientation of least resistance to horizontal movement and then is driven further by environmental forces. A second, though less common, mechanism is that of a bulldozing keel. In this case the keel is under the leading edge of the iceberg and remains in this relative position as the environmental forces continue to drive the berg forward. These mechanisms are shown in Figure 3.

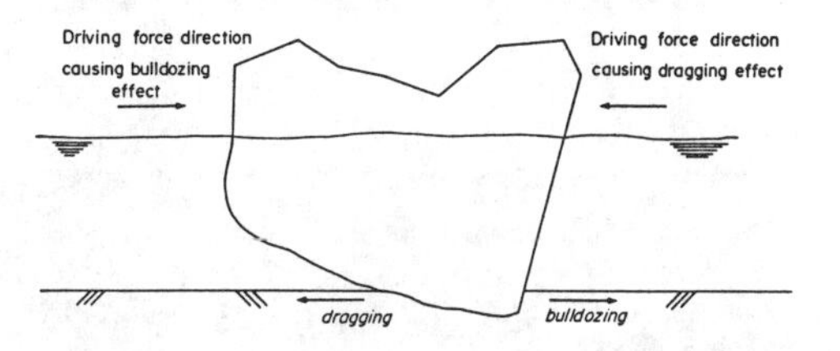

Figure 3. Ice-seabed failure mechanisms as determined by the direction of driving forces relative to keel shape.

The bulldozer mechanism is used in two published scour prediction models, by Chari (1979) and Fenco (1975), previously mentioned. The third mechanism is that of an iceberg which rolls and impacts the seabed, but remains grounded. When environmental loads become sufficiently large, as for example from storm waves or ice pack loading, the resultant inclined load produces a localized failure, called an iceberg pit, which is deeper than normal scours. An example of a 10m-deep pit on the Grand Banks is shown in Figure 4.

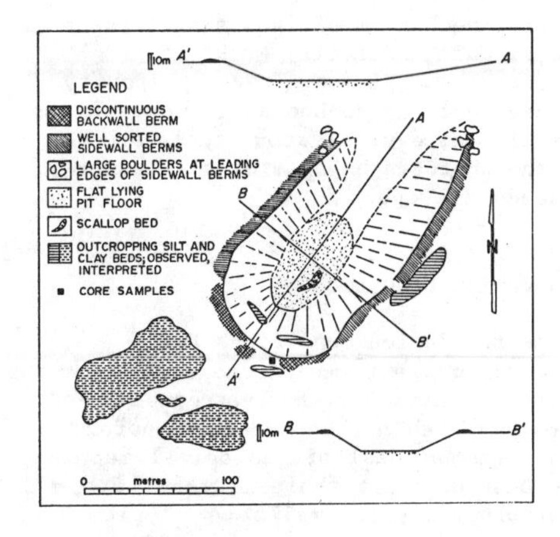

Figure 4. Schematic view of an iceberg pit based on submersible observations.

Figure 5. Examples of East Coast scours.

Figure 5A. Example of a typical heavily iceberg-scoured seabed on Saglek Bank, Labrador Shelf, at a depth of approximately 120m. Two 'chattermark' scours traverse this sidescan record, one (at left) resembles a tire track in morphology and probably represents scour by an oscillating berg whose keel remained in constant contact with the seabed. The second scour (just right of the first) is a 'crater chain' with coalesced and, in places, separated craters formed by an oscillating berg whose keel probably lifted free of the seabed in places. These two scours are superimposed over normal curvilinear scours and occasional single iceberg-generated pits.

Another mechanism of less significance is the so called "chatter mark" scour produced by an iceberg that apparently bounces off the seabed as it moves along. Examples of several scour types are shown in the photographs of side scan sonar mosaics of Figure 5.

Foundation Conditions

The foundation conditions for production platforms on the east coast, although variable, are less complex and more favourable than for the Beaufort Sea. A great deal of geological mapping has been done but there is very little geotechnical data available. Typically the geologic setting consists of an inner shelf of strong rock with very little sediment cover, then a central area consisting of a marginal trough up to 800m deep in places and an outer shelf of broad shallow bank tops parallelling the coastline (Brown 1986). With respect to foundation conditions for gravity based structures the most significant sediments are those of the Quaternary period. Brown (1986) provides an overview of the geological, geotechnical and environmental settings of the Labrador Shelf, the Grand Banks and the Scotian Shelf, drawing upon extensive marine geology studies carried out by scientists and engineers from the Atlantic Geoscience Centre. He draws attention to the fact that very little information is available for relatively large sections. The most extensive data in the public domain are for the Quaternary sediments of the Scotian Shelf.

Industry has focussed attention on

Figure 5B. Part of a sidescan record in the same vicinity and water depth as in Figure 5A. Here a normal scour is interrupted by a complex 'wallow pit', formed when the berg ceased to move forward. Slight adjustments whilst at rest formed the irregular pit outline before the berg moved forward again. Movement was from top left to bottom right.

determining foundation conditions for three prospective sites for a gravity based structure in the Hibernia field (Long 1986). This is the most extensive geotechnical data for the Grand Banks area published to date and it includes in situ test data as well as laboratory data. The Tertiary clays were shown to have an overconsolidation ratio of about two to six for the upper 50m, corresponding to a vertical loading of about 600 kPa in excess of present in situ vertical stress. The mechanism accounting for the over consolidation has not been determined. In summary, although there is a dearth of geotechnical properties data for potential production fields on the east coast, foundation conditions at the most probable sites are very good and material models will generally not be a major constraint in developing reliable foundation designs.

STANDARDS

There are at present no approved standards or codes for guiding the geotechnical aspects of investigation and design of offshore production platforms but such standards should be in place by 1988. The Canadian Standards Association (CSA) with support from the Canadian Petroleum Association (CPA) and the Canada Oil and Gas Lands Administration (COGLA) is currently developing standard S472 for foundations for offshore structures and standard S471 for environmental loads. The Committees comprising volunteers drawn from industry, government and university sectors have been working for approximately three years in developing the standards. The Foundations Committee has chosen to produce a performance standard, accompanied by a commentary to amplify or explain various clauses in the standard, rather than to produce a design code. The commentary includes recommended load and resistance factors but it is silent on reliability.

The advantage of a performance standard is that it retains a great deal of flexibility while still providing distinct guidelines and a framework for the design process. The disadvantage is that those responsible for regulatory matters have a more difficult job in that more judgement enters the selection of appropriate material and analytical models and in the prediction of the structural response to environmental and operational loads.

Given the very limited experience base, the sparce geotechnical data, the exceptionally high cost of acquiring such data, the limited information on environmental loadings and global ice strengths, it is inevitable and indeed proper that probability methods and decision theory enter the design and prediction process.

PROBABILITY AND ANALYSIS

The application of probability methods to the solution of engineering problems is well established (see for example Maes 1986) but has not been widely applied in geotechnical practice in Canada. Soil properties and loads can be characterized to produce load factors and resistance factors and exceedance probabilities can be specified to thus define the risk or probability of failure. The probability of whether or not a site will include a particular geologic feature that may result in a weak foundation soil, for example a thin weak layer, depends upon the geologic history. Whether or not that weak layer or other variations in the soil profile will be detected by a site investigation, depends upon the site investigation methods adopted, as does the probability that proper samples will be secured for testing. The probability of the tests properly reflecting the soil strength depends on a number of factors relating to the sampling, handling and testing procedures. A limited number of laboratory strength tests of high quality specimens may be of much greater value than a large amount of field data that may not be reliable such as standard penetration tests or vane tests.

Thus, applying probability theory to foundation design and performance prediction must be done in the context of geologic history, field investigation procedures, laboratory test methods and analytic methods. Soil variability which can lead to localized failures is less important than systematic errors which can lead to total failure. Probabilistic theory is particularly appropriate for estimating environmental loads from wind waves and ice.

An example is shown on Figure 6 where probability density function is plotted against load for (a) frequent loads and (b) infrequent loads. For frequent loads such as wave loading a typical specified load would be that of a 100 year return period and it would be multiplied by a factor to arrive at the design load. Infrequent loads are much different. The greatest probability density function may occur at zero - for example the annual probability of an iceberg impacting a gravity base structure on the Grand Banks. The curve is much flater and an appropriate load would be selected (say 1 in 10,000) for design and it would not be factored. Figure 7 illustrates probability density function plotted against resistance or strength. The design strength is determined by multiplying the peak resistance by a factor such that the actual resistance being less than the design value is equal to some selected small value shown here as Br (Maes 1986). Figure 8 shows a load curve A for frequent loads and a resistance curve B, with the shaded area representing the probability of exceedence of the load. If L represents the design load and R the resistance, the safety factor Fs can be conveniently determined by $Fs = L/R$. For exploration platforms a lower factor of safety would be accepted than for a production platform, simply on the basis of design life. However, as pointed out by Wu (1986), a given factor of safety, for example 1.3 has a different meaning when there is little uncertainty about resistance and load than when there is a great deal of uncertainty about either or both. Thus to be meaningful, a reliablity function is required with the factor of safety. The mean and the variance of the resistance are determined from the uncertainties associated with the foundation soil and the analysis and the mean and variance of the load are determined from the properties of the ice, wave statistics the analytic model. These in turn are used to determine the mean and variance of the factor of safety, which without such analyses, may be less accurate than required.

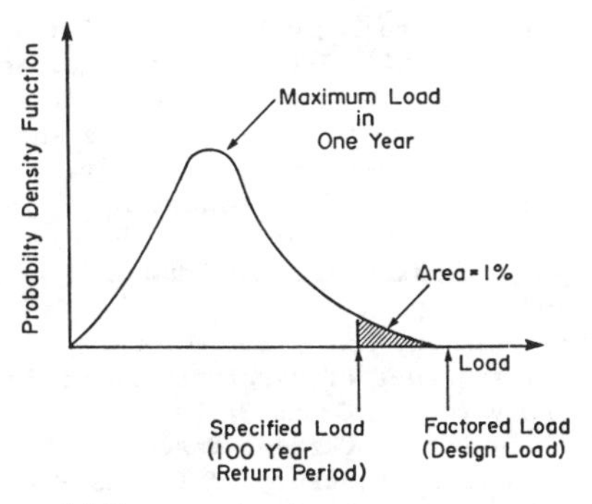

(a) Frequency Loads

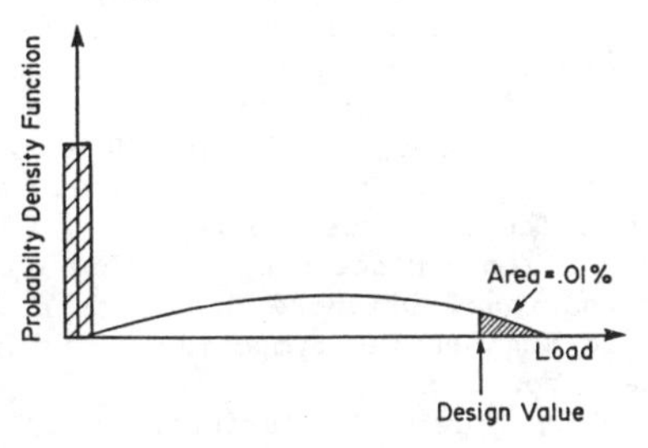

(b) Infrequent Loads

Figure 6. Probability density function vs. load.

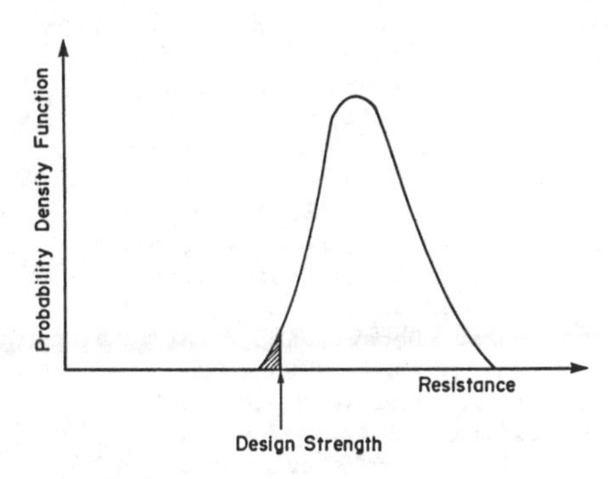

Figure 7. Probability density function vs. resistance.

CONCLUSIONS

1. The foundation conditions typically encountered in the Beaufort Sea are more complex and less favourable than those typically encountered at potential production sites on the East Coast of Canada.

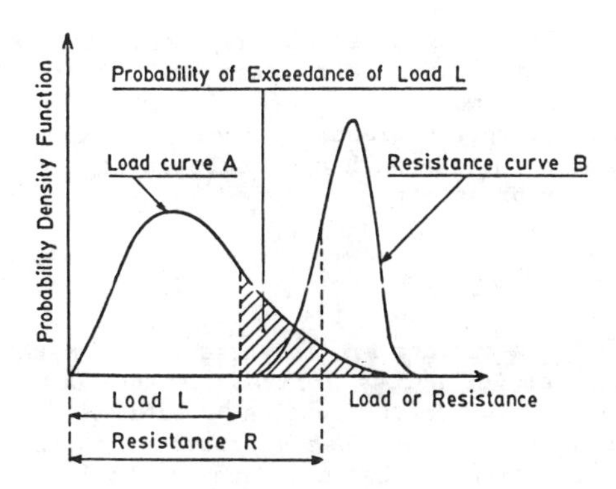

Figure 8. **Probability density function vs. load and resistance.**

2. The extensive geotechnical data base and the monitoring records from exploration platforms in the Beaufort Sea can reduce the uncertainty of material properties and results in greater reliability in computing factor of safety and in predicting performance than is currently possible for the geotechnical data base and performance records for potential East Coast production sites.

3. Ice loading may be the dominant design load for structures in both East Coast and Arctic environments. Uncertainties are related to ice properties and global ice loads. Prediction of wave loads has a higher reliability than ice loads.

4. Ice impacting the seabed is a frequent occurrence over much of the Arctic offshore region as well as the east coast. Predictive models in the public domain assume the ice to be very strong in relation to the soil. An analytical model incorporating ice strength in relation to soil strength is required but it would suffer the same limitation with respect to strength as ice loads on structures.

5. Probability methods can be used to assess reliability in mapping of features such as a layer of clay being present; the site investigation; the strength or resistance of the material; the loads to which a structure will be subjected and the analytical models.

6. Global factors of safety or factors of safety based on load factors and resistance factors are more meaningful if

accompanied by a reliability analysis. A structure with a calculated lower factor of safety may have a safer design if it has a larger reliability index than the same structure with a calculated higher factor of safety.

ACKNOWLEDGEMENTS

The authors are indebted to Dr. Mark Maes of Det norske Veritas (Canada) Ltd. for helpful discussions on probability methods. Mr. Jørn Landva assisted in the preparation of the paper and Mr. Chris Woodworth-Lynas provided the side scan sonar pictures.

REFERENCES

Barnes, P.W., Reimnitz, E., Hunter, R.E., Phillips, R.L., and Wolf, S. (Eds.), 1983. Geologic processes and hazards of the Beaufort and Chukchi Sea Shelf and Coastal Regions. U.S. Dept. Commerce, NOAA, OCSEAP Final Report 34 (1985), pp. 223-250.

Chari, T.R. 1979. Geotechnical aspects of iceberg scours on ocean floors. Canadian Geotechnical Journal.

Christian, H.A. and Morgenstern, N.R. 1986. Compressibility and Stress History of Holocene Sediments in the Canadian Beaufort Sea. Proceedings Third Canadain Conference on Marine Geotechnical Engineering, St. John's. Newfoundland, pp. 275-300.

Clark, J.I., Landva, J., Collins, W.T., and Barrie, J.V. 1986. Geotechnical aspects of seabed pits in the Grand Banks area. Proceedings Third Canadian Conference on Marine Geotechnical Engineering, St. John's, Nfld., pp. 431-455.

Clark, J.I., and Landva, J. 1986. The Controlling Factors of Ice Created Seabed Features Related to Production Systems in Canadian Cold Oceans. Proceedings Polartech '86, October 27-30, 1986, Helsinki, Finland, pp. 565-582.

Croasdale, K.R. 1985. Arctic fixed platforms for water depths greater than 50 metres. Proceedings, 1st Spilhaus Symposium: Arctic Ocean Engineering for the 21st Century, Williamsburg, VA, pp. 169-180.

Crooks, J.H.A., Jeffries, M.G., Becker, D.E., and Been, K. 1986. Geotechnical properties of Beaufort Sea clays. Proceedings, Third Canadian Conference on Marine Geotechnical Engineering, St. John's, Newfoundland, pp. 329-346.

Diemand, D. 1984. Iceberg temperatures in the North Atlantic-theoretical and measured. Cold Regions Science and Technology, Vol. 9, pp. 171-178.

Fact Sheet 1986 - The Norwegian Continental Shelf Published by the Royal Ministry of Petroleum and Energy, Norway, 98p.

Fader, G.B. and King, L.H. 1981. A reconnaissance study of the surficial geology of the Grand Banks of Newfoundland. Current Research, Part A. Geological Survey of Canada Paper 81-1A, pp. 45-56.

FENCO, 1975. An analytical study of ice scour on the sea bottom. Edited by R. Pilkington and H. Iyer. APOA Project 69.

Goff, R.D., Thomas, G.A.N., and Maddock, W. 1987. Applications of spray ice and rubble ice for Arctic offshore exploration. Proceedings of the Sixth International Offshore Mechanics and Arctic Engineering Symposium, Houston, Texas, pp. 1-8.

Hill, P.R., Moran, K., Kurfurst, P.J., Pullan, S. 1986. Physical and sedimentological properties of Nearshore Sediments in the Southern Beaufort Sea. Proceedings Third Canadian Conference on Marine Geotechnical Engineering, St. John's, Newfoundland, pp.301-328.

Hnatiuk, J. and Wright, B.D. 1983. Sea bottom scouring in the Canadian Beaufort Sea. Proceedings, 15th Offshore Technology Conference, Houston, Texas. OTC paper 4584, pp. 35-40.

Lachance, J. 1985. "Etude du comportement fragile de la glace d'iceberg et de la glace columnaire d'eau douce" M.Sc. thesis, Civil Engineering Department, Laval University, Quebec City.

Lewis, C.F.M. and Barrie, J.V. 1981. Geological evidence of iceberg grounding and related seafloor processes in the Hibernia discovery area of Grand Bank, Newfoundland. Proceedings, Symposium on Production and Transportation Systems for the Hibernia Discovery, Newfoundland Petroleum Directorate, St. John's, Nfld., pp. 146-177.

Long, L.G., Thompson, G.R., Brown, J.D., and Rivette, C.A. 1986. Hibernia site geotechnical characterization, Proceedings Third Canadian Conference on Marine Geotechnical Engineering, St. John's, Newfoundland, pp. 99-116.

Maes, M.A. (1986). Calibration of

Partial Factors in the New CSA Code for Fixed Offshore Production Structures. Environmental Protection Branch, Canada Oil and Gas Lands Administration, Technical Report No. 9.

Mitchell, D.E. 1984. Liquefaction slides in hydraulically placed sands. Proc. of 4th Int'l Symp. on Landslides, Toronto, Ontario.

Nadreau, J-P. 1987. Survey of physical and mechanical properties of icebergs. Presented at NRC Workshop on Extreme Features, November 3-5, 1986, Banff, Alberta.

Offshore Engineer. "Valhalla is sinking too". December, p.5.

Rogers, B.T., Hardy, M.D., Neth, V.W. and Metge, M. 1986. Performance modelling of the Molikpaq while deployed at Tarsiut P-45. Proceedings Third Canadian Conference on Marine Geotechnical Engineering, St. John's, Newfoundland, pp. 363-384.

Sladen, J.A., D'Hollander, R.D., Krahn, J. and Mitchell, D.E. 1985. Back analysis of the Nerlerk berm liquefaction slides. Canadian Geotechnical Journal, Vol. 22, No. 4, pp. 579-588.

Wu, T.H., Kjekstad, O. and Lee, I. 1986. Reliability analysis of foundation stability for gravity platforms in the North Sea. Proceedings Third Canadian Conference on Marine Geotechnical Engineering, St. John's, Newfoundland pp. 165-180.

Prediction and Performance in Geotechnical Engineering / Calgary / 17-19 June 1987

Some observations on the stability of structures founded on soft clays

J.H.A.Crooks
Golder Associates, Calgary, Alberta, Canada

ABSTRACT: Well established procedures have been developed for the evaluation of the stability of fills founded on soft clays. Given that many of these structures are designed with low factors of safety (< 1.5) it appears that stability procedures are reasonably successful. However, examination of the assumptions and input data used in these analyses indicates significant uncertainties related to the selection of undrained strength for total stress analyses, prediction of porewater pressure response for effective stress analyses, strengths mobilized in the crust and fill materials, and the failure mode. Because of these uncertainties, continued assessment of stability analysis procedures involving soft clays is required.

1 INTRODUCTION

The accurate prediction of the performance of structures founded on soft soils requires the selection of the appropriate soil properties for analysis. This is true regardless of the degree of sophistication of both the soil model and the method of analysis. In the absence of a comprehensive stress-strain-strength model for soils, problems involving stability and deformation are normally dealt with separately with relevant soil properties being measured in laboratory and/or field tests. Because of the differences between the boundary conditions associated with the various tests, a range of values for a single property is typically produced together with an associated range in predicted performance. Therefore, it is common practice to rely on "calibration" of a prediction method based on a particular soil property measured in a specific test, with measured field performance. The result of this process is often the development of empirical correction factors which are applied either to the input soil properties or to the results of the analyses. While this approach has some advantages (e.g. it provides a framework for quantifying experience with specific materials and classes of problems) it tends to deflect attention from the examination of actual soil behaviour. Thus empirical correction factors based on prediction - observation correlations may be limited in their application and these limitations are not always well understood. The purpose of this Paper is to examine some of the uncertainties and limitations associated with analyses to predict the stability of structures founded on soft clays.

2 TOTAL STRESS ANALYSES AND UNDRAINED STRENGTH

2.1 Background

Bishop and Bjerrum (1960) presented the results of analyses of various failures involving soft clays based on standard analytical techniques and undrained strength data obtained using a variety of test types. Their results are summarized on Figure 1 and suggest that regardless of method of strength measurement, accurate assessments of failure loads could be made. Thus the problem of predicting the stability of structures founded on clays appeared to have been solved.

Further analyses of failures by Bjerrum (1973) based on field vane strengths alone demonstrated that this was not the case. Computed factors of safety were generally greater than unity with the difference

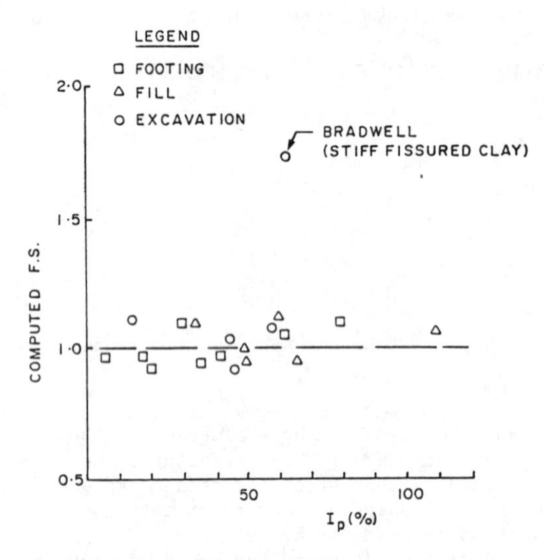

Figure 1 Computed Factors of Safety for Field Failures (Bishop and Bjerrum, 1960)

increasing with increasing plasticity (Figure 2). Bjerrum explained this difference predominantly based on the effects of strain rate and anisotropy. He proposed that vane strengths be corrected by an empirical factor to produce the "correct" factor of safety. Although this proposal has been widely adopted in practice, its validity has been frequently disputed. Other case records have been published which illustrate a much wider scatter in computed factors of safety at failure than was indicated by Bjerrum (Figure 2).

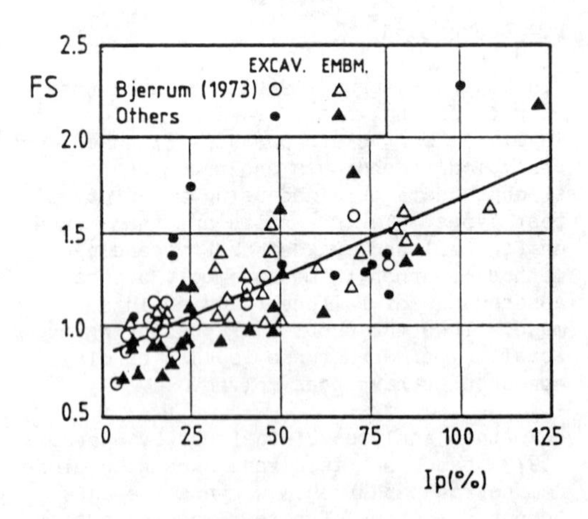

Figure 2 Computed Factors of Safety for Field Failures (Bjerrum, 1973; Aas et al, 1986).

There is ample evidence that the undrained strength of clays is dependent on the rate and direction of shearing. Graham et al (1983) report the results of a variety of laboratory strength tests at varying strain rates on different clays. Their data (Figure 3) indicate that a tenfold change in strain rate results in a strength change of typically 10-15% regardless of test type. The data also clearly indicate that the effect of strain rate does not increase with increasing plasticity. Assuming that the same trend would apply to the field vane test, it appears that one of the basic reasons for Bjerrum's correction factor (i.e. strain rate effect) is not valid.

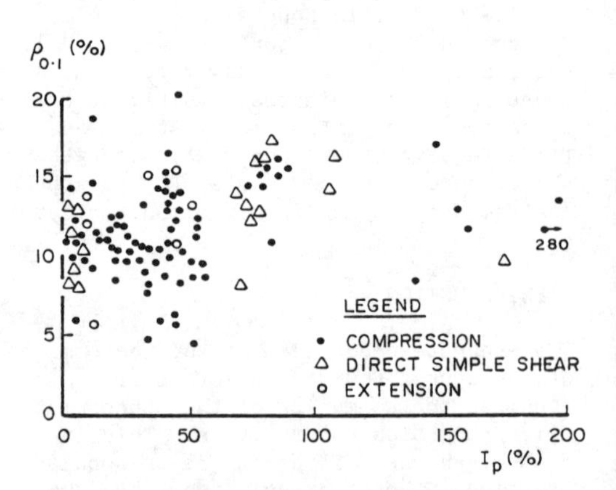

Figure 3. Effect of Strain Rate on the Undrained Strength of Clays (Graham et al, 1983)

Laboratory test data (Larrson, 1980) also clearly demonstrate the anisotropic nature of the undrained strength of clays (Figure 4). However, the degree of strength anisotropy decreases with increasing plasticity and obviously cannot explain the difference between actual and computed factors of safety reported by Bjerrum.

2.2 Interpretation of Undrained Strength in Terms of State

Like other aspects of the behaviour of soft clays, measured undrained strength depends on the boundary conditions associated with the test. These vary widely and as a result, measured undrained strengths vary.

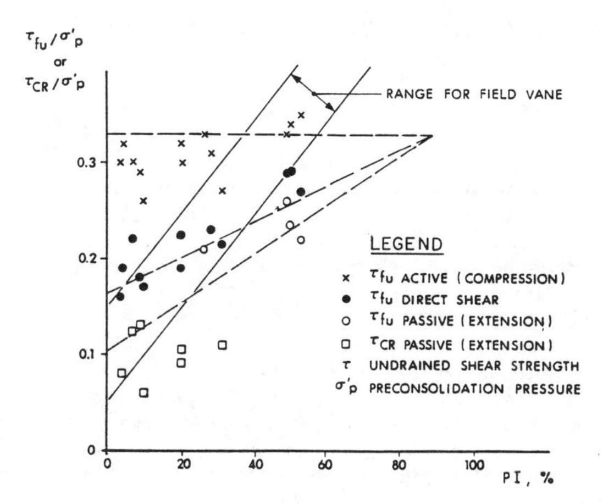

Figure 4 Undrained Strength Anisotropy in Clays (Larrson, 1980)

It is well appreciated that undrained strength is largely dependent on the state of the clay which is defined in terms of its void ratio and current stresses. State can be represented by a point in void ratio - log stress space and is quantified by reference to a behaviour which can also be represented as a line in void ratio - log stress space. The virgin consolidation line (VCL) which describes a unique behaviour of clay (i.e. the conditions following primary consolidation), is traditionally used for this purpose. Thus, current state is quantified in terms of the stress difference between the current state and the state on the VCL following loading (i.e. its over-consolidation ratio). Good correlation between behavioural properties of clays and OCR has been demonstrated by a number of researchers and forms the basis of the SHANSEP approach proposed by Ladd and Foott (1974). The normalized strength relationships shown on Figure 4 are also an indirect reflection of strength dependence on state. The critical state soil model is based on a similar concept except that the critical state line is used as a reference. Been and Jefferies (1985) have developed a similar concept for characterizing sand behaviour using the steady state line as a reference to quantify the current state of sand.

It is noted that in describing the state of a clay, void ratio can be represented by the yield stress and the definition of the in situ stress state includes both vertical and horizontal in situ effective stresses (i.e. σ'_{vo} and σ'_{ho}). OCR is commonly defined in terms of the vertical yield and effective stresses (OCRV = $\sigma'_{vy}/\sigma'_{vo}$).

Because of the boundary conditions and stress paths associated with various strength tests, it is preferable that over-consolidation be defined based on a more global description of current and yield stresses in the clay. A useful, albeit simplified, definition for over-consolidation is:

$$OCRI = (\sigma'_{vy} + 2\sigma'_{hy})/(\sigma'_{vo} + 2\sigma'_{ho})$$

Becker et al (1987) presented an interpretation of the field vane strength based on the state of the clay. Their interpretation is shown on Figure 5 in terms of OCRI. It is clearly evident that correlating field vane strength with the state of the material defined in terms of mean stresses provides a consistent interpretation of strength.

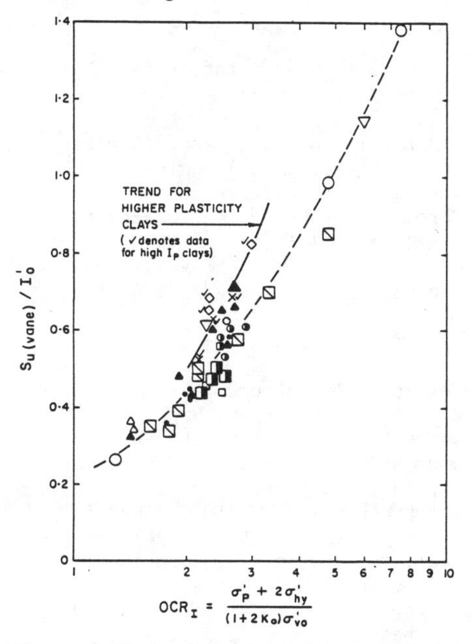

Figure 5 State Interpretation of Field Vane Strength (Becker et al, 1987)

A similar approach is being developed for cone penetration tests and the results available to date are summarized on Figure 6 in terms of OCRV. (At this stage insufficient data are available to summarize CPT data in terms of OCRI. Based on the field vane interpretation, less scatter might be expected for this relationship). Again, it is evident that normalized cone tip resistance correlates well with the state of the clay.

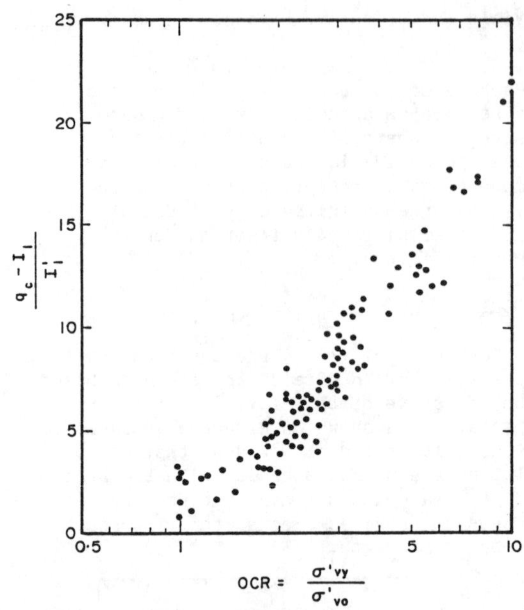

Figure 6 State Interpretation of CPT Tip Resistance

The trends in the field vane strength and cone tip resistance - OCR relationships are similar (Figures 5 and 6). Therefore a constant factor relates the two quantities. This is in stark contrast to the wide variation in N_k values reported in the literature where N_k is defined as:

$$S_u = (q_c - \sigma_{vo})/N_k$$

For example (De Ruiter, 1982) reports values of N_k which vary between 10 and 20. Except for the expectation that N_k increases with OCR there is little quantitative guidance available as to the choice of the appropriate N_k factor. Therefore unless a regional data base developed from field case records is available, this situation precludes the effective use of the CPT for determination of undrained strength based on conventional interpretation.

2.3 The Significance of K_0 in Undrained Strength Interpretation

Since the trends in the relationships between q_c and Su_v with OCR are the same, it is apparent that N_k is not dependent on over-consolidation. Instead, it is the inclusion of the in situ horizontal stress as a normalizing parameter which permits a consistent correlation between Su_v and q_c to be developed. It is noted that the data base for both the field vane and cone tip resistance include data from clay deposits which exhibit a wide range of K_o values including values as high as 3.

While it is appreciated that some uncertainties exist in the measurement of K_o, it must also be accepted that simply because it is difficult to achieve this measurement does not mean that it can be ignored (Jefferies et al, 1987).

Dependence of undrained strength on in situ stress conditions does not only apply to field tests. It is well appreciated that the results of laboratory shear tests also depend on the stresses to which test specimens are reconsolidated. Since a laboratory test is usually intended to represent an element of the soil in the field, it is generally accepted that the sample should be reconsolidated to its in situ stress state; this requires knowledge of both the in situ vertical and horizontal effective stresses (i.e. σ'_{vo} and K_o). The influence of consolidation stress history on measured undrained strength in compression is illustrated on Figure 7.

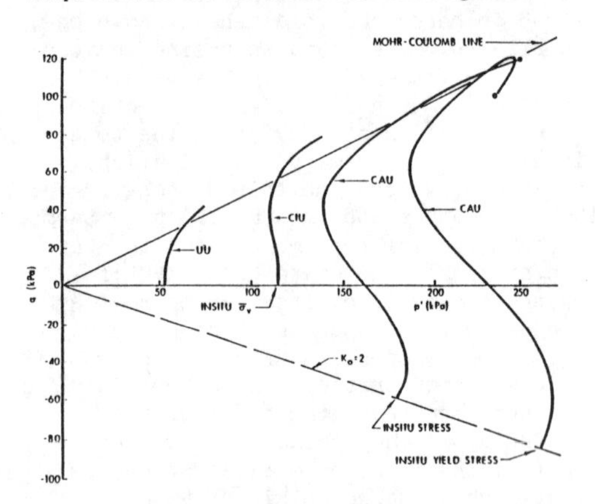

Figure 7 Dependence of Undrained Strength on Consolidation Stresses (Jefferies et al, 1985)

The conventional view of K_o is that this quantity is a function of the degree of over-consolidation of the clay and its friction angle (Ø'). Laboratory tests by Brooker and Ireland (1965) produced the widely accepted correlation between K_o, OCR and IP in which the latter is taken as a direct reflection of Ø'. However, stress history alone (i.e. OCR) is not always an adequate description of the geological processes which can affect the in situ state of stress. Jefferies et al (1987) presented data for lightly overconsolidated Beaufort Shelf clays in which extremely high K_o values were measured. They attribute the lack of correspondence between these findings and conventional expectations to geological processes which occurred during deposition. Similarly high $K_o$values were measured in the Lake Edmonton clays by Chan and Morgenstern (1986) and were attributed to post depositional processes. While these examples indicate that K_o measurements should be made routinely, it is reasonable to expect that the conventional expectation of K_o will not be seriously in error for those deposits which have not been subjected to significant geological processes other than stress history. For the purposes of discussion in this paper, then, it is assumed that the somewhat simplistic Brooker and Ireland correlation is appropriate. Thus it is assumed that the value of K_o will increase with both increasing OCR and plasticity.

2.4 Yield Envelope Interpretation of Undrained Strength

The variation in undrained strength depending on whether the clay is sheared in compression or in extension (Figure 4) can also be related to in situ stress conditions. It is generally accepted that the yield envelope for a natural clay will be approximately symmetrical about the K_o axis in s'- t space. Thus for highly plastic clays where K_o values of about 1 can be expected, the yield envelope in s'- t stress space will be approximately symmetrical about the t = 0 axis as shown on Figure 8(a). Typical effective stress paths for compression and extension tests indicate that equal strength values will be measured in extension and compression. On the other hand, for low values of K_o (i.e. low plasticity clays), the strengths in compression and extension vary significantly as shown on Figure 8(b). Thus for low plasticity clays (i.e. low K_o) stress induced strength anisotropy is significant; for high plasticity clays (i.e. high K_o) strength anisotropy is much less. This is the general trend observed on Figure 4.

2.5 Comparison of Field Vane Strength and Strength Mobilized During Field Failures

The above sections indicate the importance of in situ stresses on the magnitude of the undrained strength measured in various test types. The field vane test in particular is strongly influenced by in situ stresses. For example, it is possible to attribute the general trend of increasing field vane strengths (normalized with respect to the vertical yield stress) with increasing plasticity (Figure 4) to the fact that K_o increases with plasticity index.

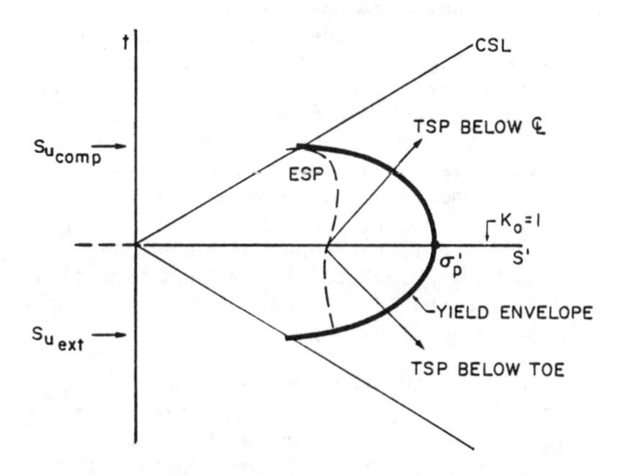

(a) High K_o, High IP

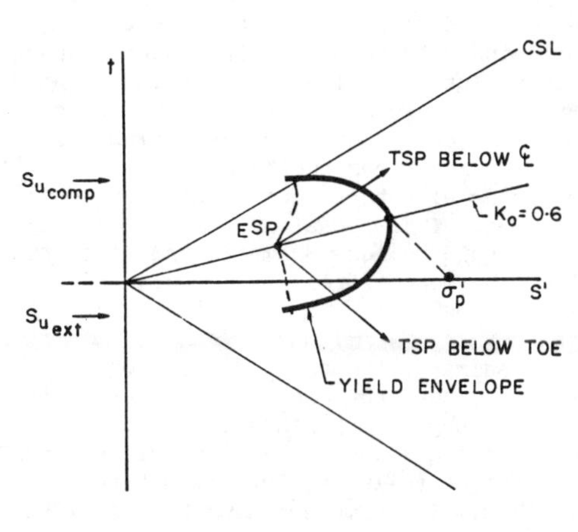

(b) Low K_o, Low IP

Figure 8 Undrained Strength Related to In Situ and Yield Stresses

To some extent, strength values back-calculated from field failures will also depend on the in situ stress state; this important aspect does not receive attention in traditional limit equilibrium analyses. For example, consider the total stress paths below the centre of fill and beyond the toe of a fill, locations where the principal stress directions remain approximately constant during loading. For these conditions, available laboratory data describing the yield envelope of clays are appropriate. Typical effective stress paths for these locations are shown on Figure 8 for both high K_o and low K_o conditions. Now the back-calculated undrained strength from field failure (Su_F) is directly related to the surface loading which can be applied. Inspection of the length of the total stress paths for the high K_o and low K_o cases (Figure 8) indicates that more load can be imposed in the former case than in the latter case before yielding occurs. Again, assuming that K_o is a function of plasticity, a higher back-calculated Su_F value will derive from failures involving plastic clays than is the case for lean clays. Thus back-calculated Su_F values from field failures should increase with increasing plasticity (i.e. K_o).

However, the dependence of back-calculated Su_F values on plasticity is likely not as great as is the case for field vane strengths. For example, the illustration of dependence of Su_F on K_o for fills (Figure 8) is based on an appreciation of yield/strength behaviour under conditions where the in situ major principal stress directions do not change during loading. There is little, if any, available information related to yield/strength behaviour where the principal stress directions during loading do not coincide with those existing prior to loading. It is possible that the situation would be less dramatic than inferred from the more extreme cases shown on Figure 8.

In summary, the above discussion suggests that the expectation that field vane strength (Su_v) should directly reflect the strength back-calculated from embankment failures (Su_F) is not well founded. It is more reasonable to expect that the two quantities should differ not because of strain rate effects but because of the different significance of in situ stresses with respect to each quantity. Further, it can be speculated that the wide scatter exhibited (Figure 2) is probably the result of K_o having different values than would be conventionally expected.

It is therefore concluded that unless the influence of the in situ stress state is included in the comparison of strength values measured in different test types and computed from back-analyses of field failures, there is little prospect of achieving a more rational appreciation of the appropriate strength for use in total stress stability analysis. Comparison of undrained strength values obtained from various test types and those back-calculated from field failures should be based on normalizing the strength data to account for the state of the clay as shown on Figure 5.

Unfortunately, the required data base is not available to facilitate such comparisons to be made at this time; information related to horizontal in situ effective and yield stresses are generally missing. However, the potential success of this approach can be anticipated based on the interpretation of strength proposed by Becker et al (1987). They recognized that the field vane strength is largely controlled by the horizontal yield stress (σ'_{hy}) in the clay and normalized Su_v with respect to σ'_{hy}. The resulting correlation with plasticity index shows a dramatically decreased dependence between the two quantities (Figure 9). Also shown on Figure 9 is the correlation between Su_F/σ'_{hy} and plasticity index; there is obviously much better agreement between the normalized Su_v and Su_F values than is the case if the vertical yield stress is used to normalize the strength values. This agreement should only be treated as a qualitative indication that strength values should be expressed in terms of the state of the clay including in situ stresses; it should not be viewed as a basis for a new generation of correction factors.

Finally, it should be appreciated that even if strength values from various test types and field failures are shown to be more consistently related based on state interpretation, it should not be expected that a unique expression describing the undrained strength of clays will emerge. This is because the boundary conditions for the various test types are different and the constitutive behaviour of various clay types will be different. Further, the stress paths imposed by different field loading geometries (e.g. fills, footings and cuts) also vary; therefore the response of the clay will vary. This is the inevitable result of attempting to characterize the complex behaviour of clay under varying loading conditions using a single (strength) parameter.

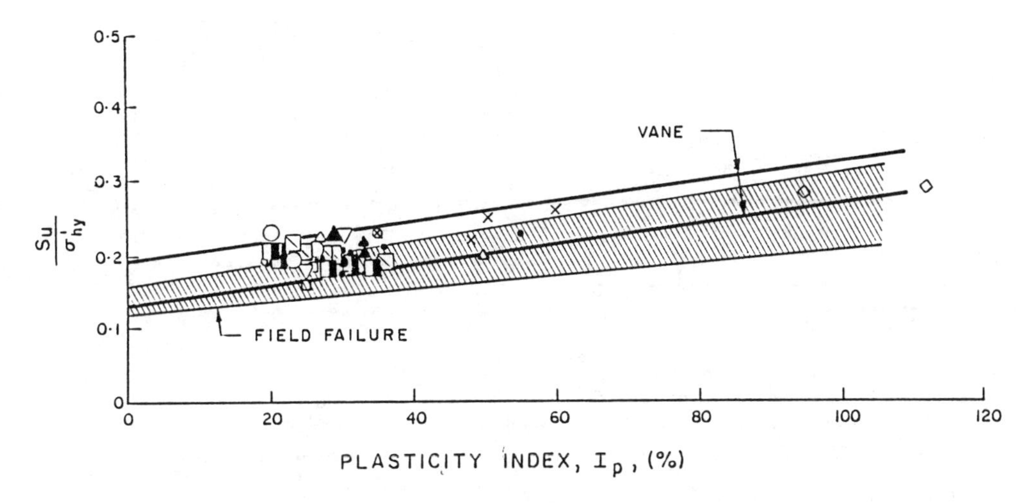

Figure 9 Field Vane and Field Failure Normalized Using Horizontal Yield Stress (Becker et al, 1987)

Nevertheless, it is anticipated that strength values expressed in terms of the state of the clay will be more consistent than is currently the case and therefore more consistent trends between these values will be evident. This will form the basis for developing more rational empirical factors which account for field loading geometry, variation in the constitutive behaviour of natural clays and varying boundary conditions between tests.

3 EFFECTIVE STRESS ANALYSIS AND POREWATER PRESSURE RESPONSE

Effective stress stability analysis is, of course, a more desirable approach to stability evaluation since the behaviour of any soil (e.g. deformation and stability) is controlled by the prevailing effective stresses. However, the major problem in the use of effective stress analysis is the accurate prediction of excess porewater pressures caused by surface loading.

Given the large number of factors which can affect porewater pressure response, it is hardly surprising that no single constitutive model can provide universally accurate porewater pressure predictions. To illustrate this, various case records are reviewed using the procedure developed by Folkes and Crooks (1985) for examining the effective stress path (ESP) behaviour in foundation clays below fills based on measured porewater pressures and computed total stresses. The shapes of the resulting effective stress paths reflect the relationship between current effective stresses and the yield envelope (YE) for the foundation clay. Yield behaviour, critical stressing and strain softening are clearly indicated by the computed ESP shapes. Although the ESP/YE approach was developed for analysis of monitored field behaviour, sufficient cases have now been examined to permit general trends in behaviour to be established. Some typical responses are summarized on Figures 10 and 11 and are discussed below.

The ESP associated with loading when the imposed shear stress is less than the shear strength can vary significantly as indicated on Figure 10. Clays with a high modulus will generate only modest excess porewater pressure during initial loading and associated ESPs will be as indicated by ESP A on Figure 10. On the other hand low modulus soils will exhibit significant porewater pressures and lie to the left of the range (i.e. ESP B) shown on Figure 10. The rate of loading, boundary drainage conditions and permeability of the foundation clay also significantly affect the ESP location on initial loading.

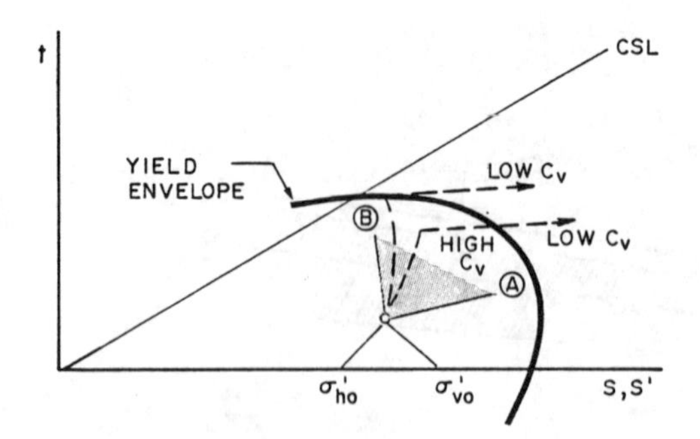

Figure 10 Effective Stress Paths During Initial Loading and Consolidation (Folkes and Crooks, 1985)

The response of a clay to loading in cases where the imposed shear stress reaches its shear strength depends on its stress-strain behaviour. ESPs computed for case records involving strain softening, plastic and dilatant soils are shown on Figure 11. In the strain softening case (Figure 11(a)), it is assumed that the vertical total stress remains constant and that the ESP moves down along the critical state line after the peak shear strength is reached. The final effective stress state is indicated by fitting the measured excess porewater pressure between these two lines. The excess porewater pressure generated by strain softening is very large as is the increase in the horizontal total stress and associated lateral deformation. Plastic behaviour is similar except that the effective stress state where the ESP first reaches the critical state line does not change; as a result, the total stress path moves to the right (Figure 11(b)) by an amount equal to the excess porewater pressure and there is an associated increase in horizontal total stress. For dilatant soils (Figure 11(c)), the ESP tracks up along the critical state line and of course, is associated with much lower porewater pressure response.

Accurate prediction of the consolidation behaviour of foundation clays is important not only for computing settlements but also for determining strength gain for staged loading and for preloading operations. In cases where the imposed shear stress does not equal the shear strength of the soil, the final effective stress state following loading will lie within the yield envelope. Dissipation will cause the ESP to move to the right in t - s' space (Figure 10). As drainage takes place, associated changes in Poisson's ratio will allow the ESP to move upward as well as to the right. The rate of consolidation will be relatively fast while the effective stress state lies within the yield envelope; however, a large decrease in C_v can occur in some materials when the ESP crosses the yield envelope. This is also observed in laboratory oedometer tests.

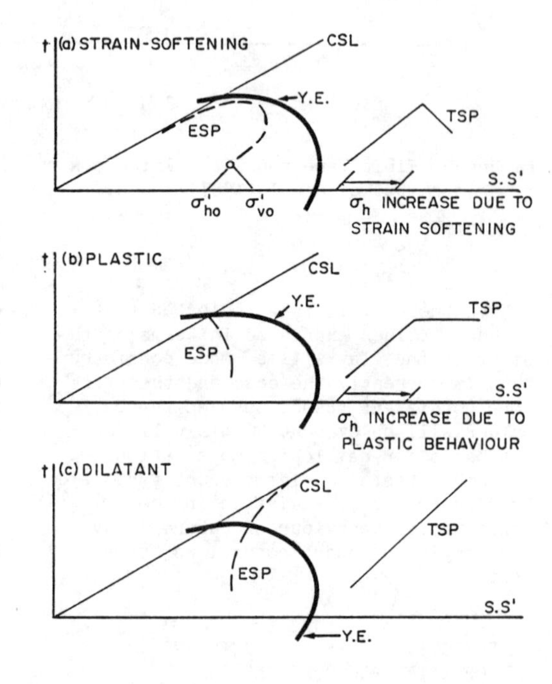

Figure 11 Behaviour of Critically Stressed Clays

For conditions where the imposed shear stress equals the shear strength of the clay, whether or not strain softening occurs, dissipation of excess porewater pressures occurs very slowly (Figure 10). Depending on the material type (e.g. sensitive soils), reduction in porewater pressure can be effectively offset by continued generation of porewater pressure due to breakdown/creep of the soil fabric. In these cases, the net effect is that there is little or no decrease in excess porewater pressure with time.

Depending on the geometry of the problem, it is also possible for excess porewater pressures to continue to increase even following completion of loading. This occurs when a significant portion of the foundation clay is critically stressed (i.e. imposed shear stress equals available shearing resistance). Notable examples of this behaviour have been reported by Crooks et al, (1984) and Becker et al, (1985 and 1986) indicating that this is not an uncommon occurrence. They attribute the continued rise in excess porewater pressure following construction to time dependent development of increased total horizontal stress. Although the final magnitude of excess porewater pressure can be predicted based on ESP/YE analysis, the time for the maximum value to develop cannot be predicted at this time. It should be appreciated that for non-strain softening soils, an increase in excess porewater pressure following completion of construction does not imply a reduction in available shearing resistance. In fact, the effective stress state (and therefore available shearing resistance) remains constant; it is the total stress and excess porewater pressure which change. Therefore rising porewater pressures following construction do not necessarily herald inevitable failure; they do however, indicate critical stressing of a significant portion of the foundation clay.

4 FAILURE MECHANISMS

In the preceding sections, the issue of evaluating the shearing resistance of foundation clays for stability analysis has been discussed. While this is obviously important, it is not the only uncertainty in evaluating the stablity of fills on soft soils. Other factors are also important as discussed below.

It is often the case in limiting equilibrium analysis that the failure surface is assumed to be circular. While this assumption is attractive from the viewpoint of the kinematics of the problem and is probably reasonable in some cases, it is not universally correct. For example Leonards (1982) presents very convincing arguments that for the specific cases he examined, failure was controlled by weak layers in the foundation material. Milligan (1972) made a similar statement. Clearly the results of analyses which do not recognize the controlling effect of weak zones or layers must be suspect. This is true in spite of the perennial difficulty of detecting and estimating the strength of weak layers together with the difficulty in modelling the more complex associated failure mechanisms.

It is noted that in most clay deposits, the strength is not constant with depth even if there are no defined weak layers present. In most cases the strength is high in the weathered surface crust and decreases to a minimum at the top of the unweathered clay before increasing again with depth. Therefore in many deposits there is a zone of lower strength clay immediately below the weathered crust. As is well appreciated, the rate of increase in strength with depth increases with increasing plasticity. Therefore the weak zone below the weathered crust is likely to be more pronounced in more highly plastic clays than in lower plasticity materials. As a result, the stability of fills on more plastic clays may be controlled by this zone and the failure surface may be more translational in these cases than rotational. It is noted that if translational-type failures are analyzed using circular failure surfaces, the computed factor of safety will be over-estimated. This may be a further reason (i.e. incorrect assumption regarding the nature of the failure surface) for Bjerrum's finding that the computed factor of safety at failure increases with increasing plasticity.

The plasticity of the foundation clay is not the only factor in controlling the location of the failure surface. For example, the thickness of the high strength crust and the fill in relation to the foundation stratigraphy will also be important. The influence of the fill/ foundation stratigraphy on failure loads was illustrated by Crooks et al (1986) and their data is summarized on Figure 12. In summary, for those situations where the fill thickness is large in relation to the thickness of the crust, higher stability numbers (N~ 8-12) are computed based on the (minimum) strength immediately below the crust. Under these conditions, failures would tend to be more translational by nature than is the case where the crust to fill thickness ratio is large. In the latter case, the failures would tend to be rotational and are associated with stability numbers (based on minimum shear strength) of about 5-6 which is the conventional expectation. While there is scatter in the data summarized on Figure 12, the general trend suggests different failure mechanisms associated with different geometries.

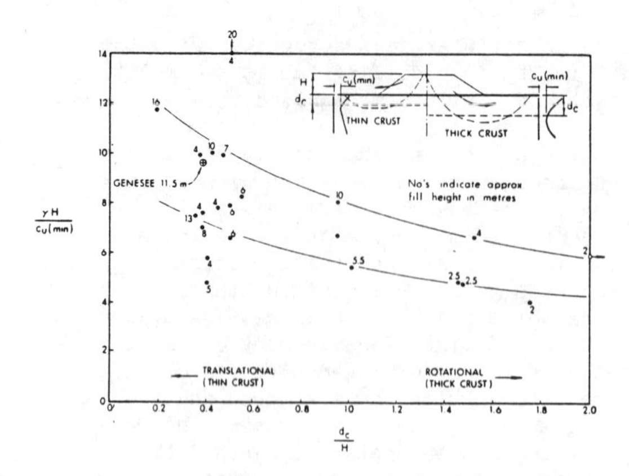

Figure 12 Influence of Fill and Foundation Stratigraphy Geometry on Failure Mode (Crooks et al, 1986)

In the back-analysis of failures, there is a tendency to assume that the limits of the failure surface which controls stability are coincident with the limits of disturbed ground (i.e. the hindmost scarp and the furthest extent of disturbance in front of the slope toe). However, it is often the case that secondary failures occur behind the main failure. Similarly, the main failure mass can affect the ground surface beyond the point where the failure surface exits; for example, where the surface crust is strong, it can be deformed as a result of simply being pushed by the failure mass. These possibilities should be taken into account in the analysis of failures since an over-estimate of the length of the failure surface will result in an over-estimate of the factor of safety.

Finally, it is noted that the selection of the shear strength parameters for the crust and fill materials has a significant influence on the results of stability analysis. The selection of these parameters is discussed by LaRochelle et al (1974). Although their recommendations for crust strength assessment is definitive for the case they examined, quantified guidelines for other cases are difficult to develop. In most cases, it is unlikely that the full strength of the fill and crust materials will be mobilized because pre-failure deformations will cause tension cracks to develop predominantly at the base of the fill where they are not obvious. Further, the degree to which fill/crust strength can be mobilized depends on the nature of the failure. For example, translational failures will tend to cause tearing of the crust and fill with low strength mobilization. Evaluation of a fill failure on Lake Edmonton clay where the pre-failure geometry and failure surface are well defined (Crooks et al, 1986) illustrates this point. The results of analyses of this failure are summarized on Figure 13 in terms of the combination of crust/fill strength and foundation clay strength required to provide a computed factor of safety of 1. It is noted that the average crust strength was about 100 kPa and the compacted clay fill strength was even higher. However, using an average crust/fill strength of only 40 kPa and a measured average strength of 20 kPa in the foundation clay would provide a computed factor of safety of unity. Regardless of the actual strength of the foundation clay, the strength mobilized in the crust and fill at failure is only a fraction of its shearing resistance in compression. This is attributed to the translational nature of the failure. For rotational types of failures, it is reasonable to expect that a higher proportion of the crust and fill strengths would be mobilized.

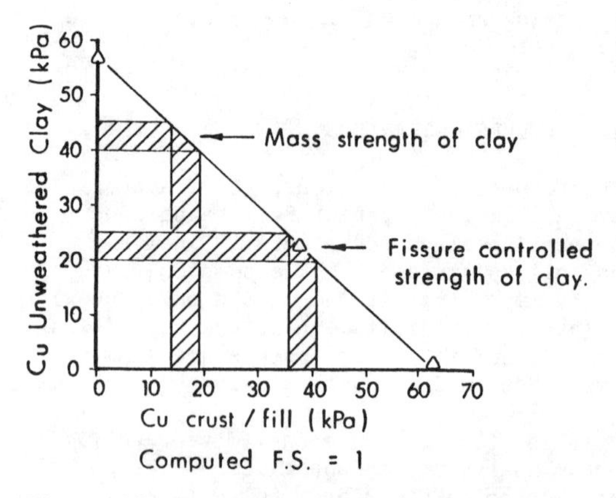

Figure 13 Total Stress Analysis of the Genesee Embankment Failure (Crooks et al, 1986).

5 CONCLUSIONS

Evaluation of the various factors which influence the analysis of stability of structures founded on soft clays indicates that there is considerable uncertainty in the selection of appropriate strength parameters and failure mechanisms. Thus despite the apparent success of current approaches to assessing stability, continued effort is required to evaluate uncertainties and to better understand the limitations of conventional limitations.

6 ACKNOWLEDGEMENTS

Many of the concepts described in this paper are the product of discussions with my colleagues over the past several years. Particular thanks are due to D.E. Becker, K. Been, M.G. Jefferies and R.M.C. Ng However, the opinions expressed in this paper should be attributed to the author alone. Thanks are also due to K. Koch, R. Kroeker and C. Gibson for their efforts in preparing this manuscript.

REFERENCES

Aas, G., Lacasse, S., Lunne, T and Hoeg, K. 1986. Use of in situ tests for foundation design on clay. Proc. of ASCE Conf. on Use of In Situ Tests in Geotechnical Engineering, Blacksburg, Virginia, June 1986, pp 1 - 30.

Becker, D.E., Jefferies, M.G., Shinde, S.B. and Crooks, J.H.A. 1985. Porewater pressures in clays below caisson islands Proc. ASCE Specialty Conf. on Civil Engineering in the Arctic Offshore, San Francisco, March 1985, pp 75 - 83.

Becker, D.E., Crooks, J.H.A., Rothenburg, L., Yam, C.C. and Bathurst, R. 1986. Behaviour of the second stage of the Gloucester test fill: prediction and performance. Submitted to NRC Workshop held in conjunction with the 39th Canadian Geotechnical Conference, Ottawa, August 1986.

Becker, D.E., Crooks, J.H.A. and Been, K. 1987. Interpretation of the field vane test in terms of in situ and yield stresses. ASTM International Symposium on Laboratory and Field Vane Shear Strength Testing, January 1987, Tampa, Florida.

Been, K. and Jefferies, M.G. 1985. A state parameter for sands. Geotechnique, 35, 2, pp 99-112.

Bishop, A.W. and Bjerrum, L. 1960. The relevance of the triaxial test to the solution of stability problems. Proc. of ASCE Research Conf. on Shear Strength of Cohesive Soils, Boulder, Colorado, June 1960, pp 437 - 501.

Bjerrum, L. 1973. Problems of soil mechanics and construction on soft clays. State of the art report Session IV, Proc. 8th Int. Conf. Soil Mech., Moscow Vol. 3, pp 111-159.

Brooker, E.W. and Ireland, H.O. 1965. Earth pressures at rest related to stress history. Canadian Geotechnical Journal, Vol. 2, pp 1-15.

Chan, A.C.Y. and Morgenstern, N.R. 1986. Measurement of lateral stress in a lacustrine clay deposit. Proceedings 39th Canadian Geotechnical Conference, Ottawa, August 1986, pp 285 - 290.

Crooks, J.H.A., Becker, D.E., Jefferies, M.G. and McKenzie, K. 1984. Yield behaviour and consolidation: Part 1. Pore pressure response. Proceedings of ASCE Symposium on Sedimendation Consolidation Models: Predictions and Validation, San Francisco, October 1984, pp 356-381.

Crooks, J.H.A., Been, K., Mickleborough, B.W. and Dean, J.P. 1986. An embankment failure on soft fissured clay. Canadian Geotechnical Journal, Vol. 23, No. 4, pp 528-540.

DeRuiter, J. 1982. The static cone penetration test - State of the art report. Proc. of the Second European Symposium on Penetration Testing, Amsterdam, May 1982, pp 389-405.

Folkes, D.J. and Crooks, J.H.A. 1985. Effective stress paths and yielding in soft clays below embankments. Canadian Geotechnical Journal, Vol. 22, No. 3, pp 375-374.

Graham, J., Crooks, J.H.A., and Bell, A.L. 1983. Time effects on the stress-strain behaviour of natural soft clays. Geotechnique, Vol. 33, No. 3, pp 327-340.

Jefferies, M.G., Ruffel, J.P., Crooks, J.H.A. and Hughes, J.M.O. 1985. Some aspects of the behaviour of Beaufort Sea clays. ASTM Special Technical Publication 883.

Jefferies, M.G., Crooks, J.H.A., Becker, D.E. and Hill, P.R. 1987. On the independence of geostatic stress from over-consolidation in some Beaufort Sea clays. Submitted to the Canadian Geotechnical Journal.

Ladd, C.C. and Foott, R. 1974. New design procedure for stability of soft clays. ASCE Journal of Geotechnical Engineering, Vol. 100, No. GT7, pp 763-786.

LaRochelle, P., Trak, B., Tavenas, F. and Roy, M. 1974. Failure of a test embankment on a sensitive Champlain clay deposit. Canadian Geotechnical Journal, Vol. 11, pp 142 - 164.

Larrson, R. 1980. Undrained shear strength in stability calculation of embankments and foundations on soft clays. Canadian Geotechnical Journal, Vol.17, pp 591-602.

Leonards, G.A. 1982. Investigation of failures. ASCE Journal of Geotechnical Engineering, Vol. 108, pp 185 - 246.

Milligan, V. 1972. Discussion on 'Embankments on Soft Ground', Proc. of ASCE Specialty Conference on Earth and Earth Supported Structures, Vol. 3, pp 41 - 48.

Prediction and Performance in Geotechnical Engineering / Calgary / 17-19 June 1987

Seismic response analysis: Prediction and performance

W.D.Liam Finn, M.Yogendrakumar & A.Nichols
Department of Civil Engineering, University of British Columbia, Vancouver, Canada

ABSTRACT: Procedures for the analysis of site response and seismic soil-structure interaction are discussed in this paper. The reliability of these procedures is investigated by a number of case histories and data from simulated earthquake tests on centrifuged models. The results show that if the input motions can be characterized properly, seismic response can be predicted adequately for engineering purposes. However the examples illustrate clearly the great uncertainty associated with the selection of input motions.

1 INTRODUCTION

The basic elements in the dynamic analysis of a soil-structure system are input motion, appropriate models of site and structure, constitutive relations for all materials present, and a stable, efficient, accurate, computational procedure. The specification of the input motion and the selection of an appropriate constitutive relation are the most difficult steps in the analysis.

Linear elastic analysis is appropriate for low levels of shaking in relatively firm ground. As the shaking becomes more intense, soil response becomes nonlinear. A great variety of constitutive relations are available for nonlinear response analysis ranging from equivalent linear elastic models to elastic-plastic models with both isotropic and kinematic hardening. An additional complication is the effect of seismically induced porewater pressures. If these become significant, the corresponding reduction in effective stresses will result in significant reductions in moduli and strength which must be taken into account. Therefore, for some problems, the simpler total stress methods of analysis are not adequate; effective stress methods must be used.

The most widely used methods for dynamic analysis are based on the equivalent linear model. Computer programs representative of this approach are SHAKE (Schnabel et al., 1972) for one-dimensional analysis (1-D) and FLUSH (Lysmer et al., 1975) for 2-D analysis. These programs perform total stress analyses only. Equivalent linear models can exhibit pseudo-resonance, an amplification of computed response that is a function of the nature of the model only. This phenomenon can lead to increased design requirements (Finn et al., 1978).

In recent years, there has been a distinct shift towards the use of nonlinear total or effective stress methods of analysis. A number of nonlinear 1-D programs are now available which give similar results for a given site (Streeter et al., 1973; Lee and Finn, 1975; Lee and Finn, 1978; Martin et al., 1978; Dikmen and Ghaboussi, 1984). A widely used program of this kind is DESRA-2 (Lee and Finn, 1978) and some field applications of this program will be discussed later.

A number of programs are also available for 2-D nonlinear dynamic effective stress analysis. The simplest kind are based on nonlinear hysteretic models of soil response using hyperbolic skeleton curves and unloading-reloading response defined by the Masing criterion (Masing, 1926). A representative program of this type is TARA-3, the third in an evolving series of TARA programs (Finn et al., 1986). This program has been subjected to critical evaluation over the last three years using data from centrifuge model tests sponsored by the U.S. Nuclear Regulatory Commission through the European Office of the U.S. Army Corps of Engineers. Some of these tests have been

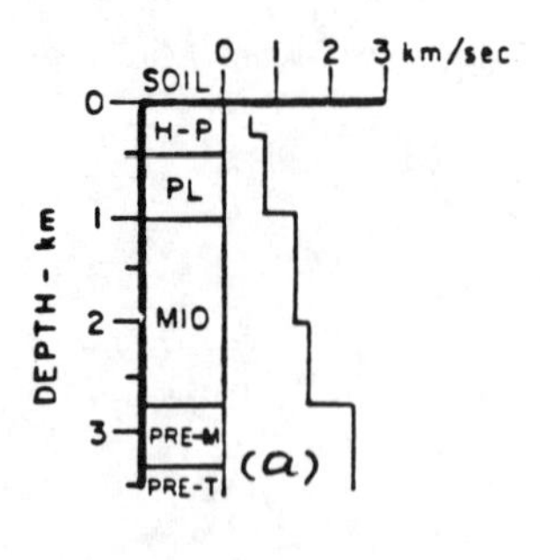

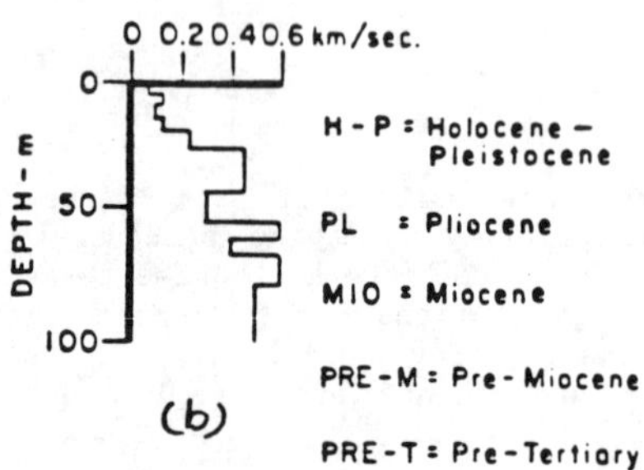

Figure 1. Geological profile of site.

described previously by Finn (1986). Some results from this study will be presented later.

2-D elastic-plastic models for dynamic effective stress analysis are generally based on Biot's equations (Biot, 1941) for coupled fluid-soil systems. However few of these have been incorporated in commercially available programs. The most widely used program of this type is DYNAFLOW (Prevost, 1981). The elastic-plastic effective stress models offer the most complete description of soil response but the properties required in some of them are difficult to measure accurately and they make heavy demands on computational time. Analyses using these models have been conducted on super computers to cut the turn around time.

2 ELASTIC RESPONSE ANALYSIS

Site response analyses are usually conducted to get site specific ground motion spectra for design, especially at soft sites. It is generally assumed that the most reliable results will be obtained if motions on local rock outcrops are available or can be estimated. These can be inputted at the rock-soil interface under the site.

An interesting example of this kind of analysis is provided by data from a Japanese site, (Ohta et al., 1977). The geological profile of the site is shown in Figure 1a for a total depth of 3.5 km. The shear wave velocities are also given for the various soil and rock formations. The top 100 m of the formation is shown in Figure 1b to a larger scale. A layer of soft material about 20 m thick with a low shear wave velocity exists at the surface. Below this layer the shear wave velocities generally exceed 400 m/s in the top 100 m of soil.

During an earthquake, accelerations were measured in the base rock at a depth of 3.5 km and at various elevations up to the surface. Elastic response analyses were conducted using the base rock motions as input. The computed accelerations at various elevations are compared with the recorded accelerations in Figure 2. The computed motions agree very well

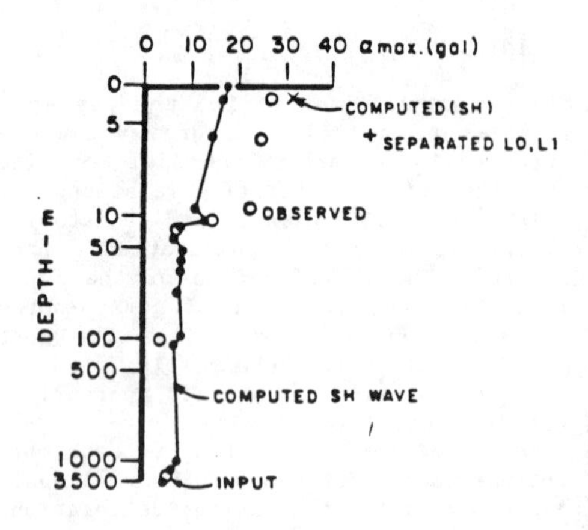

Figure 2. Computed and recorded accelerations at various elevations.

with the recorded motions except in the soft layer near the surface where the computed motions greatly underestimate the recorded motions. Decomposition of the measured ground motions at the surface showed that surface waves of the Love type existed in the soft surface layer. These waves were separated from the recorded motions as shown in Figure 3. When these motions are added to the computed surface accelerations, the computed and measured surface motions agree very closely. This example is especially

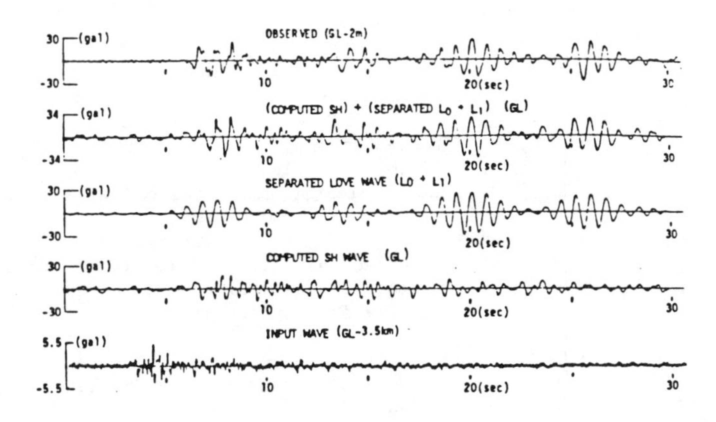

Figure 3. Computed and observed motions at surface.

interesting because motions were recorded from bedrock to surface. Therefore there was none of the usual uncertainty about base input motion. Yet the predictions of surface motions were poor. In this case the contribution to surface motions from shear waves propagating vertically were accurately predicted but there was an additional component of motion which was not represented in the modelling process, the surface wave.

This example suggests that the common practice of using rock outcrop motions as input motions to soft sites should be followed with caution. Additional evidence of difficulties with this procedure are given in the next section.

3 EQUIVALENT LINEAR ANALYSIS OF MEXICO CITY SITE

During the 1985 Mexico earthquake, ground accelerations were measured in the free field on the soft deposits of the former lake bed in downtown Mexico City and on rock and hard sites in the University area (UNAM). These data provide an opportunity to check the reliability of the usual procedures for conducting site response analyses.

Romo and Seed (1986) conducted total stress site response analyses for Mexico City sites using the SHAKE program (Schnabel et al., 1972). The sites were assumed to consist of relatively homogeneous clay deposits of various thicknesses underlain by a dense sand layer. This layer was considered the base of the site and seismic motions were inputted at this level. Following usual practice, they selected a representative record for input. They used the Pasadena record of the 1952 Kern County earthquake (M=7.6) as a representative motion and scaled it appropriately. Both acceleration and frequency were scaled so that average computed spectra for various sites matched the average spectra of the recorded motions.

A downtown site on the old lake bed known as the SCT site was analyzed using the acceleration components recorded at UNAM and the Romo-Seed model of the site. The results are shown in Figures 4 and 5 for the N90W and N00E acceleration components.

In the case of the N90W component, the spectrum of computed motions badly underestimates the spectrum of recorded motions both as to peak spectral acceleration and the range of strong response.

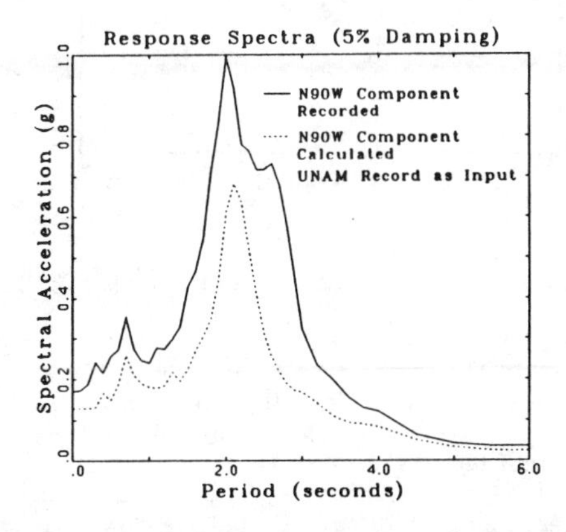

Figure 4. Recorded and computed spectra at SCT site: N90W component.

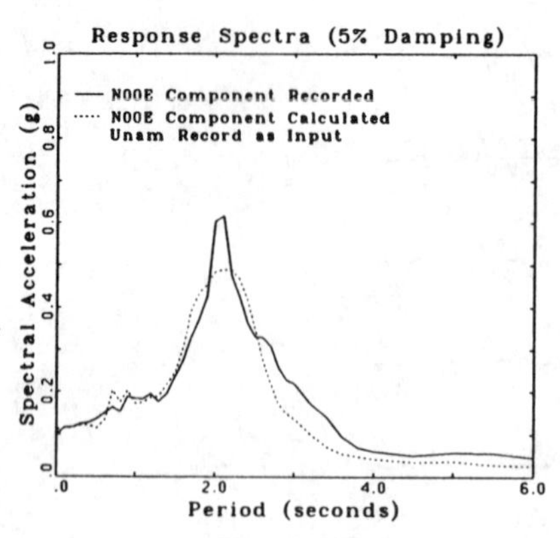

Figure 5. Recorded and computed spectra at SCT site: N00E component.

The agreement between the spectra in the case of the N00E component is better although the peak response is also somewhat underestimated. Clearly the motions at the SCT site have a much greater directional bias than those recorded at UNAM. This is clearly evident from the acceleration plot in Figure 6 which shows

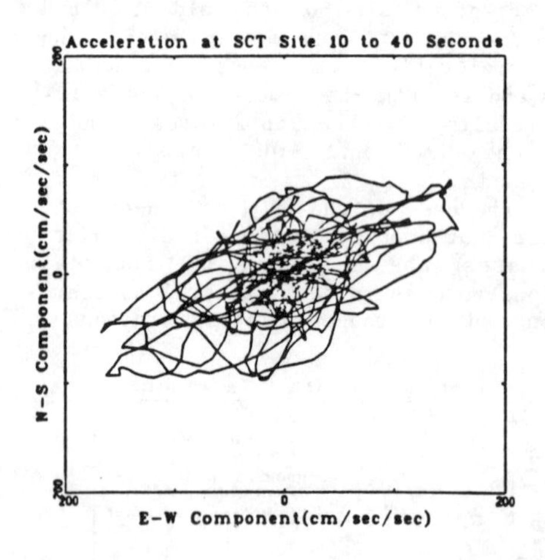

Figure 6. Recorded accelerations at SCT site.

a strong bias in the E-W direction. To match the spectra of the recorded motions the recorded N90W component should be amplified by a factor of 2 to a peak acceleration of 0.07 g before use as input motion. The N00E component should be amplified to 0.035 g an amplification of 1.25.

The simplest explanation for the discrepancy between recorded and computed surface motions is that as the motions pass from the surrounding rock basin into the lake bed, they are modified, acquiring a strong directional bias and increased amplitude.

The examples from Mexico City and Japan clearly demonstrate that selecting suitable input motions for dynamic analysis is a very difficult task. It requires a deep understanding of the factors influencing dynamic response both locally and regionally.

4 NONLINEAR DYNAMIC EFFECTIVE STRESS ANALYSES

The distinguishing characteristic of effective stress methods is the capability of predicting seismically induced porewater pressures and taking their effects into account during analysis. The DESRA-2 program has been used to evaluate site response at a number of sites in Japan (Finn et al., 1982; Iai et al., 1985) at which seismic porewater pressures were measured or liquefaction occurred. Some results from these studies are presented in the next section.

4.1 Seismic response of Owi Island test site

4.1.1 Owi Island test site

Owi Island No. 1 is an artificial island located on the west side of Tokyo Bay. A test site at the south end of the island is instrumented to record porewater pressures and ground accelerations during earthquakes. Porewater pressures are recorded by piezometers installed at depths of 6 m and 14 m. The transducer in each recorder is of the strain-gauge type with a full capacity of 200 kN/m^2. A two-component seismograph is installed on the ground surface to measure horizontal acceleration. The soil profile at the site, is shown in Figure 7. The sand layers in which the piezometers were embedded at depths of 6 m and 14 m had almost identical blow counts of N = 5. The depths from which undisturbed samples were recovered are also shown in Figure 7.

The Mid-Chiba earthquake, with a magnitude M = 6.1, shook the Tokyo Bay area on September 25, 1980. The ground shaking, due to the earthquake, was of intensity V on the Japanese Meterological Agency Scale in the Tokyo Bay area and was sufficient to develop the porewater

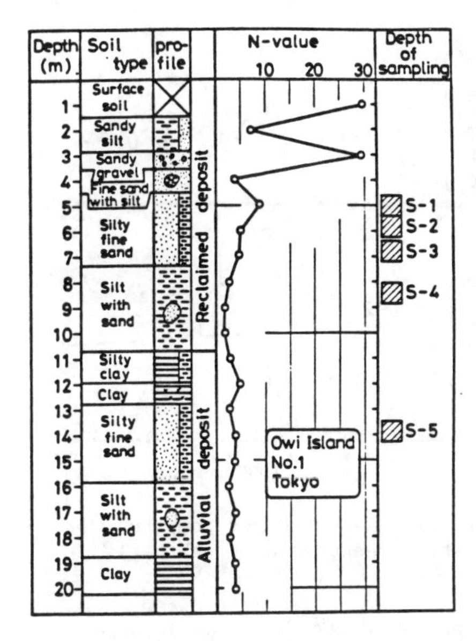

Figure 7. Soil profile at Owi Island.

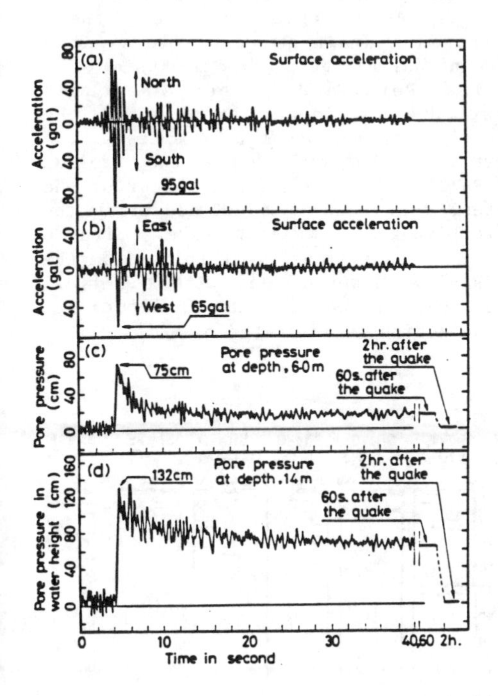

Figure 8. Recordings of accelerations and porewater pressure at Owi Island.

pressures and accelerations shown in Figure 8. The maximum horizontal accelerations at the ground surface were 95 gals in the N-S direction and 65 gals in the E-W direction. The rise in porewater pressure was 0.75 m of water in the sand layer at a depth of 6 m and 1.32 m at a depth of 14 m. Fourier spectra of the acceleration records indicate that the predominant periods of motion were 0.64 sec and 0.5 sec in the E-W and N-S directions, respectively.

All the properties required for the analysis of Owi Island No. 1 by DESRA-2 were obtained using data usually available from conventional site and laboratory investigations.

Full details of the instrumentation, recorded data, and the site investigations on Owi Island and the associated laboratory testing have been described by Ishihara et al. (1981). A more detailed description of the analysis of the site is by Finn et al. (1982).

4.1.2 Comparison of Field and Computed Responses

The recorded ground motions are shown in Figure 9a; those computed by DESRA-2 in Figure 9b. Except for minor differences in frequency and magnitude in the 8-10 s range, the computed record is very similar to the recorded motions.

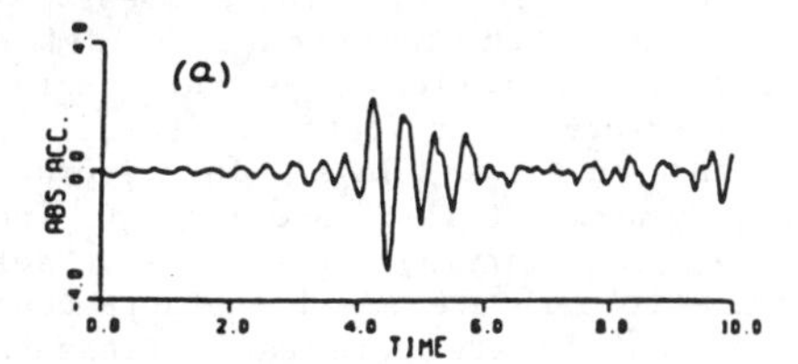

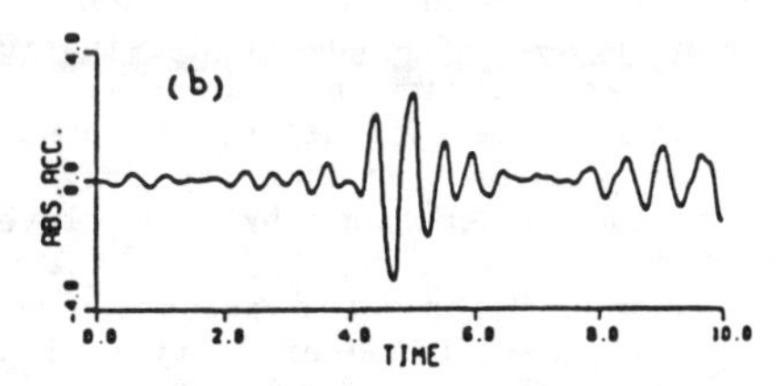

Figure 9. Recorded and computed accelerations at Owi Island.

The porewater pressures recorded at the 6 m depth are shown to an expanded scale in Figure 10a. During the low level shaking of the first 4 s, the response was elastic and porewater pressures developed in instantaneous response to

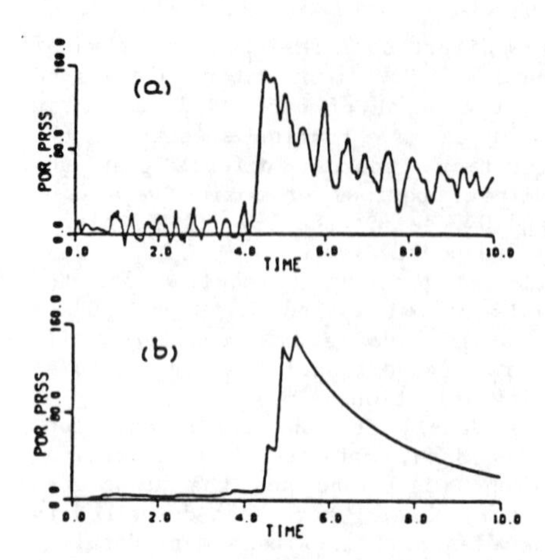

Figure 10. Recorded and computed porewater pressure at depth of 6 m.

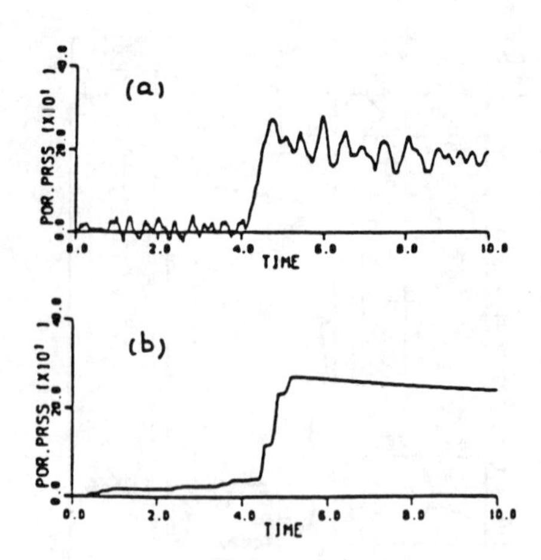

Figure 11. Recorded and computed porewater pressures at depth of 14 m.

changes in the total applied mean normal stresses. Such porewater pressures result from the elastic coupling of soil and water. With the onset of more severe shaking, plastic volumetric strains are induced and these result in the development of residual porewater pressures which are independent of the instantaneous states of stress. These pressures accumulate with continued plastic volumetric deformation. Residual porewater pressure is indicated by the steep rise and sustained level in recorded porewater pressure in Figure 10a. During shaking, the varying applied stresses continue to generate small instantaneous fluctuations in the pore water pressure which are superimposed on the larger residual porewater pressures. The gradual decay in the sustained level of porewater pressure is due to dissipation of porewater pressure by drainage. At this stage in the excitation, the dissipation of porewater pressure by drainage exceeds the generation by low level excitation.

The computed porewater pressures are shown in Figure 10b and are very similar to the recorded values. DESRA-2 computes only residual porewater pressures so there are no fluctuations in the computed pressures due to changes in instantaneous stress levels.

Recorded and computed porewater pressures for the sand layer at a depth of 14 m are shown in Figures 11a and 11b, respectively. DESRA-2 results compare very favourably with the recorded values. Dissipation of porewater pressure is negligible in the lower sand compared to the upper sand because it is capped by a clay layer instead of by pervious fill. The DESRA-2 program can take these different drainage conditions into account during the dynamic analysis.

The Port and Harbour Research Institute of Japan investigated the liquefaction potential of 6 sites at the port of Ishinomaki using the program DESRA-2 (Iai et al., 1985). Four of the sites, A,B,D and F liquefied during the 1978 Miyagi-Ken-Oki earthquake. Sites C and E did not liquefy. Results from the DESRA-2 analyses agreed with the field experience as may be seen in Figure 12. Liquefaction is indicated by $u/\sigma'_{vo} = 1$.

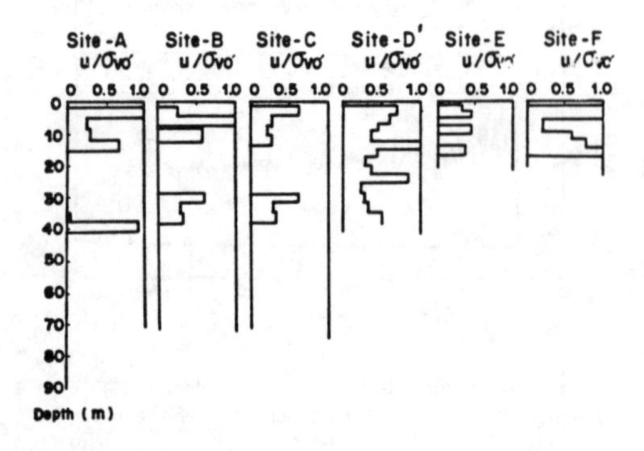

Figure 12. Computed porewater pressures at 6 Japanese sites.

5 2-D NONLINEAR DYNAMIC EFFECTIVE STRESS ANALYSIS

2-D dynamic analyses are usually conducted using equivalent linear finite element analyses in the frequency domain. There has been little verification of these methods because of a lack of adequate field data.

There are certain important phenomena in soil-structure interaction outside the scope of conventional frequency domain analysis. Typical examples are uplift during rocking, permanent deformations, the effects of seismically induced porewater pressures, hysteretic behaviour and stick-slip behaviour at interfaces between structure and foundation soils.

The program TARA-3 (Finn et al., 1986) was developed to cope with such problems. The capability of the program will be demonstrated by using it to analyze one of the NRC centrifuge tests which models the response of a heavy two-dimensional structure embedded in a saturated sand foundation to seismic excitation.

6 ANALYSIS BY TARA-3

In TARA-3, response in shear is assumed to be nonlinear and hysteretic with unloading and reloading stress-strain paths defined by the Masing criterion (Masing, 1926). The response of the soil to uniform all round pressure is assumed to be nonlinearly elastic and dependent on the mean normal effective stress. Porewater pressures during shaking are computed using the Martin-Finn-Seed porewater pressure model (Martin et al., 1975) modified to take into account the effects of initial static shear stress. Moduli and strength are continuously modified during analysis to reflect changes in the effective stress regime. A detailed description of the constitutive relations in TARA-3 is given by Finn (1985).

For analysis involving soil-structure interaction it may be important to model slippage between the structure and soil. Slip may occur during very strong shaking or even under moderate shaking if high porewater pressures are developed under the structure. TARA-3 contains slip elements of the Goodman (Goodman et al., 1968) type to allow for relative movement between soil and structure in both sliding and rocking modes during earthquake excitation.

7 MODEL STRUCTURE EMBEDDED IN SATURATED SAND

A schematic view of the model structure is shown in Figure 13. It is made from a

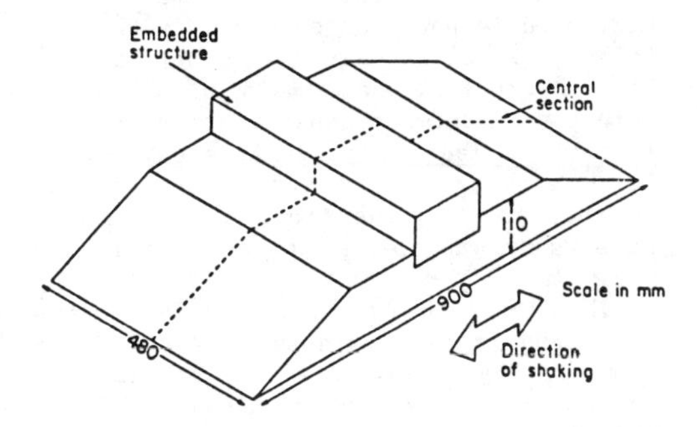

Figure 13. Centrifugal model of embedded structure.

solid piece of aluminum alloy and has dimensions 150mm wide by 108mm high in the plane of shaking. The length perpendicular to the plane of shaking is 470mm and spans the width of the model container. The structure is embedded a depth of 25mm in the sand foundation. Sand was glued to the base of the structure to prevent slip between structure and sand.

The foundation was constructed of Leighton Buzzard Sand passing British Standard Sieve (BSS) No. 52 and retained on BSS No. 100. The mean grain size is therefore 0.225mm. The sand was placed as uniformly as possible to a nominal relative density D_r = 52%.

During the test the model experienced a nominal centrifugal acceleration of 80 g. The model therefore simulated a structure approximately 8.6m high by 12m wide embedded 2m in the foundation sand.

De-aired silicon oil with a viscosity of 80 centistokes was used as a pore fluid in order to model the drainage conditions in the prototype during the earthquake. If the linear scale factor between model and prototype is N, then excess porewater pressures dissipate approximately N^2 times faster in the model than in the prototype if the same fluid is used in both. The rate of loading by seismic excitation will be only N times faster. Therefore, to model prototype drainage conditions during the earthquake, a pore fluid with a viscosity N times the prototype viscosity must be used. This viscosity was achieved by an appropriate blending of commercial silicon oils. Tests by Eyton (1982) have shown that the stress-strain behaviour of fine sand is not changed when silicon oil is substituted for water as a pore fluid. In the gravitational field of 80g, the structure underwent consolidation settlement which led to a significant increase in density under the structure compared to that in the free field. This change in density was taken into account in the analysis.

The locations of the accelerometers (ACC) and pressure transducers (PPT) are shown in Figure 14. Analyses of previous centrifuge tests indicated that TARA-3 was capable of modelling acceleration response satisfactorily. Therefore, in the present test, more instrumentation was devoted to obtaining a good data base for checking the ability of TARA-3 to predict residual porewater pressures.

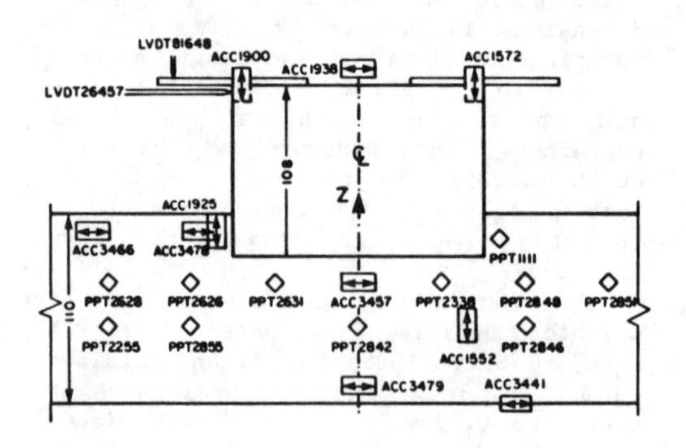

Figure 14. Instrumentation of centrifuged model.

As may be seen in Figure 14, the porewater pressure transducers are duplicated at corresponding locations on both sides of the centre line of the model except for PPT 2255 and PPT 1111. The purpose of this duplication was to remove any uncertainty as to whether a difference between computed and measured porewater pressures might be due simply to local inhomogeneity in density.

The porewater pressure data from all transducers are shown in Figure 15. These records show the sum of the transient and residual porewater pressures. The peak residual pressure may be observed when the excitation has ceased at about 95 milliseconds. The pressures recorded at corresponding points on opposite sides of the centre line such as PPT 2631 and PPT 2338 are generally quite similar although there are obviously minor differences in the levels of both total and residual porewater pressures. Therefore it can be assumed that the sand foundation is remarkably symmetrical in its properties about the centre line of the model. Hence measured and computed porewater pressures are compared only for locations on one side of the centre line of the model only, the right hand side.

8 COMPUTED AND MEASURED ACCELERATION RESPONSES

The soil-structure interaction model was converted to prototype scale before analysis using TARA-3 and all data are quoted at prototype scale. Soil properties were consistent with relative density.

The computed and measured horizontal accelerations at the top of the structure at the location of ACC 1938 are shown in Figure 16. They are very similar in frequency content, each corresponding to the frequency of the input motion given by ACC 3441 (Figure 15). The peak accelerations agree fairly closely.

The vertical accelerations due to rocking as recorded by ACC 1900 and those computed by TARA-3 are shown in Figure 17. Again, the computed accelerations closely match the recorded accelerations in both peak values and frequency content. Note that the frequency content of the vertical accelerations is much higher than that of either the horizontal acceleration at the same level in the structure or that of the input motion. This occurs because the foundation soils are much stiffer under the normal compressive stresses due to rocking than under the shear stresses induced by the horizontal accelerations.

9 COMPUTED AND MEASURED POREWATER PRESSURES

The porewater pressures in the free field recorded by PPT 2851 are shown in Figure 18. In this case the changes in the mean normal stresses are not large and the fluctuations of the total porewater pressure about the residual value are relatively small. The peak residual porewater pressure, in the absence of drainage, is given directly by the pressure recorded after the earthquake excitation has ceased. In the present

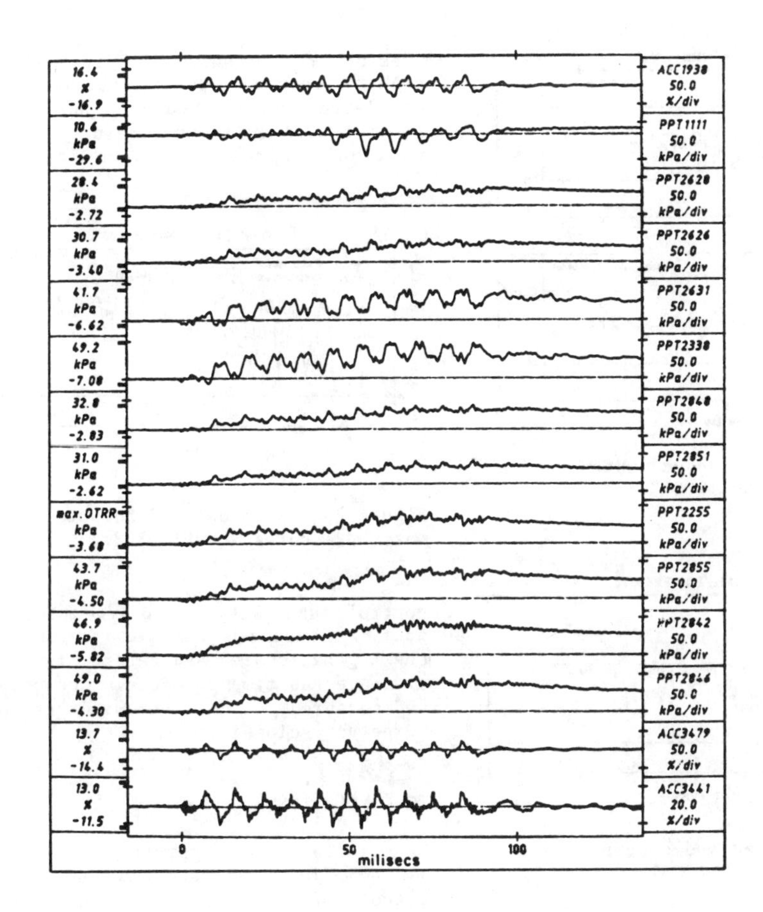

Figure 15. Partial data from centrifuge test.

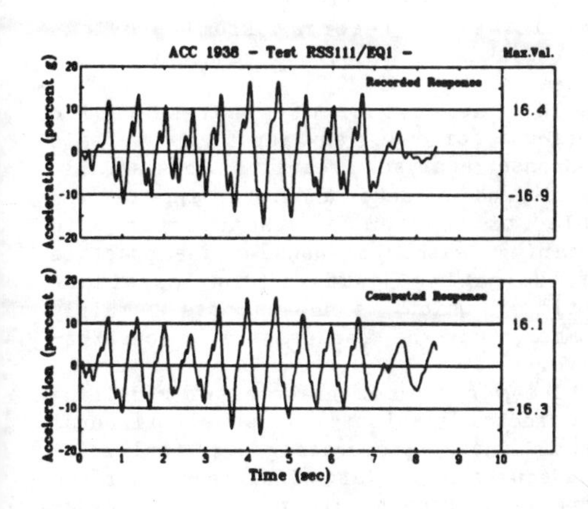

Figure 16. Recorded and computed vertical accelerations at ACC 1938.

test, significant shaking ceased after 7 seconds. A fairly reliable estimate of the peak residual pressure is given by the record between 7 and 7.5 seconds. The recorded value is slightly less than the value computed by TARA-3 but the overall agreement between measured and computed pressures is quite good.

As the structure is approached, the recorded porewater pressures show the increasing influence of soil-structure interaction. The pressures recorded by PPT 2846 adjacent to the structure (Figure 19) show somewhat larger oscillations than those recorded in the free field. This location is close enough to the structure to be affected by the cyclic normal stresses caused by rocking. The recorded peak value of the residual porewater pressure is given by the relatively flat portion of the record

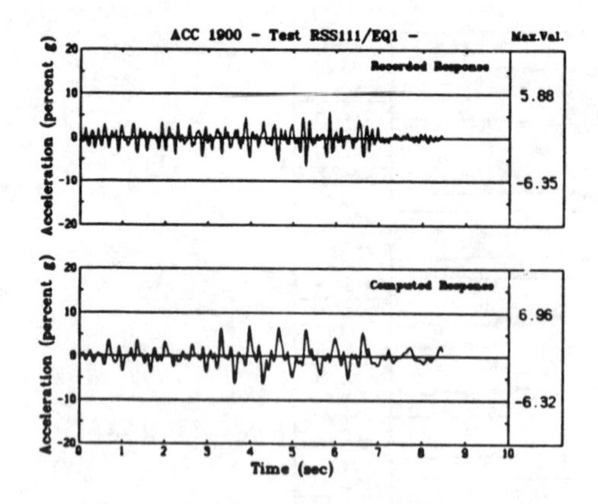

Figure 17. Recorded and computed vetical accelerations at ACC 1900.

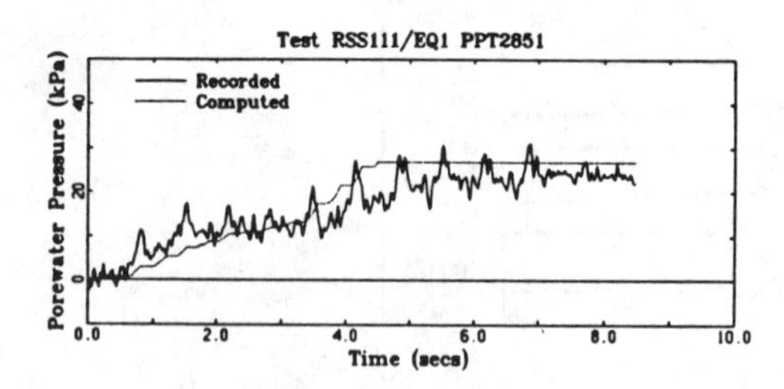

Figure 18. Recorded and computed porewater pressures at PPT 2851.

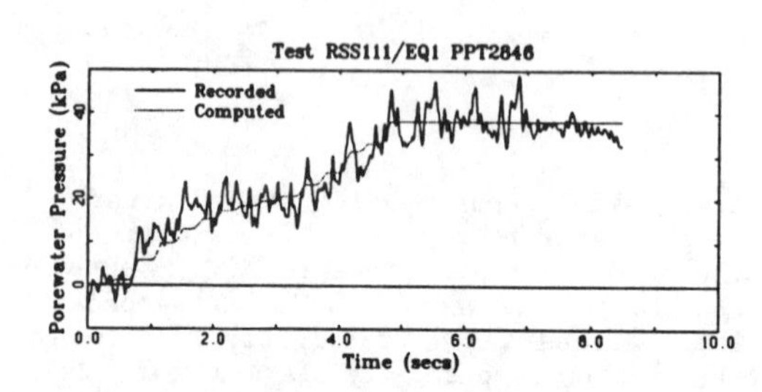

Figure 19. Recorded and computed porewater pressures at PPT 2846.

between 7 and 7.5 seconds. The computed and recorded values agree very closely.

Transducer PPT 2338 is located directly under the structure near the edge and was subjected to large cycles of normal stress due to rocking of the structure. These fluctuations in stress resulted in similar fluctuations in mean normal stress and hence in porewater pressure. This is clearly evident in the porewater pressure record shown in Figure 20. The higher frequency peaks superimposed on the larger oscillations are due to dilations caused by shear strains. The peak residual porewater pressure which

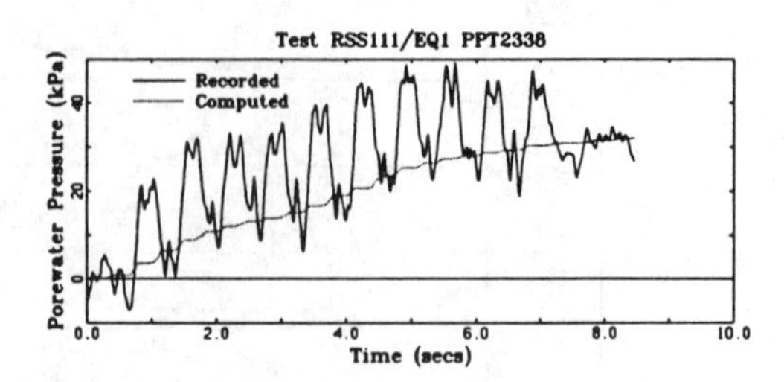

Figure 20. Recorded and computed porewater pressures at PPT 2338.

controls stability is observed between 7 and 7.5 seconds just after the strong shaking has ceased and before significant drainage has time to occur. The computed and measured residual porewater pressures agree very closely.

10 CONCLUSIONS

Many data from centrifuge tests and limited field data confirm that reliable methods of analysis are available for estimating the seismic response of sites and some soil-structure systems. The predictions of seismic response are heavily dependent on the degree to which the input motions for analysis can be estimated, especially in the absence of acceleration records from previous earthquakes at locations adjacent to the site.

The selection of representative motions for use as input in seismic response analyses requires considerable skill and a deep understanding of the role of system characteristics in defining seismic response. The practice of selecting just a few candidate motions, which is unfortunately fairly common, may be dangerously unconservative.

The very common practice of modelling the incoming seismic waves as horizontal shear waves propagating vertically is inadequate wherever significant surface waves are present. It may also be inadequate close to the epicentre.

Dynamic analysis provides a constant ordered approach to estimating the characteristics of site specific motions. It allows parametric studies to be conducted which are powerful guides to judgement. Estimating "exact" seismic response parameters is impossible; defining safe but economical design parameters with the help of an adequate supply of representative input motions for dynamic response analysis is feasible and practical.

ACKNOWLEDGEMENTS

The analyses of site response in Mexico City were conducted as part of a study for the Canadian Council on Earthquake Engineering. The development of TARA-3 was supported by the National Science and Engineering Research Council of Canada under Grant No. 1498, and the Exxon Production and Research Company. The centrifuge tests were funded by the U.S. Nuclear Regulatory Commission through the European Office of the U.S. Army Corps of Engineers. The support of these sponsors is gratefully acknowledged. Descriptions of the centrifuge model test and related figures are used by permission of Cork Geotechnics Ltd., Ireland. The text was typed by Mrs. Kelly Lamb.

REFERENCES

Biot, M.A. 1941. General Theory of Three-Dimensional Consolidation. J. Appl. Phys., 12, 155-64.

Dikmen, S.U. and Ghaboussi, J. 1984. Effective Stress Analysis of Seismic Response and Liquefaction: Theory. Journal of the Geotech. Eng. Div., ASCE, Vol. 110, No. 5, Proc. Paper 18790, pp. 628-644.

Eyton, D.G.P. 1982. Triaxial Tests on Sand with Viscous Pore Fluid. Part 2, Project Report, Cambridge University, Engineering Department.

Finn, W.D. Liam. 1985. Dynamic Effective Stress Response of Soil Structures; Theory and Centrifugal Model Studies, Proc. 5th Int. Conf. on Num. Methods in Geomechanics, Nagoya, Japan, Vol. 1, 35-36.

Finn, W.D. Liam. 1986. Verification of Nonlinear Dynamic Analysis of Soils Using Centrifuged Models. Proc., International Symposium on Centrifuge Testing, Cambridge University, U.K., 20 pp.

Finn, W.D. Liam, Iai, S. and Ishihara, K. 1982. Performance of Artificial Offshore Islands Under Wave and Earthquake Loading: Field Data Analysis. Proceedings, Offshore Tech. Conf., Houston, Texas, OTC Paper No. 4220, Vol. I, pp. 661-672.

Finn, W.D. Liam, G.R. Martin and M.K.W. Lee. 1978. Comparison of Dynamic Analyses for Saturated Sands. Proceedings of the ASCE Specialty Conference on Earthquake Engineering and Soil Dynamics. Vol. I, ASCE, New York, N.Y., pp. 472-491.

Finn, W.D. Liam, M. Yogendrakumar, N. Yoshida, and H. Yoshida. 1986. TARA-3: A Program for Nonlinear Static and Dynamic Effective Stress Analysis, Soil Dynamics Group, University of British Columbia, Vancouver, B.C.

Goodman, R.E., R.L. Taylor and T.L. Brekke. 1968. A Model for the Mechanics of Jointed Rock, J. Soil Mech. and Found. Div. ASCE, 94 (SM3), 637-659.

Iai, S., Tsuchida, H. and Finn, W.D. Liam. 1985. An Effective Stress Analysis of Liquefaction at Ishinomaki Port During the 1978 Miyagi-Ken-Oki Earthquake. Report of the Port and Harbour Research Institute, Vol. 24, No. 2, June, pp. 1-84.

Ishihara, K., Shimizu, K. and Yasuda, Y. 1981. Porewater Pressures Measured in Sand Deposits During an Earthquake. Soils and Foundation Journal, Vol. 21, No. 4, December, pp. 85-100.

Lee, M.K.W. and Finn, W.D.L. 1975. DESRA-1, Dynamic Effective Stress Response Analysis of Soil Deposits. Dept. of Civil Engineering, University of British Columbia, Vancouver, B.C.

Lee, M.K.W. and Finn, W.D.L. 1978. DESRA-2, Dynamic Effective Stress Response Analysis of Soil Deposits with Energy Transmitting Boundary Including Assessment of Liquefaction Potential. Soil Mechanics Series No. 38, Department of Civil Engineering, University of British Columbia, Vancouver, B.C.

Lysmer, J., Udaka, T., Tsai, C.F. and Seed, H.B. 1975. FLUSH: A Computer Program for Approximate 3-D Analysis of Soil-Structure Interaction Problems. Report No. EERC 75-30, Earthquake Engineering Research Center, University of California, Berkeley, California.

Martin, G.R., W.D. Liam Finn, and H.B. Seed. 1975. Fundamentals of Liquefaction Under Cyclic Loading, Soil Mech. Series Report. No. 23, Dept. of Civil Engineering, University of British Columbia, Vancouver; also Proc. Paper 11284, J. Geotech. Eng. Div. ASCE, 101 (GT5): 324-438.

Martin, P.P. and Seed, H.B. 1978. MASH - A Computer Program for the Nonlinear Analysis of Vertically Propagating Shear Waves in Horizontaly Layered Soil Deposits. EERC Report No. UCB/EERC-78/23, Univ. of California, Berkeley, California, October.

Masing, G. 1926. Eigenspannungen und Verfestigung beim Messing, Proc., 2nd Int. Congress of Applied Mechanics, Zurich, Switzerland.

Ohta, T., Niva, M. and Andoh, H. 1977. Seismic Motions in the Deeper Portions of Bedrock and in the Surface and Response of Surface Layers. Proc., 4th Japan Earthquake Engineering Symposium, Tokyo, pp. 129-136 (in Japanese).

Prevost, J.H. 1981. DYNAFLOW: A Nonlinear Transient Finite Element Analysis Program. Princeton University, Department of Civil Engineering, Princeton, N.J.

Romo, M.P. and Seed, H.B. 1986. Analytical Modelling of Dynamic Soil Response in the Mexico City Earthquake of September 19, 1985. Presented at International Symposium on Mexican Earthquake, Mexico City. To be published in proceeding.

Schnabel, P.B., Lysmer, J. and Seed, H.B. 1972. SHAKE: A Computer Program for Earthquake Response Analysis of Horizontally Layered Sites. Report No. EERC 72-12, Earthquake Engineering Research Center, University of California, Berkeley, California.

Streeter, V.L., Wylie, E.B. and Richart, F.E. 1973. Soil Motion Computations by Characteristics Method. ASCE National Structural Engineering Meetings, San Francisco, California, Preprint 1952.

Prediction and Performance in Geotechnical Engineering / Calgary / 17-19 June 1987

The prediction and performance of structures on expansive soils

D.G.Fredlund
University of Saskatchewan, Saskatoon, Canada

ABSTRACT: Numerous analytical procedures have been proposed for the prediction of heave in swelling soils. These procedures have recently been examined within the context of unsaturated soil theory. This paper describes the theory related to the swelling of soils, outlines the procedure for testing these soils in a one-dimensional oedometer, and also explains how the data should be interpreted. Two case histories are also presented.

1 INTRODUCTION

Lightly loaded structures commonly suffer severe distress subsequent to their construction. Changes in the environment around the structure result in changes in the (negative) pore-water pressure, thereby producing volume changes in the soil. Soils with a high swelling index, C_s, in a changing environment are commonly found to be highly swelling soils.

Krohn and Slosson, 1980, estimated that 7 billion dollars are spent each year in the United States as a result of damage to all types of structures built on swelling soils. Jones and Holtz (1973) pointed out that more than twice as much is spent on damage due to swelling soils as is spent on damage from floods, hurricanes, tornadoes and earthquakes. Certainly the problem is of enormous financial proportions.

The prediction of heave of light structures has probably received more attention than any other analysis associated with swelling soils. Numerous analytical procedures have been proposed in various countries. Most methods have been used to a limited extent within a restricted geographical region. Only recently has there been an attempt to embrace the different methods for predicting heave within one consistent theoretical context.

It is necessary to relate soil behavior to the stress state in the soil, in order to develop a transferable science for swelling soils. The engineer must be able to visualize volume changes in terms of appropriate stress state variable changes. The success of the practice of <u>saturated</u> soil mechanics can be attributed largely to the ability of engineers to relate soil behavior to changes in the effective stress state variable. Swelling soils are generally <u>unsaturated</u> and engineers have found it much more difficult to relate soil behavior to stress state variable changes.

The primary objective of this paper is to assist engineers in relating the volume change behavior of unsaturated, swelling soils to changes in the stress state. Specifically, the objectives can be summarized as follows:

1. to explain how past, present and future behavior of a swelling soil can be explained in terms of stress state variables. An attempt will be made to maintain a similar philosophical framework to that used in saturated soil mechanics.
2. to describe a method that can be used to predict heave. The method involves the use of one-dimensional oedometer tests. Emphasis will be placed on the interpretation of the laboratory results.
3. to briefly present two case histories involving swelling soils. The results of these studies are used to confirm the reasonableness of the proposed method.

2 STRESS STATE VARIABLES CONTROLLING BEHAVIOR

Three stresses must be measured, estimated

or predicted in order to describe the behavior of an unsaturated soil. These are the total stress, σ, the pore-water pressure, u_w, and the pore-air pressure, u_a. These variables can be combined into two independent stress state variables for unsaturated soils (Fredlund and Morgenstern, 1977). Although various combinations of independent stress variables are possible, the $(\sigma - u_a)$ and $(u_a - u_w)$ combination has proven to be most advantageous since the effects of total stress changes and pore-water pressure changes can be separated. This is beneficial both from a conceptual and analytical standpoint since pore-air pressure can generally be assumed to be atmospheric. The $(\sigma - u_a)$ term is referred to as the "net total" stress and the $(u_a - u_w)$ term is referred to as the matric suction. These stress state variables provide a smooth transition when going from the unsaturated to the saturated soil case. As the degree of saturation approaches 100 percent, the pore-air pressure and the pore-water pressure become approximately equal in magnitude. When the matric suction term goes to zero, the pore-air pressure in the $(\sigma - u_a)$ term becomes the pore-water pressure.

The independent stress state variables can be used to assist in understanding the behavior of a swelling clay deposit. Let us consider a deposit of proglacial, lacustrine origin. The present physical properties and state of stress of the clay are dependent upon stress influences subsequent to deposition. When studying a potential heaving problem, the engineer must evaluate the present state of stress in the soil and determine suitable physical properties to predict future behavior.

2.1 Stress History

Deposits in a proglacial lake are initially consolidated by the bouyant weight of the overlying sediments. The drainage of the lake and the subsequent evaporation of water over the lake sediments commences a desiccation of the underlying sediments. The term "desiccation" is used to mean the drying of the soil by evaporation and evapotranspiration. The water table is simultaneously drawn below the ground surface. The total stress on the sediments remains essentially constant, while the stress in the water phase is reduced (i.e., it becomes negative above the water table). This gives rise to an increase in effective stress and the soil consolidates. The tension in the water phase acts in all directions and as a result, there is a tendency for cracking and overall desaturation of the upper portion of the profile (Figure 1).

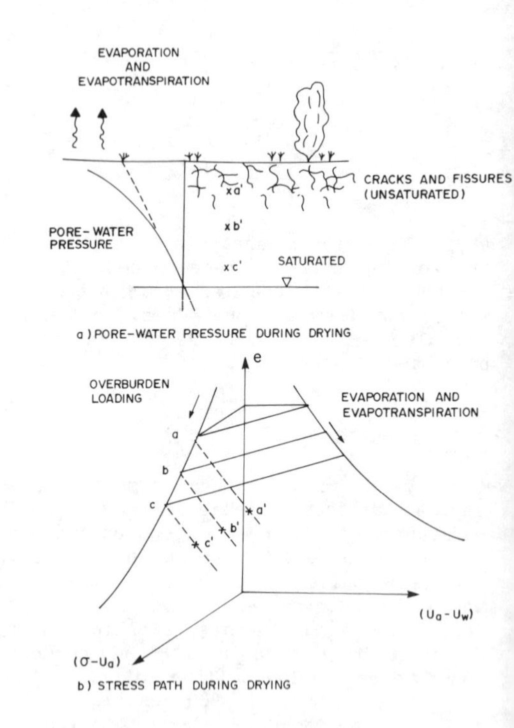

Fig.1 Stress representation after the lake sediments are subjected to evaporation and evapotranspiration.

Grass, trees, and other plants also start to grow on the surface with the net effect of further drying the soil by applying a tension to the water phase. Most plants are capable of applying 10 to 20 atmospheres of tension to the water phase prior to reaching their wilting point. A high tension in the water phase (i.e., high matric suction) means that the soil is highly desiccated. The drying results in an affinity of the soil for water (Figure 1a).

Year after year, the surface deposit is subjected to varying and changing environmental conditions. In response to these changes, the upper portion of the deposit swells and shrinks. Volume changes may extend to depths in excess of 10 feet. Environmental changes transmit a change in stress to the pore-water. The stress changes are isotropic. On the other hand, changes in total stress imposed by man are generally anisotropic. It is advantageous to separate the effects of total and pore-water pressure changes in accordance with the stress state variables involved.

Evaporation and evapotranspiration are depicted as movements in the matric suction plane, whereas loads applied to the soil structure are shown in the net total stress plane (Figure 1b). Wetting and drying due to environmental effects are visualized as changes along hysteresis loops in the matric suction plane. In arid and semi-arid regions, the natural water content gradually decreases.

Low water contents in clay deposits indicates that the soil has the potential for swell if evaporation and evapotranspiration are not permitted from the ground surface as a result of covering the area with a building, asphalt, etc.

2.2 Present State of Stress

When the soil is sampled for laboratory testing purposes, the insitu state of stress may be anywhere along either a drying or wetting portion of the void ratio versus stress relationship. Figure 2 illustrates a typical, complex stress history. In reality, the soil has undergone thousands of cycles of drying and wetting. At the point of sampling, the soil is subjected to a specific net total stress and a specific insitu matric suction.

The primary laboratory information desired by the engineer for analyzing a swelling problem is an assessment of: (i) the insitu state of stress, and (ii) the swelling properties with respect to changes in matric suction. It is necessary to develop a simple, rapid, and economical procedure to obtain the information required for solving practical swelling clay problems.

Several laboratory testing procedures are used in practice to obtain the required soils information. These generally involve the use of the one-dimensional consolidation apparatus (i.e., oedometer) and are classified as the "constant volume" and "free swell" testing procedures (Noble 1966).

The oedometer can test the soil on the total stress plane. The assumption is made that it is possible to eliminate the matric suction in the soil by immersing of the specimen in water and obtain the necessary soil properties and stress values from the total stress plane.

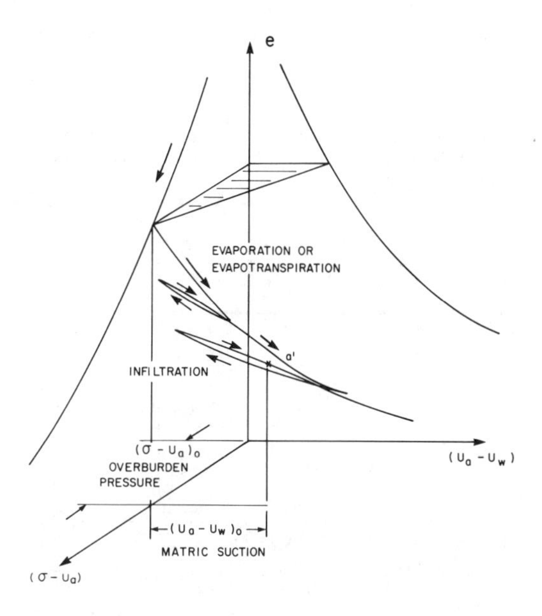

Fig.2 Stress representation when soil has undergone a complex stress history caused by drying and wetting.

Let us first consider the "constant volume" oedometer test procedure. The sample is subjected to a token load and submerged in water. As the sample attempts to swell, the applied load is increased to maintain the sample at a constant volume. This procedure is continued until there is no further tendency for swelling. The applied load at this point is referred to as the "uncorrected" swelling pressure. P_s. The sample is then further loaded and unloaded in the conventional manner.

The test results are commonly plotted as shown in Figure 3a. The actual stress paths followed during the test can be more clearly understood by use of a three-dimensional plot with the stress state variables forming abscissas (Figure 3b). An understanding of the stress paths followed during the test assist in the interpretation of the data. The void ratio and water content stress paths are shown for the situation where there is a minimum of disturbance due to sampling. Even so, the loading path displays some curvature as the total stress plane is approached. In actuality, the stress path may show even more influence from sampling (Figure 4). Engineers have long recognized the significance of sampling disturbance when determining the preconsolidation pressure for a saturated clay. However, only recently has the significance of sampling disturbance been recognized in evaluating the swelling pressure of a soil (Fredlund, Hasan and Filson 1980).

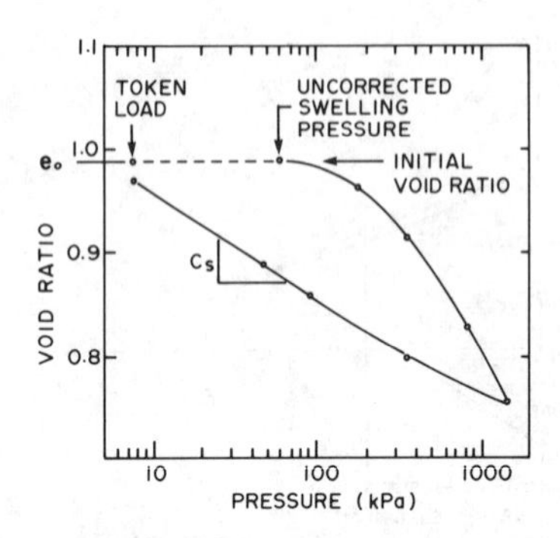

3a.) CONVENTIONAL PROCEDURE FOR PLOTTING 'CONSTANT VOLUME' OEDOMETER DATA

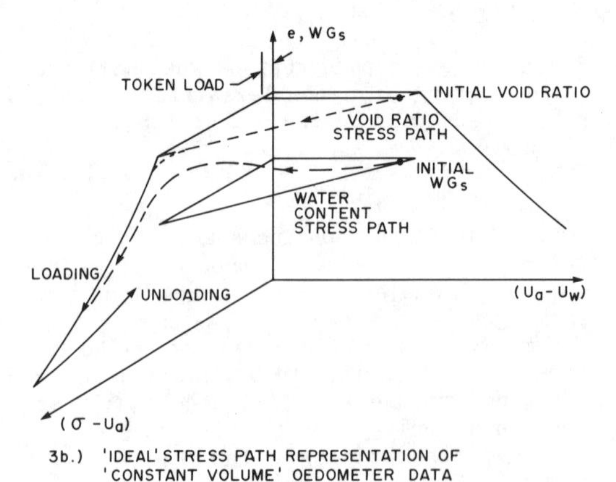

3b.) 'IDEAL' STRESS PATH REPRESENTATION OF 'CONSTANT VOLUME' OEDOMETER DATA

Fig.3 Interpretation of data from a 'constant volume' oedometer test.

Sampling disturbance causes the conventional swelling pressure, P_s, to fall well below the "ideal" or "corrected" swelling pressure, P_s'. The "corrected" swelling pressure represents the insitu stress state translated to the total stress plane. It is equal to the overburden pressure plus the insitu matric suction translated onto the total stress plane. The translated suction is called the "matric suction equivalent" (Yoshida, Fredlund and Hamilton 1982). The magnitude of the matric suction equivalent will be lower than the insitu matric suction; the difference being primarily a function of the insitu degree of saturation. The engineer needs to obtain the "corrected" swelling pressure from the oedometer test in order to reconstruct the insitu stress conditions. The procedure for accounting for sampling disturbance is discussed later in this paper.

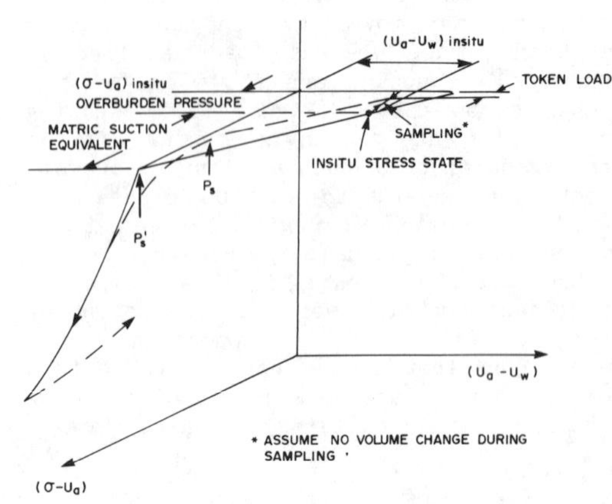

Fig.4 Actual stress path showing the effect of sampling disturbance.

The "free swell" oedometer test can also be used to measure the swelling pressure and swelling properties of a soil. The sample is initially allowed to swell freely with a token load applied (Figure 5). The load required to bring the sample back to its original void ratio is termed the swelling pressure. The stress paths adhered to can be understood from a three-dimensional plot of stress state variables versus void ratio and water content. This test has the limitation that it allows volume change and incorporates hysteresis into the estimation of the insitu stress state (i.e., swelling pressure). On the other hand, this procedure somewhat compensates for the effect of sampling disturbance.

2.3 Future Ground Movements

The prediction of future ground movements requires a knowledge of (i) the initial insitu state of stress, (ii) the swelling moduli and (iii) the final state of stress. The initial state of stress can be quantified from the "corrected" swelling pressure. The swelling moduli can be obtained from the rebound data. The final state of stress corresponding to several years after construction must be estimated on the basis of local experience. Possible final pore-water pressure profile are discussed under final boundary conditions.

For discussion purposes, let us assume that the final pore-water pressures go to zero. Figure 6 shows the stress path that would be followed by a soil element at a specific depth. Swelling would follow a path from the initial void ratio, e_o, to the final void ratio, e_f, along the rebound surface of the matric suction plane. The rebound surface can be assumed to be unique (Matyas and Radhakrishna 1968; Fredlund and Morgenstern 1976). Therefore, it is also possible to follow a stress path from the insitu stress state to the "corrected" swelling pressure and then proceed along the rebound curve in the total stress plane to the final stress condition. The advantage of the latter stress path is that the soil properties determined in the total stress plane can be used to predict total heave.

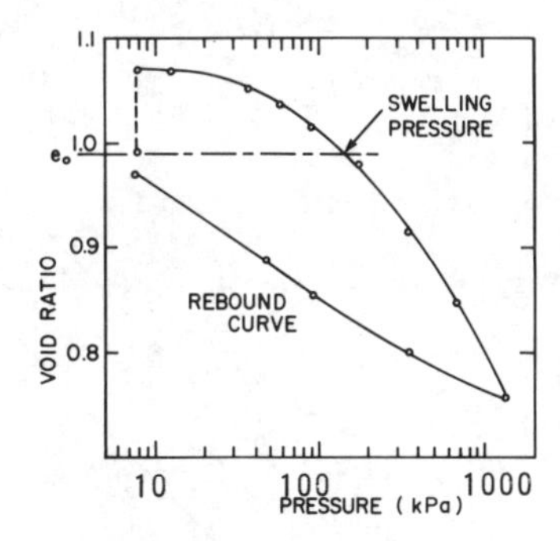

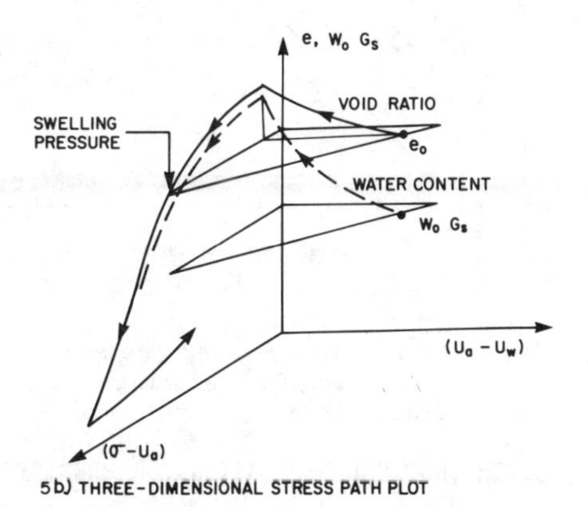

Fig.5 Stress path representation for the 'free swell' oedometer test.

The effects of excavation, replacement of soil with a relatively inert material (e.g., gravel) and loadings can also be taken into account by using appropriate moduli for loading and unloading. However, it is preferable to assume that there is insufficient time for the soil to respond to each loading and unloading, and that long term heave is in response to the net loading or unloading.

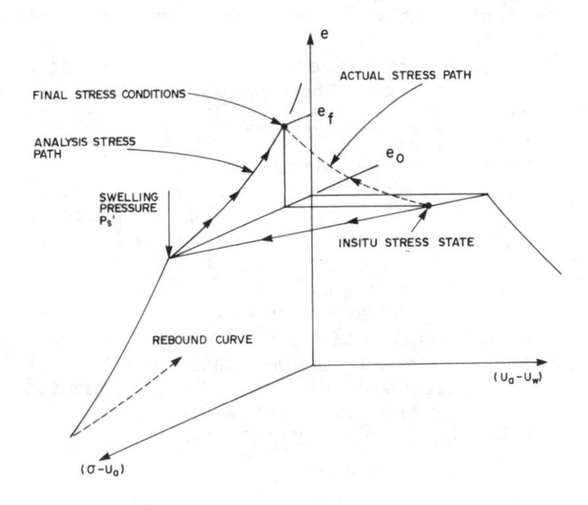

Fig.6 'Actual' and 'analysis' stress paths representing swelling of the soil.

2.4 Determination of Insitu Consolidation/Swelling Curve

When testing <u>saturated</u> clays, the laboratory oedometer test is used to reconstruct the insitu void ratio versus effective stress plot. Likewise, the laboratory oedometer test on desiccated soils can be used to construct a void ratio versus pressure plot for analysis purposes. Often the entire laboratory loading curve is on the recompression portion; not even reaching the virgin compression branch. The preconsolidation pressure of the clay may exceed the highest load applied in the laboratory. The "corrected" swelling pressure indicates the present insitu state of stress on the total stress plane. The lower, "uncorrected" swelling pressure shows the effect of sampling disturbance. Upon access to water in the field, the soil swells along the rebound curve. The laboratory rebound curve in the vicinity of the initial void ratio, e_o, must be translated upward to pass through the "corrected" swelling pressure in order to show the stress path that would be followed.

The following procedure is suggested for

obtaining the "corrected" swelling pressure. First, an adjustment should be made to the laboratory data in order to account for the compressibility of the oedometer apparatus. Desiccated, swelling soils have a low compressibility and the compressibility of the apparatus can significantly affect the evaluation of insitu stresses and the slope of the rebound curve (Fredlund 1969). Second, a correction must be applied for sampling disturbance. Sampling always increases the compressibility of a soil and does not permit the laboratory sample to return to its insitu state of stress at its insitu void ratio. Casagrande (1936) proposed an empirical construction on the laboratory curve to account for the effect of sampling disturbance when assessing the preconsolidation pressure of a soil. Other construction procedures have also been proposed (Schmertmann 1955). A modification of Casagrande's construction is suggested for finding the "corrected" swelling pressure (Figure 7).

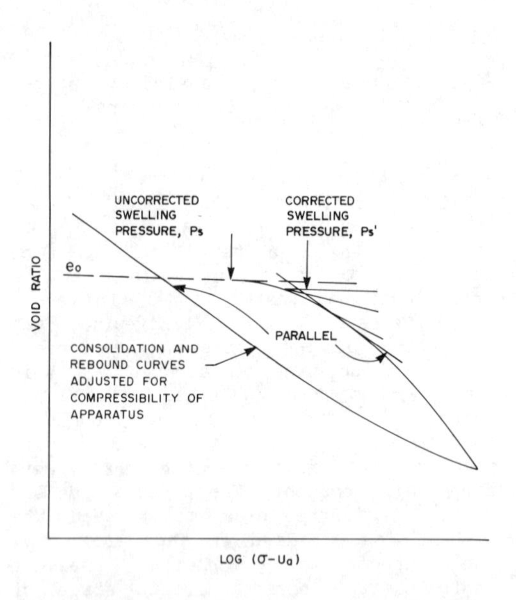

Fig.7 Construction procedure to correct for the effect of sampling disturbance.

The need for applying a correction to the laboratory measured swelling pressure is revealed in numerous ways. First, it would be anticipated that such a correction is necessary as a result of experience in determining preconsolidation pressure. Second, attempts to use the "uncorrected" swelling pressure in the prediction of total heave result in predictions which are too low. Predictions using "corrected" swelling pressure may often be twice the magnitude of those predicted when no correction is applied. Third, the analysis of oedometer results from desiccated deposits often produce results which are difficult to interpret if no correction is applied for sampling disturbance (Fredlund, Krahn and Hasan, 1980).

Figure 8 shows a comparison of "corrected" and "uncorrected" swelling pressure data from 2 soil deposits. The results indicate that it is possible for the "corrected" swelling pressures to be more than 300 percent of the "uncorrected" swelling pressures.

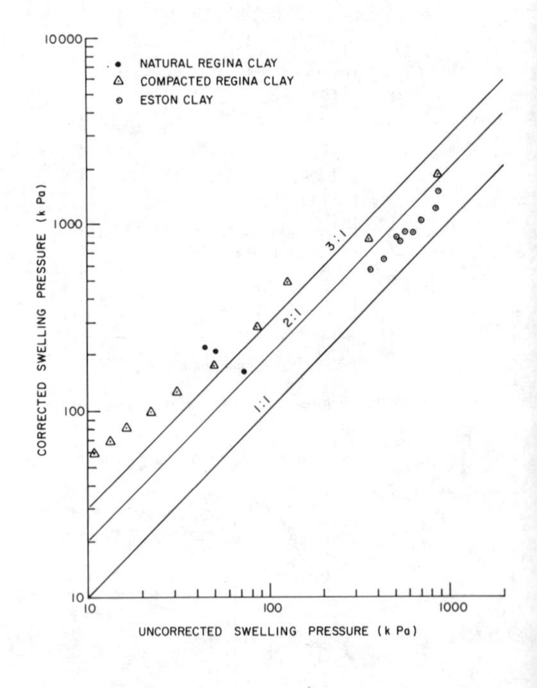

Fig.8 Change in swelling pressure due to correction for sampling disturbance.

3. THEORETICAL DERIVATION FOR PREDICTION OF HEAVE

An equation for the prediction of heave has been previously derived using the unsaturated soil theory (Fredlund, Hasan and Filson 1980). In this paper, reference is made only to the theory necessary to use the results of an oedometer test to predict total heave.

All stress paths considered are transferred to the total stress plane.

The rebound portion of the oedometer test data plotted in a semi-logarithm form, is essentially a straight line.

$$e_f = e_o - C_s \log P_f/P_o \qquad [1]$$

where: e_f = final void ratio
e_o = initial void ratio
C_s = swelling index
P_f = final stress state
P_o = initial stress state.

The initial stress state, P_o, is the sum of the overburden pressure and the matric suction transferred to the total stress plane (i.e., matric suction equivalent). The initial stress state is always equal to the "corrected" swelling pressure.

$$P_o = \sigma_v + (u_a - u_w)_e \qquad [2]$$

where: σ_v = original overburden pressure
$(u_a - u_w)_e$ = matric suction equivalent.

It is necessary to have some understanding of the "corrected" swelling pressure versus depth relationship for the deposit under consideration. The final stress state, P_f, must account for total stress changes and the final pore-water pressure conditions.

$$P_f = \sigma_v \pm \Delta\sigma - u_{wf} \qquad [3]$$

where: $\Delta\sigma$ = change in total stress due to excavation or placement of fill.
u_{wf} = estimated final pore-water pressure.

The heave in an individual soil layer can be written in terms of changes in void ratio.

$$\Delta h_i = h_i \, \Delta e/(1 + e_o) \qquad [4]$$

where: Δh_i = heave in a layer,
h_i = thickness of the layer under consideration
Δe = change in void ratio (i.e., $e_f - e_o$)

The heave in a layer in a strata can be written,

$$\Delta h_i = h_i \frac{C_s}{1 + e_o} \log P_f/P_o \qquad [5]$$

$$\Delta h_i = h_i \frac{C_s}{1 + e_o} \log \frac{(\sigma_v \pm \Delta\sigma - u_{wf})}{(\sigma_v + (u_a - u_w)_e)} \qquad [6]$$

The total heave Δh, is the sum of the heaves computed for each layer.

$$\Delta h = \Sigma \Delta h_i \qquad [7]$$

The matric suction is often a maximum near the ground surface of a deposit. This is also the zone of lowest oveburden pressure. Therefore, the ratio of P_f and P_o is most negative in this region, resulting in the largest amount of heave.

3.1 Initial and Final Pore-water Pressure Boundary Condition

The initial and final stress states must be known in order to perform a heave analysis. The initial and final total stresses can be computed using conventional total stress theory. The initial and final pore-air pressure is equal to atmospheric pressure. The need to know the initial insitu pore-water pressures is circumvented through the manner in which the laboratory oedometer test data is interpreted.

One of three possibilities provides the most logical estimation of the final pore-water pressure conditions. First, it can be assumed that the water table will rise to ground surface, creating a hydrostatic condition. This assumption produces the greatest heave prediction. Second, it can be assumed that the pore-water pressure approaches zero throughout its depth. This may be a realistic assumption; however, it should be noted that it is not an equilibrium condition. Third, it can be assumed that under long-term equilibrium conditions the pore-water pressure will remain slightly negative. This assumption produces the smallest prediction of heave. It is also possible to have variations of the above assumptions with depth. As well, there may be a limit placed on the depth to which wetting will occur. Any of the above assumptions produce similar predictions of heave in most cases. This is due to the fact that most of the heave occurs in the uppermost soil layer where the matric suction change is largest.

The choice of a final pore-water pressure boundary condition can vary from one geographic location to another depending upon the climatic conditions. Russam and Coleman (1961) related the equilibrium suction below asphaltic

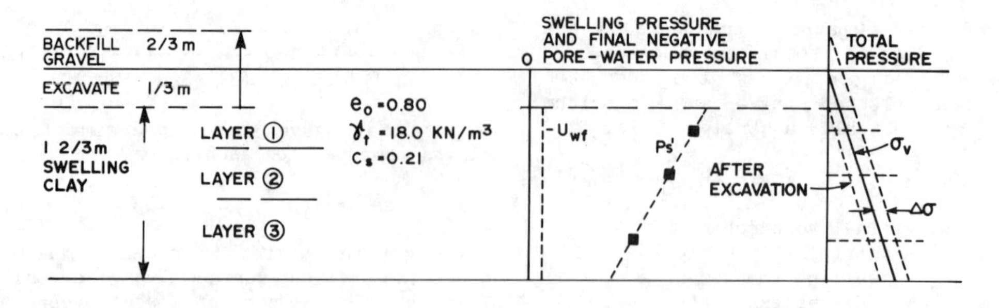

LAYER NO.	THICKNESS (mm)	INITIAL STRESS STATE: $P_o = P_s'$ (kPa)	FINAL STRESS STATE: INITIAL OVERBURDEN σ_v (kPa)	CHANGE IN TOTAL STRESS $\Delta\sigma$ (kPa)	FINAL PORE-WATER PRESSURE u_{wf} (kPa)	$P_f = \sigma_v \pm \Delta\sigma - u_{wf}$ (kPa)	Δh_i (mm)
1	333.	800.	9.0	+6.0	-7.0	22.0	60.6
2	500.	608.	16.4	+6.0	-7.0	29.4	76.7
3	833.	300.	28.4	+6.0	-7.0	41.4	83.6
						TOTAL HEAVE	= 220.9 mm

Fig.9 Calculations for Example

pavements to the Thornthwaite Moisture Index. On many smaller structures, however, it is often man-made causes such as leaky water lines and poor drainage that control the final pore-water pressure in the soil.

3.2 Example Calculations

An example problem is presented to illustrate the calculations required to predict heave (Figure 9). Let us consider a 2-meter layer of swelling soil with an initial void ratio of 0.8, a total unit weight of 18.0 kN/m^3 and a swelling index of 0.21.

Three oedometer tests were performed which show a decrease in the "corrected" swelling pressure with depth (Figure 9).

Suppose the engineering design suggests the removal of 1/3 metre of swelling clay from the surface, prior to the placement of 2/3 metre of gravel. The unit weight of the gravel is assumed as being equal to that of the clay. The 1-2/3 metres of swelling clay is subdivided into 3 strata as shown in Figure 9.

The initial stress state, P_o, can be obtained by interpolation of the laboratory data to the midpoint of each layer The final stress state, P_f, must take into account changes in the total stress and the final pore-water pressure. The final pore-water pressure is assumed to be -7.0 kPa. Equations [5] or [6] can be used to calculate the heave in each layer. The total amount of heave is computed to be 22.1 cm.

Two assumptions are made concerning the heave analysis in the Example. First, it is assumed that the independent processes of excavation and placement of the gravel fill do not allow sufficient time for equilibrium to be established. Therefore, the soil responds only to the net changes in stress. Second, the designation of a final negative pore-water pressure assumes that near saturation, the slopes of the rebound curves on the matric suction plane and the total stress plane approach the same value. This assumption is reasonable provided the final pore-water pressures are relatively small.

4 CASE HISTORIES

Two case histories are briefly presented to demonstrate that the proposed method for predicting total heave can be used with a reasonably high degree of confidence.

4.1 Slab-on-Grade Floor, Regina, Saskatchewan

In 1961, the Division of Building Research, National Research Council, undertook to monitor the performance of a light industrial building which was being constructed in north-central Regina. Details of the study have been presented by Yoshida, Fredlund and Hamilton, 1983. Instrumentation was installed to monitor ground movements at various depths below the slab. Water content changes were monitored using a neutron moisture meter probe. Undisturbed samples were taken as part of the subsurface exploration prior to the construction of the building. Constant volume oedometer tests were performed on 3 samples and the swelling pressures are shown on Figure 10. The average swelling index was 0.09.

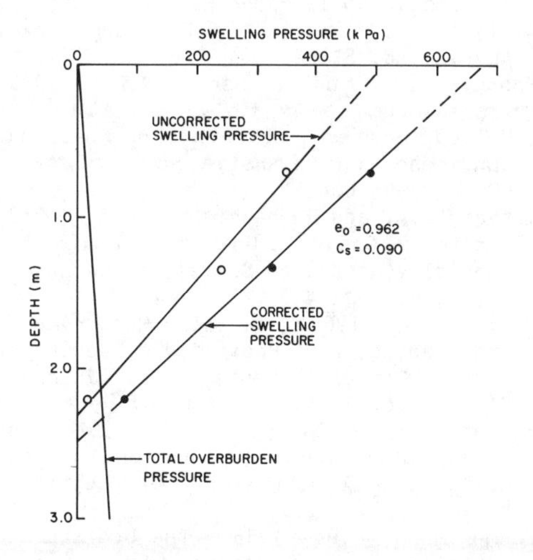

Fig.10 Swelling pressure versus depth for Regina clay (from Yoshida, Fredlund and Hamilton, 1983).

Approximately one year after construction, the owner noticed considerable cracking of the floor slab. Precise level surveys showed the maximum total heave to be 106 mm. The owner had also noticed a significant increase in water consumption (i.e., 35,000 litres). It was discovered that a leak had occurred in the hot water line beneath the floor slab, at the location of maximum heave. The leak was immediately repaired.

Heave analyses were performed using the laboratory oedometer data. Various assumptions were made concerning the final pore-water pressures. When it was assumed that the soil had become saturated and the water table rose to the base of the floor slab, the predicted heave was 141 mm. Assuming that the negative pore-water pressures were reduced to zero, gave a total heave prediction of 118 mm. Assuming a final pore-water pressure of -50 kPa, gave a total heave prediction of 66 mm. On the basis of the heave analysis, it appears that the assumption of zero pore-water pressure was probably the most realistic for this case history. It appears that further heave would likely have taken place had the leak not been repaired. The prediction of heave at various depths also showed close agreement with the actual measurements.

4.2 Eston School, Eston, Saskatchewan

Soils in the Eston area of Saskatchewan have long been known as extremely high swelling. The stratigraphy consists of approximately 7-1/2 metres of highly plastic, brown clay overlying the glacial till. Many light structures in the area have undergone serious distress. The building of particular interest is the Old Eston School constructed in the late 1920's.

The school building was constructed on concrete strip footings and a wooden basement floor was supported by interior surface concrete footings. The school was a two-storey structure with classrooms in both the lower and upper levels. The lower floor was approximately 1.2 m (4 feet) below grade. The exterior concrete walls were founded approximately 1.8 m (6 feet) below grade.

A substantial amount of heave has taken place below the interior footings. Although the performance has not been precisely recorded, the heave in one portion of the basement area has been severe. On two occasions during the history of the school, 15 cm to 30 cm (6 to 12 inches) of soil have been removed from below the interior footings. As much as 45 cm to 90 cm (1-1/2 to 3 feet) of total heave has occurred during the life of the school according to maintenance records. Large amounts of differential heaving of the floor (i.e. 15 cm or 6 inches) were measured in 1960. The school was demolished in 1967.

In 1981, a subsurface investigation was conducted adjacent to the location of the old school. Undisturbed soil samples were taken and "constant volume" oedometer tests were performed. The results are presented in Figure 11. The average natural water content throughout the profile is 25 percent. The average lastic limit is 27 percent and the .verage liquid limit is 100 percent. The average swelling index is 0.21. Due to a lack of detailed information on the soil and performance conditions of the school, it is not possible to do a precise total heave analysis. It is of interest, however, to perform an approximate analysis. Using the corrected swelling pressure from Figure 11 and assuming the negative pore-water pressures went to zero, the predicted heave would be in excess of 90 cm (3 feet).

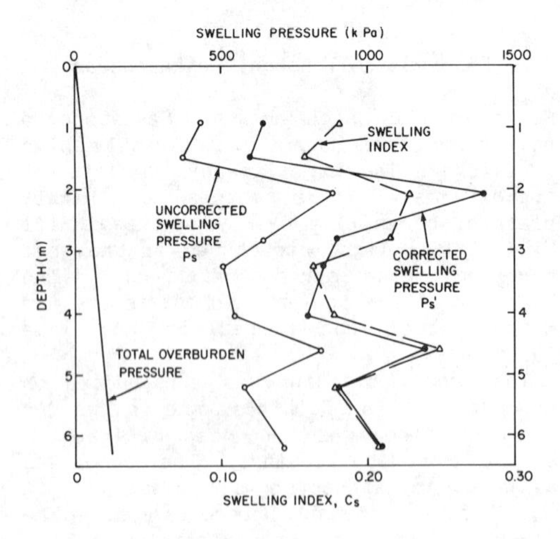

Fig.11 Swelling index and swelling pressure versus depth for Eston clay.

Further case history studies would be useful in confirming the proposed procedure for predicting total heave.

REFERENCES

Casagrande, A. 1936 The Determination of the Pre-Consolidation Load and Its Practical Significance , Discussion D-34, Proceedings of the First International Conference on Soil Mechanics and Foundation Engineering, Cambridge, Vol. III, pp. 60-64.

Fredlund, D.G. 1969 Consolidometer Test Procedural Factors Affecting Swell Properties , Proceedings of the Second Conference on Expansive Clay Soils, Texas A & M Press, College Station, TX, pp. 435-456.

Fredlund, D.G., Hasan, J.U., and Filson, H. 1980 Proceedings of the Fourth International Conference on Expansive Soils, Denver, CO, pp. 1-17.

Fredlund, D.G., Krahn, J. and Hasan, J.V. (1980) Variability of an Expansive Clay Deposit , Proceedings of the Fourth International Conference on Expansive Soils, Denver, CO, pp. 322-338.

Fredlund, D.G. and Morgenstern, N.R. 1977 Stress State Variables for Unsaturated Soils , Journal of the Geotechnical Engineering Division, ASCE, Vol. 103, No. G75, pp. 447-466.

Fredlund, D.G. and Morgenstern, N.R. 1976 Constitutive Relations for Volume Change in Unsaturated Soils , Canadian Geotechnical Journal, Vol. 13, No. 3, pp. 261-276.

Jones, D.E. and Holtz, W.G. 1973 Expansive Soils - The Hidden Disaster , Civil Engineering, ASCE (New York), August, pp. 87-89.

Krohn, J.P. and Slosson, J.E. 1980 Assessment of Expansive Soils in the United States , Fourth International Conference on Expansive Soils, Denver, CO, pp. 596-608.

Matyas, E.L. and Radhakrishna, H.S. 1968 Volume Change Characteristics of Partially Saturated Soils , Geotechnique 18(4), pp. 432-448.

Noble, C.A. 1966 Swelling Measurements and Prediction of Heave for a Lacustrine Clay , Canadian Geotechnical Journal, Vol. 3, pp. 32-41.

Russam, K. and Coleman, J.D. 1961 The Effect of Climatic Factors on Subgrade Moisture Conditions , Geotechnique, Vol. II, pp. 1-22.

Schmertmann, J.H. 1955 The Undisturbed Consolidation Behavior of Clay , Transactions, ASCE, Vol. 120, pp. 1201-1227.

Yoshida, R.T., Fredlund, D.G. and Hamilton, J.J. 1983 The Prediction of Total Heave of a Slab-on-Grade Floor on Regina Clay , Canadian Geotechnical Journal, Vol. 20, pp. 69-81.

Prediction and Performance in Geotechnical Engineering / Calgary / 17-19 June 1987

Fact and fiction in the field of vertical drainage

S.Hansbo
Chalmers University of Technology, Gothenburg, Sweden, and AB Jacobson & Widmark, Lidingö, Sweden

ABSTRACT: Suspicion thrown on the utility of drain installations for the purpose of speeding up consolidation, or on the use of a certain type of drain, may be ill-founded and based solely on whim. Thus, when due consideration is taken to the consolidation characteristics of the soil and to the specific drain behaviour, the consolidation process can undoubtedly be predicted with acceptable accuracy, irrespective of the type of drain chosen. In this paper, the most important soil and drain parameters governing the consolidation process in vertically drained soil are discussed. The reliability of design methods is discussed in the light of factual data from well-monitored test areas on clay subsoil, areas that have been under observation for more than 30 years. The utility of vertical drain installations in peat is also discussed on the basis of field tests.

1 INTRODUCTION

Precompression and vertical drains for stabilization of weak, highly compressible soils have now been used for more than 50 years. Originally only sand drains were utilized although prefabricated drains were invented and tested in a full-scale field test in Sweden only a few years after the first sand drain installation in the USA (see, for example, Hansbo, 1977). In recent years prefabricated drains have been used to an ever increasing extent whereas sand drains are loosing their share of the market. This has entailed that a great number of different makes of drains, emanating from the Kjellman cardboard wick of 1939, have appeared on the market, Fig. 1, and all claim good performance and a high draining capacity. In the jungle of drain products it may be difficult to select the best drain for the purpose, and therefore the price of the drain generally decides which drain is chosen. However, there are methods by which the properties of a specific drain can be judged with relatively good reliability, even on a laboratory scale. The best method, of course, is to base one's judgement on full-scale field tests, which may be quite costly and time-consuming.

The number of drain installations throughout the world is now very large and one would have thought that the experience gained from these installations would have eliminated any doubts about the drains not serving their purpose. But obviously this is not the case. The suspicion thrown on the use of vertical drains may derive from negative experience of some publicized failures indicating that drains have had no positive effect on the consolidation process. Undoubtedly, however, such failures are a result of erroneous design due to misinterpretation of the consolidation characteristics of the subsoil, or of bad execution.

In this paper, some results of full-scale field tests and of laboratory tests will be presented for the purpose of elucidating some points in dispute within the field of vertical drainage. Hopefully, the data presented will contribute to a better understanding of the problems involved, and ultimately, to a safer design of precompression and vertical drain installations.

2 GOVERNING THEORETICAL RELATIONS

The theoretical solution to the hydrodynamically delayed (primary) consolidation process for a vertically drained soil is either based on the assumption that the vertical strains in the soil develop freely in full accordance with the

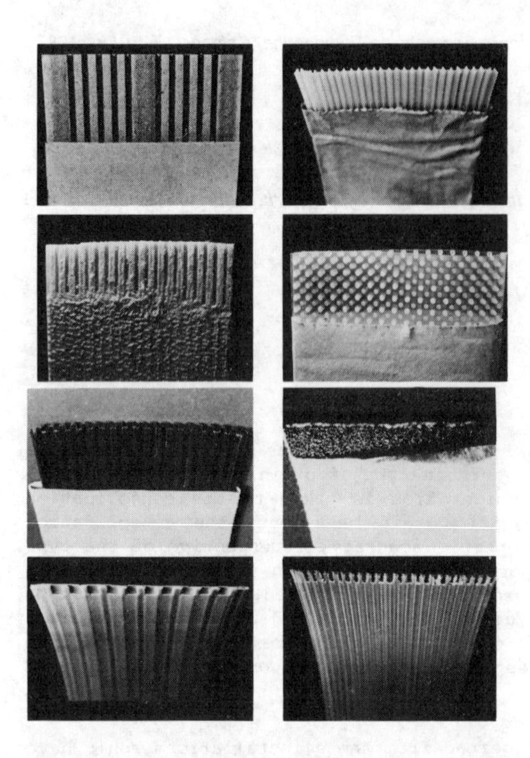

Fig. 1 - Some of the drain makes introduced on the market. From left to right, top to bottom: Cardboard wick, Geodrain, Castle Board, Alidrain, Mebradrain, Colbond, PVC and Desol.

rate of dissipation of excess pore water pressure ("free strain theory"), or on the assumption that horizontal sections in the soil remain horizontal throughout the consolidation process ("equal strain theory"). The latter assumption was shown by Barron in 1944 to lead to very nearly the same average excess pore pressure dissipation (rate of consolidation) as the former one. Barron's original solution was based on the simplified assumption of no well resistance (highly pervious drains) and no effect of drain installation on the soil properties. Later, Barron (1948) introduced also the effect of well resistance (limited drain permeability) and of disturbance by drain installation (zone of smear). Since then, many researchers have presented extended solutions taking into consideration some more sophisticated cases, e.g. time-dependent loading (Schiffman, 1958), deviation from Darcy's flow law (Hansbo, 1960), variations in soil properties during consolidation (Mesri & Rokhsar (1974), well resistance (Yoshikuni & Nakanodo, 1974), influence of well resistance and radial strain (Yoshikuni, diss. 1979 - in Japanese), viscous resistance in organic soils (Fürstenberg et al., 1983), partially penetrating drains (Runesson et al., 1985).

A large number of case records were presented at a symposium on vertical drains arranged by the Institution of Civil Engineers in London in 1982. Many interesting case records can also be found in the Proceedings of the 8th ECSMFE, Helsinki, 1983. A very extensive text book on vertical drains has been written by Magnan, Tec&Doc - Lavoisier, Paris 1983.

Although researchers may need to go into all the subtle details of the consolidation problem of a vertically drained soil, this is hardly true for engineers who work with vertical drain projects in practice. For them the important aspect is that prediction does not deviate from performance insomuch that it entails structural damage.

A very simple solution to the consolidation problem in vertically drained mineral soil, and one which is easy to apply, was outlined by Hansbo (1979) and presented in full detail by Hansbo (1981). This solution, which is based on the equal strain theory and takes into account the effect of smear and well resistance gives almost exactly the same results as the far more complicated solutions presented by Barron and Yoshikuni and Nakanodo. Possible deviations are negligible. According to this solution, the rate of excess pore pressure dissipation in a

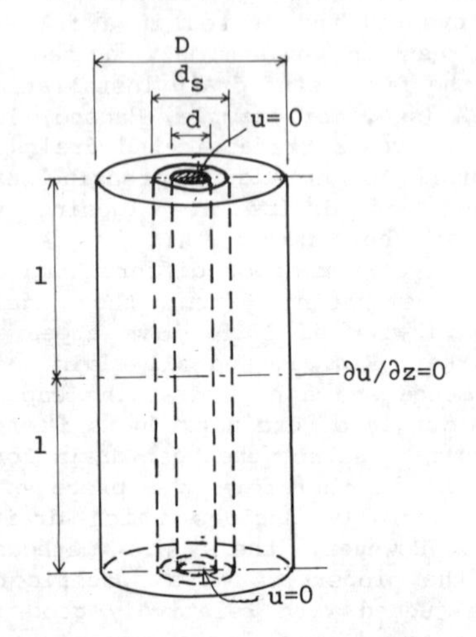

Fig. 2 - Each drain is assumed to dewater a circular cylinder with diameter D. The cross-sectional area of this cylinder is equal to that of the parallelogram formed by four neighbouring drains in two adjacent drain rows.

water saturated soil is governed by the relation (Fig. 2).

$$(\bar{u}_h/\bar{u}_o)_z = \exp(-8T_h/\mu) \qquad (1)$$

where $(\bar{u}_h/\bar{u}_o)_z$ = ratio at depth z of average excess pore pressure at time t and average initial excess pore pressure

$T_h = c_h t/D^2$ time factor

c_h = coefficient of consolidation in horizontal pore water flow

t = time of consolidation

D = diameter of soil cylinder dewatered by the drain ($D=2\sqrt{A/\pi}$, where A is the parallelogram formed by four neighbouring drains in two adjacent rows).

$\mu = \ln(n/s)+(k_h/k_s)\ln(s)-0.75+$
$+\pi z(2l-z)k_h/q_w$

n = D/d, where d = drain diameter

$s=d_s/d$, where d_s = diameter of zone of smear.

l = full length of drain if closed at bottom (half length of drain if open at bottom)

k_h = permeability (in horizontal direction) of undisturbed soil

k_s = permeability (in horisontal direction) of soil within zone of smear

q_w = specific discharge capacity of drain ($q_w=k_w A_w$, where k_w = permeability in vertical direction of the drain, and A_w = cross-sectional area of drain)

In the ideal case of negligible well resistance ($\pi l^2 k_h/q_w \to 0$) and disturbance ($s \to 1$) we have
$\mu = \ln(n)-0.75$.

The average degree of consolidation

$$\bar{U}_h = 1-\bar{u}_h/\bar{u}_o = s/s_p \qquad (2)$$

where s = settlement at time t
s_p = final primary settlement.

Vertical drains are utilized in cases where the consolidation process without drains would be very slow . Therefore, the effect of pore pressure dissipation due to pore water being squeezed out of the soil in the vertical direction in between the drains can usually be disregarded. If this is not the case, for example if the distance between horizontal pervious layers is relatively small, then the coupled effect of radial and vertical drainage can be calculated according to Carillo (1942)

$$(u/u_o)_z=(\bar{u}_h/\bar{u}_o)_z(u_v/u_o)_z \qquad (3)$$

where $\bar{u}_h$ is obtained from Eq. (1) and u_v from Terzaghi's one-dimensional consolidation theory. In practice, u_v is most easily obtained by using finite difference methods and the approach made by Helenelund (1951).

Expressing the average degree of consolidation at a certain depth z as $\bar{U}_z$, we find

$$\bar{U}_z=1-(\bar{u}/\bar{u}_o)_z=\bar{U}_{hz}+U_{vz}-\bar{U}_{hz}U_{vz} \qquad (4)$$

where $\bar{U}_{hz}$ is given by Eq. (2)

and $U_{vz} = 1-(u_v/u_o)_z$

3 SOIL PARAMETERS

Coefficient of consolidation: The most important soil parameter governing the consolidation process is the coefficient of consolidation for horizontal pore water flow (c_h). Thus, the time required to reach a certain degree of consolidation is inversely proportional to the numerical value of the coefficient of consolidation according to the relation

$$t = (\mu D^2/8c_h)\ \ln(\bar{u}_o/\bar{u})$$

or

$$t = -(\mu D^2/8c_h)\ \ln(1-\bar{U}_h) \qquad (5)$$

Therefore, a correct determination of c_h is of paramount importance for the design of the drain installation. However, in the conventional oedometer test, usually only the coefficient of consolidation for vertical pore water flow is determined. The c_v value thus obtained will be affected by sample disturbance: the higher the disturbance, the lower the coefficient of consolidation. Moreover, the existence of horizontal seams with higher permeability entails a higher c_h than c_v value. Even in apparently homogeneous soil the permeability (and thus the coefficient of consolidation) in the horizontal direction is generally much higher than in the vertical direction. Undoubtedly, choosing the c_v value found by oedometer tests for c_h may lead to a very conservative design, particularly in a case where the soil is layered.

The best way to find a reliable basis for design is to measure in-situ the permeability in the horizontal direction (k_h) by means of permeability probes developed for the purpose, and to determine, in the laboratory, the compression modulus (M) by oedometer tests. From these results, the coefficient of consolidation can be calculated: $c_h = k_h M/\gamma_w$, where $\gamma_w = g\rho_w$ is the unit weight of water. Again the influence of sample disturbance on the M value has to be considered.

Preconsolidation pressure: another extremely important soil parameter that has to be determined is the preconsolidation pressure σ'_c, or as it is also designated: the maximum past pressure p_c. This ought to be done on the basis of oedometer tests on so-called undisturbed soil samples taken at different depths. However, the soil samplers used frequently cause severe sample disturbance and the results of the oedometer tests may therefore be misleading. The disturbance will not only cause a reduction of the σ'_c value but will also lead to a possible underestimation of the compressibilty of the soil (overestimation of the M value).

The result of an oedometer test on disturbed clay is presented in Fig. 3. The conventional presentation of strain in linear scale vs. stress in logarithmic scale may lead to serious misjudgement of the preconsolidation pressure. A presentation of strain vs. effective stress in linear scales gives a much better picture of the preconsolidation stress level.

The most plausible reason why, in some cases, vertically drained areas subjected to loading have performed in a very similar way as undrained areas, indicating that drains have had no positive effect on the consolidation process, is that the stress increase due to loading is below the preconsolidation pressure. In such cases, of course, the use of drains is usually meaningless. Thus, the coefficient of consolidation of overconsolidated soil is in the order of magnitude of 10-100 times higher than that of normally consolidated soil. The consolidation time without drains is speeded up correspondingly. Moreover, the disturbance caused by drain installation may entail such a decrease of the coefficient of consolidation that the consolidation time when drains are installed will be only slightly reduced (or even unaffected).

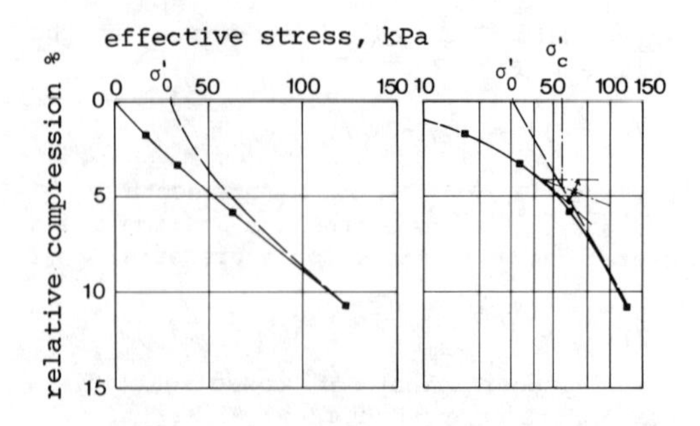

Fig. 3 - The conventional semilogarithmic plot of the oedometer curve (right) may give a false impression of the soil's preconsolidation pressure. The linear plot indicates severe sample disturbance and shows no sign of a preconsolidation pressure. The oedometer modulus $M=\Delta\sigma'/\Delta\varepsilon$, determined from the inclination of the disturbed σ'vs. ε curve, will lead to an underestimation of the settlement of normally consolidated soil.

Zone of smear: the extent and condition of the more or less remoulded zone around each drain depend very much on the installation method. Installation by careful rotary jetting or by the Dutch jet-bailer method seems to cause a minimum disturbance while installation with the aid of a closed-end mandrel, particularly if driven into the soil by pile or vibratory hammers, seems to cause a maximum disturbance. It is doubtless very difficult to tell what values of s and k_h/k_s should be inserted for calculation of the μ value in Eq. (1). Detailed investigations of the extent of the zone of smear presented by Holtz & Holm (1973) and Akagi (1976) indicate that s = 2 is a realistic and acceptable figure for drains installed by a closed-end mandrel.

Regarding the ratio k_h/k_s, this is very much dependent on the texture of the soil. One way of solving this problem is to replace k_s for k_v, i.e. to assume that horizontally oriented texture has been transformed, by remolding, into vertically oriented texture. The choice of s and k_h/k_s values has a very great influence on the time of consolidation. For example, in a case where n = 10, and the zone of smear has a diameter d_s=2d (i.e. s=2), the time required to achieve a certain degree of consolidation according to Eq. (5) will be nearly doubled if k_h/k_s=3, and more than tripled if k_h/k_s=6.

4 DRAIN PARAMETERS

The drain parameters governing the consolidation process are the drain diameter and the discharge capacity of the drain. For prefabricated drains, filter criteria also have to be considered.

Drain diameter: For sand drains, the question of which diameter to insert in Eq. (1) may seem simple to answer but is not always so. The diameter may not only be dependent on the diameter of the installation tool but also on the method of installation and on the shear strength of the clay. However, this does not create a serious problem since the influence on the consolidation process of an erroneous choice of diameter is rather limited. A misjudgement of the diameter by ±20% will change the consolidation process by only around ±5%. A more serious problem is the possible risk of necking or of other discontinuities, leading to clogging.

For prefabricated drains (including sand wicks, Fig. 4), the question of drain diameter is a matter of how the drain itself is designed. For band-shaped drains, we have to find an equivalent diameter to be inserted in Eq. (1). It can be shown (Hansbo, 1979) that a band-shaped and a circular drain have the same consolidation effect if their circumference is equal. Therefore, the equivalent diameter can be calculated from the relation

$$d = 2(a+b)/\pi \qquad (6)$$

where a = thickness of drain
b = width of drain.

Fig 4 - Sand wicks are characterized by the sand being enclosed in a "stocking" which prevents necking. Diameter usually 60 mm.

Well resistance: well resistance can entail a considerable increase in consolidation time, particularly where drains are installed to great depth. As shown by Eq. (1), the maximum influence is obtained at depths corresponding to half the length of the drains if open at top and bottom, or to the full length of the drains if closed at bottom.

As an example, for 40 m deep drains, closed at bottom, a discharge capacity of 500 m^3/year will cause a prolongation of the consolidation time by a maximum of around 10% in relation to the ideal case of no well resistance. The corresponding figure for a discharge capacity of 100 m^3/year is around 50%. In the more common cases where drains are 10-20 m deep and open at both top and bottom, the discharge capacity may be reduced to around 30 m^3/year without causing more than about 10% prolongation of the consolidation time.

The discharge capacity of sand drains will be dependent mainly on the grain size distribution of the drain sand and on the risk of siltation (silt intruding into the sand when the mandrel is withdrawn). The permeability of clean drain sand can be expected to vary from around 300 to around 30000 m/year. Choosing these extremes we find, for example, for drains 0.18 m in diameter q_w=8-800 m^3/year.

As soon as different makes of drains were introduced on the market, a somewhat infected discussion began regarding their respective advantages over other competing drains. At an early stage in these discussion the concept of open surface area was advocated by Fellenius (1978). According to this concept, the "open" part of the contact surface between filter and drain core is decisive for the efficiency of the drain. In consequence of this concept, Fellenius suggests that the diameter of a prefabricated drain should be related to that of a circular sand drain in relation to their respective relative open surface areas (for sand drain equal to the porosity of the sand). Thus, according to Fellenius, Geodrain should be given an equivalent diameter of 100 mm, Castle Board 70 mm and Alidrain 150 mm. Unfortunately, the concept of free surface area is still considered by some people to be relevant. In my opinion, it is mearly an attempt to replace the hydraulic conductivity with something else that might seem logical at first sight, but which in reality will only cause confusion. Going back to the deduction of the equations governing the consolidation process and the analysis leading to Eq. (6), it is immediately realized that this concept is not convincing. Thus, an empty cylindrical hole which represents a drain with no well resistance (one of the original theoretical assumptions) would accordingly lead to an equivalent diameter 3 times larger than that of a sand drain with 33% porosity. This, of course, cannot be true. It is the discharge capacity of the drain that matters.

A detailed discussion of the discharge capacity of different drain makes has been presented earlier by the Author (Hansbo, 1986). Fig. 5 summarizes some of the results obtained by laboratory investigations of the q_w values for different drain types. In all these investigations the drains were placed in a triaxial cell and were surrounded by soil. As can be seen, q_w decreases with increasing lateral pressure against the drain. The reason for this decrease is either that the filter sleeve is squeezed into the channel system or that the channels themselves are compressed and, consequently, diminished by the lateral soil pressure.

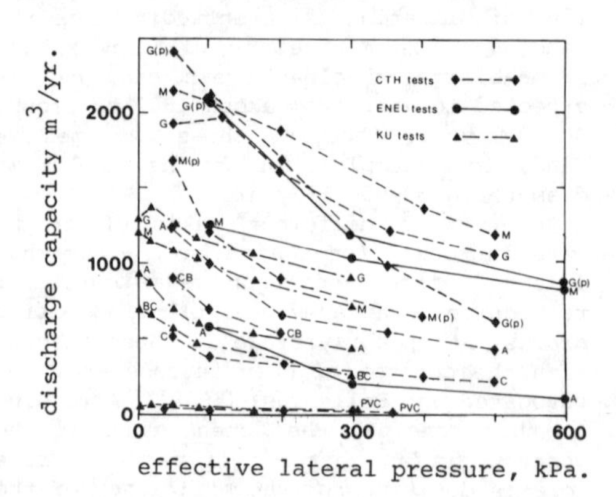

Fig 5 - Results of discharge capacity tests for different drain makes carried out on a laboratory scale. Drains enclosed in soil. In the ENEL tests (Jamiolkowski et al., 1983) the drains have been tested in natural shape (100 mm width), in the CTH tests (Hansbo, 1983) and in the KU tests (Kamon, 1984) with reduced width (40 mm and 30 mm, respectively). Legend: A=Alidrain; BC=Bando Chemical; CB=Castle Board; C=Colbond; G=Geodrain; M=Mebradrain. (p) indicates filter sleeve of paper. (After Hansbo, 1986).

Filter criteria: The filter material originally used (cardboard wick and Geodrain) was of paper. In spite of the fact that all drain installations with Geodrains with paper filter showed excellent performance, suspicion was nevertheless thrown on the use of paper as filter material, particularly by competitors who used synthetic filter material. As a result of the spread and increased acceptance of this unjustified suspicion all drains are now provided with synthetic filter sleeves, thereby making the drains more expensive than necessary.

One of the main objections against paper filter was the risk of deterioration caused by fungi or bacteria. Doubtless, the most obvious risk would exist in organic soils. A very interesting investigation of this problem was made by Prof. Wolski at Warsaw Agricultural University (Koda et al., 1986). Two types of Geodrain, one with paper filter and the other with synthetic filter, were installed inside perforated steel tubes 320 mm in diameter, to 6 m depth of which 3 m was in peat and 3 m was in slime. The tubes, containing the soil with the drains, were pulled up at different times after the insertion of the drains, Fig. 6, and taken to the laboratory. The soil with the drains in its centre was carefully trimmed to fit into a triaxial cell for determination of the discharge capacity. The results obtained are given in Fig. 7. Obviously, the synthetic filter has a better durability than the paper filter and consequently a less pronounced time effect on discharge capacity. However, after more than 1½ year in peat and slime, the drains with filter paper can be considered to perform in quite a satisfactory manner.

Fig 6 - Tube containing soil with Geodrain in centre, being pulled up.

Another filter problem frequently discussed is the range of permeability that should be required. Since the thickness of the filter sleeves generally used for drains is very small (some tenths of a mm) a very low filter permeability can be allowed. For a filter thickness of 0.2 mm it can be shown (Hansbo, 1981, and 1986) that there is no need for a higher long-term filter permeability than about 0.03 m/year, even in a case where the discharge capacity is 2000 m^3/year and the drains are placed in silt. On the other hand, too coarse-meshed filters may create a risk of the channel system becoming silted as a result of not retaining fines in the pore water.

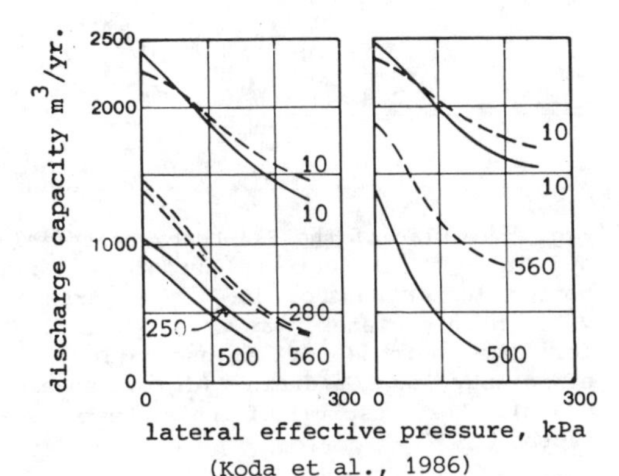

(Koda et al., 1986)

Fig. 7 - Results of discharge capacity tests carried out on Geodrains that have been pulled up after having been in peat (left) and slime (right) in-situ for different lengths of time (number of days in soil given in figure). Results obtained with filter sleeves of paper are shown with full lines and with filter sleeves of synthetics with broken lines.

Stress vs. strain characteristics: From a practical point of view it is important that prefabricated drains can withstand the hard treatment to which they are subjected under field conditions. Among other things, the integral parts of the drains cannot be allowed to fail during installation. Therefore, the filter sleeve and the core should reach about the same elongation at rupture. Moreover, the ultimate elongation should be as large as possible.

An example of the stress/strain characteristics of filter and drain core is given in Fig. 8.

Another important problem is how the drains will function in extremely compressible soil, such as peat and slime, or when the load is heavy enough to cause

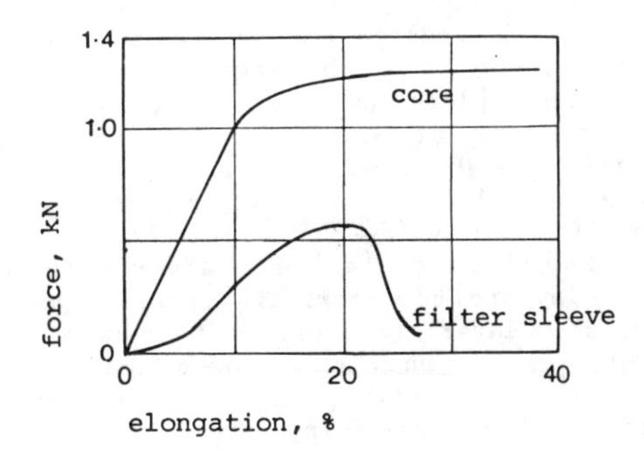

Fig. 8 - Stress vs. strain characteristics of Geodrain core and filter sleeve of synthetic material (Storalene) with permeability 270 m/year (8.5×10^{-6} m/s). Rate of strain 0.3%/min.

unusually large relative compression. The resultant folding ("buckling") of the drain may alter the discharge capacity of the drain. Brittle drain cores that break easily in bending and drain cores where the channel area will be seriously reduced in bending should be avoided. However, usually the negative effect of possible folding will be restricted mainly to the end of the consolidation process when the need for high discharge capacity is less essential.

5 RELIABILITY IN PREDICTION

Primary consolidation of clay: A great number of case records have been published showing an excellent agreement between predicted primary consolidation process and performance (see, for example, the Proceedings of 10th ICSMFE, Stockholm, 1981 and of 8th ECSMFE, Helsinki, 1983). However, difficulties in prediction are often encountered in the case of heterogeneous subsoil (see for example, Géotechnique, Vol XXXI, No 1, 1981). Moreover, when judging the performance, important differences may be found depending upon which basis for interpretation is chosen: remaining excess pore water pressure, or remaining settlement to be expected. It may seem justified to base one's judgement of the consolidation process simply on the rate of pore pressure dissipation, particularly since this is also the basis for the theoretical approach. However, it may be that the placement of the load for precompression will change the final equilibrated pore pressure distribution, thus making the interpretation quite difficult. Moreover, the consolidation process in itself may entail self-induced pore water pressures. This, for example, has been reported to have taken place in a test field at Väsby, 30 km north of Stockholm with two test areas, 15 m by 15 m, arranged in 1945-1946 (Chang, 1981). Here the soil consists of a 15 m thick clay layer underlain by bedrock intervened by a thin sand layer between the clay and the rock. The upper part of the clay layer, to a depth of 13-14 m, is postglacial, and the lower part is glacial. The average excess pore water pressure still remaining after 33 years of loading is about 38% of the initial average excess pore water pressure in spite of the fact that a load reduction of 32% was obtained due to submergence of part of the fill in the course of settlement. The remaining excess pore water pressure, corrected with regard to load reduction, is thus around 56% of the original value, which indicates an average primary consolidation degree of only 44%. Settlement observations indicate that primary settlement was still taking place 33 years after the placement of the load, The observed settlement after 33 years exceeds the calculated final primary consolidation settlement by about 25%.

Of course, from a practical point of view, expected future settlement is of greater interest than remaining excess pore water pressure. Due to the difficulties involved in the interpretation of field data, both settlements and pore pressures should be monitored.

In papers presenting case records, the agreement between prediction and performance is generally achieved by adjustment of the soil parameters to fit performance in the best possible way. This, of course, is not possible in real design. The Skå-Edeby test field represents one case record where the answer to the complete consolidation process was not known when the results were first published (Hansbo, 1960). This test field has now been under settlement observations for more than 30 years. Pore pressure observations went on for 14 years (Holtz & Broms, 1972). My own prediction of final primary settlement and of the time required to achieve a certain degree of primary consolidation can now be checked and thus be used as an indication of the reliability in design.

The Skå-Edeby test field, located about 25 km west of Stockholm, contains five circular test areas, Fig. 9. The soil consists of normally consolidated postglacial clay to a depth of 4-5 m underlain by glacial clay. The depth to firm bottom

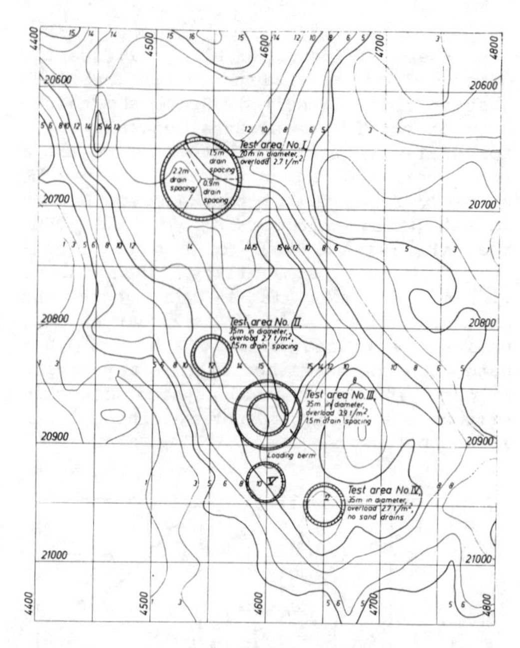

Fig. 9 - Plan of the Skå-Edeby test field and of depth from ground surface to firm bottom (after Hansbo, 1960) Test area No V, 30 m in diameter, was arranged in 1972 for the purpose of investigating the efficiency of Geodrain (drain spacing 0.9 m). For results of this test, see Hansbo (1977).

of moraine or bedrock varies from round 10 to 15 m. The geotechnical characteristics of the clay are presented in detail in earlier publications (Hansbo, 1960; Holtz & Broms, 1972), and will therefore not be repeated here. Only the parameters related to design will be given. For the 5 m thick soil layer from 2.5 to 7.5 m depth, originally selected for study in order to avoid the influence of the dry crust and irregularities in soil characteristics near firm bottom, c_v varies from 0.13 to 0.22 m^2/year. The compression ratio CR = $\varepsilon_2/\log 2 = C_c/(1+e_o)$ varies from 0.43 to 0.56. The coefficient of consolidation in horizontal pore water flow was determined in an oedometer test where drainage was allowed through a central drain (Hansbo, 1960) and was found to be c_h = 0.7 m^2/year, i.e. 3 to 5 times c_v.

A comparison between the results obtained during a loading time of 3 years and the results to be expected according to the then conventional design method (assuming validity of Darcy's law, no smear and no well resistance) showed that an acceptable agreement would be obtained by using a coefficient of consolidation c_h

of around 0.4 m^2/year for 1.5 and 2.2 m drain spacing and around 0.2 m^2/year for 0.9 m drain spacing. This would mean that 99% primary consolidation would be achieved after 2.5 years at 0.9 m drain spacing, after 5.5 years at 1.5 m drain spacing, and after 14.5 years at 2.2 m drain spacing. Here, the disturbance effect of drain installation is reflected in the smaller c_h value chosen to fit the observational data at 0.9 m drain spacing. (The best agreement between predicted and observed pore pressure dissipations was obtained when non-validity of Darcy's law was assumed.) Introducing instead a zone of smear with diameter $d_s = 2\ d$ and with permeability $k_s = 0{,}25\ k_h = 0.005$ m/ year, and using $c_h = 0.8\ m^2$/year and $q_w = 100\ m^3$/year, we find that 99% primary consolidation is reached after 2 years at 0.9 m spacing, after 6 years at 1.5 m spacing and after 15 years at 2.2 m spacing.

The actual performance is shown in Figs 10 and 11. According to these figures, primary consolidation can be considered as completed after about 2 years at 0.9 m spacing, after about 4 years at 1.5 m spacing and after about 9 years at 2.2 m spacing. Calculations performed on the same basis as previously, with the suggested values of zone of smear, discharge

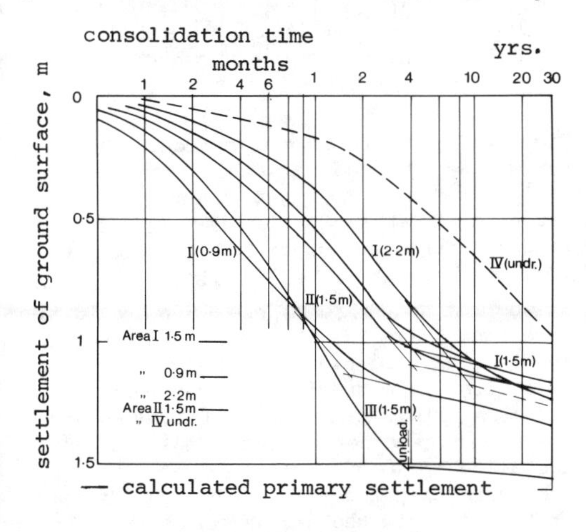

Fig. 10 - Time vs. settlement curves for test areas at Skå-Edeby. Test areas I, II and IV loaded with 1.5 m gravel fill, test area III with 2.2 m gravel fill (overload of 0.7 m gravel fill which was removed after 4 years). Primary settlement calculated with due regard to load reduction by submergence of fill in the course of settlement. Compression index modified with regard to disturbance (as suggested by Hansbo, 1960).

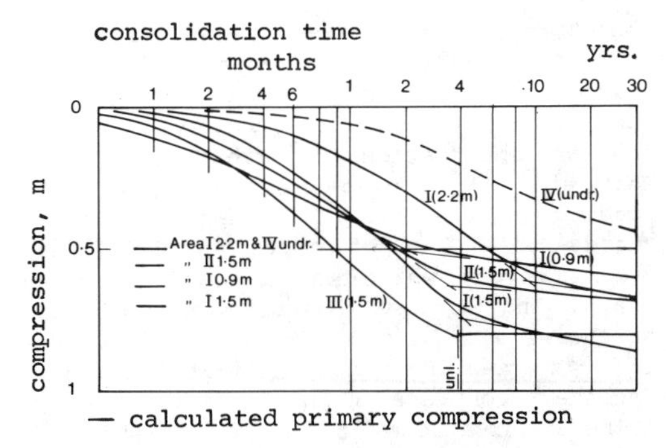

Fig. 11 - Consolidation of the 5 m thick clay layer between 2.5 and 7.5 m depth. After removal of overload in Area III, no secondary compression has taken place during a period of observation of 26 years.

capacity etc., leads to $U_h = 95\%$ after 4 years at 1.5 m spacing, and to $U_h = 94\%$ after 9 years at 2.2 m spacing. Taking into consideration the results obtained in the undrained area, U_v after a loading time of 4 years can be estimated at about 40% and after 9 years at about 60%. Using the Carillo equation we find $U_{tot} = 97\%$ in the former case and $U_{tot} = 98\%$ in the latter case.

In conclusion, we find that in the Skå-Edeby case the time of primary consolidation can be predicted with acceptable accuracy. In my opinion this can also be done in other cases of vertically drained clay provided that the consolidation characteristics have been determined in a satisfactory way.

The magnitude of total primary compression of the 5 m thick layer, located between 2,5 and 7,5 m depth, was calculated on the basis of the CR values given above. The results, which are given in Fig. 11, show a maximum deviation from the observed magnitude of about +25% at 0.9 m spacing and about -25% at 2.2 m spacing. At 1.5 m spacing a better correlation is obtained. Looking at the total primary settlement of the soil surface, Fig. 10, the agreement between prediction and performance is better.

Secondary consolidation of clay: It is well recognized that so-called secondary consolidation is not a phenomenon starting up only after primary consolidation is completed. In fact, the rate of secondary consolidation is probably most pronounced in the beginning of the consolidation process. However, with regard to the aim of this paper, the conventional approach to the problem of secondary consolidation is used here.

Looking at the results given in Figs 10 and 11 we find, as expected, that the effect of secondary consolidation is considerable. However, there is also another observation of great interest. Test area III represents a case of preloading by the use of a temporary surcharge of 0.7 m of gravel (total load 2.2 m gravel) which was removed after 4 years. The purpose of using a temporary surcharge was to find out whether or not it was possible to eliminate further secondary compression in this way.

The observations of the settlements at different depths in test area III show no tendency of secondary consolidation during the 25 years that have passed since removal of the temporary surcharge down to a depth of 7.5 m in spite of the fact that an excess pore water pressure of around 10 kPa still remained at 5 m depth 10 years after removal of the surcharge. The glacial clay layer below 7.5 m (about 6-7 m in thickness) is, however, still undergoing some secondary consolidation - around 50 mm in 25 years which from a practical point of view is negligible.

Vertical drainage of peat: In many regions, a great percentage of the land area is covered by peat. As peat is among the poorest soil types from the foundation point of view, building activity on such areas is often greatly restricted. In the event of construction being decided on the peat soil, this is either excavated or consolidated by preloading. Vertical drainage and precompression have been utilized in some rare cases with mixed results.

Due to the extremely high compressibility of peat and to the low horizontal soil resistance there is doubtless an obvious risk of necking if sand drains are used. Now that we have good prefabricated drains at hand the risk of necking can be considered to be eliminated. Instead, folding, accompanied by decreasing discharge capacity, may entail another possible problem. The general idea seems to be that vertical drainage does not contribute enough to the speeding up of the consolidation process for this investment to be worthwhile.

In order to investigate the efficiency of vertical drainage of peat soil, a test field was arranged in 1985 in a marshy area in the immediate vicinity of Chalmers University of Technology in Gothenburg, Fig. 12. Here the subsoil (Fig. 13) consists of around 5 m of more or less decomposed peat underlain by 20 m of soft, highly plastic clay on sand. The clay below a depth of 10 m contains a large amount of shells.

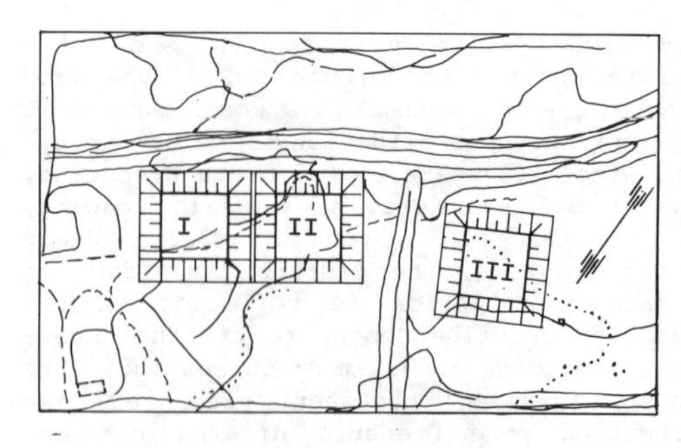

Fig. 12 - Plan of the Mossen test field in Gothenburg. Test area I provided with Geodrains, 1.5 m spacing, and test area III with Geodrains, 1.0 m spacing. Test area II undrained.

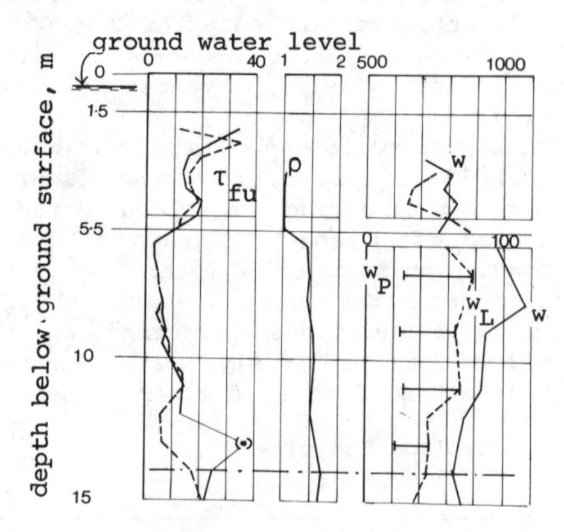

Fig. 13 - Typical geotechnical characteristics of soil at the Mossen test field. From left to right: undrained shear strength (kPa), bulk density (t/m^3) and water contents and consistency limits (%).

Two test areas, each one 15 m by 15 m, were provided with drains placed in an equilateral triangular pattern, in one case with 1.5 m spacing, in the other with 1.0 m spacing. The drains (of type Geodrain with synthetic filter sleeve) were installed to a depth of 15 m in order to make sure that the pore water squeezed out of the soil via the drains could only flow upwards in the drains (drains closed at bottom). (The aim behind this was to find out if folding at an early stage of consolidation would create a visible delay of the rate of consolidation.) A third area has no drains. Full load (1.5 m sand and gravel, including the drainage layer, 0.5 m in thickness), was placed both on the test area with 1.5 m drain spacing and the one with no drains, while only the drainage layer (0.5 m sand) was placed on the test area with 1 m spacing. Full load was placed in this case more than 1½ years after drain installation. Some of the results obtained to date are shown in Figs. 14-16.

The results are undoubtedly evidence that the drains have performed excellently and contributed to speeding up the consolidation process of the peat layer in a remarkable way. Folding does not seem to have delayed or influenced the consolidation process at all.

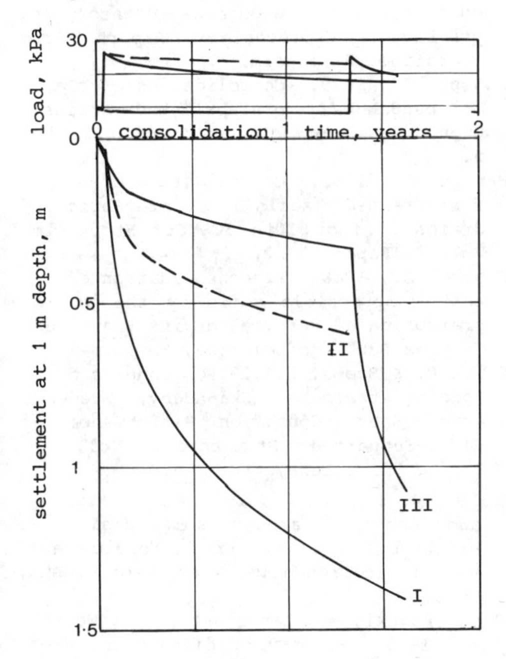

Fig. 14 - Consolidation settlement and loading conditions in centre of test areas. Full load was placed on test area III with about 1 1/3 years of delay because of actions by environmental activists.

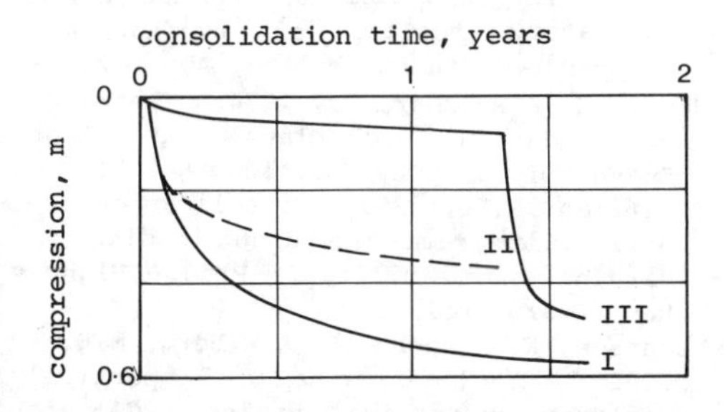

Fig. 15 - Consolidation of 4 m thick peat layer from 1.5 to 5.5 m depth. As can be seen, vertical drains have speeded up consolidation of the peat layer considerably.

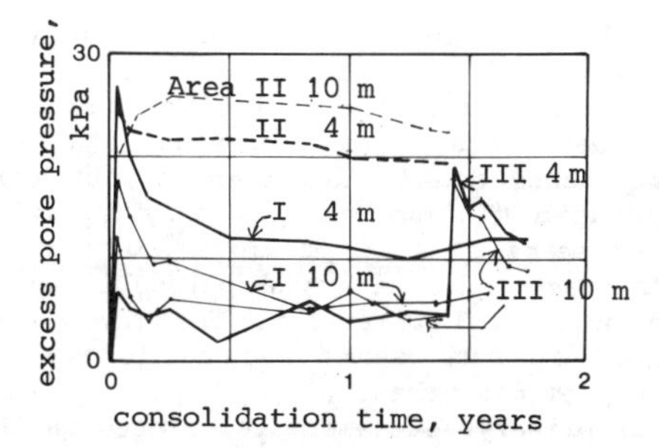

Fig. 16 - Dissipation of excess pore water pressure at 4 m and 10 m depth in centre of test areas. Results obtained in undrained area represented by broken lines.

From the compression characteristics of the peat, the compression of the 4 m thick peat layer is estimated at 0.5 m, taking into consideration the load reduction due to submergence of part of the fill. The contribution to vertical settlement attributed to observed horizontal outward displacment of the peat layer is about 0.1 m. Again prediction underestimates the real settlement.

The process of consolidation in peat cannot be predicted with any degree of accuracy by the conventional consolidation theory, which is intended for clay, (see, for example, Fürstenberg et al., 1983), and therefore, a comparison between performance and prediction by Eq. (1) is meaningless.

6 CONCLUDING REMARKS

Experience has clearly shown that vertical drains are an effective way of achieving quick consolidation of low-permeable soils, such as clay and slime, and even peat. However, drains should not be utilized unless the stress increment in the soil caused by the load exceeds the preconsolidation pressure.

For practical purposes, prediction of the settlement process can be based, with acceptable accuracy, on the theoretical hydrodynamic delay of the rate of load-induced excess pore pressure dissipation, although in reality settlement in itself may induce additional pore pressures in excess of those caused by the placement of the load. In the design of a drain installation, consideration should be taken to the fact that the permeability in horizontal pore water flow (and consequently the coefficient of consolidation) can be many times higher than in vertical pore water flow.

Well resistance can cause a serious delay in the consolidation process in those cases where the drains are installed to great depth. It is important that the disturbance due to drain installation is kept at a minimum, particularly in soils where seams of more pervious soil exist. Such seams indicate a high permeability in the horizontal direction. The favourable effect of such seams may be lost as a result of disturbance.

The primary settlement calculated on the basis of oedometer compression moduli will be affected negatively by the disturbance caused by drain installation and also by secondary settlement taking place during the primary phase of consolidation. Therefore, in settlement prediction one has to take into account the effect on the settlement of disturbance and of secondary settlement. One way of compensating for the effect of disturbance is to calculate the settlement without regard to the fact that submergence of part of the load during settlement will reduce the load intensity. Another is to increase, based on experience, the compressibility determined by the oedometer test.

Secondary consolidation settlement during the lifetime of a structure can be more or less completely avoided by using a temporary overload which is not removed until the primary and secondary settlement to be expected for the permanent load have ceased.

Prefabricated band-shaped drains of good quality have usually a much higher discharge capacity (lower well resistance) than sand drains. It is easier to ensure complete control during installation of prefabricated drains and they do not run the risk of being necked.

REFERENCES

Akagi, T., 1976. Effect of displacement type sand drains on strength and compressibility of soft clays , University of Tokyo. (Doctoral thesis)

Barron, R.A., 1948. Consolidation of fine-grained soils by drain wells , Trans. ASCE, Vol. 113, Paper No. 2346.

Carillo, N., 1942. Simple two- and three dimensional cases in the theory of consolidation of soils . J. Math. Phys., Vol. 21, No. 1.

Chang, Y.C.E., 1981. Long-term consolidation beneath the test fills at Väsby , Swed. Geot. Inst., Report No. 18.

Fellenius, B., 1978. Dimensionering av vertikala dräner - Ett diskussionsinlägg (Design of vertikal drains - Discussion). Väg- och vattenbyggaren No. 1.

Fürstenberg, A., Lechowicz, Z., Szymanski, A. & Wolski, W., 1983. Effectiveness of vertical drains in organic soils . Proc. 8th ECSMFE, Vol. 2, Pp 6.11, Helsinki.

Hansbo, S., 1960. Consolidation of clay, with special reference to influence of vertical sand drains. A study made in connection with full-scale investigations at Skå-Edeby . Svs. Geot. Inst., Proc. No. 18 (Doctoral thesis).

Hansbo, S., 1977. Geodrains in theory and practice . Geotechnical Report from Terrafigo, Stockholm.

Hansbo, S., 1979. Consolidation of clay by bandshaped, prefabricated drains , Ground Engineering, July, Vol. 12, No. 5.

Hansbo, S., 1981. Consolidation of fine-grained soils by prefabricated drains . Proc. 10th ICSMFE, Stockholm, Vol. 3, Paper 12/22.

Hansbo, S., 1986. Preconsolidation of soft compressible soils by the use of prefabricated vertical drains , Ann. des Travaux Publ. de Belgique, No. 6.

Holtz R. & Broms, B., 1972. Long-term loading tests at Skå-Edeby, Sweden , Proc. Spec. Conf. on Performance of Earth-Supported Structures, Vol. 1, Purdue Univ. Lafayette, Ind.

Holtz, R. & Holm, G., 1973. Excavation and sampling around some drains at Skå-Edeby, Sweden , Proc. Nordic Geot. Meeting in Trondheim, Norw. Geot. Inst., Oslo.

Jamiolkowski, M., Lancelotta, R. & Wolski, W., 1983. Summary of discussion, Proc. 8th ECSMFE., Vol. 3, Spec. Session 6.

Kamon, M., 1984. Function of band-shaped prefabricated plastic board drain . Proc. 19th Jap. National Conf. on SMEF.

Koda, E., Szymanski, A. & Wolski, W., 1986. Laboratory tests on Geodrains-Durability in organic soils , Seminar on Laboratory testing of prefabricated band-shaped drains, Milano, April 22-23.

Mesri, G. & Rokshar, A., 1974. Theory of consolidation for clays , J. Geot. Engng. Div., ASCE, Vol. 100, No. GT 8.

Schiffman, R.L., 1958. Consolidation of soil under time-dependent loading and varying permeability , Proc. Highway Res. Board, Vol. 39.

Runesson, K., Hansbo, S. & Wiberg, N.E., 1985. The efficiency of partially penetrating vertical drains . Géotechnique 35, No. 4.

Yoshikuni, H. & Nakanodo, H., 1974. Consolidation of soils by vertical drain wells with finite permeability , Jap. Soc. SMFE, Vol. 14, No. 2.

Prediction and Performance in Geotechnical Engineering / Calgary / 17-19 June 1987

Cold regions predictions using centrifuge testing

Andrew N.Schofield
Cambridge University, UK

Geotechnical Centrifuge Operations

Since February 1975 more than 85 different model test series have been undertaken on the Cambridge 10 metre Geotechnical Centrifuge involving Over 950 tests.

Studies of soil-structure interaction have included piles tested with cyclical loads; laterally loaded piles in dry and saturated quartz sand; laterally loaded piles and pile groups in saturated calcareous sand; axially and laterally loaded piles in normally consolidated and overconsolidated clay; axially loaded driven piles in saturated quartz sand. Tests of footings on saturated granular materials have included axially loaded shallow circular flat footings and eccentrically loaded shallow circular spud footings. There have been tests of flexible oil tanks on overconsolidated clay, tests of gravity walls retaining dry quartz sand, of reinforced earth retaining both dry quartz sand and clay, of multi-anchored bulkheads retaining dry quartz sand, of cantilevered T-walls retaining dry quartz sand, and of diaphragm walls in stiff overconsolidated clay. There have been tests of buried pipes and flexible culverts in dry quartz sand.

The problem of earthquake hazard mitigation has led to study of earthquake effects on piles and on shallow strip footings in dry quartz sand, and on towers resting on the surface of dry and saturated quartz sand. There have been tests of earthquake effects on gravity and on cantilever walls retaining dry quartz sand, slopes in overconsolidated clays, on dry and submerged quartz sand embankments and circular islands, embankment dams and coastal dykes.

Slope stability tests have included quartz sand slopes destabilised by groundwater surges, flowslides in undisturbed quick clay, slips in overconsolidated plastic clays, tests of staged embankment construction on soft clay and on soft ground adjacent to piled foundations, and spill-through bridge abutments incorporating piled foundations. There have been tests of void migration in dam cores, and of breaching of tailings dams, studies of the excavation and long term performance of lined tunnels in dry sand, unlined tunnels and tunnel headings in saturated silt and overconsolidated clay, unlined trenches and axisymmetric and rectangular shafts in overconsolidated clay.

As the scope of centrifugal model testing has widened, hazardous waste disposal studies have included pollution migration from engineered trenches and landfill sites, and ocean bed disposal of hot cylindrical canisters in soft normally consolidated clay. Preliminary studies of thaw induced settlement have involved movement of a trap door below a model pipeline. These tests were to have been followed by studies of warm pipelines buried in the cold sea floor above permafrost, with heat transfer and salt transfer both affecting the thaw rate. The fall in the price of oil has meant that this research has had to be funded by earnings on other contracts, but although progress is delayed there is clear potential for tests to model and correctly predict these and other various cold regions processes.

Test facilities in Cambridge

The 10 m beam operates with swinging platforms on which a model at a working radius of about 4 m can be tested at an acceleration of up to 125 g. There has been a series of tests in which a pool of salt water of 750 mm diameter had cold gas poured over its surface and sea ice has formed on the surface. There was doubt about the structural safety of the cold gas generator and that test series was discontinued in 1986.

A new 2 m diameter drum centrifuge has been built to be capable of eventual operation at 500 g. It had been hoped to form ice in this centrifuge before the end of 1986, but that is now expected in April 1987. The pool surface exposed to cold gas is increased to an area of 6000 x 1000 mm^3, and with the increased acceleration this will eventually be equivalent to a cold region of area 3 km x ½ km. In June 1987 it should be possible to show progress with this facility.

A third refrigerated centrifuge capable of operation at -20°C has recently been purchased for general laboratory use but with a view to use for some cold regions studies. It is a bench-top machine with a windshielded swing-out rotor that can apply 3000 g to 4 buckets each of ¾ litre capacity. At a scale of 1:3000 the 3 litre capacity gives a total prototype equivalent volume of $3 \times 10^{-3} m^3 \times 3000^3 = 81$ million m^3. Each bucket has an internal volume of about 90 mm dia x 120 mm height, which at 3000 g is equivalent to a prototype volume of 270 m dia x 360 m height. The cold region model test capability of this machine has still to be demonstrated, but should begin to be known by June 1987.

The sea ice experiments which were interrupted at the end of 1986 will continue with collaboration between the University Departments of Engineering and of Earth Sciences. In addition to this link between the University Geotechnical Centrifuge Centre and the Scott Polar Research Institute, the location of the Centre next to the British Antarctic Survey will make it possible to have access to the UK main cold regions research facilities. At present cold regions centrifuge research is funded by earnings in other regions. More rapid progress is anticipated when appropriate funds are generated, but in the interim period facilities and techniques are being prepared in readiness for future developments.

Prediction and performance

The scale factor for transient heat conduction is found simply from the heat equation to be the square of the model scale. In a bucket of fresh water permafrost under a salt water pool at 3000 gravities if the combined salt and heat transfer processes follow the heat equation the advance of thawing into the permafrost in one day of continuous refrigerated centrifuge flight could be regarded as a model for the 3000 x 3000 days = 25,000 years of the present interglacial period. Similarly, one week of continuous flight of the drum centrifuge at 500 g will model a period of 500 x 500/52 = 4800 years, which corresponds to the period in which the onset of the next glaciation is expected to occur. There are civil engineering requirements for prediction of hazardous waste migration over such periods. In centrifuge model tests the relative density of water with different salt concentrations is seen to introduce a strong gravitational component in models.

The influence of effective stress on the yielding of soil was the reason for the emphasis placed on geotechnical centrifuge model testing at Cambridge over the past twenty years. Fears were expressed at the outset of the research that the effect of creep, or of particle size in soil, might make the models incapable of interpretation; the fears were exaggerated. In the case of ice and of frozen soil similar fears are now expressed; once again it is appropriate to advance into the research programme in stages, with verification wherever possible.

It should be possible to model ice-ice and ice-structure interaction in cold oceans with waves. A study is proposed of stresses in icebergs, of the influence of air bubbles and effects during calving or capsizing, of the scouring of sea beds by icebergs and of damage to the sea bed caused by wallowing or capsizing icebergs. Explosive fragmentation of model icebergs will be possible. In sea ice studies a first objective is punch indentation, and then pressure ridge formation. Grounded rubble field and pressure ridge keel interactions with a sea bed will be studied. The motion of large vessels through unbroken sea ice with pressure ridges, or through channels, or through brash ice, will all be modelled.

Future opportunities

The opportunities for research are varied. Operational needs differ in the Arctic and Antarctic. On land in the Arctic there are long term field trials in progress, or planned, which could be supplemented by parametric studies in centrifuge model test series. If spray ice construction in a cold ocean above subsea permafrost, with movement of vessels and of pack ice, can all be modelled, then a rise in the oil price could lead to opportunities for research in the offshore Arctic.

The opportunities for research will depend on availability of funds. The underlying academic problem is to determine the scope of various alternative analyses of theory of elasticity and plasticity, of diffusion and dispersion, of heat and of salt transfer, of yielding and of fracture. In centrifuge modelling our aim is to study the effect of changes in pressure within the depth of a body and of differences of density between different constituents of the body, on the solution of boundary value problems, for geotechnical materials in all regions of the earth's surface.

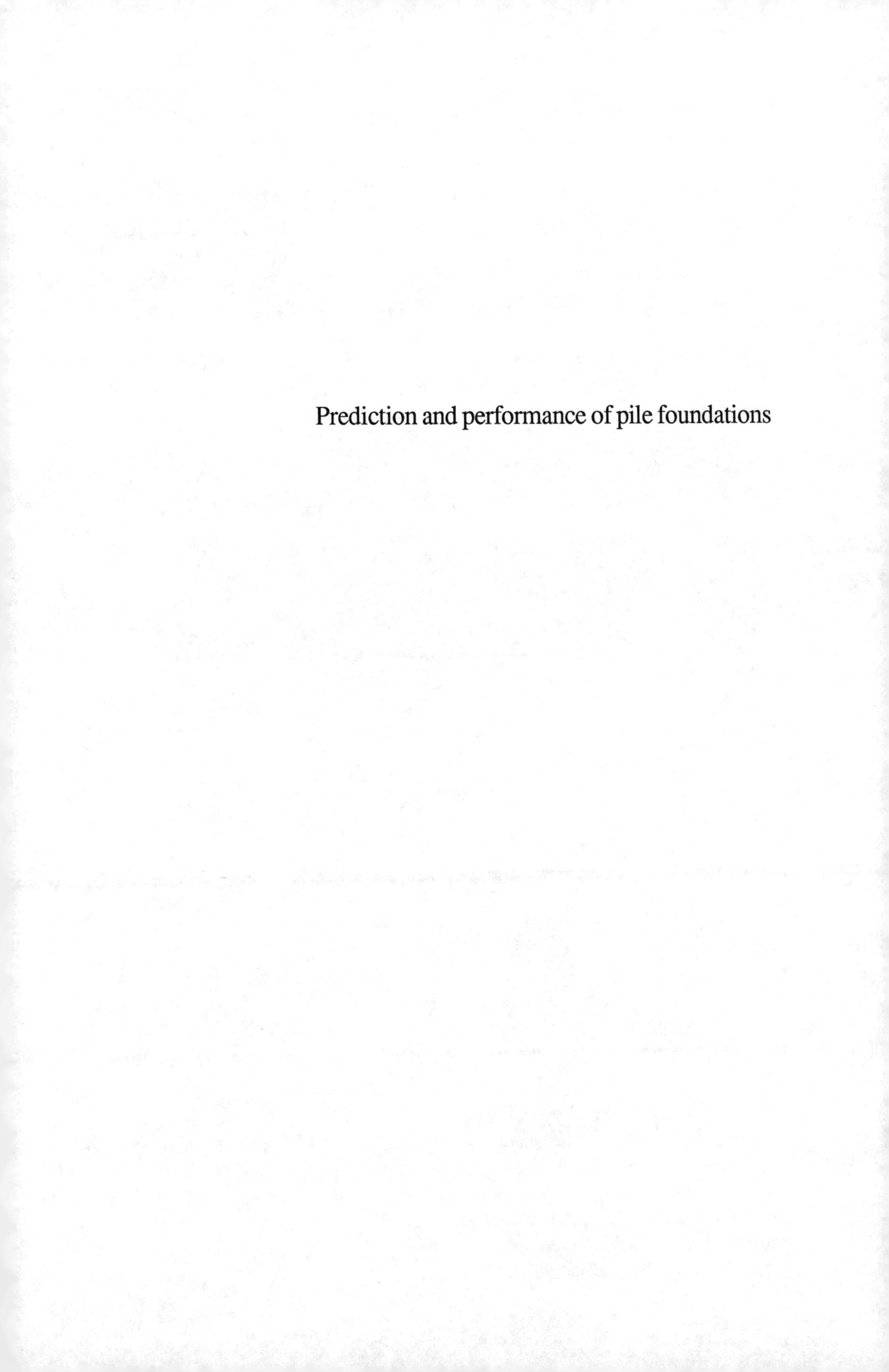

Prediction and performance of pile foundations

Prediction and Performance in Geotechnical Engineering / Calgary / 17-19 June 1987

Prediction of pile foundation response under static lateral loads and dynamic loading using the transfer matrix approach

D.J.Cook & V.Chandrasekaran
National Building Technology Centre, North Ryde, Sydney, Australia

ABSTRACT: The response characteristics of soil-pile systems subjected to lateral loads have been analysed. The analysis is performed using the transfer matrix approach utilising modal analysis and determinant search procedures. For a massless pile with flexural stiffness and mass lumped at the top the fundamental mode response is identical to that of laterally loaded piles. Existing solutions compare identically with the results presented. Response of piles embedded in granular, clayey and layered soils have been discussed and the methods of analysis explained. The static load behaviour and the dynamic response for the fundamental mode and the various factors of the soil-pile system which control them have been discussed.

1 INTRODUCTION

There is an increased awareness amongst the engineering profession of the importance of foundation support conditions in controlling the behaviour of structures during earthquakes. In this context, pile foundations are one of the most frequently used foundation types to resist the lateral load components of earthquake loading.

Furthermore, the problem of laterally loaded piles is of particular interest in connection with drilling platforms, defence installations as well as structures subjected to high wind loads.

The problem of a laterally loaded pile is closely related to the problem of a beam on an elastic foundation. However, the uniqueness in the present case is that all the external forces and moments applied to the pile-soil systems are introduced by virtue of the boundary conditions at the top of the pile, whereas, the effect of loading is felt along the entire length of the pile.

This paper describes part of a continuing research programme dealing with response characteristics of soil-pile systems subjected to static and dynamic loading. In our approach it is envisaged that it is possible to provide unique solution techniques to static as well as dynamic loadings.

Solutions to the problem of piles subjected to lateral loads embedded in soils have been discussed by Matlock and Reese (1960). They considered the soil modulus variations to be of a general type and proposed methods to evaluate non-linear effects. Nondimensional solutions are obtained on the basis of dimensionless analysis. Solutions to piles embedded in layered soils have been determined by Davisson and Gill (1963) using analog computer techniques.

2 SOIL PARAMETERS

Fig. 1(a) shows a fully embedded pile subjected to a system of loads at the ground surface. The vertical load is Wp, M the moment and Q the horizontal load. The pile will deflect under the applied load and as shown in Fig. 1(b), the soil reaction, w, at any depth x, is normally expressed as a linear function of the deflection at that point, $w = k.y$. Fig. 2 describes the probable shape of w versus deflection y for a point along the embedded length of the pile. This relationship is taken to be linear by considering either the tangent or secant modulus. This results in the possibility of handling this phenomena by considering the pile as a beam on closely spaced elastic springs, ie., the Winkler model.

The variations of the subgrade modulus with depth have been illustrated in Fig. 3. For normally loaded clays and silts, and for granular soils, k varies linearly with depth. Fig. 3(b) shows the variations in the case of granular soils, where nh is the constant of horizontal subgrade reaction (Terzaghi 1955). Herein, layered soil systems have been analysed in which the ratio of the modulus of the upper layer, K_c and that of the lower layer, K_o, have been incorporated. The cases of stiff upper layers and soft upper layers in relation to the bottom layers have been considered. Fig. 4 schematically describes the various cases considered in the present study. Crust depths of 0.10; 0.50; 1.0; 1.50 have been considered. The ratios of soil modulus at the top, K_c and that at the bottom, K_o which were considered include K_c/K_o = 0.0; 0.25; 0.5; 3.0; 5.0; 15.0; and 25.0. Finally, solutions for the cases of piles embedded in uniform deposits of granular soils and normally loaded clays have also been obtained.

3 SOIL PILE INTERACTION AND LUMPED MASS MODEL

In the mathematical model used, the Winkler reaction offered by the soil is descretised springs connected at desired points. In order to achieve this, the soil reaction is assumed to be a distributed loading intensity and at the division points the reactions are lumped. The steps involved in the descretisation of soil springs in the case of piles embedded in soils of the type described in Fig. 3(a) and 3(b) have been described in Fig.5. A similar approach has been used in the case of layered soils.

3.1 Components of the model

In Fig.6 the descretised model of the actual soil-pile system is shown. The mass M_t lumped at the top includes the safe carrying capacity of this isolated pile, m_r, the lumped mass at the division point r, and M_n, the lumped mass at the nth. or last division point.

K_1, the linear spring having a stiffness K_1 is attached to the mass M_t and has an immovable support. Similarly, spring K_r corresponds to mass at r and K_n the n th. mass. The analysis requires the adoption of the realistic end conditions of the form shown in the figure.

4 THE CONCEPT

Considering the free vibration characteristics of the pile to be of the form, Fig.7a, then

$$Y = X\ (x) \sin pt$$

then

$$EId^4y/dx^4 - \rho A(p^2x) + kx=0$$

The various components of this equation are well known. In Fig.7a the forces acting on a displaced pile have been illustrated. Considering the rotational and translational displacements and the equilibirium of the segments, we have

$$V = -\sigma AG\ dYs\ /\ dx$$

$$M = EI\ d^2\ Y_b\ /\ dx^2$$

$$dM/dx = V - \rho I_p{}^2\ \theta b$$

$$dV/dx = mp^2y - k.y$$

$$Y = Y_b + Y_s$$

where,

- V is the shear force due to shear deformations
- σ ratio of the average shear stress on a section to the product of the shear modulus and the angle of shear at the neutral axis, defined as the shape factor, 1.1 for circular sections.
- I_p moment of inertia of the section
- M bending moment in a section
- m the mass of an element of the section
- Y_s deflection due to shear deformation
- Y_b deflection due to bending deformation
- θ_b rotation due to bending deformation
- ρ mass density of the material
- Y the total deflection due to bending and shear
- G shear modulus of the structural material
- EI flexural stiffness
- A area of cross section
- p the natural frequency of vibration in a mode.

Consider three mass locations and a section drawn through the mass point 1, (Fig. 7b). A finite change in shear force occurs at each mass which is equal to the algebraic sum of the inertia force of the mass and the soil reaction (spring reaction). Each of these quantities are dependent on the deflection of the mass point.

Therefore, we have:

$$\Delta V = mp^2y - ky$$

assuming that quantities V_o, M_o, θ_{bo}, Y_{bo} and Y_{so} are known to the left of the section and for a division length $x = (\Delta x)_1$ then,

$$V_1 = V_o + M_o p^2 Y_o - kY_o$$

$$M_1 = M_o + V_1(\Delta x)_1 - (\rho I)p^2 x \theta_{bo}$$

$$Y_{s1} = Y_{so} - \left(\frac{V}{\sigma AG}\right)_1 (\Delta x)_1$$

Now, the bending moment M at any distance x, from the left side of the section, o is:

$$M = M_o + \frac{M_1 - M_o}{(\Delta x)_1} \cdot x$$

$$\text{Slope} \quad \theta b = \frac{1}{(EI)_1} \int M \, dx + B$$

$$= \frac{1}{(EI)_1} \quad M_o x + \frac{M_1 - M_o}{(\Delta x)_1} \cdot \frac{x^2}{2} + \theta_{bo}$$

and deflection

$$Y_b = \int \theta \, dx + B$$

$$\frac{1}{EI} \left(M_o \cdot \frac{x^2}{2} + \frac{(M_1 - M_o \cdot x^3)}{(\Delta x)_1 \; 6}\right) + \theta_{bo}\, x + Y_{bo}$$

For a distance $x = (\Delta x)_1$

$$\theta_{b1} = \frac{(\Delta x)_1}{EI} \left(\frac{M_o}{2} + \frac{M_1}{2}\right) + \theta_{bo}$$

and

$$Y_{b1} + \frac{(\Delta x)_1}{EI} \left(\frac{M_o}{3} + \frac{M_1}{6}\right) (\Delta x)_1$$

$$+ \theta_{bo} \quad (\Delta x)_1 + Y_o$$

4.1 The transfer matrix solutions

Now, instead of the example of the three lumped mass spring systems, if we consider the discretised model of the entire soil-pile system, similar recursion relationships in a general form for the mass location, say 'j', in terms of the adjacent mass locations (j-1) and (j+1) can be written (Fig 8),

$$\begin{bmatrix} Y \\ \theta \\ M \\ V \end{bmatrix}_{j+1} = \begin{bmatrix} 1 & \left(\frac{1_j - 1_j^3 \rho I p^2}{GEI}\right) & (1_j^2 / 2EI) & \left(1_j / 6EI - \frac{\sigma 1_j}{GA}\right) \\ 0 & \left(1 - \underline{\rho I p^2 1_j^2}\right) & (1_j / EI) & (1_j^2 / 2EI) \\ 0 & (-\rho I 1_j p^2) & 1 & 1_j \\ (m_j p^2 - k_j) & 0 & 0 & 1 \end{bmatrix} \begin{bmatrix} Y \\ \theta \\ M \\ V \end{bmatrix}_j$$

This equation may be expressed in a generalised manner as shown below:

$$\{Z\}_{j+1} = [U]_j \quad \{Z\}_j$$

The column matrices

$$\{Z\}_{j+1} \text{ and } \{Z\}_j$$

are the state vectors and contain the state variables which give the information concerning the response characteristic of the soil-pile system at each station point. The response quantities for the lateral loading and hence the lateral displacements of the pile are Y, θ, M, and V.

The square matrix [U] is known as the transfer matrix at station, j.

The elements of the transfer matrix depend on the soil-pile characteristics and the frequencies of vibrations.

5 THE SOLUTION PROCEDURE

With the known boundary conditions at the two ends of the piles, it is possible to start either at the pile top (or pile bottom) and proceed to satisfy the boundary conditions at the other end. Thus starting from point 1, the pile top, the following equations relating the state vectors at the successive stations can be written:

$$\{Z_2\} = [U]_1 \{Z_1\}$$

$$\{Z_3\}=[U]_2\{Z_2\}=[U_2][U]_1\{Z_1\}$$

Finally on reaching the bottom the state vectors at the top and bottom are found to be related as:

$$\{Z_n\}=[U]_{n-1}[U]_n \cdot\cdot [U]_2[U]_1 Z_1\}$$
$$= [P] \{Z_1\}$$

The matrix [P] is the overall transfer matrix formed by taking the products of all the intermediate transfer matrices in the order indicated. It relates the vectors at the two end points at which the boundary conditions are known. These conditions fix values of some of the elements in each of these state vectors, and hence to provide constraints on the elements of the matrix [P]. The imposed conditions can only be satisfied by the natural frequency, p of the system.

5.1 The modal analysis and determinant search

The steps involved in the determination of the natural frequencies of vibrations as well as the response quantities namely the deflection Y, rotation θ, bending moment M and shear force S, are as follows. As shown in the equation, the state vector at any division or mass point, say the j+1, is given in terms of the state vector at station 1. The matrix [U] and hence the product matrix [P] contain elements which are dependant on the natural frequencies. In the pile problem, it is convenient to start from station 1 or the pile top.

The state variables Y and θ at the pile top are unknown, particularly for the pile top free to rotate conditions. A similar situation is also encountered at the pile bottom.

Therefore, by resorting to modal analysis, unit values can be assigned to the quantities Y and θ in turn and the boundary condition at the other end may be satisfied. This results in the following equation:

$$\begin{bmatrix} Y \\ \theta \\ M \\ V \end{bmatrix}_n = \begin{bmatrix} C_{11} & C_{12} & C_{13} & C_{14} \\ C_{21} & C_{22} & C_{23} & C_{24} \\ C_{31} & C_{32} & C_{33} & C_{34} \\ C_{41} & C_{42} & C_{43} & C_{44} \end{bmatrix} \begin{bmatrix} Y \\ \theta \\ M \\ V \end{bmatrix}_1$$

By satisfying the boundary conditions at the pile bottom would result in the equation as below

$$\begin{bmatrix} C_{31} & C_{32} \\ C_{41} & C_{42} \end{bmatrix} \begin{bmatrix} Y_t \\ \theta_t \end{bmatrix} = \begin{bmatrix} 0 \\ 0 \end{bmatrix}$$

However, in the above equation, Y_t and θ_t have to be non-zero unless the pile is in a state of rest equilibrium. Therefore, the condition for Y_t and θ_t to be non-zero is for the determinant to approach zero.

$$\begin{vmatrix} C_{31} & C_{32} \\ C_{41} & C_{42} \end{vmatrix} = 0$$

It is apparent that the process depends on correct estimation of the natural frequencies at different modes of vibration. Therefore, with a low starting value of a trial frequency and proper interpolation techniques, the exact natural frequency at which the determinant will be zero is easily assessed.

Having determined the natural frequencies it is possible to start from the pile top and proceed to the pile bottom to determine the required modal quantities.

6 ANALYSIS OF PILES SUBJECTED TO LATERAL LOADS

Suitable computer programmes were developed to evaluate the response characteristics of piles subjected to lateral loads. In the mathematical model, Fig.6, the mass lumped at the top includes the safe carrying capacity considering individual action. The balance of the (other) masses are assigned zero values, that is, the pile section is considered to be massless but to possess flexural stiffness, EI. The free vibration analysis of such a model results in a situation in which the system is subjected to an inertia force equivalent to mp^2y at the top.

Following the transfer matrix techniques together with the modal analysis and the determinant search steps, the natural frequency of the system is easily evaluated.

For the case of a single mass lumped at the top this frequency is the fundamental frequency, with a mode participation factor of 1.0.

Having determined the fundamental frequency, the modal quantities, deflection Y, rotation θ, bending moment M and shear S at each division points are evaluated. This defines the entire response of the soil-pile system subjected to a lateral load at the top or the ground surface.

7 SOLUTIONS FOR PILES EMBEDDED IN SOILS IN WHICH THE SOIL MODULUS REMAIN CONSTANT OR VARIES LINEARLY WITH DEPTH

In the case of laterally loaded piles, it has been a practice to define the following non-dimensional parameters:

depth coefficient $Z = x/T$
maximum depth coefficient $Z_{max}=L/T$
soil modulus function $K_x=n_h.x$

where,
x, is any depth
L, the embedded length of the pile
T, the relative stiffness factor (with units of length).

In the case of soils in which the soil modulus varys linearly with depth, the relative stiffness factor T = $\sqrt[5]{(EI)/n_h}$, where n_h is the constant of horizontal subgrade reaction with units of FL^{-3}

In the case of piles embedded in soils in which the soil modulus remains constant with depth, Fig. 3b the relative stiffness takes the $R = \sqrt[4]{(EI)/k_x}$ where K_x is the soil modulus in FL^{-2} units.

As discussed previously, one of the steps in the modal analysis consists of displacing the pile with an arbitrary displacement at the top. Evaluation of the fundamental frequency with a zero determinant results in the assessment of the response quantities.

The response quantities, deflection Y(I), rotation θ(I), bending moment M(I) and shear S(I) at any station point are successively determined.

As a result of such modal analysis using transfer matrix solutions, it was possible to define the following coefficients for various modal quantities.

For piles embedded in soils in which soil modulus remain constant with depth the following coefficients were evaluated:

frequency coefficient

$$F_f = p \cdot \sqrt{M_t/KR}$$

deflection coefficient

$$A_{yc} = (EI/M_t p^2)R^3))*Y(I)$$

slope coefficient

$$A_{\theta c} = (EI/(M_t p^2)R^2)* \theta(I)$$

moment coefficient

$$A_{mc} = M(I)/(M_t p^2)$$

shear coefficient

$$A_{sc} = S(I)/(M_t p^2)$$

For piles embedded in soils in which the soil modulus varies linearly with depth: ($k_x = n_h. x$), only the frequency coefficient was found to be different i.e.,

$f_f = p \cdot \sqrt{M_t / n_h T^2}$. The remainder of the other coefficients were found to be the same as for the constant soil modulus case, except that the relative stiffness factor R is replaced by a relative stiffness factor T. The relative stiffness factor T and have been defined previously.

The results of the response analysis using transfer matrix procedures are given in Fig. 9 to Fig. 13. It is emphasised that these results were obtained by an entirely different method to that of the widely adopted procedures of Matlock and Reese (1960). However it can be seen from the figures that the coefficients of deflection and bending moment have the same values and trends as determined by the previous studies. The following observations from the present results can be made:

1. The variations of deflections and bending moments are dependent on the relative stiffness factor and Z_{max}.

For various pile section details, EI soil modulus values, the maximum values of deflection and bending moment are only functions of the relative stiffness factor and Z_{max}.

2. From the deflection coefficient variations, Fig 11 and Fig 12, it can be seen that pile lengths below $Z_{max} = 2$ produce only rigid body movements, whereas $Z_{max} =3$ can be considered to be an intermediate length. The embedded lengths of piles with $Z_{max} = 5$ or more, behave as an infinitely long pile. The maximum bending moment values increase with Z_{max} but more or less remain the same when the pile length results in $Z_{max} = 5$ or more.

3. The uniqueness of the results are seen in Fig 13, where the variations of the frequency factor with Z_{max} are shown. For various pile section details and soil modulus values, for similar variations of soil modulus with depth, unique frequency factor values exist and depend only on Z_{max}, (that is a function of embedded length, EI and soil modulus). It is interesting to note that the frequency factor and hence the fundamental frequencies increase with an increase in Z_{max} and become constant after $Z_{max} = 5$ and above. Short pile ranges have larger time periods or are less stiff compared to long pile ranges.

Also it can be seen that for long pile ranges, an increase in K results in lesser time periods or stiffer systems. A pile embedded in softer soils would have larger time periods than the one in stiffer soils. For a given pile length, an increase in flexural stiffness would result in a decrease in time periods and hence an increase in stiffness.

With the increase in EI, relative stiffness values would increase and thus offset Z_{max} values and this could be significant in border line cases of intermediate lengths with Z_{max} =3 to 4. For Z_{max} = 5 or more, an increase in EI, results in lesser time periods.

The above conclusions are true for piles embedded in clayey as well as granular soils of the type shown in Fig 3a and 3b.

8 RESULTS OF ANALYSIS OF PILES EMBEDDED IN LAYERED SOILS

As mentioned previously, the dynamic response characteristics of piles embedded in layered soils were determined to understand the behaviour under static lateral loading conditions. Recapitulating Section 2 and Fig 3, the analyses were performed on piles embedded in layered soils with layer thickness (crust depths) of 0.1, 0.5, 1.0 and 1.5 metres. The ratios of the modulus top to bottom layers (K_c/K_o) considered include 0, 0.25, 0.5 ,3.0, 5.0, 10.0, 15.0 and 25.0.

A representative example of some of the results have been presented in Figs. 14 to Fig 20. Previous investigators (Davisson and Gill 1963) have provided solutions to piles in layered soils using analog computer techniques. Furthermore, they have only analysed piles of characteristic length Z_{max} = 4.0. In this paper only the results of free head piles are shown and the balance of the results are available elsewhere (Chandrasekaran 1987).

It is emphasised that these solutions for the static lateral loading case are based on the transfer matrix approach and modal response analysis involving determinant search procedures.

Furthermore, the variations of frequency factors and the maximum values of deflection coefficients, A_y and bending moment coefficients, B_m have been shown only for the maximum length factor, Z_{max}, of the pile. Variations with depth or x/R though available cannot be shown here due to space limitations. Since the designer is interested in maximum values this presentation is justified. The frequency factors and other coefficients are based on the soil modulus of the bottom layer Ko and not of the crust.

Based on these results some of the most significant observations are as follows:

1. With an increase in Z_{max}, the frequency factor increases and the time periods decreases. However, no significant changes occur after Z_{max} = 5 or more. With an increase in K_c/K_o, the time periods tends to decrease sharply.

2. As can be seen in Figs 15 and 18, there is a significant magnification of the deflection as the modulus of the surface layer is increased. As seen in Figs 16 and 19, the same is also true for maximum bending moment.

3. The results for crust depths of 0.1 and 1.0 meters show the significance of crust depth. For an increase in crust depth and a decrease in layer coefficients, the effect on the maximum values of deflections and moments are dramatic. This can be seen in Figs 15 and 18 and Figs 16 and 19 for K_c/K_o = 0.0 ,0.25 and 0.5. For smaller crust depth, there is not a pronounced effect on layer coefficient values below unity. However, the reverse is true for greater crust depths.

4. When the layer coefficient values are more than 5 there is a severe damping of the maximum bending moment and deflections especially for greater crust depths.

5. In Figs 20 and 21, variations of crust depths with frequency factors have been shown for Z_{max} =2 and 10. These two curves provide one of the most significant insights to pile response whether under static lateral loading or dynamic loading conditions. Fig 20 correspond to K_c/K_o =0.25 and Fig 21, K_c/K_o=15. When the layer coefficients are less than unity, with an increase in crust depth there is a sharp increase in time periods or reduction in fundamental frequencies and hence the overall stiffness of the entire system. There is a marked difference in stiffness and time periods for short and long pile ranges.

However the reverse is true for a layer coefficient of K_c/K_o=15. With an increase in crust depth, the time periods decrease and the fundamental frequency increase. The difference in time periods between short and long pile ranges are not that significant.

6. The crust depth and the modulus of the upper layer in comparison to the lower layer is important in the case of laterally loaded piles. Very small upper layer thickness and its characteristics could govern the overall behaviour.

7. The dynamic response analysis to any earthquake loading can be assessed by using the response coefficients and the fundamental period.

9 CONCLUSIONS

The present investigation has shown that a transfer matrix approach and modal analysis with determinant search methods can be used effectively to evaluate the response characteristics of pile-soil systems. Solutions to pile response for piles embedded in soils in which soil modulus remain constant with depth and varies linearly with depth have been provided. Also for the first time response characteristics of piles embedded in layered soils for long and short pile ranges have been provided for various layer coefficients and crust depth values. The dynamic response for the fundamental mode for a massless pile with mass lumped at the top has been discussed. Also it has been shown that the static lateral load response is identical to that of the discussed special situation under dynamic conditions. It has been shown that it is very effective to follow generalised procedures to obtain static and dynamic pile responses without changing the general approach.

10 REFERENCES

1. Chandrasekaran V 1987 Static and Dynamic Analysis of Pile Foundations, Report, National Building Technology Centre, North Ryde, Sydney, Australia

2. Davisson, M.T. & Gill H.L. 1963 Laterally Loaded Piles in Layered Soil Systems Proc ASCE Jnl of Soil Mechanics and Foundation Dn, pp 63-93

3. Matlock H & Reese L.C 1960 Generalised Solutions for Laterally Loaded Piles Proceedings ASCE, Jnl of Soil Mechanics and Foundation Division, pp 64-94

4. Terzaghi.K 1955 Evaluation of Coefficient of Subgrade Reaction, Geotechnique, Vol 5 pp 297-326

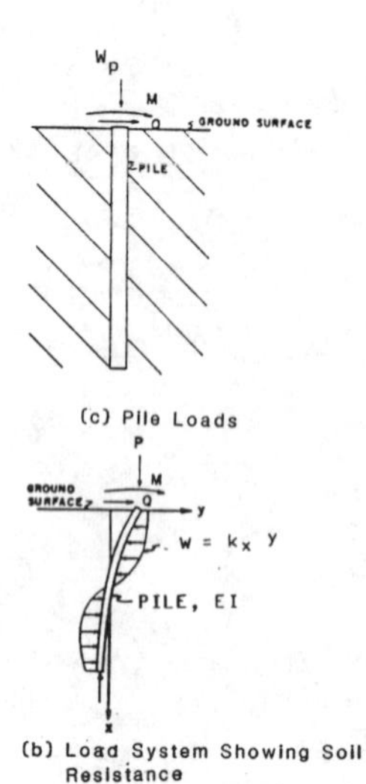

(c) Pile Loads

(b) Load System Showing Soil Resistance

Fig. 1: Pile Loads and Soil Resistance

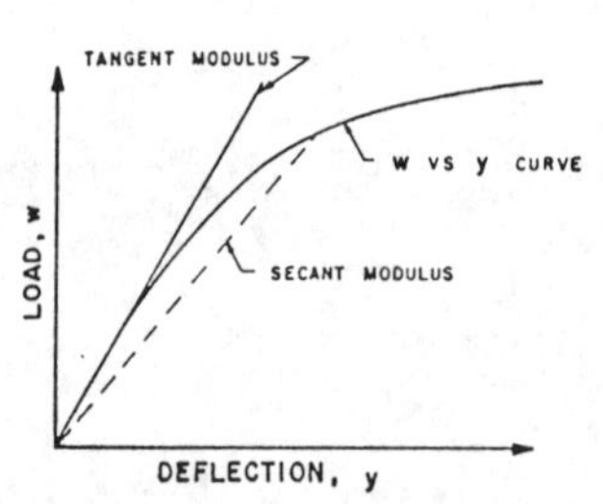

Fig. 2: Typical Soil Reaction-Deflection Curve

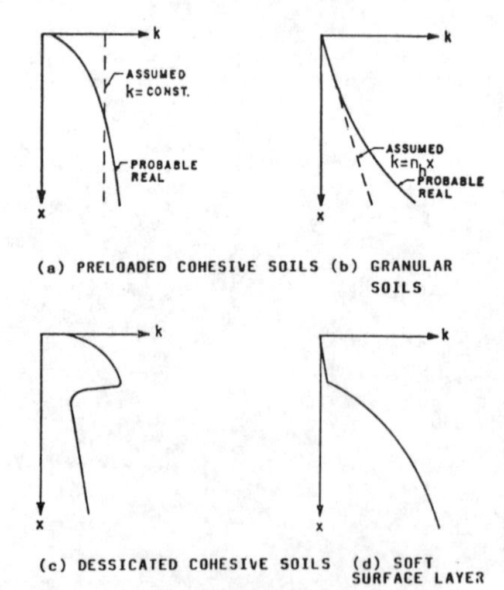

(a) PRELOADED COHESIVE SOILS (b) GRANULAR SOILS

(c) DESSICATED COHESIVE SOILS (d) SOFT SURFACE LAYER

Fig. 3: Variation of Subgrade Modulus with Depth

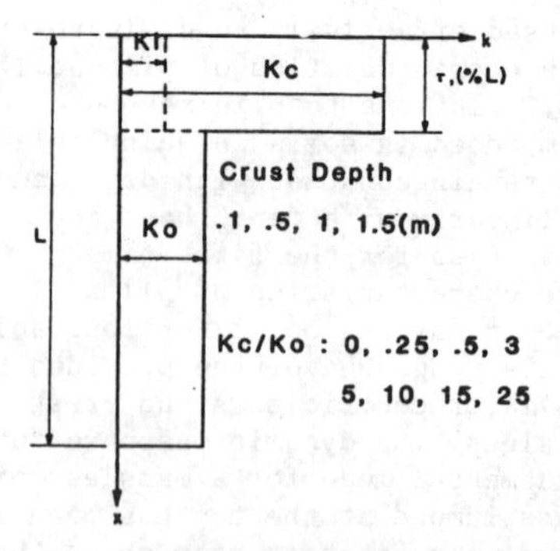

Fig. 4: Subgrade-Modulus Variations used in Analysis

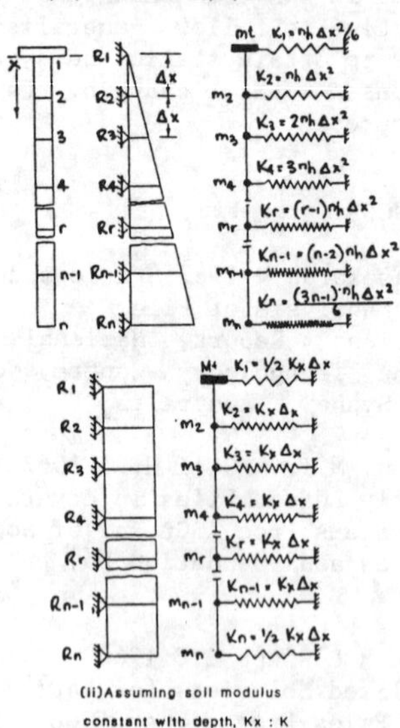

Fig. 5: Discretisation of soil pile interaction effects

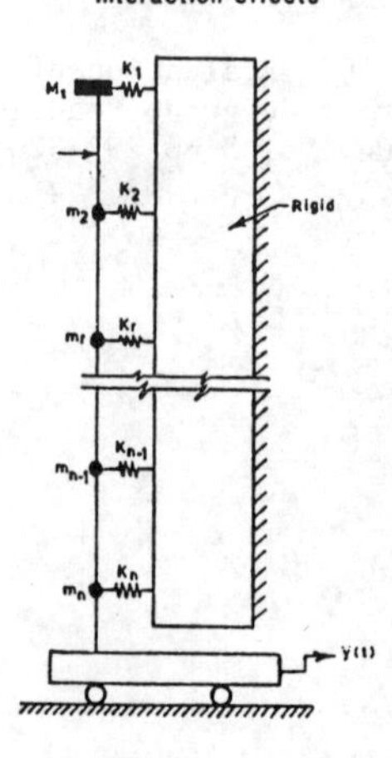

Fig. 6: Mathematical model

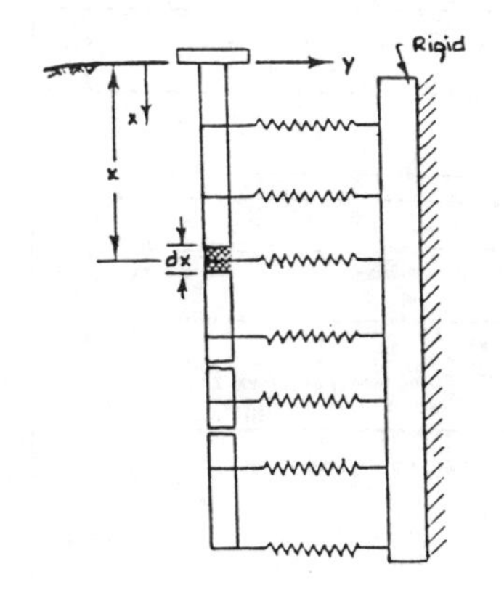

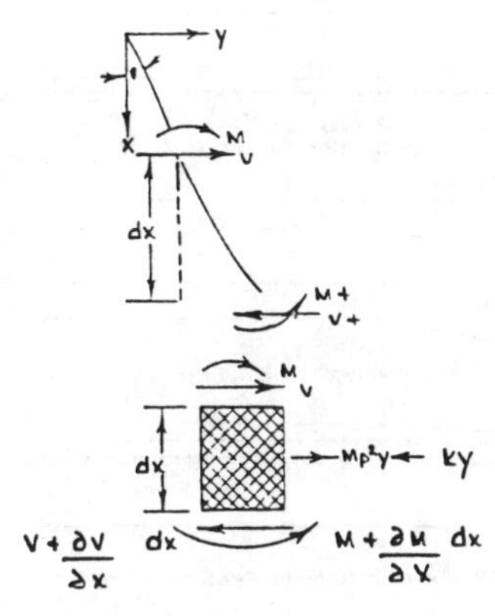

Fig 7a: Elastic and Inertia Forces on an Element

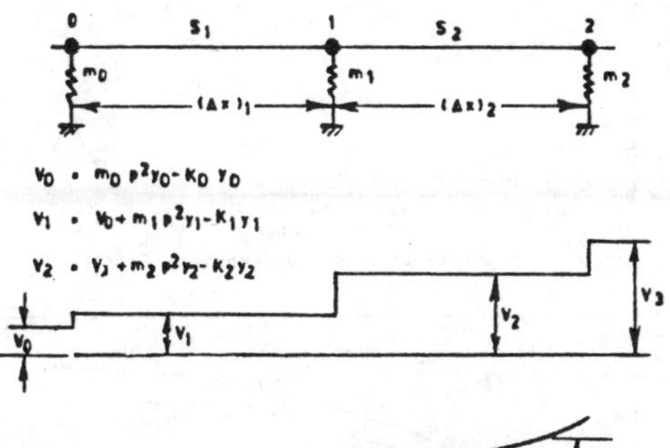

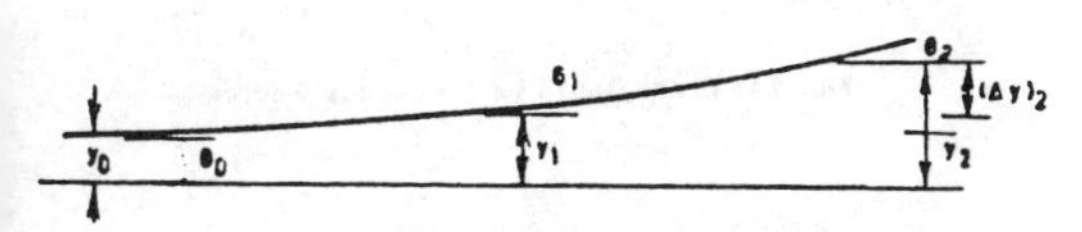

Fig. 7b Transfer Operation

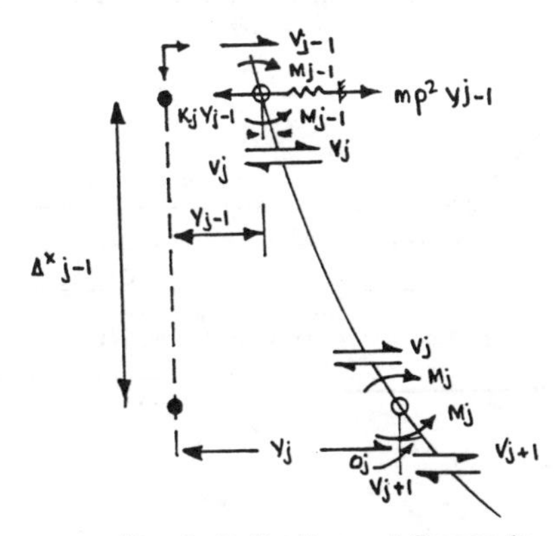

Fig. 8: Deflection and Forces in a Segment

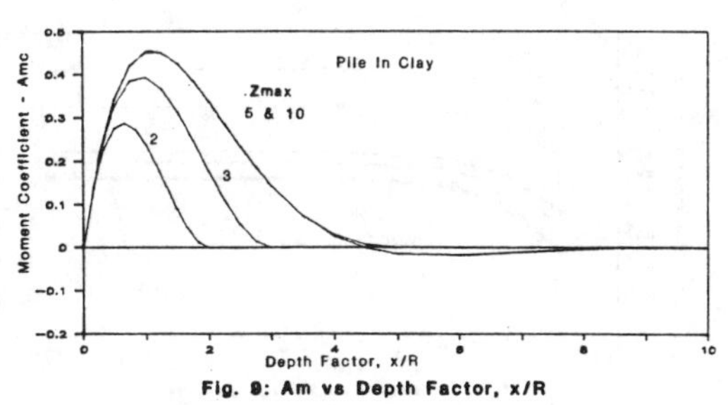

Fig. 9: Am vs Depth Factor, x/R

Fig. 10: Am vs Depth Factor, x/T

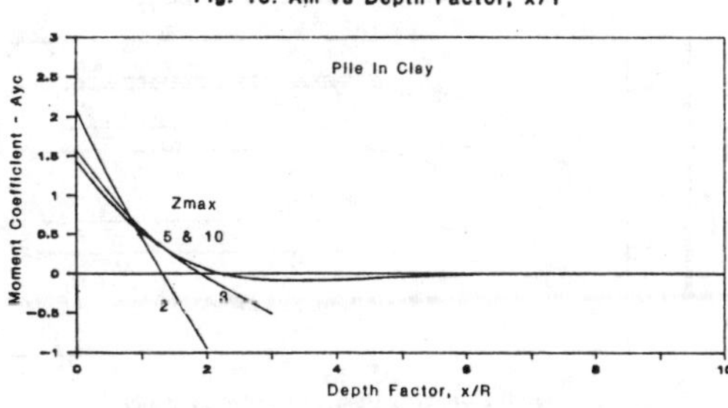

Fig. 11: Ayc vs Depth Factor, x/R

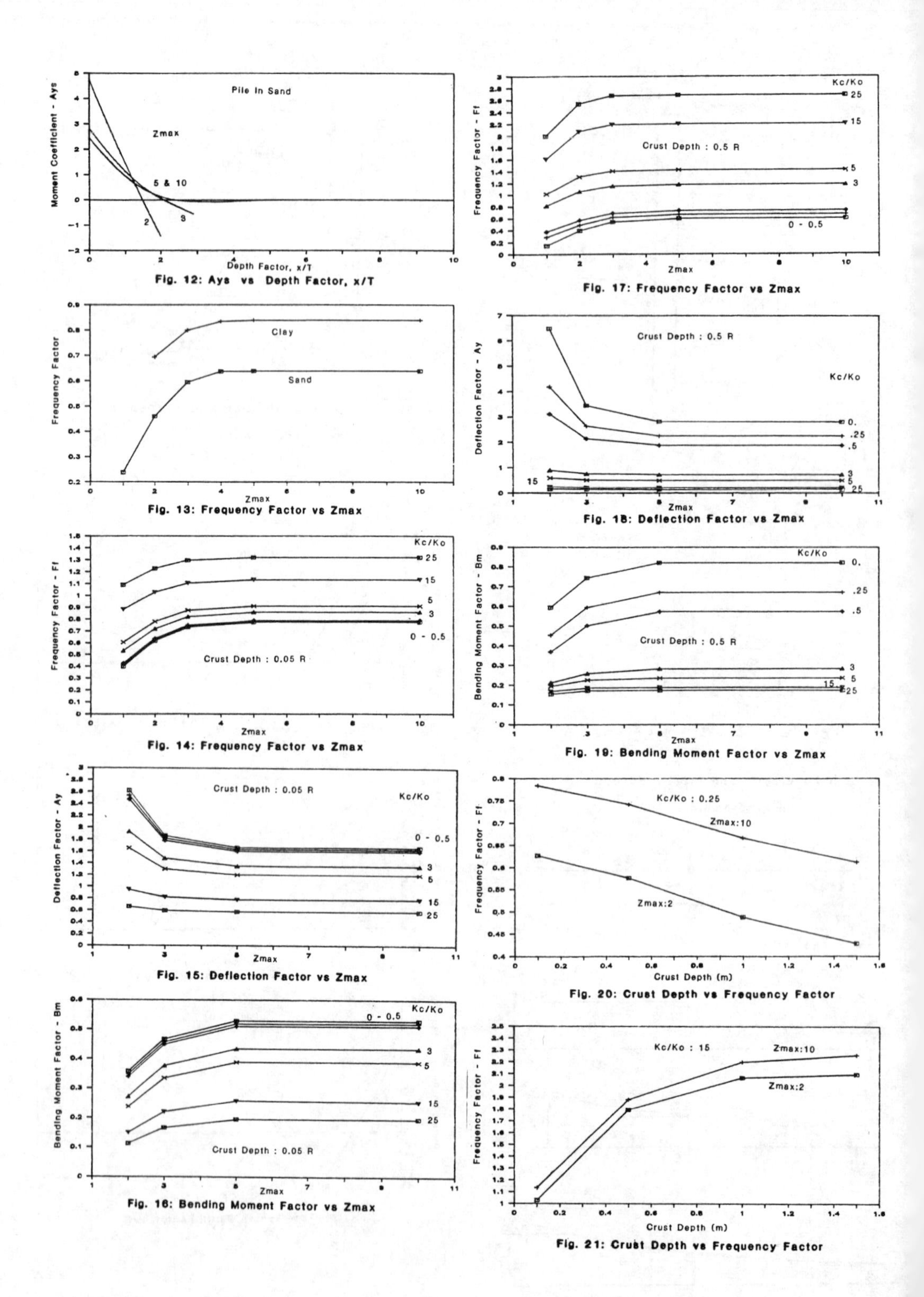

Fig. 12: Ays vs Depth Factor, x/T

Fig. 13: Frequency Factor vs Zmax

Fig. 14: Frequency Factor vs Zmax

Fig. 15: Deflection Factor vs Zmax

Fig. 16: Bending Moment Factor vs Zmax

Fig. 17: Frequency Factor vs Zmax

Fig. 18: Deflection Factor vs Zmax

Fig. 19: Bending Moment Factor vs Zmax

Fig. 20: Crust Depth vs Frequency Factor

Fig. 21: Crust Depth vs Frequency Factor

Prediction and Performance in Geotechnical Engineering / Calgary / 17-19 June 1987

Friction mobilization F-y curves for laterally loaded piles from the pressuremeter

Trevor D.Smith
Portland State University, Oreg., USA

ABSTRACT: A discussion is presented, and field evidence given, to confirm both the existence and importance of lateral shear drag on the pile side under lateral load. The relationship between the pressuremeter and the pile side friction mobilization rates is given in elasticity. Torsion testing shows that depth effects influence the ultimate side shear which can be mobilized in the same way as vertical shear mobilization. Existing pressuremeter vertical shear rules are modified to calculate horizontal side friction. A tentative recommended procedure is given to formulate elastic-plastic F-y curves which when added to Q-y curves produces the pile's P-y curve. The procedure is applied to a rigid shaft in clay.

INTRODUCTION

It is common geotechnical design practice to recognize the presence and importance of soil - structure friction effects. In axially loaded piles, the magnitude of the shaft friction is of fundamental importance and probably reaches peak shear at very small movements. In retaining wall design the wall-soil friction effects have long been recognized and were incorporated into modified Active and Passive coefficients. This same friction is also present in the horizontal loading of piles and rigid shafts mobilized to oppose the direction of movement.

The calculation of lateral load P-y curves and the use of the finite difference method has been the subject of much research and discussion. Despite the wide variety of methods available to formulate these curves, no methods based on conventional laboratory testing, and only one method of at least seven pressuremeter methods, Briaud (1986 a), makes any attempt to quantify the friction contribution.

The pressuremeter methods are increasingly being recognized for improving the quality of P-y curve data by virtue of the tool's versatility and sound theoretical base. There is wide agreement that the soil response around the pressuremeter probe is directly analogous to the soil's behavior ahead of the pile. However, some researchers convert the pressuremeter pressure by magnification factors based solely on plasticity, Robertson et al. (1986), and dispute the existence of any shear, while others, Briaud, Smith, Meyer (1983), Smith (1983), accommodate the theoretical stress and shear distributions to formulate pressure contributions (Q-y curves) and side shear contributions (F-y curves) independently, Fig. 1.

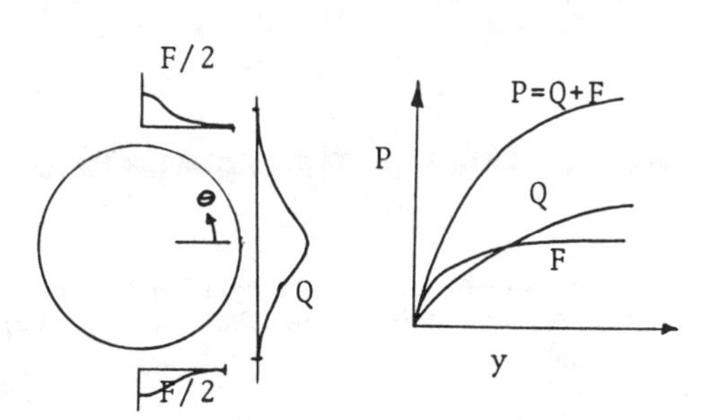

Fig. 1. Q-y and F-y contributions.

Many questions with respect to side shear mobilization rates need to be addressed; do the elements to the side of a pile and around a pressuremeter correspond?, are there any depth effects (which significantly reduce front pressures)?, and what role is the most effective for the pressuremeter?

THEORETICAL MOBILIZATION RATES AND ULTIMATE VALUES

Baguelin, Frank and Said (1977) from a single Airy function reported the theoretical distributions of front pressures and side shears around a translating disc in an elastic medium.Despite the limitation of soil modeling in elasticity, it can be stated that any assumptions of uniform pressures across the pile face and uniform shears along the sides are far from the truth. In elasticity, the contribution of shears and pressures in resisting pile movements are equal. As real soils approach failure, we have long recognized that passive type pressure require large movements to deliver the full soil capacity but peak shear requires very little.

Using the expressions given by Baguelin et al. for pile displacement, y, and side shear τ, and applying them to the element shown in Fig. 2, the mobilization rate is given by:

$$\frac{\tau}{y} \quad \frac{(1-\upsilon)G}{R(.75-\nu)\left(3-\frac{1}{(3-4\upsilon)^2}\right)} \qquad (1)$$

where:

R = pile radius

for elasticity assumptions.

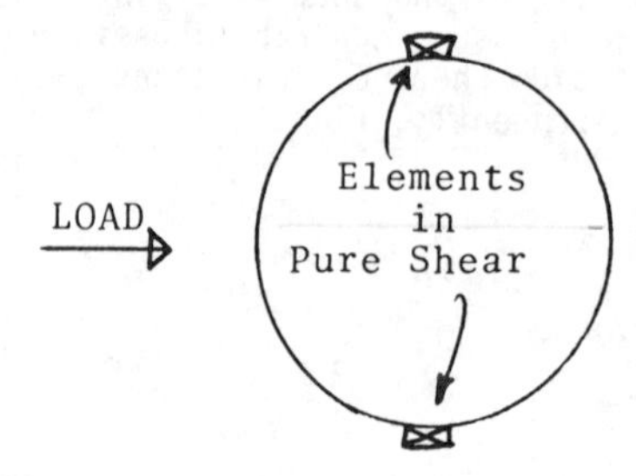

Fig. 2. Pile elements under pure shear.

If typical values of Poisson's ratio, from 0.3 to 0.5, are introduced into Eq. (1) then we have:

$$\frac{\tau}{y} \simeq \frac{.5G}{R} \text{ to } \frac{G}{R} \qquad (2)$$

Turning to the expressions which govern the soil cavity elements around the pressuremeter in elasticity, given by Baguelin, Jezequel and Shields (1978), we find:

$$\frac{\tau}{\Delta r} = \frac{2G}{r} \qquad (3)$$

where:

r = initial soil cavity radius
Δr = increase in soil cavity radius

Of additional particular interest is the soil response around a pile subject to pure side shear when under an applied torque, Fig. 3. Like the pressuremeter

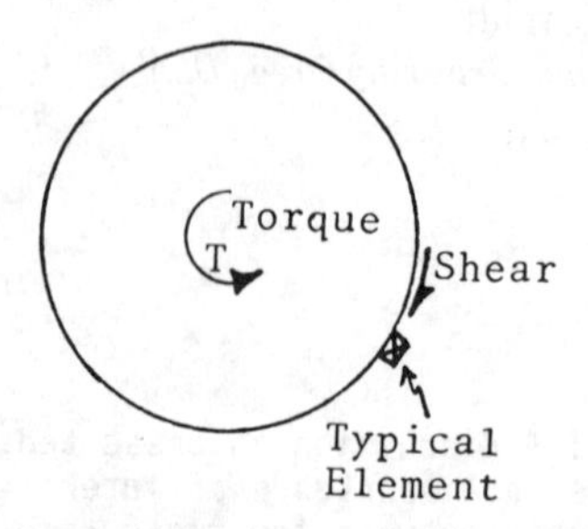

Fig. 3. Pile under torsion.

the boundary conditions are axially symmetric and from the work of Randolph (1981) for rigid piles it is found:

$$\frac{\tau}{v} \simeq \frac{2G}{R} \qquad (4)$$

where:

v = circumferential movement (length)

which is the same as the pressuremeter. A rational way to compare Eqs. 2, 3 and 4 is to rearrange the mobilization rate such that the denominator is radial strain (i.e. movement/radius). At equal radial strains the torsion pile and pressuremeter correspond exactly and both show mobilization rates two to four times the laterally loaded pile. In all cases, the mobilization rate is inversely proportional to radius. It should be now recalled that conventional prebored pressuremeter tests underestimate the intact shear modulus, G, by the $1/\alpha$ values, Menard (1975), which gives a range ofreduction factors of 1 to 3. Hence, the use of Eq. 3, to calculate the pile's mobilization rate, with G values measured too low, produces a result of the correct order, Briaud, Smith and Meyer (1983).

In plasticity from a limit analysis Poulos and Davis (1980) illustrated very well the contribution from side friction to soil reaction P on pile sections of different width to depth ratios. For ratios of 1, (i.e. square or circular) the presence of full soil friction increased the soil reaction, P, by 50% above that for a smooth pile wall. They further indicated that as the section depth increased above the pile width, the F contributions rapidly grew as the Q contribution decreased.

MEASURED MOBILIZATION RATES AND ULTIMATE VALUES

Two groups of investigations give insight into mobilization rates and ultimate values of the pile side friction: firstly, full scale pile load tests which are instrumented to record soil pressures and/or shears, and secondly, pile testing under torsion.

Of the first group, most of the instrumented piles reported incorporated strain gauges, and/or slope indicator tubes, which enable direct calculation of soil reaction, P, but not the two components Q and F, (Fig. 1). However, the series of 3 rigid shafts in stiff clay reported by Coyle and Bierschwale (1983), and later studied by Smith and Ray (1986), do show mobilization rates for pressures. In the latter analysis Smith and Ray, from horizontal equilibrium, reported that the final shaft had 84% of its soil reaction, H, from side shear, F, at a factor of safety of 8 to ultimate load. This is illustrated in Fig. 4.

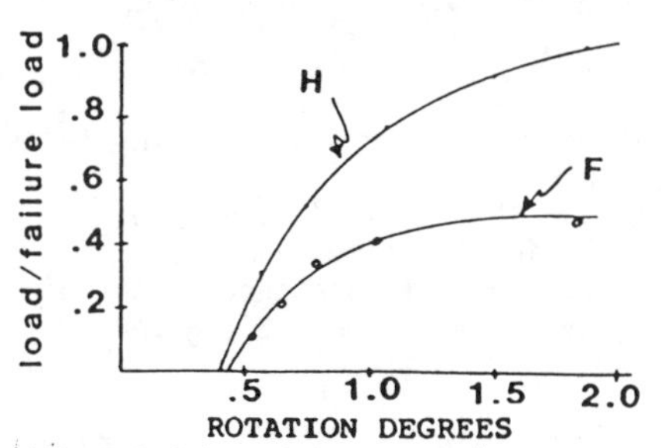

(after Smith and Ray, 1986)
Fig. 4. Horizontal equilibrium for rigid shaft.

This confirms the earlier point that friction is mobilized very quickly and may reach its ultimate value at working load level displacements. Smith and Ray further concluded that predictions of soil to pile adhesions based on pressuremeter tests were within 5%, but predictions based on unconfined compression test results were low by over 30%.

A review of model pile testing under torsion was reported by Smith and Slyh (1986) together with results from testing in dense sand, stiff clay and cemented sand. It clearly showed that ultimate lateral shear followed the same phenomenon as vertical shear; a distinct depth effect in clays and high K values, approaching Kp for driven piles, in sand. Their conclusion suggests that existing design rules to calculate vertical side friction for axially loaded piles may be suitable for ultimate lateral friction also.

THE USE OF PRESSUREMETER METHODS FOR VERTICAL SHEAR

A review of pressuremeter methods for vertical pile settlement and capacity,together with an evaluation based on 51 load test results, was presented by Briaud (1986 b). For settlement predictions at a factor of safety of 2.5 to the predicted ultimate vertical capacity (when friction would be expected to dominate), it was concluded the predicted settlements compared favorably with those measured. Settlements in the data base ranged from 0.6 mm to 15 mm and averaged 2.5 mm or about 0.1 inches. It is of interest to recall that the peak vertical movement to reach shear capacity in the wave equation of pile driving (soil quake) is generally taken as 0.1 inches. Settlements in this study were calculated using a vertical shear transfer rate, f/w of:

$$\frac{f}{w} = E_r/2(1+\upsilon)(1+L_n(L/2R))R \qquad (5)$$

where:

E_r = the deformation modulus from the reload cycle on a prebored test.
L = pile embedded length

and was taken from the method proposed by Baguelin, Frank and Jezequel (1982). Since $E_r = 2(1+\upsilon)G_r$, Eq.(5) reduces to

$$\frac{f}{w} = \frac{G_r}{(1 + L_n(\frac{L}{2R}))R} \qquad (6)$$

and if applied directly to the lateral shear mobilization rate the length, L, would be the equivalent pile depth over which the shear, τ, is acting. Based on the theoretical distributions shown in Fig. 1, Smith (1983) reported the shape factor for circular piles and square piles as 0.79 and 1.76 respectively when used with the diameter (or width) D. Since D = 2R, Eq.(6) becomes, for lateral shear transfer:

$$\frac{\tau}{y} = 0.64\,\frac{G_r}{R} \text{ to } 1.3\,\frac{G_r}{R} \qquad (7)$$

for square to circular piles, respectively. The similarity to the expression in elasticity given by Eq. (2) is remarkable. Recalling the relationship given by Briaud, Smith and Meyer, and now dividing by y:

$$\frac{F}{y} = \frac{\tau}{y} \times SF \times 2R \tag{8}$$

then substituting from Eq. (7) for $\frac{\tau}{y}$ we have:

$$\frac{F}{y} = \alpha \times G_r \times SF \tag{9}$$

where:

α = 1.28 and 2.6 for square and circular piles respectively
and hence, by further substitution of the shape factors, SF, given above:

$$\frac{F}{y} = 2.25\,G_r \text{ and } 2.0\,G_r \tag{10}$$

for square to circular piles.
For practical application this could be simplified to:

$$\frac{F}{y} = 2.0\,G_r \tag{11}$$

for all piles.

RECOMMENDED F-y PROCEDURE

From the review of both elastic and plastic pile theories, full scale load testing, model pile torsion testing and existing vertical shear pressuremeter design rules, the following summary can be made:
(1) All evidence confirms the existence of substantial side shear on laterally loaded piles.
(2) From a direct analogy to vertically loaded piles, the prediction of shear mobilization curves (F-y) is of paramount importance for correct calculation of the governing P-y curve.
(3) It follows from (2) then, that the evaluation of installation disturbances and initial stresses against the pile wall must be taken into account.
(4) Comparison of Eqs. (7) and (2) confirms a rational link between vertical shear transfer and horizontal shear transfer does exist. Torsion testing suggests this link is also present at ultimate conditions by equal ultimate shear transfer in the vertical and horizontal directions.

Based on this review, the following tentative procedure is suggested for elastic-plastic F-y curves:

(1) From prebored pressuremeter tests, establish the reload shear modulus, G_r, in addition to the initial cycle shear modulus, G_o, Fig. 5.

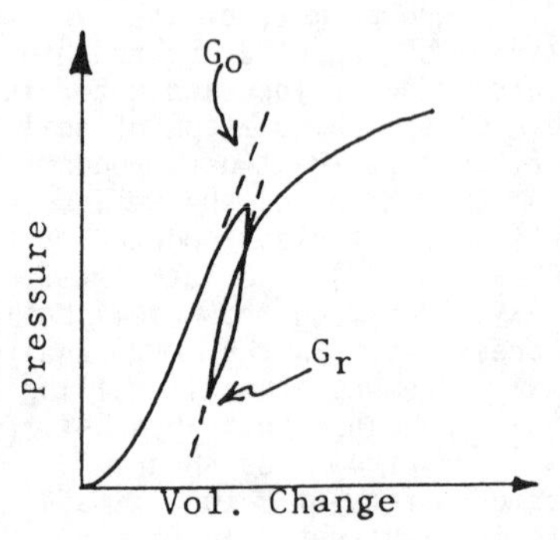

Fig. 5. Pressuremeter initial and reload cycles.

(The initial modulus from self-boring tests may also be used.)
(2) Establish the net limit pressure, P_L^*, profile for the pile.
(3) Calculate the ultimate shear reaction, F_f using the shape factors, SF, given by Smith from:

$$F_f = \tau_f \times SF \times 2R \tag{12}$$

where:

SF = 0.79 circular piles
SF = 1.76 square piles

with τ_f calculated from existing pressuremeter design rules. A method which accommodates at least pile type, installation and soil type should be used.
(4) Calculate the mobilization rate using Eq. (11) with reload shear modulus, G_r:

$$\frac{F}{y} = 2G_r \tag{11}$$

For disturbance effects, reduce this rate as appropriate.

This procedure is illustrated in Fig. 6.

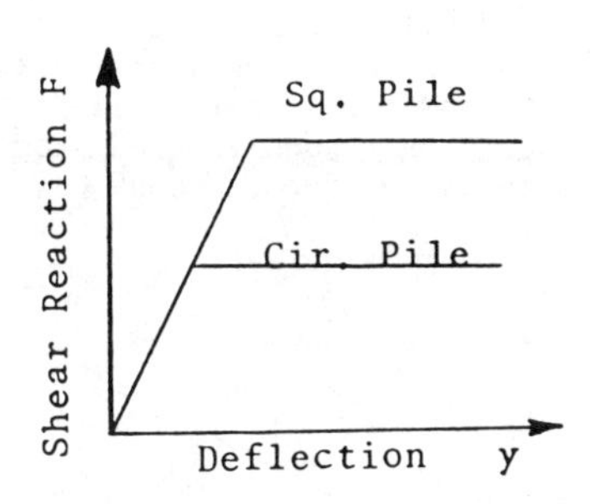

Fig. 6. Schematic F-y curves.

EXAMPLE F-y AND P-y FORMULATION

To illustrate the above procedure, the F-y curve is constructed for the drilled shaft quoted earlier, Coyle and Bierschwale (1983), which contained earth pressure cells. The P-y curve is assembled using the Q-y procedure recommended by Briaud, Smith and Meyer (1983) with critical depth effects, Smith (1983), applied only to the Q-y curve. The concrete shaft was 0.914 m (3 feet) in diameter and 6.1 m (20 feet) long. Since the shaft was drilled and the pressuremeter tests were performed in a prebored hole, the pile and pressuremeter disturbances should be equal. Comparisons between lateral load and surface deflection responses from three pressuremeter P-y techniques were presented for the shaft by Briaud, Smith and Meyer (1983).

F-y curve at 4' depth

Smith (1983) reports the following

E_r = 23.4 MPa
P_L^* = 620 kPa

The most recent pressuremeter based design rules to calculate vertical side friction, and proposed directly for horizontal shear, are based on Laboratoire des Ponts et Chaisses work and are reported by Briaud et al.(1985)
For stiff clays at P_L^* = 620 kPa
τ_f=60kPa
Therefore from Eq. (12)

$$F_f = \tau_f \times SF \times 2R = 60 \times 0.79 \times .914 = 43.5 \text{ kN/m.}$$

and the mobilization rate is given by Eq.(11).

$$\frac{F}{y} = 2Gr$$

since $Gr = \frac{Er}{2(1+u)}$

then $\frac{F}{y} = \frac{Er}{(1+u)} = \frac{23.4}{1.33} = 1.76$ MPa

The maximum shear is then attained at a pile displacement of

$$y = \frac{43.5}{17600} = 2.47 \text{ mm}$$

The friction F-y curves from the pressuremeter tests at 1.2 m (4 feet), 2.74 m (9 feet), 3.6 m (12 feet) and 4.5 m (15 feet) are shown in Fig. 7.

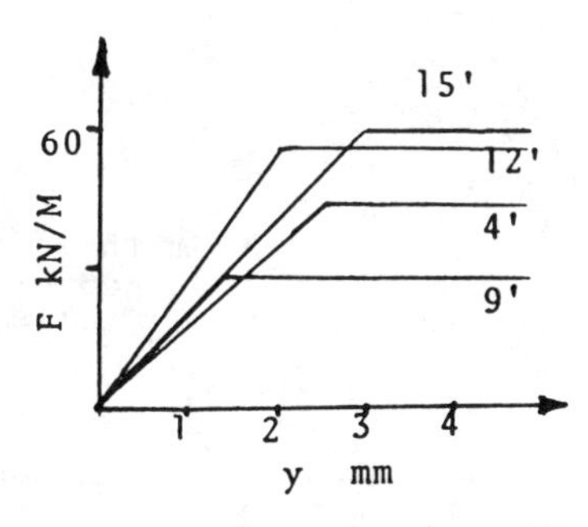

Fig. 7. Calculated F-y curves.

Again, it is of interest to note the shaft movement to mobilize peak lateral friction of the order 0.1 inches.
Construction of the P-y curve can then be accomplished by the addition of the F-y and Q-y contributions. Following Briaud, Smith and Meyer (1983), the prediction of Q-y curves uses the entire non-linear expansion curve from the pressuremeter:

$$Q = P^* \times B \times SQ$$

where:

P^* = net pressuremeter pressure
B = pile front width or diameter
SQ = shape factor for pressure
= 1.0 square piles
= 0.75 circular piles

at a pile movement given by

$$y = \frac{\Delta r}{r} \times R$$

where:

Δr = increase in soil cavity radius at P*
r = soil cavity initial radius
R = Pile radius, (or $\frac{B}{2}$)

Prediction of the lateral load response for the shaft using the F-y curves in Fig. 7 and Q-y curves from Briaud, Smith and Meyer is shown in Fig. 8.

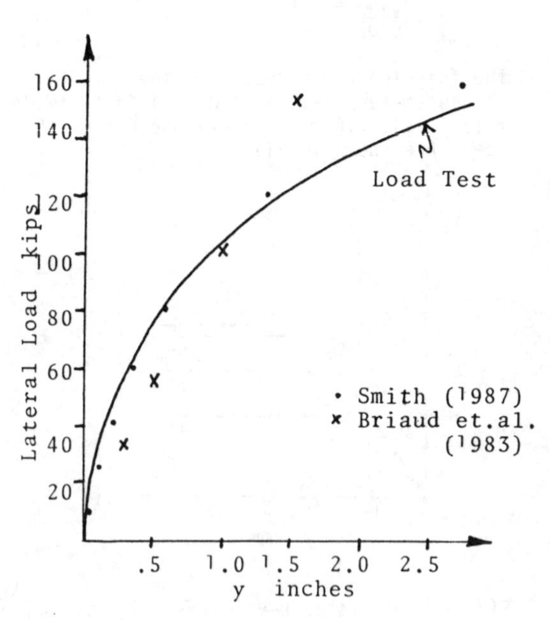

Fig. 8. Lateral load response for the shaft.

When compared to the predictions reported by Briaud, Smith and Meyer (1983), the improved non-linearity in the shaft's behavior is directly a result of the F-y improvement.

CONCLUSIONS

Based on both elastic and plastic limit equilibrium approaches, there is extensive theoretical evidence to support the presence of lateral side friction on a horizontally loaded pile. From measured field results, including pressure cell data and model pile testing in torsion, the lateral shear phenomena and vertical shear phenomena are closely related, in both mobilization rates and ultimate values. It further appears that direct use of existing ultimate vertical shear rules can be used which take account of installation methods, soil types, and pile types. Due to the highly local nature of lateral friction, the prevelant conditions at the pile/soil interface should be recognized and proper judgement used in the formation of F-y curves.

REFERENCES

Baguelin, F., Frank, R., and Said, Y.U. 1977. Theoretical Study of Lateral Reaction Mechanism of Piles. Geotechnique.

Baguelin, F., Jezequel, J.F., and Shields, D.H. 1978. "The Pressuremeter and Foundation Engineering", Rockport, Mass. Trans Tech Publications.

Baguelin, F., Frank, R., and Jezequel, J.F. 1982. "Parameters for Friction Piles in Marine Soils", Austin, Second Int. Conf. in Numerical Methods for Offshore Piling.

Briaud, J.L. 1986a. "Pressuremeter and Deep Foundation Design", The Pressuremeter and its Marine Application: 2nd Intl. Symposium. ASTM STP950

Briaud, J.L. 1986b. "Pressuremeter and Foundation Design", Proceedings of ASCE Speciality Conference on "Use of In Situ Tests in Geotechnical Engineering". Virginia Polytechnic.

Briaud, J.L., Anderson, J.S., Tucker, L.M. and Coyle, H. 1985. "Measured and Predicted Response for 100 Piles: Part I", Research Report 4981-1, Civil Engineering, Texas A&M University.

Briaud, J.L., Smith, T.D., and Meyer, B. 1983. "Laterally Loaded Piles and the Pressuremeter: Comparison of Existing Methods", ASTM STP 983.

Coyle, H.M. and Bierschwale, M. 1983. "Design of Rigid Shafts in Clay for Lateral Load", Journal of Geotechnical Engineering Division, ASCE, pp 1147-1164.

Menard, L., 1975. "The Menard Pressuremeter: Interpretation and Application of the Pressuremeter Test Results to Foundations Design", Sols - Soils No. 26

Poulos, H. and Davis E. 1980. "Pile Foundation Analysis and Design", New York, John Wiley and Sons.

Randolph, M.F. 1981. "Piles subjected to Torsion", Journal of Geotechnical Engineering Division, ASCE, pp. 1095-1112.

Robertson, P.K., Hughes, J.M.O., Campenella, R.G., Brown, P. and McKeown, S. 1986, "Design of Laterally Loaded Piles Using the Pressuremeter", The Pressuremeter and Its Marine Application: 2nd Intl. Symposium, ASTM STP 950, ASTM.

Smith, T.D. 1983. "Pressuremeter Design Method for Single Piles Subjected to Static Lateral Load", Dissertation submitted to Graduate College, Texas A&M University, in partial fulfillment for Doctor of Philosophy. Vol. I and II.

Smith, T.D. 1985. "Lateral Shear Transfer on Model Piles", Research Report to Portland State University Graduate Office, Grant 90-050-5801 TSA.

Smith, T.D. and Ray, B. 1986. "Shear Mobilization on Laterally Loaded Shafts", Proceedings of the Symposium on Geotechnical Aspects of Stiff and Hard Clays, ASCE Spring Convention. Seattle. April 7-11.

Smith, T.D. and Slyh, R. 1985. "Lateral Friction Mobilization Rate for Laterally Loaded Piles", Edmonton. Proceedings of the 38th Canadian Geotechnical Conference.

Smith, T.D. and Slyh, R. 1986. "Side Friction Mobilization Rates for Laterally Loaded Piles from the Pressuremeter", The Pressuremeter and Its Marine Application: 2nd Intl. Symposium, ASTM STP 950.

Behavior of two piles in clay subjected to cyclic lateral loads

James H.Long
University of Illinois, Urbana-Champaign, USA

Lymon C.Reese
University of Texas, Austin, USA

ABSTRACT: The behavior of a pile subjected to repetitive lateral loads can be influenced significantly by 1) the accumulation of progressive permanent strains, 2) the reduction of soil modulus and strength due to cyclic loads, 3) the effect of water entering fissures within the soil and reducing the strength of the soil, and 4) the removal of soil particles (scour) due to water entering and exiting a gap that forms along the side of the pile. Results of full-scale pile-load tests conducted near Manor, Texas and near Sabine, Texas were investigated. Tests in the laboratory were performed to determine the stress-strain behavior of the soils when subjected to cyclic loads and to determine the effect of water above the ground surface. From results of the tests, it was apparent that the reduction of soil resistance measured during the lateral-load test at Manor was affected significantly by the presence of water.

1 INTRODUCTION

In many cases, foundations of a structure must be designed to resist horizontal forces applied by the action of wind, waves, current, traffic, impact, and earthquakes. These loads are cyclic in nature. Structures, such as offshore platforms, transmission-line poles, bridges, and buildings are founded on piles that must be designed to resist the effects of cyclic lateral load.

The behavior of a pile subjected to repetitive lateral loads depends on many factors such as: the magnitude and characteristics of the load, the material and geometric properties of the pile, and the response of the soil to cyclic loading. In addition, of possible importance is the influence of water entering and exiting the gap that may form between the pile and soil during cyclic loading.

Lateral-load tests with full-sized piles were conducted to measure the response of piles in soft clay (Matlock, 1970) and stiff clay (Reese, et al., 1974). Reductions in soil resistance were observed during the cyclic lateral-load tests. Reductions in the resistance provided by the stiff clay were significantly greater than observed for the soft clay. A research program was initiated to investigate the results of these two tests and provide explanations for the differences that were observed.

The original test sites were located and undisturbed samples of soil were taken near each site. Specimens of soil were tested in the laboratory to identify the behavior of the soil subjected to cyclic loads and its resistance to erosion by water. Results of these laboratory tests help to identify important aspects of soil behavior during cyclic lateral loading of piles, and provide a basis for the selection of appropriate recommendations for predicting the behavior of such piles.

2 BEHAVIOR OF PILE AND SOIL DURING CYCLIC LOAD

Shown in Fig. 1 is a simple model to illustrate the response of a soil and pile to the application of a cyclic lateral load at the head of a pile. In this model, the pile is embedded in cohesive soil, and the pile is long and vertical. The free surface of water is above the ground surface; therefore, water can enter gaps that form between the pile and the cohesive soil. The horizontal load at the head of the pile varies in magnitude sinusoidally with time at a frequency slow enough that the effects of inertia are

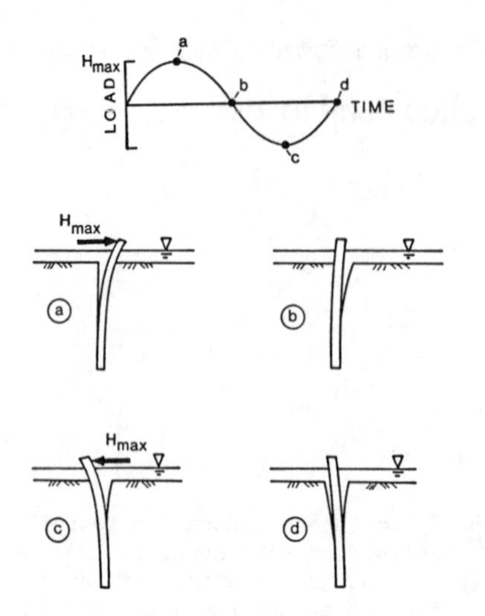

Fig.1 Simplified Illustration of the Response of Pile in Clay due to Cyclic Load

minimal. The behavior of the pile is described for each quarter-cycle of load. The influence of further cycles are mentioned.

The magnitude of lateral load varies from zero to a maximum horizontal load H_{max} during the first quarter-cycle. The head of the pile rotates and translates as the load increases to H_{max} (Fig. 1a). If deflections of the pile are large enough, a gap between the soil and pile-wall will form along the back of the pile and water will flow into the gap. The velocity at which the water flows is dependent on the rate of deformation of the pile, and the geometric characteristics of the gap.

The lateral load decreases from a value of H_{max} to zero during the second quarter-cycle. The head of the pile returns to a position similar to its original configuration and forces out water between the pile and soil (along the back of the pile). The wall of the pile and soil may separate along the front of the pile due to permanent deformations induced within the soil-mass during the first quarter-cycle of loading (Fig. 1b). Water enters the newly-formed gap along the front of the pile at a velocity proportional to the rate of change of the geometry of the gap. As the magnitude of load approaches zero, the pile returns to a configuration similar to its original shape, but displaced in the direction of the lateral load H_{max}.

The direction of the lateral load, H_{max}, and the corresponding deflection of the pile head are reversed during the third quarter-cycle. The gap along the face of the pile closes while the gap along the opposite face of the pile opens (Fig. 1c) as the magnitude of horizontal load increases to H_{max}.

The responses of the pile and soil during the fourth quarter-cycle are similar, but opposite in direction, to the responses described during the second quarter-cycle. At the end of the fourth quarter-cycle, with no load imposed on the head of the pile, gaps between the soil and pile may occur on both sides of the pile (Fig. 1d).

The effect of further cycles on the maximum horizontal deflection of the pile and bending moments within the pile depend upon the accumulation of permanent strains in the soil, the degradation of mechanical properties (strength, modulus) of the soil, and the amount of soil removed by the action of water entering and exiting the gap. The results of full-scale lateral-load tests conducted in soft clay and stiff clay were studied to better understand the effect of cyclic load on pile behavior.

3 DESCRIPTION OF TESTS CONDUCTED IN SOFT CLAY

Lateral-load tests were conducted in a site located south of Sabine Pass, Texas (Matlock, 1970) along the west bank of the Sabine river. The soil profile consists of soft, slightly overconsolidated marine clay with occasional thin layers of fine sand. The clay, classified as CH according to the Unified Soil Classification System, and had a natural water content near its liquid limit. The undrained shear strength averaged 14.4 kPa for the depth of soil that significantly influenced the behavior of the pile. Sensitivity of the soil was between 2 and 3.

An open-ended pipe pile, 324 mm in diameter, was driven into the clay. Special care was taken to minimize disturbance of the soil within the top few pile diameters. Water was ponded above the ground surface to simulate conditions of a submerged pile.

Static lateral-load tests were conducted by applying a lateral load and recording deflection. Cyclic lateral-load tests were performed by applying a lateral load in one direction (major direction) and then by applying a load of 8.9 kN in the direction opposite to the major load. Each cycle of load required approximately 20 seconds. Cyclic loads for each load level were discontinued after it was determined that the effect of cycling would effect pile behavior insignificantly. Between 100 and 300 cycles of load were applied.

The load-deflection relationships measured during the static and cyclic lateral-load tests are shown in Fig. 2. Observations were made of fluid entering and exiting the gap between the soil and pile while the pile was loaded cyclically. No significant scour was observed along the gap between the pile and soil during the application of cyclic load in the major direction. However, after the major load was applied, the lateral load was released and the pile returned rapidly to a position close to its original position. Velocity of the fluid exiting the gap was therefore higher along the backside of the pile. A slight amount of scour was observed along the backside of the pile. More details regarding these tests are presented elsewhere (Long, 1984; Matlock, 1970; and Matlock and Tucker, 1961).

4 DESCRIPTION OF TESTS CONDUCTED IN STIFF CLAY

In 1966, lateral-load tests on piles embedded in stiff clay were conducted at a site near Manor, Texas (Reese, et al., 1975). The soil profile consisted of a medium stiff, overconsolidated clay of high plasticity. The clay was jointed and fissured with some of the fissures having lighter colored, softer clay within them.

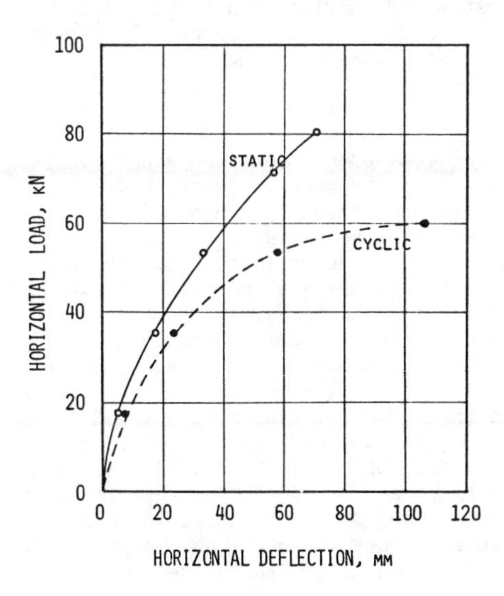

Fig.2 Load-Deflection Relationship for Pile Load Test in Soft Clay

The clay had a natural water content near the plastic limit. The undrained shear strength of the clay varied linearly from near zero at the ground surface to about 335 kPa at 4 m below the ground surface.

Two open-ended pipe piles 641 mm in diameter were driven into the stiff clay and water was ponded above the ground surface.

One of the piles was loaded statically in a manner similar to the test at the soft clay site. The other pile was loaded cyclically by applying a major load in one direction and a load of smaller magnitude in the opposite direction. The magnitude of minor load was 30 percent of the major load. Each cycle of load required 20 to 30 seconds and approximately 100 cycles were applied for each level of load. The load-deflection relationships measured during the static and cyclic lateral-load tests are shown in Fig. 3.

A considerable volume of soil was observed to be removed along the gap between the pile and soil as the pile deflected during cyclic loading. An opening in the soil, approximately 100 mm in diameter and 600 mm in depth, was observed in the soil to the side and slightly behind the pile at the conclusion of cyclic loading. Further details

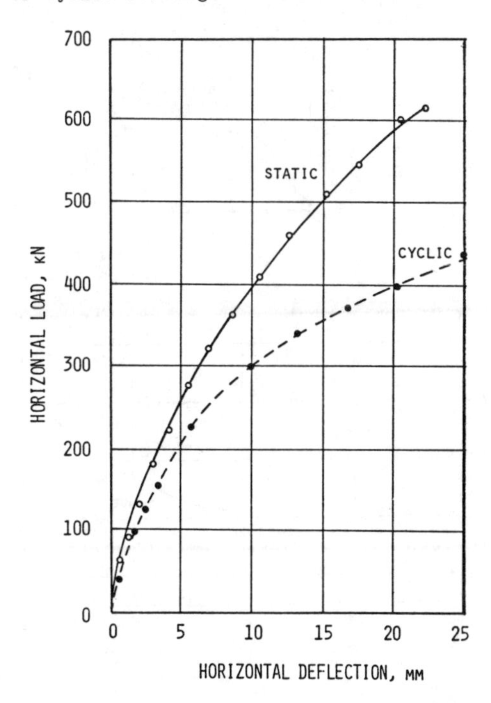

Fig.3 Load-Deflection Relationship for Pile Load Test in Stiff Clay

concerning the lateral-load tests conducted near Manor are given by Long (1984), Reese, et al. (1975), and Reese, et al. (1968).

5 EFFECT OF CYCLIC LOAD ON BEHAVIOR OF PILES

The effect of cyclic loads upon the behavior of piles at Manor and Sabine are illustrated in Figs. 2 and 3. A simple measure of the effect of cyclic loads is the ratio of deflection at the end of cyclic loading to the deflection during static loading, Y_c/Y_s. Results from the lateral-load tests near Sabine and Manor are illustrated and compared by plotting the ratio Y_c/Y_s versus the static deflection Y_s (Fig. 4). Two important observations can be made regarding the effect of cyclic loads and are 1) as the static deflection, Y_s, increase, the ratio, Y_c/Y_s, increases, and, 2)for a given static deflection, Y_s, the ratio, Y_c/Y_s, is much greater for the Manor tests than for the Sabine tests.
The results show that the relative loss of soil resistance during cyclic loading was much greater during tests at Manor than during tests at Sabine.

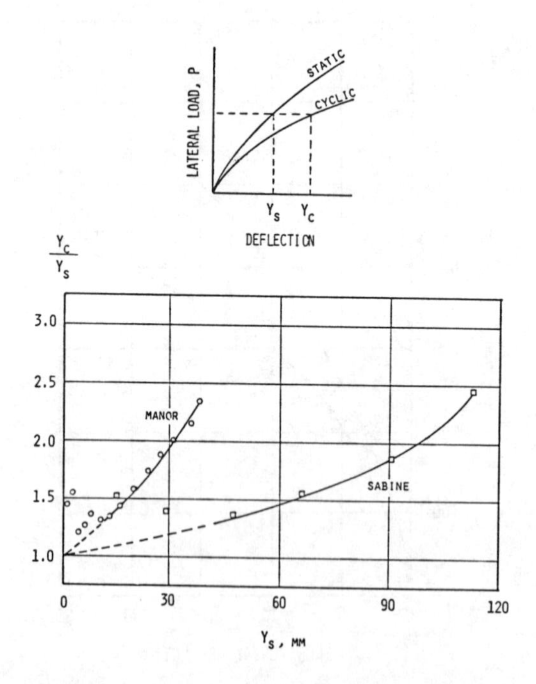

Fig.4 Effect of Cyclic Loads on Deflection Ratio, Y_c/Y_s

Four important phenomena that appear responsible for the reduction in soil resistance during cyclic load are 1) the accumulation of progressive permanent strains, 2) the reduction of soil modulus and strength due to cyclic loads, 3) the effect of water entering fissures within the soil and reducing the strength of the soil, and 4) the removal of soil particles along the side of the pile due to water entering and exiting.

Laboratory and field tests conducted during the original research did not include tests to determine the effects of cyclic load on soil behavior nor the effect of water at the ground surface.

6 FIELD INVESTIGATION AND LABORATORY TESTS

6.1 field investigation

In 1981, a program was initiated to return to sites where lateral-load tests were conducted and obtain undisturbed samples. These samples were used to study the response of soil to cyclic loads and the resistance of the soil to scour or erosion.

The Sabine test site is located south of Sabine Pass, Texas, along the west bank of the Sabine River. It was not possible to obtain samples at the exact location of the original site, therefore, soil specimens were obtained from a site approximately 70 m from the original site. The original site for the Manor test was located and samples of soil were taken within a few meters of where the lateral-load tests were performed.

6.2 cyclic triaxial tests

Controlled-deformation, undrained cyclic triaxial tests were conducted on isotropically-consolidated specimens to determine the stress-strain characteristics of soil subjected to cyclic loads. Several cycles of load, usually greater than 100, were applied to each triaxial specimen.

The effect of cyclic loads on soil behavior is shown in Fig. 5 as a plot of the secant modulus of the soil versus the number of cycles. The slope, t, of a line passing through the points is a measure of the effect of cyclic loads. Idriss et al (1978) first introduced the t-parameter to express the effect of cyclic loads on reduction in value of secant modulus. The following equation can be used:

$$E_N = E_O N^{-t} \quad (1)$$

where: E_N = secant modulus at any cycle, N,
E_O = secant modulus at the first cycle of loading,
N = number of cycles of loading, and
t = degradation parameter.

The value of the t-parameter increases with in the magnitude of cyclic strain, and that relationship is discussed herein.

Triaxial tests were performed on specimens of cohesive soil from both the Sabine and Manor sites. Plots of t versus axial strain for Sabine and Manor clays are shown in Fig. 6 along with the relationship determined by Idriss et al. (1978). It can be seen that the t-parameter increases with the magnitude of cyclic axial strain. The curves found by Idriss plot slightly above the range of values found for the Manor soil. The values of the t-parameter for a given value of cyclic strain are generally higher for the Sabine soils than for the Manor soil. This result is opposite than suggested from the results of full-scale lateral-load tests because the effect of cyclic load was found to be more severe for load tests conducted near Manor than load tests conducted near Sabine.

6.3 immersion tests

A factor that may also play a significant role in the reduction of soil resistance is the removal of soil along the side of the pile due to the movement of water in and out of the gap that may form between the pile and soil. The soil properties, pile geometry, pile deflection, and rate of loading play important roles in determining the loss of soil resistance (due to scour).

Several conditions must be met for scour to occur. These conditions are 1) the soil must be erodible, 2) the pile must be displaced horizontally so that permanent strains are induced in the soil and a gap between the pile and soil will occur upon unloading, 3) the soil must possess enough strength to allow the gap to remain open for some depth, and 4) the lateral load must be applied at a rate that will cause the velocity of the water to be high enough to scour the soil.
If any of these conditions is absent, scour will not occur.

Very little scour was observed during the tests at Sabine Pass, while a significant amount of scour was observed during the tests at Manor. Velocity of fluid entering and exiting the gap was estimated for the tests conducted near Manor and

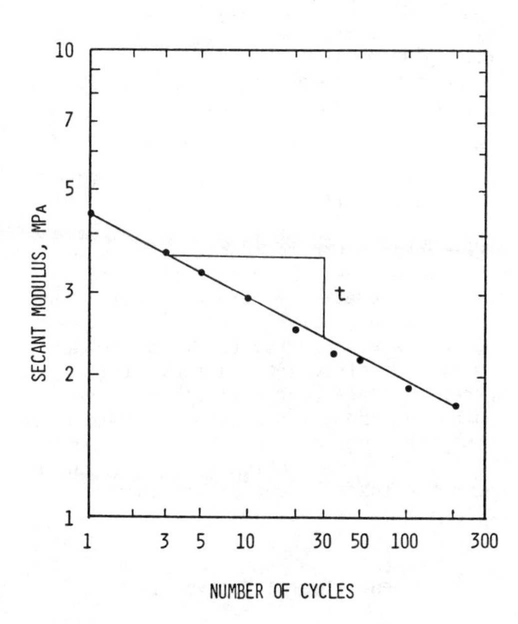

Fig.5 Variation of Secant Modulus with Number of Cycles

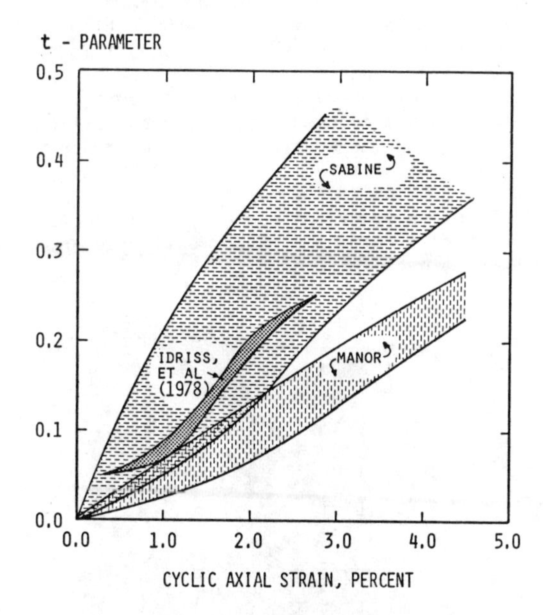

Fig.6 Comparison of t-Parameters

Sabine. Velocities were calculated to be slightly greater for the tests conducted near Manor; however, considering the short time required to test the piles (less than 24 hours), the difference in velocities was considered not to be an important factor contributing to scour.

A simple test was devised to illustrate the effect of access to free-water along a possible gap between pile and soil. An undisturbed specimen of soil was immersed in water and observations were made and recorded with time. This test, termed herein as an "immersion test", is similar in principle to the crumb test described by Sherard, et al. (1976).

Immersion tests showed that presence of free water appeared to play an important role in the behavior of the soil. Upon immersion of the undisturbed soil specimen of Manor soil, the sides of the specimen began to roughen. Within 1 minute, the entire surface of soil was rough and small flakes of soil, about 3 to 10 mm in diameter, were observed to separate from the specimen and settle slowly to the bottom of the glass container. Flaking continued for 2 days; however, the rate of flaking decreased with time. A photograph of a specimen originally smooth and cylindrical after 4 hours of immersion is shown in Fig. 7. The removal of soil particles by the process of the soil flaking results in the rough surface shown, and also provides an explanation for the scour observed during the field tests near Manor.

After completing the immersion test, the specimen was removed from the water-filled container. The specimens from the Manor site broke readily into blocks with sides corresponding to natural fissure planes within the specimen. In additional tests on other specimens of Manor soil, blocks of soil separated and fell from the top

Fig.7 Photograph of Stiff Fissured Clay Four Hours After Immersion in Water

and sides of the soil specimen. The sides of these blocks corresponded to natural fissure planes in the soil. It appears water entered the fissures and softened the soil during the immersion test.

Specimens of soil from Sabine were also immersed in water and observed. During the first 10 minutes after immersion, particles of fine-sand were seen settling to the bottom of the glass container. These particles originated from the many thin fine-sand seams within the soil specimen and left small pockets along the sides of the specimen. No further changes in the sides of the specimen were observed during the additional 48 hours. The effect of immersion on the visual appearance was minor.

6.4 discussion

Several phenomena are responsible for reducing the soil resistance during a cyclic, lateral-load test in cohesive soils. For the tests performed at Sabine Pass and Manor, the effects of undrained cyclic strain and presence of water above the ground surface were investigated. The difference in the reduction of soil resistance observed during the two field tests can not be explained from the results of cyclic triaxial tests. However, the response of the Manor soil when immersed in water and the presence of fissures within the soil appear to account for the differences observed.

7 RECOMMENDATIONS FOR DETERMINING CYCLIC BEHAVIOR OF PILES

Three sets of recommendations are made for determining the behavior of piles subjected to repetitive lateral loads. The first option involves conducting a field test and would provide the most relevant and site specific information. The second option involves performing laboratory tests, and the third option involves soil identification and some laboratory tests. The third recommendation is the option in which the least amount of confidence resides.

7.1 recommendation 1: full-scale test

A full-scale, repetitive, lateral-load test in which the pile geometry, soil profile, and loads are identical or similar to the conditions expected in design would provide information most relevant to the design of a pile subjected to cyclic loads. A carefully controlled lateral-load test can be conducted, and the effect of cyclic loading on the pile behavior can be measured by observing the deflection and rotation of the pile head. Laboratory tests can be performed to identify the effect of cyclic loads upon the behavior of the soil and the susceptibility of the soil to scour. The results of these tests can be used to explain behavior observed during the lateral-load test, and conclusions can be drawn regarding the relationship between soil behavior measured in the laboratory and pile behavior measured in the field.

7.2 recommendation 2: laboratory tests

In many cases, full-scale lateral-load tests are too expensive; therefore, a second option involves performing laboratory tests to identify and define the static, cyclic, and erosional properties of the soil. The results of these laboratory tests can be compared with the results of the cyclic and erosional tests presented herein and conclusions regarding the similarities can be drawn. For example, if data from cyclic triaxial tests and erosional tests appear similar to those found from the tests conducted on the soil from Sabine, then the soft clay analysis seem most appropriate (Matlock, 1970). However, if the soil is stiff, fissured and readily breaks up into small pieces when submerged, or if the soil is easily scoured, then recommendations for analysis based on the Manor tests are appropriate (Reese, et al., 1974).

7.3 recommendation 3: soil identification and limited testing

This set of recommendations represents the minimum investigation that should be conducted to predict the behavior of piles due to repetitive lateral loading. Laboratory tests should be conducted to identify the static properties of the soil. Thus, no tests are conducted to determine specifically the effect of cyclic loads. Erosional characteristics of the material should be identified either by visual inspection to identify the degree of fissuring and jointing, or by performing a laboratory test such as the immersion test.

If the soil appears readily erodible, stiff and fissured, then the recommendations proposed for stiff clay (Reese, et al.), 1974) are more appropriate. If the clay is not readily erodible, recommendations provided by Matlock (1970) are more appropriate.

8 CONCLUSIONS

The following conclusions can be drawn based upon comparisons between results of laboratory tests and the field tests conducted in stiff fissured clay (Manor) and in soft clay (Sabine):

1. Results from undrained cyclic triaxial tests indicate that degradation of soil resistance should have been greater for soft clay than for stiff clay; however, from results of full-scale pile tests the opposite was observed. This illustrates that phenomena other than undrained mechanical degradation are important.
2. The effect of the gap that forms between the pile and soil can be very important, and can contribute to the pile behavior by allowing water to enter and exit the gap.
3. Fissures in the soil and the response of the soil with free access to water appear responsible for significant rates of scour observed at the stiff clay (Manor) site. Water entering and exiting the gap provided a mechanism by which water was introduced to the soil and soil particles were carried away. No significant scour was observed during lateral-load tests near Sabine, nor during the immersion tests performed in the laboratory.
4. Significant degradation of soil resistance for piles in stiff fissured clays may occur due to water entering the fissures and softening the soil.

9 ACKNOWLEDGEMENTS

The research for this paper was supported by a grant from the U.S. Department of Interior, Minerals and Management service, in cooperation with the Army Corps of Engineers, Waterways Experiment Station. Valuable data from full-scale lateral-load tests were provided by Shell Development Company.

REFERENCES

Hetenyi, M., Beams on Elastic Foundation, The University of Michigan Press, Ann Arbor, 1946.

Idriss, I. M., R. Dolbry, and R. D. Singh, "Nonlinear Behavior of Soft Clays During Cyclic Loading," Journal of the Geotechnical Engineering Division, ASCE, Vol. 104, No. GT12, 1978, pp. 1427-1447.

Long, J. H., "The Behavior of Vertical Piles in Cohesive Soil Subjected to Repetitive Horizontal Loading, Dissertation, University of Texas at Austin, 1984.

Matlock, H., "Correlations for Design of Laterally Loaded Piles in Soft Clay," Offshore Technology Conference, Proc.-Paper #1204, Vol. I, 1970, pp. 577- 593.

Matlock, H., and R. L. Tucker, "Lateral-Load Tests of an Instrumented Pile at Sabine, Texas," Report to Shell Development Company on research conducted by Engineering-Science Consultants, Austin, Texas, 1961.

Reese, L. C., and Matlock, Hudson, "Nondimensional Solutions for Laterally Loaded Piles with Soil Modulus Assumed Proportional to Depth," Proceedings, Eighth Texas Conference on Soil Modulus and Foundation Engineering, Special Publication No. 29, Bureau of Engineering Research, The University of Texas, Austin, Texas, September, 1956.

Reese, L. C., W. R. Cox, and F. D. Koop, "Lateral-Load Tests of Instrumented Piles in Stiff Clay at Manor, Texas," unpublished report to Shell Development Company, Houston, Texas, 1968.

Reese, L. C., W. R. Cox, and F. D. Koop, "Field Testing and Analysis of Laterally Loaded Piles in Stiff Clay," Offshore Technology Conference, Proc. Paper #2312, 1975, pp. 671-690.

Sherard, J. L., L. P. Dunnigan, and R. S. Decker, "Identifying Dispersive Soils," Journal of the Geotechnical Engineering Division, ASCE, Vol. 102, No. GT1, 1976.

NOTATION

The following symbols are used in this paper:

E_0 = secant modulus during first cycle of load;

E_N = secant modulus at Nth cycle of load;

N = number of cycles;

P = lateral load at pile load;

t = degredation parameter;

Y = deflection of pile load at ground surface;

Y_S = deflection of pile head at ground surface for static load;

Y_C = deflection of pile head at ground surface after many cycles of load.

Prediction and Performance in Geotechnical Engineering / Calgary / 17-19 June 1987

Prediction of drilled shaft displacements under repeated axial loads

John P. Turner
University of Wyoming, Laramie, USA

Fred H. Kulhawy
Cornell University, Ithaca, N.Y., USA

ABSTRACT: Large-scale models of drilled shaft foundations were constructed in medium-dense deposits of dry uniform sand and tested under static and repeated axial loads. The test results indicate that two-way repeated displacements can be predicted by a log-linear relationship between the displacements and the number of loading cycles. This relationship is shown to be applicable for variable as well as constant levels of repeated loading. Guidelines are presented for predicting displacements of drilled shafts in sand as a function of the level of repeated loading and number of loading cycles, and for limiting displacements to avoid decreases in axial load capacity.

1 INTRODUCTION

Drilled shaft foundations supporting many types of structures are subjected to repeated applications of axial load. Transient loads induced by wind, ice, construction and maintenance activities, etc. are capable of subjecting foundations to repeated compression and uplift forces. In the case of lightweight structures, such as electrical power transmission structures, these transient loads often exceed the dead loads used for static design.

Compared to sustained static loads, repeated applications of axial load to deep foundations normally result in increased displacements. For static axial loading, elastic methods or the load transfer (t-z) method commonly are used to predict movements. For repeated loading, no generally accepted method substantiated by field performance is currently available. The data which are available from documented field repeated load tests are limited and insufficient to evaluate the influence of soil properties, foundation type and geometry, and loading characteristics (Turner, et al., 1987). In particular, no documented laboratory or field repeated axial load tests have been conducted in which movements of drilled shaft foundations are evaluated. This paper describes an experimental study of displacements of drilled shafts under repeated axial loading and the development of an empirical method for predicting the displacements of shafts in sand.

2 TESTING PROGRAM

Large-scale models of drilled shaft foundations were constructed and tested at Cornell University using a large-scale facility for load testing of deep foundations. This facility includes an in-floor testing pit, 2.9 m deep and 2.1 m in diameter, in which medium-dense deposits of dry uniform sand were carefully prepared. Details of the experimental methods, equipment, materials, and instrumentation are described by Turner and Kulhawy (1987). Seven concrete model shafts, 1.2 m deep and 150 mm in diameter, were cast in place in the prepared deposits and tested under static and repeated axial loads. Table 1 summarizes the loading conditions for each test.

The repeated loading parameters are defined in Figure 1. The magnitude of repeated load is the maximum value of two-way loading applied on a repeated basis. For example, a magnitude of repeated load of $\pm$ 2.0 kN consists of 2.0 kN in uplift followed by 2.0 kN in compression, and so on. The level of repeated load is defined as the ratio of the magnitude of repeated load minus the shaft weight, $(Q-W)_{repeated}$, to the (average)

Table 1. Summary of conditions for axial load tests on model drilled shaft foundations

Test Number	Loading Mode (1)	Magnitude of Repeated Load (kN)	Level of Repeated Loading, LRL (percent)(2)	Number of Cycles, N
1	SU	--	--	1
2	SU	--	--	1
3	TR	± 1.43	±16	1000
4	TR	± 2.00	±26	200
"	TR	± 2.67	±38	111
5	TR	± 2.45	±35	216
6	SU	--	--	1
7	TR	± 2.00	±26	100
"	TR	± 2.45	±35	70
"	TR	± 2.23	±30	110
	TR	± 2.45	±35	10

(1) SU = static uplift; TR = two-way repeated loading

(2) LRL = level of repeated loading defined by Equation 1

static uplift capacity minus the shaft weight, (Q_u-W), expressed as a percentage:

$$LRL = [(Q-W)_{repeated} / (Q_u-W)] \times 100\% \quad (1)$$

The average static uplift capacity determined from test shafts 1, 2, and 6 was 6.1 kN. The magnitude of two-way cyclic displacement, z_{cyc}, is defined as the difference between the minimum and maximum displacements for a single loading cycle, as illustrated in Figure 1.

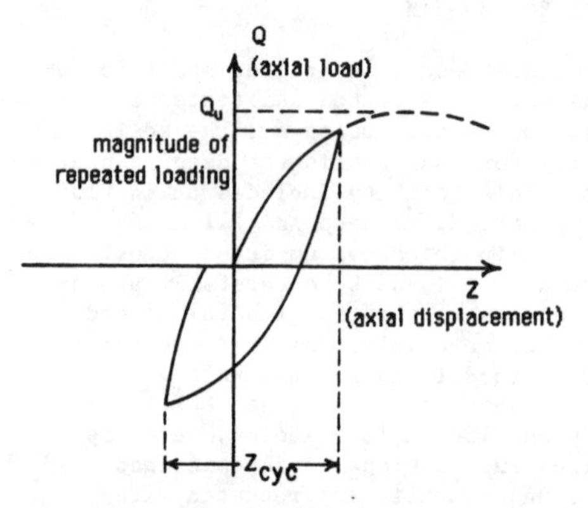

Figure 1. Repeated loading parameters.

3. EFFECTS OF CYCLIC DISPLACEMENTS ON FOUNDATION CAPACITY

Several researchers, including Poulos (1981, 1985) and Bea (1984), have suggested that cyclic degradation of deep foundation capacity is influenced strongly by the magnitude of cyclic displacements. Turner, et al. (1987) reviewed the available data from field and laboratory repeated axial load tests on driven piles, drilled shafts, and soil anchors and noted that repeated loads producing axial displacements greater than about 5 percent of the foundation diameter or greater than 50 percent of the displacement at static failure generally result in capacity reductions. The results of numerous laboratory shear test studies on soils also indicate that cyclic degradation of soil strength and stiffness are fundamentally dependent on cyclic shear strain amplitude.

Laboratory repeated axial load tests on small to large-scale model drilled shaft foundations in loose, medium, and dense deposits of sand, including the tests described herein to evaluate displacements, were conducted by Turner and Kulhawy (1987) to evaluate the effect of cyclic displacements on shaft uplift capacity. A capacity change coefficient, M, was defined as the ratio of foundation capacity after repeated loading to capacity before repeated loading. The results are presented in Figure 2 and indicate that an approximately linear relationship exists between two-way cyclic displacements and capacity reductions. The model shafts were subjected to 100 cycles of two-way repeated loading before being loaded statically to failure in uplift, except for those which failed in less than 100 cycles. This linear relationship appears to be independent of soil density, shaft geometry (D/B), and model scale, indicating that the magnitude of cyclic displacement is a fundamental parameter in determining capacity changes under repeated loading. To utilize this observation in practice requires some means of predicting the displacements.

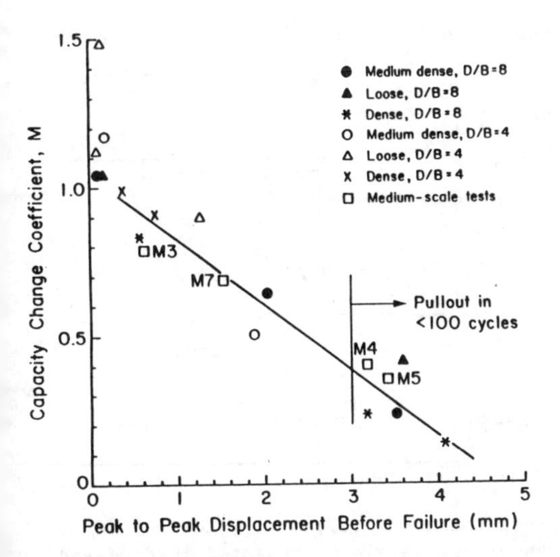

Figure 2. Effects of cyclic displacements on foundation uplift capacity.

4. TEST RESULTS

Figure 3 is a plot of the two-way cyclic displacements, z_{cyc}, versus the logarithm of the number of loading cycles, N, for different levels of repeated loading. Initially, the curves are linear, but with increasing number of cycles, the curves for shafts loaded at a LRL of 35 and 38 percent become more steep as failure is approached. For the smallest level of repeated loading (LRL = 16 percent), the curve is linear up to about 100 cycles, then begins to curve upward slightly, indicating that failure may eventually occur at large numbers of cycles ($N>10^3$), even for relatively low levels of repeated loading. Because each curve consists of a linear portion followed by a nonlinear portion, it is difficult to interpret the results in terms of a single expression relating displacements, number of cycles, and level of repeated loading. Note, however, that the magnitude of cyclic displacement below which the z_{cyc} versus log N curves are linear corresponds to the displacement below which repeated loading has little effect on uplift capacity, and above which uplift capacity decreases occur. In Figure 2, the threshold level of cyclic displacement beyond which capacity reductions occur is in the range of 0.4 to 0.6 mm. From Figure 3, once the displacements exceed approximately 0.4 to 0.6 mm, the plots begin to curve upward. This threshold magnitude of cyclic displacement is related to the mechanism of sand caving beneath the tip during the uplift portion of the loading cycle. Apparently, once the cyclic displacement becomes about 0.4 mm, sand grains begin to move under the tip, having two effects. First, it causes a decrease in the horizontal stress near the tip of the shaft, reducing side resistance and resulting in the capacity decreases observed in Figure 2. Second, it initiates a process of cyclic creep, causing the rate of displacement to increase and the side resistance to be progressively degraded, resulting in the upward curving plots of z_{cyc} versus log N in Figure 3. A rational method for design therefore would be based on limiting the displacements to prevent soil caving and cyclic creep.

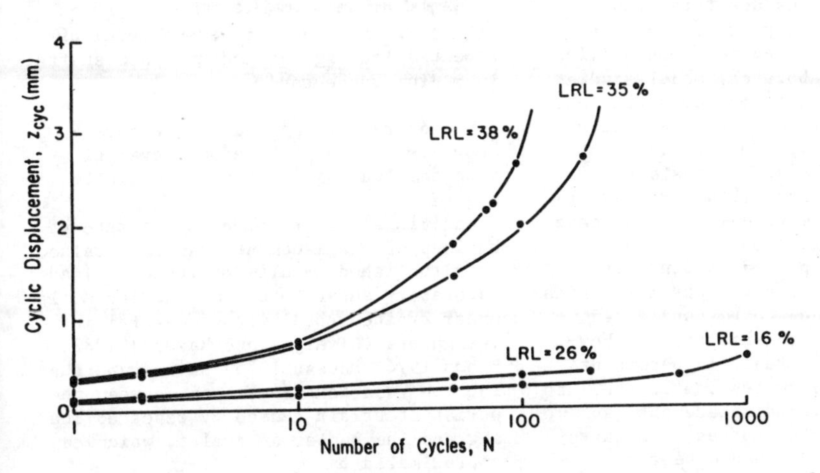

Figure 3. Two-way cyclic displacement versus log number of cycles for model shafts under two-way repeated axial loading.

Table 2. Values of the parameter B from repeated load tests on model deep foundations

Type of Foundation	Type of Loading (1)	Soil Type	Relative Density D_r (percent)	B	Reference
model helical anchor	RU	uniform fine sand	67	0.26-0.40	Clemence
model plate anchor	TR	fine to medium sand	77	0.23	Hanna
model plate anchor	RU	fine to medium sand	77	0.13-0.16	Hanna
model drilled shafts	TR	medium sand	43-50	0.22-0.31	this study

(1) RU = one-way repeated uplift; TR = two-way repeated loading

4.1 Analysis of test results

A semi-logarithmic curve of z_{cyc} versus N which is linear can be described by an equation of the form:

$$z_{cyc} = A + B \log N \tag{2}$$

in which the parameter A is equal to z_{cyc} for N = 1 and the parameter B is the slope of the semi-log curve. A simple approximate fit of the data within the linear range in Figure 3 yields values of B between 0.22 and 0.31, with B increasing for increasing level of repeated load. To predict the displacements caused by repeated loading, it would be useful to know what typical values the parameter B can assume for a range of conditions. As noted by Turner, et al. (1987), data from repeated axial load tests on deep foundations are scarce. Besides this study, which includes the only available data from repeated axial load tests on drilled shafts, only two laboratory model studies on anchors in dry sand provide sufficient data to compare values of the parameter B. Table 2 summarizes the available information from the anchor tests of Clemence and Smithling (1984), Hanna, et al. (1978), and this study. In each case, the permanent displacement after each cycle of two-way repeated loading was plotted versus log number of cycles, which is another difference to be considered in comparing results with this study. However, the rate of increase in permanent displacement should correspond to the rate of increase in cyclic displacement, so that the slopes of the curves, and therefore the values of B, should be comparable. Analysis of the data reported by Clemence and Smithling (1984) indicates that the lower values of B correspond to the lower levels of repeated loading and lower values of anchor prestress. Analysis of the results of Hanna, et al. (1978) show that the value of B increases for two-way loading compared to one-way uplift loading. As noted already, the values of B from this study increase with increasing

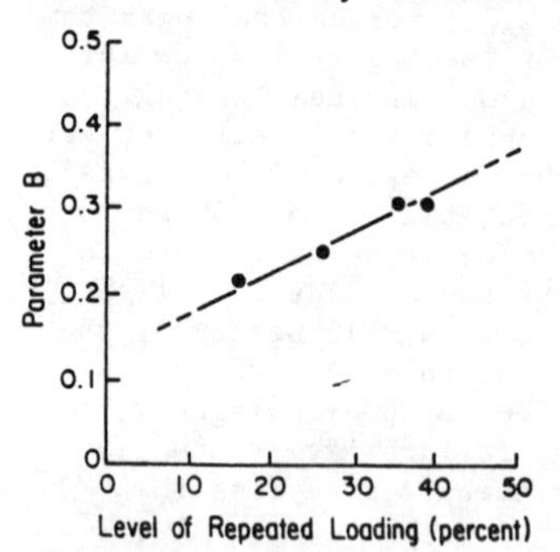

Figure 4. Parameter B versus level of repeated loading for medium-scale shafts in medium dense sand.

level of repeating loading. Figure 4 shows the value of B versus level of repeated loading for the tests plotted in Figure 3.

Additional information on the rate of buildup of displacements can be obtained from published results of repeated load laboratory shear tests on granular soils under drained conditions. Several invesigators (Diyaljee and Raymond 1982, Brown 1974, Moussa 1975) have found that a log-log relationship exists between the permanent strain caused by repeated loading and the number of cycles, which can be approximated by:

$$\varepsilon_p = aN^b \tag{3}$$

Table 3. Values of the parameter b from laboratory repeated load shear tests on granular soils.

Test Type (1)	Soil Description (2)	b	Reference
TX	Syndenham sand	0.09	Diyaljee and Raymond
TX	Ottawa sand	0.03-0.07	Diyaljee and Raymond
TX	subgrade sand	0.12	Diyaljee and Raymond
TX	dolomitic railway ballast	0.14	Diyaljee and Raymond
TX	well-graded crushed granite	0.15-0.20	Brown
SS	rounded quartz sand	0.05-0.25	Moussa

(1) TX = triaxial test, one-way repeated loading; SS = box-type simple shear, two-way repeated loading

(2) Soil densities were not reported, although in each case the soil was reported as being compacted

Table 3 summarizes in which ε_p = permanent strain after the Nth cycle and a and b are determined from the experimental data. The parameter a is equal to the permanent strain remaining after the first cycle while b is the slope of the log-log curve.reported values of the parameter b for several soils and test conditions. Diyaljee and Raymond (1982) reported values of b for three different sands and for coarser railroad ballast, all tested under one-way repeated loading in a triaxial apparatus. The value of b ranged from 0.03 to 0.12 for sand and was slightly higher (0.14) for the ballast. They also reported that b appears to be independent of confining pressure. Analysis of the results of Brown (1974) yields values of b ranging from 0.15 to 0.20 for crushed granite, also in one-way triaxial loading. Moussa (1975) reported an average value of b = 0.16 for sand tested under two-way loading in a simple shear device. He noted that b appears to be independent of density and cyclic shear stress level. Further examination of the results shows that b actually varied between 0.05 and 0.25, with the higher values corresponding to higher levels ofcyclic shear stress level. Values of the parameter b from laboratory shear tests generally are lower than the values of B obtained from tests on model foundations for one-way repeated loading. The tests of Moussa (1975) provide the only results obtained from two-way loading and, at the higher levels of repeated shear stress, the values of b (0.25) are within the range of B obtained from the tests of this study (B = 0.22 to 0.31) and the two-way repeated load tests of Hanna, et al. (1978) (B = 0.23). A preliminary assessment would indicate that values of b obtained from two-way repeated load tests in simple shear can be used to estimate the value of B. Further research on this hypothesis is required.

From the above information, it appears that the slope of the curve of displacement versus log number of cycles depends primarily on the level of repeated loading, the loading mode (one-way or two-way), and the soil type. For a given soil, B increases with increasing level of repeated loading and is greater for two-way loading than for one-way loading. From the tests of this study and data from other studies on model anchors, a value of B = 0.30 is recommended for predicting the cyclic displacements caused by two-way loading of drilled shafts. This recommendation is subject to modification if additional laboratory or field testing provide reliable data for other test conditions. In particular, field repeated load testing of drilled shafts would provide much needed information required for confirming the results of laboratory model tests.

Evaluation of the parameter A requires that the cyclic displacement for the first cycle of loading be predicted. Evaluation of model tests by Turner and Kulhawy (1987) indicate that uplift displacement accounts for 70 to 80 percent of the two-way displacement on the first cycle. Therefore, a good estimate of the initial two-way cyclic displacement can be obtained from the uplift displacement under static loading.

Several methods exist for predicting uplift displacements during static loading. The most reliable information is obtained from a field load test for the foundation and site conditions to be encountered; however, such data usually are not available. Analytical prediction techniques include elastic methods (Poulos and Davis, 1980) and the load transfer (t-z) method. With elastic methods, it is assumed that the soil-foundation system behaves as a linearly elastic medium, which may be a reasonable assumption for predicting small displacements within the range of interest for determining the parameter A. The load transfer method is a numerical procedure for evaluation the shear stress-displacement behavior along the side of a deep foundation. Callanan and Kulhawy (1985) evaluated in detail the procedures for predicting drilled shaft uplift displacements. In applying the load transfer method to model tests of Stewart and Kulhawy (1981) on drilled shafts in the same sand used in this study, they showed that a good t-z correlation resulted by using the shear stress (t) parameters from direct shear tests with constant normal stress and displacement (z) parameters from constant volume direct shear tests. The use of the load transfer method with this modification to the direct shear test data is recommended for drilled shafts in sand.

4.2 Response under variable level of repeated loading

The repeated loads transmitted to a drilled shaft supporting a transmission line tower subjected to wind forces are not of constant magnitude, like the model tests described thus far. Constant magnitudes of repeated load are useful, however, for determining the levels of repeated load leading to capacity changes and failure, and for isolating the effects of other variables on shaft response. Once that has been accomplished, load testing at variable magnitudes of repeated loading can be used to test the resulting hypotheses and conclusions. Shaft 7 was tested at several different levels of repeated loading to determine whether the cyclic displacements could be predicted and to test the hypothesis that those displacements control capacity changes. The test conditions and results are summarized in Table 1.

Stewart (1986) performed variable-amplitude repeated load triaxial tests on well-graded railroad ballast. He noted that when the level of repeated loading was increased beyond any past maximum value, the permanent strain continued to increase. If the level of repeated loading was reduced below any maximum past value, negligible increases in permanent strain occurred with increasing cycles. A similar observation was made regarding the two-way cyclic displacements during testing of shaft 7 under variable levels of repeated loading. During the first loading sequence at a LRL = 26 percent, z_{cyc} reached 0.31 mm after 100 cycles. The LRL was increased to 35 percent and z_{cyc} increased to 1.36 mm after an additional 70 cycles. The LRL then was decreased to 30 percent. During the first load cycle at a LRL of 30 percent, z_{cyc} decreased to 1.16 mm, then increased to 1.22 mm after an additional 100 cycles, a relatively small increase. The LRL was again increased to 35 percent and, after 10 cycles, z_{cyc} was 1.59 mm.

The observed relationship between cyclic displacement, level of repeated loading, and number of loading cycles for shaft 7 can be determined from the data plotted in Figure 3, and then be re-plotted with the predicted loading path in Figure 5, as

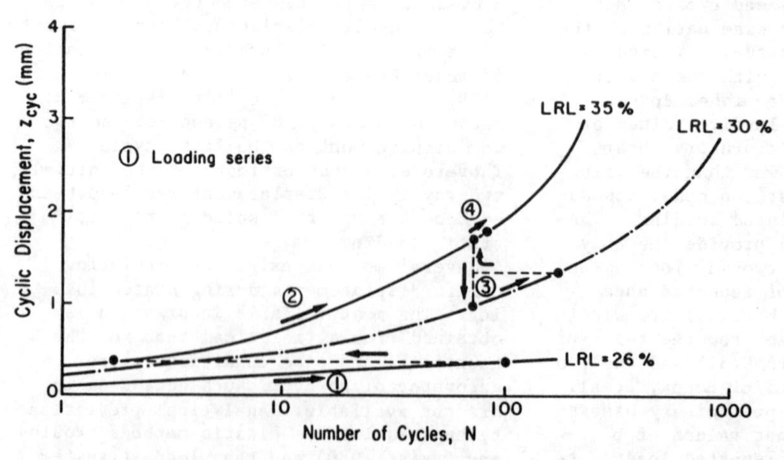

Figure 5. Predicted z_{cyc} versus log number of cycles relationship for shaft 7.

Table 4. Predicted and measured values of two-way cyclic displacement for shaft 7.

Loading Series	Level of Repeated Loading (percent)	Number of Cycles	(z_{cyc})predicted (mm)	(z_{cyc}) measured (mm)
1	± 26	100	0.37	0.31
2	± 35	70	1.69	1.36
3	± 30	100	1.30	1.22
4	± 35	10	1.77	1.59

follows. The initial loading series at a LRL of 26 percent should follow the z_{cyc} versus log N curve already established for that level, yielding z_{cyc} = 0.37 mm at 100 cycles. For the next series at LRL of 35 percent, the shaft starts out with z_{cyc} = 0.37 mm, so the load path moves horizontally to the LRL = 35 percent curve, yielding an equivalent number of cycles, N_{eq}, of 2. An additional 70 cycles of LRL of 35 percent brings the loading path to z_{cyc} = 1.69 mm at N_{eq} = 72. If the LRL is reduced, the cyclic displacement also decreases. An approximation of z_{cyc} at the lower level is made by dropping vertically to the curve for the lower LRL. The next loading sequence for shaft 7 was at a LRL of 30 percent, for which no previous test data were obtained, but which is extrapolated in Figure 5 between the curves for LRL of 26 and 35 percent. On the first load cycle at LRL of 30 percent, z_{cyc} should decrease to 0.95 mm. After an additional 100 cycles, z_{cyc} should increase to about 1.30 mm at N_{eq} = 172. Finally, the LRL is increased to 35 percent again. If the cyclic displacement at the lower LRL had exceeded the maximum past displacement, then the loading path would move horizontally to the path for the higher level. However, since z_{cyc} at the lower level (1.30 mm) did not exceed the previous maximum value (1.69 mm), the loading path moves to the previous maximum displacement on the curve of the new level of repeated loading. For shaft 7, the loading path moves to the curve for LRL of 35 percent at z_{cyc} = 1.69 mm and N_{eq} = 1.71. An additional 10 cycles at LRL of 35 percent increases z_{cyc} to 1.77 mm. The measured and predicted values of cyclic displacement are summarized in Table 4 and show close agreement.

At the end of the four series of repeated loading, shaft 7 was tested in uplift to failure, yielding a value of 0.68 for the capacity change coefficient, M, for two-way cyclic displacement before failure of 1.59 mm. These data were plotted in Figure 2 and are in excellent agreement with the relationship established using the results of tests at constant levels of repeated loading.

The experimental data obtained from shaft 7 under variable levels of repeated loading provide information on several aspects of drilled shaft behavior. First, the load-displacement response of a drilled shaft is determined largely by the maximum level of repeated loading and therefore by the maximum previous two-way cyclic displacement. Reducing the level of repeated loading has little effect on the subsequent behavior. Second, the uplift capacity after repeated loading is determined by the maximum past two-way cyclic displacement and is independent of the sequence of loading for variable levels of repeated loading. And third, if the relationship between cyclic displacements and number of loading cycles can be established or predicted for constant levels of repeated loading, then the displacements at variable levels of repeated loading can be predicted and used as a tool for designing against capacity decreases and failure under repeated axial loading.

Future research should concentrate on developing prediction curves relating cyclic displacements, levels of repeated loading, and number of loading cycles for a wider variety of conditions. If possible, such predictions should be based on repeated axial load field tests.

5 SUMMARY AND CONCLUSIONS

This study has demonstrated that degradation of deep foundation capacity and load-displacement response under repeated axial loading depend primarily on the magnitude of cyclic displacements. Repeated axial load tests conducted on model drilled shaft foundations showed that a semi-logarithmic relationship exists between the two-way cyclic displacements and the number of loading cycles. For a given level of repeated loading, two parameters are required to describe the relationship. One is the cyclic displacement on the first loading cycle,

A, which can be estimated using conventional elastic or t-z methods of predicting foundation movements. The second, B, is the slope of the curve of displacements versus log number of cycles. The results of this study, plus the limited data available from repeated load tests on anchor foundations were evaluated and it was found that values of the parameter B vary over a relatively narrow range for deep foundations in sand, and that generally B increases with increasing level of repeated loading and is greater for two-way than for one-way repated loading. Design values of B are recommended for limiting cyclic displacements of deep foundations in sand.

The results of a repeated axial load test using variable levels of repeated loading, which is more representative of field loading conditions, showed that cyclic displacements for irregular loading histories can be predicted accurately based on relationships established for constant levels of repeated loading.

6 ACKNOWLEDGEMENTS

This paper is based on research sponsored by the Electric Power Research Institute at Cornell University under EPRI 1493-4. Vito J. Longo was the EPRI project manager. The support of EPRI is gratefully acknowledged. The text was typed by B. Powell and figures were drafted by A. Avcisoy.

7 REFERENCES

Bea, R.G. 1984. Dynamic response of marine foundations. Keynote for Ocean Structural Dynamics Symposium '84, Corvallis, OR.

Brown, S.F. 1974. Repeated load testing of a granular material. J.Geotech. Eng.Div. (ASCE) 100: 825-841.

Callanan, J.F. & F.H.Kulhawy 1985. Evaluation of procedures for predicting foundation uplift movements. Elec. Power Res. Inst. Rpt. EL-4107.

Clemence, S.P. & A.P.Smithling 1984. Dynamic uplift capacity of helical anchors in sand. Proc. 4th. Australia-NZ Conf. Geomech., Perth: 88-93.

Diyaljee, V.A. & G.P.Raymond 1982. Repetitive load deformation of cohesionless soil. J. Geotech. Eng. Div. (ASCE) 108: 1215-1229.

Hanna, T.H., Sivapalan, E. & A.Senturk 1978. Behavior of dead anchors subjected to repeated and alternating loads. Ground Eng. 11: 34-45.

Moussa, A.A. 1975. Equivalent drained-undrained shearing resistance of sand to cyclic simple shear loading. Geotechnique 25: 485-494.

Poulos, H.G. 1981. Cyclic axial response of single pile. J. Geotech. Eng. Div. (ASCE) 107: 41-58.

Poulos, H.G. 1985. Some recent developments in pile design. Guest lecture, 3rd Indonesian Natl. Conf., Jakarta.

Poulos, H.G. & E.H.Davis 1980. Pile foundation analysis and design. New York: Wiley.

Stewart, H.E. 1986. Permanent strains from cyclic variable-amplitude loadings. J. Geotech. Eng. (ASCE) 112: 646-660.

Stewart, J.P. & F.H.Kulhawy 1981. Experimental investigation of the uplift capacity of drilled shaft foundations in cohesionless soil. Niagara Mohawk Power Corp. Contract Rpt. B-49(6).

Turner, J.P. & F.H.Kulhawy 1987. Experimental analysis of drilled shaft foundations subjected to repeated axial loads under drained conditions. Elect. Power Res. Inst. Rpt. on Research Project 1493-4 (in press).

Turner, J.P., Kulhawy, F.H. & W.A. Charlie 1987. Review of load tests on deep foundations under repeated loading. Electric Power Research Inst. Report on Research Project 1493-4 (in press).

Supporting a rock layer during construction

T.J.Kaderabek, D.Barreiro & M.Call
KBC Consultants Inc., Miami, Fla., USA

ABSTRACT: Construction of a 12-level office building in Coral Gables, Florida required a 16-foot deep vertical excavation be constructed within 3 feet of an existing 3-level structure. The 3-level structure was supported on shallow foundations.

The 3-level building was estimated to fail in 1 of 2 ways: a shear failure of the shallow foundations through the limestone or the entire limestone beam failing in catilever action. A reinforced concrete beam installed near the bottom portions of the rock formation was the underpinning system used

The completed underpinning system was monitored with surveying equipment, a telltale monitoring device accurate to 1/1000 of an inch and through crack monitors on the existing 3-level structure. Movement of 1/10 of an inch was recorded during the construction process. Anticipated settlement of the Miami Limestone was expected to be on the order of 1/4 inch.

1 PROJECT DESCRIPTION

Construction of a 12-level office building in Coral Gables, Florida involved 2 levels of below grade basement construction on a site where new construction was coincidental with the property line. The tower portion of the project was supported on pressure grouted auger piles while the garage area used shallow foundations. The working pile capacity was 100 tons per pile for piles installed about 50 feet below street grade.

Street or site grade before construction was about elevation +13 feet MSL. The stabilized groundwater level is located at about elevation +3 feet MSL. Well points were used to depress the groundwater level to elevation -3 feet in the parking area and -8 feet beneath the tower. Well point water was disposed of via discharge wells and adjacent storm drains.

An existing 3-level structure was located 3 feet from the proposed 16-foot deep vertical excavation in the northeast portion of the site. The project owners requested to build basement walls in contact with the property line as close to the existing 3-level structure as possible.

This paper discusses field procedures, lab tests, and design methods which were used in association with construction of a 16-foot deep vertical excavation 3 feet away from a 3-level concrete block structure which was supported on shallow foundations.

2 SUBSURFACE CONDITIONS

Prior to the time of basement excavation 3 borings were performed near the property line in the northeast portion of the site. The borings utilized rock coring in the near-surface Miami Limestone which recovered rock core having a diameter of about 3.9 inches. The Miami Limestone encountered in these 3 borings had a thickness ranging from 7 to 10 feet.

A prior exploration at this site used standard penetration test borings which revealed sandy conditions beneath the Miami Limestone. The top portions of the sand beneath the Miami Limestone were reported to be loose in consistency. (SPT N-valves less than 10 blows per foot.)

3 LABORATORY TESTS

The following tests were performed on representative rock specimens: unconfined compressive strength, splitting tension, modulus determination, and punching shear.

The unconfined compressive strength tests were performed on core samples which had been trimmed and capped. A core length to diameter ratio approaching 2 was attempted. Five compressive strength tests were performed and yielded test results ranging from 23 to 35 ksf, with an average value of 29 ksf.

Modulus determinations for the limestone cores were obtained during the unconfined compressive strength testing. Micrometer dial gauges were attached to the testing machine platens in order to record stress-strain curve during the loading process. Soil modulus was approximated from the slope of the straight line portion of the stress-strain curve. The modulus of elasticity values averaged 14,500 ksf.

Splitting tension tests were performed to approximate the limestone's tension characteristics. Rock core specimens with a length to diameter ratio approaching one were used for this test procedure. The uncapped rock core was failed in diametral compression or with the load applied across the diameter of the sample. An analytical calculation results in splitting tension values. Tests results averaged 6 ksf.

Sixteen punching shear tests were performed. Punching shear tests were performed by forcing a 2-inch diameter plunger through the rock disc. The rock disc had a diameter of about 3.9 inches and was 2 inches thick. The rock disc was supported on a circular steel member which had a 2-1/16 inch diameter center hole. A hydraulic machine was used to shear the center portion of the rock disc. The results are as follows: average shear strength 12.5 ksf, average deflection at shear failure 0.08 inch, average residual strength 9.9 ksf.

4 ENGINEERING PROPERTIES

Graphical construction of a Mohr envelope suggests the Miami Limestone has an average ultimate shear value of about 11 ksf. This compares favorably with the punching shear tests results which had an average shear value of 12.5 ksf.

The Miami Limestone was estimated to have an average ultimate unconfined compressive strength of about 29 ksf and an average ultimate tension value of about 6 ksf.

5 APPLIED FORCES

Construction of 2 levels of basement required an excavation to elevation -3 feet MSL from the current grade of +13 feet MSL. This excavation was made 3 feet from the face of an existing 3-story structure which was supported on shallow foundations. Considering that this structure is supported on shallow foundation results in two modes of failure: the rock mass shearing as a result of adjacent excavation and removal of the sand beneath the Miami Limestone formation resulting in cantilever failure of the Miami Limestone beam.

Structural loading was estimated for the 3-story reinforced concrete structure. An estimated line load was calculated along the 80-foot length of building. The line load was estimated to be 7.7 kips per lineal foot for dead and live load. The load was concentrated 3 feet from the property line and was assummed to span 20 feet in an east-west direction. In other words, an applied load of 7.7 kips per lineal foot was assummed to act at each end of the 20-foot span. A footing width of 2 feet was assummed along with a 1-foot depth of embedment.

The Miami Limestone formation was to remain vertical during construction. Using the applied load of 7.7 kips per lineal foot, and a 45 degree failure plane translates into a factor of safety against shear failure of about 6.

Once basement excavation proceeded below about elevation +4 feet MSL, sand from beneath the Miami Limestone began to slough into the excavation, thereby permitting the Miami Limestone beam to deflect as a cantilever. Excavation to the proposed -3 feet MSL elevation would result in exposing about 8 feet of sand beneath the Miami Limestone. The sand sloughed into the excavation on an angle of repose or a slope of about 1:2 (H:V). However, it was expected the excavation process would result in stress relief to a slope of about 2:1 (H:V), which results in a 16-foot cantilever span. It is for this cantilever condition that forces were estimated.

The force necessary to support the Miami Limestone at the property line, at elevation +5 feet MSL was evaluated as a propped cantilever. The limestone beam spans 16 feet with a fixed reaction at one end and a simple support at the property line. The two reactions support load from the weight of the limestone and the 3-story structure. The limestone was conservatively estimated to weight 150 pounds per cubic foot. The building load of 7.7 kips per lineal foot was applied 3 feet east of the property line. Evaluation of this free body diagram suggests a reaction at the support system of 13.1 kips per lineal foot along the 80-foot length of building. It was estimated that the Miami Linestone had a factor of safety against shear failure, to resist the 13.1 kips per lineal foot load, ranging from 2 to 6, depending upon the analytical method used to estimate forces.

6 DESIGN CONCEPTS

A support system was required at the property line which would resist deflection and be compatible with site characteristics. Since auger pilings were being used to support the tower area, they were incorporated into an underpinning scheme because of their 100-ton working capacity and lack of vibration during installation. Auger piles were reinforced with steel wide flange sections (W8x24). Details of the reinforced concrete beam are shown in Figures 1 and 2.

The reinforced concrete beam was designed to resist bending and shear due to an applied load of 13.1 kips per lineal foot applied in a vertical direction at elevation +5 feet MSL. Load factors were applied, resulting in a total design load of 22 kips per lineal foot of beam. Using applicable codes results in a 14-inch wide beam with a depth of 20 inches.

Auger piles were spaced 10 feet on centers to transfer applied loads to a deeper limestone strata(elevation -37 feet). Bending forces in the design were minimal with shear forces controlling the final design. Angle brackets were welded to the wide flange beam in order to develop the shear force in the channel section. The purpose of these angle brackets was to permit some degree of pile misalignment both vertically and from pile rotation.

The project owners requested the upper level basement encroach on the underpinning system to provide for additional parking space. This encroachment resulted in the limestone block ABCD in Figure 3 being removed. Removing this rock mass was made possible by placing concrete backfill behind the retaining system thereby permitting force to be transferred through the concrete backfill and into the auger piles. Backfill tubes were used for this grout placement. Concrete backfill was placed in two 4-foot thick lifts in order to be in contact with the bottom side of the limestone.

Figure 4 is a sketch of the final underpinning system with the limestone block ABCD removed. Volclay panels were used as a waterproofing system between basement walls and the surrounding soil or rock condition.

7 CONSTRUCTION MONITORING

Monitoring during construction consisted of elevation surveys with surveying equipment, crack monitors, and a telltale monitoring device.

Elevation surveys were accurate to 0.001 feet and recorded no settlement around the perimeter of the structure or the excavation.

Crack monitors were placed across hairline cracks in the stucco of the existing 3-level structure. Crack monitors are accurate to about 1/2 millimeter. Crack monitors did record fluctuations in crack width depending on moisture and temperature conditions.

The telltale monitoring device consisted of a steel pipe grouted into the limestone formation at elevation -37 feet. This steel pipe was surrounded with plastic pipe in order to isolate it from

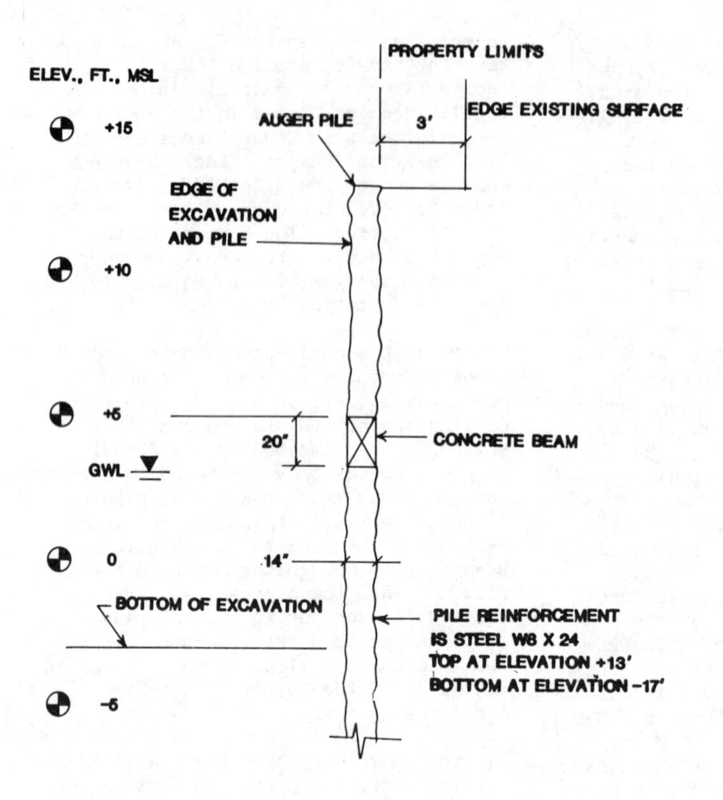

Figure 1. Section (looking north) at retaining system near existing 3-level building.

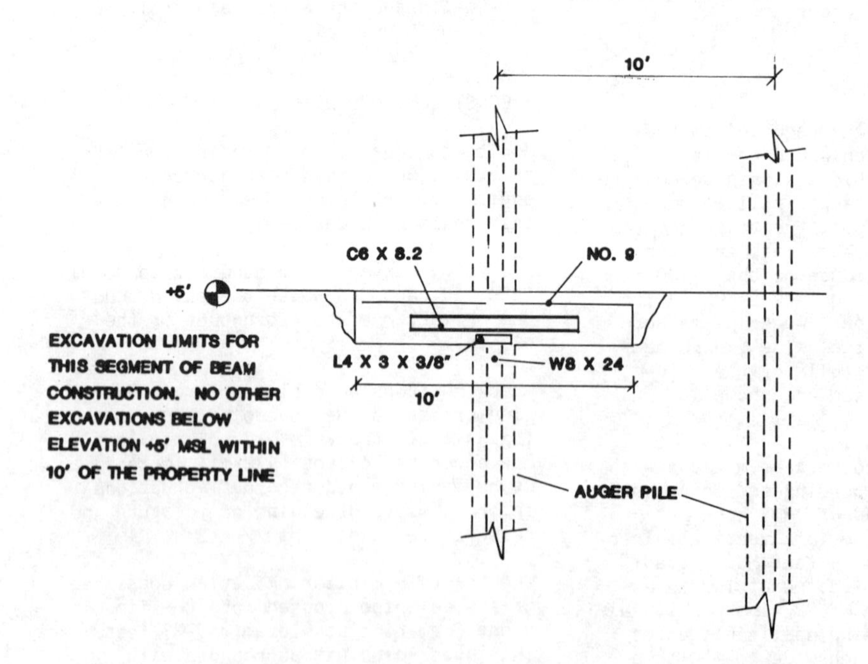

Figure 2. Front view of concrete beam, looking east.

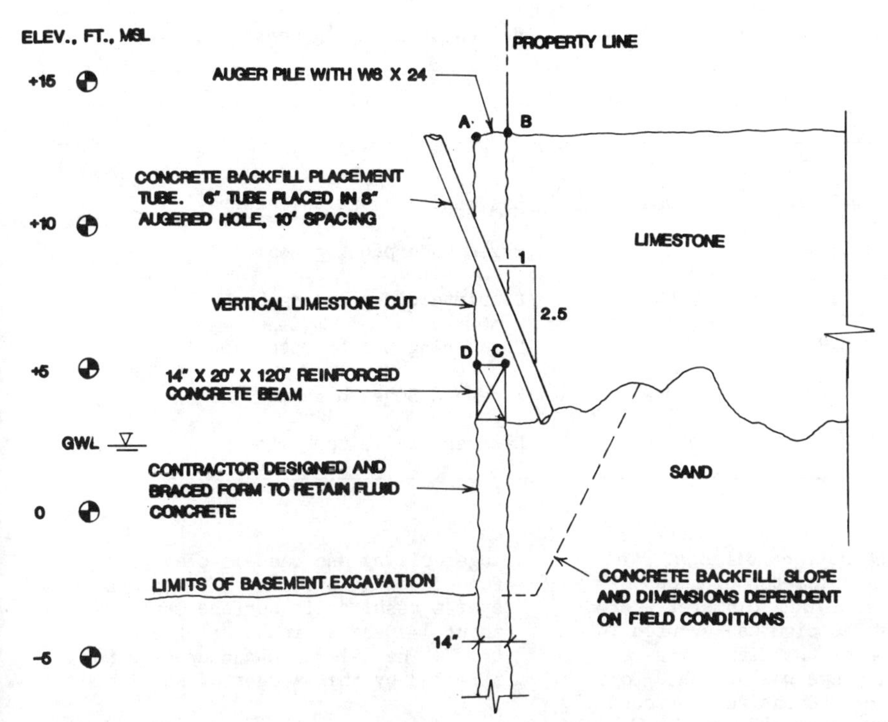

Figure 3. Section (looking north) before concrete backfill placement, with limestone block ABCD intact.

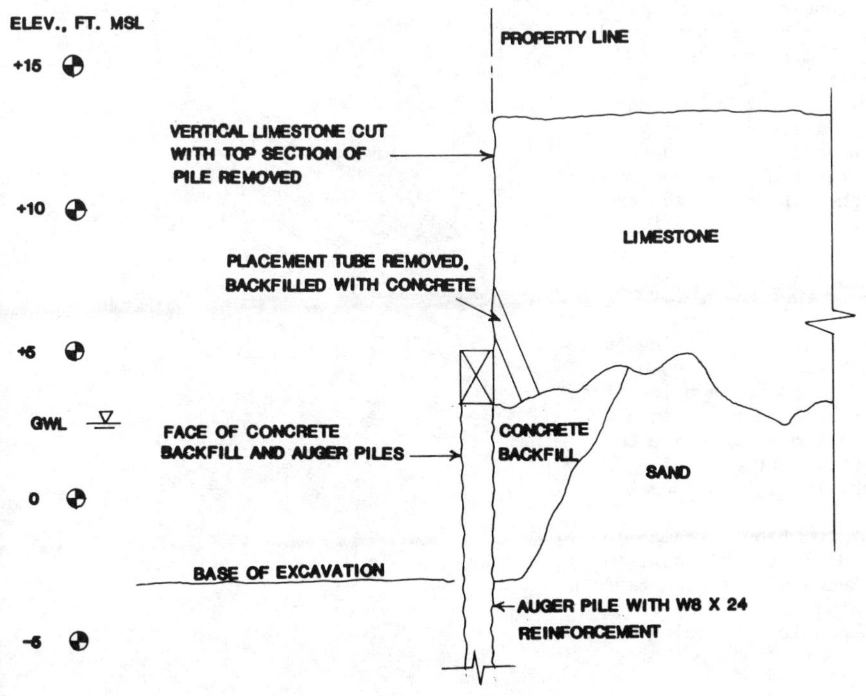

Figure 4. Section (looking north) after concrete backfill placement with limestone block ABCD removed.

Table 1. Recorded Settlement at Property Line Measured During Construction.

Date	Settlement, Inches	Construction Activity
10-31-84	0.024	Basement Excavation Underway
11-12-84	0.038	Begin Underpinning Beam
11-19-84	0.041	Complete Underpinning Beam
3-8-85	0.059	Dewatering System Operational
6-10-85	0.082	Excavate Beneath Beams
8-27-85	0.090	Basement Walls Complete

the surrounding soil conditions. This PVC sleeve had a diameter of about 3 inches. Next a 12-inch diameter, 2-foot long, piece of PVC pipe was grouted into the top portion of the Miami Limestone. A dial gauge was attached to the 12-inch diameter PVC casing to record differential movement between the Miami Limestone and the steel bar grouted into the 50-foot deep limestone layer. As the Miami Limestone moved down, differential settlement readings between the upper and lower limestone was possible. The telltale monitoring device had an accuracy of about 0.001 inch. Data from the telltale monitoring device is presented in Table 1. The data recorded indicates the limestone settled about 1/10 of an inch during construction.

8 SUMMARY

Construction of a 12-level office building in Coral Gables, Florida resulted in construction of 2 basements below street level. The 2 basements required a 16-foot deep excavation 3 feet from the face of a 3-level apartment building supported on shallow foundations.

A subsurface exploration revealed 10 feet of cemented limestone underlain by loose sands. Removing the sand beneath the limestone would cause limestone deflection and possible building distress.

An underpinning system was developed which supported the limestone beam using auger piling and cast-in-place reinforced concrete beams. The underpinning system resulted in surface settlement of about 1/10 of an inch during construction. The 3-level structure was not affected by this amount of settlement.

Effect of pile driving on group efficiency

R. Janardhanam
University of North Carolina, Charlotte, USA

ABSTRACT: The traditional methods of evaluating the bearing capacity and group efficiency of piles, especially those driven in sand, are not satisfactory. Static formulae based on the assumption that the soil conditions remain unchanged by pile driving are very conservative. This study investigates the effects of pile driving on surrounding soils and the resulting influence on pile group efficiency. The factors considered are soil strength properties, pile geometry, and pile driving resistance. Regression analysis is used to analyze the field data to determine the functional relationship and the degree of importance of the parameters. All existing efficiency formulae are seen to underestimate the pile group capacity in case of loose and medium dense sands.

1 INTRODUCTION

Ever since engineers began using pile foundations, they have attempted to develop rational methods to predict its behavioral performance and, in particular, it's load carrying capacity. A critical review of the literature indicates that the traditional methods of evaluating the bearing capacity of piles, especially those driven in sand, are not satisfactory. Poor correlations have been observed between the ultimate bearing capacity determined by various driving formulae and the field pile load test results. Static formulae based on the assumption that the soil conditions remain unchanged by pile driving are very conservative. The action of driving a pile into sand creates a zone of compacted soil in the vicinity of the driven pile. The factors on which pile design should be based must depend upon the properties of the soil after the piles have been driven and these soil properties are extremely difficult to determine.

To increase the understanding of the behavior of friction piles in sand, much research has been carried out by numerous investigators and many analysis methods have been proposed. Most methods include friction angle of sand as the primary parameter for evaluating bearing capacity. Very few methods include parameters such as pile size, depth of penetration, and relative density of sand.

Most of the laboratory and field investigations carried out and reported have been on single piles. Research findings from these investigations have thrown considerable light on the uncertainties in accessing the contributing factors. However, these may not be applied to the behavior of pile groups driven in sand. Furthermore, in the laboratory studies, the pile models were "driven" by jacking them into the soil medium, thus omitting the impact and vibrational effects induced by the hammer impact found in actual driving.

Improvement is needed in the currently used methods of evaluating the bearing capacity of pile groups driven in sand. This study intends to use field data of pile driving in sand and pile load tests to determine the functional relationship and degree of importance of factors such as pile geometry, soil property, and pile group geometry in developing improved procedures for predicting the bearing capacity of pile groups.

2 EXISTING METHODS AND THEIR LIMITATIONS

Most widely used static formula method to evaluate the ultimate axial bearing capacity (Q_u) can be expressed as

$$Q_u = Q_p + Q_s = q_o A_p + f_s A_s \quad (1)$$

where Q_p = ultimate point load, Q_s = ultimate side load, q_o = ultimate point resistance, f_s = ultimate unit side resistance, A_p = area of pile point and A_s = area of pile shaft. The commonly used equations for q_o and f_s are

$$q_o = p_o N_q \text{ , and } f_s = K\ p \tan \phi \quad (2)$$

where p_o = effective overburden pressure at the pile point level, $N_q = f(\phi)$ bearing capacity factor, p = average effective overburden pressure along the pile shaft, $\tan\phi$ = coefficient of friction between pile and soil and K = a lateral earth pressure coefficient.

Here p_o and p are calculated without considering any stress change that may occur in the soil mass during pile penetration. An examination of a summary of the range of values for N_q for different values of the angle of internal friction, according to different theories, shows that there are major variations in N_q from one theory to another. Furthermore, the main problem when evaluating unit side resistance is the determination of the magnitude of the lateral earth pressure coefficient K. When a pile is driven in loose or medium dense sand below a critical depth, the sand is displaced only in the horizontal direction and the horizontal pressure against the pile is then equal to K times the vertical effective overburden pressure. Since the pile wall pushes against the sand and the horizontal movement is large, it is theoretically possible for the magnitude of K as high as the passive earth pressure coefficient. However, experimental evidence indicates that the magnitudes of K can be less than one and may even be as low as the active earth pressure coefficient.

The variation of predicted values of N_q and K is so wide that the choice of one theory over another is a very difficult decision to make. The need for a better understanding of the failure mechanisms and the development of a correct theory has resulted in a number of laboratory and field investigations. The findings of some investigators confirm the concept that relative density increases by approximately 100% at the face of pile shaft and reduces to the initial value at 6 to 8 diameters distance. The rate of change is reported to be parabolic. A similar trend is observed for horizontal displacement and porosity.

When piles are installed in groups, the pressures induced by the individual units in the surrounding soil combine and the compaction of soil is increased. Overlapping of the individual compaction zone theoretically occurs whenever the spacing of individual piles is less than approximately 7 or 8 times the diameter. If the pile spacing is sufficiently close, the ultimate load capacity of the group may exceed the summation of the individual piles capacity and the group efficiency may be greater than one. Suggestions are made to extend the findings of single pile performance to pile groups by the method of superposition. It may be meaningful in elastic zone but not in plastic zones. There is need for research to develop a convenient, logical, reasonable, and reliable procedure for determining group efficiency of piles driven in sand.

3 OBJECTIVE

The primary objective is to study the effects of driving piles in sand on their load carrying capacity. The secondary objective is to develop new correlations relating soil properties such as relative density, friction angle of soil, pile properties and pile group geometry with pile group load carrying capacity. Driving stress at the top of piles is measured to determine the load capacity of piles driven to bearing in soils using wave equations. A similar concept is used, where the driving record of the piles will be used to determine the average group capacity.

4 METHODOLOGY

The important factors influence the degree of compaction of soil are (1) mode of compaction (M), (2) physical properties of sand, namely initial relative density (D_{ri}), (3) the pile-soil interaction (δ), (4) pile size (B), and (5) spacing of piles (S) - (pile group geometry). The degree of compaction (C_d) can now be expressed as a function

$$C_d = F(M, D_{ri}, \delta, S, B) \quad (3)$$

This in turn governs the load carrying capacity (Q_u) of pile groups and can be expressed as

$$Q_u = F(C_d, , K, \delta) \quad (4)$$

The remaining three variables are tied to the degree of compaction attained. Variables like M, S, δ and B are maintained constant. Field results (pile driving logs) from projects where the same kind of hammer (diesel hammer: KOBEK 13) is used to drive piles of the same type (precast concrete) and size (14 x 14 inch square) in a group of equal spacing (2.5 feet) is used for analysis.

Pile driving operations are presently very systematized, well-controlled, and consistent. Therefore, pile driving data can be considered fairly trustworthy as the standard penetration test results. The number of hammer blows required to drive a pile through one foot at a particular depth (R) is hereafter known as pile driving resistance value at that depth. The R value can be treated as an index property of the soil. Field measurement of changed properties around the driven pile is feasible but very complex and likely misleading. Thus, the significant influence of generated stresses due to pile driving on the soil medium is achieved here in an indirect fashion in terms of R values.

The N values before pile driving and R values of the first pile driven at a site can be related as both of them refer to the virgin state of the ground. Data is drawn from a specific job site in the western beaches of Florida. The subsoil condition site indicates that the upper stratum, of 6±ft. in general, contains loose to medium dense cohesionless sands. Below this stratum, fine sands with variable amounts of organic material is encountered to a depth of about 20±ft. This weak stratum is underlain by medium dense cohesionless sands to the terminated depths of all the borings. Groundwater level is ranging from 2.5 to 4 feet below the existing ground surface. Ten standard penetration tests have been conducted at this site and they have been terminated at a depth of 30 ft. In this job site, a total of 493 piles have been driven either as single piles or in groups of 2, 3, 4 and 5 piles. Piles are driven up to practical refusal at 25 blows per foot. When piles are driven in sand, all the piles in a group do not reach the same depth of penetration. The general trend of varied projected length of the piles in a group is an indication of an increase in the supporting strength of the soil. Here additional blows are given to drive the piles in a group to reach the same depth of penetration. Extra blows given is correlated as an increase in supporting strength of the soil; in other words, the increase in pile group capacity.

5 DATA ANALYSIS

Statistical methods are used to remove the outliers in the field data. Figure 1 shows the variation of weighted average of N and R values with depth. The relation of R and N values is determined as

$$R = 14.7N + 13.3 \qquad (6)$$

The sequence of pile driving has significant influence on the number of blows required to drive subsequent piles to the same level. Figure 2 shows the change in numbers of blows as piles are driven sequentially in a group. With the R values known at different depths for individual piles in a group, the maximum loads they can carry individually are determined using the Janbu's formula

$$P_u = \frac{e_h E_h}{K_u S} \qquad (7)$$

where $K_u = C_d\left[1+\sqrt{1+\frac{\lambda}{C_d}}\right]$

where $C_d = 0.75+0.15\frac{W_p}{W_r}$ and $\lambda = \frac{e_h E_h L}{AE_s^2}$

Here $E_h = W_r \times h$, S = penetration/blow, $e_h = 0.85$, W_r = weight of ram, W_p = weight of pile, A = area of cross section of pile, L = Length of pile, E_s = Modulus of elasticity of pile.

Figure 3 shows the load carrying capacity of precast concrete piles, according to Janbu formula. The total load a group can carry is the sum of the maximum loads the individual pile in a group can carry. Figure 4 shows the maximum load carrying capacity of three different pile groups. The results show that all the parameters considered here are approximately linear functions of the pile's embedded depth. These parameters are analyzed to obtain the degree of association between these variables.

The correlation is put in a mixed polynomial of the following form

$$Y=\alpha_0+\alpha_1X_1+\alpha_2X_2+\alpha_3X_3+\alpha_4X_4+\alpha_5X_5+\alpha_6X_6 \qquad (8)$$

The coefficients of α_1, α_2, α_3, α_4, α_5, and α_6 are the constants to be determined by regression analysis. X_1, X_2, X_3, X_4, X_5, and X_6 are the known independent variables and Y is the dependent variable, namely the load carrying capacity (Q_u-lb). The independent variables are: X_1 = number of piles in the group (n), X_2 = depth of pile embedded in soil (D-feet), X_3 = number of hammer blows per foot (R), X_4 = the bearing capacity factor (N_q), X_5 = unit weight of soil (γ-pcf) and X_6 = shear strength of soil (f_s-psf). Substituting the variables, the equation is changed to the form

$$Q_u = \alpha_0 + \alpha_1 n + \alpha_2 D + \alpha_3 R + \alpha_4 N_q + \alpha_5 \gamma + \alpha_6 f \qquad (9)$$

The values of variables N_q, γ and f_s are determined from laboratory test results. The relation between bearing capacity factor (N_q) and depth (D) is estimated to be

$$N_q = 7D - 123.2 \qquad (10)$$

The shear strength (f_s) of soil at different depths is found to vary as well

$$f_s = 280D - 4396 \qquad (11)$$

The regression analysis yields the values of the coefficients for a best fit and the new formula to determine the load carrying capacity of pile groups is established. The coeficients are $\alpha_0 = -4.5 \times 10^5$, $\alpha_1 = 3.5 \times 10^5$, $\alpha_3 = -3.2 \times 10^3$, $\alpha_5 = -1.5 \times 10^3$ and $\alpha_6 = 3.5 \times 10^2$. α_2 and α_4 are insignificant. A table of correlation coefficents is shown in Table 1.

6 VERIFICATION

The validity of this equation established is tested by back-prediction technique. A set of data from a site of similar subsoil conditions is selected. The value of load carrying capacity for a group of four piles is calculated and compared with the predicted values as shown in Figure 5. The percentage of difference between the two values is found to be about 8%. The group capacity of three different pile groups is determined by the various efficiency formulae presently used and also by the proposed method and shown in Figure 6. It is seen that all existing efficiency formulae underestimate the pile group capacity in case of loose and medium dense sands. This leads to over design of foundations and increased costs.

7 CONCLUSIONS

Field and laboratory test data are used to develop a method of determining load carrying capacity of pile groups driven in sand. Driving a pile into a loose or medium dense sand causes the soil around and between the piles to become highly compacted. The important factors governing the load carrying capacity of piles are relative density, pile size, pile group geometry, and frictional properties of soil. Existing efficiency formulae underestimates the pile group capacity. The proposed method can be used to determine the realistic load carrying capacity value resulting in an economical foundation design. The proposed method needs to be future verified as additional data becomes available.

8 REFERENCES

Bergdahl, U., and Wennerstrand, J., "Bearing Capacity of Driven Friction Piles in Loose Sand," Geotechnique, 1976, Vol 9, pp. 355-360.

Coyle, H. M., and Castello, R. R., "New Design Correlations for Piles in Sand," ASCE Journal of Geotechnical Engineering, 1981, Vol. 107, pp. 965-986.

Hanna, T. H., "Model Studies of Foundation Groups in Sand," Geotechnique, 1963, Vol. 13, pp. 334 -351.

Janbu, N., "Static Bearing Capacity of Friction Piles," Sixth European Conference on Soil Mechanics, 1976, Vol. 1, pp. 479-488.

Meyerhoff, G. G., Compaction of Sands and Bearing Capacity of Piles," Proceedings of ASCE, 1959, Vol. 85, pp. 1-18.

Thompson, C. D., and Thompson, D. E., "Influence of Driving Stresses on the Development of High Pile Capacities," ASTM STP 470, 1979, pp. 562-577.

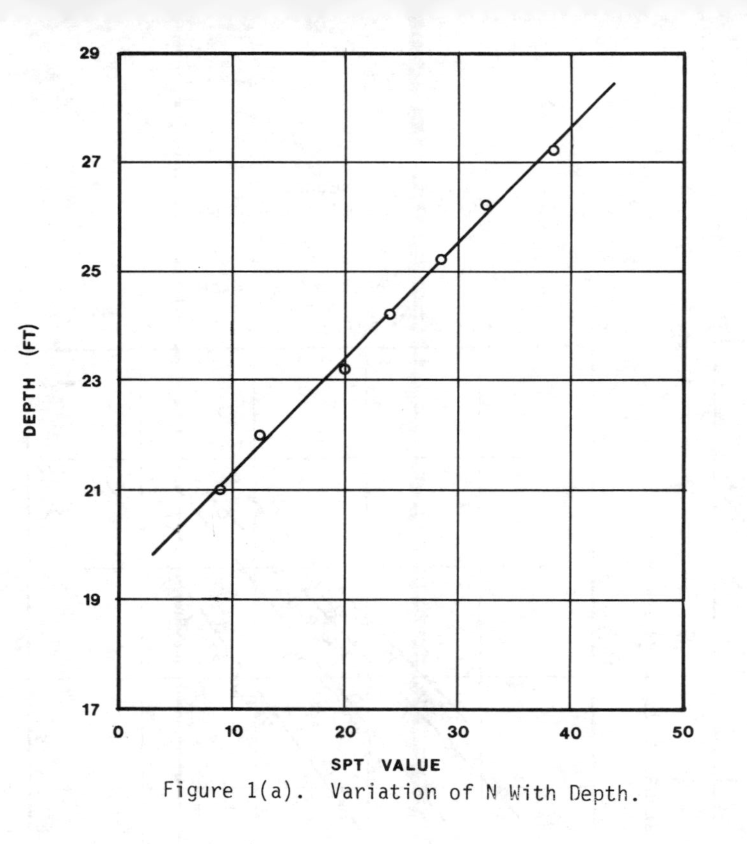

Figure 1(a). Variation of N With Depth.

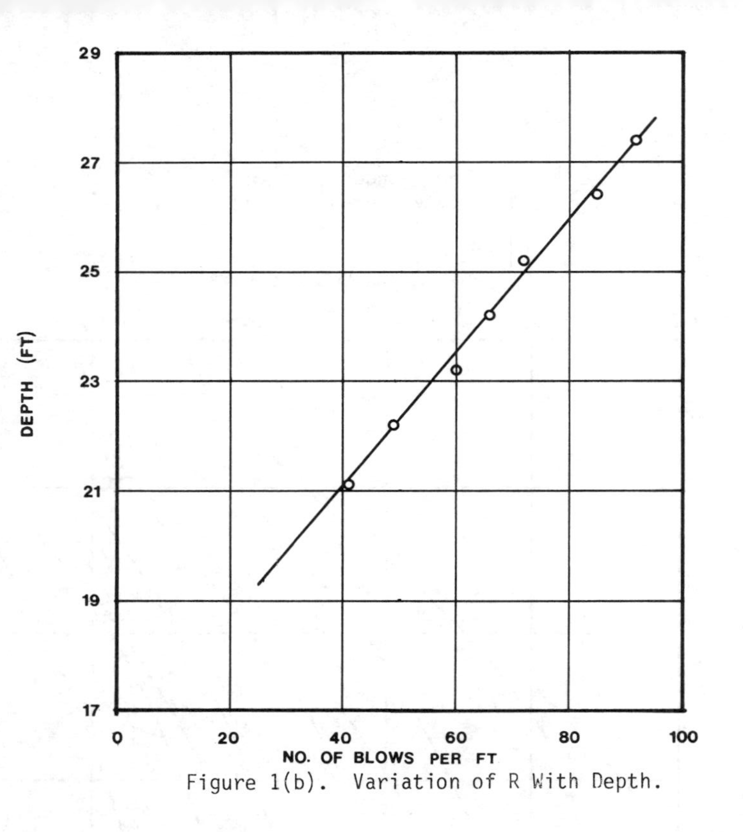

Figure 1(b). Variation of R With Depth.

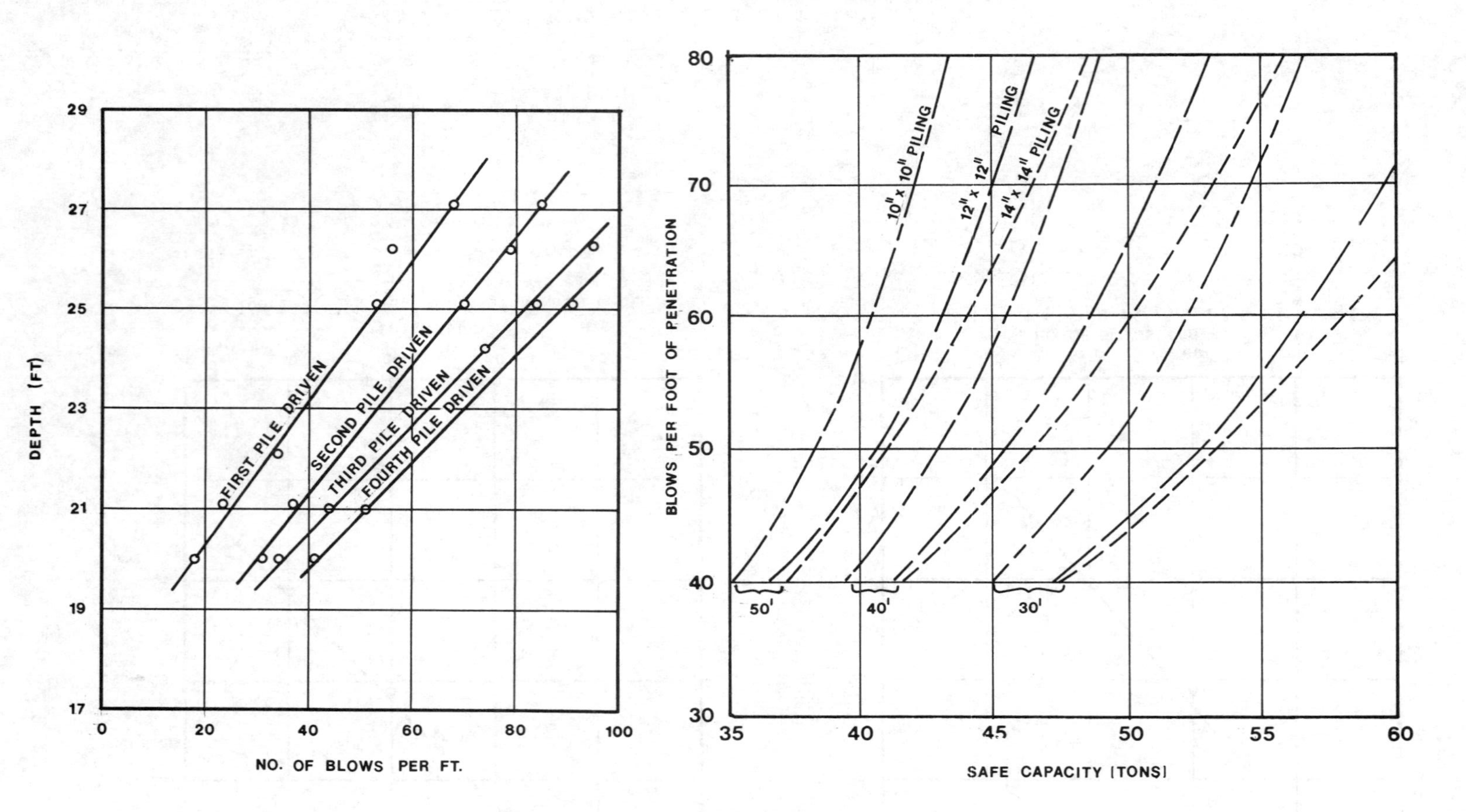

Figure 2. Change in Number of Blows Due To Sequence of Driving

Figure 3. Load Carrying Capacity of Piles (Janbu Formula).

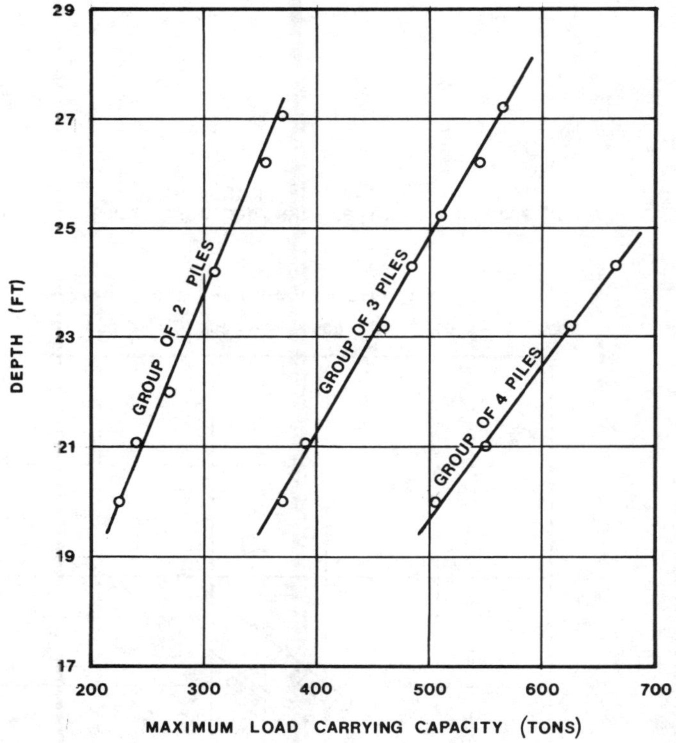

DEPTH (FT)

29 27 25 23 21 19 17

200 300 400 500 600 700

GROUP OF 2 PILES

GROUP OF 3 PILES

GROUP OF 4 PILES

MAXIMUM LOAD CARRYING CAPACITY (TONS)

Figure 4. Relation Between Depth and Load Carrying Capacity of Pile Groups.

TABLE 1.

Correlation Coefficients of Different Variables.

	n	D	B	N_q	γ	f_s
n	1.0	.19	.27	.20	.11	.20
D	.19	1.0	.98	.99	.91	.99
B	.27	.98	1.0	.98	.89	.98
N_q	.20	.99	.98	1.0	.91	.99
γ	.11	.91	.89	.91	1.0	.91
f_s	.20	..99	98	.99	.91	1.0

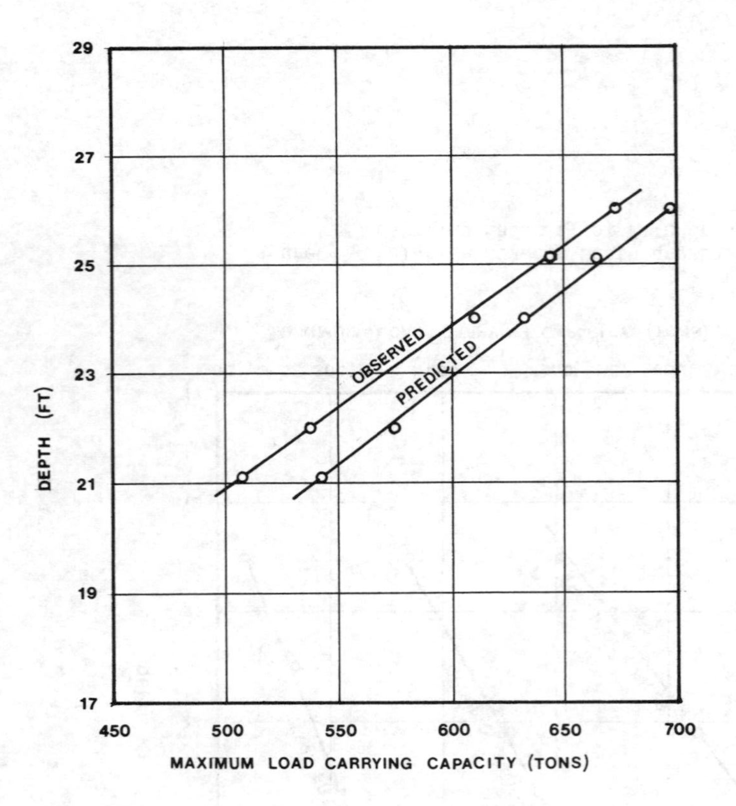

Figure 5. Comparison Between Predicted and Observed Values of Load Carrying Capacity.

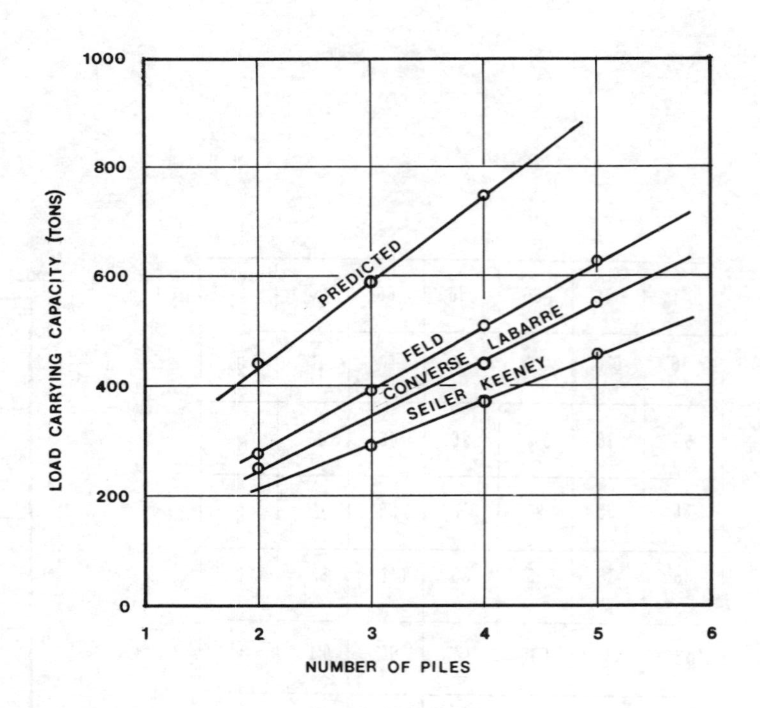

Figure 6. Load Carrying Capacity of Pile Groups By Different Methods.

Design and performance of large diameter cast in place pile groups subjected to lateral loads

S.McKeown
Golder Associates, Calgary, Alberta, Canada

J.I.Clark
Memorial University of Newfoundland, St.John's, Canada

ABSTRACT: The Olympic Oval on the University of Calgary Campus is founded on groups of large diameter cast in place concrete piles. The foundation system was required to support very large lateral loads (up to 6600 kN) with differential movements between adjacent supports of less than 10 mm. The design methodology for the project included the following:

- A detailed investigation program using a cone penetrometer and self boring pressuremeter, supplemented by boreholes.

- A lateral load testing program on different size piles both individually and in groups.

- Modelling based on the pile load test results to extrapolate the results to areas of slightly different soil conditions.

Due to the unique nature of the structure and the magnitude of the loads on the foundation system, an ongoing series of surveys have been carried out to monitor both initial foundation response to loading and long term creep. Comparisons of predicted to observed movements are presented and an assessment of the effect of group effects and varying soil conditions is made.

1 INTRODUCTION

The Olympic Oval facility is a multi-purpose enclosed sports complex recently completed on the campus of the University of Calgary. It will be used as the venue for the speed skating events during the 1988 Winter Olympic Games and as a field house after the games.

The roof of this structure spans 199 m by 88 m and is the largest clearspan prestressed concrete supported roof in the world. The design involved the use of intersecting long span shallow arches supported at 18 m centres. The loads are transferred to the foundation system by a series of buttresses at an angle of 35 to 45 degrees from the horizontal. The magnitude of the resultant horizontal loading is up to 6600 kN per support with the dead load comprising 82 to 84 percent of the total load at the higher load supports. Ninety percent of the dead load is from the concrete girders supporting the roof structure. A planview of the roof and a section of a typical support buttress are shown in Figure 1 and 2 respectively.

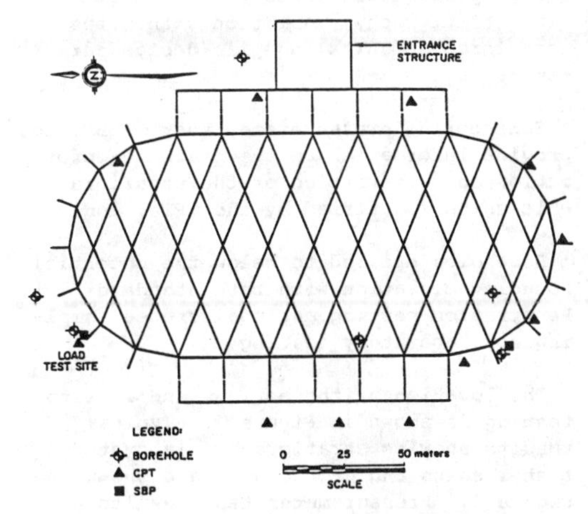

Figure 1 - Building Layout and Investigation Program

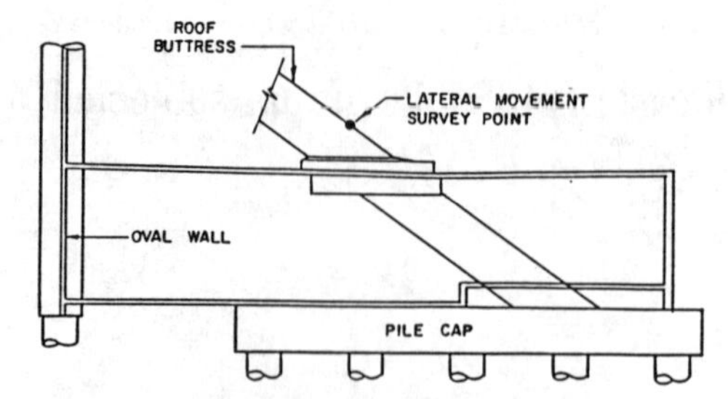

Figure 2 Typical Buttress Design

This paper describes the methodology employed to design the foundation system and the performance of the foundations under initial dead loads. The results of full scale lateral load tests have been described earlier but are summarized nerein for completeness (Clark et al 1985).

2 SITE DESCRIPTION AND STRATIGRAPHY

After an initial borehole investigation program to determine the upper soil conditions and the elevation of the water table, a study was made of alternative foundations appropriate to the structure and site conditions. It was concluded that large diameter cast in place concrete piles were economically and technically the best solution. An alternative system of tie supports commonly used for arch roof structures was precluded primarily due to the width of the structure. The final investigation program, then, was tailored towards the design and consisted of the following components:

- Cone penetration tests (CPT) to determine stratigraphy, variation within the stratigraphic units and pile design parameters,

- Self boring pressuremeter tests (SBP) to provide information on the pressure-deformation characteristics of the stratigraphic units identified by the CPTs, and,

- Boreholes drilled to below the potential founding elevation with both Standard Penetration testing and undistrubed sampling for laboratory testing.

The location of the borings and in situ testing is shown in Figure 1. Typical results showing stratigraphy, laboratory test results and CPT results are shown in Figure 3. Pressuremeter test results at the same location are shown in Figure 4. An example of the soil stratigraphy, as

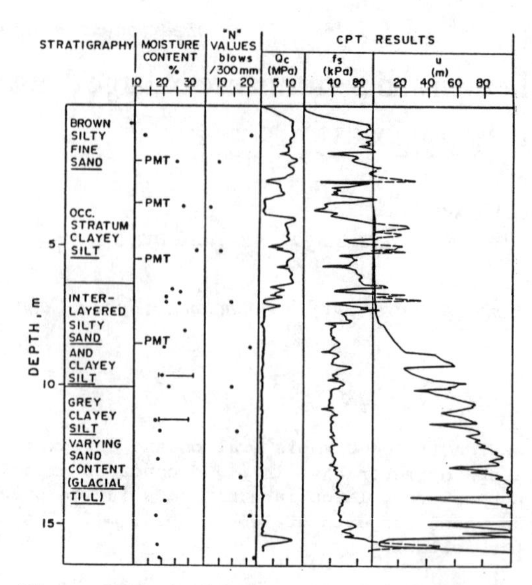

Figure 3 Typical Laboratory and CPT Test Results

indicated by the CPT tip resistance is shown in Figure 5 for a section down the west side of the structure.

The soils at this site were laid down by glacial and periglacial processes. Glaciers invaded the area at least twice during the last glaciation and left the bedrock and preglacial gravel covered with glacial drift. During the last advance, the area was inundated with the result that lacustrine sediments were deposited over the till. These lacustrine sediments have been subjected to erosion and weathering since deposition and the upper layers shows aeolian features such as cross bedding. The section shown in Figure 5 indicates the following stratigraphic sequence:

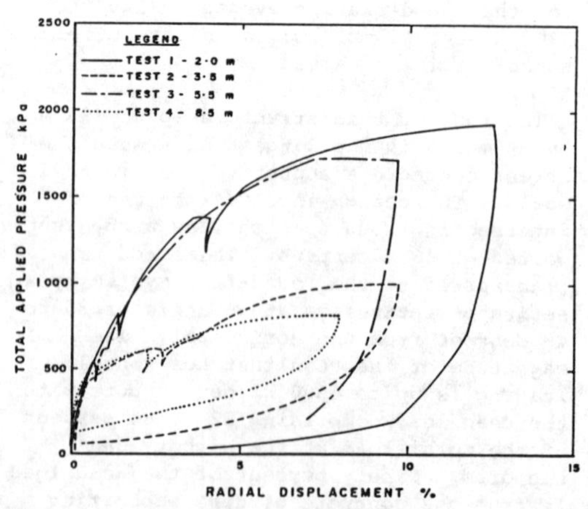

Figure 4 Pressuremeter Test Results

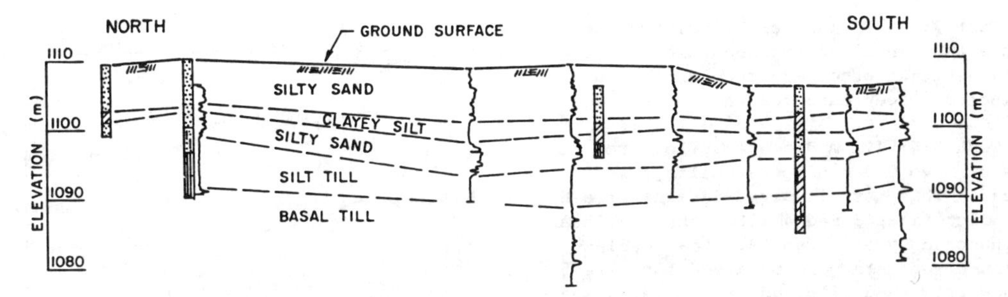

Figure 5 Soil Stratigraphy - West Side

· Surficial silty sands.
· A relatively thin clayey silt stratum
· Compact to dense sand grading into an ablation till.
· A clayey silt ablation till that has some silt and sand lenses.
· A very stiff heterogeneous basal till.

The cone penetration tests, pressuremeter tests and soil borings indicate that the surficial silty sands are denser at the northern and eastern side than for the south and west sides. Standard Penetration Test "N" values were typically 15 to 25 in the more competent areas and 10 to 15 in the western and southern portions of the site.

3 PILE LATERAL LOAD TEST PROGRAM

Owing to the expense of installing piles and pile caps, it was decided to test piles in a location where they could be used after the tests were completed as part of the main structure. The loads are less at the single arch supports at the end of the oval and hence, one of these locations, at the northwest corner of the oval (Figure 1), was selected for the test site.

The configuration of the test site and pile caps is shown in Figure 6. The piles were cast in place reinforced concrete having steel reinforcing cages comprising

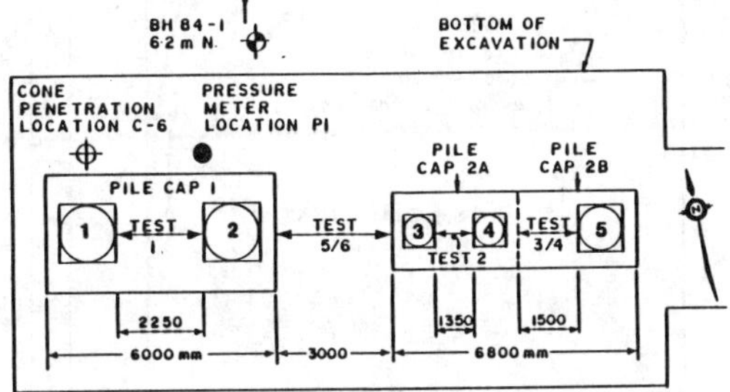

Figure 6 Pile Load Test Configuration

1.25% of the total cross sectional area. All piles were installed to a depth of 15 m below grade and were belled. The piles were laid out to allow free headed tests by jacking one pile against another and then fixing the piles by casting a heavily reinforced concrete cap which would then be jacked against an adjacent cap.

The piles were loaded laterally using a 9 MN capacity calibrated jack. In order to provide a suitable jacking surface, all piles were capped for the free headed tests. These caps extended 600 mm above the pile cutoff elevation at ground surface. They were square with a width equal to the pile diameter and each had a 600 mm square steel plate 5 mm thick cast in it to act as a bearing plate. For the fixed head tests, the piles and jacking caps were cast into 1.5 m deep pile caps of dimensions shown in Figure 7. Reinforcing steel was extended from the piles into the pile cap.

The objectives of the pile lateral load test program were:

1. To establish fixed and free head pile capacities for piles ranging in diameter from 0.9 to 1.5 m.

2. To assess group effects.

3. To provide information on pile deflected shape in order to provide a basis for structural design of the piles.

4. To provide a model for design in areas where soil conditions vary from the test site.

The program established to meet these objectives was:

Test 1: 1.5 m diameter piles, free head. This test involved jacking Piles 1 and 2 against each other and recording load-deflection-creep data for each pile.

Test 2: 0.9 diameter piles, free head. This test involved jacking Piles 3 and 4 against each other and recording load-deflection-creep data for each pile.

Test 3/4: 1.2 m diameter pile, free head; two 0.9 m diameter piles fixed head. This test involved jacking Pile 5 against Piles 3 and 4 after they had been connecetd by Pile Cap 2a. Load-deflection-creep data were obtained for pile 5 as a free head pile, and for piles 3 and 4 as fixed head piles.

Test 5/6: Group Test: Group 1: two 1.5 m diameter piles encased in Pile Cap 1. Group 2: 1.2 m diameter pile and two 0.9 m diameter piles encased in Pile Cap 2 (2a and 2b). Load-deflection-creep data were obtained for each pile group in addition to information on vertical movements, twist and rotation of the caps. A creep test of one week duration was undertaken as part of this test.

The deflection of the piles was measured at discrete load intervals using dial gauges. In addition, the deflection-time (creep) behaviour was also recorded while the load was held constant. The deflected shapes of the piles was measured by placing a slope indictor casing down the middle of each pile. The dial gauges were positioned so that twist and tilt of the piles and pile caps could be recorded.

4 TEST RESULTS

The test results are described in detail by Clark et al (1). Typical lateral load-deflection curves for free head piles (Test 1) and fixed head piles (Test 5/6) are shown in Figures 7 and 8 respectively.

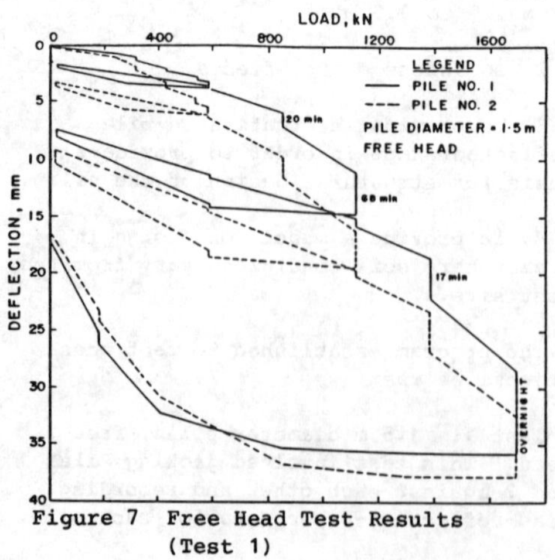

Figure 7 Free Head Test Results (Test 1)

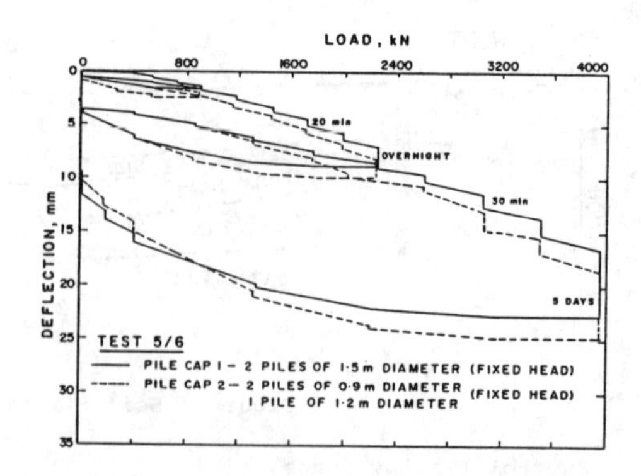

Figure 8 Free Head Test Results (Test 5/6)

Typical deflected shapes of these piles are shown in Figures 9 and 10. A summary of the load per pile at 10 mm deflection versus the pile diameter is shown in Figure 11. Although a graph such as that shown in Figure 11 is dependent on testing procedures, and particularly on the creep the results presented are considered to be reasonably representative since the creep deformation accounts for a relatively small proportion of the total deformation at the loading range under consideration.

Creep deformation at various loading increments was observed to be generally linear with the logarithm of time. A typical example of this is shown in the "long term" (one week duration) test of fixed head piles shown in Figure 12.

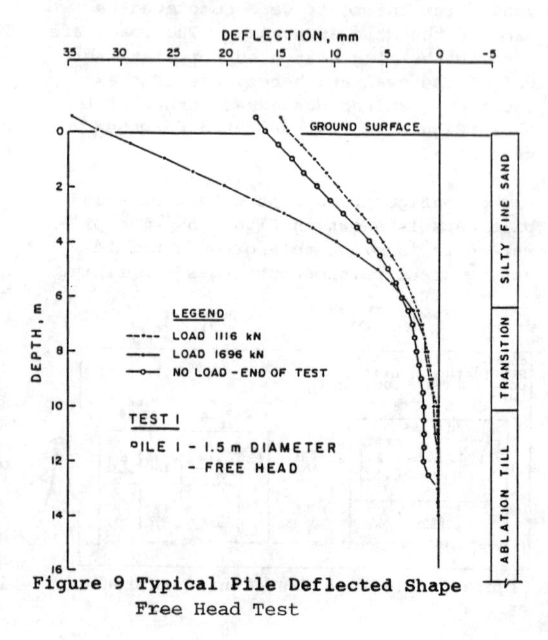

Figure 9 Typical Pile Deflected Shape
Free Head Test

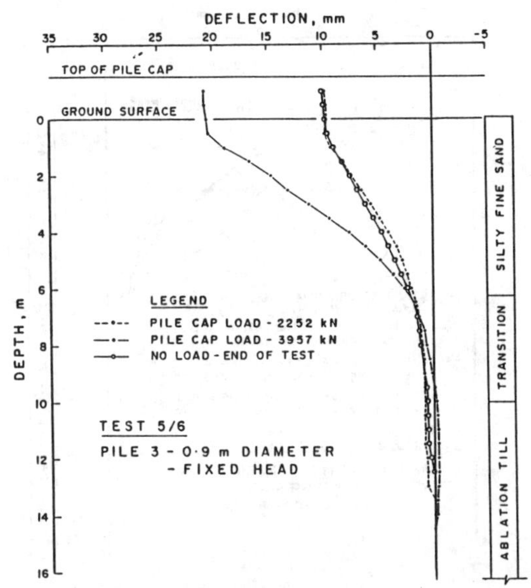

Figure 10 Typical Pile Deflected Shape Fixed Head Test

Summaries of the observed creep rates versus the normalized pile load are shown in Figure 13. The normalizing load (i.e.- that causing 10 mm deflection) was chosen because it is the designated allowable short term deflection.

Differential movements between piles and pile groups may be caused by non-uniform loading, variable soil conditions, and construction related differences. The free head tests indicate 5 mm differential deflection for the 1.5 m diameter piles (Figure 7) and 3 mm differential deflection for the 0.9 m diameter piles when the largest pile deflection was 10 mm.

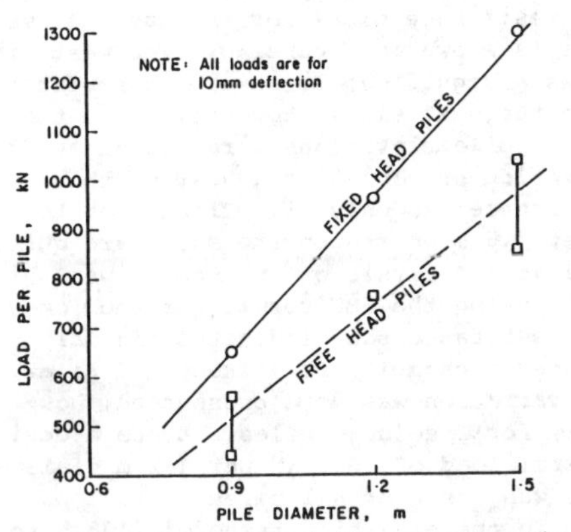

Figure 11 Pile Loads at 10 mm Deflection

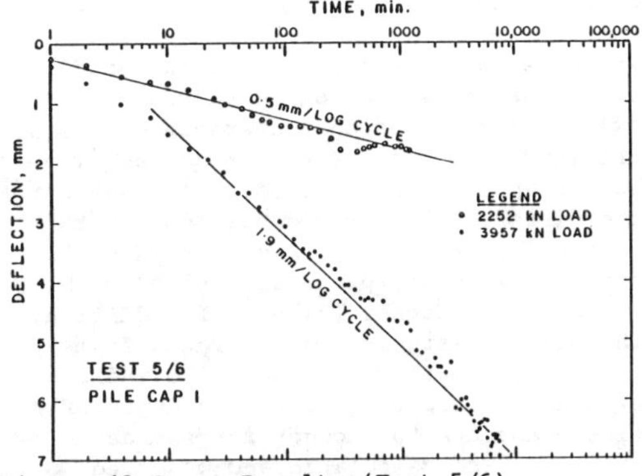

Figure 12 Creep Results (Test 5/6)

It was determined by the structural consultants that the roof structure would readily tolerate a differential movement of 5 mm between adjacent supports but that a 10 mm differential movement would be the maximum tolerable without inducing unacceptable stresses in the arches. Hence, the design load capacity was selected on the basis of an allowable immediate lateral deflection of 10 mm and a total allowable deflection of 15 mm for a combination of immediate deflection and creep. Since the load capacity is roughly proportional to the diameter (see Figure 12) and the cost is proportional to the diameter squared, it is economic to consider relatively smaller diameter piles. However, due to uncertainty about large group effects and the cost of larger pile caps, piles of 1.2 m diameter were adopted for all major lateral load bearing foundations.

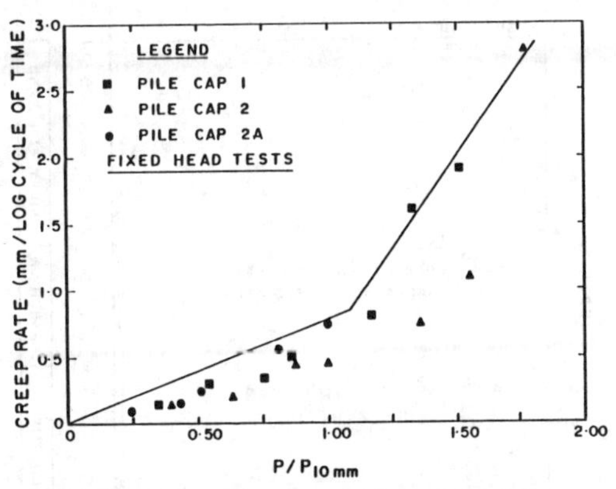

Figure 13 Creep Rate vs Normalized Load

5 PILE LOAD TEST ANALYSIS

The pile load tests were back analyzed using the computer program LATPILE (Reese 1977) and soil response characteristics as obtained from the SBP test results. The soil response was modified to account for the different boundary conditions in the pressuremeter loading (radial strain) to the pile loading (plane strain) by methods suggested by Hughes (1979). In addition, the soil resistance was pro-rated from zero at the surface to the value indicated by the pressuremeter test at a depth of 2 pile diameters to account for surface effects. The calculated and observed deflected shapes for a 0.9 m diameter pile (using the effective moment of inertia for a cracked section) are shown in Figure 14.

The correspondence is good at the design deformation of 10 mm. A discrepancy was observed at large deformations (25 mm), however, with the analytical model under-predicting pile deformation. This is thought to be due primarily to creep deformation during the load test which was not modelled during the short term pressuremeter test. This creep was more predominant at higher loads.

For fixed head tests, both surface and group effects must be considered. The latter is a function of the soil type, pile spacing, and shape of the group. This group effect can be taken into account by reducing the soil stiffness and ultimate resistance. A comparison of the observed pile deflected shape versus the predicted shape using full soil stiffness, 66 precent of full soil stiffness, and 33 percent of full soil stiffness is shown in Figure 15. The 66 percent curve agrees well with the observed results.

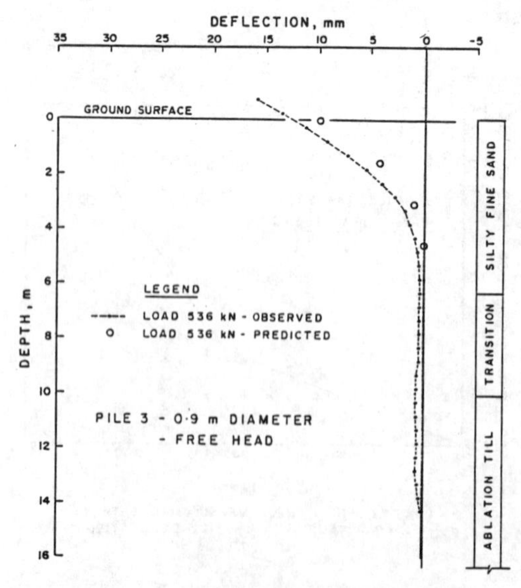

Figure 14 Free Head Test - Calculated vs Predicted Deformation

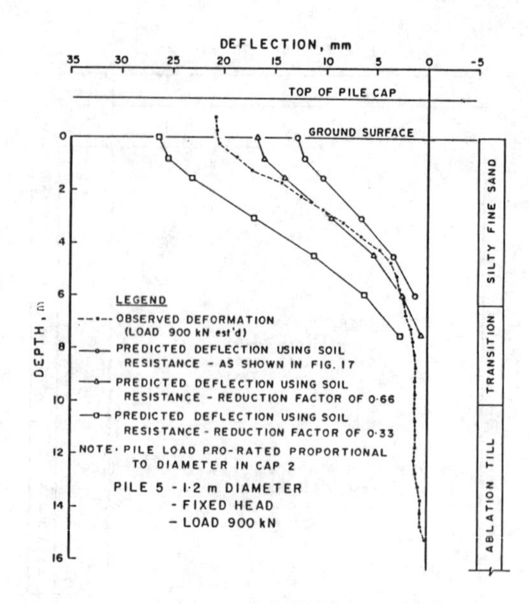

Figure 15 Free Head Test Calculated vs Predicted Deformations

6 DESIGN OF FOUNDATION SYSTEM

The variation in density of the surficial sand can be an important factor in the design of laterally loaded pile foundations for a structure as large as the Olympic Oval. A comparison of typical CPT tip resistance plots for the east and west side is shown in Figure 16. The west side shows a significantly lower tip resistance than the east side. However, since the top of pile elevations were set at an elevation of 1103.5 to 1103.9 m, this difference was only significant for the upper 1.5 m whereupon the soils are quite similar. Analysis of the looser west side soils using the SBT result for the lower tip resistance sand indicated minimal change in capacity for fixed head piles. The variation was more pronounced, however, for free head piles. Hence a design lateral load of 1000 kN per 1.2 m diameter pile was used for all piles where tops were in the elevation range of 1103.5 to 1103.9 m.

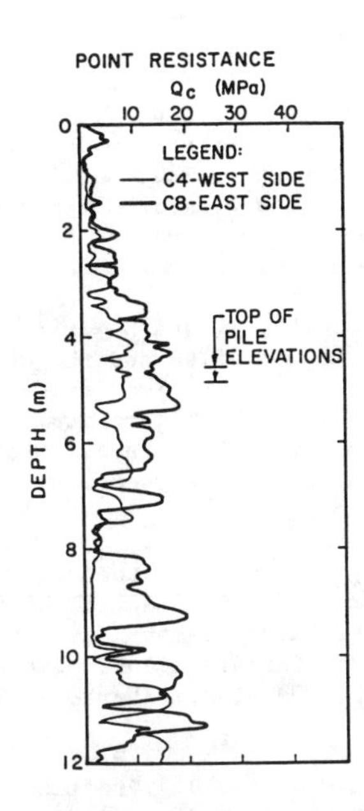

Figure 16 Comparison of Typical East and West Side SPT Results

Typical pile configurations on the west side of the structure are shown in Figure 17. The dead load per buttress was calculated to be 5400 kN with approximately 85 percent of this coming on the foundation system as soon as the construction scaffolding system was released from under the roof girders. Piles not in the pile caps were tied to the main lateral load bearing piles by thickened floor slabs and, in some cases, by tension ties. The live load was designed to be taken by the compacted backfill against the exterior walls of the structure. This soil was not placed until the dead load was on the structure so that the walls would not have to be designed for passive earth pressures.

The pile groups are very large with up to 6 piles per group. Pile group effects were a concern since the largest group tested was three in a line. This, however, was countered by the fact that the pile caps would be buried by typically 2 to 3 m of compacted soil and hence surface effects would not have the impact they had for the pile load tests. Using elastic analysis to estimate group effect (Poulos), it was computed that a five to six pile group would be expected to deflect 25 to 40 percent more than a three pile group at the same load per pile. The LATPILE computer model predicts a reduction in deflection by 25 - 30 percent

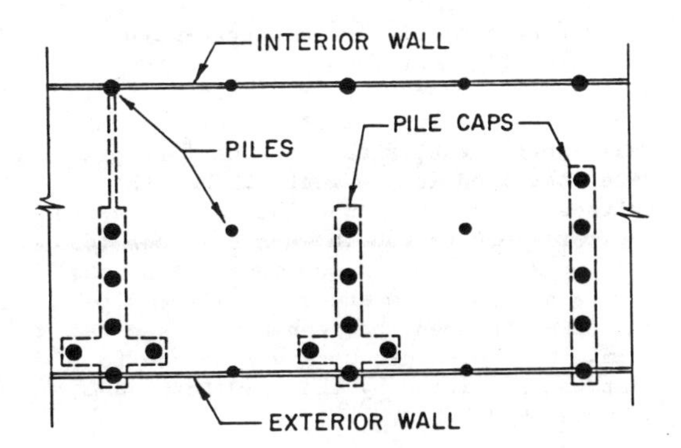

Figure 17 Typical Pile Groups - West Side

due to the absence of surface effects. Hence, group and surface effects are offsetting and the 1000 kN design load was used for larger buried pile groups.

7 PERFORMANCE OF THE STRUCTURE

A total of four surveys have been made over a three month period to quantify the outward deformation of the buttress on the east and west side of the oval. The buttress on the curves at each end were not surveyed. The survey measurements were taken on the backs of the buttresses at a location shown in Figure 2. The survey dates were June 14 (immediately before the first scaffolding was lowered), July 2 (immediately after the final scaffolding was lowered), and July 7, 1986. Deformations recorded on these dates are presented in Table I.

TABLE I
DEFORMATION SURVEY RESULTS

	Buttress Deformation	
	July 2	July 7
WEST SIDE	4	3
(North to South)	5	4
	11	11
	8	9
	7	8
	6	6
	6	7
EAST SIDE	3	2
(North to South)	3	3
	0	0
	1	1
	5	5
	-1	-1
	-2	-2

The west side of the structure moved significantly more than the east side. This is consistent with the stiffer soil conditions and the effect of the entrance hall, on the east side although more piles were installed in the west side at the buttress locations. The load per pile was typically 650 to 800 kN when the roof load was imposed. The maximum movement on the west side was 11 mm and the maximum differential movement between buttresses was 6 mm. The movements were very close to what was predicted by the load test and computer model.

Long term monitoring for creep deformations is currently being carried out. Although the creep deformations are expected to be very small and within the accuracy of the survey techniques, it is hoped that, with sufficient time and readings, trends and creep rates will emerge.

8 CONCLUSIONS

The results of the pile lateral load test and LATPILE analytical computer model sucessfully predicted the performance of the Olympic Oval structure when the dead load was imposed on the foundation system. The analytical model was a useful tool in back calculating pile load test results although allowances had to be made for surface and pile group effects. The use of the cone penetrometer and self-boring pressuremeter provided high quality stratigraphic and soil property data that was instrumental in the design of the structure.

9 ACKNOWLEDGEMENT

The authors would like to acknowledge the help of Mr. B. Lester of the Simpson Lester Goodrich Engineering Partnership who was responsible for the structural design of the Oval, Mr. L. Eibner of the University of Calgary for his support and input into the pile load test program, and Mr. I. Duncan of the University of Calgary for permission to publish the results. Special thanks are due to Mr. M. Chapman of the University of Calgary survey engineering program for organizing the survey program for building deformation.

10 REFERENCES

Briaud, J., Smith, T. and Meyer, B., 1983. Design of Laterally Loaded Piles using Pressuremeter Test Results, Symposium on the Pressuremeter and its Marine Applications, Paris, May 1983.

Briaud, J., Smith, T. and Mayer B., 1984. Laterally Loaded Piles and the Pressuremeter: Comparison of Existing Methods, ASTM Special Technical Publication STP 835, Laterally Loaded Deep Foundation: Analyses and Performance.

Clark, J.I., McKeown, S., Lester, W.B. and Eibner, L.J. The Lateral Load Capacity of Large Diameter Concrete Piles. 38th Canadian Geotechnical Conference, Edmonton, Alberta 1985.

Hughes, J.M.O., Goldsmith, P.P. and Fendall, H.D.W., 1979. Predicted and Measured Behaviour of Laterally Loaded Piles for the Westgate Freeways, Melbourne, Victoria Geomechanics Society, Australia.

Poulos, H. and Davis, E., 1980. Pile Foundation Analysis and Design, Chapter 8. John Wiley and Sons.

Reese, L.C., 1977. Laterally Loaded Piles: Program Documentation, Journal of Geotechnical Engineering Division, ASCE, Vol. 103, GT4. April 1977, pp. 287-305

Prediction and Performance in Geotechnical Engineering / Calgary / 17-19 June 1987

Prediction of ultimate pile capacity from load tests on bored and belled, expanded base compacted and driven piles

R.C.Joshi
University of Calgary, Alberta, Canada

H.D.Sharma
Fluor Canada Ltd., Calgary, Alberta, Canada

ABSTRACT: On many projects pile load tests are carried out as a part of the field investigation program, and on the basis of the results of these full scale pile load tests, pile capacities are established. One of the criteria for establishing allowable pile capacities is to estimate the ultimate pile load capacity and then to apply an appropriate safety factor. Prediction of the ultimate pile capacity from pile load tests is, therefore, an important aspect of pile design. This paper first reviews various pile load test methods and the available ultimate load prediction or interpretation methods. These interpretation methods have then been applied on five pile load test data. These test piles consisted of two bored and belled piles, two expanded base compacted piles and one driven steel H pile. Applicability of these load test interpretation methods for these different pile types has finally been discussed.

1 PILE LOAD TEST METHODS

Many pile load test methods have been used by practicing engineers and researchers and are reported in many publications (ASTM, D 1143-81, Broms, 1972, Butler and Hoy, 1977, Fellenius, 1975 and 1980, Mohan et al, 1967, New York State DOT, 1974, Swedish Pile Commission, 1970, Weele, 1957, Whitaker, 1957 and 1963 and Whitaker and Cooke, 1961). From the available numerous test methods the four described in Table 1 can be identified as the basic test methods.

As shown in Figure 1 the SM tests and SC tests are the slowest tests. Figure 2 provides a comparison of typical load-deformation behavior for the above four test methods. This figure exhibits that the shape of CRP test is well defined and agrees well with the QM test before the failure is reached. The SM test method is commonly used in North America because it is simple, most engineers are familiar with it, interpretation based on gross and net settlements can be made easily and it can furnish a rough estimate of the expected pile settlement under working load.

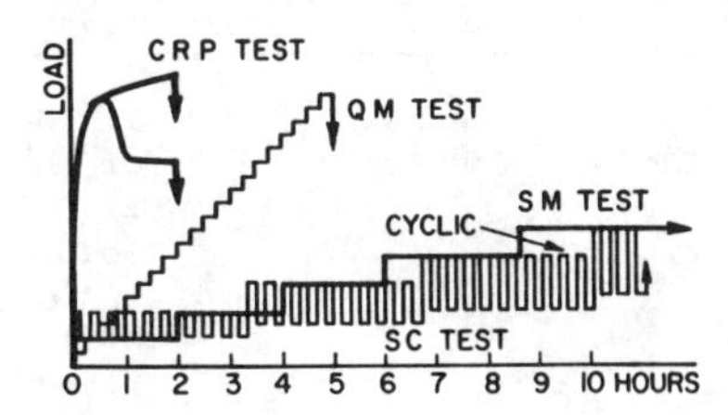

Fig.1 Comparison of required time for test methods (Fellenius, 1975)

2 PILE LOAD TEST DATA INTERPRETATION METHODS

The ultimate failure load, for a pile which is stronger than the soil, is defined as the load when the pile plunges or the settlements occur rapidly under sustained load. Plunging, however, may require large movements which may exceed the acceptable range of the soil-pile system. Other definitions of failure consider arbitrary settlement limit such as, the pile is considered to have failed when the pile head has moved 10% of the pile end diameter or the failure load is at the

Table 1. Pile load test methods.

Method	Code Agency	Procedures	Applicability and Limitations
1) Slow Maintained Load Test Method (SM Test)	ASTM D-1143 (standard)	1) Load the pile in eight equal increments to 200% of the design load. 2) Maintain each increment until the rate of settlement has decreased to 0.01 in/hr. 3) Maintain 200% load for 24 hrs.	1) Commonly used as a part of site investigation prior to installing contract piles and writing specifications. 2) Time consuming (30-70 hrs. or more).
2) Quick Maintained Load Test Method (QM Test)	New York State Dept. of Transportation Federal Highway Administration ASTM D-1143-81 (optional)	1) Load the pile in 20 equal increments to 300% of the design load. 2) Maintain each load for a period of 15 min. with readings taken every 3 minutes.	1) Fast and economical (typically 3-5 hrs.) 2) Represents more nearly undrained conditions. 3) Cannot be used for settlement estimation.
3) Constant-Rate of Penetration Test Method (CRP Test)	Swedish Pile Commission New York State Dept. of Transportation ASTM D-1143-81 (optional)	1) The pile head is forced to settle at 0.42 inch/min. 2) The force required to achieve the penetration is recorded. 3) The test is carried out to a total penetration of 2-3".	1) Fast (2 to 3 hrs) and economical. 2) Yields good load-deformation plot. 3) It is of particular value for friction piles.
4) Swedish Cyclic Test Method (SC Test)	Swedish Pile Commission	1) Load the pile to 1/3 of the design load. 2) Unload to 1/6 the design load. Repeat this loading and unloading cycle 20 times 3) Increase the load by 50% higher than the item (1) & then repeat as item (2). 4) Continue until failure is reached.	1) Cycling changes the pile behavior so it is different than the original pile. 2) It is time consuming. 3) Only recommended for special projects.

intersection of the initial tangent to the load-movement curve and the tangent to or extension of the final portion of the curve. All these definitions are judgemental. A failure definition should be based on some mathematical rule and should provide a repeatable value. This value should be independent of scale effects and individual opinion.

This section provides a brief review of the following nine interpretation methods. These methods are then applied for different pile load test data presented in the next section. These methods are:

1. Davisson Method (1972)
2. Chin Method (1970, 1971)
3. De Beer Method (1967) or De Beer and Wallays (1972)
4. Brinch Hansen 90% criterion (1963)
5. Brinch Hansen 80% criterion (1963)
6. Mazurkiewicz Method (1972)
7. Fuller and Hoy Method (1970)
8. Butler and Hoy Method (1977)
9. Vander Veer Method (1953)

Table 2 summarizes the procedure and the applicability of these various methods. Fellenius (1980) used the

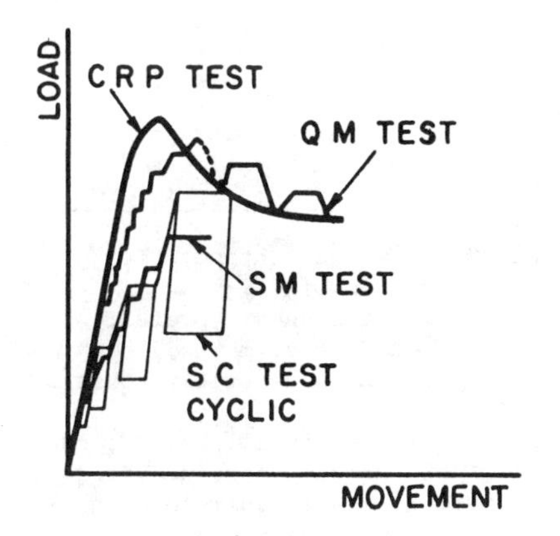

Fig.2 Comparison of load-movement behavior for test methods (Fellenius,1975)

load test results of a 12 inch (305 mm) concrete pile installed through 60 ft (18 m) of sensitive clay, 10 ft (3 m) of clayey silt and 6 ft (1.8 m) of silt. The pile was tested by CRP method six weeks after driving. Table 3 provides the failure loads interpreted by various methods, for this pile.

Interpretation by the methods, listed in Table 3, shows that the Davisson method provided the lowest failure load while the Chin method yielded the highest failure load. The Fuller and Hoy, Brinch Hansen's 90% criterion, Vander Veen and Mazurkiewicz Methods provided failure loads closer to each other and the pile also appeared to have failed around these load levels.

The criterion one prefers to select for a particular project depends mainly on one's past experience. At this time it is difficult to recommend or select a criterion that can be applied for all soil types, pile types and test methods. In order to arrive at one or a few recommended interpretation methods it is necessary to establish a set of reference cases which have been analyzed by these methods. In the following sections these nine methods will be applied on pile load test data for two bored and belled piles, two compacted expanded base (Franki) piles and a driven steel H pile. All of these piles were tested by the Slow Maintained (SM) Test Method.

3 APPLICATION OF INTERPRETATION METHODS FOR BORED AND BELLED CONCRETE PILES

Two bored and belled concrete piles, installed at locations where soil conditions primarily consisted of glacial clay till overlying the Paskapoo Formation clay shale and siltstone bedrock, were tested under axial compression loads. Figures 11 and 12 exhibit the pile dimensions, soil profile and load-movement curves for these piles. These two piles are identified as TP-1 and TP-2. Load tests were carried out by the SM Test method as per ASTM D 1143-81. Further information on these piles and load test details are provided by Sharma et al (1984).

The various failure load interpretation methods were used for the load-movement data obtained for these two piles. Table 4 provides a summary of the failure loads interpreted by these methods. The results show that the De Beer method yields the lowest failure load, followed by Brinch Hansen's 90% criterion. This method could not be applied for TP-1 because the shape of the curve was such that P_u and Δ_u gave twice the movement of pile head as obtained for 90% of P_u; and, therefore, these values could not be found on the curve.

The Brinch Hansen's 80% criterion as well as Vander Veen and Chin methods yielded failure loads at or higher than the test failure loads. The Mazurkiewicz method could not be applied for these load-movement data because the curves are not approximately parabolic which is the basic assumption of this method. The Fuller and Hoy method gave interpreted failure loads close to the test failure loads. The Davisson method and the Butler and Hoy method predicted failure loads closer to each other for both the load test data and provided conservative estimates of failure loads both for TP-1 and TP-2.

4 APPLICATION OF INTERPRETATION METHODS FOR EXPANDED BASE COMPACTED (FRANKI TYPE) PILES

Two expanded base compacted (Franki types) piles, TP-3 and TP-4, were load tested by the SM Test Method as per ASTM D 1143-81. Soil conditions at pile TP-3 location consisted of 0 to 25 ft (7.6 m) of high plasticity clay, 25 ft (7.6 m)

Table 2. Load test result interpretation methods.

Method	Procedure	Applicability and Limitations
1) Davisson	1) Draw the load-movement curve as shown in Figure 3. 2) Obtain elastic movement $\Delta = PL/AE$ of the pile. 3) Draw line OA based on equation in item 2. 4) Draw offset line BC where $x = 0.15 + D/120$ inches. 5) The failure load is then at the intersection of BC with load-movement curve, i.e. point C.	1) It was intended for driven piles. 2) It's use is preferred for QM test method. 3) It is generally conservative. 4) The limit line, BC, can be drawn before starting the test; therefore it can be used as one of the acceptance criteria for for proof-tested contract pile.
2) Chin	1) Draw the Δ/P versus Δ plot as shown in Figure 4. 2) The failure load or Pult is then equal to $1/C_1$. Figure 4 explains all the terms. (It assumes that the load-movement curve is approximately hyperbolic).	1) It is applicable for both the QM and SM tests provided constant time increments are used. 2) One should be careful in selecting Chin's straight line which may not begin to appear until the test load has passed Davisson's limit value. 3) May not provide realistic failure value for tests carried out as per ASTM standard method because it may not have constant time load increments.
3) De Beer	1) Plot load and movement on logarithmic scales as shown on Figure 5. 2) These values then fall on two approximate straight lines. 3) The failure load is defined as the load which falls at the intersection of these two straight lines.	1) It was originally proposed for slow tests e.g. SM tests.
4) Brinch Hansen's 90% criterion (Swedish Pile Commission)	1) Plot load-movement curve, Figure 6. 2) Find the load P_u and Δ_u that gives twice the movement of the pile head as obtained for 90% of the load P_u, where P_u is the failure load.	1) It is applicable to CRP tests irrespective of the soil type.
5) Brinch Hansen's 80% criterion	1) Plot $\sqrt{\Delta}/P$ and Δ as shown in Figure 7. 2) The failure load is determined as follows. $P_u = \frac{1}{2\sqrt{c_1 c_2}}$, $\Delta_u = \frac{c_2}{c_1}$	1) It is applicable for both the quick & slow tests e.g. QM tests and SM tests. 2) Agrees well with the failure test values. 3) Cannot be plotted in advance of tests.

	All the terms are defined in Figure 7. (It assumes that the load-movement curve is approximately parabolic).	
6) Mazurkiewicz	1) Plot load-movement curve as shown in Figure 8. 2) Choose a series of equal pile head movements and draw vertical lines to intersect on curve. Then draw horizontal lines from these intersection points or curve to intersect load axis. 3) From intersection of each load draw 45° line to intersect with next load line. 4) These intersections fall approximately on a straight line; the intersection of which on load line is the failure load.	1) It also assumes that load-movement curve is approximately parabolic; therefore values are close to 80% criteria (item 5 above). 2) When drawing the lines through intersection some disturbing freedom of choice is normally used.
7) Fuller & Hoy	1) Plot load-movement curve as shown in Figure 9. 2) Find the failure load P_u on the curve where the tangent on load-movement curve is sloping at 0.05 in/ton.	1) Applicable to QM test. 2) It penalizes the large piles because they will have larger elastic movements & 0.05 in/ton will therefore occur sooner.
8) Butler & Hoy	1) Use the load-movement curve plotted in Figure 9. 2) The failure load is then the intersection of the 0.05 in/ton slope line with either the initial straight portion of the curve or the line parallel to rebound curve or the elastic line.	1) Applicable to QB test.
9) Vander Veen	1) Choose a value of failure load Pult 2) Plot ℓ_n (1-P/Pult) against the movement for various load P values. 3) When the plot becomes a straight line this represents correct Pult as shown in Figure 10.	1) Time consuming calculations are required.

Note: Figures 3 through 10 inclusive are taken from Fellenius (1980).

to 37 ft (11 m) of silt and 37 ft (11 m) to 60 ft (18 m) of sand. Figure 13 shows the load-movement curve and soil conditions for this pile. Installed shaft diameter for this pile ranged between 18 inches (450 mm) to 20 inches (500 mm). Load test data for test pile TP-4 as reported by Nordlund, 1982, are shown in Figure 14. This pile was 22 inches (560 mm) in diameter and was installed through fine sandy silt to silty sand.

Table 5 provides a summary of failure loads for both TP-3 and TP-4 interpreted by various methods. For TP-3, the De Beer method provides a very low failure load. However, for TP-4 this method yielded a much higher failure load. The Davisson and the Butler and Hoy methods provided slightly lower or close to test failure loads. Fuller and Hoy's method gave interpreted failure loads close to the test failure loads. All other methods generally yielded failure loads higher than the test failure loads. As in TP-1, Brinch Hansen's 90% criterion could not be applied for TP-3 because of the shape of the load-

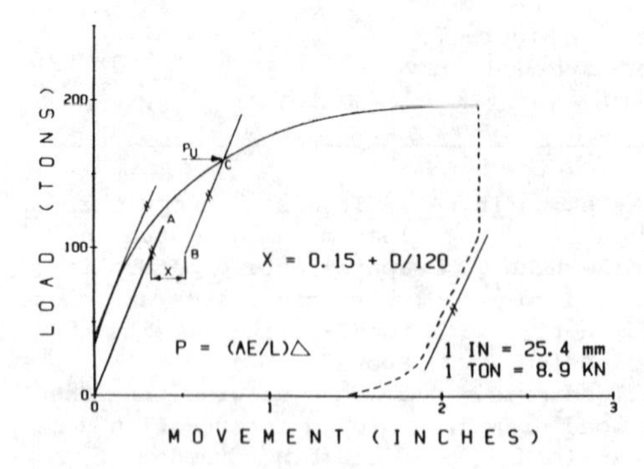

Fig.3 Davisson's method of load test interpretation

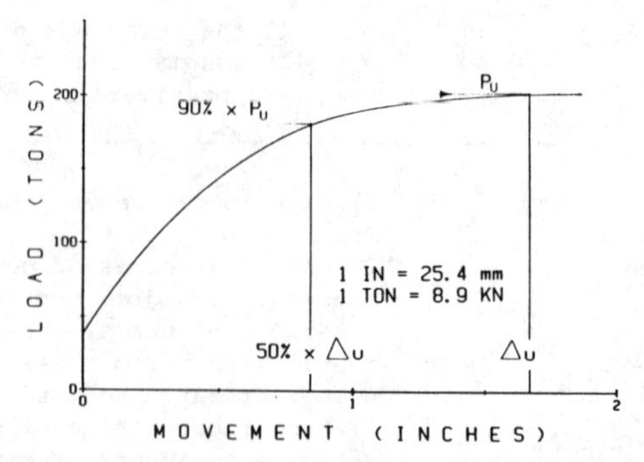

Fig.6 Brinch Hansen's 90% criterion method of load test interpretation

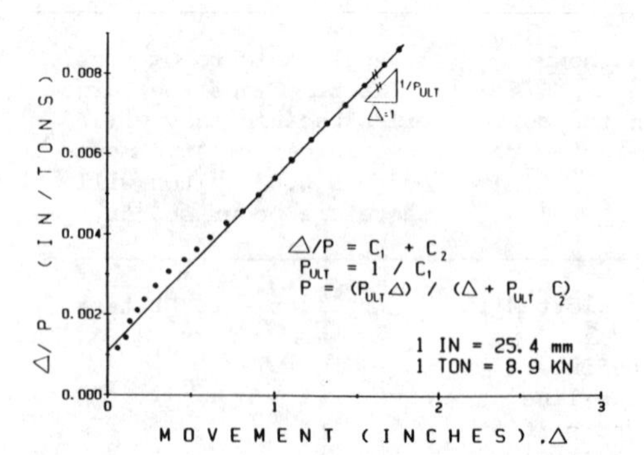

Fig.4 Chin's method of load test interpretation

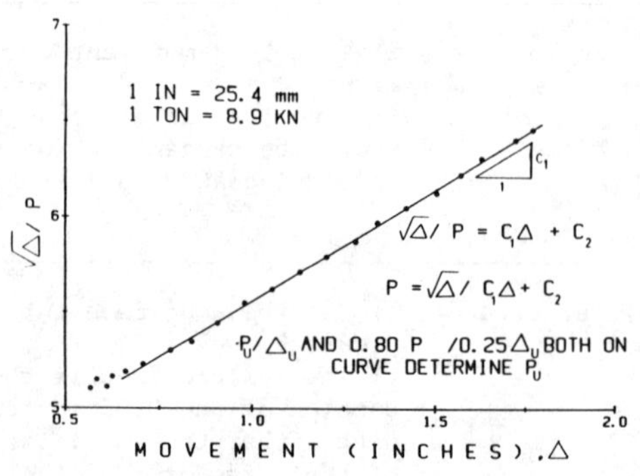

Fig.7 Brinch Hansen's 80% criterion method of load test interpretation

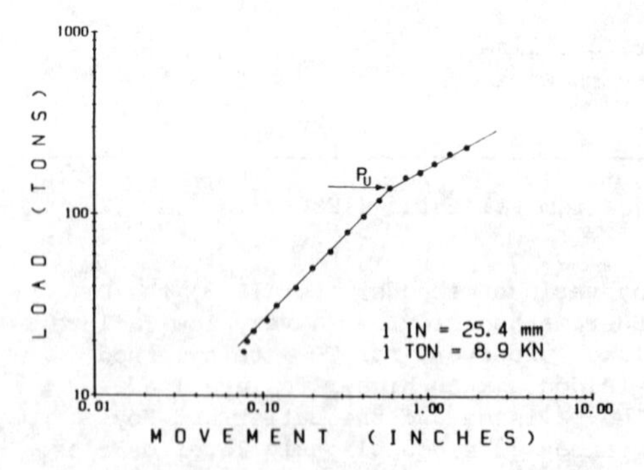

Fig.5 De Beer's method of load test interpretation

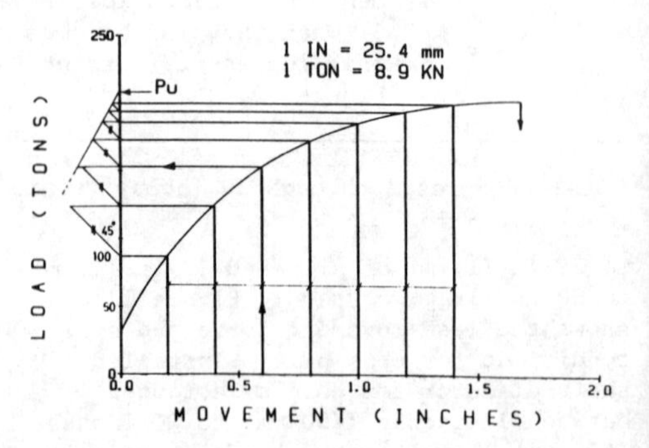

Fig.8 Mazurkiewicz's method of load test interpretation

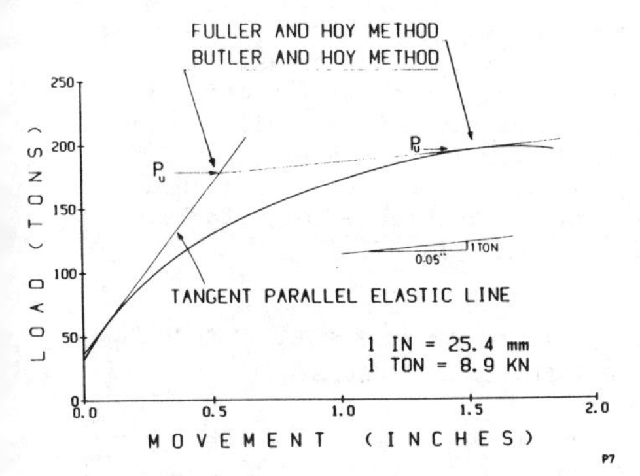

Fig.9 Fuller and Hoy's and Butler and Hoy's methods of load test interpretations

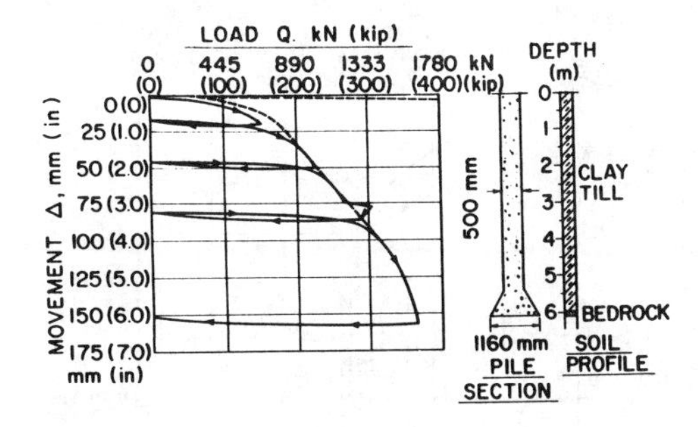

Fig.11 Axial compression pile load test (TP-1)

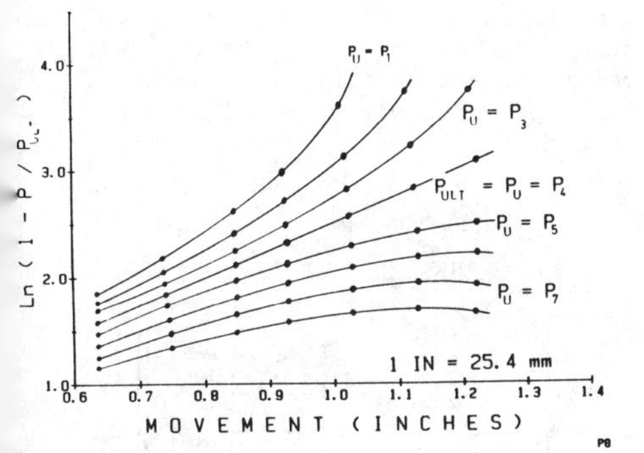

Fig.10 Vander Veen's method of load test interpretation

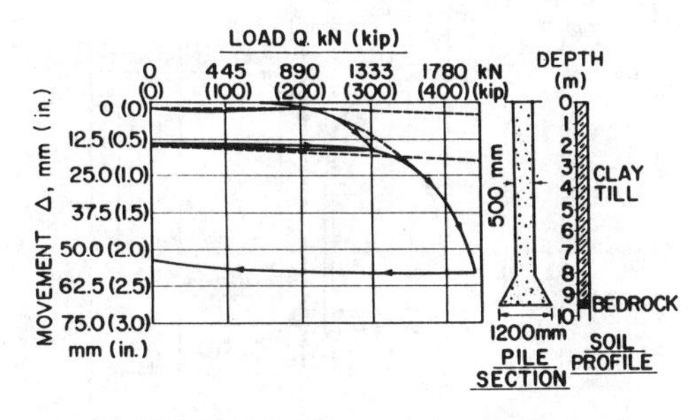

Fig.12 Axial compression pile load test (TP-2)

Table 3. Failure loads interpreted by various methods on a concrete driven pile.

Method	Failure Load, P_u
Davisson's	262 kips
Butler & Hoy	370 kips
De Beer	372 kips
Fuller & Hoy	406 kips
Brinch Hansen's 90% criterion	410 kips
Vander Veen	410 kips
Mazurkiewicz	416 kips
Brinch Hansen's 80% criterion	422 kips
Chin	470 kips

Note: The load-movement curve indicated that the test failure load was at around 412 kips.

Table 4. Failure loads interpreted by various methods for bored and belled concrete piles.

Method	Failure Load, P_u TP-1	TP-2
De Beer	103 kips	182 kips
Brinch Hansen's 90% criterion	-	250 kips
Davisson	138 kips	304 kips
Butler & Hoy	162 kips	356 kips
Fuller & Hoy	315 kips	425 kips
Vander Veen	404 kips	428 kips
Mazurkiewicz	-	-
Brinch Hansen's 80% criterion	448 kips	452 kips
Chin	484 kips	511 kips

Note: TP-1 test showed failure at approximately 355 kips and TP-2 test showed failure at approximately 430 kips.

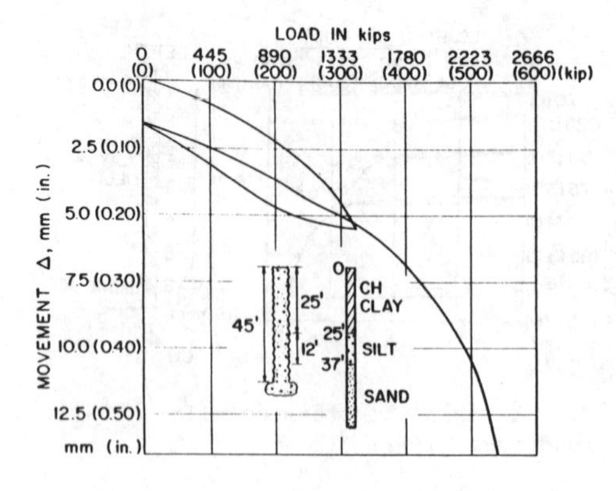

Fig.13 Load-movement curve for TP-3

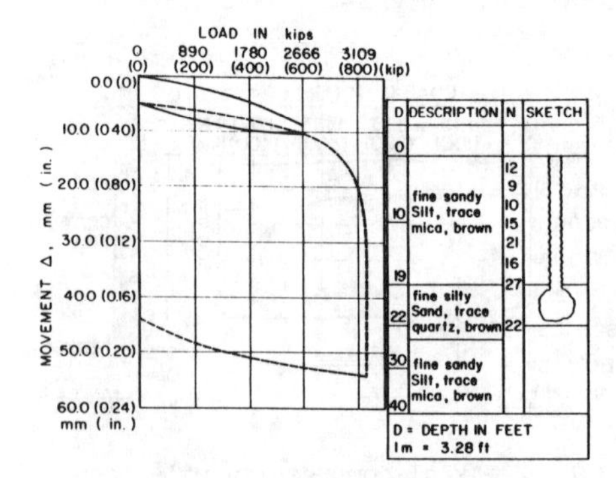

Fig.14 Load-movement curve for test pile TP-4

Table 5. Failure loads interpreted by various methods for expanded base compacted (Franki Type) piles

Method	Failure Load, P_u TP-3	TP-4
De Beer	223 kips	740 kips
Davisson	540 kips	650 kips
Butler & Hoy	530 kips	800 kips
Fuller & Hoy	540 kips	830 kips
Vander Veen	562 kips	787 kips
Mazurkiewicz	580 kips	845 kips
Brinch Hansen's 90% criterion	-	845 kips
Brinch Hansen's 80% criterion	750 kips	846 kips
Chin	783 kips	949 kips

Note: TP-3 test showed failure at approximately 540 kips and TP-4 test showed failure at approximately 840 kips.

movement curve. In general, it appears that the Davisson, Butler and Hoy and Fuller and Hoy methods provided failure loads in the vicinity of failure loads obtained from the field tests of piles TP-3 and TP-4. The Davisson method, however, interpreted a lower failure load for test pile TP-4.

5 APPLICATION OF INTERPRETATION METHODS FOR DRIVEN STEEL H PILES

Pile load test results on a steel H pile, TP-5, reported by Bethlehem Steel (1970) are presented in Figure 15. The pile was BP 10 x 57 lb and was driven through 7.5 ft (2.3 m) thick clay over 3 ft (1 m) of sand, underlain by 10.5 ft (3.2 m) thick soft clay over 6 ft (1.8 m) thick sand and gravel. The pile was driven to rock to a final resistance of 11 blows per inch with a No. 1 Vulcan steam hammer.

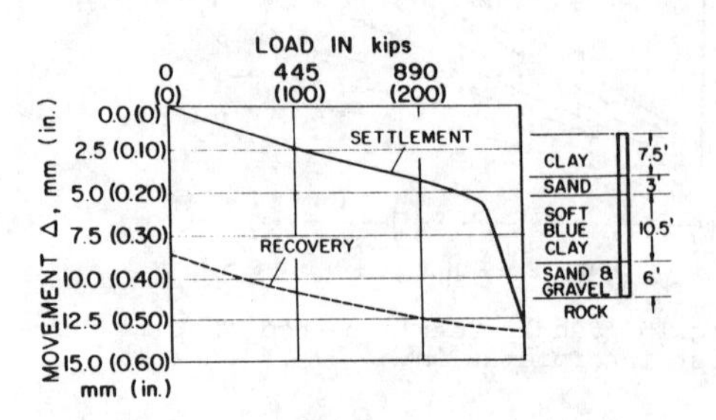

Fig.15 Load-movement curve for driven steel H-pile

Table 6 provides a summary of failure loads for TP-5 interpreted by various methods. This table indicates that the scatter in failure loads interpreted by various methods is smaller than obtained for the other piles (TP-1, TP-2, TP-3 and TP-4). Failure loads obtained by the De Beer, Vander Veen, Butler and Hoy and Davisson methods were lower than the test failure load of 280 kips (1245 kN). Brinch Hansen's 80% criterion, and the Chin and Mazurkiewicz methods gave failure loads higher than the test failure load. Brinch Hansen's 90% criterion and the Fuller and Hoy method predicted failure loads closer to the actual failure loads.

Table 6. Failure loads interpreted by various methods for driven steel H piles (TP-5).

Method	Failure Load, P_u
De Beer	247 kips
Vander Veen	258 kips
Butler & Hoy	268 kips
Davisson's	270 kips
Brinch Hansen's 90% criterion	276 kips
Fuller & Hoy	276 kips
Brinch Hansen's 80% criterion	288 kips
Chin	292 kips
Mazurkiewicz	300 kips

Note: TP-5 showed failure at test load of approximately 280 kips.

6 CONCLUSIONS

Based on nine different interpretation methods of five load-movement curves obtained using the SM test method, the following conclusions can be drawn.

1. The Davisson method always predicts conservative values of failure loads. The Chin method invariably yields failure loads that are higher than the actual failure loads. The Fuller and Hoy method always gives failure loads that are the best approximation of test or actual failure loads. The length to diameter ratio of the piles analyzed here varied between 12 to 32. This conclusion should be checked for longer piles.
2. For bored and belled concrete piles the De Beer method, Brinch Hansen's 90% criterion, as well as the Davisson, and Butler and Hoy methods yield very conservative failure loads. Brinch Hansen's 90% criterion and the Vander Veer and Chin methods yield too high failure loads. The Fuller and Hoy method provides a reasonable estimate of failure loads for these piles.
3. For expanded base compacted piles the De Beer method provides very conservative estimate of failure loads. It appears that the Davisson, Butler and Hoy, and Fuller and Hoy methods provide reasonable approximations for failure loads. Other methods, in general, yield too high failure loads.
4. For a driven H-pile the De Beer method provides the lowest estimate of failure load. Brinch Hansen's 80% criterion, and the Chin and Mazurkiewicz methods provide too high failure loads. Other methods provide reasonable approximations of failure loads; Brinch Hansen's 90% criterion and the Fuller and Hoy methods predict the failure loads similar to the failure test loads for this pile.
5. The above conclusions should be further checked by carrying out similar analyses for longer piles, other pile types, soil conditions and load test methods.

7 REFERENCES

ASTM Designation 1986. D 1143-81 Standard Method of Testing Piles Under Static Axial Compressive Load. Annual Book of ASTM Standards, Vol. 04.08, pages 239-254.

Bethlehem Steel 1970. Steel H. Piles. Bethlehem Steel Corporation. Bethlehem, Pa, January, pages 64.

Brinch Hansen, J. 1963. Discussion, Hyperbolic stress-strain response. Cohesive soils. ASCE, J. SMFD, Vol. 89, SM4, pp. 241-242.

Broms, B.B. 1972. Bearing Capacity of Cyclically Loaded Piles. Preliminary Report No. 44, Swedish Geotechnical Institute, Stockholm, Sweden.

Butler, H.D. & Hoy, H.E. 1977. Users manual for the Texas quick-load method for foundation load testing. Federal Highway Administration, Office of Development, Washington, 59 pp.

Chin, F.K. 1970. Estimation of the ultimate load of piles not carried to failure. Proc. 2nd Southeast Asian Conf. on Soil Engineering, pp. 81-90.

Chin, F.K. 1971. Discussion, Pile tests. Arkansas River Project. ASCE, J. SMFD, Vol. 97, SM6, pp. 930-932.

Davisson, M.T., 1972. High capacity piles. Proceedings, Lecture Series, Innovations in Foundation Construction, ASCE, Illinois Section, 52 pp.

De Beer, E.E. 1967. Proefondervindelijke bijdrage tot de studie van het grensdraag vermogen van zand onder funderingen op staal, Tijdshrift der Openbar Werken van Beigie Nos 6-67 and 1-, 4- 5-, 6-68.

De Beer, E.E. & Wallays, M., 1972. Franki piles with overexpanded bases. La Technique des Travaux, No. 333, 48 pp.
Fellenius, B.H. 1975. Test load of piles and new proof testing procedure. ASCE, Jr. of Geotechnical Engineering Division, Vol. 101, No. GT 9, Sept. pp. 855-869.
Fellenius, B.H. 1980. The analysis of results from routine pile load tests. Ground Engineering, September pp. 19-31.
Fuller, F.M. & Hoy, H.E. 1970. Pile load tests including quick-load test method conventional methods and interpretations. HRB 333, pp. 78-86.
Mazurkiewicz, B.K. 1972. Test loading of piles according to Polish regulations. Royal Sw Acad of Engng Sciences Comm on Pile Research. Report No. 35, Stockholm, 20 pp.
Mohan, D., Jain, G.S. & Jain, M.P. 1967. A new approach to load tests. Geotechnique Vol. 17, pp. 274-283.
New York DOT 1974. Static load test manual. N.Y. DOT Soil Mech. Bureau, Soil Control Procedure SCP4/74, 35 pp.
Nordlund, R.L. 1982. Dynamic formula for pressure injected footings. ASCE Geotechnical Engineering Division, Vol. 108, No. GT3, pp. 419-437.
Sharma, H.D., Sengupta, S. and Harron, G. 1984. Cast-in-place bored piles on soft rock under artesian pressures. Canadian Geotechnical Journal, Vol. 21, No. 4, pp. 684-697.
Swedish Pile Commission 1970. Recommendations for pile driving test and routine test loading of piles. Royal Sw. Acad. of Engng. Sciences. Comm. on Pile Research, Report No. 11, Stockholm, 35 pp.
Vander Veen, C. 1953. The bearing capacity of a pile. Proc. 3rd ICSMFE, Zurich, Vol. 2, pp. 84-90.
Weele, A.F. 1957. A method of separating the bearing capacity of a test pile into skin friction and point resistance. Proceedings, 4th International Conference on Soil Mechanics and Foundation Engineering, Vol. 2, London, England, pp. 76-80.
Whitaker, T. 1957. Experiments with model piles in groups. Geotechnique, London, England, Vol. 7, pp. 147-167.
Whitaker, T. 1963. The constant rate of penetration test for the determination of the ultimate bearing capacity of a pile. Proceedings, Institution of Civil Engineers, Vol. 26, London, England, pp. 119-123.
Whitaker, T. and Cooke, R.W. 1961. A new approach to pile testing. Proceedings, 5th International Conference on Soil Mechanics and Foundation Engineering, Vol. 2, Paris, France, pp. 171-176.

Soil improvement

Prediction and Performance in Geotechnical Engineering / Calgary / 17-19 June 1987

Prediction of bond between soil and reinforcement

G.W.E.Milligan & E.P.Palmeira
University of Oxford, UK

ABSTRACT: One important aspect of the design of reinforced soil structures concerns the interaction between soil and reinforcement under direct sliding and pull-out conditions. Values of bond for sheet and strip reinforcement materials, including geotextiles, lie within a reasonably narrow range and may be easily measured in appropriate tests. Suitable tests for pull out of polymer and metal grid reinforcements are more difficult to perform accurately. Methods of predicting results are compared with experimental values and the importance of the relationship between grid geometry and soil particle size established. The significance of the results for the design of reinforced soil structures is

1 INTRODUCTION

The use of soil reinforcement is now accepted practice in geotechnical construction; the technique may be used in retaining walls, embankments, road pavements, foundations etc. Many different types of reinforcing material have been used, including metal, polymer and glassfibre strips, geotextile sheets, and metal and polymer grids. Design methods generally follow similar principles to those applied to unreinforced soil construction, but a number of additional factors need to be considered. These involve in particular:

(1) failure of reinforcement, especially in the long term for which corrosion of metal and creep of polymer reinforcement may become significant;

(2) the interaction mechanisms by which stresses are transferred between soil and reinforcement and vice versa.

A typical example is shown in Figure 1. The stability of the retaining wall is confirmed by checking possible critical failure mechanisms; the tensile forces in the horizontal reinforcing elements increase the forces resisting the movement of the soil block under consideration. For mechanism 1, failure may occur if each reinforcing element either breaks at point A or pulls out of the ground along the length AB, whichever occurs first. For mechanism 2, the only effect of the reinforcement is to modify the sliding resistance along the length CD.

This paper is primarily concerned with th interaction between soil and reinforcement which resists the pull-out type of failure; it is analogous to the bond between concret and reinforcement in reinforced concrete, and is therefore referred to as bond. A clear distinction is made between bond and the resistance to failure along a plane containing reinforcement (such as CD), here referred to as direct sliding resistance. The latter is generally easy to evaluate an will be considered briefly first.

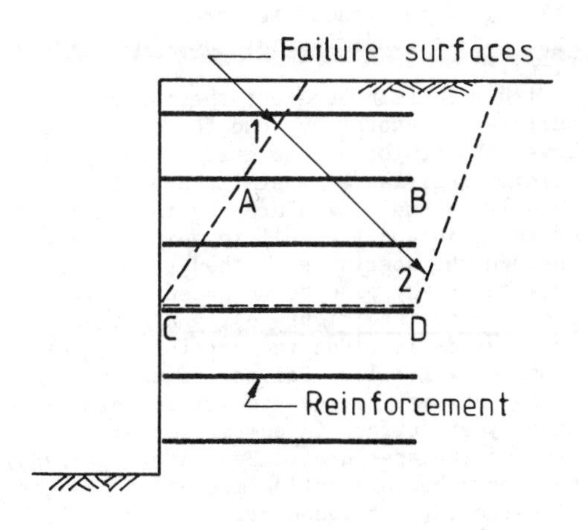

Fig. 1 Mechanisms of failure in a reinforced soil structure

2 DIRECT SLIDING RESISTANCE

The interaction mechanisms between soil and reinforcement in direct sliding are shown in Figure 2. With sheet reinforcement, relative movement is resisted by shear stresses at the interface given by:

under undrained conditions,
$\tau = a = \alpha c$ where $\alpha \leq 1$ (1)
under drained conditions
$\tau = \sigma_n \tan\delta = \sigma_n \beta \tan\phi$ where $\beta \leq 1$ (2)

With flat strip reinforcement, sliding along the reinforcement is controlled by Equations 1 and 2, and by the full shear strength of the soil between strips. If the strips are ribbed to improve bond, then sliding is resisted by a combination of shear on the horizontal surface and bearing against the rib. If sufficient resistance is generated against the ribs, it will be easier for the soil to fail along a plane outside the ribs, generating the full shear resistance of the soil.

Values of α and β are not easily predicted, but are easily measured by direct shear tests of soil over the surface of the reinforcement, as shown schematically in Figure 3; results of many such tests are reported in the literature. For normally smooth metal surfaces both α and β typically lie in the range 0.4 to 0.65. For geotextile sheets both α and β are normally in the range 0.7 to 1.0. However β may be as low as 0.4, the value varying with stress level, type of geotextile, size and shape of soil particles and the degree of saturation of the soil. For any particular application it is best to assume a reasonable value for preliminary design and then measure the value in direct shear tests under appropriate conditions at the final design stage.

With grid reinforcement the failure surface will normally tend to form at the level of the top of the grid members, with sliding resistance generated over the relatively small plan area of the material of the grid and the soil to soil contact through the apertures of the grid. This resistance may also be measured in direct shear tests (see Figure 3), and for most metal grids is close to that for the soil along ($\alpha = \beta = 1$). Polymer grids such as Tensar SR2 may have a significant plan area of polymer surface in comparison with the area of the apertures. Jewell et al (1984) show how the interaction mechanism changes with the ratio between grid aperture width and soil D_{50} size. Very large soil

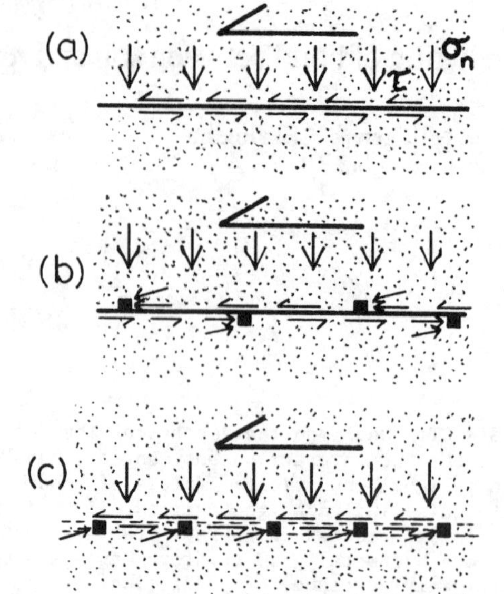

Fig. 2 Interaction in direct sliding:
(a) sheet or strip reinforcement
(b) ribbed strip reinforcement
(c) grid reinforcement

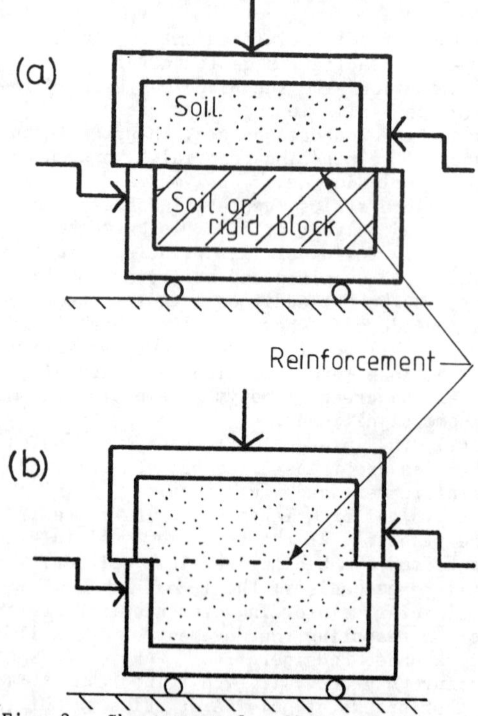

Fig. 3. Shear test for direct sliding resistance:
(a) sheet or strip reinforcement
(b) grid reinforcement

particles may not be able to lock into the grid apertures, and will tend to slide over the polymer surface, with a value of β of about 0.6. When the aperture width to D_{50} ratio is about 3 to 30 the soil particles interlock effectively with the grid and β is in the range 0.9 to 1.0. With fine-grained soils the failure surface has more freedom to follow the surface of the polymer and β falls to about 0.8.

3 BOND RESISTANCE

Bond is also generated by friction, bearing and adhesion between soil and reinforcement, but the relative movements are different from those occurring in direct sliding (see Figure 4). Nevertheless, for the soil on either side of strip or sheet reinforcement the interaction between soil and reinforcement is the same as in direct sliding, and bond forces may be determined from the same values of α or β. These values may alternatively be measured directly in pull out tests (see Figure 5). The results of such tests may be significantly affected by boundary conditions; however, provided these are properly controlled or accounted for, the results obtained from direct shear and pull out tests agree well.

With grids, bond is induced to a small extent by shear along the surface of the longitudinal members of the grid, but mainly by bearing of transverse members of the grid against the soil (Figure 4). No soil to soil resistance is developed as there is no relative movement of the soil on either side of the grid. This mechanism cannot be produced in a direct shear test, and bond for grids may only be measured in pull out tests; the different mechanisms occurring in the various situations have been clearly demonstrated by Dyer (1986) using a photelastic technique, while the importance of boundary and other effects in pull out tests is shown by Palmeira and Milligan (1987).

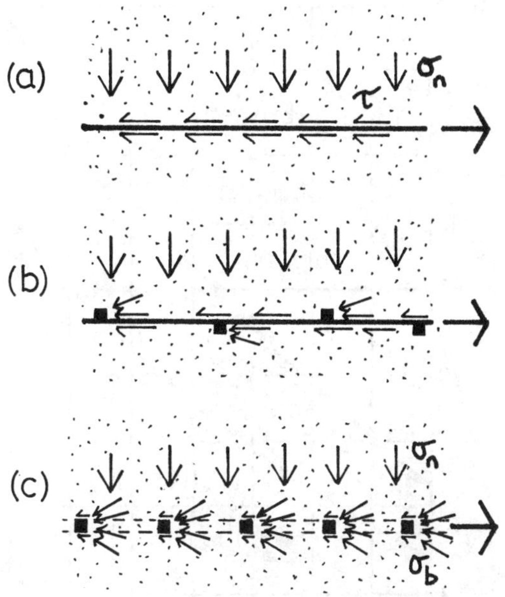

Fig. 4 Interaction in bond:
(a) sheet or strip reinforcement
(b) ribbed strip reinforcement
(c) grid reinforcement

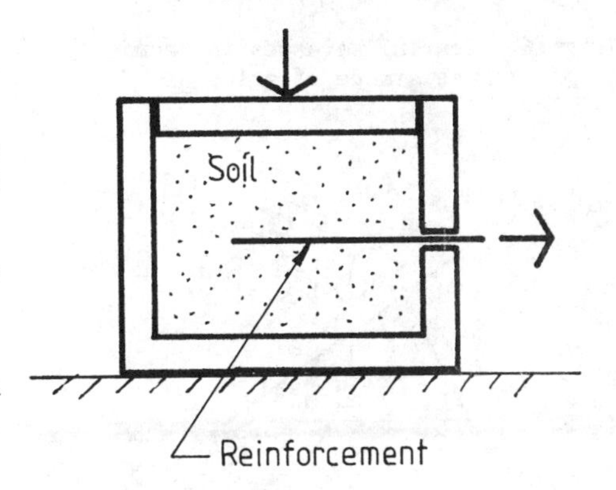

Fig. 5 Pull out test, for all types of reinforcement

3.1 Bond under drained conditions

Because of the difficulty of performing accurate pull out tests on real grid reinforcement, Jewell et al (1984) put forward a method of calculating bond for grids under drained conditions, based on the bearing stresses acting on the transverse members. Two curves are plotted relating bearing stress σ_b to the normal stress on the plane of the reinforcement σ_n as a function of soil friction angle (see Figure 6). These curves appear to provide upper and lower bounds to experimental data from tests on grids and anchors (see Jewell et al 1984 and Palmeira and Milligan 1987), the results of finite element analyses by Rowe and Davies (1982), and the empirical relationship suggested by Jones (1984) for the anchors in anchored earth. The lower curve is obtained from the relationship

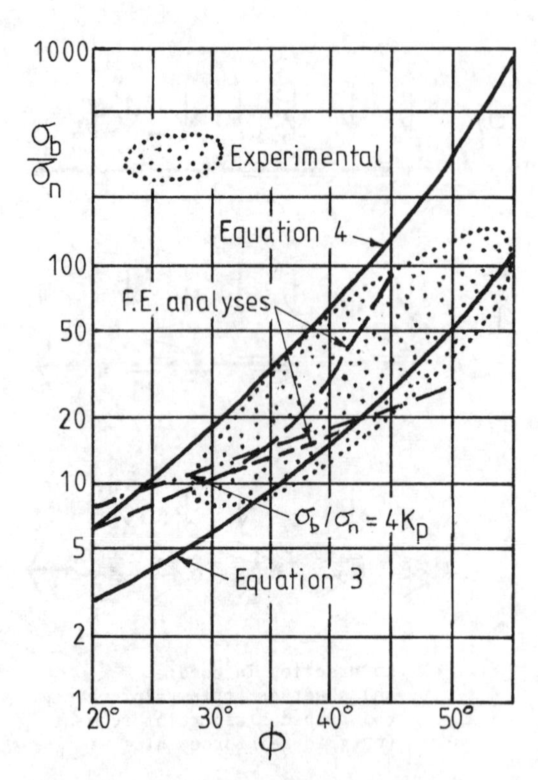

Fig. 6 Bearing stresses for bond resistance of grids

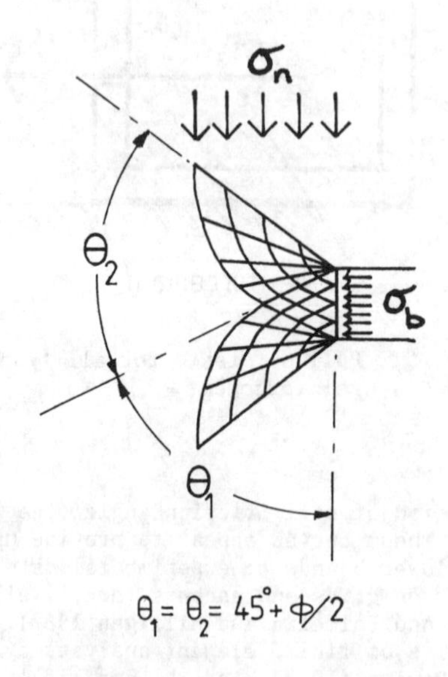

Fig. 7 Stress characteristics for calculation of bearing stresses

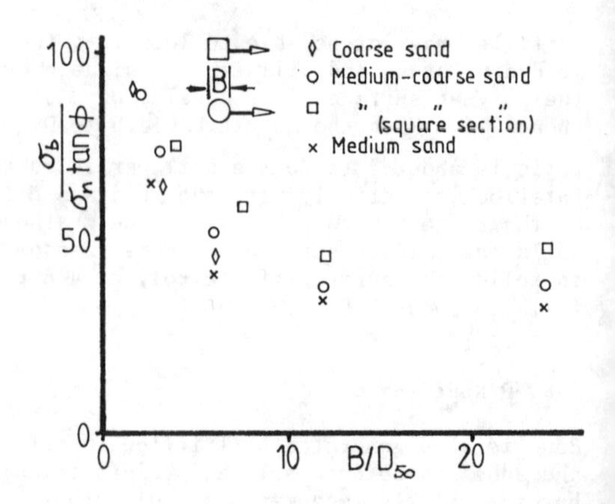

Fig. 8 Variation of bearing stress with B/D_{50} ratio

$$\frac{\sigma_b}{\sigma_n} = e^{(\frac{\pi}{2} + \phi)\tan\phi} \tan(\frac{\pi}{4} + \frac{\phi}{2}) \qquad (3)$$

which is based on the stress characteristics shown in Figure 7. The upper curve is obtained from the normal characteristic field for a foundation rotated to the horizontal, giving the relationship

$$\frac{\sigma_b}{\sigma_n} = e^{\pi \tan\phi} \tan^2(\frac{\pi}{4} + \frac{\phi}{2}) \qquad (4)$$

These expressions assume that the soil may be treated as a continuum, but for many grids in coarse-grained soils the diameter B of the grid transverse members may be relatively small in relation to the average particle size of the soil; in this case the assumption of a continuum is obviously not valid. Figure 8 presents data from Palmeira and Milligan (1987) showing that the ratio σ_b/σ_n increases significantly when B/D_{50} is less than about 15. This effect may explain some of the scatter of reported experimental results, and suggests that the lower bound curve given by Equation 3 may provide safe design values when B/D_{50} exceeds 15.

This analysis is appropriate for a single bearing member. A grid presents a succession of bearing members to the soil and if they are close together they will interact with each other. Figure 9 defines a degree of interference which for relatively large numbers of grid members (n) decreases approximately linearly from unity when the spacing, S, is zero (grid members coinciding) to zero when S/B is about 50.

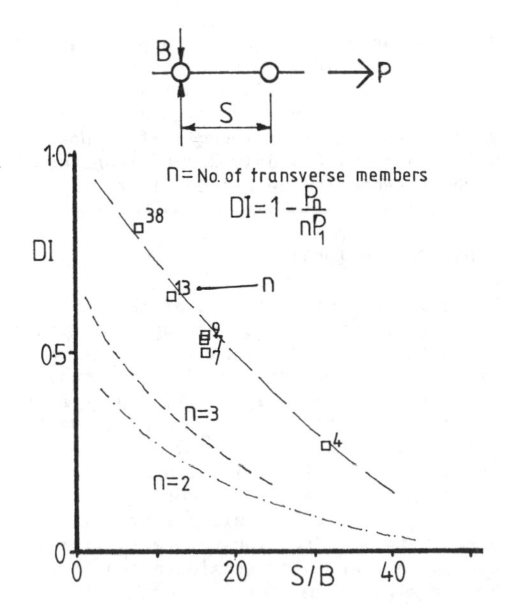

Fig. 9 Effect of interference between grid transverse members

These results were obtained with uniform sands, and it is possible that different results would be obtained with very coarse grained or well graded soils. However they suggest that for most grids in use at present, which have S/B much less than 50, the predicted bond behaviour will be markedly affected by interference.

If the average soil particle size is very large compared with the grid apertures, interlock of the soil into the grid may not occur and bond will be much lower. Milligan (1982) compared the relative performance of a number of grids with smooth and rough solid sheets in direct shear tests on reinforced soil samples (see Figure 10).

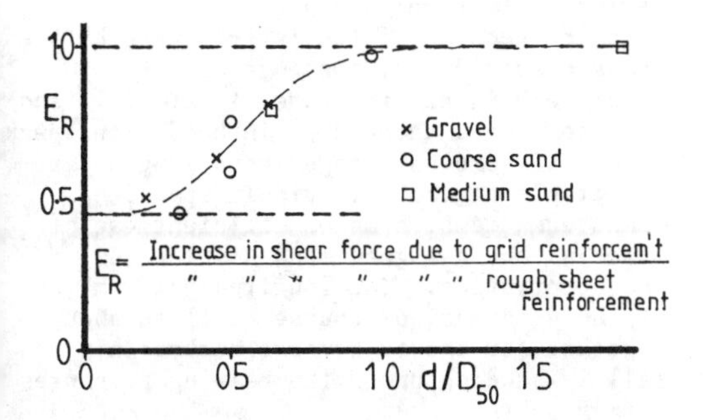

Fig. 10 Bond between grids and soil with large particles

With the grid aperture size d about equal to the soil D_{50} size there appeared to be sufficient interlock for the grid to perform as a fully rough sheet, but as the aperture size became relatively smaller the performance approached that of a smooth sheet.

3.2 Bond under undrained conditions

Similar arguments may be applied to the reinforcement of clay under undrained conditions, except that adhesion will replace friction in the calculation of sliding shear, and bond strength in bearing will depend on the undrained shear strength (Ingold 1984)

$$\sigma_b = N_c \, c \qquad (5)$$

where c is the undrained shear strength and N_c a bearing capacity factor. For appropriate values of N_c reference may be made to Randolph and Houlsby (1984); bearing capacity factors for a deeply buried cylindrical member vary with adhesion, a, from about 9 to 12, with about 10.5 for typical values of a/c. An alternative calculation based on cavity expansion suggests that in stiff compacted clay fills a value of N_c of about 7.5 may be appropriate. The limited experimental evidence available seems to support these figures and indicates that interference is not a significant factor for grids of typical geometries with $S/B > 5$.

3.3 Stress-strain behaviour

The analyses above give no indication of the strains necessary to develope the predicted bond stresses. These strains may impose a design limitation if distortions of the structure are not to be excessive when extensible reinforcing materials are used. Figure 11 shows some typical load-displacement curves for pull out tests using 500 mm long by 1000 mm wide samples of full-scale materials of various types under different normal stresses (Palmeira 1987). The steel grids gave a very stiff response under both low and high stresses, while the stiff geotextile reached its limiting bond stress at a displacement of about 15 mm. The weak and extensible geotextile could only be tested at low stresses without failing in tension before developing its limiting bond stresses (see Figure 11a). These were reached at a displacement of

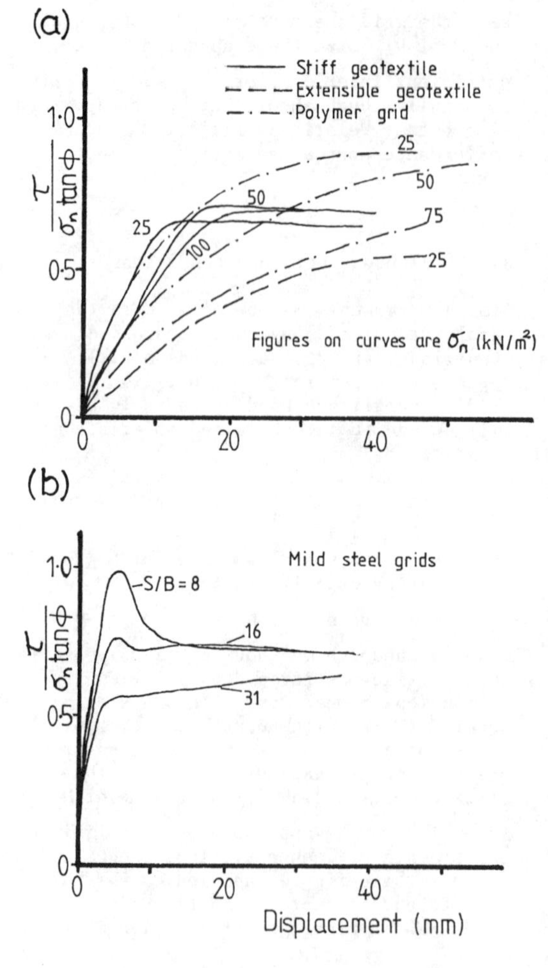

Fig. 11 Stress-displacement plots for pull out tests.

about 40 mm. The lower limit compared with the stiffer geotextile may be a function of different surface properties, but progressive failure is almost certainly a contributory factor. By the time bond has been mobilised along the length of the sample the soil has been strained past its peak shear strength at the leading end; at no stage is the peak strength mobilised at all points along the length of the reinforcement. Bond prediction for such material should be based for safety on critical state values of friction angle for the soil. The polymer grid exhibited a stiff response at the low stress level, but was relatively extensible at the loads induced at the higher stresses. All the materials shown here failed in bond in a ductile manner which would be preferable to sudden failure due to reinforcement breaking. Steel grids with transverse members relatively close together showed a marked drop in bond strength after peak, while those with widely spaced transverse members showed no peak (see Figure 11b)

4 DESIGN CONSIDERATIONS

The principal significance of the above analyses is that the bond length necessary to mobilise the full tensile strength of geotextile and polymer grid reinforcement is found to be relatively short; typically in the range 0.1 to 1.3 m for reinforcement with working loads of 5 to 50 KN/m at 2-3 m depth in a granular soil with an internal friction angle of 35°. Under such circumstances, the reinforcement will usually fail in tension before failing in bond. However, bond considerations may become significant for reinforcement at shallow depths, in lower quality soils, and with very strong metal grid or thin strip reinforcement. With very extensible reinforcement large strains will have to be developed in the structure for the reinforcement to become effective, and appropriate critical state soil parameters should be used in design.

In some situations, perhaps for instance in designing against extreme earthquake loading, it may be considered preferable to ensure that failure always occurs first by limited ductile failure in bond rather than by overstressing of the reinforcement. Strong reinforcement with deliberately low bond, such as metal or polymer bars or strips, would then be used.

On the other hand, the reinforcement with the best bond capability is a 'rough' geotextile sheet or a grid with relatively thin ($<15D_{50}$) transverse members sufficiently closely spaced for the grid to perform effectively as a rough sheet. However, because of the interference between transverse members, these grids make relatively inefficient use of material, and seem to show a sharp drop in bond after peak load. If somewhat longer bond lengths are acceptable a grid with widely spaced transverse members makes efficient use of material and has good load-displacement characteristics. The longitudinal members of the grid must of course be close enough together for the transverse members not to fail in bending under the bearing pressures acting on them.

ACKNOLEDGEMENTS

E.M. Palmeira is a research student at Oxford University, supported by the Brazilian government. Apparatus for his experimental work was constructed with the aid of grants from the Department of Engineering Science and Science and Engineering Research Council.

REFERENCES

Dyer, M.R. 1985. Observations of the stress distribution in crushed glass with applications to soil reinforcement. D. Phil. thesis, University of Oxford.

Ingold, T.S. 1984. Discussion on session on Introduction to polymer grids. Proc. Symp. on Polymer grid reinforcement in civil engineering. London: Thomas Telford.

Jewell, R.A, Milligan, G.W.E., Sarsby, R.W. and Dubois, D.D. 1984. Interaction between soil and geogrids. Proc. Symp. on Polymer grid reinforcement in civil engineering London: Thomas Telford.

Jones, C.J.F.P. 1984. Earth reinforcement and soil structures, p.78. London: Butterworths.

Milligan, G.W.E, 1982. Some tests on the relative effectiveness of grid reinforcements for granular soils. University of Oxford, Dept. of Engineering Science, O.U.E.L. report No.1441/82.

Palmeira, E.M. 1987. D.Phil. thesis, to be submitted to the University of Oxford in 1987.

Palmeira, E.M. and Milligan, G.W.E. 1987. Scale and other factors affecting the results of pull-out tests of grids buried in sand. University of Oxford, Dept. of Engineering Science, O.U.E.L. report.

Randolph, M.F. and Houlsby, G.T. 1984. The limiting pressure on a circular pile loaded laterally in cohesive soil. Geotechnique 34, No.4: 613-623.

Rowe, R.K. and Davis, E.H. 1982. The behaviour of anchor plates in sand. Geotechnique 32, No.1: 25-41.

Inextensible geomesh included in sand and clay

R.R.Al-Omari, H.H.Al-Dobaissi & B.A.Al-Wadood
Building Research Centre, Baghdad, Iraq

ABSTRACT: The behaviour of geomesh reinforced soil is strongly influenced by the interlocking mechanism between soil particles and the apertures of the mesh. Triaxial compression tests are performed on 100 mm diameter sand samples using either perforated steel discs or plastic geomesh as reinforcing materials. Perforations with constant diameter are made systematically into each steel disc. The perforation diameter ranges from 3 mm to 14.5 mm. The area of holes to the area of solid is kept constant in each disc so that the only variable in the tests will be the ratio of the aperture size of the steel mesh to the grain size of the sand. The results have clearly indicated that the improvement in strength reachs an optimum at a certain aperture size after which a sharp fall occurs. Tests using strong plastic mesh indicated that the strength of reinforced samples is slightly affected by the value of sand density. Triaxial tests are also conducted on 50 mm diameter clay samples, in the undrained condition, with the inclusion of layers of plastic mesh. Although, past published results showed a large decrease in the strength when metal foil was used, the present results are encouraging.

1 INTRODUCTION

1.1 Geomesh reinforced sand

Materials like metal strips or geotextiles have been widely used as inclusions in the construction of different soil structures. McGown et al (1978) defined the term "relatively inextensible inclusions" as those having in-situ rupture strains less than the maximum tensile strains in the soil without inclusions, under the same operational stress condition. If such inclusions are suitably located within the sand mass, their effect will be to strengthen or reinforce the sand assuming the friction between the soil and the inclusions is adequate.

Ingold and Miller (1983) pointed out that the improvement in strength is due to lateral deformation control, arising from shear stresses mobilized on the soil-reinforcement interface. A rough reinforcement would restrict the lateral deformations causing large shear stresses to mobilize which increase the strength of the loaded mass.

In the case of a geomesh reinforced sand, the inhibition to lateral deformations would mainly be due to the interlocking between the sand particles and the apertures of the mesh. The strength enhancement would be affected by the mechanism of interlocking which is in turn affected by the ratio of aperture diameter of the mesh ,T, to the diameter of sand particles , D .

Milligan (1982)and McGown et al (1985) realized the effect of the ratio T/D on the interlocking between sand and reinforcement, nevertheless, they did not investigate this effect.

There are special charts ,made by some manufacturers, which can be used in the design of geomesh and geogrid reinforced soil structures. However, these charts are based on their products and they do not specify the grain size or the relative density of the fill. This presses the design engineers to condition their design on a certain product and introduces points of uncertainty in the design.

The triaxial apparatus has been used by many authors to investigate the behaviour of reinforced sand. A description of the previously conducted work is **reported** in Ingold (1982). Values of the improvement factors obtained from the triaxial tests may not quantitatively correspond to the improvement factors in the full scale reinforced soil structures. However, these factors can help to understand the effects of the inclusion type on the sand behaviour and to predict the quality of modifications expected in the full scale soil structures.

1.2 Geomesh reinforced clay

Cohesive soils are yet not utilized in reinforced soil construction for reasons which are explained in Jewell and Jones (1981). Ingold and Miller (1983) reported that at present, the soil backfill used to construct a reinforced soil structure is almost without exception a granular material.

Ingold and Miller (1982) investigated the strength behaviour of clay samples into which discs of either aluminium foil or porous plastic were inserted. The samples were consolidated and then cut into a number of thick discs of equal height. Each sample was then reassembled with a disc of inclusion being introduced between soil discs, thereby producing a multi-reinforced sample. The reassembled samples were tested either unconfined unconsolidated undrained or triaxial consolidated undrained. The results indicated that a serious reduction occurs in the strength of clay samples due to the insertion of the "reinforcing" materials.

However, Jewell and Jones (1981) reported an improvement in the strength of overconsolidated kaolin clay tested undrained in the shear box and reinforced using a single plane of grid oriented at 45 from the vertical direction. The grid was introduced into the slurry and the clay was then consolidated.

1.3 The present work

In the present work the effect of varying the ratio T/D on the modifications in sand properties and the strength behaviour of plastic mesh reinforced sand are investigated. A study of the undrained strength of mesh reinforced clay is also presented. The investigation is carried out using the triaxial apparatus.

2 SOIL PROPERTIES

2.1 Sand

The sand used is sorted out from Karbala sand deposits located at the western part of Iraq. It has a particle size ranging from 0.425 mm to 1.18 mm with uniformity and curvature coefficients of 1.69 and 0.89 respectively. It is uniform containing more than 60% of coarse fraction of sand size. The particle size such that 50% of the particles are smaller than that size, D_{50} , is 0.74 mm. The value of the specific gravity is 2.75. The maximum and minimum porosities are 45% and 35% respectively.

2.2 Clay

The clay used is kaolin powder of liquid and plastic limits of 35% and 27% respectively. All the material passes the 75 μm BS sieve and it has a clay fraction of 30%. The effective shear strength parameters C' and ϕ' , determined using the triaxial tests, are 17 kN/m^2 and 30° respectively.

3 REINFORCEMENT PROPERTIES

3.1 Sand reinforcement

To investigate the effect of varying the ratio T/D on the improvement in strength, an inextensible (strong) reinforcement is chosen so that when inserted in a sample it will cause a significant enhancement. Steel discs, 2 mm thick, were used. The coefficient of interface friction between Karbala sand and the steel was about 75% of the sand alone. Perforations with constant diameter, T, were made systematically into each steel disc as shown in Fig.1. The diameter values used were 0.0,3.0,5.0,10.0,and14.5 mm. A value of T larger than 14.5 mm was not used due to the limited scale of a triaxial specimen. The perforation centres were located on the perimeter of circles drawn inside the disc. The diameter of the smaller circle is C=2(T+W) where W is the clear spacing between each two perforations. The diameters of the next circles are 2C,3C,..etc. If the perforation diameter of a disc is T,and the corresponding diameter of a second disc is kT ,then the number of perforations made in the second disc is X/k^2 wher X is the number of perforations in the first disc. This ensures that the area of perforations to the total area of the disc is kept constant (about 35%) in each disc so that the only variable in a test will be the ratio of aperture (perforation) size of the steel mesh to the grain size of the sand.

A strong plastic mesh ,Netlon CE 121, with aperture size of 8x6 mm was also used in the investigation. It has a maximum tensile strength of 7.68 kN/m and the extension at maximum load is 20.2%. The extension at 50% of maximum

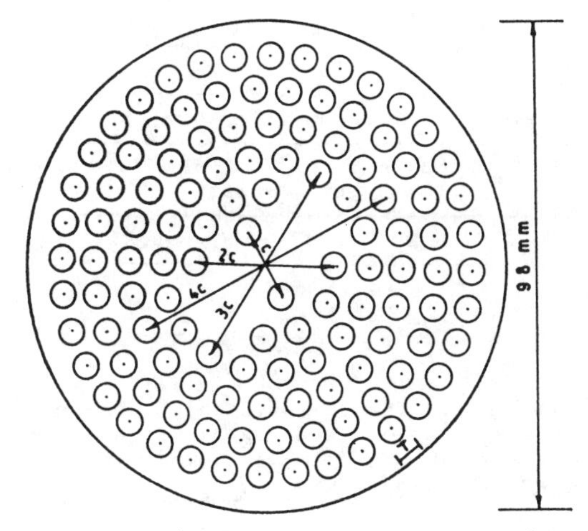

Figure 1. Typical design of a steel mesh (T = 5 mm).

load is 3.2%.

3.2 Clay reinforcement

The material used to reinforce the clay is Netlon CE 121.

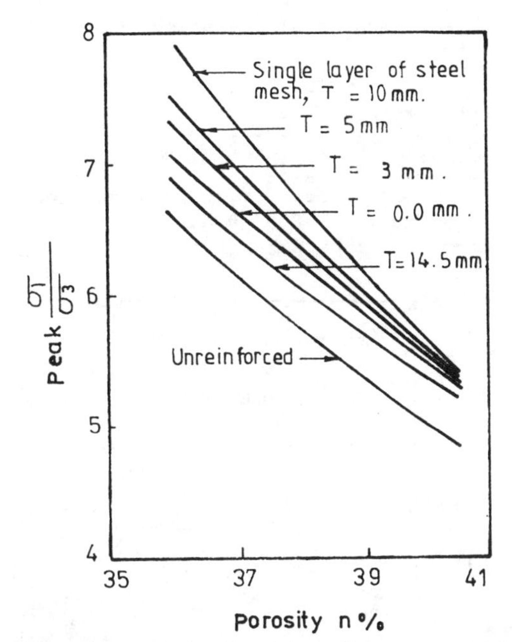

Figure 2. Relationships between peak stress ratio and sand porosity using a single layer of steel mesh.

4 TESTS ON REINFORCED SAND

Drained triaxial tests on 100 mm diameter x 200 mm high dry unreinforced and reinforced samples were performed using different sand porosities for each individual case. The range of porosities was obtained using a vibrating table. The cell pressure for all the tests was 138 kN/m² and the rate of strain was 1.3 mm/min.

The relationships between the sand porosity, n, and the peak stress ratio, σ_1/σ_3, for the sand alone and for each of the samples reinforced using a single layer of steel mesh are plotted in Fig.2 where σ_1 and σ_3 are the vertical and horizontal boundary stresses respectively. Smooth curves are fitted using an exponential regression analysis. Table 1 presents the number of points (tests) used to obtain each relationship and the corresponding determinating coefficient, r^2, of the regression analysis.

The inclusion of the unperforated steel disc (T=0.0) improved the strength. However, the results indicated that the perforations made in the discs caused alterations in the strength behaviour depending on the diameter, T, of the perforations. The alteration is a further improvement when T is less than 14.5 mm and a reduction in the improvement when T is equal to 14.5 mm.

To highlight the effect of T/D on strength improvement, the relationship between T/D_{50} and peak σ_1/σ_3 is plotted in Fig.3, for the practical range of porosities, using values from the fitted curves. The strength gradually increased with increasing T/D_{50} and the maximum occured at T/D_{50} = 13.5 . Beyond this value the strength dropped. Additional tests using two layers of reinforcing discs with the spacing between them equal to one third the initial hight of the sample were conducted. Tests using T/D_{50} values of 0.0, 4.0, 6.7, 13.5, and 19.5 were carried out and the corresponding porosities were 0.392, 0.389, 0.394, 0.391, and 0.388 respectively. Thus, the average porosity is 0.391 and the maximum deviation from the average is 0.003 . The results, drawn in Fig.3, confirmed the same behaviour. Actually, the slope of the curve between T/D_{50} = 0.0 and T/D_{50} = 13.5 is expected to be very high if the coefficient of interface friction between the mesh material and the sand is lower than that of the present case, and the area of holes to the total area is larger.

The relationships between stress ratio and axial strain for the sand alone and for the sand reinforced using two layers of steel mesh are demonstrated in Fig.4 .

Table 1. Number of tests and values of r^2.

Single layer of steel reinforcement											
Unreinforced		T= 0.0 mm		T= 3.0 mm		T= 5.0 mm		T= 10.0 mm		T= 14.5 mm	
No. of tests	r^2	No. of tests	r^2	No. of tests	r^2	No. of tests	r^2	No. of tests	r^2	No. of tests	r^2
9	0.9	8	0.66	8	0.81	7	0.76	8	0.87	7	0.71

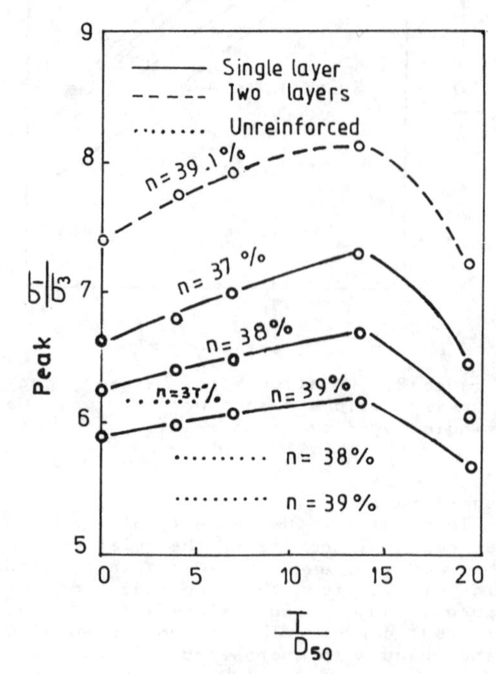

Figure 3. Variation of peak stress ratio with the dimensionless factor T/D_{50}

The relationships between the peak stress ratio and the porosity obtained from tests on plastic mesh reinforced samples are plotted in Fig.5. The results indicated that the strength of the plastic mesh reinforced samples is less dependent on the sand porosity than the strength of the steel mesh reinforced samples. This may be attributed to the large interlocking bond developed between the sand and the mesh ($T/D_{50} \simeq 9.5$) which caused the failure to be characterized by the yield of the mesh itself, and for any porosity tested. Actually, the yield was noticed after each test in the middle of the mesh. These findings are important in construction as it shows that for a certain required strength, the employment of such materials saves the efforts to compact the sand to high densities as far as the strength is concerned. The relationship between stress ratio and axial strain is compared in Fig.4 with those obtaind using the steel mesh.

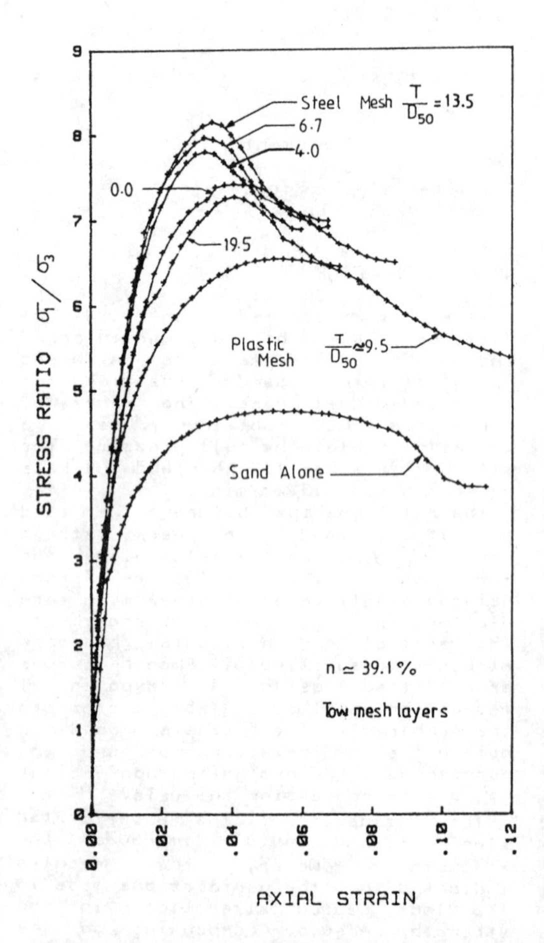

Figure 4. Relationships between stress ratio and axial strain for reinforced and unreinforced sand samples.

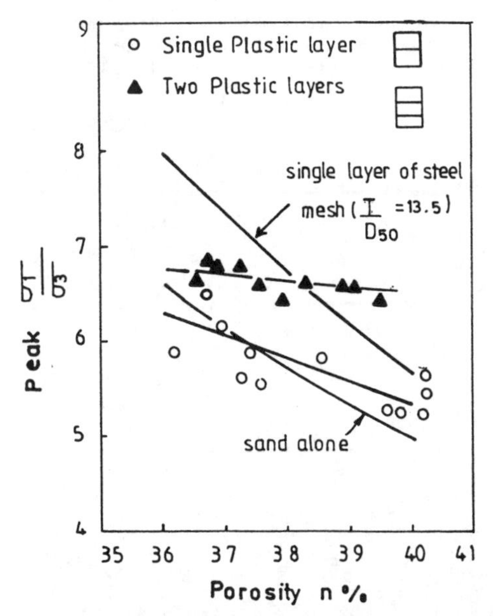

Figure 5. Relationships between peak stress ratio and sand porosity using a single and two layers of plastic mesh.

5 TESTS ON REINFORCED CLAY

Samples of 50 mm diameter were consolidated from a slurry. The intial consolidation was performed under zero lateral strain condition, in a triaxial mould, using a maximum vertical stress of 95 kN/m² . Each sample was then isotropically consolidated at a certain cell pressure which was reduced to its third value prior to shear in order to maintain the overconsolidation ratio at a constant selected value of 3 . In both stages the sample was left till no further change in the volume was noticed and the pore pressure dissipated. In the reinforced samples, the reinforcement was introduced into the slurry and the intial consolidation was then carried out.

The main object was to find out the effect of including strong mesh in the clay on its undrained strength. Four series of consolidated undrained tests were conducted using 50 mm diameter × 100 mm high samples. In each series the cell pressures prior to shear were 67,133,200,267,and 333 kN/m². The clay was unreinforced in the first series whereas a single layer ,two layers,and three layers of mesh were used in the second, third , and fourth series respectively.

The test results are expressed in terms of a strength ratio, S, which is defined as the ratio of the deviator stress of the reinforced sample to the deviator stress of the unreinforced sample.

The results are presented in Table 2. The mesh has generally improved the strength and the value of S is modified by increasing the number of reinforcing layers. The results are encouraging when compared with the results of Ingold and Miller (1982). According to their results the number of porous plastic layers in a kaolin sample, tested unconfined, has to be increased to 6 in order to restore the undrained strength to its unreinforced value. The drop in strength was even larger in the case of aluminium foil reinforcement using the unconfined test and the consolidated undrained triaxial test.

The deviator stress-axial strain curves of samples sheared at a cell pressure of 67 kN/m² are shown in Fig.6.

Table 2. Values of strength ratio,S.

No. of layers	Cell pressure kN/m²				
	67	133	200	267	333
1	1.23	1.04	0.97	1.23	0.94
2	1.63	1.21	1.25	1.07	1.23
3	1.98	1.70	1.38	1.22	1.51

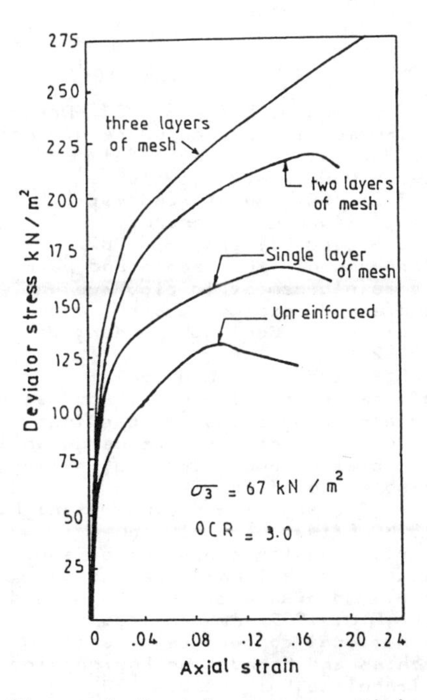

Figure 6. Deviator stress-axial strain curves of reinforced and unreinforced clay samples.

6 CONCLUSIONS

1. The strength enhancement in mesh reinforced sand depends on the mechanism of interlocking which is in turn affected by the ratio of aperture size,T, to particle diameter, D.
2. The strength of mesh reinforced sand increases to a maximum at $T/D_{50} \simeq$ 13.5 and then drops beyond this value.
3. The use of plastic mesh with the appropriate stiffness and the appropriate aperture size makes the strength of the sand approximately independent of the sand density.
4. The existance of layers of strong mesh in slightly overconsolidated clay improves the undrained strength and the improvement depends on the number of layers.

ACKNOWLEDGEMENTS

The work is conducted in the Building Research Centre of the Scientific Research Council, Iraq. The results is published according to the permission of the Director General of the centre. The authors acknowledge the help of Mr. Imad AL-Obaidi and Mr. Yahya Nazhat. The assistance of Ms. Asma Hashim, laboratory assistant, is gratefully acknowledged.

REFERENCES

Ingold,T.S.,and Miller,K.S.1983. Drained axisymmetric loading of reinforced clay. Journal of Geotechnical division,ASCE, No.7,Vol. 109 .

Ingold,T.S. 1982. Reinforced earth. Thomas Telford Ltd, London.

Ingold,T.S.,and Miller,K.S. 1982 .The performance of impermeable and permeable reinforcement in clay subject to undrained loading .Quarter Journal of Engineering Geology, London, Vol. 15:201-208 .

Jewell,R.A. and Jones,C,J.1981. Reinforcement of clay soils and waste materials using grids. Proceedings of 10th International conference on soil Mechanics and Foundation Engineering, Stockholm, Vol.3:701-706 .

McGown,A.,Andrawes,K.Z.and AL-Hasani,M. M. 1978. Effect of inclusion properties on the behaviour of sand. Geotechnique, 28, No.3:327-346 .

McGown,A.,Andrawes,K.Z.,Hytiris,N.,and Mercer,F.B. 1985. Proceedings of 11th international conference on soil Mechanics and Foundation Engineering, San francisco,Vol.3:1735-1738 .

Milligan,G.W.1982. Some scale model tests to investigate the use of reinforcement to improve the performance of fill on soft soil. Quarter Journal of Engineering Geology, London, Vol.15:209-215 .

Prediction and Performance in Geotechnical Engineering / Calgary / 17-19 June 1987

Finite element study of pilot test at Changi Airport

S.A.Tan, G.P.Karunaratne, S.L.Lee & V.Choa
National University of Singapore, Kent Ridge

ABSTRACT : The Changi International Airport in Singapore is constructed on a 645 hectare hydraulic sandfill reclaimed from the sea. Prior to its construction, a pilot test was conducted to study the performance of Geodrains in accelerating consolidation of the underlying thick marine clays. This paper describes a detailed finite element study of a test area GD3 (Geodrain spaced at 2.1 m x 2.1 m) using a CAM-Clay constitutive model and a 2D, plane strain finite element consolidation program. The effects of vertical drains were approximated by a soil with an enhanced vertical permeability. Soil parameters for the model were deduced from available boreholes and laboratory data, with the appropriate initial conditions and parameters calibrated to give consistent deformation compared to field behavior. Results of analysis show reasonable agreement with field performance overall. Pore pressure profile across the surcharge width demonstrate the back pressure effect from the pore pressure in the untreated reclamation which is still underconsolidated.

1 INTRODUCTION

A major reclamation project was undertaken by the Port of Singapore Authority in 1976. A total of 645 hectares of land was reclaimed for the construction of the new international airport at Changi in the Republic of Singapore. Soil investigation prior to the reclamation indicated that the second runway is partly on hard ground of sandstone and shale and partly on thick deposits of soft marine clay. The marine clay exists in depths up to about 40 m below the sea bed and is deposited mainly in deep channels. Long term settlements caused by the 6.5 m of hydraulic fill and the runway pavement load were estimated to be up to 2 m. In order to minimise the problem of uneven settlement and heavy maintainence on the runway, soil treatment was necessary for the runway and its associated high speed turnoffs and taxiways (Choa et al, 1979, 1981).

2 PILOT TEST AREA

A pilot test was carried out to compare the performance and feasibility of several soil improvement techniques such as vertical sand drains, prefabricated drains, surcharge and dynamic consolidation. Detailed description of the pilot test has been described by Choa et al (1979). On the basis of the pilot test a combination of Geodrains and surcharge was chosen as the appropriate method of soil treatment. Approximately 275,000 sq m of marine clay with an average depth of 18.5 m were treated.

The pilot test area for the Geodrains is shown in Fig. 1. Section A-A shows the profile across the pilot test area in a typical marine clay valley with new and old marine clays sandwiching a silty clay transition zone and overlying cemented sand. Numerous sandy lenses in the new marine clay and large pockets of sand in the form of old river terraces exist on either side of the old marine clay valley. A surcharge of 4.75 m of sand was placed on 6.5 m of hydraulic sand fill after the installation of drains. Based on the assumption of $C_h = 3\ C_v$ (lab) from Choa et al, (1979) it was decided to adopt a square grid pattern of Geodrains with spacing of around 2.5 m. Geodrains were then installed with three different spacings 3.2 m, 2.6 m and 2.1 m on square grids in areas GD1, GD2 and GD3 respectively, for the pilot test study.

For the finite element study, attention

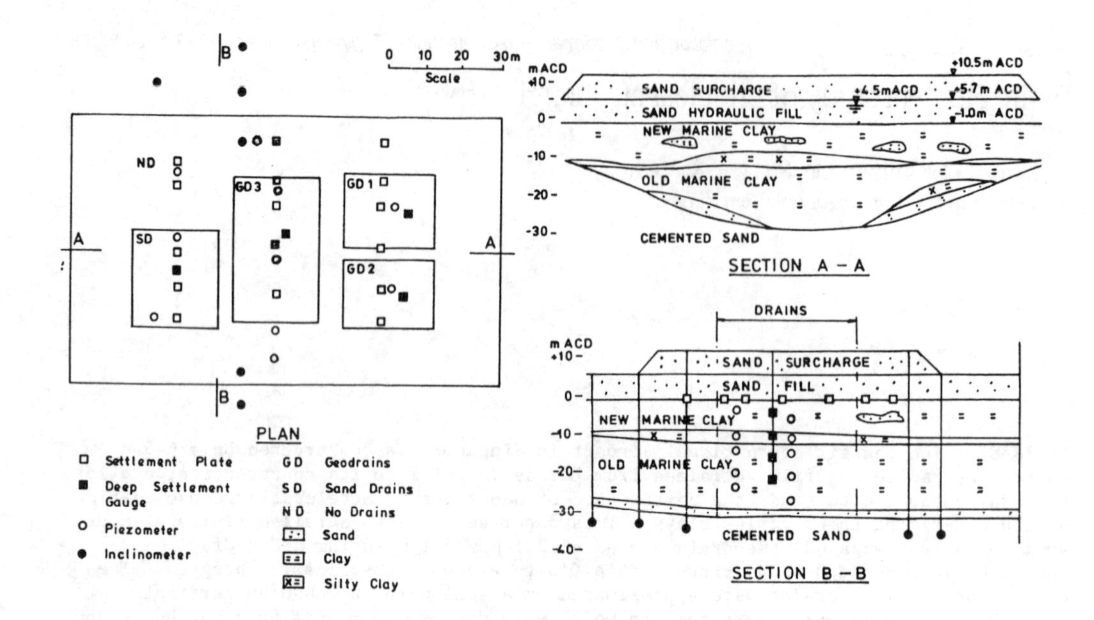

Fig. 1 Pilot test area

is focussed only on area GD3, with Geodrains spacing of 2.1 m. Section B-B shows the approximate cross section of GD3 estimated from the available boring logs, together with the locations and types of instrumentation installed at the site. Of primary importance for comparison to the finite element analysis are the instruments located on the centreline of GD3. These include surface settlement plates S83, S82 and S81, deep settlement gauges D81 to D84 at the centre of GD3, piezometers C81 to C86 at the centre of GD3, piezometers C87 to C92 at the edge of GD3, piezometers C250 to C253, outside GD3 and close to the edge of the surcharge width.

3 FINITE ELEMENT PROGRAM

The finite element program used in this study is called CON2D, developed at the University of California, Berkeley by Duncan et al (1981). CON2D is a plane strain finite element program with a Cam-Clay constitutive model for analysis of consolidation in saturated or partly saturated earth masses. It has been successfully used to analyze consolidation in embankment dams during and after construction (Duncan et al, 1981), and in foundation soils under surface loads imposed by fills, buildings and oil storage tanks (D'Orazio and Duncan (1984). The program treats the coupled problem of deformation and fluid flow, and can be used to calculate deformations and pore pressures under undrained, partly drained to fully drained conditions. Details pertaining to the finite element code and the governing equations behind the code are given by Duncan et al (1981).

4 SOIL PARAMETERS FOR THE GD3 TEST SITE

Fig. 2 shows the soil properties profile of GD3 obtained from boreholes BH 219 and BH 348. Borehole BH 219 is made after reclamation and before drain installation and surcharging. Borehole BH 348 is made at the end of the surcharging period. Detailed laboratory tests results from BH 219 and BH 348 are obtained from the Changi Soil Improvement Works reports and Choa's thesis (1984).

From the depth profiles of soil properties, it is clear that GD3 consists of three distinct layers, that is a young marine clay and an old marine clay sandwiching a transition layer of stiff clay. The boundaries of these layers are estimated as -1.5 mACD to -9.5 mACD for the young marine clay, -9.5 mACD to -12.5 mACD for the stiff clay, and

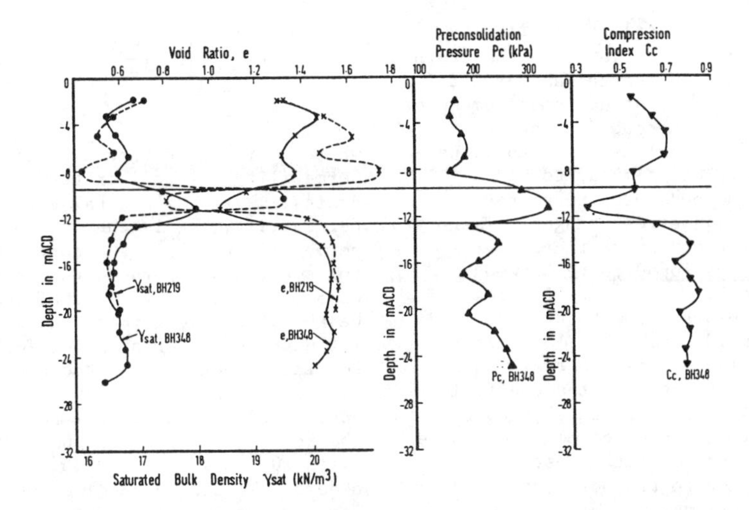

Fig. 2 Soil properties profile in area GD3

-12.5 mACD to -29.5 mACD for the old marine clay. For the finite element analysis, initial insitu stress condition is estimated from the known (γ sat) initial bulk density and (e_o) void ratio profiles obtained from BH 219, and known pore pressure profiles with depth obtained from piezometers C81 to C86 at the centre of GD3 and C87 to C92 at the edge of GD3, 10 months after reclamation but before drain installation. Values of (K_o) lateral earth pressure coefficient at rest, are estimated from empirical relationships (Brooker and Ireland, 1965 and Ladd et al, 1977) relating K_o to the values of (ϕ'), friction angle based on effective stresses and (OCR), the overconsolidation ratios.

The Cam-Clay parameters are estimated in the following way. Detailed conventional oedometer tests were performed for the undisturbed samples of BH 348, giving the variation of (C_c) compression index, (P_c) preconsolidation pressure with depth (see Fig. 2). Lambda (λ) is estimated as 0.8 times the compression index (C_c) (Duncan et al, 1981). The factor 0.8 allows for differences in compressibility between triaxial and one-dimensional conditions. Kappa (κ) is estimated to be 0.8 times the recompression index (C_r), provided C_r is available from the rebound phase of the 1D consolidation tests. Where C_r is not available, κ is estimated as approximately 0.1 times λ. With K_o values estimated previously, the isotropic preconsolidation pressure (p'_c) can be determined from the 1D preconsolidation pressure (P_c) as

$$p'_c = \frac{(1 + 2K_o)}{3} P_c \tag{1}$$

Using the values of ϕ' obtained from isotropically consolidated triaxial compression tests with pore pressure measurements, the values of (M) the slope of the critical state line can be calculated as

$$M = \frac{6 \sin \phi'}{3 - \sin \phi'} \tag{2}$$

The Poisson's ratio, ν, can be obtained from stress-strain relationships of soil in confined compression as

$$\nu = \frac{K_o}{1 + K_o} \tag{3}$$

The most difficult soil parameters to ascertain were the horizontal and vertical permeabilities. Since prior analyses only considered undrained loading and the settlement analysis was made by the classical Terzaghi theory, permeability values were not required. As no insitu permeability tests were carried out the procedure adopted was to back analyse the early part of the consolidation settlement in the pilot test area with various horizontal and vertical permeabilities. It was assumed that by the time the drain installation and surcharge was carried out, about 10 months after commencement of the reclamation, the coefficient of permeability for vertical flow would have decreased to values similar to those obtained in the oedometer tests for the various operative stress ranges. The average coefficient of vertical permeability was found to be between 0.3×10^{-9} to 0.7×10^{-9} m/s for low effective stresses below 80 kPa and between 0.1×10^{-9} to 0.3×10^{-9} m/s for higher effective stresses between 80 kPa to 300 kPa. The ratio of the coefficients of permeability for horizontal flow to vertical flow was found to be about three by comparison of oedometer test results from samples cut in the horizontal and vertical directions.

The effects of vertical drains are approximated by a soil with an enhanced vertical permeability. To evaluate the equivalent permeability the average degree of consolidation of the real soil is equated to the average degree of consolidation of the equivalent soil. From Choa (1984), the equivalent vertical permeability k'_v can be expressed in terms of the coefficient of permeability in the natural soil k_h and k_v in the horizontal and vertical directions respectively as

$$k'_v = \frac{32}{\pi^2 \mu_s} \left(\frac{h}{D}\right)^2 k_h + k_v \qquad (4)$$

where k'_v is the enhanced vertical permability for soil treated with vertical drains, k_v is the natural vertical permeability, k_h is the natural horizontal permeability, μ_s is the correction factor for smeared soil around the vertical drain, h is the maximum drainage path and height of soil cylinder, and D is the influence diameter of a vertical drain. For the computation of the equivalent vertical permeability for drains the coefficient of horizontal permeability of the remolded zone was assumed to be about 0.1×10^{-9} m/s, a value obtained from oedometer tests on remolded specimens. The other assumptions were that the drain diameter was 62 mm for the Geodrains used (Hansbo, 1981), the equivalent diameter of the soil cylinder was 1.13 times the drain spacing in a square grid and the diameter of the smeared zone was equal to the diameter of a circle with an area equal to twice the cross-sectional area of the mandrel, whose diameter was 170 mm. From the preceding discussion, Table 1 shows the range of values of the pertinent parameters used in the finite element analysis.

Table 1. Soil Parameters for GD3 used in CON2D

Material	Surcharge and Hydraulic Fill	Young Marine Clay	Stiff Clay	Old Marine Clay
γ_{sat} (kN/m^3)	17.0	16.5	18.0	16.8
e_o	0.8	1.4 to 1.7	0.8 to 1.2	1.5 to 1.6
Average e_o	0.8	1.6	1.0	1.55
K_o	0.5	0.54	0.6	0.58
λ	0.003	0.52	0.38	0.62
K	0.0023	0.052	0.038	0.062
M	1.6	1.0	1.3	1.1
ν	0.33	0.35	0.37	0.37
k_h m/s	in drained	1.0×10^{-9}	1.0×10^{-9}	1.0×10^{-9}
k_v m/s	zone	8.0×10^{-9}	8.0×10^{-9}	8.0×10^{-9}
k_h m/s	outside	1.5×10^{-9}	1.5×10^{-9}	1.5×10^{-9}
k_v m/s	drained zone	0.5×10^{-9}	0.5×10^{-9}	0.5×10^{-9}

5 THE FINITE ELEMENT MESH, BOUNDARY CONDITIONS AND INSTRUMENT LOCATIONS

The plane strain finite element mesh consisted of 99 elements and 315 nodes. The mesh describes only the compressible foundation soils and the hydraulic sand fill, but not the constructed sand surcharge. The load applied by the surcharge above the hydraulic sand fill was modeled as a progression of surface loads from 0 to 70 kPa in half a month. The mesh represented only the northern half of the GD3 profile, assuming symmetry at the centreline of the area.

The marine clays and stiff clay were modelled using eight node isoparametric elements with a pore pressure degree of freedom at each node. For the hydraulic sand fill, 5 node isoparametric elements were used, without pore pressure degree of freedom since fully drained behavior is assumed for these materials. Because of the high stiffness and high permeability of the cemented sands, the bottom boundary of the mesh was assumed rigid and free draining. The right hand vertical boundary was chosen to be twice the surcharge width, with horizontal deformations restrained, but vertical deformations unrestrained.

Fig. 3 shows the finite element mesh with the locations of various field instruments drawn in. These instruments are the surface settlement plates S81 to S83, deep settlement gauges D81 to D84 at the centre of the GD3 area, piezometers C81 to C86 near the centre of GD3, piezometers C87 to C92 at the edge of GD3, piezometers C250 to C253 outside of GD3 and close to the edge of the surcharge width, and piezometers C254 to C256 outside the surcharge width but within twice the surcharge width from the centreline.

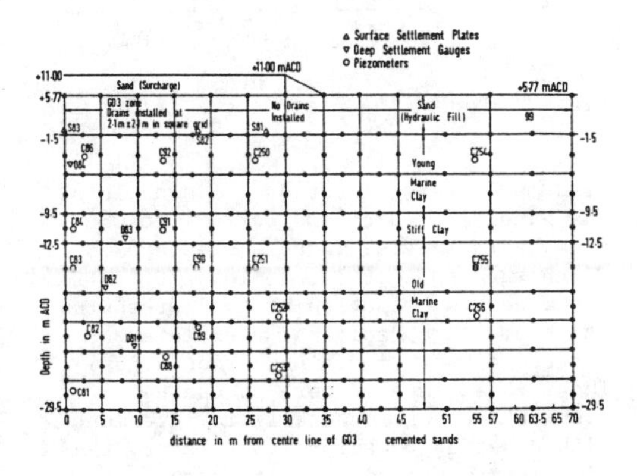

Fig. 3 Finite element mesh of GD3 with instruments location

6 RESULTS OF ANALYSIS AND DISCUSSION

The comparison between the measured surface settlements (S81 to S83) and their CON2D predictions throughout the surcharge period is presented in Fig. 4. The prediction for S83 the center vertical settlement compares very well with field measurements. The predictions for S82 and S81, the off-center vertical settlements are less than measured values by about 15% at the end of the surcharging period.

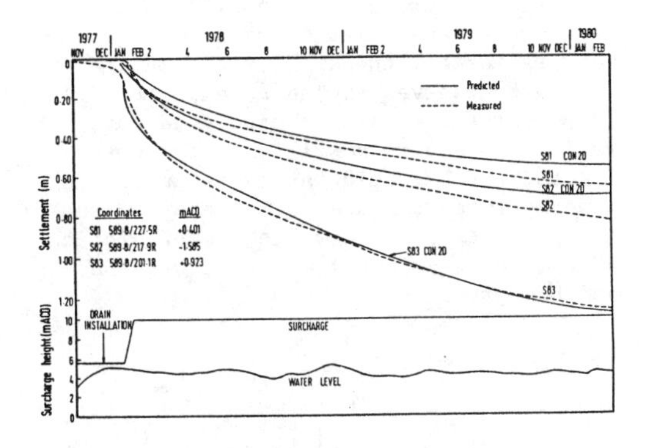

Fig. 4 Surface settlements in GD3

The measured vertical settlements of the deep gauges (D81 to D84) and the CON2D predictions of their response are shown in Fig. 5. The predicted performance for S83 and D84 agrees very well with the measured behavior of their vertical settlements. But the settlements of D81 to D83 are clearly underpredicted by the theory. If we observe the locations of D81 to D84,

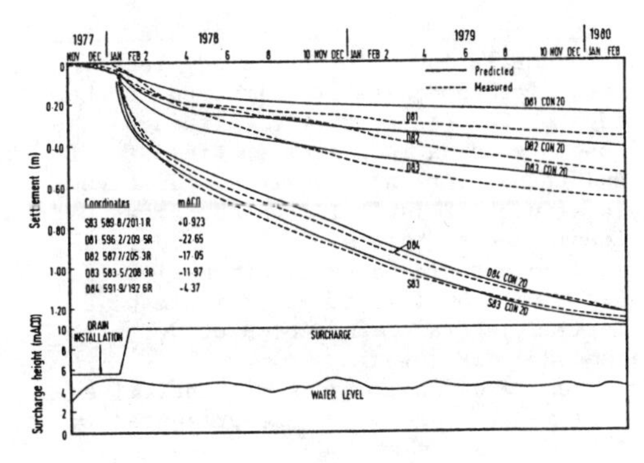

Fig. 5 Deep settlements in GD3

these indicate that this is due to an overprediction of the relative settlements between D83 and D84 (bottom of new marine clay layer), and an underprediction of the relative settlements between D81 and the cemented sand (bottom of old marine clay layer). But relative settlements between D81 and D82 and D82 and D83 seem to be correctly predicted.

The above points were is noted when we observe the void ratio changes with depth from the time before surcharge is applied to the end of surcharge period. The measured and predicted changes in void ratios with depth between the beginning and end of the surcharge period is presented in Fig. 6. Large changes in void ratios for the compressible young marine clay and relative small changes in void ratios for the stiffer old marine clay are observed as can be expected. In Fig. 6, the form of the predicted void ratio change agrees well with the observed values, but the void ratio change in the lower part of the new marine clay is much overpredicted, as shown in the prediction of the relative settlements between D83 and D84 previously.

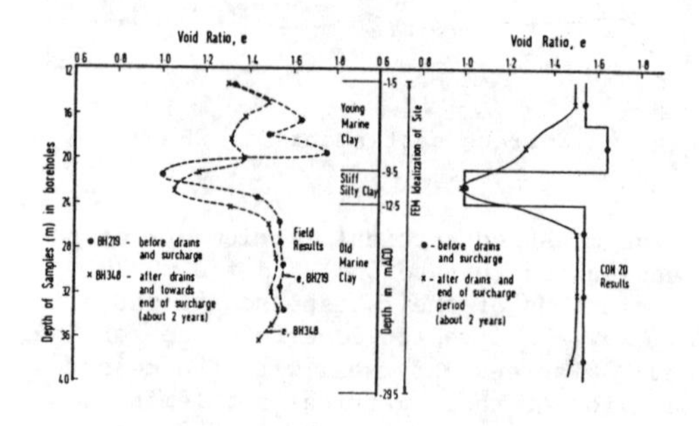

Fig. 6 Comparison of void ratio profile between FEM and field measurements

Inclinometers records though available, were not included in this study because these inclinometers are situated at the edge of GD3 (see Fig. 1). Although absolute comparison between field results and theory were not possible, relative magnitudes of lateral movement show consistent behaviors. CON2D predicted a maximum outward lateral deformation of 0.2 m whereas inclinometers indicate a maximum lateral deformation of 0.1 m in the same direction.

The field measurements of centreline piezometers C81 to C86 are presented in Fig. 7. C85 is excluded because it indicated no pore pressure response,

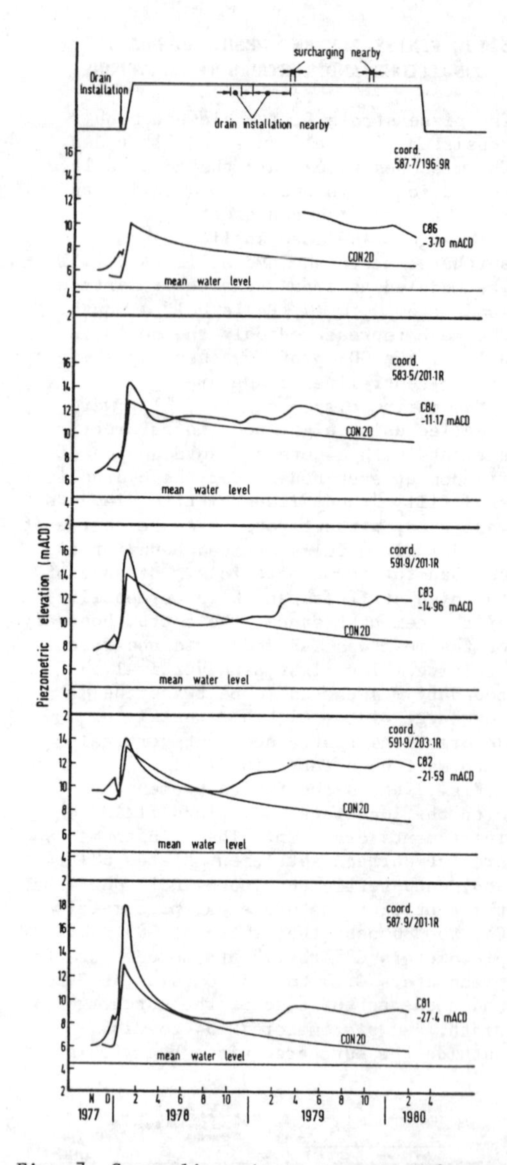

Fig. 7 Centreline piezometers in GD3

presumably being located in a sand lens. The general predictions for C81 to C86 shows a rapid rise in pore pressures upon surcharge application, followed by a continuous dissipation of pore pressures which are more rapid for piezometers located closer to drainage boundaries. The initial rate of pore pressure dissipations are generally in agreement with observed behavior. But later, the piezometers showed pore pressure stagnation, followed by porepressure generation subsequently, probably due to drain installation and additional surcharging nearby. These behaviors are not predicted by the present elasto-plastic soil model and the plane strain analysis.

The field measurements of the piezometers (C88 to C91) at the edge of the drained area and their corresponding predictions are shown in Fig. 8. Generally, the agreement between theory and field observation in this case are better than for the centreline piezometers. Field observations diverge from theory where pore pressure stagnation and generation are observed in C88 to C90.

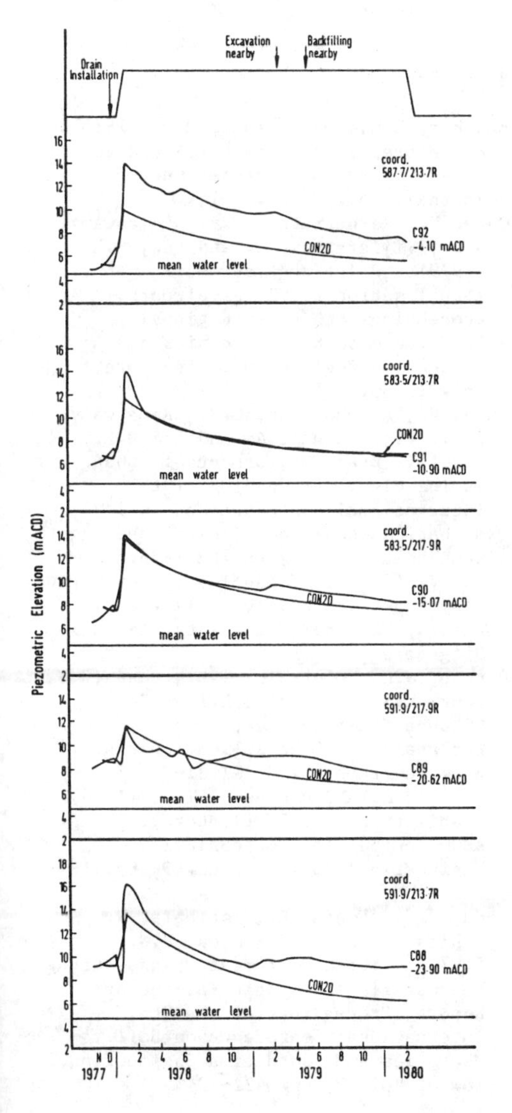

Fig. 8 Edge of drains piezometers in GD3

Piezometers C250 to C253 were installed 17 months after surcharge application to check on the pore pressure behavior in GD3 (see Fig. 9). High undissipated pore pressures were observed in C252 and C253. Observed pore pressures by C250 and C251 are reasonably well predicted by CON2D.

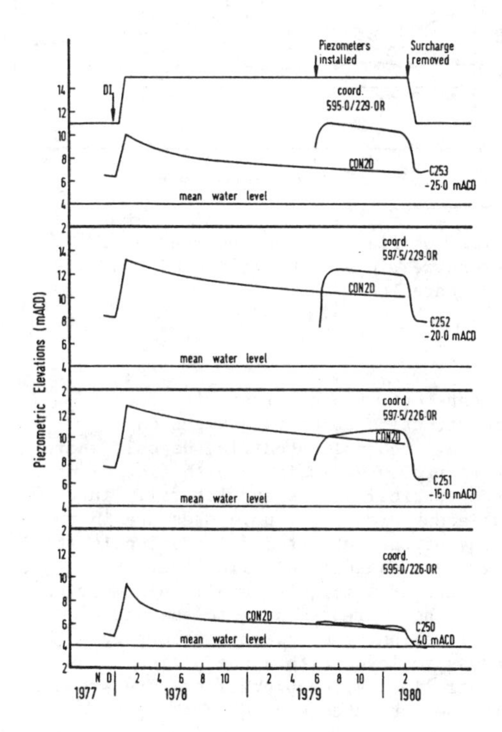

Fig. 9 Edge of surcharge width piezometers in GD3

The pore pressure variation with distance from the taxiway (GD3) centreline at different elevations 18 months after surcharge application are shown in Fig. 10. The field observations showed that after 18 months of surcharge, the excess pore pressure within the drained zone was 40 to 50 kPa. The excess pore pressure increased to around 60 to 80 kPa under the surcharge overwidth just beyond the drained zone and fell to a value of 20 kPa about 20 m from the toe of the surcharge (Choa et al, 1981). The pore pressure predictions from CON2D showed the same form as the field observations, clearly illustrating the spreading of pore pressure from the clay under the surcharge overwidth into the drained zone and also the back pressure from the clay under the reclamation beyond the surcharge

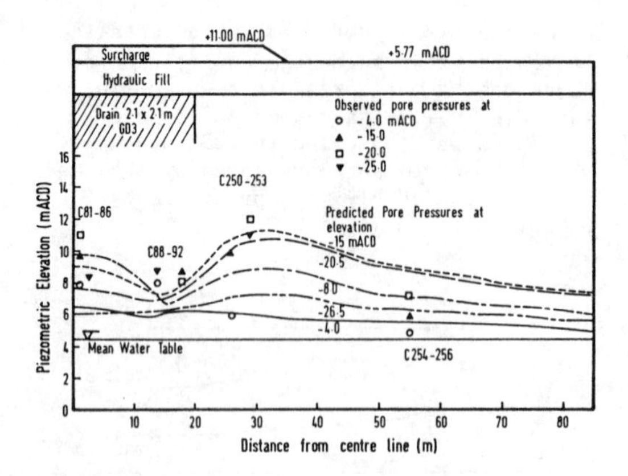

Fig. 10 Predicted vs observed pore presurres in area GD3 (18 months after surcharge)

embankment. In contrast with the construction of an embankment on a normally consolidated clay deposit where pore pressure dissipate with flow along the vertical drains and laterally into adjacent areas, the pore pressure in the clay of the untreated reclamation is still high due to underconsolidation. Since the treated area is quite small in comparison with the untreated area in the whole reclamation, the lowest stable pore pressure level within the drained area is controlled by the prevailing pore pressure in the untreated area.

7 CONCLUSIONS

(1) A 2D plane strain finite element model with an elasto-plastic soil constitutive model can be used to describe reasonably well the behavior of recently reclaimed land with drains and surcharge.

(2) Vertical drains in a 2D plane strain analysis can be accounted for by using an enhanced equivalent vertical permeability in the drained zone.

(3) The soil parameters must be carefully calibrated based on field and laboratory measurements. Well defined initial conditions and variations of compressibilities and permeabilities with changing void ratios must be accounted for before reasonable predictions can be obtained.

(4) Influence of close neighbouring activities, like drains installations and surcharging, can greatly complicate the pore pressure response of a test site making correct interpretations and predictions very difficult.

(5) Design and analysis of soil improvement works for recently reclaimed fills using drains and surcharge must take the back pressure effect due to underconsolidation of the reclaimed land into account. The back pressure effect can be predicted by a proper 2D analysis of the reclamation.

References

Brooker, E.W. and Ireland, H.O. (1965). Earth pressures at rest related to stress history. Canadian Geotechnical Journal, Vol. 2, pp. 1-15.

Choa, V., Karunaratne, G.P., Ramaswamy, S.D., Vijiaratnam, A. and Lee, S.L. (1979). Pilot test for soil stabilisation at Changi Airport. Proceedings 6th Asian Regional Conference on Soil Mechanics and Foundation Engineering, Singapore, Vol. 1, pp. 141-144.

Choa, V., Karunaratne, G.P., Ramaswamy, S.D., Vijiaratnam, A. and Lee S.L. (1981). Drain performance in Changi marine clay. Proceedings 10th International Conference on Soil Mechanics and Foundation Engineering, Stockholm, Vol. 3, pp. 623-626.

Choa, V. (1984). Consolidation of Marine clay under reclamation fills. Ph.D. thesis, National University of Singapore.

Duncan, J.M., D'orazio, T.B., Chang, C.S., Wong, K.S. and Namiq, L.I. (1981). CON2D a finite element computer program for analysis of consolidation. U.C. Berkeley report #UCB/GT/81-01.

Duncan, J.M. AND D'orazio, T.B. (1984). Stability of Steel Oil Storage Tanks. ASCE Journal of Geotechnical Engineering, Vol. 110, No. 9, pp. 1219-1238.

Ladd, C.C., Foote, R., Ishihara, K., Schlosser, F., and Poulos, H.G. (1977). Stress-deformation and strength characteristics. State-of-the-Art Report, Proceedings of the 9th International Conference on Soil Mechanics and Foundation Engineering, Tokyo, Vol. 2, pp. 421-494.

Prediction and Performance in Geotechnical Engineering / Calgary / 17-19 June 1987

Embankment for interchange constructed on soft ground applying new methods of soil improvement

Masami Fukuoka
Department of Civil Engineering, Science University of Tokyo, Japan
Masaaki Goto
Kiso-Jiban Consultants, Tokyo, Japan

ABSTRACT: A high embankment was to be constructed on very thick soft ground without causing harmful displacement at the ground surface near the toe of the embankment. Two new methods to stabilize the foundation were proposed. The first was to place steel reinforcing bars at the bottom of the embankment. The other was to drive cement stabilized piles using dry cement powder as stabilizer into the soft ground. The precision of predicting displacement was improved by conducting soil surveys, building test embankments and analyzing divergencies of predicted values from values observed in the test embankments. Displacement was thus successfully minimized. For predicting the displacement of the ground surface near the toe. The finite element method (FEM) was found to be the best method available.

1 INTRODUCTION

A high embankment on soft ground was contemplated at Hayashima Interchange, at the north end of the expressway connecting Honshu and Shikoku. Though the area at the toe of the embankment was mainly used for rice fields, significant ground surface displacement at the toe could not be tolerated. For improving the soft ground, dry jet mixing method (DJM), i.e. driving cement stabilized columns into the soft ground, and the steel bar reinforcing method, i.e. placing steel bars with anchor plates at the bottom of embankment, were adopted. These methods were used on a small scale at Muchiki about 20 km south of Hayashima Interchange. The results of the test were used in the construction work at Hayashima.

2 OUTLINE OF MUCHIKI EMBANKMENT

Figure 1 shows cross sections of the embankments at Muchiki. The east and west parts of the soft ground at embankment section A were stabilized by the DJM and by sand compaction piles, respectively. Sand piles were driven into the soft ground and steel bars with anchor plates were placed at the bottom of embankment section B. Properties of the soils are described in Fig. 2. Triaxial compression tests and consolidation tests were also performed. The soil constants used for design were as follows.

For soft clays: Unit weight γ is 17 kN/m^3, unconfined compressive strength qu is 20 kN/m^2 from the ground surface to the depth of 3m, and $\{20+6.6(z-3)\}$ kN/m^2 below the depth of 3m. Preconsolidation pressure σ_p' and effective overburden pressure σ_{v0}' increase with depth, as shown in Fig. 2. Coefficient of consolidation c_v is 3.3 x 10^{-7} m^2/s. Rate of strength increase means increase of apparent cohesion intercept $c_u = q_u/2$, when a pressure $\Delta\sigma$ is applied by an embankment. Then, $\Delta c_u = m\Delta\sigma$.

For an embankment of weathered granite: Unit weight γ is 20 kN/m^3. Cohesion c is 10 kN/m^2. Angle of internal friction ϕ' is 30°. For FEM analysis, the following soil constants were used in addition to the above soil constants. Unit weight of submerged soil for As layer γ' is 10 kN/m^3. Poisson's ratio ν and coefficient of earth pressure K for As layer and embankment are 0.3 and 0.43, respectively. Poisson's ratio ν and coefficient of earth pressure for As layer K are 0.33 and 0.50, respectively. Coefficient of deformation E for clay is derived from the coefficient of volume change m_v and the Poisson's ratio ν. Coefficient of deformation E for sand is obtained by the formula E = 7N (where, N is the N-value of the SPT). The

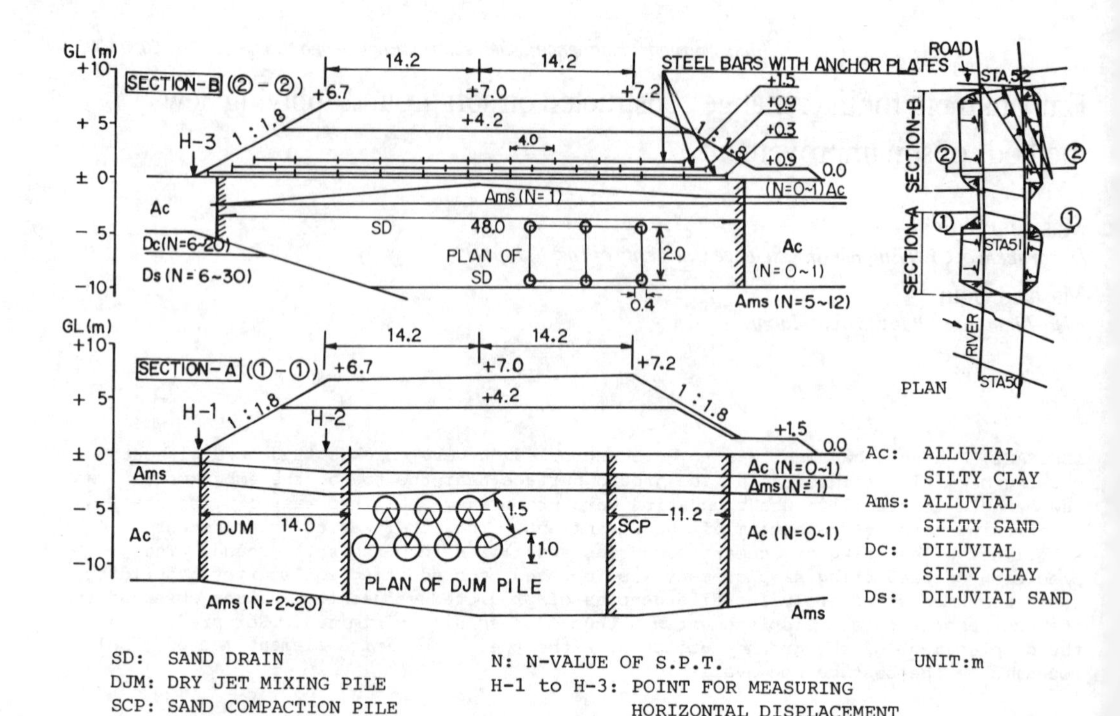

Fig. 1 Plan and cross sections of Muchiki embankment

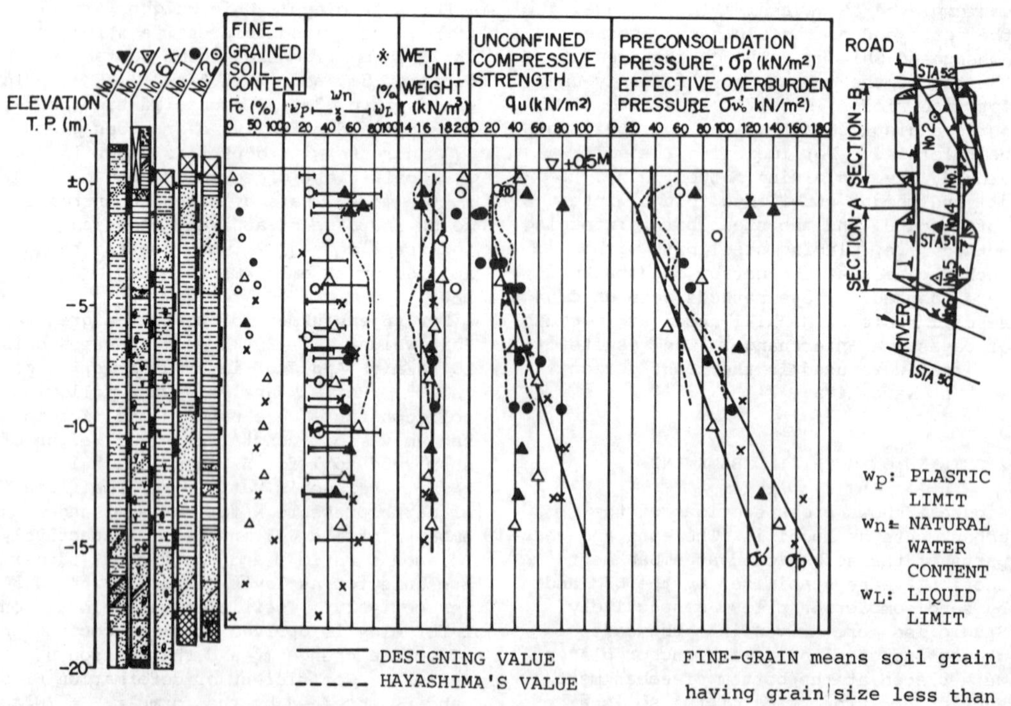

FINE-GRAIN means soil grain having grain size less than 7.5 x 10^{-5}m in diameter

Fig. 2 Soil map of sections A and B at Muchiki

results of pressuremeter tests were used for reference. Coefficient of deformation E for embankment is 10 MN/m^2.

3 DESIGN OF DJM PILES AND PREDICTION OF DEFORMATION AT SITE

DJM method is driving cement mixed clay columns or piles into soft ground to improve its bearing capacity. Dry cement powder is blown out of a hole at the tip of the mixing wing. Diameter of the pile is 1m. Compressive strength of the cement mixed clay at 28 days is about 400 kN/m^2, when 1N of cement is mixed with $1m^3$ of clay. The piles were spaced at intervals of 1.5m in a triangular pattern. Their length was 15m, and the width of improved area was 14m. Average cohesion of the improved portion is 80 kN/m^2. Safety factor Fs, according to circular arc analysis, is 1.34. According to the FEM analysis, horizontal displacement at the toe of the embankment is 8cm and settlement 2cm. The values for untreated soft soil were displacement 20cm and settlement 25cm.

4 DESIGN OF STEEL REINFORCING BARS WITH PLATE ANCHORS AND PREDICTION OF DEFORMATION AT SITE

In 1963 Fukuoka used steel reinforcing bars with plate anchors to strengthen an embankment of 40m in height on soft ground stabilized with sand compaction piles. Sand piles were used at Muchiki instead of sand compaction piles to avoid noise pollution during construction. Sand piles of 40cm in diameter were driven at intervals of 2m and to depths of 8 - 13m in pattern to square. Anchor bars were inserted only at the bottom of Muchiki embankment as it was lower than the 1963 embankment. Deformed steel bars of 22mm in diameter having steel bearing plates 250 x 300 x 9mm in dimension were placed at intervals of 50cm in horizontal direction and 60cm in vertical direction. As it was very hard to predict tensile forces acting on the bars, three layers of steel bars were provided. Therefore, the number of bars per unit length of embankment was 6. Extraction tests of bars were performed at the construction site. Extraction strength and extraction force at 1cm displacement were 120 - 140 and 70 kN/m, respectively. The following methods were used for predicting maximum tensile forces on the bars.

(1) Circular arc analysis.
Figure 3(a) is used for obtaining tensile forces on reinforcing bars.

$$T = \frac{1}{h}\{W \cdot d - \frac{1}{F}\Sigma(s_u \cdot \Delta\ell)\} \qquad (1)$$

where, s_u is shear strength, F is factor of safety.
(2) Method of using earth pressures on assumed vertical walls (Fig. 3(b)).
The horizontal earth pressure on the assumed vertical wall at the center of the embankment is $P_{A1} = 1/2\, K\gamma H^2$. Where, K is coefficient of earth pressure, γ is 20 kN/m^3, and H is 8m. Substituting these values into the formula, P_{A1} is 256 kN/m. The horizontal earth pressure on the vertical wall at the same position in the soft foundation calculated using Boussinesq's theory is P_{A2} = 387 kN/m. The coefficient of earth pressure at rest and passive earth pressure on the vertical wall in the soft foundation at the toe of the embankment are assumed to be 0.5 and 1.0, respectively. The increment of earth pressure by the construction of the embankment P_3 is calculated as the difference between the passive pressure and at rest pressure. Total shear stress along the bottom of the soft clay layer is obtained by the following formula.

$$S = \frac{1}{F}\Sigma(s_u \cdot \Delta\ell) \qquad (2)$$

where, S : total shear stress, s_u : unconfined compressive stress divided by 2, F : factor of safety, $\Delta\ell$: element of length between center and toe of the embankment.

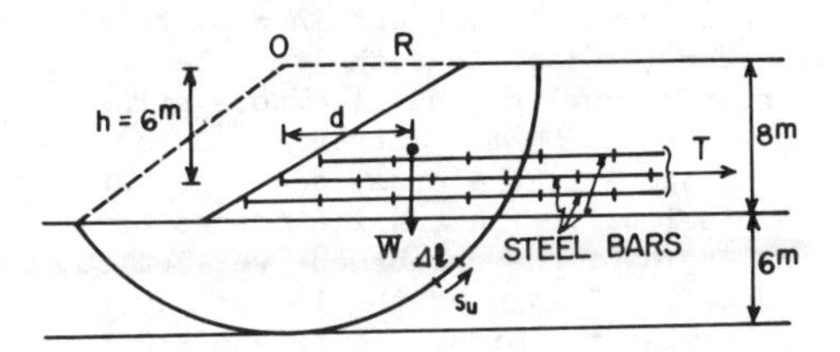

(a) Circular arc analysis

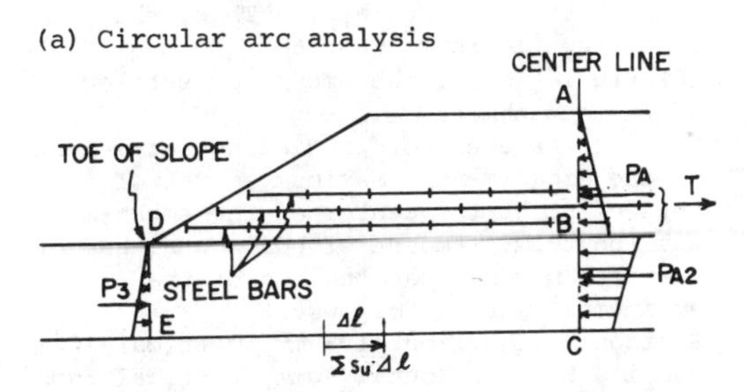

(b) Earth pressure method

Fig. 3 Method of estimating tensile force of reinforcing steel bars

Tensile force on the bars T is calculated by the following formula.

$$T = P_{A1} + P_{A2} - P_3 - S \qquad (3)$$

where, T is 358 kN/m as the result of calculation.

(3) Estimation by observations at Ebetsu Testing Embankment, Japan Road Public Corporation.

The Japan Road Public Corporation tested an embankment of 8m in height over soft ground of 10m in thickness. Two layers of steel strips were laid at the bottom of the embankment. The thickness and width of the strips were 2.3mm and 140mm, respectively. Horizontal and vertical spacings were 0.5 and 1.0m, respectively. Maximum tensile stress was observed on the strip of the lower layer at the center of the embankment. This stress was greater than the yielding stress of 3 MN/m^2. The tensile stress on the strip of the lower layer reached the yielding stress at the end of testing. Analyzing this case record of Ebetsu, 386 kN/m of tensile force was obtained for the case of Muchiki.

(4) Model test.

A model test of 1/10 scale was performed. The tensile force on the steel bars was calculated as 292 kN/m for the case of Muchiki.

(5) FEM analysis.

The soil constants used were given above. The steel bars were assumed to be bar elements resisting the deformation of the embankment by axial tensile force. The following constant was obtained by trial and error in order that the tensile stress was nearly equal to the allowable stress. Then, E_sA = 25 MN/m. Where, E_s is Young's modulus and A is the cross sectional area of a steel bar. The tensile forces per unit length of the embankment were calculated as 112.1 kN/m at the lowest layer, 50.8 kN/m at the middle layer and -10.6 kN/m at the highest layer. Thus, the total tensile force was 162.9 kN/m.

(6) Prediction of the amount of settlement of the embankment.

The FEM was used for predicting settlement of the embankment. Maximum amount of settlement at the center of the embankment was about 1m. Amount of horizontal and vertical movement at the toe of the embankment was as follows.

Section A : horizontal 80mm, vertical 20mm
Section B : horizontal 90mm, vertical 50mm

Prediction could be made by referring to the Ebetsu test embankment. Settlement at the center of the Ebetsu embankment was 2.9m. Horizontal displacement was 40cm at the right toe and 15cm at the left toe. The largest horizontal displacement under the ground was about 70cm. This large amount of displacement was caused chiefly by the existence of peat layer 4 - 5m in thickness. Therefore, the Ebetsu record seemed too large to apply to the Muchiki embankment.

5 TIME SCHEDULE FOR FILL WORK

The fill work started in January, 1984, and reached 4.2m in height at the end of March. The speed of fill work was 4.7m/day. The fill work was stopped for 122 days, from April 1 to July 31. The height of the embankment was increased by 2.8m, from 4.2 to 7.0m, in the 44 days from August 1 to September 13. The speed of fill work was 6.4 cm/day. The fill work was stopped for 57 days, from September 13 to November 9. Surcharge was placed on top of the embankment to a height of 40cm for section A and 20cm for section B.

6 SETTLEMENT OF EMBANKMENT AT THE CENTER

The middle of the section A was untreated. Settlement reached 124cm at the end of construction and 172cm 406 days after completion. Predicted settlement was 141cm. Therefore, the observed amount was 122% to the predicted one. The middle of section B with sand drains and steel bars settled by 68cm at the end of construction and 86cm 406 days after completion. The observed value was 82% of the predicted value of 95cm. The differences between the observed and predicted values are due to the lack of prediction methods and to sampling at locations other than the point of settlement measurement.

7 DISPLACEMENT AT THE TOES OF EMBANKMENT

Horizontal displacement measured at the toe of the embankment at section A is illustrated in Fig. 4(a). Comparing the observed value to the predicted value by FEM analysis, the ratio between them was as large as 190%. Horizontal displacement measured in the DJM pile inside the pile group is shown in Fig. 4(b). Comparing the observed value to that predicted by FEM analysis, the ratio between them was 190%. Judging from the curve of horizontal displacement, the pile was broken near the ground surface. Cement was not mixed with soil fully there, and this might have been the cause of the breakage of piles.

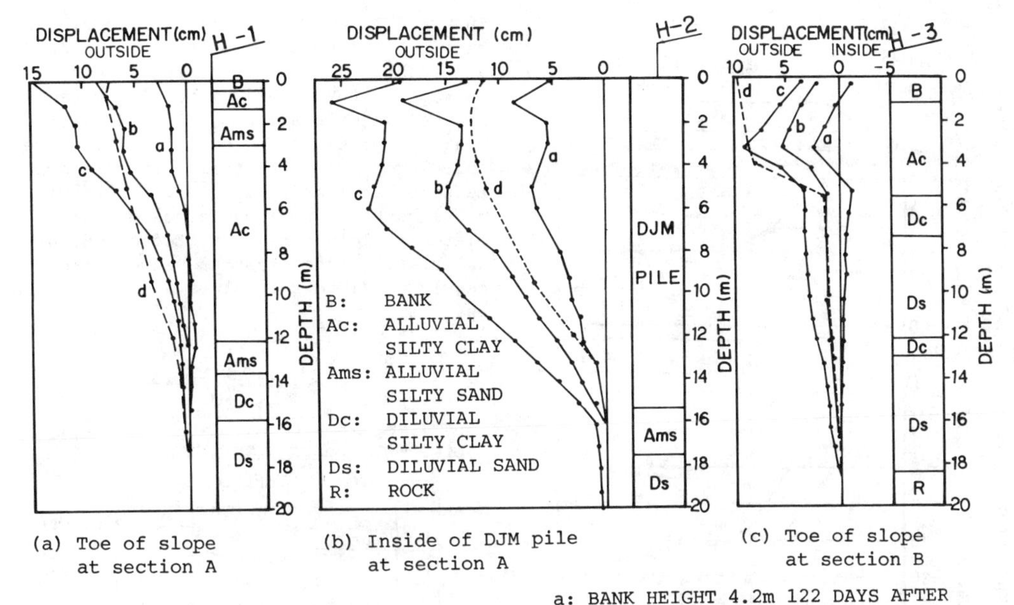

(a) Toe of slope at section A

(b) Inside of DJM pile at section A

(c) Toe of slope at section B

a: BANK HEIGHT 4.2m 122 DAYS AFTER
b: BANK HEIGHT 7.0m
c: BANK HEIGHT 7.0m 193 DAYS AFTER
d: PREDICTED BY FEM (BANK HEIGHT 7.0m)

Fig. 4 Horizontal displacement

The horizontal displacement of the ground at the toe of the embankment at section B was also predicted by the FEM. The displacement at the ground surface was 3cm, but the predicted value was 10cm (Fig. 4(c)). The sum of the tensile force of steel bars at the center of the embankment was much larger than the predicted value. This may be a cause of the difference between the predicted ground surface displacement and observed. Settlement at the toe of embankment was 4.3cm, compared to the predicted value of 10cm.

8 TENSILE FORCES ON REINFORCING BARS

The results of calculation using different methods were given in the previous paragraph for design. Based on engineering judgement, three layers of deformed steel bars of 22mm in diameter were provided with horizontal spacing of 50cm. Allowable tensile strength per bar was 69.7 kN, and for 6 bars, 418 kN/m. Values predicted by FEM and by Ebetsu test embankment were 163 kN/m and 386 kN/m, respectively. Figure 5 shows the measured tensile stresses on steel bars. The total tensile forces of steel bars at the center of the embankment was 426 kN/m, which is much larger than the predicted value. Tensile force of the bars in the lower layer was the largest among the tensile forces of bars in the three layers. The ratio of the tensile force on the lower layer to that on the middle layer was about 2.

This ratio is similar to the result of FEM analysis. The tensile forces acting on the bars at the upper layer were negative when the height of the embankment was low, and positive when the height became high. Total tensile force can be obtained by conventional circular arc analysis, but this analysis cannot provide the proportion of tensile forces in each layer of the bars. It is necessary to use FEM for obtaining the tensile forces on bars in each layer. The E_SA-value of a steel bar was assumed to be 25 MN/m at the first stage of prediction ((5) FEM analysis), but this assumption was not very correct. We should have used on E_SA-value of 22mm deformed steel bar, for which E_SA was 160 MN/m. Considering relative displacement between an anchor plate and surrounding soil, the E_SA-value was assumed to be 130 MN/m. Tensile force calculated using this E_SA value at an embankment height of 3.9m agreed with the measured value. The calculated values were given in Fig. 5 at the embankment heights of 4.2m and 7.0m. There was little difference between the calculated and observed values at an embankment height of 7.0m. As the measured yielding strength of the steel bars was 35 kN/m^2, the observed tensile

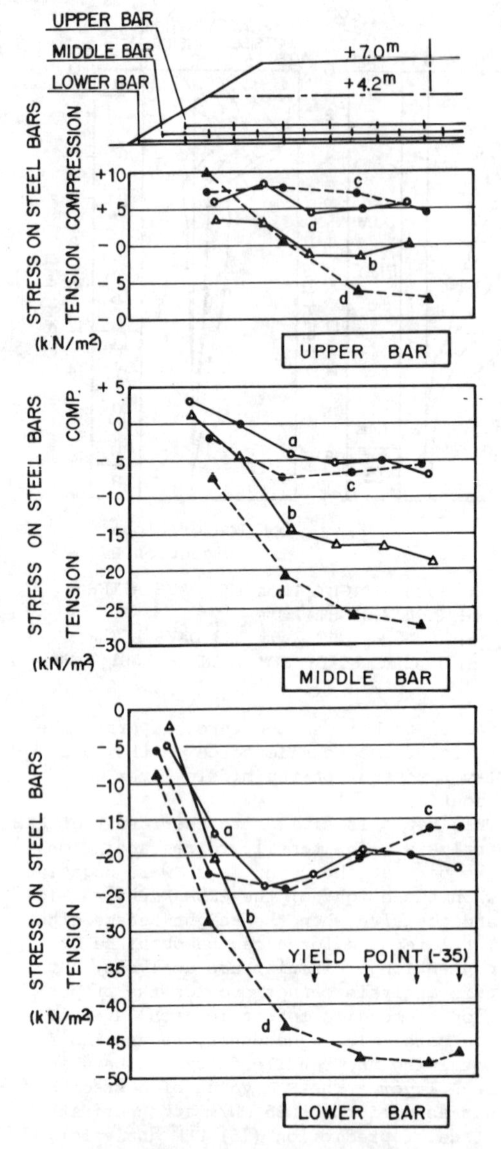

a: BANK HEIGHT 4.2m 122 DAYS AFTER
b: BANK HEIGHT 7.0m 193 DAYS AFTER
c: PREDICTED BY FEM (BANK HEIGHT 4.2m)
d: PREDICTED BY FEM (BANK HEIGHT 7.0m)

Fig. 5 Stresses on steel bars

stress was higher than the yield value. The steel bars were elongated at the yield stress without breaking. The steel bars seemed to be in the condition stated above, and they were serving as strengthening bars for the soil. It was felt that this condition should be improved in the future. It was very difficult to predict bearing forces acting on the anchor plates at the beginning. As the result of extraction test, maximum bearing capacity of the anchor plate was 120 - 140 kN. The bearing forces acting on the plates were calculated from the tensile forces on the tensile bars, and they are given in Fig. 6. Though the bearing forces on the plates could not be predicted based on calculation, the measured values were nearly equal to the values surmised by the authors.

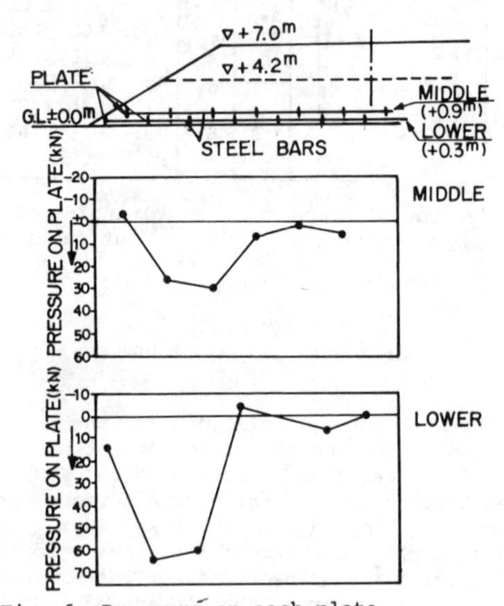

Fig. 6 Pressure on each plate

9 EMBANKMENT AT HAYASHIMA INTERCHANGE

The difference between the observed and predicted values of the Muchiki test embankment was investigated fully before constructing the embankment at the Hayashima Interchange. Points of improvement for the methods of prediction and design are stated as follows.

(1) Design.

The number of layers for the reinforcing bars, three for the Muchiki embankment was reduced to 2. Three layers were used at Muchiki because the prediction was thought to be not very accurate. Therefore, it was thought that breakage of the lower layer bars was quite possible. In that case, the upper layer bars could work instead of the lower layer bars. As a result of the embankment test, it was found that the lowest layer of bars yielded, but did not break. It was thought that the upper layer of bars could have been ommitted by improving the methods of design. The dimension of the anchor plates was 225 x 225 x 9mm, which was smaller than the plates used in the Muchiki embankment. Smaller plates were used because the excessive tensile force could not act upon the bars. An extraction test at the real construction site was performed, and the dimension was decided based on the test. A large settlement of embankment causes a deformation in an inverted arch shape at the base of the embankment. The ratio of the arch to the original straight line of the bars of the embankment is relatively large. At first, the bars are placed in a plane under the embankment, and then bent with the formation of the inverted arch. Soil can be extended but the stiff bars cannot be. Therefore, relative displacement occurs between the soil and the bars. As a result, a large tensile force appears in the bars. The original purpose of the usage of the bars is to prevent the horizontal expansion of the base of the embankment, which is caused by the lateral deformation of the soft ground. To enable the bars to perform their purpose, the formation of too large inverted arch must be prevented. It was decided to place the bars on a low embankment formed like a low positive arch. By the fill operation, the bars in a shape of low arch become straight. They form an inverted arch according to the settlement of the embankment. Thus, the bars could be used effectively to prevent lateral movement of the ground near the toe of the embankment.

(2) Method of calculation.
Calculation by FEM was improved based on the tests on the Muchiki embankment. Another improvement was made at Hayashima. Considering the relative displacement between the anchor plate and the surrounding soil, a special joint element was adopted for the reinforcing bars. The constants for the joint element were provided by an extraction test at the site.

(3) Method of construction.
As the upper part of the DJM piles were broken at Muchiki, the mixing machine was improved so the clay and cement could be mixed ideally. The arrangement of the DJM piles was changed so tensile force did not act on the inner rows of piles. An intact wall of 2 - 5m in thickness was formed at a portion near the shoulder of the embankment.

Total area, total volume of embankment, and maximum height of the Hayashima Interchange were 200,000 m^2, 1,200,000 m^3 and 11m, respectively. Depth of soft clay layer was about 12m to the west and 5m to the east. Vertical drains were installed under the middle of the embankment. DJM piles and gravel compaction piles with reinforcing bars were placed under the slope of the embankment to prevent displacement of the ground surface near the toe of the embankment. Figure 7 shows the arrangement of reinforcing bars at Hayashima Interchange.

Fig. 7 Arrangement of reinforcing bars

(4) Prediction and performance.
The construction work of the embankment is now in progress. Therefore, the predicted values cannot be compared with the observed values at the end of construction. The predicted values by FEM analysis is described in Fig. 8. But these values are calculated assuming that the settlement at 50% consolidation of the embankment with 10.1m in height is nearly equal to the measured settlement. Settlements and tensile stresses on the lower bar agree fairly well. Recorded stresses on the upper bar does not agree with the predicted value. This may be caused by placing bars on a low embankment formed like a low positive arch. Observed maximum displacement was 6cm, but the predicted value was 13cm. Height of embankment was lower than the predicted, and tensile stresses on the upper bar near the toe of embankment were larger than the predicted. This may be a cause of difference.

Figure 9 represents the embankment with DJM piles. Maximum displacement is nearly equal to the predicted value, but observed maximum displacement occurred at the top of DJM piles, while this was predicted to occur at depth of 2.5m. The strength of DJM pile was better than predicted. This may be a cause of the difference.

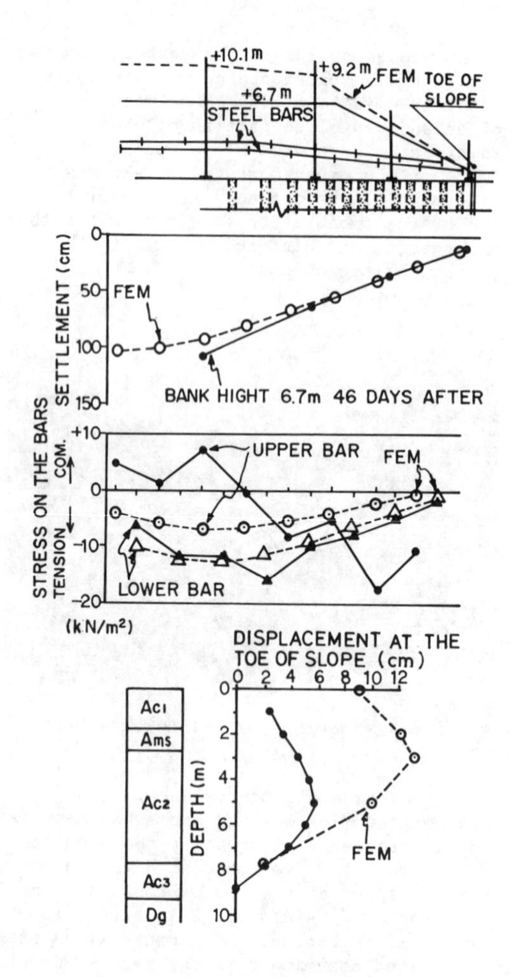

Fig. 8 Cross section of embankment with GCP with reinforcing bars

10 CONCLUSIONS

To treat the soft ground under a high embankment, DJM piles and sand compaction piles with reinforcing steel bars were used to prevent deformation at the toe of an embankment. These treatments were adopted on a small scale at Muchiki, and the predicted results were compared with the observed. FEM was used to analyze tensile forces acting on the reinforcing steel bars at the bottom of the embankment. Bar elements for the reinforcing bars were used. The EsA-value, which is the product of Young's modulus Es and cross sectional area A of the steel bars, was assumed to be 20 MN. When construction was half complete, this value was found to be too small, so it was raised to 120 MN. But a little error was obtained at the end of the construction. The position, spacing and method of construction for the DJM piles, number of layers, dimensions of anchor plates, and diameter of steel bars were changed based on the experiment at Muchiki. Joint elements were used in addition to bar elements for FEM analysis, and nearly satisfactory result for estimating displacement was attained. The construction work proceeded safety without causing serious deformation at the ground surface near the toe of the embankment.

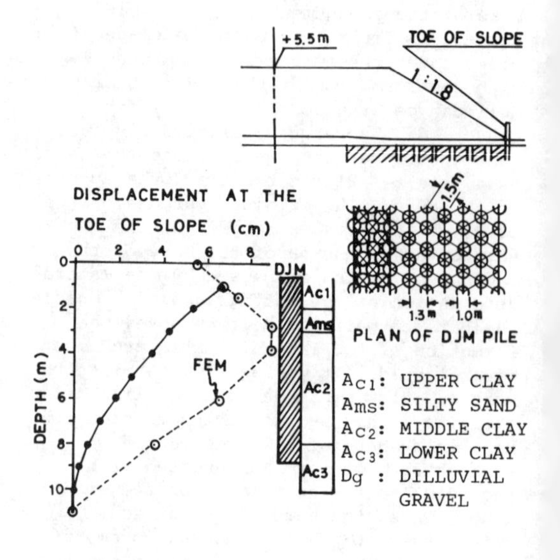

Fig. 9 Embankment with DJM piles

11 ACKNOWLEDGEMENT

This paper was prepared based on a report on the construction work by the Hayashima Interchange of the Honshu-Shikoku Bridge Authority. The authors express their sincere appreciation to the engineers of the Kurashiki Construction Office of the Authority and the committee members of Technology Center for National Land Development related to the construction work.

REFERENCES

H. OHTA, R. MOCHINAGA and KURIHARA. 1980. Investigation of soft foundations with surface reinforcement. Proc. 3rd Australia - New Zealand Geo-mechanics Conference.

Prediction and Performance in Geotechnical Engineering / Calgary / 17-19 June 1987

Stabilization of clayey soils with lime, cement and chemical additives mixing

S.Chandra
Division of Building Materials, Chalmers University of Technology, Göteborg, Sweden

ABSTRACT: Laterite clay soils have been a major building material for the rural poulation. These soils are abundant and cheap but unfortunately they are not very durable as building materials. Some mixtures with lime, cement and chemical additives were made. Their mechanical strength and water absorption were tested. It is seen that some of the properties of clayey soils such as low strengths and high water absorption, which do not encourage their use as building materials, can be improved thereby making them more durable.

1 INTRODUCTION

Laterite clay soils have been a major building material for the rural populations in many developing countries. The factors that make these soils so popular must be their availability and cheapness. Unfortunately, these soils are not very durable as building materials mainly due to lack of requisite strength. Besides this, in very wet climates, they absorb a huge amount of water which reduces their strength and results in cracking when drying (Hammond 1973).

It is known that soils stabilized with Portland cement and lime have improved properties as building material and have been used in low cost housing in some developing countries (United Nations 1964). However this, did not, receive, widespread acceptance. Polymers, both natural as well as synthetic, have been used in stabilization of clays for a long time (Ensninger & Gieseking 1942, Emerson 1956, Anon 1950 & 1966, Lambe 1951 & 1955) but they are expensive since large amounts are necessary. The idea is, that maybe the combination of polymers as admixture along with lime or cement with clay and sand will produce a durable composite material of strength. According to this idea, a systematic investigation was done on cement-clay-sand and lime-clay-sand mixes. Influence of chemical additive especially on the property of water absorption, is studied.

2 MATERIALS AND METHOD

Two series were made for this investigation. Mortar specimens were prepared by mixing three parts by weight of sand (0-4 mm) and one part of a mixture of plastic clay (moisture content 38%) and cement for series (I) to be called cement-clay-sand series. The other uses one part by weight of a mixture of plastic clay and lime to be called lime-clay-sand series (II). It was not possible to keep the water content constant for the mixes investigated due to the variations in lime and cement contents. But the consistency was kept constant and was plastic. Prisms, 4x4x16 cm, were made from the mixtures of both series. In series I, two additives were used (1) 1% Melment and (2) 0.5% Plastiment A40.

The specimens thus made were cured under water for 7 and 28 days' strength tests. The composition and strengths are shown in Table 2. In the case of lime (series II), the additive used was Mowiton M 370. The specimens made were cured in a climate room at 20°C temperature and 55% RH. The composition and strengths of the specimens are shown in Table 3.

The effect of moisture on earth buildings has been studied in detail (Hammond 1973, Clifton & Davies 1979). It has been shown that the compressive strength decreases rapidly with increase of moisture content and that with a moisture content of only

Table 1

Composition of clay	%
SiO_2	40-50
$Al_2(SiO_5)(OH)_4$	30-40
Na Al Si_3O_8	
Ca $Al_2Si_2O_8$	10-20
K Al Si_3O_8	

10% the decrease in strength could be as high as 84% of the dry strength of the same specimens. Various attempts were made to reduce the water absorption of the specimens by mixing small quantities of certain chemicals in the mixtures as well as by brushing the surfaces of the specimens with some solutions after they were demoulded and dried. It was found that by brushing with a solution made of 40% butylstearate & 60% Vanolene, the water absorption of all the specimens was significantly reduced. The results are shown in Tables 4 & 5 for cement and lime series respectively.

When the specimens were "water proofed", the effect on their strengths was also investigated. The specimens treated with the solution mentioned above were dried at 105°C until weight was constant. These specimens were put in water for 30 minutes and their flexural and compressive strengths were measured. The results are shown in Tables 6 & 7 for cement and lime series respectively.

3 DISCUSSION OF RESULTS

3.1 Strengths

Series I (cement:clay:sand mixes).

Two things are distinct from Table 2: (1) The dependence of strength on cement content and, (2) general increase in strengths of all the specimens with chemical additives.

It is seen that both Melment and Plastiment A40 are effective when used in specimens of low cement content. For example, for specimens (4:6:30) with Melment the strength increased by 128.5% and with Plastiment A40 the strength increased by 185.7%. Their effectiveness generally decreased with increase in cement content. It is also noted that Melment is more effective than Plastiment A40 in the case of high cement content (9:1:30). The specimens with Melment had increased in strength by as much as 12% both in flexural as well as in compression, whereas those with Plastiment A40 had shown loss in strength. This depends upon the interaction of chemical additives used with clay and cement. This is outside the scope of this paper and will not be discussed here.

Series II (lime:clay:sand mixes).

Unlike the specimens stabilized in Portland cement:clay mixes where the increase in cement content had a significant effect on the strengths, the changes in strengths with increase in lime content are moderate (Table 3). Besides, due to the swelling of specimens with high lime content, it may not be wise to use too much lime. All the specimens made with the addition of Mowiton M 370 showed remarkable increases in strength especially after 28 days. In some cases more than a 400% increase in strength was recorded. As in the case of cement:clay mixes, here the increase in strength also may be due to the stabilized structure obtained because of the interaction of the polymer with lime and clay (Chandra et al 1981).

3.2 Water absorption

Series I (cement:clay:sand mixes).

Two sets of specimens were made for the water absorption test: (1) treated with butyle stearate and vanolene and, (2) not treated.

It is seen from Table 4 that the untreated specimens were almost saturated after just 2 hours in water. On the other hand, the water absorption on treated specimens was very slow and even after 48 hours the specimens were far from saturation. Further untreated specimens showed that the water absorption is a function of cement content, with the highest absorption in the specimens of low cement content. No such correlation could be noted in the treated specimens.

Series II (lime:clay:sand mixes).

As in series I, the same tendency is shown by the treated specimens (Table 5), i.e. the absorption is very slow in the beginning and even after 48 hours the specimens were far from saturation. In untreated specimens, unlike series I,

Table 2 Composition and strengths of cement:clay:sand mixes

Composition	Flexural - 7 day (MPa)			Compressive - 7 day (MPa)		
	No additive	1% Melment	0.5% SIKA Plastiment A40	No additive	1% Melment	0.5% SIKA Plastiment A40
4:6:30	0.7	1.6	1.6	2.66	5.2	7.6
5:5:30	1.3	2.3	2.3	5.06	7.69	10.7
6:4:30	1.4	3.1	3.3	7.0	11.2	18.9
7:3:30	2.80	3.92	3.6	11.34	16.4	19.4
8:2:30	3.4	-	3.8	14.9	-	18.6
9:1:30	5.3	5.72	4.7	25.9	29.2	21.9
	Flexural - 28 day (MPa)			Compressive - 28 day (MPa)		
4:6:30	1.15	1.75	2.1	4.0	7.0	9.3
5:5:30	1.8	2.6	2.6	6.75	9.70	14.7
6:4:30	2.65	3.3	4.2	9.87	14.0	24.7
7:3:30	3.62	4.45	3.9	15.69	20.46	25.8
8:2:30	4.5	-	4.1	18.88	-	23.8
9:1:30	5.6	6.2	5.0	29.1	32.66	23.4

Cement - Standard Portland cement supplied by Cementa AB, Sweden
Melment - Melamine formaldehyde based admixture
Plastiment A40 - A product of AB Svenska SIKA

Table 3 Compositions and strengths of lime:clay:sand mixes

Composition	7-day strengths (MPA)				28-day strengths (MPa)			
	Flexural		Compressive		Flexural		Compressive	
	No additive	With 10% Mowiton M 370	No additive	With 10% Mowiton M 370	No additive	With 10% Mowiton M 370	No additive	With 10% Mowiton M 370
3:7:30	0.45	0.6	1.1	1.3	0.62	1.3	1.65	3.8
4:6:30	0.53	-	1.3	-	0.78	-	2.1	-
5:5:30	0.7	1.5	1.43	4.1	0.85	3.45	2.1	8.75
6:4:30	0.8	1.3	1.7	4.3	0.88	4.00	2.2	11.5
7:3:30	1.2	3.1	2.3	7.2	1.1	4.2	2.82	9.56
8:2:30	0.93	2.4	2.19	6.1	1.1	6.1	2.44	14.4
10:-:30	1.2	-	3.2	-	1.0	-	3.3	-

Lime - supplied by Cementa AB, Skövde, Sweden
Mowiton M 370 - supplied by Perstorp, Sweden

there was no clear dependence on lime content. The water absorption was more or less the same.

3.3 Effect of moisture on strength

Series I

It is seen from Table 6 that the strengths of untreated specimens after 30 minutes in water were substantially decreased. The decrease is from 23.7% for 4:6:30 mix to 69.5% for 9:1:30 mix. This is due to the hydrophobic character imparted to the surface of the specimens by the chemical solution used for brushing. Because of this, the water could not penetrate the specimens and the strength was not influenced.

Table 4 Water absorption of treated and untreated cement:clay:sand specimens

Specimen composition	Water absorption (%) at hours shown below											
	1/2	1	2	3	4	8	9	22	24	26	48	96
4:6:30 treated		1.48	1.9	2.2	2.5	3.5	3.76	7.28	7.7	8.19	11.9	15.17
4:6:30	16.6	16.86	16.9	16.9	16.96	17		17.3	17.3			
5:5:30 treated		1.18	1.46	1.72	1.89	2.70	2.90	4.7	4.9	5.2	6.9	9.26
5:5:30	13.48	14.9	15.0	15.09	15.09	15.1		15.3	15.3			
6:4:30 treated		1.2	1.68	2.0	2.3	3.2	3.4	5.8	6.1	6.57	8.88	12.00
6:4:30	10.7	12.6	13.69	13.8	13.9	13.95		14.1	14.1			
7:3:30 treated		0.71	0.81	0.97	1.0	1.18	1.26	1.87	1.9	2.03	2.56	3.5
7:3:30	9.25	11.2	12.2	12.4	12.4	12.4		12.57	12.57			
8:2:30 treated		0.68	0.82	0.92	1.0	1.28	1.38	2.08	2.16	2.28	3.0	4.28
8:2:30	6.66	8.48	10.1	10.40	10.46	10.5		10.56	10.6	10.6	10.6	
9:1:30 treated		0.58	0.62	0.66	0.78	0.97	1.0	1.5	1.5	1.63	2.1	3.25
9:1:30	6.2	7.80	9.0	9.2	9.25	9.3		9.37	9.4	9.4	9.5	

Table 5 Water absorption of treated and untreated lime:clay:sand specimens

Composition	Water absorption (%) at hours indicated below					
	1	2	3	23	24	48
4:6:30 treated	1.18	1.45	1.67	4.64	4.75	8.26
4:6:30 untreated	13.8	13.88	13.99	14.08	14.1	14.22
5:5:30 treated	0.82	0.95	1.1	2.2	2.2	2.82
5:5:30 untreated	14.2	14.38	14.4	14.61	14.61	14.72
6:4:30 treated	0.98	1.09	1.35	2.4	2.4	2.94
6:4:30 untreated	13.9	14.0	14.08	14.23	14.28	14.39
7:3:30 treated	1.19	1.40	1.60	2.92	2.92	4.65
7:3:30 untreated	14.09	14.15	14.26	14.43	14.37	14.5

Table 6 Strengths of treated and untreated specimens after 30 minutes in water (cement:clay:sand series)

Specimen composition	Flexural (MPa)		Compressive (MPa)	
	Treated	Untreated	Treated	Untreated
4:6:30	1.0	0.35	9.0	2.13
5:5:30	1.9	0.8	14.3	4.75
6:4:30	3.1	1.7	23.44	9.28
7:3:30	5.1	3.0	29.38	16.72
8:2:30	6.8	3.6	40.47	22.97
9:1:30	7.5	5.4	40.94	28.44

Table 7 Strengths of treated and untreated specimens after 30 minutes in water (lime:clay:sand series)

Composition	Flexural (MPa)		Compressive (MPa)	
	Treated	Untreated	Treated	Untreated
4:6:30	0.9	0.5	2.72	1.10
5:5:30	0.8	0.6	2.97	1.34
6:4:30	0.8	0.7	3.34	1.50
7:3:30	0.8	0.8	3.59	1.84

Series II

It is seen from Table 7, that the strengths of untreated specimens depended upon the lime content and were lower than the corresponding values of the treated specimens. In compression for example, the strength of the untreated specimens ranged from 40% for 4:6:30 to 51% for 7:3:30 mix of the values of the treated specimens.

4 CONCLUSION

This study has shown that some of the properties of clayey soils, such as low strength and high water absorption, which do not encourage their use as building material can be improved, thereby making them more durable. More work is to be done to make field tests and to test the influence of other climatic conditions on these mixtures.

ACKNOWLEDGEMENTS

The practical work was done by Robert Kwadjo, who was a guest scientist from Ghana. The manuscript is prepared by Annika Palmdin.

REFERENCES

Anon, 1950. New soil stabilization method developed using calcium acrylate, Roads and Streets, 93, p. 51.

Anon, 1966. Chemicals promote soil stabilization. Chem. Eng. News, p. 80.

Chandra, S, Flodin, P and Berntsson, L 1981. Interaction between calcium hydroxide and styrene-methacrylate polymer dispersion, 3rd Int. Congr. Polymers in Concrete, Koriyama, Japan.

Clifton, J.R. & Davies, F.L. 1979. Mechanical properties of Adobe, US department of commerce, NBS Tech. Report 996.

Emerson, W.W. 1956. A comparison between the mode of actions of organic matter and synthetic polymers in stabilizing soil crumbs. J. Agric. Sci. 47, p. 350.

Ensninger, E & Gieseking, J.E. 1942. Resistance of clay adsorbed proteins to proteolytic hydrolysis. Soil Sci. 53, p. 205.

Hammond, A.A. 1973. Prolonging the life of earth buildings in the tropics, Building Research and Practice, p. 154-164.

Lambe, T.W. 1951. Stabilization of soils with calcium acrylate. Boston Soc., Civil Eng. 38, p. 127.

Lambe, T.W. 1955. Chemical stabilization can make construction materials of weak soils. Eng. News Record, 155, p.41.

United Nations, 1964. Soil cements - its use in buildings.

Prediction and Performance in Geotechnical Engineering / Calgary / 17-19 June 1987

Ground improvement by vibro dynamic compaction techniques

M.M.Rahimi & P.M.Bayetto
Research House, Norwood, S.A., Australia

ABSTRACT: Vibro dynamic compaction is a ground improvement technique, using commonly available piling equipment; seems especially advantageous in treating inner urban pugholes and heterogeneous fill materials. The technique is capable of compacting insitu fill up to a required depth. Due to plastic waves induced by a drop hammer, surrounding soil will be compacted to an effective radius. Then, through the introduction of gravel aggregate, a high quality control stone column increases bearing capacity and reduces the settlement of the foundation soils. The energy input by V.D.C. to compact a layer of 9m thickness was 1500-2000 KNm/m^2 and carried out within 5-10 min./m^2. For light loads like one to two story domestic houses, performance of the stone column and treated soil has an elastic behaviour. Calculation and prediction of settlements based on an elastic solution, finite layer and finite element approach has a good agreement with large scale plate load test results.

1 INTRODUCTION

Ground treatment by means of stone columns were first used in 1830 by French military engineers according to Hughes (1974). Stone columns were then forgotten until the early 1930's when the fundamental concept of vibroflotation was developed in Germany by Johann Keller. The dynamic compaction process was devised by a French Engineer Louis Menard (1975) and was first used in France around 1970.

The past decades have witnessed a substantial growth in the use of compaction techniques and insitu soil treatment. The amount of work and current research on ground treatment techniques shows that the process and analysis is still in the developmental stage.

Extensive state-of-the-art reports have been prepared by Pilot in Bangkok (1977), by Mitchell in Stockholm (1981), by Eggestad in Helsinki (1983) and a general report by Schlosser et al in Brighton (1979).

The aim of this paper is to present a stone column installation technique developed in Research House and known as Vibro Dynamic Compaction (V.D.C.). The technique has been applied in two different sites for housing development, at Mira Monte and Brompton, in Adelaide, south Australia. The first site was filled with quarry material from the quarry site, brick and rubble in some areas. The second one was filled with gravel sand, ash, cinders, clayey silt, brick fragments and traces of glass, wood, metal, foundry sand and building rubble. The analysis of the test results and settlement calculation carried out at the Mira Monte site where generally the fill material was clean clayey sandy gravel, and does not contain rubbish or deleterious materials is reliable for theoretical analysis and interpretation of test results.

The observed settlement of a large scale plate load test has been compared with several methods developed for predicting the settlement of treated ground including analytical solution and the finite element and finite layer analysis.

2 CONCEPT AND ASPECT OF V.D.C.

Tamping compaction of soil is the oldest type of deep compaction. Lewis (1957) wrote "full scale compaction tests have shown that impact compaction is one of the most promising methods of compacting soil in the field" (Smoltczyk, 1983).

This method is fundamentally the same as the well known Proctor method, though with greatly increased tamper and drop height.

The concept of vibro dynamic compaction (V.D.C.), is an extension of earlier work done in the field of ground improvement techniques. The work is based on the technique of dynamic compaction after Louis Menard and on the technique of vibroflotation, after Johann Keller.

The aspect of V.D.C. is the use of compacted stone to fill each hole created by dynamic compaction. In this respect, the treatment has similarity to the installation of stone columns.

Vibroflot methods of installing stone columns are open to many criticisms. The ground around the stone columns remains largely uncompacted. Thus the stone columns are best viewed as piles. For satisfactory performance in fill, the stone columns need to extend down to near the base of the fill. Although the quality of the stone column is often measured in terms of the quantity of stone used, this may be misleading. Frequently, the diameter of the stone column may be large at ground level, but rapidly tapers off with depth, giving an overall "carrot" shape to the column.

In addition to these criticisms, the large quantity of water that is generally used to aid penetration of the vibroflot generates difficult site conditions. At worst, it is possible for significant quantities of mud to become entrained in the stone column, severely reducing the effectiveness of the treatment. We have to mention that the range of the treatable soils with vibroflotation process is restricted to clean sands and fine gravels (Bell, 1975).

Quality control of dynamic compaction of Menard is notoriously poor. The vibrations produced by the impacts are relatively large and may prohibit the employment of the technique of dynamic compaction in urban areas. In general it has been found that dynamic compaction method of Menard, especially suited to large open sites, open sites on reclaimed land, fills and alluvial soils (Menard, 1975 and Hansbo, 1978).

The V.D.C. technique seems to overcome many of the individual shortcomings of dynamic compaction and the use of stone columns. Quality control of dynamic compaction is achieved by specifying precise depths of penetration of the compacting ram. The hole created is filled with good quality stone fill which is then compacted. The high compaction energy used in forming the column will give considerably greater compaction of the surrounding soil than is the case with vibroflot installation of stone columns.

3 PROCESSES OF V.D.C. TECHNIQUES

The V.D.C. process consists essentially of driving or "punching" holes below the surface of the site using a well defined and closely spaced grid pattern. The holes are driven by a hammer-drop technique with the addition of water to assist in the reduction of inter-granular friction and densification of collapsing soils. For each site a comprehensive site investigation is undertaken and a minimum specified depth of hole is calculated for that particular site on the basis of the mechanical properties of the foundation soils, be they naturally occurring or "fill" soils. The hole punching process continues, after the minimum depth has been reached and until the required "set" is achieved for the particular foundation soil being treated. The design "set". and energy input for each site is specified and based upon the individual characteristics of the foundation soils being treated. The most important point in dynamic compaction is the optimization of energy input. The designer should carefully examine which relative density he really needs before the appropriate grid distance is chosen.

Various stages of the installation procedure of stone columns are presented in Figure 1-a. When the compaction is completed and the required set has been achieved, the hole created by the hammer is back-filled in stages with granular material and compacted by the hammer drop technique, up to 0.6m below the surface. Granular material was graded from 20-100mm and is chosen to have a Los Angeles abrasion resistance of less than 28 such that it will be hard enough so as not to be pulverized by the impacting hammer blows. To ensure uniform compaction throughout the pile length, check tests by set measurements are carried out at different stages. The control during construction of stone columns is mainly concerned with:- checking the depth of compaction, the quantity of backfill material, the consumption of water, the rate of penetration of the hammer, the final set and the total time required to complete one compacted column. The surface layer is then compacted by vibrating roller.

4 MECHANISM OF V.D.C.

When the soil in the first stage and the backfill gravel in the second stage of the V.D.C. process are being compacted in the hole, a plastic zone is formed creating plastic waves which are beyond the elastic limit (Sinitsin, 1983). Under the hammer,

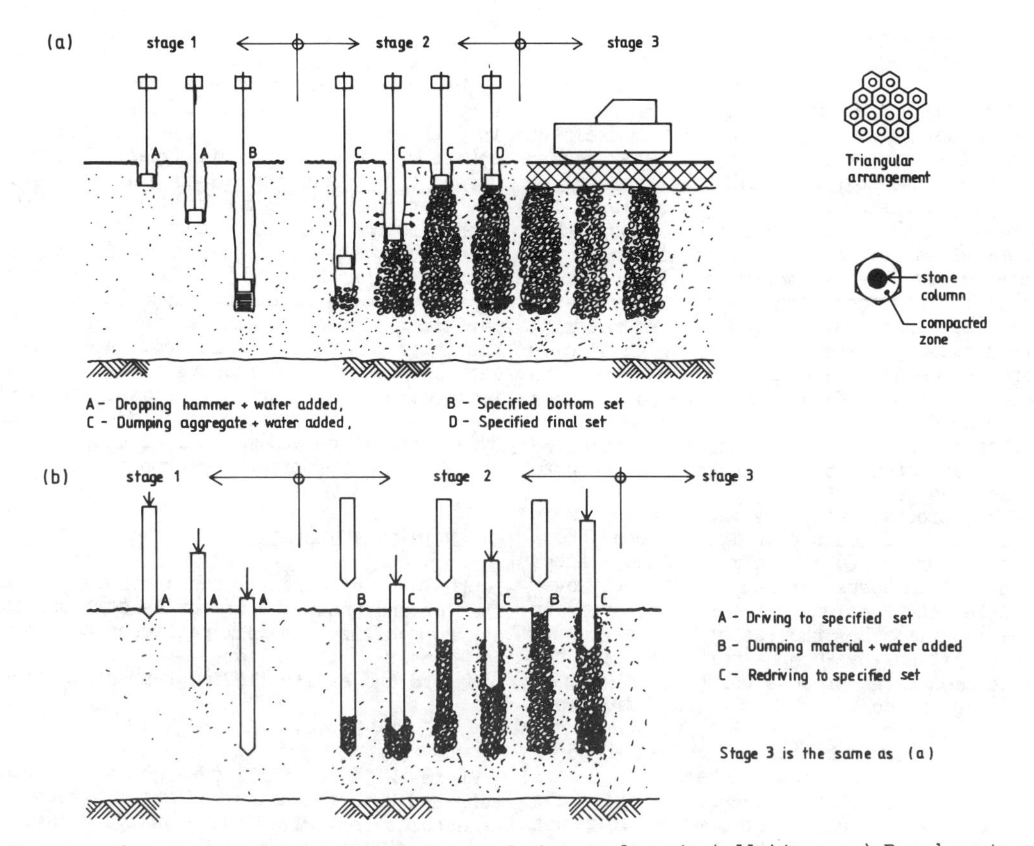

Fig. 1 The V.D.C. technique - Sequence of stone column installation - a) By dropping hammer, b) By driving casing

both elastic and plastic waves are created and these propagate with different velocities. These waves are reflected from bedrock and interaction between these waves and their reflection changes the pressure of the soil on the contact plane between falling hammer and soil, Figure 2.

If $\sigma < \sigma_A$ the elastic waves propagate in the soil.

Where $\sigma_A < \sigma < \sigma_B$ there are elastic and plastic waves.

When $\sigma > \sigma_B$ compaction waves arise and propagate with high velocity. In fact compaction energy is transmitted in time and distance without transport of matter. This wave motion is transmitted radially through the foundation soil thereby inducing lateral soil displacement and radial compaction. A circular zone of compacted foundation soil is thus created around each rock pile so formed.

Clearly, the depth and width of the plastic region depends on the mechanical properties of the soil. At the same loading intensity, the larger ϕ is the less deep extends the plastic zone, and cohesion reduces the extent of the plastic zone considerably (Jumikis, 1969). Therefore plastic waves in cohesive soil are negligible, and compaction of surrounding soil by V.D.C. is not effective.

A very similar aspect has been reported by Brandl et al (1977) and Mitchell (1981).

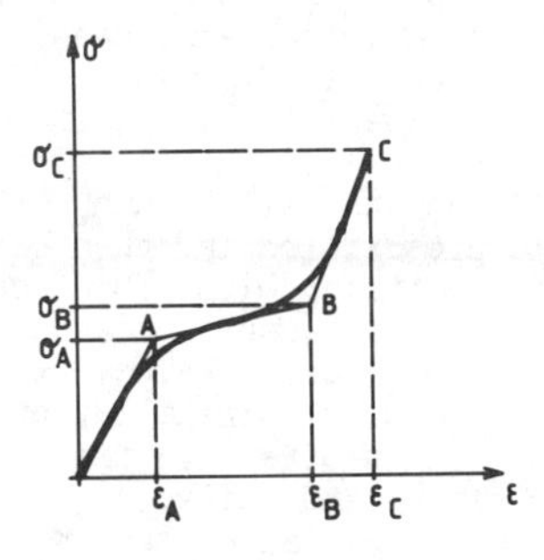

Fig. 2 Axial deformation of confined compactable soil

5 PERFORMANCE OF STONE COLUMN INSTALLATION BY DRIVING A CASING

After using a drophammer technique for punching holes and compacting backfilled material in a few sites, it has seemed that, stone columns could be performed by driving a casing with a diesel hammer. The holes are formed by driving a casing with a conical shoe to a specified depth. The casing is then withdrawn and the hole is filled in stages with clayey, sandy gravel material and water added. The material used has a grading of 0 - 75mm with 8% clay and PI = 33%. The casing is redriven to get the required set, the material is then displaced laterally. The process is repeated several times until the required set has been achieved in top 0.5 - 0.8m below the ground level.

The process was very slow taking more than 1 to 2 hours to complete one stone column or 20 to 40 min/m . Also, after plate load tests carried out directly over single stone column, it has been found out that while the process of stone column installation by drophammer reduced the settlement, the process using diesel hammer driving a conic shaped casing did not. This could be due to soil type, low energy input by diesel hammer and convex shape of the contact surface, where the soil compaction is not achieved, because the plastic zone, and plastic waves did not form. Mainly in the top two metres the soil did not compact but loosened. Therefore, all the calculations and analysis in this paper are concentrated on the Mira Monte site and based on the drophammer techniques.

Similar observation has been made by several author's i.e. Harder et al (1984) for densifying the silty sand soil. The holes were pre-drilled and then backfilled with sand by Jetting with vibroflot, where it is not the usual case in the vibrofloatation method. They found out that the stone column installation by this method was not an effective treatment for densifying silty sand soils.

Thornburn (1975) after twelve years of experience reported that the vibro-replacement method of ground treatment is not recommended for cohesive soils and organic silts, because the horizontal vibrations are effectively damped within a relatively small radius from the centre of vibration. McKenna et al (1975) using vibro replacement method for strengthening of clayey alluvium, found out that stone columns had no effect on the amount of rate of settlement. After Broms (1983) this is due to contamination and plugging of stone columns by clay slurry during the installation.

The common and possible reason why the stone columns did not reduce the settlement in these cases are the method of construction and the soil types. In many cases the plastic zone and plastic waves are not formed, and the ground around the stone column remained uncompacted. The stone column bulges under the applied load because the bearing and deformation behaviour of the columns depends on the lateral resistance of the surrounding soil. The surrounding soil is not compacted, lateral resistance remains unchanged and improvement in ground treatment is not significant. Therefore, according to soil type, condition and mechanical properties, there is an appropriate treatment.

6 DEPTH OF INFLUENCE

Depth of influence is the depth to which the ground under the hammer is stressed so as to mobilize the soil particles to a new and denser configuration.

Menard and Broise (1975) proposed that:-

$$D = \sqrt{W\ H}$$

Where W is the falling weight in metric tons and H is the height of drop in metres.

Leonards et al (1980) analysed seven cases and concluded:-

$$D = 0.5\sqrt{W\ H}$$

was more appropriate, and Lukas (1980) concluded that:-

$$D = (0.65 \text{ to } 0.80)\sqrt{W\ H}$$

P.W. Mayne (1984) collected over more than 120 sites treated with dynamic compaction at the ground surface and produced a graph relating the depth of influence and energy per blow of the hammer. The range of variation is given by:

$$D = (0.3 \text{ to } 0.8)\sqrt{W\ H}$$

and with a conservative value of:

$$D = 0.5\sqrt{W\ H}$$

Clearly, the depth of influence should depend on other factors in addition to the impact energy. Soil type might be expected to be the most important (Mitchell, 1981). Also parameters such as surface area of weight, the number of blows, grid spacing, initial soil condition and ground water level (Mayne, 1984), have the most effect on depth of influence.

Charles (1980) proposed:

$$D = 0.4d\,[\,E/A\ 1/d\ 1/Cu\,]^{0.5}$$

where E/A is the total energy applied per unit area, d is diameter of the hammer and Cu is undrained shear strength of the soil.

A more comprehensive approach, which makes allowance for soil type and strength and the diameter of the falling weight can be formulated using plastic analysis.

The thickness of compacted material under the hammer can be assumed as a cylindrical shape with linear distribution of shear stress along the boundary. Soil under the hammer is in plastic equilibrium, and for the purpose of simplification the elastic deformations and the effect of third dimensions are ignored. With the application of the theory of plasticity and plastic failure based on Prandtl's theory, after L'Herminier (1967), the depth of influence is calculated as in Figure 3.

$$\tau = \frac{P \sin \varphi}{\tan(\pi/4 + \varphi/2)} \exp[-(\pi/2 - \varphi)\tan\varphi]$$

or:

$$\tau = P\,f(\varphi)$$

P is applied pressure and τ is shear stress

$$(p-q)\,\pi d^2/4 = \pi D d\,\tau/2$$

$$D = (P - q)\;d/\,2\tau$$

with:

$$q = [\,\gamma(h + D) - U]\,s_2\,s_2' + Cu \cot\varphi\,(s_2 - 1)$$

$$s_2 = \tan^2(\pi/4 + \varphi/2)\exp[\pi \tan\varphi]$$

$$s_2' = (1 + \sin\varphi)^{-1} \exp[(\pi/2 - \varphi)\tan\varphi]$$

U is pore water pressure.

Using all above equations it can be deduced:

$$D = d\,\frac{P - [(\gamma h - U)\,s_2\,s_2' + Cu \cot\varphi\,(s_2 - 1)]}{\gamma d\,s_2\,s_2' + 2\,P\,f(\varphi)}$$

Some interesting results are presented in Table 1, where prediction of the maximum depth of influence with current methods has been compared.

For the site under study, the energy input and soil parameters are:

W = 3.8t, H = 8m, d = 0.45m, Cu = 43-70 kPa and φ = 1.9-2.1 .

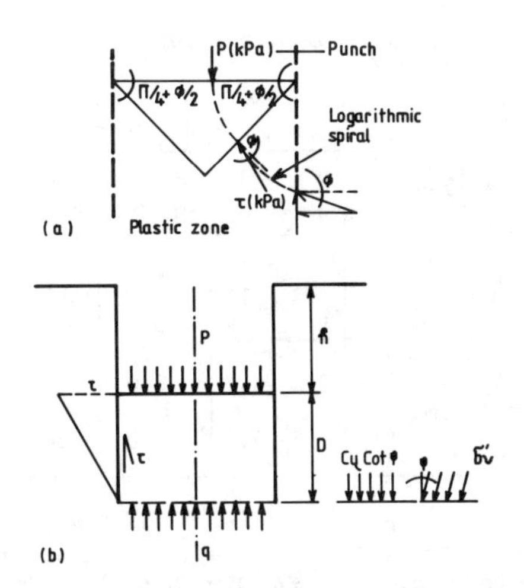

Fig. 3 Stress transfer/stress distribution

Under the pressure applied by the hammer to the ground surface and ignoring pore water pressure, the depth of influence is:

$$D = 5.5\text{m}$$

This is the minimum depth of compaction and the required set won't be achieved before this depth has been reached. The depth of treatment is the sum of the depth of compaction and the depth of influence.

It is difficult to assess how much increase in undrained shear strength would have occurred just under the hammer, but after the treatment insitu tests performed at the site show the average Cu at this depth could be around 100-150 kPa. Assuming φ = 4 the depth of influence is 2.1 to 1.5m and depth of treatment is 7 to 7.5m which shows good agreement with after treatment testing.

In the consolidated and moderately consolidated areas, which have been lying below the overburden pressures due to stockpiling for some forty years, the average undrained shear strength in these areas was 100-150 kPa; the depth of influence is:

$$D = 3.5 \text{ to } 4.5\text{m}$$

where the depth of treatment will be 5-6m.

7 LIMITS OF PROCESS

The most important question to be considered is the suitability of the method for cohesive soils as there is a plasticity

Table 1. Heavy tamping records.

Author (1)	Minkov		Harti-kainen	Gambin		Cleud	Koponen	Maly-shev	Johnson	Charles (5)	
Soil Type	Loess		Sand + silt	Sand + silt (4)		Silty sand	Rockfill	Sand + clay	Sandfill	Boul. clay	Soft Allu.
W (t)	4	15	12	40	170	18	12.5	10	15	15	15
H (m)	5	10	12	10	25	28	10	15	25	20	20
D (m) anticipated	1.5	7	6-7	24	40	12	10	8	12	5-6	-
D (m) (2) predicted	2.2	6.1	6.0	10	33	11	5.6	6.2	9.7	8.7	8.7
(3)	1.5	7.2	7.0	24	42	12	6.4	7.2	10.8	5.9	8

(1) - First author only, (2) - Mayne (1984), (3) - Plastic analysis, assuming hammer base 2m x 2m , $\phi = 4^\circ$, and under the water $\phi = 1\text{-}2^\circ$, U = 0, Cu = 0, (4) - under water, (5) - The average Cu for boulder clay and soft alluvial recorded is 50 and 25 kPa respectively.

limit which makes the application of V.D.C. prohibitive.

In soft cohesive soils, compaction of the soil mass is not possible. Silty soils with low water content and low plasticity can be compacted. In organic soils, the state of decomposition has to be contemplated and a conservative design must be considered. Single boulders with sizes not exceeding 500mm do not interfere with the process, but bigger boulders in surrounding soil will stop the propagation of plastic waves, and prohibit any compaction beyond this level.

8 INSITU TESTS BEFORE AND AFTER TREATMENT

A series of plate load tests and SPT boreholes were conducted before V.D.C. treatment. The load was kept constant until settlement had ceased. The load was then released to zero and one more load cycle was performed to around 200 kPa and while under load the borehole was flooded with water. The collapsing behaviour of the soil was measured and then loaded to failure. The initial response of the plate load test before flooded, gave a stiffness of about 33 kPa/mm, and a shear modulus of 2.7 MPa for the silty sand fill area. Plate load test and SPT results are presented in Figures 4 and 5 respectively.

The effectiveness of the Vibro Dynamic Compaction treatment has been assessed by plate load tests conducted on treated areas, and also by carrying out SPT tests at the centres of the triangular grid of compaction points. Plate load tests (300mm diameter) which were conducted directly over a compaction point, showed dramatic improvement in both bearing capacity and soil stiffness, with bearing pressures as high as 1400 kPa being achieved for settlements no greater than 40mm.

SPT tests conducted in areas that had been treated by V.D.C. showed significant increases in blow count, particularly in the upper 4.5m - 6m. In almost all cases, blow counts exceeded 20. At depths greater than 6m the SPT results reverted to values typical for the untreated fill material.

A plate test was also conducted on a larger scale (see Figure 6). A plate 2.55m by 3.46m, spanning three compaction points in an area of silty/sandy fill, was loaded up to 144 kPa. The measured settlement was 10.6mm. For this size of plate, the stiffness of 13.6 kPa/mm implies an effective shear modulus for the soil of 13.4 MPa, some 5 times greater than before treatment by the V.D.C. process.

Two different methods of V.D.C. have been specified depending on whether the upper fill material is brick and rubble or silty/sandy fill. In areas of brick fill, the compaction is carried out to a penetration depth of at least 6m before backfilling with compacted stone. "Chance" cavities or other weak areas at those depths are catered for by stipulating final sets at 6m penetration that must not exceed 60mm/blow for a 8 m drop of 3.8 t hammer. In areas of silty/sandy fill, the method of V.D.C. specifies penetrations of at least 3m, before backfilling with stone. Again, final sets of 60mm/blow for the same energy input at 3 metre penetration depth controls any weak areas.

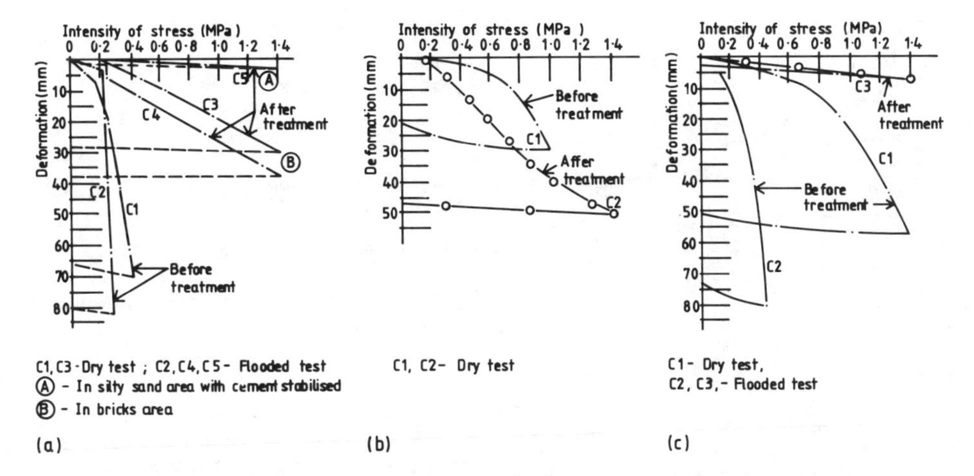

Fig. 4 Plate load tests a) At Mira Monte, treated by dropping hammer, b) At Brompton Site I, treated by driving casing, c) At Brompton Site II, treated by dropping hammer

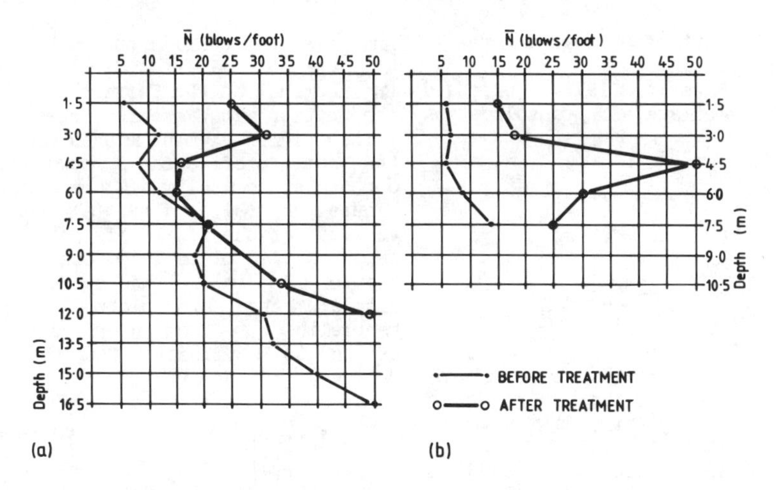

Fig. 5 Average of SPT values before/after treatment by dropping hammer, a) Mira Monte silty sand area, with 3m stone column, b) Brompton Site IV, with 5.5m stone column

The arrangement of stone columns is a regular triangular pattern with 2m spacing. The average volume of stone in each hole is $2.2m^3$ for 3 metre deep and 4.5 m^3 for 6 metre deep stone columns. These amount are 4 - 5 times the volume of the initial hole created by the drop hammer. The diameter of stone column is 0.75 - 0.95m. The SPT tests carried out after treatment confirm that the compaction is effective even at the midpoint between stone column centres.

In a silty-sand area a less consolidated zone up to 9 metres depth was identified during the process. This is an advantage of V.D.C. By controlling all the sets one is capable of finding any abnormalities, like soft spots, cavities etc. In this area a mixture of stone column depths is used, with half the number of columns penetrating to a depth of 9 metres, the rest penetrating to 3 metres. Each column is finally capped by a cement stablized layer, above which rolled compacted fill is used at the surface.

9 SETTLEMENT ANALYSIS

The total settlement of the treated ground with partially penetrating stone columns could be given by:-

$St = S1 + S2$

St = total settlement

S1 = the settlement of the treated layer

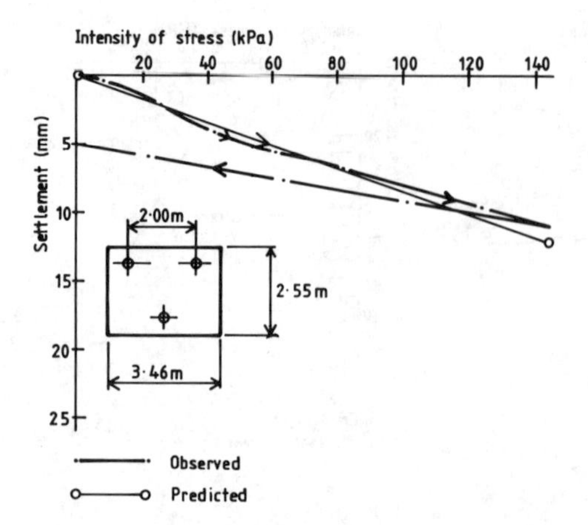

Fig. 6 Large scale plate load test with prediction of settlement

S2 = the settlement of untreated layer under the stone column.

Baumann and Bauer (1974) proposed equations for immediate and consolidation settlement:-

$$S1 = \frac{h}{Es} Ps$$

$$S2 = \sum_0^h \frac{\Delta \sigma}{Es'} \Delta Z$$

Where h = the depth of the stone column, Ps = the load shared by the soil; Es = modulus of surrounding soil, Es' and $\Delta\sigma$ = the drained modulus of deformation and increase in vertical stress at depth Z below the foundation.

Assuming the minimum diameter of stone column 0.75m and modulus of elasticities of soils and columns as given in Figure 7, the shared load between pile and soil, after Baumann et al, for plate size 2.55m x 3.46m under 100 kPa is calculated:-

Ps = 23 kPa, and Pp = 539 kPa
and
St = 8.06mm

Rao (1985) proposes the coefficient of equivalent volume compressibility of the composite mass, consisting of soil treated with stone columns as follows:-

$$m_{equi} = [\alpha Ep + (1-\alpha)Es]^{-1}$$

Where α = Ap/A and Ap = Cross - sectional area of pile after installation and A = the total area of the footing base.

The total settlement of the plate load test under 100 kPa became:-

St = 7.95mm

Using an elastic solution proposed by Balaam et al (1983):-

St = 8.07mm

Finite layer and finite element analysis show that:

St = 8.17mm

The measured value was 8.12mm. This close agreement between different solutions could be somewhat fortuitous.

10 FINITE LAYER AND FINITE ELEMENT ANALYSIS

Analysis was carried out by Dr. J. Small at the School of Civil and Mining Engineering at the University of Sydney in conjunction with Research House.

The finite element meshes used to model the soil profile for 3 metre and 9 metre depth stone columns are presented in Figures 7 (a) and 8 (a) respectively. Results of the finite element analysis for large scale plate load test are shown in Figure 7 (c) where the vertical displacements are plotted along the centreline r = 0 and outer radius r = 1.05m of the soil cylinder. The loading of 100

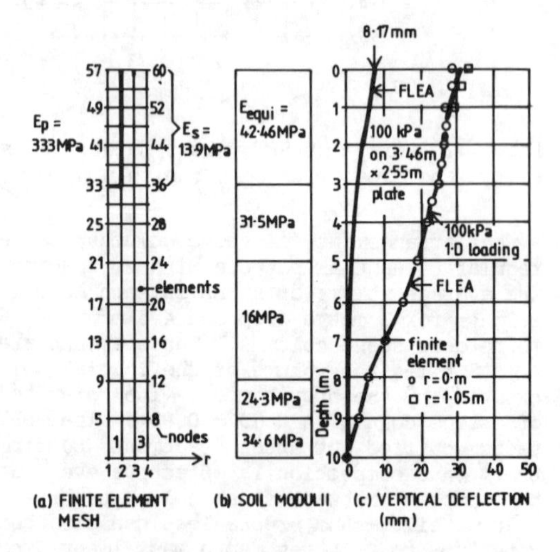

Fig. 7 Prediction of settlements for large scale plate test. Comparison of finite element and finite layer calculations for one-dimensional loading

TABLE 2

Case	Site Conditions between 6m - 9m	Soil Properties	Deflections under 15m x 14m rectangular loading of 20kPa (mm)			Deflections under 1-d loading of 20 kPa (mm)
			Centre	Edge	Δdiff	
3m Cemented stone column Finite Layer results	Consolidated	Expected	5.8	1.8	4.0	6.9
	Moderately Consolidated	Expected	8.3	2.9	5.4	11.1
		Worst Case	9.6	3.3	6.3	12.5
	Less Consolidated	Expected	11.9	4.3	7.6	16.3
		Worst Case	14.1	5.7	8.4	22.6
9m Stone Column Finite Element results	Less Consoldiated	Expected	-	-	-	11.0
		Worst Case	-	-	-	15.7

All results include 1.6m of fill over columns with modulus E = 40 MPa

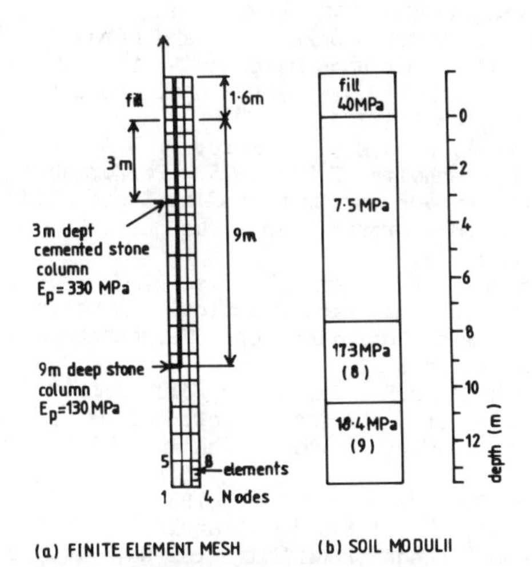

Fig. 8 Finite element mesh used for analysis of 9m deep stone column and 3m deep cemented column

kPa applied to the surface is a one dimensional loading.

In order to determine the settlement of surface loadings which are say rectangular in shape, it was decided to use a finite layer analysis. This was considered possible as it was observed from the finite element analysis that the stiff stone column carried most of the applied load, while the soil around it carried very little.

The results of the finite layer programm FLEA with a one dimensional surface loading, are compared with the finite element results. The agreement, as may be seen from Figure 7 (c), was extremely close.

In each area (treated with 3 metre and 9 metre stone columns) a 15m x 14m uniform loading of 20 kPa was applied and final deflections including consolidations were calculated at the centre and edge of the loading area. The results presented in Table 2 were obtained using the finite layer method. The results presented for the 9 metre deep column were obtained from a finite element analysis.

11 CONCLUSIONS

The analysis of the performance of treated ground and insitu testing showed that treatment of cohesionless soils or slightly cohesion soils which do not exist in their densest state, or fill soil by the V.D.C. technique was the most suitable and economic for the site being treated.

Ground improvement, using the V.D.C. technique is capable of increasing the bearing capacity and decreasing the compressibility of natural soils or fills. The new techniques make available for use sites which, due to poor ground conditions, would not previously have been considered economic to develop.

Fill material like very soft cohesive soils, domestic refuse, local authority debris, highly organic material and unstable industrial wastes are less likely to be suitable for ground stabilization by the V.D.C. process.

It has been found out that while the process of stone column installation by drophammer reduced the settlement, the process using a diesel hammer driving a conic shaped casing did not.

A theoretical approach based on a simplified plastic analysis for predicting the depth of influence of heavy tamping is presented. The proposed approach makes allowance for soil type and strength, pore water pressure and diameter of the falling weight and is in good agreement with the site results and the other emprical methods.

It has been shown that finite layer method as well as finite element method were capable of modelling the correct behaviour of soil treated by stone columns.

The results of settlements calculated from the elastic solution, finite element, finite layer and other current methods is in good agreement with the results of the large scale plate load test.

12 REFERENCES

Balaam, N.P., Poulous, H.G. 1983. The Behaviour of Foundations Supported by Clay Stabilized by Stone columns. Proc. 8th Eur. Conf. SMFE Helsinki. pp 199-204.

Baumann, V. and Bauer G.E.A. 1974. The Performance of Foundations of Various Soils Stabilized by the Vibro-Compaction Method. Can. Geotéch. J. Vol. 11. pp 509-530.

Bell, F.G. 1975. Methods of Treatment of Unstable Ground. London Newnes Butterworths

Brandl, H. and Sadgorski, W. 1977. Dynamic Stresses in Soils Caused by Falling Weights. Proc. 9th Int. Conf. SMFE Tokyo Vol.2 pp 187-194.

Broms, B.B. 1983. Soil Stabilization. Proc. 8th Eur. Conf. SMFE Helsinki, pp 1289-1301.

Charles, J.A., Burford, D. and Watts, K.S. 1981. Field studies of the effectiveness of dynamic consolidation. Proc. 10th Int. Conf. SMFE Stockholm. Vol. 1. pp 617-622.

Datye, K.R. and Nagaraju S.S. 1981. Design Approach and Field Control for Stone Columns. Proc. 10th Int. Conf. SMFE Stockholm. Vol.1. pp 637-644.

Eggestad, A. 1983. Improvement of Cohesive Soils. Proc. 8th Eur. Conf. Helsinki. pp 991-1007.

Greenwood, D.A., Thomson, G.H. 1984. Ground Stabilization: Deep Compaction and Grouting. Thomas Telford Limited, London.

Hansbo, S. 1978. Dynamic Consolidation of Soil by Falling Weight. Ground Engineering, July pp 27-30.

Harder, L.F., Hammond, W.D., Ross, P.S. 1984. Vibrofloation compaction at Thermalito Afterbay. J. Geot. Eng. ASCE (110), GT1, PP 57-70.

Hughes, J.M.O. and Withers, N.J. 1974). Reinforcing of soft Cohesive Soils with Stone Columns. Ground Engineering, May, pp 42-49.

Jessberger, H.L. and Beine, R.A. 1981. Heavy Tamping; Theoretical and Practical Aspect Proc. 10th Int. Conf. SMFE Stockholm, May, pp 695-699.

Jumikis, A.R. 1969. Theoretical Soil Mechanics. Van Nostrand, Reinhold, pp.194-230.

Leonards, G.A., Cutter, W.A. and Holtz R.D. 1980. Dynamic Compaction of Granular Soils. J. Geot. Eng. ASCE (106), GT1, pp 35-44.

L'Herminier, R. 1967. Mecanique des soils et des chaussees. C.H.E.C. Paris XVI.

Lukas, R.G. 1980. Densification of Loose Deposits by Pounding. J. Geot. Eng. ASCE (106), GT4, pp 435-446.

Mayne, P.W., Jones J.S., and Dumas, J.C. 1984. Ground Response to Dynamic Compaction. J. Geot. Eng. ASCE (110), GT6, pp 757-774.

McKenna, J.M., Eyre, W.A. and Wolstenholme, D.R. 1975. Performance of an Embankment Supported by Stone Columns in Solf Ground. Geotechnique 25, NO.1, pp 51-59.

Menard, L. and Broise, Y. 1975. Theoretical and Practical Aspects of Dynamic Consolidation. Geotechnique 25, No.1, pp 3-18.

Mitchell J.K. 1981. Soil Improvement State-of- the-Art Report. Proc. 10th Int. Conf. SMFE Stockholm, Vol. 4, pp 509-565.

Rao, B.G. and Ranjan G. 1985. Settlement Analysis of Skirted Granular Piles. J. Geot. Eng. ASCE Vol. 111 No. 11, Nov. pp 1264-1283.

Schlosser, F. and Juran, I. 1979. Design Parameters for Artificially Improved Soils. Proc. 7th Eur. Conf. SMFE Brighton Vol. 5, pp 197-225.

Sinitsin, A.P. 1983. Improving the Bearing Capacity of Two-Layered Soil. Proc. 8th Eur. Conf. SMFE Helsinki, pp 307-308.

Smoltczyk, U. 1983. Deep Compaction State-of- the-Art Report. Pro. 8th Eur. Conf. SMFE Helsinki, pp 1105-1116.

The University of Sydney 1986. Prediction of settlements for Mira Monte development site. School of Civil and Mining Engineering, Investigation Report No. S584.

Thorburn, S. 1975. Building Structures Supported by Stabilized Ground. Geotechnique 25, No.1. pp 83-94.

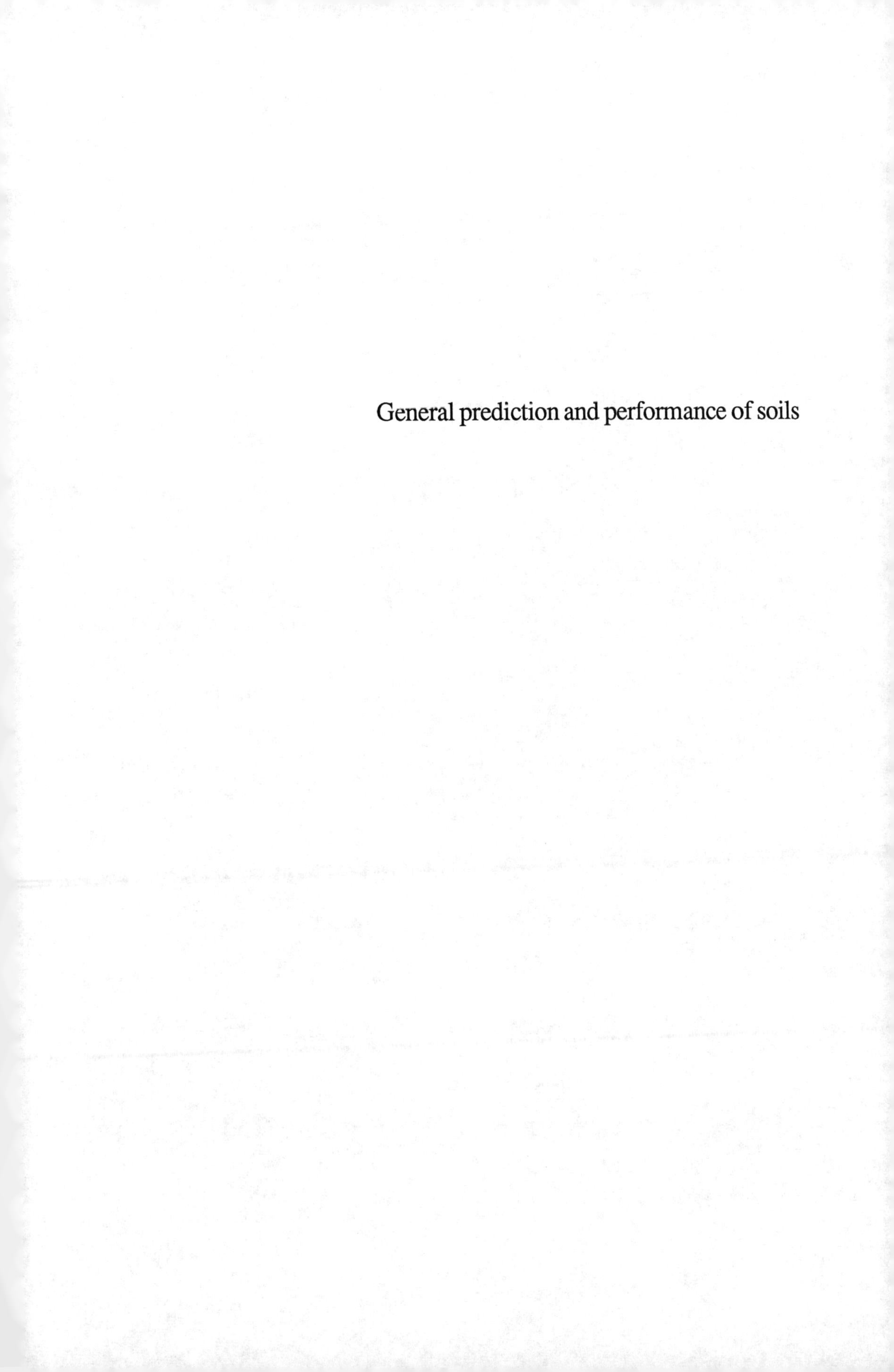

General prediction and performance of soils

Prediction and Performance in Geotechnical Engineering / Calgary / 17-19 June 1987

Influence of coarse particles on compressibility of soils

B.R.Srinivasa Murthy, T.S.Nagaraj & Bindumadhava
Department of Civil Engineering, Indian Institute of Science, Bangalore

ABSTRACT : In nature,soil consists of soil constituents varying in size from colloidal to coarse sand. The available methods to predict the compressibility of soils with liquid limit as a parameter do not account for coarse particles. Liquid limit and oedometer tests have been carried out on soils using sand and glass particles as admixtures. The test results have been compared with the generalised flow line and compressibility equation respectively to bring out the influence on physico-chemical interactions. It has been concluded that coarse particles will only reduce liquid limit linearly and correspondingly affect the comressibility behaviour. The compressibility of fine grained soils having coarse particles can be predicted using modified liquid limit.

1 INTRODUCTION

In nature, soils consist solid constituents of varying size from colloidal range to coarse sand and gravel particles. This wide range has been conveniently divided to form two groups of soils viz., fine grained and coarse grained. Though both fine grained and coarse grained soils have much in common as the members of the same family of particulate materials, they exhibit dissimilar and different behaviours. The basic difference in the behaviours of these two constituent groups make the prediction of compressibility behaviour of natural soils which may contain materials of both the groups, a complex exercise.

Several attempts, both empirical and rational, have been made in the past to model and predict the compressibility of fine grained soils using their liquid limit as a parameter. The liquid limit of a soil is determined only on soil fraction finer than 425 µm. Hence, this parameter alone cannot adequately account for the presence of particles coarser than 425 µm. This forms a limitation of the predictive methods when applying them to natural soils containing particles coarser than 425 µm.

It is attempted in this paper to examine the influence of coarse particles on both liquid limit and compressibility behaviour of fine grained soils. Further, it is attempted to formulate a method to predict the compressibility of such natural soils using their liquid limit values, determined from a standard method.

2 BACKGROUND INFORMATION

It is more than clear that in coarse grained soils the intrinsic interparticle electrical forces are of negligible magnitude and do not influence the mechanical behaviour of these soils. This is because the dimensions of the particles are themselves much larger than the distances through which the physico-chemical interaction forces can act. Further, the magnitude of gravitational forces far exceeds the surface physico-chemical interaction forces resulting in large contact stresses between the particles. In contrast, in fine grained soils the interparticle electrical forces, controlled by the physical and chemical states of the soil influence the mechanical behaviour and in particular the compressibility behaviour. The interparticle electrical forces are due to osmotic repulsion and van der Waals attraction.

Nagaraj and Srinivasa Murthy (1983) have shown from micro mechanistic considerations of soil behaviour that the influence of the interacting electrical forces can be effectively accommodated at the macro level by using liquid limit of soil as a parameter. This is because the liquid limit reflects the physico-chemical potential of the soil and at their liquid limit water contents all soils have the same magnitude of physico-chemical potential. This has resulted in generalising the compressibility behaviour of uncemented saturated fine grained soils in the form (Nagaraj and Srinivasa Murthy, 1985, 1986a and 1986b)

$$e/e_L = 1.122 - 0.2343 \log p \quad (1)$$

for normally consolidated soils and

$$e/e_L = 1.122 - 0.188 \log p_c - 0.0463 \log p \quad (2)$$

for over consolidated soils.

In the above equations the terms are defined as follows,

e is the void ratio at a consolidation pressure p in kPa,
e_L ($=w_L G$) is the void ratio at liquid limit and
p_c is the preconsolidation pressure in kPa.

Further, a number of empirical compression index equations which have liquid limit or natural water content as a parameter, are in vogue. It has been shown by Nagaraj and Srinivasa Murthy (1986a) that all these forms of equations can be related to generalised behaviours defined by the above two equations.

The above discussions clearly indicate that the compressibility of saturated fine grained soils can be predicted using the liquid limit of the soil. The majority of the soils in their natural states contain particles coarser than 425 µm. The implications of extending these predictive methods to such soils need further examination.

3 BASIC CONSIDERATIONS

Liquid limit is an index property of fine grained soils. It reflects the physico-chemical potential of these soils. Generally there is an agreement that the liquid limit depends on the base exchange capacity and the specific surface of the fine grained soil (Farrar and Coleman, 1967). The addition of coarse particles will reduce the specific surface proportionately. This indicates a possible linear reduction of liquid limit of a soil with the addition of coarse particles. Seed et al (1963) and Lupini et al (1981) have indicated such a linear reduction of liquid limit of clays with the addition of sand. However, in their investigation only sand particles finer than 425 µm could be used, to account for the specifications in liquid limit determination. Hence, it becomes logical to examine the influence of particles coarser than 425 µm on liquid limit, before extending the method to predict compressibiltiy of natural soils.

Nagaraj and Jayadeva (1981) have shown from micro - mechanistic consideration of soil behaviour that the flow curves of all the soils can be generalised in the form:

$$(w/w_L) = a - b \log N \quad (3)$$

for percussion test and

$$(w/w_L) = a + b\,D \quad (4)$$

$$(w/w_L) = a + b \log D \quad (5)$$

for cone penetration test.

where,
w is the water content corrosponding to N number of blows required to close the standard groove or that at a cone penetration of D mm.
w_L is the liquid limit of soil corresponding to 20 mm penetration or 25 blows.

Now it can be regarded that if the physico-chemical forces are controlling the behaviour of a soil, its flow line can be generalised in the form of above equation. Hence, it is attempted in this investigation to examine the possibility of generating and generalising the flow curves for fine grained soils with particles coarser than 425 µm. In the standard Casagrande percussion testing method, it is very difficult to cut the groove in soils containing coarser particles. In this investigation the cone penetrometer, wherein the coarser particles can be directly used, has been used to generate the flow curves.

For a soil in which the physico-chemical interaction forces control the behaviour, it has been indicated earlier that the compressibility behaviour of a saturated uncemented soil can be generalised in the form of equation (1) and (2). Hence, it is attempted in this investigation to generate the e vs log p plots of fine grained soils with particles coarser than 425 µm.

4 EXPERIMENTAL PROGRAMME

To bring out the effects of coarse particles on both the liquid limit and compressibility behaviour of fine grained soils, liquid limit and oedometer experiments were carried out with the parametric variations as indicated in Table 1. Three fine grained soils with a wide variation in liquid limit values have been selected. For comparison, their liquid limit and compressibility characteristics have been determined for fractions finer than 425 µm, which has been considered as soil without coarse fractions. To bring out the possible influence of coarse particles and their surface frictional characteristics two different types of coarse particles having extreme surface fractional characteristics have been selected. They are:

1. cuboidal and angular crushed quartzite of uniform size 1 mm.
2. spherical and smooth uniform glass particles of size 1 mm.

The different percents of coarse particles added as admixtures have been varied upto 40%. The liquid limit of the soil with an admixture has been taken as that of the water content at 20 mm penetration. The oedometer tests have been carried out from an initial water content equal to their respective liquid limits. The results have been presented in the form of graphs in Figs. 1 through 3.

5 DISCUSSIONS OF TEST RESULTS

Figure 1 indicates the flow lines of soils with different percent of admixtures. In the standard method of liquid limit determination, all the combinations of solid and admixture ought to have the same liquid limit, since, fractions coarser than 425 µm are removed. However, their inclusions resulted in different flow curves (Fig.1). It is also observed that these flow lines are not generalisable

Table 1 Nomenclature for soils and admixtures along with the liquid limit used in the experimental programme.

Sl No.	Soil	liquid limit
1	A (Red earth)	41.8
2	B (Black cotton soil 1)	68.2
3	C (Black cotton soil 2)	92.0
4	A: Glass = 80:20	34.0
5	A: Sand = 80:20	34.4
6	A: Glass = 60:40	26.8
7	A: Sand = 60:40	27.0
8	A: Glass : Sand = 60:20:20	27.0
9	B: Glass = 80:20	54.5
10	B: Sand = 90:10	58.0
11	B: Glass = 60:40	41.9
12	B: Sand = 60:40	42.0
13	B: Glass : Sand = 60:20:20	42.0
14	C: Glass = 80:20	75.0
15	C: Sand = 80:20	72.5
16	C: Glass = 70:30	63.5
17	C: Sand = 60:40	55.8
18	C: Glass : Sand = 60:20:20	55.8
19	A : B = 50 : 50	52.5
20	A : C = 50 : 50	63.0
21	B : C = 50 : 50	77.5

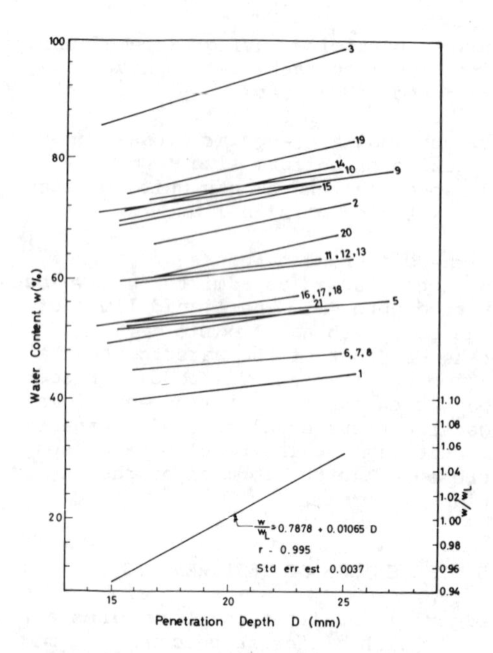

Figure 1. Experimental flow curves.

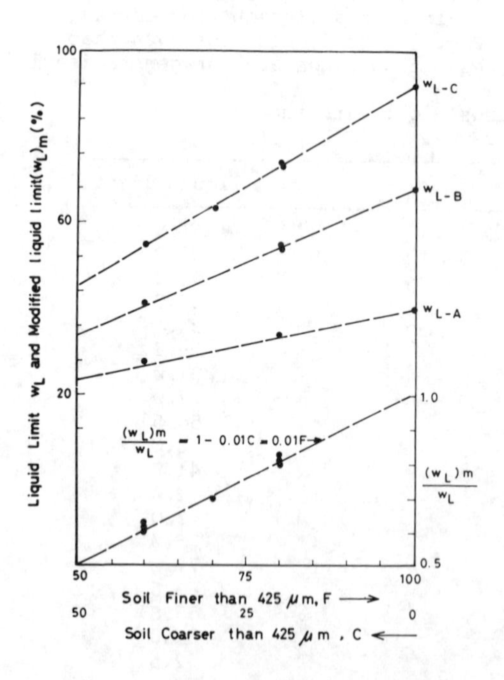

Figure 2. Modified liquid limit vs percent coarse particles.

using constant value of liquid limit. This indicates that the presence of coarse particles will decrease the value of liquid limit.

Figure 2 indicates the variation of liquid limit with percent coarse particles for all soils. This has resulted in linear law of variation similar to the observation made by Seed et al (1963). Further, for identical percentages of the two types of coarse particles the resulting flow lines are identical indicating no influence of shape and surface frictional characteristics of soil on their liquid limit. This phenomenon can be explained in terms of floating matrix concept. In this it is assumed that the coarse particles will only float in the matrix of fine grained soils resulting in only dilution of overall property. The quantification of diluation can be made in terms of linear variation of the liquid limit in the form,

$$(w_L)_m = w_L\ (F/100) \quad (6)$$

where,

$(w_L)_m$ is the modified liquid limit of the soil having fraction coarser than 425 μm of (100 - F) %

F is the percent fine passing 425 μm sieve

w_L is the liquid limit of the soil on fraction finer than 425 μm

In the above equation, the limiting value of (100 - F), can be defined by the resulting minimum value of the modified liquid limit of 25%. This is because, for all liquid limit less than 25%, there could be a physical frictional interaction between the coarser particles. It is further examined whether this modified liquid limit would serve as an adequate parameter to predict the compressibility behaviour of fine grained soils with particles coarser than 425 μm. Figure 3 indicates e vs log p plots generated from one dimensional consolidation tests on soils having various percentages of coarse particles (Table 1). All tests have been started from their respective modified liquid limit. Also indicated in Figure 3 is the generalised e/e_L vs log p plot for all the e vs log p plots, which has resulted in a single line within the limits of permissible error. This mode of generalisation could not have been possible by taking the standard liquid limit value of each soil.

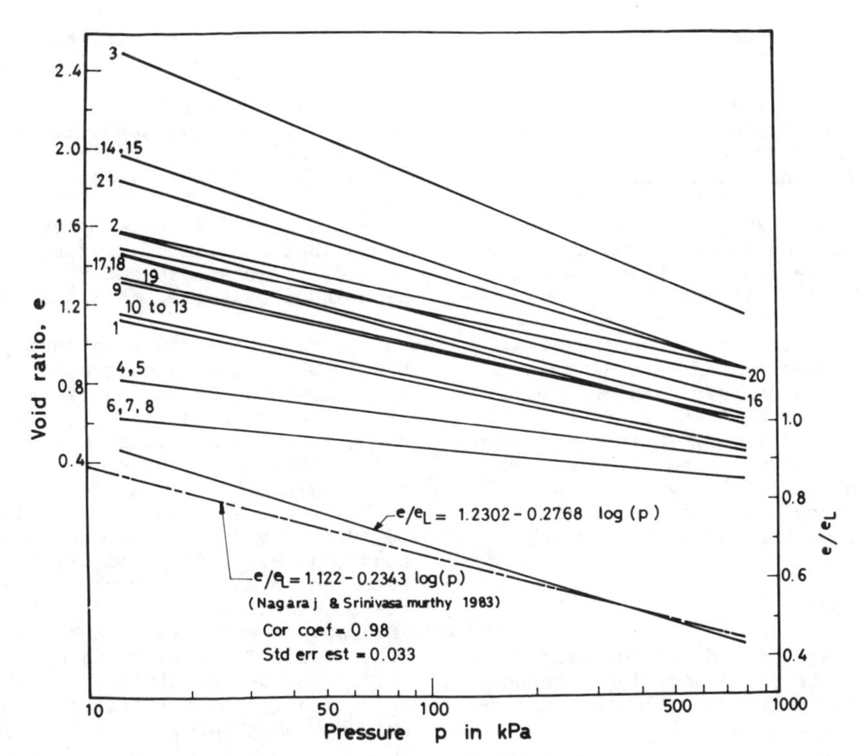

Figure 3. e vs log p and e/e_L vs log p relationship

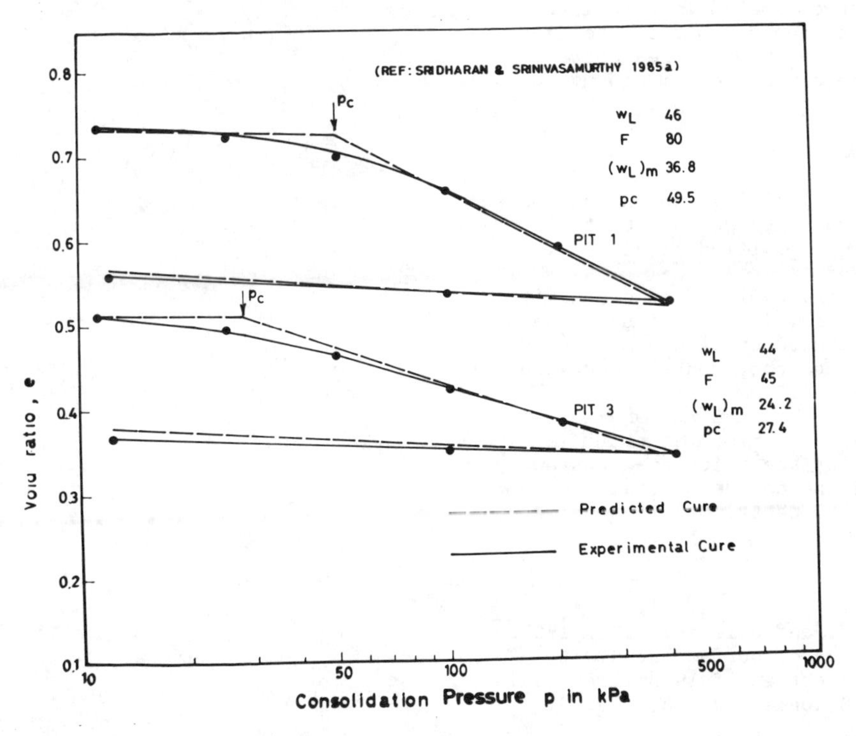

Figure 4. Prediction of e vs log p curve

For the size, shape and percent coarse particles used in this investigation, it implies that even with the presence of coarse particles the compressibility behaviour is controlled by the physico-chemical interaction rather than by the frictional interaction. From the above discussions it is clear that the compressibility of soils having coarse particles can be predicted using any of the predictive models, while modifying the liquid limit as per the linear law of equation 6. This aspect has been examined for predicting the compressibility of overconsolidated natural soils with particles coarser than 425 µm. Using the equation 1 and 2 and the modified liquid limit, it is clear from Figure 4 that the prediction can be made quite accurately, which is not a possibility by using the standard liquid limit of the soil.

6 CONCLUSIONS

From the basic considerations, experimental results and discussions presented in this paper the following conclusions can be brought out:

1. For the shape, size and percent of coarse particles used in this investigation, the liquid limit of the fine grained soil with particles coarser than 425 µm can be determined using linear law.

$$(w_L)_m = w_L\ (F/100)$$

However, the limiting percent of coarse particles could be defined by the resulting minimum modified liquid limit value of 25%.

2. Using the modified liquid limit the compressibility behaviour of saturated uncemented soils can be predicted, even when they contain particles coarser than 425 µm.

3. Generally the percent of coarse particles used in this investigation, and the surface frictional characteristics of the coarse particles do not influence the behaviour of the soil.

REFERENCES

Farrar, D.M. and Coleman, J.D., 1967. Correlation of surface area with other properties of 19 British soils. J. Soil Science Soc. 18:118-124.

Lupini, J.F., Skinner, A.E. and Vaughan, P.R., 1981. The drained residual strength of cohesive soils. Geotechnique. 32:181-213.

Nagaraj, T.S. and Jayadeva, M.S., 1981. Reexamination of one point method of liquid limit determination. Geotechnique. 31:413-425.

Nagaraj, T.S. and Srinivasa Murthy, B.R., 1985. Prediction of the preconsolidation pressure and re-compression index of soils. ASTM,GTJ. 8:199-203

Nagaraj, T.S. and Srinivasa Murthy, B.R., 1986a. A critical re-appraisal of compression index equations. Geotechnique. 36:27-32.

Nagaraj, T.S. and Srinivasa Murthy, B.R., 1986b. Prediction of compressibility of over consolidated soils. ASCE. GT Divn. 112:484-488.

Seed, H.B., Woodward, R.J. and Lundgren, R., 1963. Prediction of swelling potential for compacted clays. Trans. ASCE. 128:1443-1477.

Sridharan, A. and Srinivasa Murthy, B.R., 1985. Consultancy report No. CP.33/122/85-156, Civil Engineering Department. Indian Institute of Science, Bangalore.

Prediction and Performance in Geotechnical Engineering / Calgary / 17-19 June 1987

Dynamic reactions of composite soil medium

H.T.Chen
National Central University, Chungli, Taiwan

M.L.Liu
Southern Asia Engineering College, Chungli, Taiwan

ABSTRACT: Under the assumption of plane strain, the two-dimensional boundary element method is used to compute the dynamic reactions of a composite soil medium composed of a cylindrical inner zone and an outer infinite zone. It is found that the mass of inner zone has significant effect on the dynamic soil reactions. For a weak inner zone, the mass of that zone decreases the values of real part and increases those of imaginary part (radiation damping), as compared with the results of massless inner zone; also, as the size of inner zone increases, significant discrepancies are observed between the two sets of results.

1 INTRODUCTION

An important step in the dynamic soil-structure interaction analysis using substructure approach (Roesset 1980) is the determination of foundation stiffnesses. Over the past years, several methods have been proposed to compute the dynamic stiffnesses of foundations of arbitrary shape embedded in, or on the surface of, an elastic layered halfspace or soil stratum of finite depth (Gazetas 1983). Among them was the one specifically for embedded foundations, which replaces the soil along the side walls of foundation by equivalent Winkler springs (Novak & Sachs 1973). Under the assumption of plane strain, the dynamic stiffnesses of these springs, which are complex functions of frequency, have been computed analytically in terms of Bessel functions for a circular disk (Novak, Nogami & Aboul-Ella 1978) and by boundary element method for disks of arbitrary shape (Chen 1984). Both results were obtained for the case where the bond between the perimeter of the disk and the surrounding homogeneous, linearly viscoelastic soil medium is perfect; this is due to the fact that for most cases the dynamic stiffnesses of foundations are computed assuming perfect bond between soil and foundation walls and homogeneous soil with linearly viscoelastic behavior in the horizontal direction.

However, in practical situation, the bond is seldom perfect and separation may occur over the contact area between soil and foundation walls; furthermore, the soil may experience high level of strain in the vicinity of the foundation, and the soil medium may not be homogeneous in the horizontal direction due to, for example, backfill. A rigorous solution to account for these effects is difficult. In order to approximate these situations, the analytical solutions in terms of Bessel functions for the dynamic stiffnesses of equivalent springs corresponding to a circular disk of unit length surrounded by a composite soil medium of two different properties were presented (Novak & Sheta 1980); the nonlinear effects can then be approximated by introducing reduced shear modulus and increased material damping ratio for the inner zone which is the one immediately adjacent to the disk. However, to avoid certain mathematical difficulties, the inner zone is assumed to be massless. The effect of mass of inner zone on the dynamic soil reactions of this composite medium for axial and rocking motions has been investigated (Chen 1985). It was found in that study that a term in the analytical solution for rocking motion was left out (Novak & Shetá 1980; Novak 1985). In this paper the effect is studied for horizontal and torsional motions.

2 FORMULATION

The composite soil medium studied by Novak and Sheta (1980) is shown in Fig. 1, which is also used in this study. It is a rigid massless, circular body of infinite length embedded in a composite soil medium: an inner zone of cylindrical shape and an outer infinite zone; each zone is separated by a circular interface and is assumed to be homogeneous, isotropic and linearly elastic with material damping of hysteretic type (independent of frequency). Assuming a plane strain problem, the dynamic soil reactions are computed by treating the body as a disk of unit length and imposing unit harmonic displacements (or rotations) along the perimeter of the disk; only the steady-state solutions are considered. The dynamic soil reactions thus computed are the dynamic stiffnesses of the equivalent springs.

Due to the nature of the problem, a two-dimensional boundary element method is employed to compute the dynamic soil reactions. The use of this method requires only the discretization of the boundaries and allows one to consider boundaries of arbitrary shape rather than being restricted to circular shape as studied by Novak and Sheta (1980); In addition, the mass of inner zone can also be included in the solutions. It is known that neglecting the body forces, one can obtain the following relation for the boundary displacements and boundary tractions.

$$Hu=Qt \tag{1}$$

where u and t are the vectors of boundary displacements and boundary tractions, respectively, and the matrices H and Q are obtained through the integration of fundamental solutions (Banerjee & Butterfield 1981). The fundamental solutions for the current study have been presented (Cruse & Rizzo 1968; Cruse 1968).

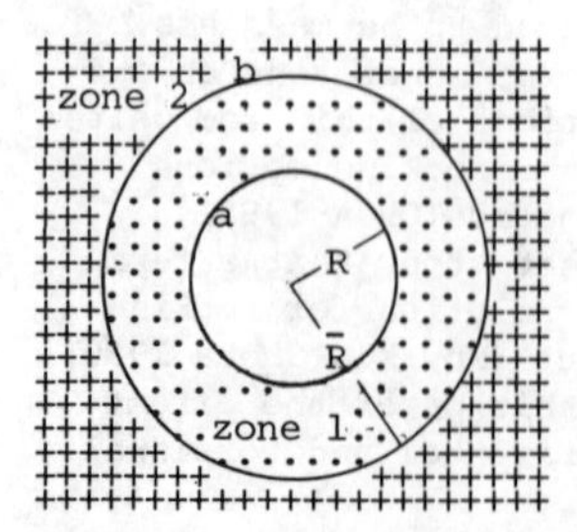

Fig.1 Composite soil medium

For zone 1, Eq. (1) is partitioned according to the boundaries as

$$\begin{bmatrix} H_{aa}^{(1)} & H_{ab}^{(1)} \\ H_{ba}^{(1)} & H_{bb}^{(1)} \end{bmatrix} \begin{bmatrix} u_a \\ u_b \end{bmatrix} = \begin{bmatrix} Q_{aa}^{(1)} & Q_{ab}^{(1)} \\ Q_{ba}^{(1)} & Q_{bb}^{(1)} \end{bmatrix} \begin{bmatrix} t_a \\ t_b \end{bmatrix} \tag{2}$$

For zone 2, the relation is simply as shown below.

$$H_{bb}^{(2)} u_b = Q_{bb}^{(2)} t_b \tag{3}$$

In the expressions above, the superscript denotes the zone and the subscript denotes the boundary. Along the boundary b, compatibility requires that

$$u_b^{(1)} = u_b^{(2)} \tag{4}$$

and

$$t_b^{(1)} = -t_b^{(2)} \tag{5}$$

By substituting these relations into Eqs. (2) and (3), the following result can be obtained through certain manipulations.

$$(H_{aa}^{(1)} - K_{ab} M_{bb} H_{ba}^{(1)}) u_a^{(1)} =$$

$$(Q_{aa}^{(1)} - K_{ab} M_{bb} Q_{ba}^{(1)}) t_a^{(1)} \tag{6}$$

where

$$M_{bb} = (Q_{bb}^{(1)} + H_{bb}^{(1)} (H_{bb}^{(2)})^{-1} Q_{bb}^{(2)})^{-1}$$

and

$$K_{ab} = Q_{ab}^{(1)} - H_{ab}^{(1)} (H_{bb}^{(2)})^{-1} Q_{bb}^{(2)}$$

Eq. (6) relates only the displacements and the tractions along the boundary a where the unit harmonic displacements or rotations are specficied. The computed tractions are then used to obtain the dynamic stiffnesses of the equivalent springs.

3 RESULTS AND DISCUSSIONS

Based on the previous formulation, different parametric studies have been made for the composite soil medium (Chen 1986). In the following discussions, the parameters with subscript 1 pertain to the inner zone, and those with subscript 2 are for the outer zone. Both the disk and the interface between the two zones are circular, and their radii are denoted by R and $\bar{R}$, respectively. The dimensionless stiffnesses, which are complex functions of frequency, are expressed in terms of real and imaginary parts and are plotted with respect to dimensionless frequency a_0, the product of frequency of excitation and the characteristic length of the body (radius for circular shape) divided by the shear wave velocity of soil in outer zone. The dimensionless stiffnesses are obtained by dividing by G_2 and G_2R^2 for the horizontal stiffness and the torsional stiffness, respectively, where G is the shear modulus of soil. The Poisson ratio for both zones is 0.4 and the material damping ratio is 0.05 for the outer zone; in addition, the following ratios are used for all the cases: G_1/G_2 =0.5 and D_1/D_2 =2, where D is the material damping ratio.

The boundary element solutions are first verified by comparing with the analytical solutions presented by Novak and Sheta (1980) for m_1/m_2=0, i.e., massless inner zone. The results are shown in Figs. 2 and 3. Excellent agreement is observed for the torsional stiffness. However, there are differences between the two solutions for the horizontal stiffness. In the real part, the agreement between the two solutions is excellent for a_0 less than 1, while differences increase with increasing a_0. On the other hand, the Novak's solution is always greater than the boundary element solution for the imaginary part; the differences also increase with increasing a_0. The reason for such differences may be due to the assumption adopted in the Novak's solution which considers the interface between the two zones to remain circular during the motion. Such assumption imposes constraints to the system, causing the stiffness obtained to be larger than the boundary element solution which allows the interface to deform freely.

The effect of mass of inner zone on the dynamic stiffnesses of soil is studied considering four different mass ratios

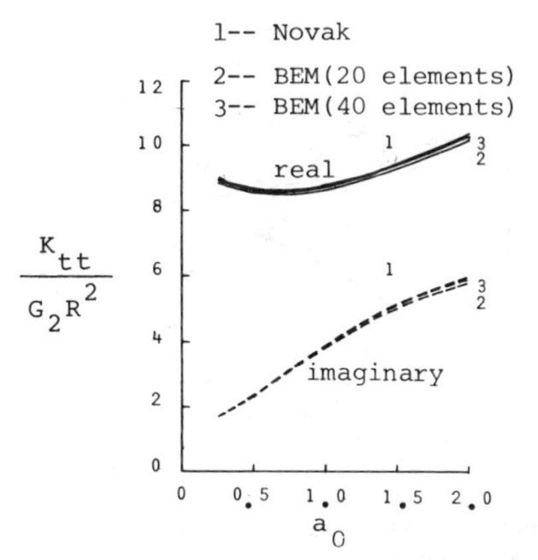

Fig.2 Comparison of BEM and analytical solution for torsional stiffness

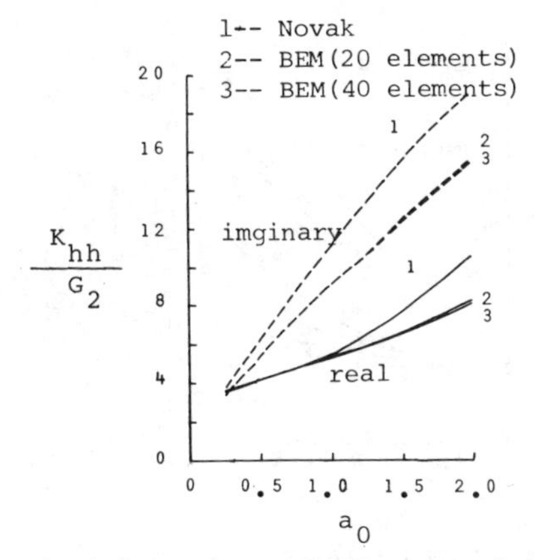

Fig.3 Comparison of BEM and analytical solution for horizontal stiffness

(m_1/m_2) 0, 0.125, 0.5 and 1. The results are shown in Figs. 4 and 5 for the horizontal stiffness and the torsional stiffness, respectively, where curves 1, 2, 3 and 4 represent the results for mass ratios 0, 0.125, 0.5 and 1, respectively; for the horizontal stiffness, curve 5 is plotted from Novak's solution. It can be seen that for a given a_0 the values of the real part of each stiffness decrease with increasing mass ratio, while the opposite trend is observed for the imaginary part. On the other hand, the differences between the results of zero mass ratio and nonzero mass ratio increase as the dimensionless frequency increases, the relative differences being smaller for the imaginary part than for the real part. In general, the effect of mass of inner zone can be neglected for a_0 less than 0.5 for the real part and for a_0 less than 1

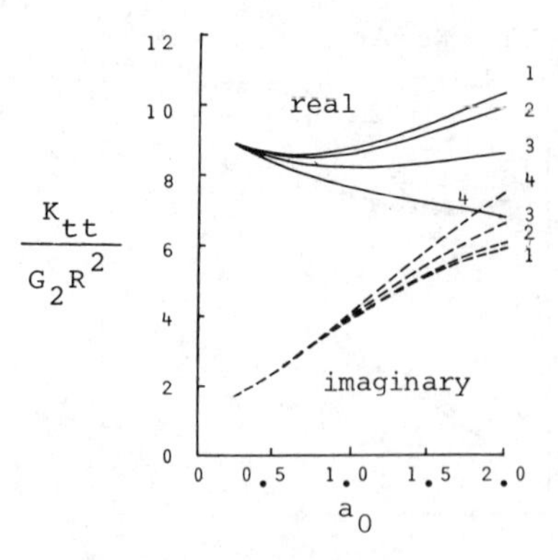

Fig.4 Effect of mass ratio on torsional stiffness

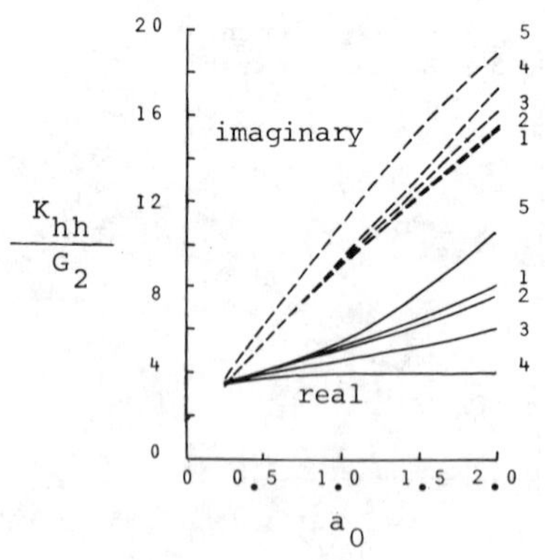

Fig.5 Effect of mass ratio on horizontal stiffness

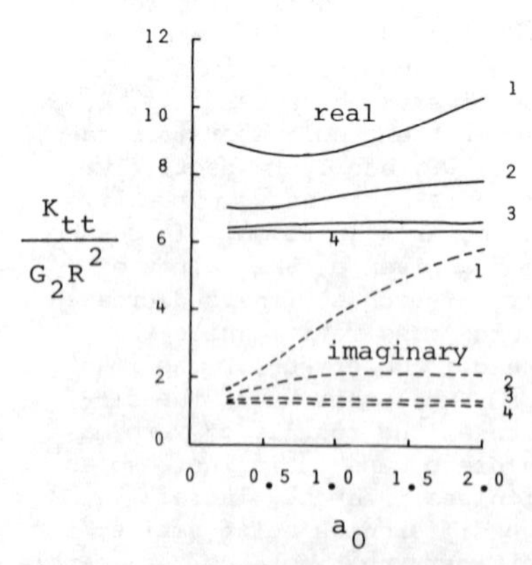

Fig.6 Effect of radius ratio on torsional stiffness for $m_1/m_2=0$

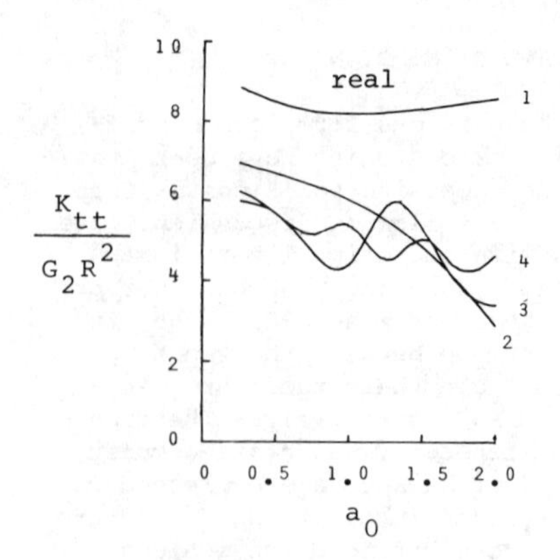

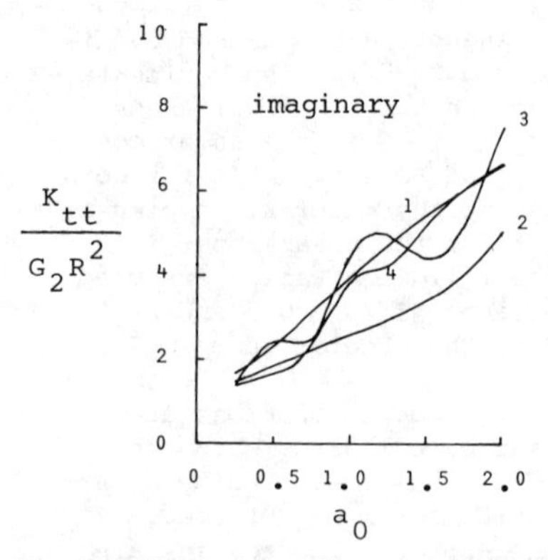

Fig.7 Effect of radius ratio on torsional stiffness for $m_1/m_2=0.5$

for the imaginary part; in addition, the differences between the results of mass ratios 0 and 0.125 are not significant over the range of a_0 considered.

The effect of size of inner zone on the dynamic soil stiffnesses is investigated considering radius ratios $(\bar{R}/R)$ 1.25, 2, 4 and 8 for the cases of mass ratios 0 and 0.5 which are represented by curves 1, 2, 3 and 4 in the figures, respectively. Figs. 6 and 7 are the results for the torsional stiffness, and the results for the horizontal stiffness are shown in Figs. 8, 9 and 10 with Fig. 9 being plotted from Novak's solution. From these figures, it can be observed that the behavior is quite dependent on the mass of inner zone. When the mass of inner zone is considered, the curves fluctuate with decreasing amplitude as the size of inner zone increases. However, for massless inner zone, the

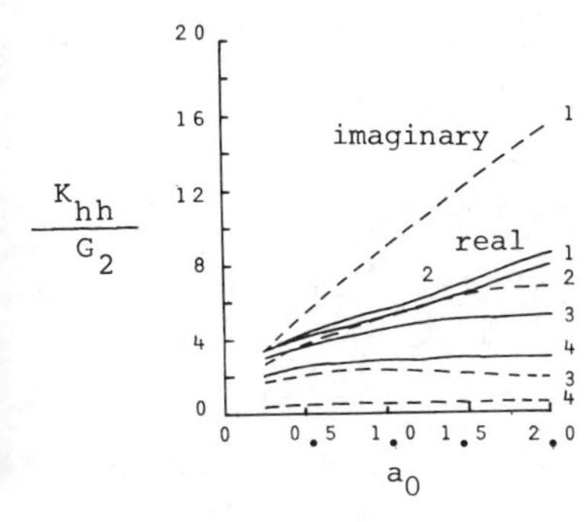

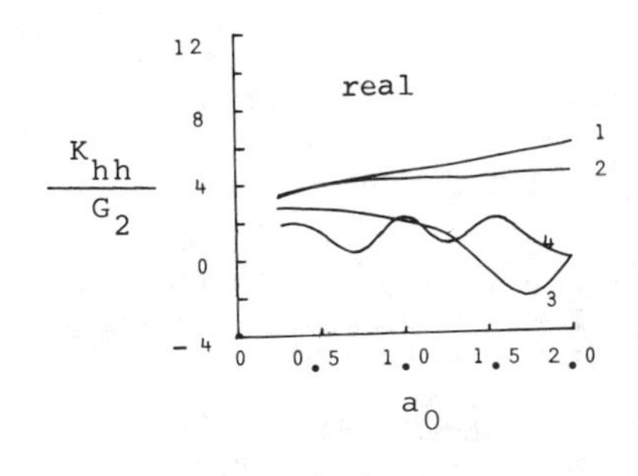

Fig.8 Effect of radius ratio on horizontal stiffness for $m_1/m_2=0$

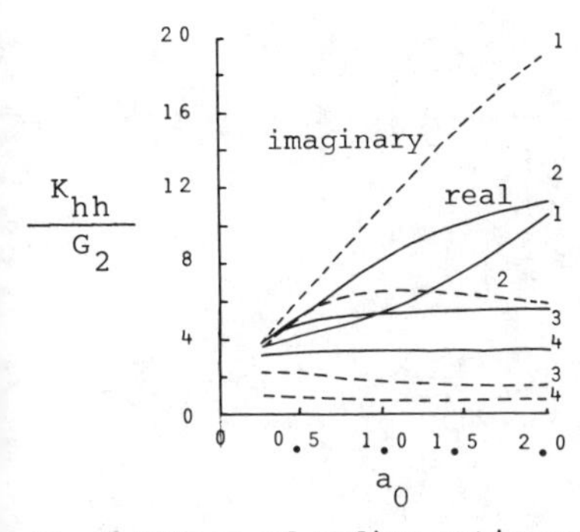

Fig.9 Effect of radius ratio on horizontal stiffness for $m_1/m_2=0$ (Novak's solution)

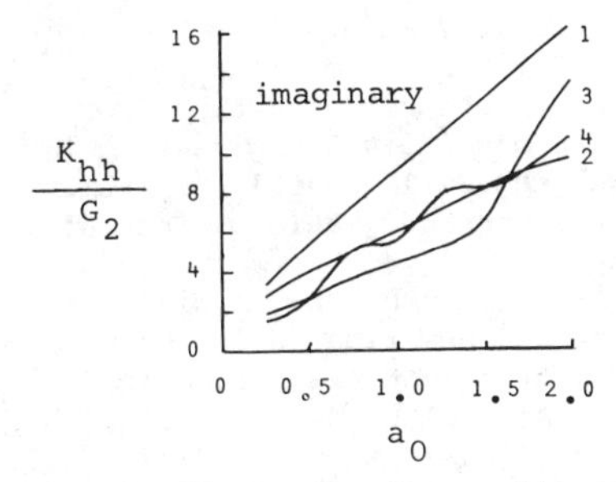

Fig.10 Effect of radius ratio on horizontal stiffness for $m_1/m_2=0.5$

values of the real part and the imaginary part decrease with increasing inner zone with no fluctuations except for the real part of the horizontal stiffness which shows the result of $\bar{R}/R=2$ is larger than that of $\bar{R}/R=1.25$. Novak's solution also shows the same trend except that the curves of radius ratios 1.25 and 4 cross each other at about $a_0=1$. This kind of difference may be attributed to the reflection of waves on the boundaries of the disk and two soil zones. Such reflection is neglected for not considering the mass of inner zone.

4 CONCLUSION

Under the assumption of plane strain, the dynamic reactions (or dynamic stiffnesses) of a composite soil medium composed of two different material properties are computed using two-dimensional boundary element method for horizontal and torsional motions. The disk and the interface between the two media are circular, and the inner zone is assumed to be weaker than the outer zone. The conclusions drawn from this study are as follows.

1. For a given set of properties, the differences in the dynamic stiffnesses for the cases with and without mass of inner zone increase as dimensionless frequency increases. Also, at a given dimensionless frequency, the values of real part decrease and those of imaginary part (radiation damping) increase as the mass of inner zone increases.

2. When the mass of inner zone is not considered, smooth curves are obtained as the size of inner zone increases; however, with the mass of inner zone included in the solution, the curves

fluctuate with decreasing amplitude.

The above conclusions indicate that the mass of inner zone has significant effect on the dynamic soil reactions, and can not be neglected for certain situations.

REFERENCES

Banerjee, P.K. & R.Butterfield 1981. Boundary element methods in Engineering science. London: McGraw-Hill.

Chen, H.T. 1984. Dynamic stiffness of nonuniformly embedded foundations. Geotechnical Engineering Report GR84-10, Department of Civil Engineering, The University of Texas at Austin.

Chen, H.T. 1985. Dynamic reactions of inhomogeneous soil for plane strain case. Proceedings of the ROC-Japan Joint Seminar on Multiple Hazards Mitigation, Vol. 1, National Taiwan University, Taipei, R.O.C.: 279-293.

Chen, H.T. 1986. Effects of inhomogeneity of soil on dynamic soil reactions for plane strain case. Civil Engineering Studies NCU-ENG-CE 12-86-001, Department of Civil Engineering, National Central University.

Cruse, T.A. & F.J.Rizzo 1968. A direct formulation and numerical solution of the general transient elastodynamic problem, I. Journal of Mathematical Analysis and Applications, Vol. 22: 242-259.

Cruse, T.A. 1968. A direct formulation and numerical solution of the general transient elastodynamic problem, II. Journal of Mathematical Analysis and Applications, Vol. 22:341-355.

Gazetas, G. 1983. Analysis of machine foundation vibrations: state of the art. Soil Dynamics and Earthquake Engineering Vol. 2, No. 1: 2-42.

Novak, M. & K.Sachs 1973. Torsional and coupled vibrations of embedded footings. Earthquake Engineering and Structural Dynamics, Vol. 2, No. 1: 11-33.

Novak, M., T.Nogami & F.Aboul-Ella 1978. Dynamic soil reactions for plane strain case. Journal of the Engineering Mechanics Division, ASCE, Vol. 104, No. EM4: 953-959.

Novak, M. & M.Sheta 1980. Approximate approach to contact effects of piles. In M.W. O'neill & R. Dobry (eds.), Dynamic response of pile foundations: analytical aspects, p. 53-79. New York: ASCE.

Novak, M. 1985. Personal correspondance.

Roesset, J.M. 1980. A review of soil-structure interaction. Seismic Safety Margins Research Program, Project III-soil-structure interaction, Lawrence Livermore Laboratory.

Prediction and Performance in Geotechnical Engineering / Calgary / 17-19 June 1987

Predicting the hydraulic conductivity of compacted clays

Kingsley Harrop-Williams
BDM Corp., McLean, Va., USA

ABSTRACT: A theoretical relationship is developed between the hydraulic conductivity and the easily measured dry unit weight and moisture content of a compacted clay liner. The development assumes that due to the low permeability of compacted clays, the soil never attains complete saturation and the line of optimum is below the zero air voids curve. The analysis then follows from the point of view of flow through unsaturated porous media.

1 INTRODUCTION

The suitability of a clay soil for the prevention of seepage depends on the hydraulic conductivity (permeability) of the soil. Experiments have shown that this property of the soil is affected by the type and effort of compaction. Also, for a given level and type of compaction, it generally reaches a minimum value on the wet side of optimum moisture content and at about maximum dry unit weight. This suggests that any given clay can be compacted to minimize the permeability.

It is, however, very difficult to monitor construction activities by directly measuring permeability insitu, or with insitu samples tested in a laboratory, due to the length of time required for testing. Therefore for real time control of clay placement to achieve a desired permeability, other parameters must be measured and the permeability implied. The most frequently used technique for the control of earth placement is to monitor the moisture content and dry unit weight and correlate them to the desired property. This is typically done for strength and displacement characteristics, and to a lesser extent for permeability. The obvious advantage for such an approach is real-time monitoring. Direct monitoring can be performed as part of the compaction process to provide immediate feedback regarding the acceptability of the compacted clay. If the permeability can be predicted directly from the moisture content and dry unit weight, the monitoring of the construction of clay-lined containment ponds will be greatly enhanced.

In this paper, a theoretical expression for predicting the permeability of compacted clays as a function of its dry unit weight and moisture content is derived. In particular it shows that this relationship is general for any permeant not interacting with the soil. It identifies the physical properties that are inherent to the fluid and soil, and the role they play in the prediction process. The result shows excellent agreement with experimental data for varying levels and types of compaction.

2 HYDRAULIC CONDUCTIVITY AND SATURATION

The pore water in clays can be thought of as travelling through a tortuous narrow channel in which capillary pressure develops. The equilibrium of forces in a full channel requires that the capillary pressure at full saturation be expressed by the equation (Wu 1976):

$$\pi R^2 P_b = 2\pi R\, \sigma \cos\beta \qquad (1)$$

where R is the radius of the channel and $\sigma\cos\beta$ is the surface tension.

In unsaturated flow the radius R is replaced by the hydraulic radius $R_h = R/2$. Also, the degree of saturation in a channel is proportional to the wetted area, and to the square of the hydraulic radius. Hence, for a proportionality constant C, the capillary pressure at partial saturation becomes:

$$P_c = \sigma \cos\beta / (S/C)^{1/2} \quad (2)$$

The change in capillary pressure with saturation is obtained by differentiating this equation, and is:

$$dP_c / dS = -P_c / (2S) \quad (3)$$

Imposing the boundary condition $P_c = P_b$ at $S = 1$, and solving equation (3), gives the degree of saturation as:

$$S = (P_b / P_c)^{\lambda} \quad (4)$$

where $\lambda = 2$. The substitution of equations (1) and (2) into equation (4) gives:

$$R_h = RS^{1/\lambda}/2 \quad (5)$$

Equation (4) was observed experimentally by Brooks and Corey (1966), who called P_b the bubbling pressure (air pressure needed to force air through an initially water saturated sample).They found $\lambda=2$ for typical porous media, λ less than 2 for soils with well developed structures, and λ greater than 2 for sands.

The general velocity on which permeability is based is $v = v_a A_f / A_t$, where A_t is the cross-sectional area of the soil, A_f is its wetted area, and $v_a = k_a i$, where (Wu 1976):

$$k_a = R_h^2 \gamma_f / (C_s \mu T^2) \quad (6)$$

Here i is the hydraulic gradient, γ_f is the unit weight of the fluid, μ is its viscosity, T is the toruosity of the channel, and C_s is a shape factor that depends on the cross-sectional shape of the channel. For circular cross-sections $C_s = 2$, and for square ones $C_s = 1.78$. The ratio A_f/A_t is equal to the ratio of liquid volume to soil volume, which is the product of the porosity, n , and the degree of saturation S. Further, Wyllie and Gardner (1958) showed that $T = T_1/S$, where T_1 is the tortuosity at complete saturation. Combining all of these observations with equation (5), we get v= ki, where the permeability is:

$$k = Bn(\gamma_f/\mu)S^{\alpha} \quad (7)$$

for

$$\alpha = (3\lambda + 2)/\lambda \quad (8)$$

and

$$B = R^2/(4C_s T_1^2) \quad (9)$$

The parameters α and B are grain size distribution and channel geometry constants, respectively.

3 PREDICTION FOR COMPACTED CLAYS

During compaction, the degree of saturation of a soil can be expressed in terms of the moisture content w, the specific gravity of the solids G, the unit weight of water γ_w , and the dry unit weight of the soil γ_d as

$$S = \gamma_d wG/(G\gamma_w - \gamma_d) \quad (10)$$

Also one can express the porosity as

$$n = 1 - \gamma_d/(G\gamma_w) \quad (11)$$

Hence, from equations (7), (10) and (11), the permeability, k, is defined in terms of the compaction variables γ_d , and w.

Experimental comparison of the general form of equation (7) for varying levels and methods of compaction was made using the data provided by Mitchell et al (1965) on a compacted Silty Clay. Regression curves using equation (7) on their data for varying compaction efforts and methods of compaction are shown in Figures 1 and 2, respectively. Here the permeant was water, and it was assumed that G = 2.7 (a typical value for clays). It was also observed that, due to the range of dry density shown in the compaction curves, the porosity is approximately constant. Hence, we can treat $K = \ln(Bn\gamma_f/\mu)$ as a constant. For the 30 lb, 46 lb and 50 lb compaction effort curves shown in Figure 1, the values of K were -25.15, -19.80 and -20.15, respectively. The corresponding values of α were -96.13, -28.86 and -18.60, with correlation coefficients of 0.94, 0.90 and 0.97, respectively. For the kneading and static compaction method curves shown in Figure 2, the values of K were -21.41 and -17.40, respectively. The corresponding values of α were -36.11 and -14.34, with correlation coefficients of 0.99 and 0.98, respectively.

DISCUSSION

In equation (7) the primary parameters dependent on the fluid are its viscosity and unit weight. This is true for steady state flow only. Before steady state is reached some fluids cause chemical exchanges with the soil that may affect the shape and tortuosity of the pore channels. Thus, altering the structure of the clay. The structure of the clay is one of the most

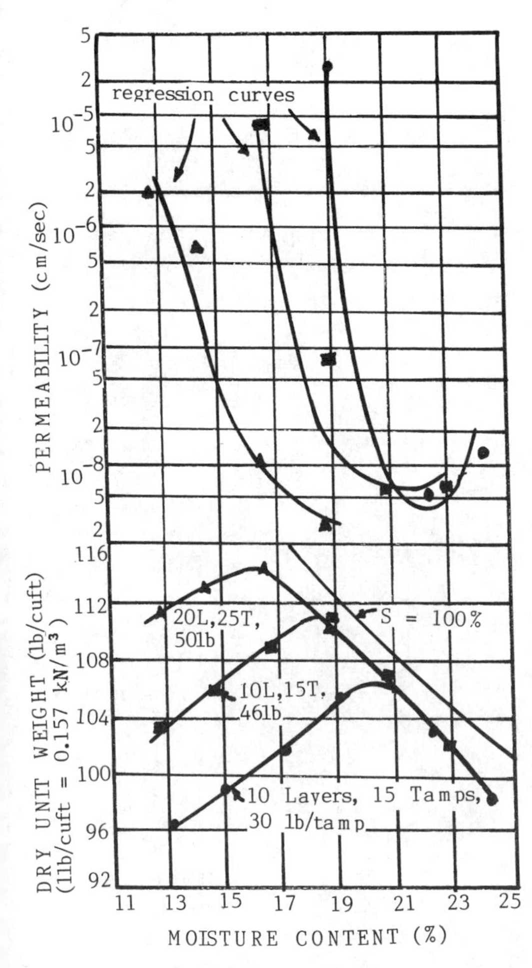

Fig.1 Effect of compaction effort (Data from Mitchell et al. 1965)

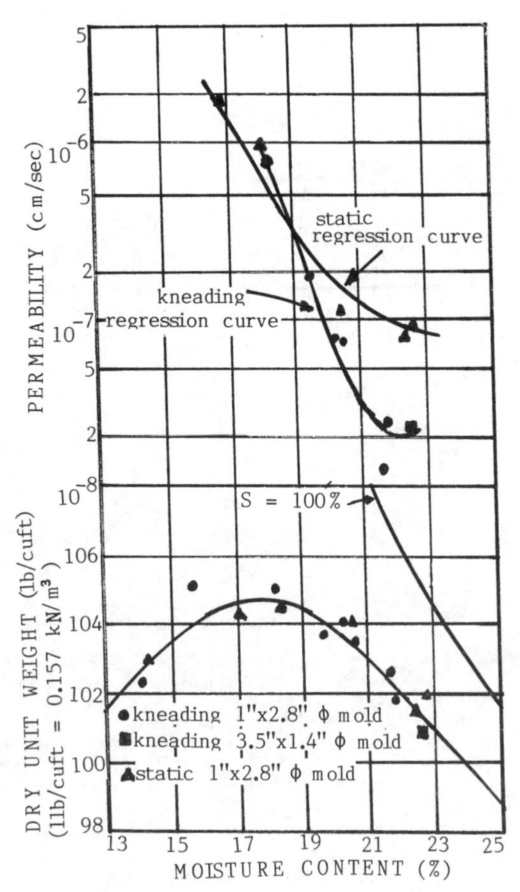

Fig.2 Effect of method of compaction (Data from Mitchell et al. 1965)

influential parameters on its permeability. It is represented here by parameters B and α , and is seen to depend greatly on the compaction effort and method of compaction.

The developed relationship between permeability and the easily measured dry unit weight and moisture content allows clay liners to be tested as they are being built. The empirical structure-specific parameters B and α, being sensitive to compaction effort and method of compaction, should be obtained by regression beforehand using laboratory or field determined permeability values of field compacted soils. With equation (7) and a fast moisture content and dry unit weight measuring device like the nuclear densometer, immediate permeability values can be obtained throughout the liner. Hence direct monitoring of construction is possible and increased sampling facilitated.

REFERENCES

Brooks, R. H., & A. T. Corey 1966. Properties of porous media affecting fluid flow. ASCE J. of Irrig. & Drain. 92,IR2.

Mitchell, J. K., D. R. Hooper, & R. G. Campenella 1965. Permeability of compacted clay. ASCE J. of Soil Mech. & Fdn. Eng. 91,SM4: pp 41.

Wu, T. H. 1976. Soil mechanics. Boston: Allyn & Bacon.

Wyllie, M. R., and G. H. Gardner 1958. The generalized Kozeny-Carman equation, a novel approach to problems of fluid flow. World Oil Prod. Sect. 210-228.

Downslope movements at shallow depths related to cyclic pore pressure changes

J.P.Burak & K.D.Eigenbrod
School of Engineering, Lakehead University, Thunder Bay, Ontario, Canada

ABSTRACT: The instrumentation of a flat slope affected by shallow slow ground movements indicated a relationship between ground movements and seasonal fluctuating pore pressures in the slope. From a triaxial testing program of undisturbed samples from the slope quantitative information on the rate of lateral deformations due to drained cyclic pore pressure changes was obtained. Laboratory results and field observations correlate reasonably well.

INTRODUCTION

Shallow, slow ground movements have been regularly observed on relatively flat slopes of less than 10° (for example, Skempton and DeLory, 1957; Williams, 1966). It has been suggested on one hand, that movements of this kind may result in slope failure (Yen, 1969); or alternatively that they continue for long periods without leading to failure (Kojan, 1967).

Slow soil movements in surficial slope materials have been related to heaving of the soil normal to the slope during freezing, followed by vertical settlement during thawing (Williams, 1966; Washburn, 1973). The process will clearly result in net downslope movements. However Williams (1966) showed that the observed rates of movement could not be caused by this process alone, and additional mechanisms

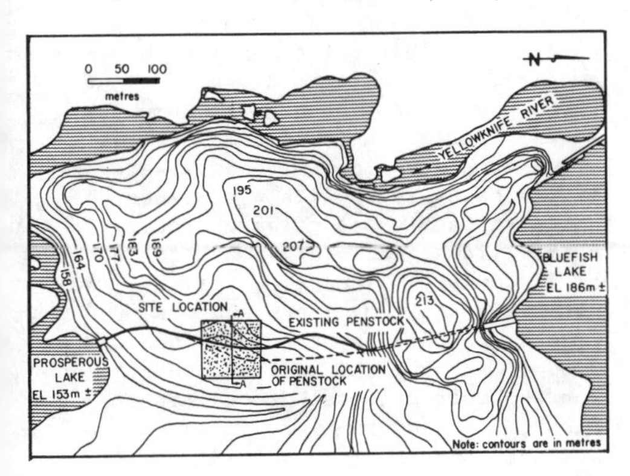

Fig. 1 General Site Plan

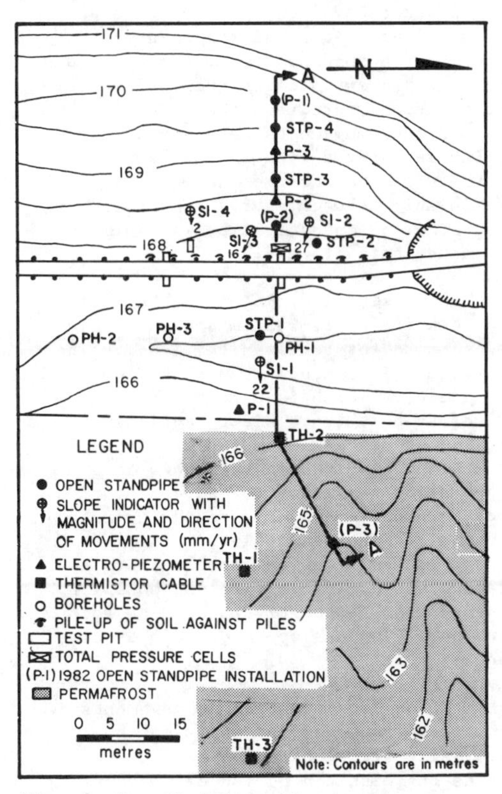

Fig. 2 Detailed Site Plan

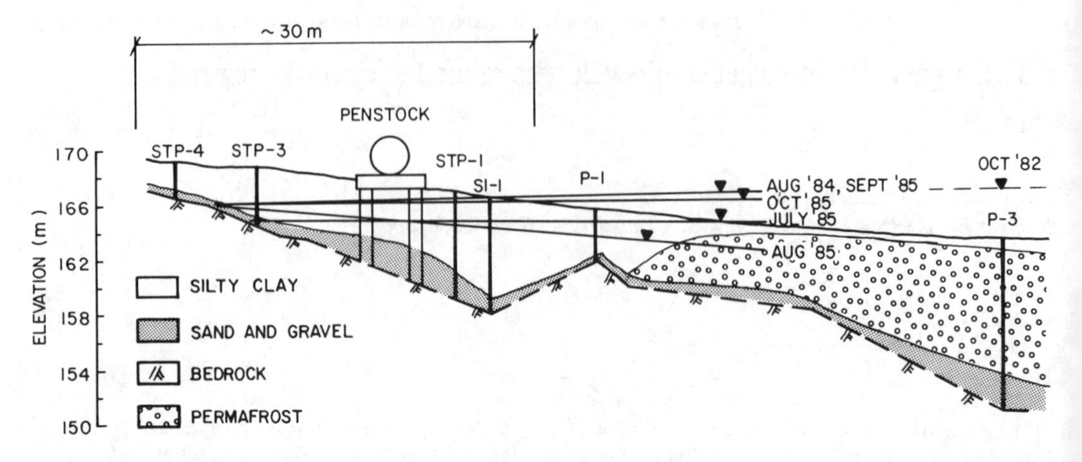

Fig. 3 Slope Section A-A (See Fig.'s 1 and 2) with observed piezometric levels.

are required to explain the higher rates that are commonly observed. McRoberts and Morgenstern (1974) and Tavenas and Leroueil (1981) suggested that slow shallow ground movements could be related to fluctuating pore-water pressure conditions caused by groundwater level changes in the slope, or by elevated porewater pressures that accompany a thawing front. However, due to insufficient data no quantified correlations were reported. Thus, when slow ground movements were encountered which interfered with an engineering structure at a location north of Yellowknife (Burak, 1986) a detailed geotechnical investigation was initiated to study this process.

The site is located on the Yellowknife River System approximately 50 km north of Yellowknife, N.W.T. The structure is a wooden penstock 2100 mm in diameter which is part of the Bluefish Lake Hydro Electric Plant. The penstock carries water from the south end of Bluefish Lake down to a generating station situated at the northernmost corner of Prosperous Lake (Fig. 1). The penstock runs in a north-south direction along a south east facing slope. In its present position the penstock is partly located on rock and partly on a slope underlain by clay and silt.

Slow continuous slope movements have occurred over the years on the clay-silt slope, necessitating on two occasions within a period of 50 years, the relocation and replacement of a section of the penstock.

Along the original location as indicated on Fig. 1 the penstock was based largely on a gravel embankment. Possibly due to thawing permafrost below the embankment the penstock experienced excessive deformation, which on one occasion caused a rupture of the penstock. Thus in 1968, the penstock was relocated to its present position. The upper portion was founded on a rock ledge, that was blasted into the side of a bedrock outcrop. Along the clay portions of the slope timber pile foundations (300 mm in diameter) were selected, carying square timber pile caps which act as a foundation system for the penstock. The pile caps and the associated piles support the penstock at approximately 2.7 metre centres and elevate the penstock above ground. The foundation depths of the piles are not known.

Twelve years later in 1980 excessive tilting of the piles became a concern, because large distortions of the penstock were leading to numerous leaks. During an inspection at that time, downhill slope movements were indicated by soil which was pushed up against a series of piles on their uphill sides, while a gap had formed on their downhill sides (see Fig. 2). No other signs of movements such as open cracks, or bulging were indicated on the slope. A stability analysis undertaken for the most critical ground water conditions and with very conservative strength parameters resulted in a factor of safety well above unity, indicating that the slope was not in a state of limit equilibrium.

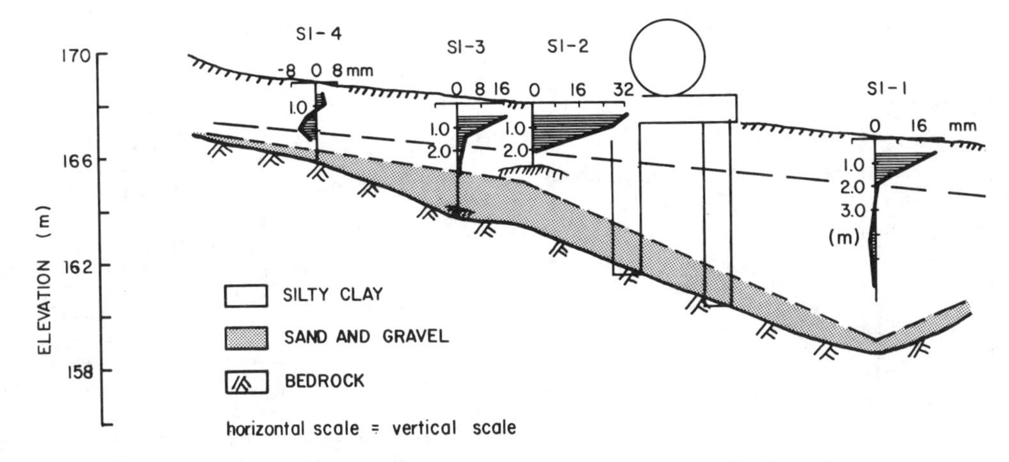

Fig. 4 Slope section A-A with observed lateral deformations (mm/year)

Eventually a portion of the pipe and its supports had to be replaced in 1983. The wooden pipe in this area, was substituted by a steel section which was bridged over the moving slope portion. In order to reduce the span of the pipe-bridge one of the bridge piers had to be located in the centre of the problem area as shown on Fig. 2.

SITE DESCRIPTION AND SURFICIAL GEOLOGY

The surficial geology of the general area is characterized by the effect of several advances of the laurentide ice sheet, which exposed the bedrock surface in most of the area (Aspler 1984). Local sequences of tills resting on glacio fluvial material are a result of basal melt-out. Following deglaciation sands and gravel were deposited in some of the valleys. The area was then covered by glacial Lake McConnell which deposited silts and clays in topographic lows , but sands on topographic highs. Accordingly, the topography of the area is characterized by smoothly rounded bedrock hills that are separated by valleys which generally are infilled with clay, silt and sand deposits.

The study site is sloped at an average angle of 7° towards a marshy area which is connected to Prosperous Lake. The stratigraphy of the slope is characterized by a silty gravelly sand zone above the bedrock surface which in turn is overlain by soft preglacial silty clays (See Fig. 3).

In the area downhill of the old penstock alignment massive ground ice of 5 meter thickness was found at a depth of less than 2 metres. At the base of the ground ice artesian ground water conditions were detected in the underlying unfrozen sand zone. In the upper slope portions permafrost was not indicated.

The wooden penstock pipe has been leaking continuously since its construction contributing relatively warm water to the slope throughout the year. There is also evidence that surface water that accumulates higher up in the slope in a trough like area is flowing towards the study slope.

FIELD INVESTIGATION

The Bluefish Lake site was investigated by a total of twenty boreholes, two test pits, and various field instruments. The location of the boreholes and instruments is shown in Fig. 2. Drilling of the holes for the piers of the new bridge supports was inspected in March of 1983. At this occasion the sand layer below the clay deposit was found to be unfrozen carrying considerable amounts of water.

In order to monitor the water pressure in the sand layer at the base of the clay deposit a total of 10 piezometers were installed in this zone, at the locations shown in Fig. 2. Two types of piezometers were selected: three electro-piezometers in order to obtain continuous readings and seven stand-pipes for control of the electrical instruments.

In order to monitor slope

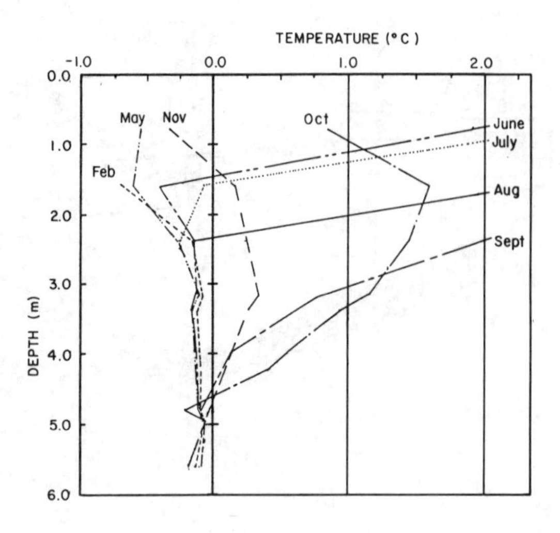

Fig. 5 Ground temperature at location TH-2.

movements four slope indicator casings were installed in 1984 at locations shown in Fig. 2, with the base anchored in the sand layer which was considered not to be affected by lateral movements.

Further, two 250 mm diameter total pressure cells were installed against a foundation pier of the penstock at depths of 1.45 m and 1.7 m below the ground surface to measure the soil presssures exerted on the pier due to the ground movements.

A total of three thermistor strings was installed at the locations shown on Fig. 2 in order to obtain the ground temperature profile as well as the uphill extent of the permafrost in the lower section of the slope. The observation period for most instruments was between the fall of 1984 and the summer of 1986. Some of the instruments were read manually and some with an automatic field data logger. The observed data are summarized on Fig.s' 3, 4, 5, 6 and 7.

On Fig. 3 piezometric levels are plotted for various points of time on a slope section, indicating artesian conditions in the sand layer below the clay deposit for the lower slope portions during late summer and fall. The lowest piezometeric levels were observed in July, on average approximately 2 meters below the maximum levels. It should be pointed out that only the standpipe piezometers provided data because the electrical piezometers stopped operating after the first winter.

The slope indicator readings as summarized on a slope section on Fig. 4, show that only the upper two meters of the clay deposit were affected by deformations, with the movements gradually increasing from this depth toward the ground surface. There was no indication of a concentrated shear zone. The deflections are referred to the first set of readings obtained in August 1984, soon after installation. The average rates of movements during the two year observation period varied between 2 mm/year for SI-4 in the upper slope, 10 mm/year for SI-3, 25 mm/year for SI-2 (both in the upper centre of the slope) and 22 mm/year for SI-1 in the lower centre of the slope.

The maximum lateral deformations observed at the various slope indicator installations are plotted on the site plan on Fig. 2. Limits and lateral directions of the earth movements can be recognized.

The ground temperatures recorded at location TH2 (see site plan Fig. 2) are plotted versus depth on Fig. 5. Thermistor TH-2 showed an active zone of almost 5 m thickness, indicating that this installation is located at the upper boundary of the permafrost feature. Thermistor installations TH-1 and TH-3, both located well in permafrost, indicated an active zone of approximately 1.5 m thickness.

Fig. 6 represents a composite plot of all field data. The deformations observed near the ground surface are plotted for all slope indicator installations, showing a generally slower rate of movements during the summer months.

The piezometric levels plotted versus time indicate a small peak in June and a major peak in August/ September.

The plots of the total pressure readings,as shown on the composite diagram indicate a total of three peaks for the upper cell during the observation period, whereas, a total of two peaks could be recorded for the lower loadcell. The pressures varied for the upper load cell between a low of

16.4 KPa and highs in excess of 55 KPa . For the lower load cell the pressures ranged between 6.0 KPa and more than 55 KPa (Unfortunately, the data logger was not set sufficiently, to read the full range of the pressure values).

The ground temperatures for thermistor TH-2 at 1.6 m and 0.8 m depths are plotted versus time, indicating that the ground temperatures at the depths of the total pressure cells were never below -1°C, which may suggest that the peaks in total pressures could be possibly related to pore pressures developing during freezing and thawing at these levels.

From the composite plot it may be assumed that the slope movements are related to seasonal fluctuating pore water conditions, caused by the ground water level changes as recorded in the underlying sand zone, and possibly also to elevated pore water pressures that accompany a thawing or a freezing front (McRoberts and Morgenstern 1974).

LABORATORY TESTS

Undisturbed block samples of the clay were obtained from approximately 1.80 m depth in a pit adjacent to the concrete pier (See Fig. 2). The soil from this location was a light brownish-grey, low plasticity silty clay with occasional ice-rafted pebbles. The liquid limit from air-dried samples was 43%, the plasticity index 18% and the clay fraction 73%. Occasional rootlets produced a noticeable smell. The clay was soft, with an average natural water content of 31%. A nuggetty structure with 2-3 mm spacing necessitated considerable care during specimen preparation. The clay is inactive and not susceptible to swelling.

Effective strength parameters of $c' = 12$ KPa and $\phi' = 31^\circ$ were determined in routine consolidated undrained triaxial tests with pore pressure measurements (CIU-tests). Preconsolidation pressures of 200 250 KPa were interpreted from the consolidation stages of these tests.

In order to investigate the correlation between lateral movements and fluctuating pore water conditions an additional special testing program was undertaken:

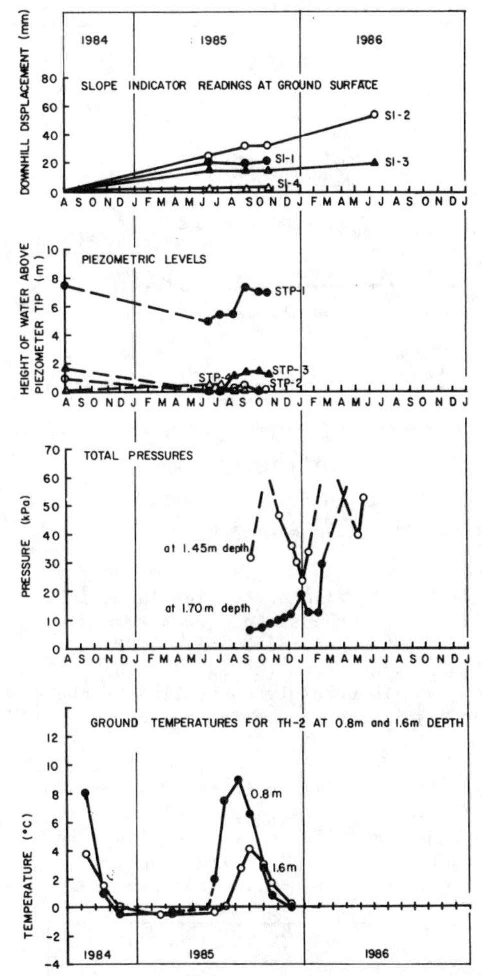

Fig. 6 Composite plot of observed data.

Triaxial specimens were tested by cyclically varying the porewater pressure by an amount Δu, and measuring the resulting strains per cycle (Eigenbrod, Burak and Graham, 1987). The specimens were initially anisotropically consolidated with normal and shear stresses corresponding to those in the moving mantle (See Fig. 7). Drainage was permitted throughout the testing. This procedure represents changes that can occur in a natural slope from (a) seasonal groundwater level changes, and (b) elevated pore-water pressures that accompany thawing.

After 60 to 100 cycles, the porewater pressure was systematically increased to the value Δu_f at which

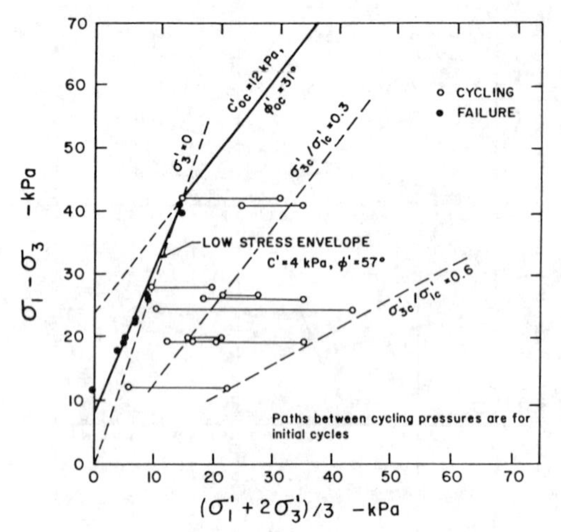

Fig. 7 Drained triaxial tests with cyclically varying pore pressures

the samples failed. This occurred for average effective stress levels of less than 10 KPa on a steep, low-stress envelope, approximately c' = 4 KPa, $\phi' = 57^o$, as shown in Fig. 7. The envelope is probably controlled by the nuggety macrostructure of the clay.

The strains-per-cycle were approximately linear in the range 30-100 cycles. As a first approximation they have been modelled as varying linearly with the ratio $\Delta u/\Delta u_f$ almost up to failure at $\Delta u/\Delta u_f = 1.0$ (See Fig. 8).

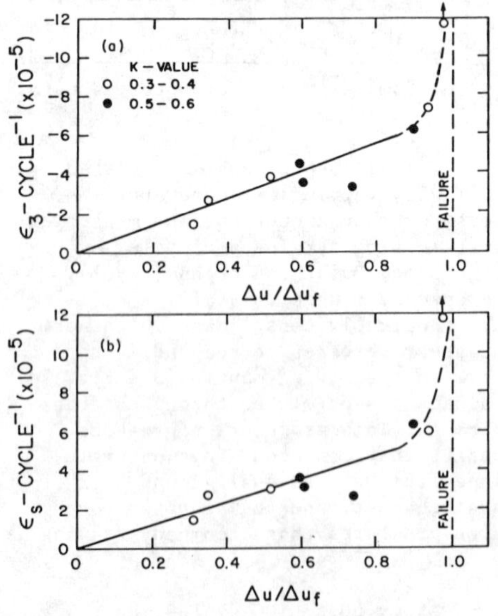

Fig. 8 a) lateral strains; b) shear strains related to pore water pressure ratio $\Delta u/\Delta u_f$

ANALYSIS OF DATA

In order to relate the deformations observed for triaxial test-conditions with plane strain field conditions a correction factor of 1.25 was applied to the laboratory deformation data (Burak 1986).

The fact that in the triaxial setting vertical and horizontal directions are principal stress directions, whereas in the slope they are not, was neglected since the slope inclination is only 7° and the correction would be very small.

In order to relate the linear correlation between strains per cycle and ratio $\Delta u/\Delta u_f$, presented in Fig. 8, the pore pressure changes required for failure of the slope must be known. The field Δu_f - values were determined from a stability analysis of the surface layer of the slope with strength parameters for low effective stress values of $\phi' = 57^o$ and c' = 4 KPa. The resultant average pore pressure changes Δu_f for failure of the slope portion under investigation can be expressed as a function of the depth "z" below the ground surface: $\Delta u_f = 19.1 \times z - u_o$ (Burak, 1986). u_o is the lowest pore pressure encountered in the slope at the depth "z". In this analysis u_o was set to zero.

Pore pressures measured in piezometer STP-1, and deformations recorded at slope indicator installation SI-1, were considered to represent the average values for the slope, as these instruments are located approximately in the center of the moving slope area. It was further assumed that the same pore pressures measured in the sand zone are also valid in the clay. Thus, the maximum pore pressure variations are equivalent to 2.0 m head of water or 20 KPa.

Pore pressure changes due to freezing or thawing were not observed in this study. Pore pressures developing during thaw of an active zone were reported from other locations by McRoberts and Morgenstern (1974) and Chandler (1972) and are summarized on Fig. 9. Data were obtained, however, only for shallow depths of less than 1.0 m. Assuming that the pore pressures at 1.0 m depth are representative for the 2 m thick soil mantle, and that these pore pressures are in excess of zero, the pore pressures changes are equivalent to 1.2 m head of water or 12 KPa.

No field records exist to the authors' knowledge on pore pressures developing in a soil ahead of a freezing front even though measurements in the laboratory have been obtained (McRoberts and Morgenstern, 1974). For simplicity sake they were considered in the analysis to be equal to thaw pore pressures .

The maximum lateral deformations due to lateral strains per cycle $\Delta u/\Delta u_f$ at location SI-1 were computed by referring the strain ε_r near the ground surface to the length of the moving slope extending from its upper boundary to the location of SI-1, which is approximately 30 m (See Fig.'s 2 and 3). The maximum lateral deformation due to shear strain ε_s per cycle $\Delta u/\Delta u_f$ can be computed by integrating the shear strains over the depth of the 2 m soil layer (See Burak, 1986).

Thus the lateral deformation due to water level variations equivalent to Δu = 20 KPa, range between 3.0 and 4.5 mm/cycle depending on the interpretation of the data.

The total lateral deformation due to thaw pore pressure are approximately of the same magnitude.

From the composite plot of the field observations on Fig. 6 it can be deduced that a total of 3 pore pressure cycles may occur per year. Thus the total deformations per year are between 9.0 and 14.0 mm/year.

The displacement component related to heaving during freezing and subsequent settlement during thaw (Williams, 1966) can be estimated to approximately 8 mm/year if frost heave of approximately 100 mm/year is considered. Including this component the total lateral displacement per year can be predicted to range between 17.0 and 22.0 mm/year which is close to the average yearly displacement observed in the field.

The analysis of the laboratory results suggest that there will be a slow transition from non-moving to irregulary or cyclically-moving material. The magnitudes of the

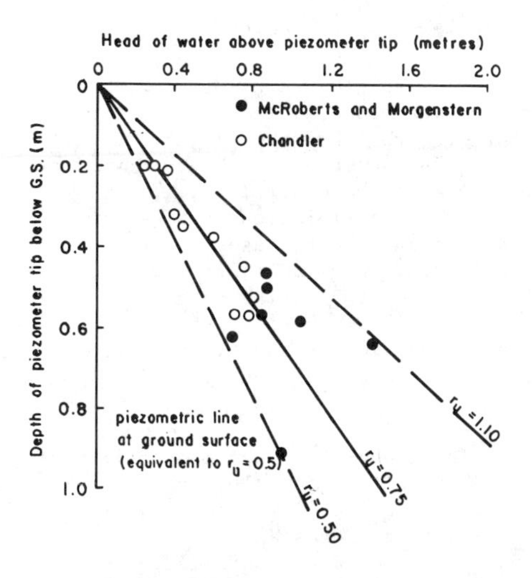

Fig. 9 Pore pressures observed during thaw of an active zone

movements depend on the size of the porewater pressure changes at any given depth and the number of cycles of pore water pressure change in the period under consideration. This conclusion agrees with the field observation that no clear boundary between moving and non-moving portions of the slope could be recognized: The extent of the moving slope portion can be recognized only by the slope indicator readings (See Fig. 2) and by the soil which was piled up on the uphill side of the old timber foundations as shown on Fig. 2.

The cyclic pore pressure tests were undertaken at conditions equivalent to lateral earth pressure coefficients between 0.3 and 0.6, which implies that some lateral unloading occurs, resulting in K-values lower than would be expected for at-rest conditions. Lateral unloading was assumed to be due to continuous deterioration of ice-rich permafrost lower down the slope. There is some indication from the laboratory test data that specimens consolidated with higher values of K showed lower average straining rates. The data is limited and somewhat scattered at this stage, thus not permitting any quantitative evaluation. A quantitative correlation between the lateral stress conditions and the lateral extent of the slope movements is further restricted by the fact that only little information exists at present about the lateral extent of the ice rich permafrost down the slope.

CONCLUSIONS AND RECOMMENDATIONS

Slow shallow downslope movements are not very spectacular events, but can be of concern, when they interfere with engineering structures. These downslope movements can be controlled, if quantitative correlations between the movements and their causes can be established. From the instrumentation of the slope at Bluefish Lake a correlation between slow, shallow downslope movements and fluctuating pore-water pressure conditions is indicated.

In a laboratory testing program of clay samples from the slope the correlation between cyclically varying pore water pressures and lateral deformation could be confirmed. The lateral deformations predicted on the basis of the quantitative correlations between number and magnitude of pore pressure cycles and strains are in the same order of magnitude as the lateral deformations observed in the field. Even though reasonably good agreement was demonstrated, it is based on rather crude assumptions which need further confirmation.

The field data as well as the laboratory data suggest that lateral downslope movements are also related to lateral unloading which at the Bluefish Lake site is believed to be caused by continuous deterioration of ice rich permafrost lower down the slope. At the present stage however the data is too limited to permit a quantitative correlation.

Thus, in order to improve the suggested correlations and to strengthen the conclusions additional data will be collected, both in the field and in the laboratory.

ACKNOWLEDGEMENTS

The research was funded by Operating and Equipment Grants from NSERC. Con Mine, Yellowknife, permitted access to the site at Bluefish Lake and provided logistical support. The authors acknowledge valuable discussions with Dr. J. Graham and Dr. W.D. Roggensack.

LIST OF REFERENCES

Aspler, L.G. 1984. Surficial Geology, Permafrost and Related Engineering Problems, Yellowknife, N.W.T. Internal Report, D.I.A.N.D., Yellowknife, N.W.T.: 119-135.

Burak, J.P. 1986. Downslope movements at shallow depth related to drained cyclic pore pressure changes. M.Sc. thesis, University of Manitoba, Winnipeg, Manitoba. Canada.

Chandler, R.J. 1972. Periglacial mudslides in Vestspitsbergen and their bearing on the origins of fossil 'solifluction' shears in low angled clay slopes. Q.J. Engineering Geol. 5: 223-241.

Eigenbrod, K.D., Burak, J.P., and Graham J. 1987. Deformations due to drained cyclic pore pressure changes and failure at low stress levels for a natural clay. (to appear in the Canadian Geotechnical Journal, May 1987).

Kojan, E. 1967. Mechanics and rates of natural soil creep. Proceedings of the 5th Annual Engineering Geology and Soils Engineering Symp., Pocatello, Idaho, Idaho Department of Highways: 233-253.

McRoberts, E.C. and Morgenstern, N.R. 1974. The stability of thawing slopes, Canadian Geotechnical Journal 11: 447-469.

McRoberts, E.C. and Morgenstern, N.R. 1974. Pore water expulsion during freezing. Canadian Geotechnical Journal. 12: 130-138.

Tavenas, F. and Leroueil, S. 1981. Creep and failure of slopes in clays. Canadian Geotechnical Journal 18: 106-120.

Washburn, A.L. 1973. Periglacial processes and environments. p170-191. New York: St. Martin's Press.

Williams, P.J. 1966. Downslope soil movement at a sub-Artic location with regard to variations with depth. Canadian Geotechnical Journal 8: 191-203.

Yen, B.C.Y. 1969. Stability of slopes undergoing creep deformation. Journal of Soil Mech. and Found. Division, Proc. of ASCE: 1075-1096.

Prediction and Performance in Geotechnical Engineering / Calgary / 17-19 June 1987

Predicting performance of granular pavements

R.B.Smith & W.D.Yandell
Department of Main Roads, Parramatta, N.S.W., Australia

ABSTRACT: Most methods for the prediction of the performance of granular pavements assume that the behaviour of the constituent materials is elastic whereas the behaviour is more likely to be elasto-plastic. The mechano-lattice analysis used in this study models a rolling wheel and can predict chanellised rutting and the development of residual stresses because it allows for elastoplastic behaviour. The paper briefly describes the method and uses it to predict rutting and stress distributions in granular pavements. Predicted results showed good agreement with the measured results.

1 INTRODUCTION

The mechano-lattice method of analysis has been successfully used to predict pavement performance (Smith & Yandell 1986a, 1986b, 1986c, 1987a, 1987b; Yandell 1971, 1983, 1984,1985). The model utilises the elasto-plastic behaviour of each of the pavement and subgrade materials. Bound and unbound material characteristics can be simulated.

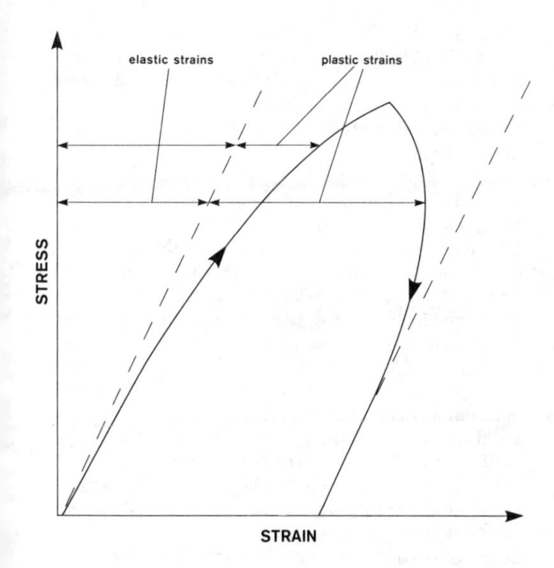

Figure 1. Axial stress-strain hysteresis loop of a hypothetical soil (after Yandell 1971).

Various aspects of the model have been described in the references cited above so only an outline will be provided.

Figure 1 shows the axial stress-strain hysteresis loop of a hypothetical elasto-plastic soil. Figure 2 shows the elastic components of the axial load versus deflection hysteresis loop during cyclic loading of the model. The axial behaviour is considered to be to that of an elastic element that has a greater modulus when unloading (E'_U) than when loading (E'_L). The plastic factor is e/f and is equal to the damping factor (ξ) which is the area in the loop divided by the hatched area under the loading line (Figure 2).

A diagram of a face of one of the mechano-lattice units is given in Figure 3. The unit simulates the behaviour of a material with a particular rigidity, plasticity, and Poisson's ratio. The volume cross and the shear cross are free to rotate and enable the shear and volumetric behaviour to be independent. The left-hand cross permits the shear elements to be activated only when shear deformation is taking place, whereas the right-hand cross permits the volume elements to be activated only during volume changes. The complex case of a rolling wheel moving in one direction, on an elastoplastic material is solved by connecting a number of units at their joints. The rolling wheel effect is represented by displacements to the central upper surface joints conforming to the radius of the wheel. Figure 4 shows a longitudinal section through the pavement.

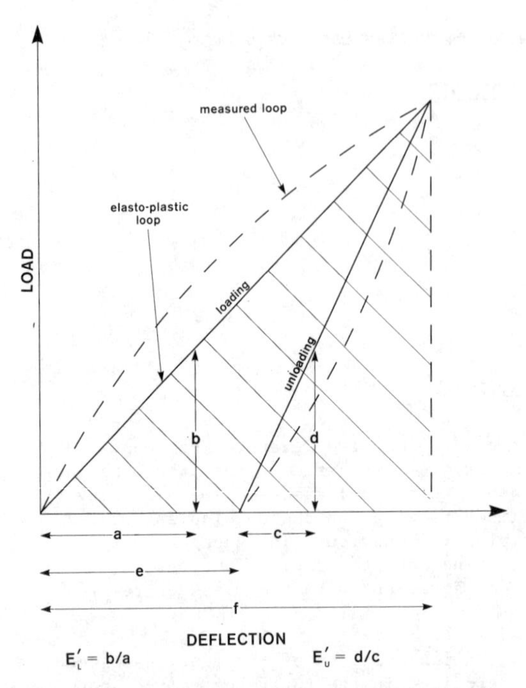

Figure 2. Simplification and partition of an axial elastoplastic load-deflection hysteresis loop from a measured loop (after Yandell 1971).

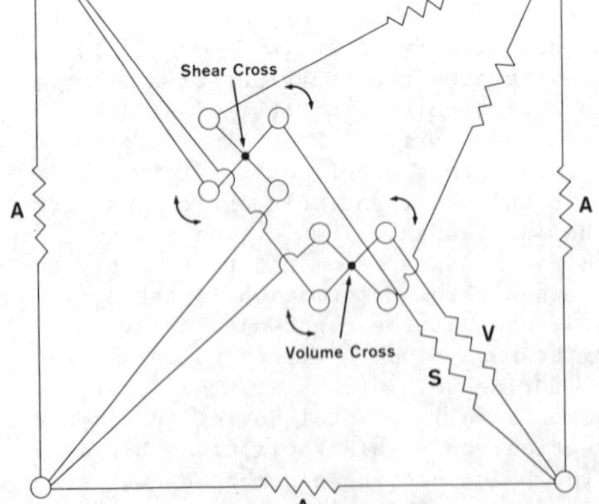

Figure 3. A unit of the lattice structure simulating the behaviour of the soil (after Yandell 1971).

The units on the extreme left-hand side of Figure 4 (shown by broken lines) represent the initial conditions before a particular wheel pass. Elastic theory is used for predicting the shape of each unit as it arrives at the simulating region from the residual no-load condition well forward of the "present" load. Subsequent change in unit shape causes the elements to change in length and change their element load. A similar action occurs when the "wall" of units moves another plane closer to the load. The sequential movement of the wall of units from left to right is followed mathematically in terms of the load-deflection history of each element. The load-deflection history is followed by calculating changes in length and load with the aid of stiffness factors. The hysteresis loop at the bottom of Figure 4 was formed by plotting sequential load-deflection values through positions A, B, C, D, E, F, and G of a vertical element in one particular unit. A similar, but more complex, task is performed by the computer after each cycle of element-load calculations. Calculations are performed until convergence which was generally between 1500 and 2000 computation cycles but has since been markedly reduced to 300. A total of 32 load deflection histories are followed for each of the 1980 mechano-lattice units.

2 THE TEST TRACK STUDY

This study was originally reported by Yandell (1982) and an overview included in Smith and Yandell (1986b). It is included here as it was the first crushed rock pavement analysed using the mechano-lattice analysis.

Yandell (1982) used performance results originally reported by Sparks (1970) for two pavements placed in the Sydney University Miniature Test Track (re-located to University of New South Wales in late 1986). The test track consists of a 250 mm loaded pneumatic tyre repeatedly rolling in one direction along a test bed 4.87 m long and 0.9 m wide. Yandell used the data from Bays 3 and 4 which consisted of dolerite crushed rock mixed with 10 per cent clay. The prediction had to take into account the fact that the wheel was smaller than standard, had a much smaller load, and was not torque driven. A further complication was the fact that whilst repeated loading triaxial tests are preferred for analysis of pavements, only repeated bearing test results were available from the original study. These

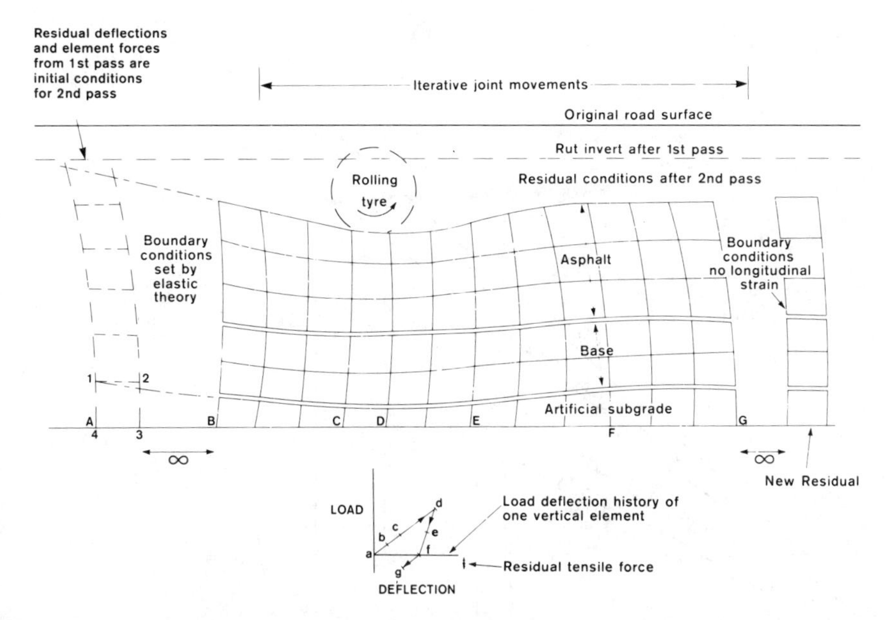

Figure 4. Longitudinal section of the three-dimensional pavement analysis showing the boundary conditions, and computational process (after Yandell 1983).

results were utilised for life prediction purposes.

The results for measured and predicted rut depth are given in Table 1. These results were obtained from Figure 8 of Yandell's paper. As well Yandell analysed the rut shape across the pavement after 14 630 passes. The rut prediction for Bay 3 was of a similar shape to the rut profile measured by Sparks but of smaller magnitude. The predicted values for Bay 4 were closer. The predictions of horizontal movement of the surface were lower but reasonably close to the measured values. Yandell considered that this prediction of rutting in a single layer pavement resting on a rigid foundation was surprisingly accurate in view of the fact that only five load repetitions were used in the four original material tests. This work was also briefly discussed by Yandell (1984).

In view of the accuracy of these predictions the mechano-lattice should be be able to provide fairly accurate predictions of the performance of in-service granular pavements.

Table 1. Predicted and measured rut invert depths (after Smith and Yandell 1986b).

Number of Wheel-passes	Bay 3 Predicted (mm)	Bay 3 Actual (mm)	Bay 4 Predicted (mm)	Bay 4 Actual (mm)
100	0.38	0.50	0.60	0.75
300	0.48	0.72	0.67	0.97
800	0.53	0.88	0.80	1.13
1 800	0.58	1.00	0.90	1.28
3 800	0.67	1.13	1.03	1.43
14 630	0.88	1.25	1.28	1.53

3 ANALYSIS OF AN IN-SERVICE PAVEMENT

3.1 General

Because of space limitations, and in line with the theme for this symposium, the emphasis in this section is on prediction of field performance. The discussion is not exhaustive but details are given of the information required for input and the types of information that can be obtained from the program. It might be noted that at this stage all output from the mechano-lattice program is in terms of tabulated data. Graphical data are prepared by hand.

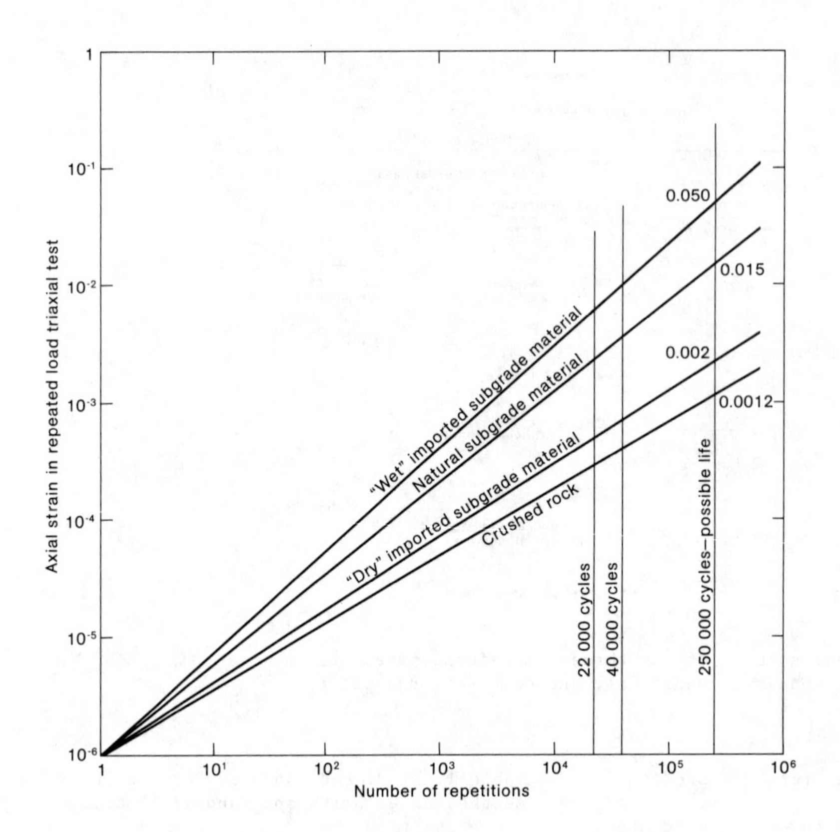

Figure 5. Axial strain in repeated load triaxial test versus number of repetitions of axial load (after Smith & Yandell 1986a).

The information which follows was obtained from an instumented pavement section. A crushed rock pavement was placed over a uniformly weak clay subgrade (200 mm thick) which had been "imported" because the natural in-situ subgrade was considered to be too variable. The imported subgrade had CBR values between 2% and 4% after soaking for 4 days. The crushed rock pavements was 100 mm thick.

Full details of the repeated loading characteristics are given by Smith and Yandell (1986a) and are summarised in Figure 5. The base was tested at a moisture content of 4.5%, the natural subgrade at a moisture content of 11.6%, whilst the "imported" subgrade was tested at two moisture contents, ie 9.2% ("dry" condition) and 14% ("wet" condition). The moisture contents and relative compaction were close to the field values. For the "imported" subgrade the higher ("wet") moisture content is considered to be the appropriate moisture content. Further discussion will be restricted to predictions based on the wet condition.

3.2 Predicted stresses in the pavement

In each case several longitudinal sections were prepared for the 1st calculation rolling pass. During preparations of the sections for the 3rd calculation rolling pass it was obvious that the essential difference between most of the stress distribution diagrams during the 1st calculation rolling pass and during the 3rd calculation rolling passes was the stress level rather than the manner in which the stresses were distributed.

The example presented here is the development of transient longitudinal stress.

The effect of the number of passes is very significant. Pockets of tensile stress developed at the bottom of the base layer as the number of rolling passes increased. Similarly the tension zone at the bottom of the base layer tripled in magnitude and increased in size with an increase in the number of rolling passes. The stresses (compression) in the imported subgrade layer became more diverse with

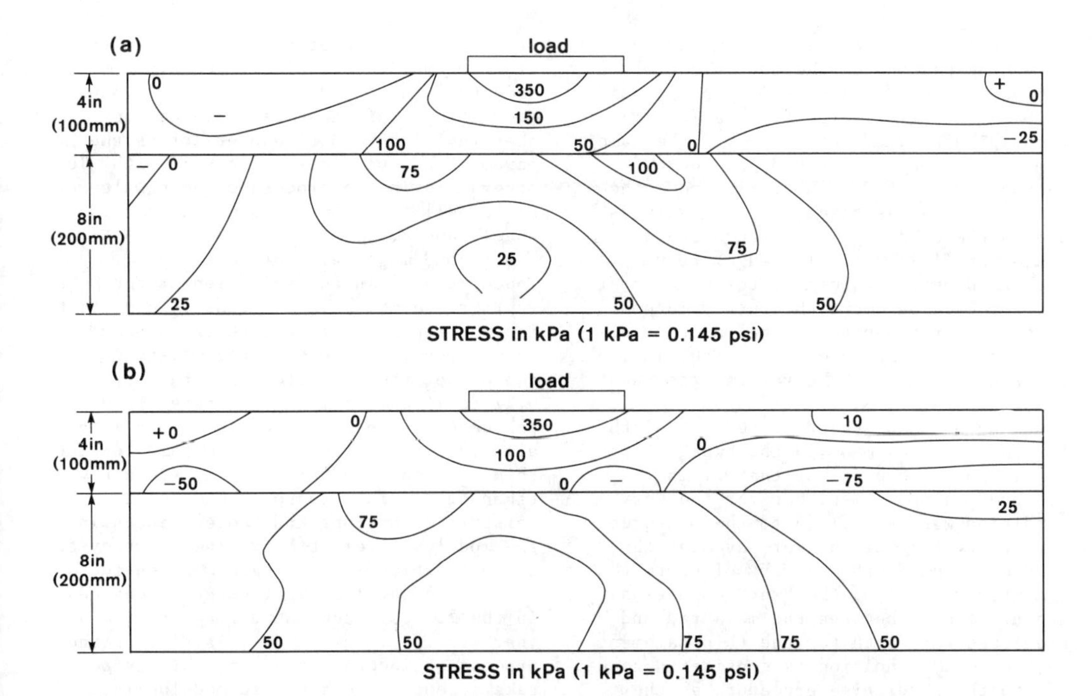

Figure 6. Predicted transient longitudinal stress during (a) 1st and (b) 3rd rolling passes with imported subgrade in "wet" condition. Longitudinal section through the load.

the 3rd rolling pass. There was a doubling of the stress developed at the bottom of the imported subgrade layer during the 3rd rolling pass. The sections under the load (Figure 6 - longitudinal section) indicate that a tensile stress area will develop away from the load with compression closer to and under the load.

Because of model limitations it is not possible to determine from the results the predicted stress pattern in the natural subgrade.

3.3 Predicted and measured rut depth

The rut depth predictions for the thin section after the 3rd Calculation Rolling Pass is given in Figure 7. The figure also shows the predicted residual rut surfaces at the top of the base and the top of the imported subgrade. As can be seen the predicted surface of the base is identical to that at the top of the imported subgrade. The maximum difference in the change in level between the layers was predicted to be 0.03 mm. In most cases the values were identical which means that the layers would move in unison and not independently.

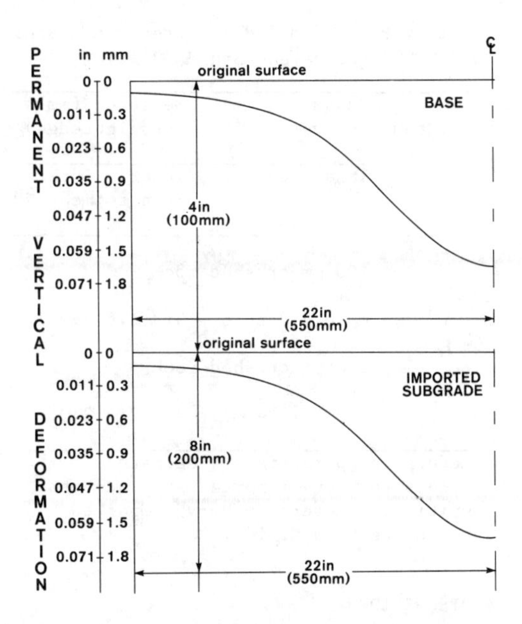

Figure 7. Predicted residual rut surfaces after 3rd rolling pass with imported subgrade in "wet" condition.

The predicted and measured rut depths after the passage of the equivalent of 250 000 standard axles are given in Table 2. Data are also included for a 250 mm crushed rock pavement built over the same "imported" subgrade. Rut depths were measured using two methods. In one method the rut depth was based on field surveys and in the other was measured using a four-foot (1.2 m) straightedge. Both methods provided values of the same order but for each pavement the survey method gave lower rut depths in the inner wheel-paths than in the outer wheelpath. In the straightedge method the reverse was the case.

As can be seen the predicted rut depth for the thin pavement in the "wet" condition is three times that of the measured value whilst that for the "dry" condition was very close to the measured condition. For the thicker pavement the predicted rut depth was very close to the measured value (for the "wet" condition). The difference between the measured and predicted rut depth for the thin pavement in the "wet" condition is considered to be due to the prediction procedure as the field in-service moisture content is considered to have been close to the "wet" condition.

Table 2. Predicted and measured rut depths at 250 000 Equivalent Standard Axles.

Predicted (mm)	Measured (mm) (survey)		Measured (mm) (straightedge)			
	Inner+	Outer+	Inner+ mean	sd	Outer+ mean	sd
	100 mm crushed rock					
20* / 6@	0 to 2	2 to 6	3.3	1.6	2.4	1.3
	250 mm crushed rock					
3*	0	4	3.9	2.4	2.0	1.0

+ Refers to appropriate wheelpath.
* Refers to "wet" imported subgrade test condition. @ Refers to "dry" imported subgrade test condition.

4 DISCUSSION

It is clear that predictions are a good first approximation, especially as only one set of input data was used for each prediction. Whilst the mechano-lattice analysis simulates channelised rutting the program available at the time of the study assumed that the wheel tracked along the same path thus overestimating the amount of rutting. This results from the fact that vehicles tend to wander across the pavement and not follow the same path. The program is in the process of being altered by Yandell to allow for simulation of wheel wander.

A second major area of potential inaccuracy is in the selection of the appropriate moisture and density conditions for the repeated loading tests. As was shown above, a change in the moisture content of the "imported" subgrade from 9.2% to 14.0% resulted in a three-fold increase in the predicted rut depth with all other variables being the same for the thin pavement. In this case no account was taken of possible seasonal changes in moisture conditions and hence changes in the modulus of elasticity. The accuracy of the prediction in this case was considered only to be marginally affected by changes in the moisture content during its two year service life. Yandell is also in the process of incorporating the ability to take account of changes in modulus into the analysis program.

5 CONCLUSIONS

The prediction of development of transient and permanent stress and strain regimes in the pavement aids our understanding of the performance of pavements. The mechano-lattice method of analysis allows us to perform such analyses. In this case the predicted rut depths are considered to be reasonable.

The prediction ability of the mechano-lattice method of analysis can be improved by the incorporation of features which allow for predictions to take account of changes of modulus over time and spread of wheelpasses across the pavement.

ACKNOWLEDGEMENT

The authors wish to thank the Mr G.B. Fisk, Commissioner for Main Roads, N.S.W., for permission to publish this paper.

REFERENCES

Smith, R.B. & W.O. Yandell 1986a. Prediction of surface deformation in a full scale test pavement. Computers and Geotechnics 1:23-41.

Smith, R.B. & W.O. Yandell 1986b. The use of mechano-lattice analysis in

prediction of pavement performance. Australian Road Research 16:10-17.
Smith, R.B. & W.O. Yandell 1986c. Prediction of performance of a granular pavement using mechano-lattice analysis. In G.N. Pande & W.F. van Impe (eds) Numerical Models in Geomechanics, p. 577-586. Redruth, Cornwall: Jackson.
Smith, R.B. & W.O. Yandell 1986d. Predicted performance of two crushed rock pavements in initial service life. Proc 13th Aust. Road Research Board/5th REAA Conf. 13(5):193-205.
Smith, R.B. & W.O. Yandell 1987a. Investigation of bound and unbound behaviour in road pavements using mechano-lattice analysis. Proc. 2nd Int. Conf. and Short Course on Constitutive Laws for Engineering Materials: Theory and Application, Tuscon 5th-10th Jan 1987 (in press).
Smith, R.B. and W.O. Yandell 1987b. Predicted and field performance of a thin full depth asphalt pavement placed over a weak subgrade. Proc. 6th Int. Conf. Structural Design of Asphalt Pavements, (in press).
Sparks, G.H. 1970. Development and use of a machine for examining the behaviour of pavement structures under the action of moving wheel loads. M. Eng. Sci. Thesis, Univ. Sydney.
Yandell, W.O. 1971. Prediction of the behavior of elastoplastic roads during repeated rolling using the mechano-lattice analogy. Highway Research Record 374:29-41.
Yandell, W.O. 1982. Measurement and prediction of forward movement and rutting in pavements under repetitive wheel loads. Transportation Research Record 888:77-84.
Yandell, W.O. 1983. The use of mechano-lattice analysis to investigate relative plastic behaviour. Proc. Int. Conf. on Constitutive Laws for Engineering Materials: Theory and Applications, Tuscon, Arizona, p. 375-381.
Yandell, W.O. 1984. Mechano-lattice prediction of pavement performance. Proc. 4th Conf. on Asphaltic Pavements for Southern Africa, Capetown 1:201-215.
Yandell, W.O. 1985. New method of simulating layered systems of unbound granular material. Transportation Research Record 1022:91-98.

Evaluation of dilatancy theories of granular materials

A.M.Hanna & H.Youssef
Concordia University, Montreal, Quebec, Canada

ABSTRACT: During the last two decades, several attempts were made to correlate the stress-strain-volume change behavior of granular materials. In this respect, Rowe [1962, 1969] proposed a relationship which is based on evaluation of the energy loss during shear, and most recently Bolton [1986] presented a simplified model for the dilatancy which is based on new definitions for the angle of shearing resistance and relative dilatancy index.

The purpose of this paper is to examine these theories utilizing the experimental results deduced from plane-strain and triaxial compression tests on sands. The results of this study will shed light on the behavior and the mechanism of cohesionless material, during shearing.

BACKGROUND

Rowe [1962, 1969] presented the dilatancy theory which is based on evaluating the energy losses during shear as given in Eqn.[1]:

$$R = K \cdot D \qquad [1]$$

where; R is the principal stress ratio $[\sigma_1/\sigma_3]$ and σ_1, σ_3 are the major and minor principal stresses respectively, and D is the dilatancy rate, furthermore:

$$D = 1 - \frac{d\varepsilon_v}{d\varepsilon_1} \qquad [2]$$

where $d\varepsilon_v$ and $d\varepsilon_1$ are the volumetric and axial strain increments respectively, and K is a coefficient given by:

$$K = \tan^2 \left[45° + \frac{\phi_f}{2}\right] \qquad [3]$$

ϕ_f is defined as Rowe's frictional angle, which accounts for dilatancy of the material to its shear strength. Rowe indicated that for triaxial testing $\phi_\mu \leq \phi_f \leq \phi_{cv}$ where for plane-strain testing $\phi_f \cong \phi_{cv}$, where: ϕ_{cv} is defined as the angle of shearing resistance at constant volume change and ϕ_μ is the angle of friction between the sand particles. Cagout [1934] and Bishop [1954] have established relationships between ϕ_μ and ϕ_{cv} in triaxial compression tests; these are:

$$\tan \phi_{cv} = \frac{\pi}{2} \tan \phi_\mu \qquad \text{Cagout [1934]} \qquad [4]$$

and

$$\sin \phi_{cv} = \frac{15 \tan \phi_\mu}{10 + 3 \tan \phi_\mu} \qquad \text{Bishop [1954]} \qquad [5]$$

Horne [1965] presented a relationship between ϕ_μ and ϕ_{cv} which is given in Fig. [1]. Horne [1965] and Rowe [1969] have found that, the upper and lower limits for dilatancy rate D are:

$$1 \leq D \leq 2 \qquad [6]$$

Recently Bolton [1986] has presented a modified dilatancy theory, based on the "saw blades model", Fig. [2]. In this theory, he correlated the angle $[\phi_{p.s.}]$ to the dilatancy angle $[\Psi_{max}]$ as follows:

$$\phi_{p.s.} = \phi_{cv} + \Psi_{max} \qquad [7]$$

where:

$$\sin \Psi_{max} = \frac{-d\varepsilon_v}{d\varepsilon_1 - d\varepsilon_3} = \frac{-\left[\frac{d\varepsilon_1}{d\varepsilon_3}\right]_{max} + 1}{\left[\frac{d\varepsilon_1}{d\varepsilon_3}\right]_{max} - 1} \qquad [8]$$

After examining Eqn. [7] using the available experimental data, Bolton [1986] has modified this equation to the following one, Eqn. [9], which was found to be in good agreement with Rowe's [1962] stress-dilatancy theory as shown in Fig. [3]. The modified equation is in the form:

$$\phi_{p.s.} = \phi_{cv} + 0.80\ \Psi_{max} = 5°\ I_R \quad [9]$$

where I_R is the relative dilatancy index which is given in Eqn. [10]:

$$I_R = I_D\,[\,10 - \ln p'\,] - 1 \quad [10]$$

I_D is the relative density, furthermore:

$$I_D = \frac{e_{max} - e}{e_{max} - e_{min}} \quad [11]$$

and p' is the mean principal stress; furthermore:

$$p' = \frac{1}{3}\,[\sigma'_1 + \sigma'_2 + \sigma'_3\,] \quad [12]$$

For triaxial strain conditions, Bolton [1986] presented the following formula:

$$\left[\phi_{max} - \phi_{cv}\right]_t = 3°\ I_R \quad [13]$$

Also, Bolton [1986] indicated that for both types of testing (triaxial and plane-strain compression), the following equation is valid:

$$\left[-\frac{d\varepsilon_v}{d\varepsilon_1}\right] = 0.30°\ I_R \quad [14]$$

where $\left[-\frac{d\varepsilon_v}{d\varepsilon_1}\right]$ is the rate of volumetric to axial strains.

The relationship between Ψ_{max}, I_D and the stress parameter p' [at constant volume change rate is not stressed to cause particle crushing] is given by:

$$\Psi_{max} = A\ I_D \ln\left[\frac{p'_{crit}}{p'}\right] \quad [15]$$

where A is a constant.

In an effort to evaluate the above mentioned theory, an extensive experimental program was carried out on three types of sands using triaxial compression and plane-strain apparatus.

Soil Tested and Testing Procedures:

The soil used for this investigation is an air dry uniform silica sand. The physical properties of these sands are presented in Table [1]. The triaxial tests were performed on cylindrical samples of 76 mm in length and 38 mm in diameter. The testing procedure is similar to the one described by Bishop and Henkel [1978] and Bowles [1978]. The results are summarized in Table [2] and typical results are presnted in graphical form in Figure [4].

The plane-strain tests were carried out on presmatic samples having nominal dimensions of 92 mm, 38 mm, 75 mm in width, length and height respectively. Figure [5] shows a schematic drawing of the horizontal projection of the plane-strain apparatus, as well as an overview of the experimental set-up. As can be seen from this figure, that the two parallel ends of the samples are fixed to provide zero lateral strain [$\varepsilon_2 = 0$] and the intermediate principal stress [σ_2] was measured by means of pressure transducers. Test results are summarized in Table [3] and typical results are presented in graphical form in Fig. [6].

ROWE'S [1962, 1969] STRESS-DILATANCY THEORY

The present experimental results were analysed to determine the angle of shearing resistance deduced from triaxial and plane-strain tests. In case of triaxial tests the angles of shearing resistance ϕ_t were determined using the classical approach of Mohr-Coulomb criteria, while for plane-strain tests, both Bishop [1954] and Mohr-Coulomb approaches were used to determine the angle $\phi_{p.s.}$ Bishop approach is in the form:

$$\cos\,[\phi_{p.s.}] = \left[\frac{2\,\sigma_2}{\sigma_1 + \sigma_3}\right]_{\text{at failure}} \quad [16]$$

and Mohr's-Coulomb approach [Rowe 1969, and Bolton 1986]

$$\sin\,[\phi_{p.s.}] = \left[\frac{\sigma_1 - \sigma_3}{\sigma_1 + \sigma_3}\right]_{\text{at failure}} \quad [17]$$

It should be mentioned here that, the two approaches yielded the same angle of friction [$\phi_{p.s.}$] [see Table 3].

The angles of shearing resistance at constant volume change ϕ_{cv} were determined from Fig. [7] which show the angle of shearing resistance [ϕ_{max}] versus the rate of volumetric change

[$\frac{d\varepsilon_v}{d\varepsilon_1}$]. The values of [$\phi_{cv}$] were taken as the values of [ϕ_{max}] at [$\frac{d\varepsilon_v}{d\varepsilon_1} = 0$]. It is of interest to note that the values of [ϕ_{cv}] were also determined from [Fig. 8] which show the angle [ϕ_{max}] versus the relative prosity [n_r]. The values of ϕ_{cv} were taken as the values of ϕ_{max} at $n_r = 0$, at which the sand is at its loosest state and the rate of volumetric change tends to be equal to zero. The two methods revealed almost the same values. The values of the angle of particles friction ϕ_μ were determined using all the methods described in the background section. A summary of these results is given in Table [4]. The calculated values, using Horne [1965] and Rowe [1969], showed good agreement with the measured values by Rowe [1969], for material similar to our sand type B used in this investigation. It should be mentioned here that Rowe [1969] indicated that the values of ϕ_{cv} for plane-strain conditions are slightly higher than those for triaxial compression. These can be depicted from the results of the present investigation.

The values of Rowe's frictional angle [ϕ_f] were determined using Eqn. [3] for triaxial and plane-strain test results of the present investigation. These values are also given in Tables [2 and 3], respectively, and represented in graphical form in Fig. [9]. It can be seen from these figures that for triaxial tests the values of ϕ_f have the limits of $\phi_\mu \leq \phi_f \leq \phi_{cv}$, while for plane -strain tests the values of ϕ_f remain almost constant and equal to the ϕ_{cv} value. Furthermore, this Figure [9 a and b] show the shear strength components as described by Rowe [1962, 1969]; mainly frictional sliding, particle rearrangements and orientation and interlocking.

Although it can be noticed from Fig. [9] that the values of these angles ϕ, ϕ_{cv} and ϕ_μ are independent of the applied minor principal stress [σ_3], it is quite evident that the above f inding will hold only for low values of [σ_3]. In case of higher [σ_3] particles crushing is expected. This will be introduced as the fourth parameter [particle crushing] beside: friction, orientation, and interlocking governing the shear strength of granular material.

Figure [10] shows the results of Lee and Seed [1967] presented in the form of Mohr's-Coulomb diagram. This figure shows that the extrapolation of the envelope at low confining pressure does not represent the actual measured envelope.

Figure [11] shows the values of the dilatancy rate D deduced from the present investigation and other test results [Cornforth 1964 and Rowe 1962, 1969]. It can be seen from this Figure [11], that the values of D vary between one and two, which confirm Horne [1965].

BOLTON [1986] STRESS-DILATANCY THEORY

The present test results were analysed to determine all the paramteres mentioned in equations [9, 13, 14 and 15]. A summary of this analysis is given in Table [5]. It should be mentioned here that the angle Ψ_{max} can be determined graphically from Mohr's circle of strain as demonstrated in Figure [12]. It can be seen from this Table [5] that the angle of Ψ_{max} increases due to an increase of the relative prosity [n_r], furthermore, it varies within the range of 0.80° to 20.53°.

The results of this analysis are plotted in a graphical form [Figures 13 and 14] together with other experimental data compiled by Bolton [1986] [Tabel 6]. It can be seen from these figures that the present results confirm Bolton's dilatancy theory.

CONCLUSION

Experimental investigation was conducted on the behavior of granular material in the two and three dimensional shear failure. The purpose of this investigation was to examine the dilatancy theories presented by Rowe [1962, 1969] and Bolton [1986]. The following conclusions can be drawn:

1- Values of angles of shearing resistance [ϕ_{cv}] can be determined experimentaly as demonstrated in figure [7] and [8]. It should be noticed that the values of ϕ_{cv} vary between 32° and 37° for all types of soil Table [4].
2- Values of angles of particle friction [ϕ_μ] can be determined from Horne [1965] as shown in Fig. [1].
3- Values of [ϕ_{cv} and ϕ_μ] are independent from the applied minor stress σ_3 for low σ_3 values while at higher values of σ_3 particle crushing will reduce the dilatancy and consequently will affect the values of ϕ_{cv} and ϕ_μ [Lee and Seed 1967].
4- Values of ϕ_{cv} and ϕ_μ are slightly higher for plane-strain tests compared to triaxial tests.
5- Bolton [1986] presented empirical relationships taking into account the effect of the relative density and confining pressures respectively. These empirical relationships showed good agreement with the present test results.

ACKNOWLEDGEMENTS

The financial assistance of the Natural Sciences and Engineering Research Council of Canada is gratefully acknowledged.

REFERENCES

Bishop A.W. [1954]. Correspondence on Shear Characteristics of a Saturated Silt Measured in Triaxial Compression. Géotechnique 4, No.1, pp.43-45.

Bishop A.W. [1972]. Shear Strength Parameters for Undisturbed and Remolded Soils Specimens. In Stress-Strain Behavior of Soils, pp. 3-58 [R.H.G. Parry, Editor] London; Foulis.

Bishop A.W., and Henkel D. [1978]. The Measurements of Soil Properties in the Triaxial Test. Fourth Edition, Edward Arnold Ltd., London, England.

Bolton M.D. [1986]. The strength and Dilatancy of Sands. Géotechnique 36, No. 1, pp. 65-78.

Caquot A. [1934]. Equilibrium of Solids with Internal Friction. Gauthier-Villars, Paris.

Cole E.R.L. [1967]. Soils in the S.S.A.Ph.D. Thesis. Cambridge University.

Cornforth D.H. [1964]. Some Experiments on the Influence of Strain Conditions on the Strength of Sand. Géotechnique 14:2, pp. 143-167.

Hanna A.M. and Massound N. [1981]. Interlocking of Granular Materials in Two and Three Dimensional Shear Failure. Proceedings of the 8th Canadian Congress of Applied Mechanics, Moncton, N.B., Vol. 2,pp. 239-240.

Hanna A.M., Massound N. and Youssef H. [1987]. Prediction of Plane-Strain Angles of Shearing Resistance from Direct Shear Test Results. The 11th Canadian Congress of Applied Mechanics [CANCAM'87], Canada.

Hanna A.M., Massound N. and Youssef H. [1987]. Prediction of Plane-Strain Angles of Shearing Resistance from Triaxial Shear Test Results. The International

Symposium of Prediction and Performance in Geotechnical Engineering. The University of Calgary, Alberta, June 17/19, 1987, Canada.

Heller A. [1981]. Verschiebungen Starrer und Elastischer Gründungskörper in Sand bei Monotomner und Zyklischer Belastung.Veröff. Inst. Bodenmech. Felsmech. No. 90, 1-127.

Horne M.R. [1963]. The Behavior of an Assembly of Rotund, Rigid Cohesionless Particles. Parts I and II. Proc. R. Soc. A 286, pp. 62-91.

Lee I.K. and Seed H.B. [1967]. Drained Strength Characteristics of Sands. J. Soil Mech. Fdns. Div. Am. Soc. Civil Engrs., 93, SMG, pp. 117-141.

Massoud N. [1981]. Shear Strength Characteristics of Sands. M.Sc. Thesis; The Department of Civil Engineering, Concordia University, Montreal, Quebec, Canada.

Rowe P.W. [1962]. The Stress-Dilatancy Relation for Static Equilibrium of an Assembly of Particles in Contact. Proceedings R. Soc. A. 269-500-527.

Rowe P.W. [1969]. The Relation between the Shear Strength of Sands in Triaxial Compression, Plane Strain and Direct Shear. Géotechnique 19, No. 1, pp. 75-86.

Wood C.C. [1958]. Shear Strength and Volume Change Characteristics of Compacted Soil Under Conditions of Plane-Strain. Ph.D. Thesis; University of London.

LIST OF SYMBOLS

R = Principal stresses ratio [σ_1/σ_3]

D = Rowe's dilatancy rate $[1 - \frac{d\varepsilon_v}{d\varepsilon_1}]$

$d\varepsilon_1$ = Axial strain increment

$d\varepsilon_v$ = Volumetric strain increment

K = Rowe's frictional parameter

ϕ_f = Rowe's friction angle which accounts for dilatancy of sands

n = Prosity

n_r = Relative prosity

ϕ_μ = Angle of frictional sliding

ϕ_{cv} = Angle of shearing resistance corresponding to constant volume change

I_R = Bolton's relative dilatancy index

I_D = Relative density

e = Void ratio

σ_1 = Major principal stress

σ_3 = Minor principal stress

σ_2 = Intermediate principal stress

p = Mean normal stresses [1/3 [$\sigma_1 + \sigma_2 + \sigma_3$]]

$[\phi_{max}]_{T.}$ = Angle of shearing resistance deduced from triaxial compression tests.

$[\phi_{max}]_{P.S}$ = Angle of shearing resistance deduced from Plane-strain compression tests.

τ = Shear stress

Ψ_{max} = Bolton's definition for maximum angle of dilatancy

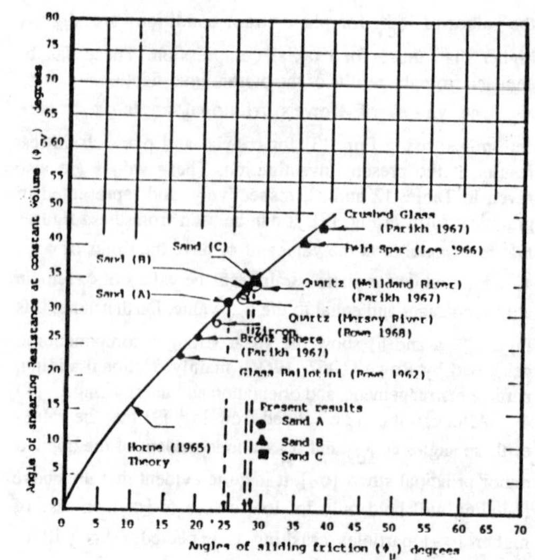

FIG. 1 Theoretical and experimental relations between (ϕ_μ) and (ϕ_{cv}) (After Rowe 1969, with change)

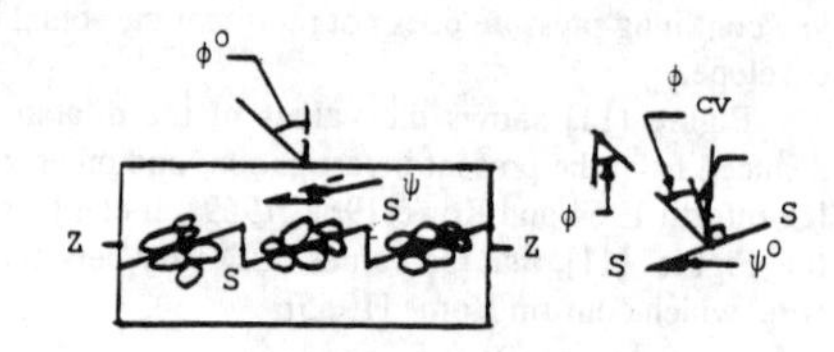

FIG. 2 Bolton's (1986) Saw Blades Model of Dilatancy (After Bolton 1986)

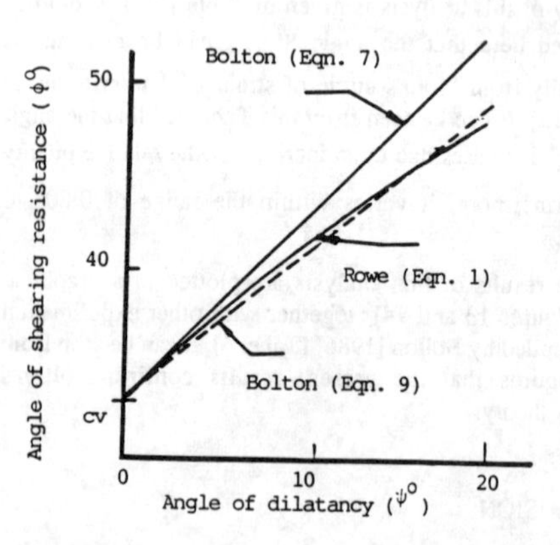

FIG. 3 Rowe's (1962) and Bolton (1986) Stress-Dilatancy (After Bolton 1986)

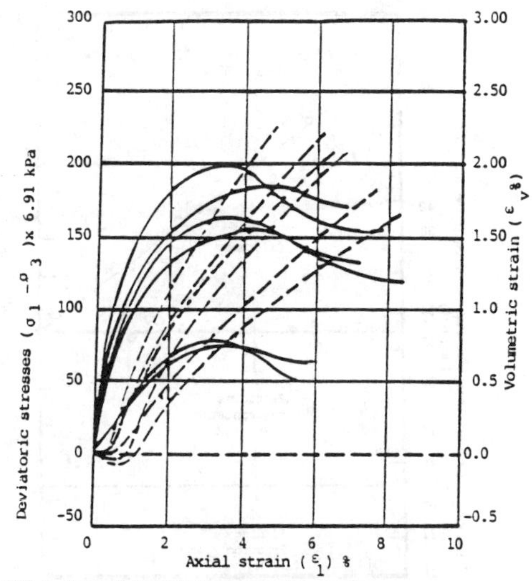

FIG. 4 Typical triaxial compression results

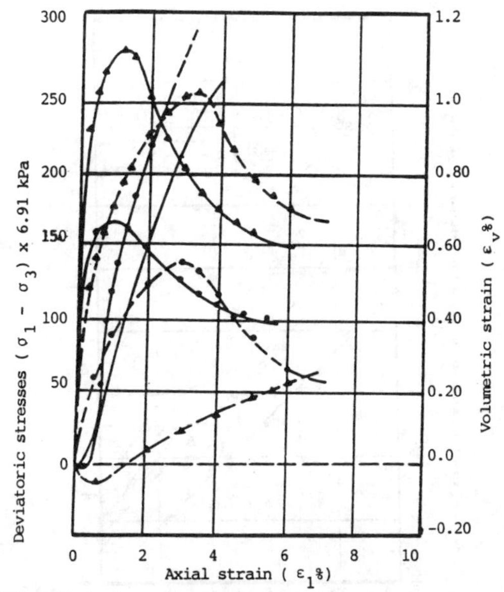

FIG. 6 Typical results of plane-strain tests

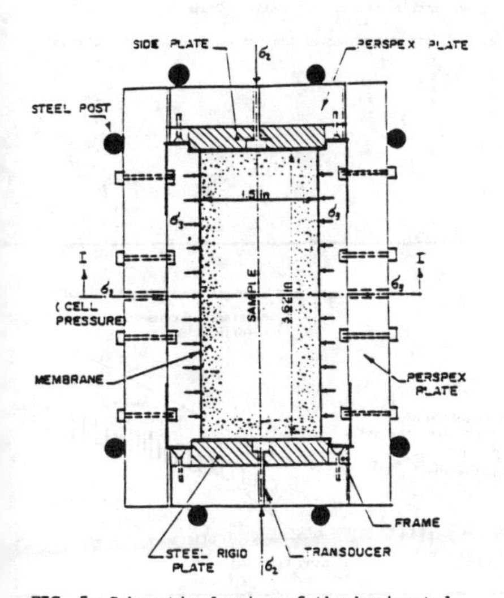

FIG. 5a Schematic drawing of the horizontal plan of the plane-strain apparatus

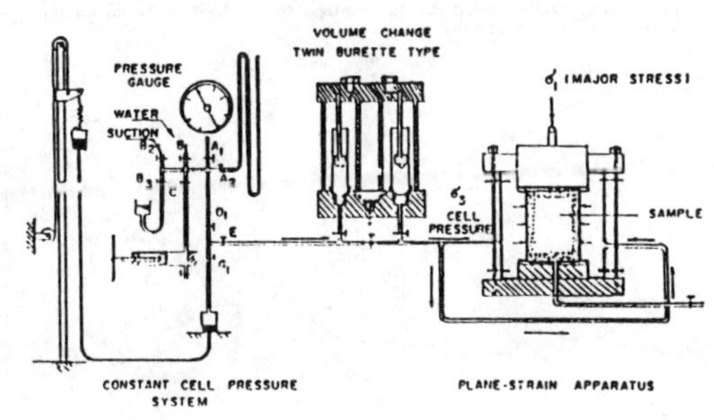

FIG. 5b Experimental layout of the plane-strain experimentation

FIG. 5 Plane-strain apparatus and the associated measurement instrumentation

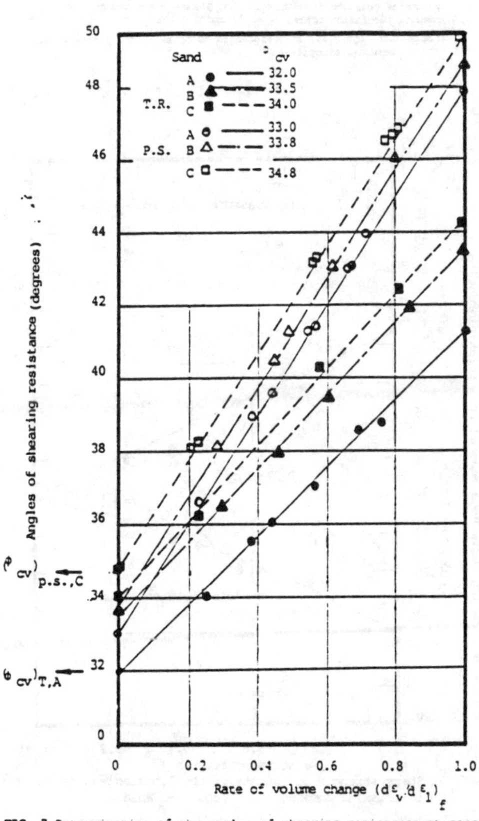

FIG. 7 Determination of the angles of shearing resistance at constant volume (ϕ_{cv}) for the tested cohesionless soil for both triaxial and plane-strain compression (confining pressure=172 and 344 kPa)

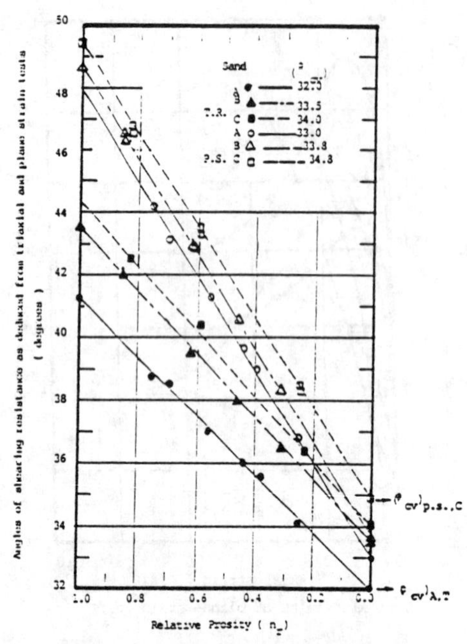

FIG. 8 Comparison between angles of shearing resistance at constant volume as obtained from triaxial and plane-strain compression results (Confining pressures = 172 and 344 kPa).
(Note: T.R. and P.S. are references to triaxial and plane- strain results respectively)

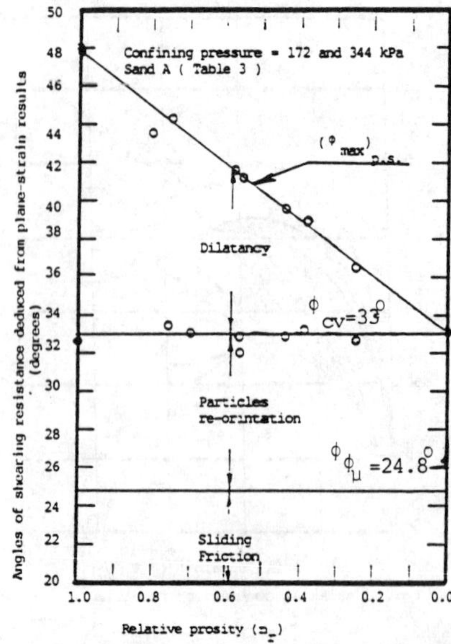

FIG. 9b Shear strength components of the cohesionless sands tested in plane-strain conditions (Sand A- Table 3)

FIG. 9 Shear strength components of the tested cohesionless material

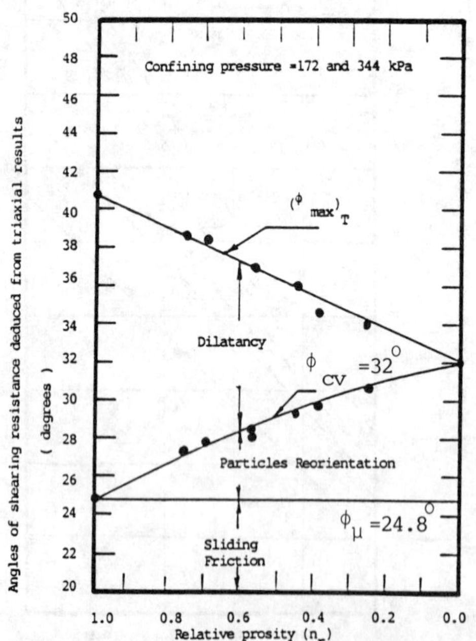

FIG. 9a Shear strength components of the cohesionless sands tested in triaxial compression (Sand A - Table 2)

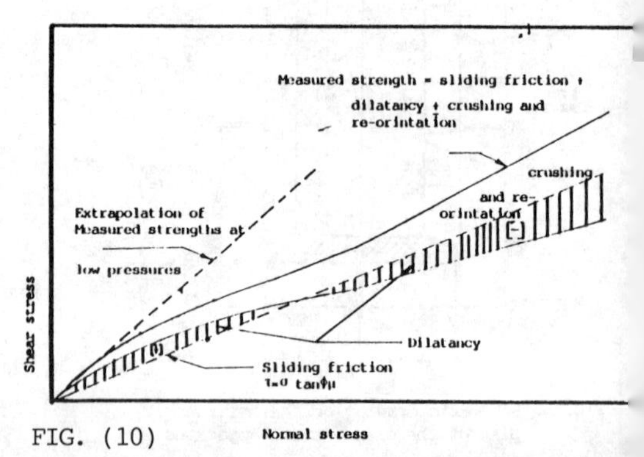

FIG. (10)

Schematic illustration of contribution of sliding friction, dilatancy and crushing to the measured Mohr's envelope for drained tests on sands. (After Lee and Seed 1967)

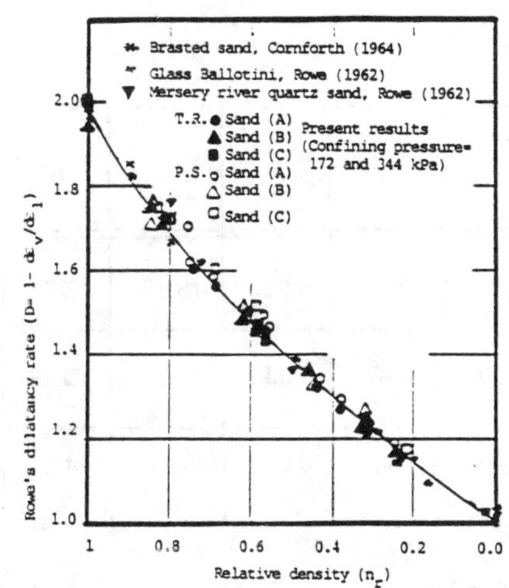

FIG. 11 Variation of the dilatancy rate (D) with the relative prosity (n_r) for triaxial and plane-strain compression results (Note T.R. and P.S. are triaxial and plane-strain results)

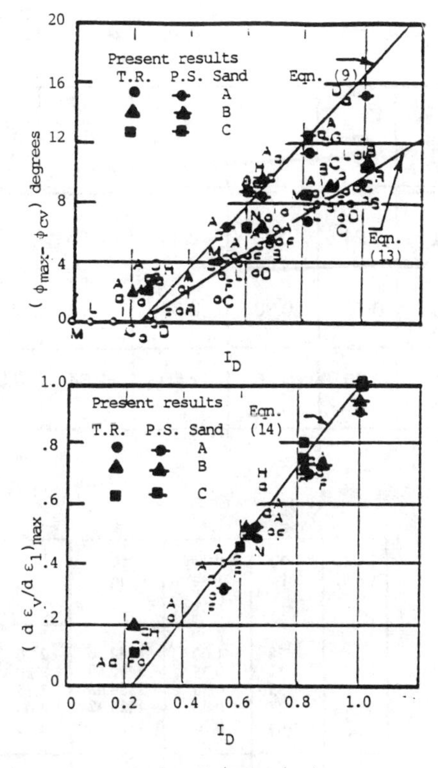

FIG 13 Vartiation of ($\phi_{max} - \phi_{cv}$) and ($d\varepsilon_v/d\varepsilon_1)_{max}$ with the relative density (I_D) for both triaxial and plane-strain compression (Data partially compiled by Bolton 1986). (Plane strain, Triaxial)

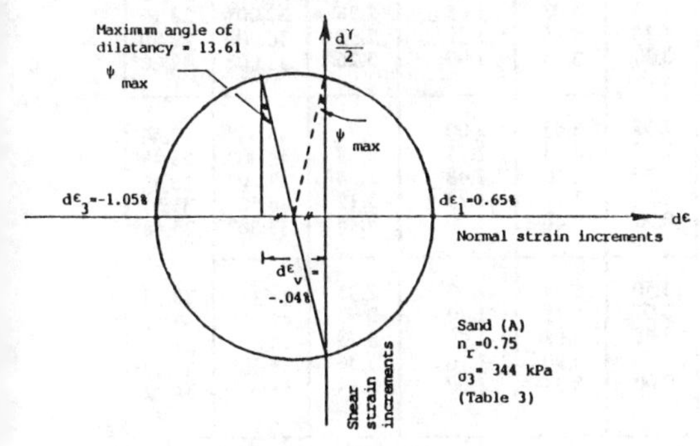

FIG. 12 Mohr's Circle for strain utilizing the present results (Table 3)

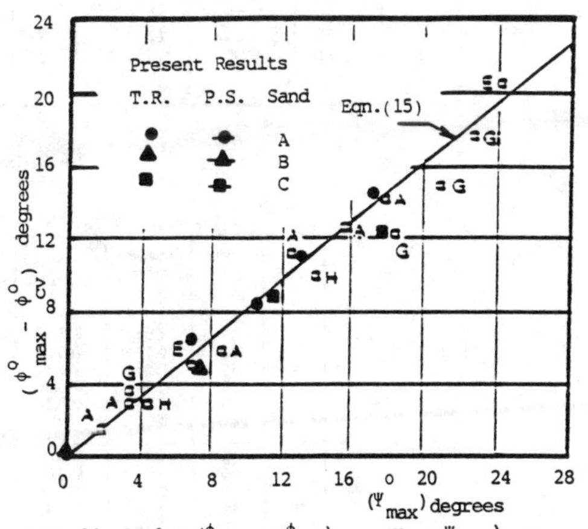

FIG. 14 Angles ($\phi_{max} - \phi_{cv})_{p.s.}$ vs. ($\Psi_{max})_{p.s.}$ (Data partially compiled by Bolton 1986)

TABLE [1]
Physical Properties of the Tested Sands [Cohesionless Soil]

Type of Sands	D_{60}	D_{10}	C_u	G_s	e_{max}	e_{min}	n_{max}	n_{min}	γ_{max}	γ_{min}
A	0.24	0.10	2.40	2.65	0.80	0.40	0.44	0.29	18.75	15.81
B	0.70	0.30	2.33	2.63	0.90	0.50	0.47	0.33	17.77	15.49
C	0.70	0.35	2.00	2.64	0.95	0.40	0.49	0.29	17.28	14.34

TABLE [2.a]
Triaxial Compression Results and Analysis
[σ_3 = [illegible] KPa]

SAND TYPE	Prosity [n]	Relative Prosity $[n_r]$	σ_1 KPa	$d\epsilon_1$ %	$d\epsilon_v$ %	$R = \sigma_1/\sigma_3$	$D' = [1 - \frac{d\epsilon_v}{d\epsilon_1}]$	K = R/D	$[\phi^\circ_t]_{max}$	ϕ°_f
A	28	1.00	1682.16	2.70	-2.70	4.89	2.00	2.45	41.33°	24.85°
	32	0.75	1490.89	3.20	-1.92	4.34	1.60	2.71	38.70°	27.45°
	35	0.56	1384.48	3.85	-1.73	4.03	1.45	2.78	37.00°	28.10°
	37	0.44	1325.72	4.00	-1.28	3.85	1.32	2.92	36.00°	29.30°
	44	0.00	1121.44	4.20	0.00	3.26	1.00	3.26	32.00°	32.00°
B	33	1.00	1867.92	2.55	-2.55	5.43	2.00	2.72	43.55°	27.54°
	35	0.85	1734.55	3.50	-2.65	5.04	1.75	2.88	42.00°	28.98°
	38	0.62	1546.18	3.95	-1.90	4.50	1.48	3.04	39.50°	30.33°
	42	0.31	1354.62	4.80	-1.15	3.95	1.24	3.19	36.50°	31.51°
	46	0.00	1190.24	5.50	0.00	3.46	1.00	3.46	33.50°	33.48°
C	29	1.00	1943.60	3.50	-3.50	5.65	2.00	2.83	44.37°	28.54°
	32	0.82	1772.46	4.38	-3.07	5.15	1.70	3.03	42.50°	30.25°
	36	0.59	1610.55	5.50	-2.48	4.68	1.45	3.23	40.40°	31.82°
	42	0.24	1342.28	9.00	-1.44	3.90	1.16	3.36	36.30°	32.77°
	46	0.00	1217.76	9.30	0.00	3.54	1.00	3.54	34.00°	34.02°

TABLE [2.b]
Triaxial Compression Results and Analysis
[σ_3 = 172 KPa]

SAND TYPE	Prosity [n] %	Relative Prosity $[n_r]$	σ_1 KPa	ϵ_1 %	ϵ_v %	$R = \sigma_1/\sigma_3$	$D = [1 - \frac{d\epsilon_v}{d\epsilon_1}]$	K = R/D	$[\phi^\circ_t]_{max}$	ϕ°_f
A	28	1.00	841.10	2.55	-2.55	4.89	2.00	2.45	41.33°	24.85°
	33	0.69	740.25	3.40	-1.90	4.30	1.56	2.76	38.50°	27.91°
	35	0.56	690.51	3.55	-1.53	4.01	1.43	2.80	37.00°	28.27°
	38	0.38	648.88	3.85	-1.04	3.77	1.27	2.97	35.50°	29.75°
	40	0.25	609.50	3.95	-0.60	3.54	1.15	3.08	34.00°	30.65°
	44	0.00	560.72	4.00	0.00	3.26	1.00	3.26	32.00°	32.00°
B	33	1.00	934.00	2.50	-2.50	5.43	2.00	2.72	43.55°	27.54°
	35	0.85	870.00	2.55	-1.91	5.06	1.75	2.89	42.00°	29.07°
	38	0.62	773.10	2.67	-1.25	4.50	1.47	3.05	39.50°	30.41°
	40	0.46	723.83	2.80	-0.99	4.21	1.35	3.12	38.00°	30.97°
	42	0.31	677.38	2.95	-0.68	3.94	1.23	3.20	36.50°	31.59°
	46	0.00	595.12	3.06	0.00	3.46	1.00	3.45	33.50°	33.48°
C	29	1.00	971.90	2.00	-2.00	5.65	2.00	2.83	44.37°	28.54°
	32	0.82	888.41	4.06	-2.92	5.17	1.72	3.01	42.50°	30.08°
	35	0.59	806.30	5.38	-2.45	4.68	1.46	3.21	40.40°	31.66°
	42	0.24	671.14	7.20	-0.95	3.90	1.13	3.45	36.30°	33.41°
	46	0.00	608.88	8.00	0.00	3.54	1.00	3.54	34.00°	34.02°

TABLE [3.a]
Plane-Strain Compression Results and Analysis
[σ_3 = 344 KPa]

SAND TYPE	Prosity [n]	Relative Prosity [n_r]	σ_1 KPa	σ_2 KPa	e_1 %	ε_v %	R = σ_1/σ_3	D = $[1 - \frac{de_v}{de_1}]$	K = R/D	$[\phi^\circ_{p.s.}]^*_{max}$	ϕ°_f
A	28	1.00	2332.32	600.00	0.50	-0.45	6.78	1.90	3.57	48.00°	34.22°
	32	0.75	1918.61	583.40	0.65	-0.45	5.58	1.70	3.28	44.10°	32.20°
	35	0.56	1679.53	571.04	0.70	-0.35	4.88	1.50	3.25	41.30°	32.00°
	37	0.44	1551.32	563.19	0.81	-0.28	4.51	1.34	3.37	39.60°	32.85°
	44	0.00	1166.16	531.28	1.84	-0.00	3.39	1.00	3.39	33.00°	33.00°
B	33	1.00	2414.88	602.21	0.50	-0.48	7.02	1.95	3.60	48.65°	34.42°
	35	0.85	2204.15	595.12	0.60	-0.42	6.41	1.70	3.77	46.90°	35.50°
	38	0.62	1806.00	577.92	0.61	-0.32	5.25	1.52	3.45	42.90°	33.41°
	42	0.31	1466.53	557.28	0.90	-0.23	4.26	1.25	3.41	38.30°	33.13°
	46	0.00	1207.44	535.45	2.60	-0.00	3.51	1.00	3.51	33.80°	33.80°
C	29	1.00	2518.08	605.94	1.25	-1.38	7.32	2.08	3.52	49.40°	33.89°
	32	0.82	2204.15	595.12	1.44	-0.32	6.41	1.80	3.56	46.80°	34.20°
	36	0.59	1854.08	580.33	1.75	-0.39	5.39	1.54	3.55	43.40°	34.09°
	42	0.24	1471.30	557.62	2.90	-0.46	4.28	1.24	3.69	38.40°	35.00°
	46	0.00	1259.04	540.36	5.94	-0.00	3.66	1.00	3.66	34.80°	34.81°

* Calculated using Eqns. [16 and 17].

TABLE [3.b]
Plane-Strain Compression Results and Analysis
[σ_3 = 172 KPa]

[illegible] TYPE	[illegible] [n]	[illegible] Prosity [n_r]	[illegible] KPa	[illegible] KPa	[illegible] %	[illegible] %	[illegible] σ_1/σ_3	[illegible] $[1 - \frac{d\varepsilon_v}{d\varepsilon_1}]$	[illegible] R/D	[illegible] max	[illegible]
A	28	1.00	1166.16	299.78	0.48	-0.50	6.78	2.04	3.32	48.00°	32.48
	33	0.69	920.20	289.83	0.60	-0.22	5.35	1.62	3.39	43.20°	33.00
	35	0.56	842.75	285.69	0.65	-0.36	4.90	1.55	3.16	41.40°	31.30
	38	0.38	755.22	280.19	0.85	-0.25	4.39	1.29	3.40	39.00°	33.10
	40	0.25	679.49	274.51	1.20	-0.04	3.95	1.19	3.32	36.60°	32.48
	44	0.00	583.08	265.64	1.80	-0.05	3.39	1.03	3.29	33.00°	32.26
B	33	1.00	1207.74	301.12	0.45	-0.47	7.02	2.04	3.44	48.65°	33.34
	35	0.85	1069.88	295.36	0.55	-0.45	6.22	1.81	3.44	46.30°	33.34
	38	0.62	899.65	288.79	0.58	-0.35	5.23	1.60	3.23	42.80°	32.10
	40	0.46	813.67	283.97	0.85	-0.37	4.73	1.44	3.56	40.60°	34.15
	42	0.31	733.26	278.64	1.25	-0.36	4.25	1.29	3.55	38.30°	34.10
	46	0.00	603.72	267.72	2.50	-0.05	3.51	1.02	3.44	33.80°	33.34
C	29	1.00	1259.04	302.65	1.22	-1.28	7.32	2.05	3.57	49.40°	34.22
	32	0.82	1100.18	297.49	1.40	-1.09	6.40	1.78	3.60	46.50°	34.39
	35	0.59	920.06	289.82	1.65	-0.99	5.35	1.59	3.34	43.20°	32.63
	42	0.24	728.52	278.30	2.35	-0.42	4.24	1.21	3.59	38.20°	34.35
	46	0.00	629.52	270.04	5.10	-0.10	3.66	1.02	3.59	34.80°	34.35

* Calculated using Eqns. [16 and 17].

TABLE [4]
Evaluation of Angle of Friction Sliding by Several Approachs

TYPE of SANDS	ϕ°_{cv} Present Study Measured		$[\phi^\circ]_\mu$ measured Rowe [1962]	ϕ°_μ Horne [1965] Figure [1]	ϕ°_μ Eqn.(1)	$[\phi^\circ]_\mu$ Bishop Eqn. (2)	$[\phi^\circ]_\mu$ Rowe [1962, 1969] Stress-Dilatancy Theory Eqn. (3)
	Triaxial	Plane-Strain					
A	32.0	33.0	24.80°	–	21.70°	21.56°	24.80°
B	33.5	33.8	27.50°	26.00°	22.85°	22.48°	27.50°
C	34.0	34.8	28.50°	–	–	22.80°	28.50°

TABLE [5]
Analysis of Triaxial and Plane-strain Compression Results Utilizing Bolton's Approach
[σ_3= 344 KPa]

SAND TYPE	n_r	e	I_D eq. 11	ϕ^o_{cv} measured	TRIAXIAL COMPRESSION RESULTS						PLANE-STRAIN RESULTS				
					ϕ_t	$[\phi^o_t - \phi^o_{cv}]$	$3 I_R$	$0.3 I_R$	$-\frac{d\varepsilon_v}{d\varepsilon_1}$	D_t	$\phi^o_{P.S.}$	$[\phi^o_{P.S.} - \phi^o_{cv}]$	$-\frac{d\varepsilon_v}{d\varepsilon_1}$	$D_{P.S.}$	$\Psi^o_{P.S.}$
A	1	0.40	1.0	T.R. 32°	41.33	9.33	7.20	0.72	1.00	2.00	48.0	15.00	0.90	1.90	18°
	0.75	0.47	0.82	P.S. 33°	38.70	6.70	5.65	0.57	0.60	1.60	44.1	11.10	0.70	1.70	13.61°
	0.56	0.54	0.65		37.00	5.00	3.90	0.39	0.45	1.45	41.3	08.30	0.50	1.50	11.54°
	0.44	0.59	0.53		36.00	4.00	2.95	0.295	0.32	1.32	39.6	06.60	0.34	1.34	07.11°
	0.00	0.80	0.00		32.00	0.00	-3.00	-0.30	0.00	1.00	33.0	00.00	0.00	1.00	00.00°
B	1	0.50	1.00	T.R. 34°	43.55	10.05	6.93	0.69	1.00	2.00	48.65	09.75	0.95	1.95	18.10°
	0.85	0.54	0.90	P.S. 34.8°	42.00	8.50	6.25	0.63	0.75	1.75	46.90	08.20	0.70	1.70	05.56°
	0.62	0.64	0.65		39.50	6.00	3.73	0.37	0.48	1.48	42.90	09.10	0.52	1.52	08.00°
	0.31	0.72	0.20		36.50	3.00	2.13	0.21	0.24	1.24	38.30	04.50	0.25	1.25	07.00°
	0.00	0.90	0.00		33.50	0.00	-3.00	-0.30	0.00	1.00	33.80	00.00	0.00	1.00	00.00°
C	1	0.41	1.00	T.R. 33.5°	44.37	10.37	6.55	0.66	1.00	2.00	49.4	09.57	1.08	2.08	20.53°
	0.82	0.47	0.82	P.S. 33.8°	42.50	8.50	4.89	0.49	0.70	1.70	46.8	12.00	0.80	1.80	18.20°
	0.59	0.56	0.59		40.40	6.40	2.65	0.27	0.45	1.45	43.4	08.50	0.57	1.57	09.97°
	0.24	0.72	0.24		36.30	2.30	-0.52	-0.05	0.16	1.16	38.4	03.60	0.24	1.24	09.17°
	0.00	0.82	0.00		34.00	0.00	-3.00	-0.03	0.00	1.00	34.8	00.00	0.00	1.00	00.00°

TABLE [6]
Data on Sands
[Partially compiled by Bolton, 1986]

Identification [SAND]		Name	d_{60} mm	d_{10} mm	e_{min}	e_{max}	ϕ_{cv}	Reference
[1]	[A]	Brasted river	0.29	0.12	0.47	0.79	32.6	Cornforth (1964, 1973)
[2]	[B]	Limassol marine	0.11	0.003	0.57	1.18	34.4	Cornforth (1973)
[3]	[C]	Mersey river	~ 0.20	~ 0.10	0.49	0.82	32.0	Rowe (1969)
								Rowe & Barden (1964)
[4]	[D]	Monterey no. 20	~ 0.30	~ 0.15	0.57	0.78	36.9	Marachi, Chan
								Seed & Duncan (1969)
[5]	[E]	Monterey no. 0	~ 0.50	~ 0.30	0.57	0.86	37.0	Lade & Duncan (1973)
[6]	[F]	Ham river	0.25	0.16	0.59	0.92	33.0	Bishop & Green (1965)
[7]	[G]	Leighton Buzzard 14/25	0.85	0.65	0.49	0.79	35.0	Stroud (1971)
[8]	[H]	Welland river	0.14	0.10	0.62	0.94	35.0	Barden *et al.* (1969)
[9]	[I]	Chattahoochee river	0.47	0.21	0.61	1.10	32.5	Vesic & Clough (1968)
[10]	[J]	Mol	0.21	0.14	0.56	0.89	32.5	Ladanyi (1960)
[11]	[K]	Berlin	0.25	0.11	0.46	0.75	33.0	De Beer (1965)
[12]	[L]	Guinea marine	0.41	0.16	0.52	0.90	33.0	Cornforth (1973)
[13]	[M]	Portland river	0.36	0.23	0.63	1.10	36.1	Cornforth (1973)
[14]	[N]	Glacial outwash sand	0.90	0.15	0.41	0.84	37.0	Hirschfield & Poulos (1964)
[15]	[P]	Karlsruhe medium sand	0.38	0.20	0.54	0.82	34.0	Hettler (1981)
[16]	[R]	Sacramento river	0.22	0.15	0.61	1.03	33.3	Lee & Seed (1967)
[17]	[S]	Ottawa sand	0.76	0.65	0.49	~ 0.8	30.0	Lee & Seed (1967)
18	[A]	Silica sands	0.24	0.10	0.40	0.44	32.00	Hanna & Youssef (1987)
19	[B]	Silica sands	0.70	0.30	0.50	0.47	33.50	
20	[C]	Silica sands	0.70	0.35	0.40	0.49	34.00	

Prediction and Performance in Geotechnical Engineering / Calgary / 17-19 June 1987

Influence of groundwater table fluctuation on performance of bases and subgrades

Sabry A.Shihata, Zaki A.Baghdadi & Ahmed M.Khan
King Abdulaziz University, Jeddah, Saudi Arabia

ABSTRACT: Water table levels fluctuate in areas where no storm and/or sewage systems exist causing significant variations in the moisture contents of the base and subgrade materials that lead to a major deterioration in the performance of pavements. In this paper, strength characteristics and basic properties of the soil are examined. The effect of the degree of saturation in the range of 50 to 100%, on the pavement deflection and resilient modulus of the soil, are studied in the repétitive undrained triaxial tests with lateral pressures in the range 100-140 kN/m^2.

1 INTRODUCTION

In developing areas without sufficient storm and/or sewage systems, waste waters are disposed of locally and the storm water is left to percolate through the ground. This recharge raises the existing water table near or to the ground surface if there is a shallow impermeable layer or if the ground water table is already high. The problem also exists when the rate of discharge of the waste water is greater than the soil can dissipate. This causes significant variations in the moisture content of the soil which is used as a subgrade for the pavement of the street network and also on the base coarse as well, depending on the material used.

Rise of the water table affects adversely the performance of street pavements, if not taken into consideration, it causes premature failure of the pavement surface manifested as extensive alligator cracking and rutting. These types of distress are typical of the affected areas. Increasing moisture content in the soil reduces its resilient modulus and increases permanent strains.

The effects of degree of saturation and compaction on the elastic modulus and permanent strain of the predominant soil in a badly affected area, by the rise of ground water table, was studied using strain controlled repetitive undrained triaxial tests. This is the first stage of a more comprehensive program which is going on to study the problem thoroughly.

Utilization of static triaxial testing for the determination of the resilient modulus and permanent deformation under repeated loading was done earlier by other investigators and many cases are cited in the literature.

It was possible, using the elastic properties as determined in this study, to assess the influence of the rise of water table on the deflection of the pavement surface and the vertical strain over the depth from which permanent strain and rutting potential may be evaluated.

2 REVIEW OF LITERATURE

In this review of literature two parameters which are used extensively to assess the performance of bases, subbases and subgrades, are discussed: resilient modulus, Er and permanent strain, ε_p. The resilient modulus has been used for evaluating the dynamic response of materials in the laboratory. It is defined as the ratio of repeated axial deviatoric stress, σ_d, to the recoverable or resilient axial strain ε_r

or $E_r = \frac{\sigma_d}{\varepsilon_r}$

Studies have indicated that resilient deformations generally stabilize after 200 load repetitions and hence the Er value is usually computed at this repetition level.

Thompson and Robnett (1979) conducted an intensive study on the resilient properties of fine-grained soils (A and BC horizons).

They found that plots of Er - σ_d relations typically displayed a "breaking point" deviator stress where there was a substantial change in slope. The resilient moduli at the breaking point Er_i, for all soils tested were found to occur at a stress level around 41 kN/m^2 (6 psi). Several factors were found to affect the resilient properties of the soils studied. Some of these factors are:

- soil properties; such as liquid and plastic limits, group index, silt content, clay content, specific gravity and organic content.
- shear strength; strong positive correlations with Eri for the static stress-strain data, unconfined compressive strength and static modulus of elasticity.
- degree of saturation; strong correlation with Eri. For soils substantially wet of optimum, high degrees of saturation and low Eri are characteristic regardless of level of compaction.
- compaction effects; the affect of increased compaction was not significant for A horizon soils but was significant at all stress levels for the BC horizon soils.

Rada and Witczak (1981) conducted a comprehensive study on the results of 271 tests on granular materials. Their analysis indicated an inverse correlation between k_1 and k_2 (constants in $Er = k_1 \theta^{k_2}$) exists for the global class of all granular materials, where θ is the bulk stress, and k_1 and k_2 are regression constants. Several factors affect the resilient moduli of granular soil, some of these are:

- stress state
- degree of saturation: Values of k_1 and k_2 change with the degree of saturation depending on the material under consideration.
- degree of compaction (density): It results in an increase in the k_1 term while the k_2 value remains essentially constant.
- gradation: Tests showed that the influence of percentage passing sieve # 200 varies depending on the type of material considered.

Numerous studies including those mentioned above utilize regression techniques for predicting Er. Others used more complex models and finite element procedures in analysing the nonlinear response of the granular materials, Brown and Pappin (1981).

Permanent strain and hence the rut depth in granular layers are interdependent and should be controlled because of their significant influence on the safety and performance of the supporting system. As reported in the literature, some of the important factors that affect the magnitude of permanent strain in granular materials subjected to repeated loading may be summarized as follows:

- number of loading cycles: A considerable amount of the total permanent strain takes place after the first few cycles and it increases with increasing magnitude of deviator stress, Olowoker (1975) and Brown (1974). A linear relationship has also been shown between permanent strain and the logarithm of the number of cycles up to 10000 cycles, by Lentz and Baladi (1980).
- state of stress: Deviator stress and confining stresses are significant in determining the amount of permanent strains. They should be considered simultaneously in the analysis, Thompson (1979) and Borksdale (1972).
- density: Increasing density leads to decrease in permanent strain, Kalcheff (1974).
- amount of fines (passing # 200): Borksdale (1972) has found that the relation between plastic strain and amount of fines is stress dependant. At low stress levels, the amount of fines had no effect, while at higher stress levels the rate of increase in plastic strain decreased as the amount of fines is more than 10%.
- degree of saturation: Haynes and Yoder (1963) have shown that increasing the degree of saturation from about 70% to 95% resulted in large plastic strain increments.

Hyperbolic functions and constitutive equations with parameters obtained from static triaxial tests have been used to predict permanent strains, e.g., Lentz and Baladi (1980, 1981).

3 EXPERIMENTAL PROGRAM

The experimental program was carried out on a silty sand collected from Jeddah area. The identification and classification properties of this soil are summarized in Table 1.

Fig.1 shows its grain size distribution and Fig. 2 shows compaction characteristics as obtained from modified AASHTO test specifications.

The experimental part of the study included determination of the grain size distribution (sieve analysis and hydrometer), (Fig. 1) compaction (Fig. 2) and unconsolidated undrained triaxial tests applying repeated loads.

Table 1. Properties of the soil tested

Location:	Prince Sultan Street, Jeddah, Saudi Arabia
Color	: light brown
Liquid Limit	= 28%
Plastic Limit	= 21%
Plascicity Index	= 7
Shrinkage Limit	= 18%
Specific gravity	= 2.64
Classification groups: AASHTO	= A-2-4(0)
Unified	= SM
Max. grain size	= ½" (14mm)
D_{10}	= .02mm
C_u	= 70
C_z	= 1.43
% Clay	= 5
% -#200	= 18

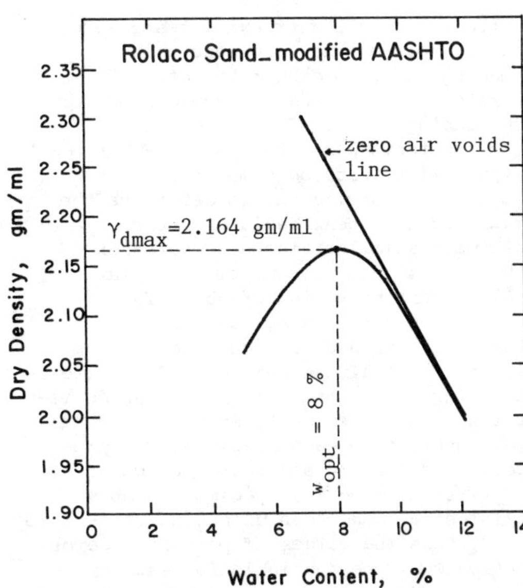

Fig.2 Dry density-water content relationships.

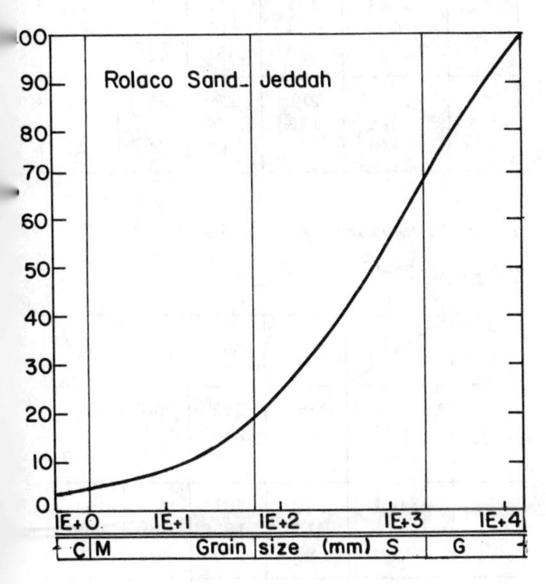

Fig.1 Particle size distribution of test sand.

In order to explore the material characteristics over a range of degrees of compaction and saturation, it was planned to test at values of relative compaction of 90%, 95% and 100%. The degrees of saturation were selected to be 50%, 75%, 90% and 100%. For each degree of saturation and relative compaction, the amounts of water and solids were worked out for preparation of triaxial test specimens. The various moisture contents for different degrees of saturation and compaction are listed in Table 2.

Table 2. Moisture contents for various degrees of saturation and relative compaction.

Dry Density gm/ml	Relative Compaction %	S=50%	S=75%	S=90%	S=100%
1.948	90	6.7	10.0	12.0	13.5
2.024	95	5.7	8.6	10.3	11.5
2.164	100	4.0	6.2	7.4	8.3

The triaxial specimens were 50 mm dia and 100 mm high; the diameter of the specimens was selected keeping in view the maximum aggregate size. The dry material for the known dry density was accurately weighed, to which pre-determined amounts of distilled water were added, before thorough mixing. The specimens were statically compacted.

Strain controlled triaxial tests were carried out on the specimens using loading and unloading rates of 1.25mm/min., the rate which is usually recommended for UU loading tests.

As the tests were of the controlled strain type, trial tests on similar specimens were first carried out to determine the magnitudes of peak deviator stresses and ultimate axial strains. It was observed that the specimens showed plastic bulging and some kind of shear failure plane when approaching failure. It was therefore decided to load specimens up to 15% axial strain as recommended in literature for the determination of peak deviator stress. Five repetitions of loading were applied to each specimen. The procedure of loading and reloading consisted of first loading the specimen to about 1% axial strain and then unloading it to zero load, thus the values of permanent deformation would be available for each cycle. For the next cycles, the axial strain levels were increased to 2%, 3%, 4% and 5% and each time, the specimens were unloaded to zero deviator stress. After having completed the 5 cycles, the loading was continued until the axial strain was 15% and the test was discontinued. A typical stress-strain curve is shown in Fig. 3.

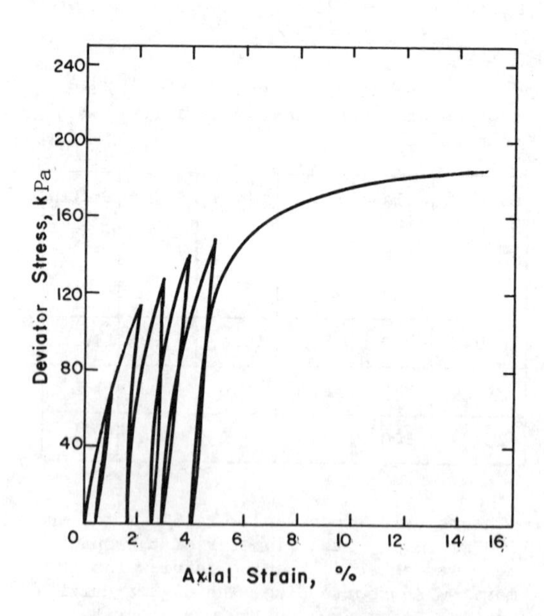

Fig.3 Typical deviator stress-strain relationship from triaxial test.

The tests were carried out on identical specimens with confining pressures of 100 kN/m^2 and 140 kN/m^2.

4 TEST RESULTS

The elastic modulus, deviator stress and permanent strain at each strain level were determined. Corresponding values at a strain level of $9.8x10^{-3}$ are given in Tables 3, 4 and 5. Effect of the degree of saturation on the elastic modulus is

Table 3. Elastic modulus in MPa at $\varepsilon_T = 9.8x10^{-3}$ (1MPa = 145 psi)

σ_3 KPa	γ_d \ S	50%	75%	90%	100%
100	2.164	181	147	71	16
	2.024	154	107	45	12
	1.948	74	53	43	8.7
140	2.164	211	157	109	64
	2.024	153	74	55	38
	1.948	144	63	24	15

Table 4. Deviator stress in KPa at $\varepsilon_T = 9.8x10^{-3}$

σ_3 KPa	γ_d \ S	50%	75%	90%	100%
100	2.164	596	383	200	96
	2.024	370	213	104	68
	1.948	222	148	65	8.7
140	2.164	740	501	305	194
	2.024	414	185	109	98
	1.948	287	126	24	30

Table 5. Permanent strain at $\varepsilon_T = 9.8x10^{-3}$ $x 10^{-3}$

σ_3 KPa	γ_d \ S	50%	75%	90%	100%
100	2.164	6.5	7.2	7.0	3.9
	2.024	7.4	7.8	7.5	3.9
	1.948	6.8	7.0	8.3	8.8
140	2.164	6.3	6.6	7.0	6.8
	2.024	7.1	7.3	7.8	7.2
	1.948	7.8	7.8	9.1	7.8

shown in Fig. 4 using a confining pressure of 100 KPa. Fig. 5 shows the effect of the deviator stress and degree of saturation on the permanent strain. Fig. 5

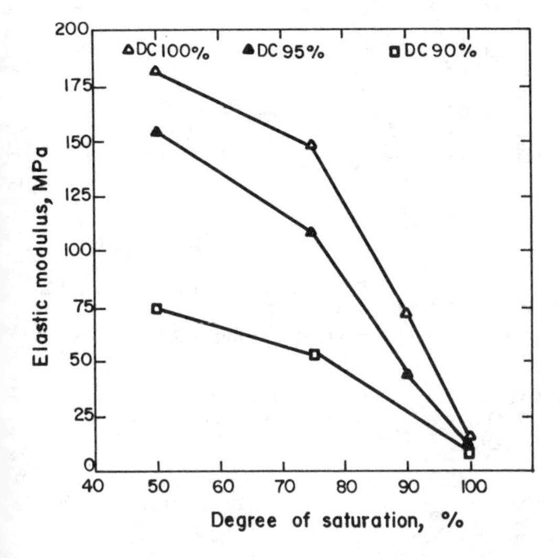

Fig.4 Effect of Degree of Saturation on elastic Modulus.

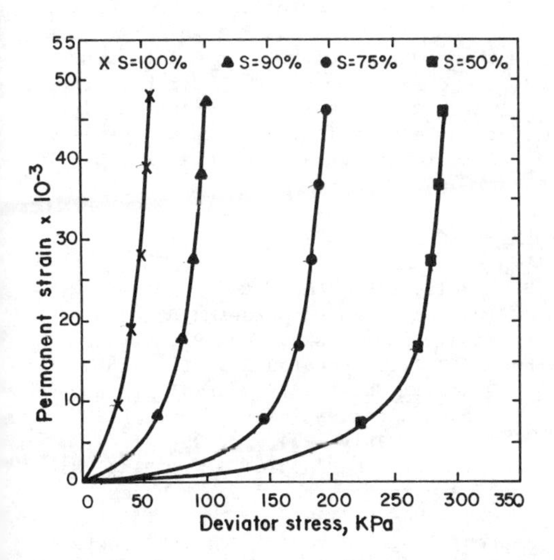

Fig.5 Effect of Deviator Stress and degree of Saturation on permanent strain.

also shows that for each degree of saturation there exists a deviator stress at which the rate of increase in the permanent strain becomes infinite. This deviation stress may be called the critical stress. Effect of the degree of saturation and compaction on this critical stress is shown in Table 6 for a confining pressure of 100 KPa. The pavement was analysed under a load of 81.7 kN applied through dual wheels, the tyre pressure was taken as 0.57 MPa.

Table 6. Effect of degree of saturation and dry unit weight on the critical deviator stress. (σ_3 = 100 KPa)

γ_d \ S_r	50%	75%	90%	100%
2.164	845	638	441	246
2.024	370	213	104	68
1.948	222	148	65	26

5 DISCUSSION & ANALYSIS OF RESULTS

To study the effect of water table on the performance of street pavements, the actual distribution of degree of saturation with depth was determined. The water table was 2.7 meters below the ground surface. The actual distribution is shown in Fig. 6, the shaded area indicates the extent of evaporation. Using this distribution, hypothetical profiles of the distribution of degree of saturation with depth assuming different levels of water table are also given in Fig. 6.

The elastic moduli of the soil at several depths and for different levels of water table are obtained from Figs. 6 and 5.

Theoretical analysis using elastic layered program (ELSYMS) was done. The pavement consisted of an asphaltic top layer 50 mm thick with a modulus of elasticity of 4138 MPa. An aggregate base 200 mm thick with elastic modulus of 552 MPa was considered. Properties of the surface and base coarse were kept constant. The subgrade was taken to a depth of -2500mm.

The effect of the level of water table on surface deflection is shown in Fig. 7. The figure shows that the rise of the water table causes larger deflections of the pavement surface and that there is a critical depth of the water table above which the rate of increase of surface deflection becomes excessive.

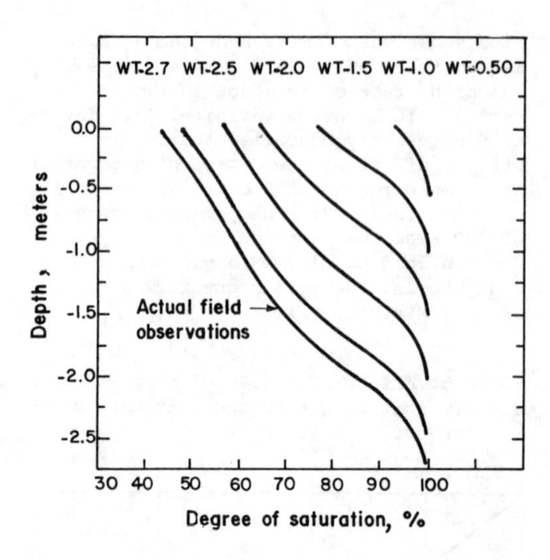

Fig.6 Variation of the degree of saturation with depth.

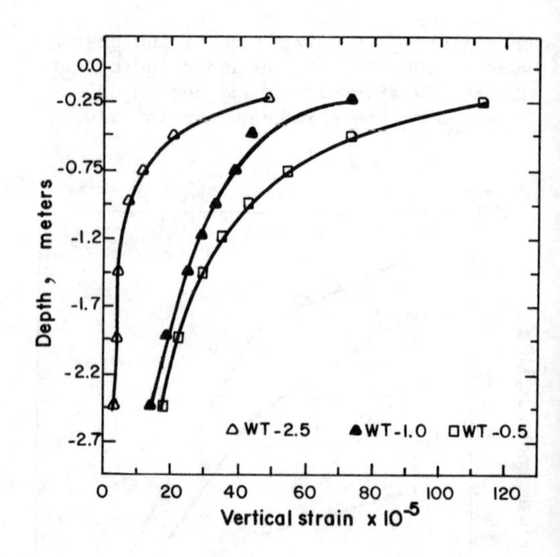

Fig.8 Effect of the level of water table on the variation of vertical strain with depth.

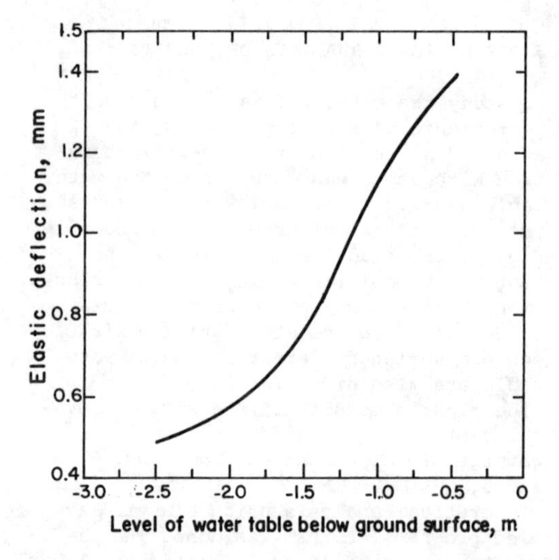

Fig.7 Effect of the level of water table on the deflection of the pavement surface.

Effect of the rise of water table on vertical strains from which rutting potential of the pavement surface is estimated is illustrated in Fig. 8. The figure shows that the rise of the water table causes higher strains over greater depths which in turn means higher potential for rutting.

CONCLUSIONS

In light of this study, the following conclusions may be drawn:

1. Strain controlled repetitive undrained triaxial tests may be used to evaluate the resilient modulus and permanent strain of the type of soil tested and it is in general agreement with the literature.
2. The critical degree of saturation for the soil tested was 75% which is lower than the 85% cited in the literature.
3. There exists a critical depth of the water table, based on surface deflection below which, the pavement will deteriorate at an excessive rate.

REFERENCES

Borksdale, R.D., "Repeated Load Test Evaluation of Base Course Materials," Final Report, GIT, May 1972.

Brown, S.F., "Repeated Load Testing of A Granular Material," Journal of the Geotechnical Engineering Div., ASCE, No. GT7, July 1974, pp.825-841.

Brown, S.F., and Pappin, J.W., "Analysis of Pavements with Granular Bases," TRB, TRR810, 1981, pp.17.

Haynes, J.H. and Yoder, E.J., "Effects of Repeated Loading on Gravel and Crushed Stone Base Course Materials Used in the

AASHO Road Test, "HRR 39, 1963, pp. 82-96.
Kalcheff, I.V., "Characteristics of Graded Aggregates As Related To Their Behaviour Under Varying Loads And Environments," NCSA, March 1974.
Lentz, R.W. and Baladi, G.Y., "Prediction of Permanent Strain of Sand Subjected to Cyclic Loading," TRB, TRR 749, 1980, pp. 54-58.
Lentz, R.W and Baladi, G.Y. "Constitutive Equation for Permanent strain of Sand Subjected to Cyclic Loading," TRB, TRR 810, 1981, pp. 50-54.
Olowekerc, D.O., "Strength and Deformation of Railway Ballast Subjected to Triaxial Loading," M.S. Thesis, Queen's University, Kingston, Ontario, 1975.
Rada, G. and Witczack, M.W., "Comprehensive Evaluation of Laboratory Resilient Moduli Results For Granular Materials," TRB, TRR 810, 1981, pp. 23-32.
Thompson, M.R., "Repeated Triaxial Testing Program Traprock Ballast Gradation Effects Study," UIUC., May 1979.
Thompson, M.R. and Robnett, Q.L., "Resilient Properties of Subgrade Soils," Final Report, UIUC., June, 1976.

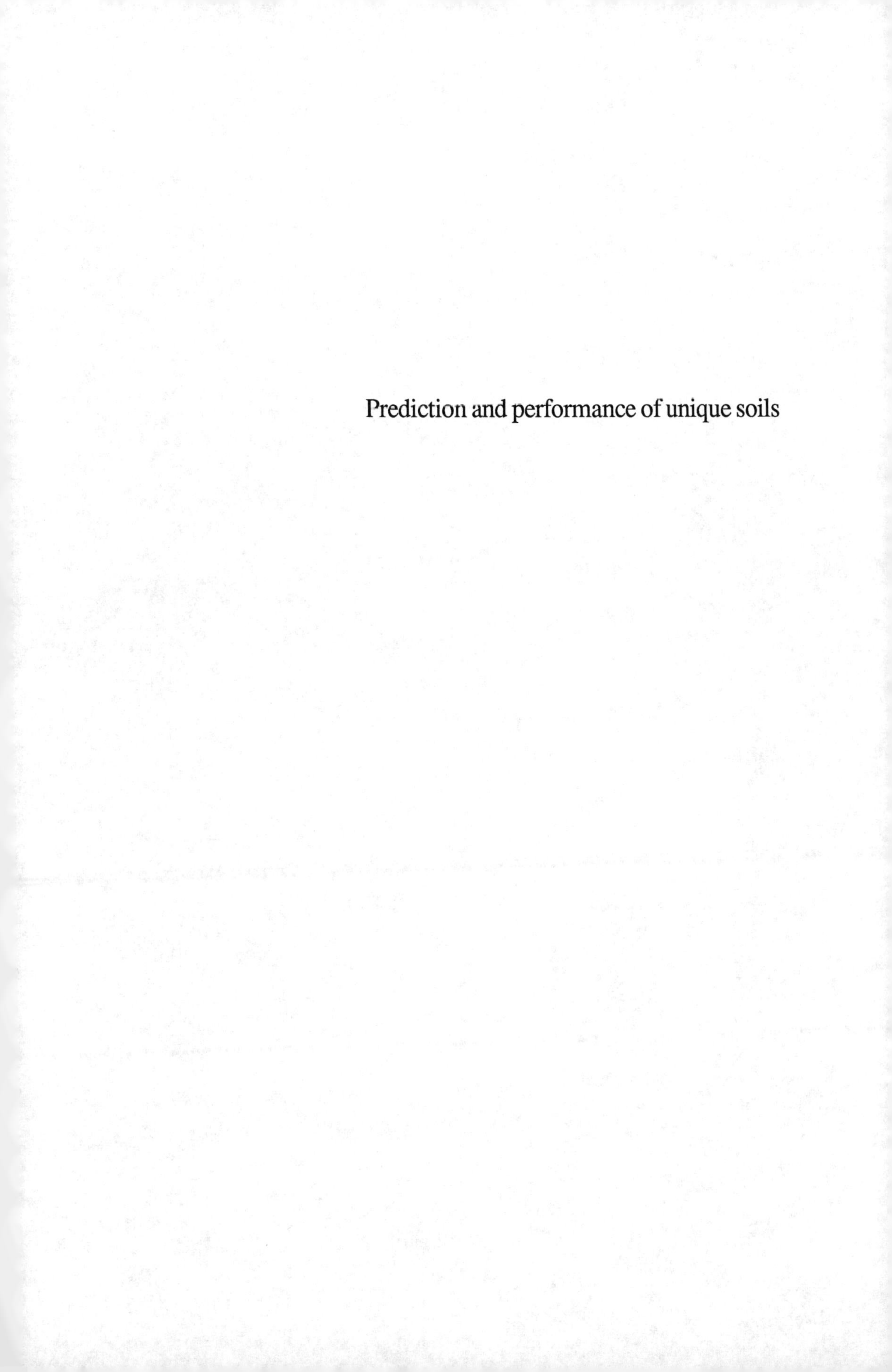

Prediction and performance of unique soils

Roadway embankment construction through a muskeg area

Murthy Pariti
Alberta Transportation Department, Edmonton, Canada

ABSTRACT: Alberta Transportation constructed a local road recently across a deep muskeg area. Because of the considerable depth of muskeg and height of fill, special construction procedures were adopted in terms of (a) using a granular drainage layer on top of the muskeg, and (b) stage construction of fill. This paper presents the details of the project and settlement observations and compares with estimates based on data observed elsewhere.

INTRODUCTION

During 1984-85, Alberta Transportation undertook construction of a major portion of a local road called Wolf Lake Road which runs in a southerly direction off Highway 16, just west of the Highway 32 junction. In that process, part of the roadway alignment had to cross a deep muskeg area at NE 24-47-14-5. The grade line design calls for an average fill height of 12 metres in that section. Based on preliminary soil survey information, the Regional District staff identified a possible concern for the stability of fill through the muskeg area and requested from the Department's Geotechnical Section a detailed evaluation of the site conditions in relation to the proposed grade construction.

It was also made known to the Geotechnical section that (a) if necessary, the department does not have any objection to spreading the grade construction through the muskeg area over a period of two construction seasons, (b) the work will be done by the Departmental forces on a day-labour basis, and (c) availability of granular material for the purpose of any drainage medium, if required, is within reasonable haul distance.

INVESTIGATION

From an observation of the site conditions and the airphoto of the project site (Figure 1), it generally appears that the muskeg formation is along an old river valley, which is about 200 metres in width. A total of four test holes were drilled on either side of the centre line. Each test hole was located slightly beyond the possible toe location at the corresponding station. Also exploratory test holes were drilled at locations where installation of settlement plates was planned. The test holes drilled indicated variable depth of muskeg, ranging anywhere from 4 to 8 metres, at different points.

Underlying the muskeg, finely suspended silty sand exists to a depth of 5-8 metres below which sandstone/shale bedrock was identified.

RECOMMENDATIONS MADE

Because of the availability of relatively good granular material within reasonable haul distances and flexibility in proceeding with construction scheduling, the construction procedure recommended incorporated the following salient features:

1. Using a drainage blanket of about 1.5 m thickness on top of the muskeg.

2. Truck hauling of fill material for the initial 2 to 3 metre height of fill placement.

3. Stage construction with a 2-3 week waiting period between any two stages in a particular area.

4. Fill placement in the outer areas first and then proceeding towards the centre line.

Figure 1. Airphoto of the muskeg site.

This procedure was considered helpful to minimize the potential risk for a muskeg failure, as the toe areas' muskeg may gain some strength due to consolidation. A typical cross-section indicating the above suggested features is shown in Figure 2.

CONSTRUCTION AND INSTRUMENTATION MONITORING

Placement of the drainage layer was initiated in April 84 while the muskeg area was still in a frozen condition. Subsequent lifts of fill material continued until November 84. The height of fill that was constructed in this manner was about 6 metres before the work was shut down for the winter. Placement of the remaining height of fill required to reach the design grade line was resumed in October 85 and was completed during the same month.

Information regarding (a) the depth of muskeg along the centre line and (b) the average height of grade line as constructed, is shown in Figure 3. A total of six settlement plates were installed, three on either side of centre line close to the shoulder location, so that any construction traffic would not knock them down. Figure 4 shows the progress chart of fill placement and its settlement profile with time at a typical location.

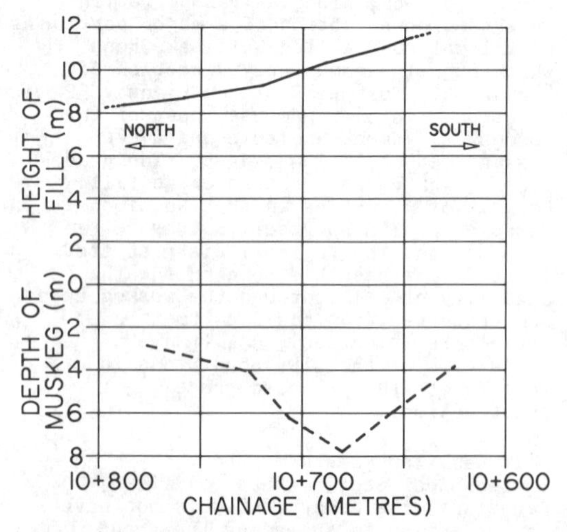

Figure 3. Grade line height and muskeg depth information.

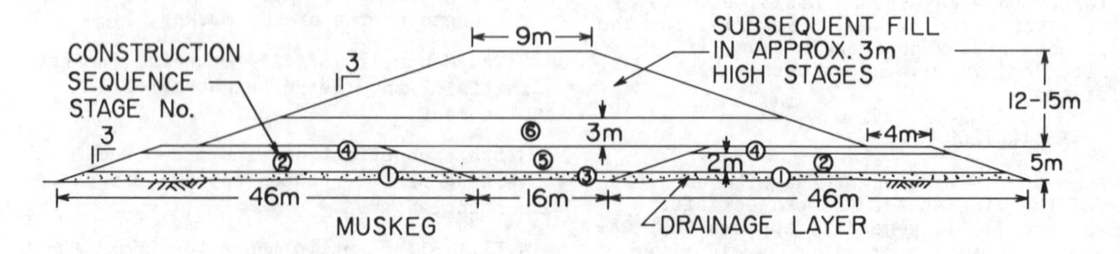

Figure 2. Suggested embankment construction procedure.

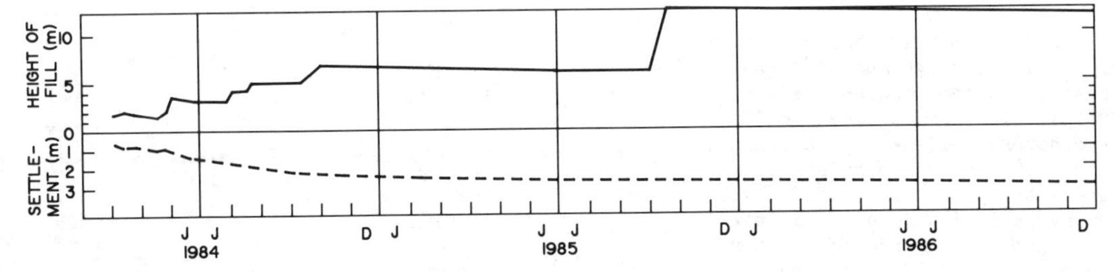

Figure 4. Fill placement and settlement monitoring at Station 10+645.

DISCUSSION

Although there was about a 2-3 week waiting period between any two successive lifts of fill placement, the settlement monitoring clearly showed that settlements had not levelled off close to the end of each waiting period. (Figure 4). No doubt, the settlements levelled off with time after the construction was suspended for winter of 1984 and after the final construction stage fill material placement was completed in October 1985.

The information of settlement observed at each settlement plate location has been compiled in Table 1, which also shows settlement values estimated from information based on field data observations elsewhere (Millions, 1977). In general, the actual settlements observed in this project are approximately 78% of the estimated values (Figure 5).

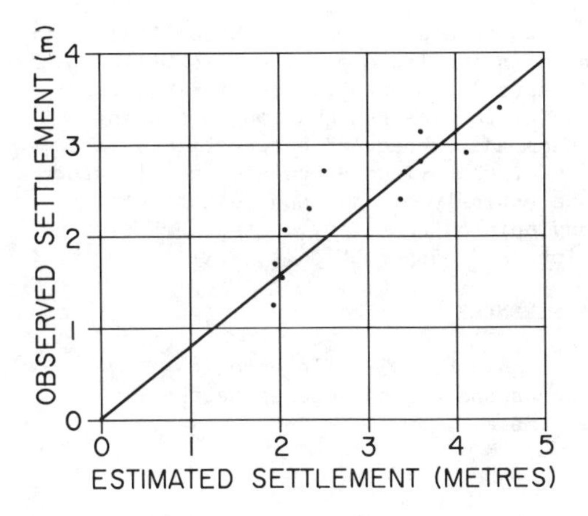

Figure 5. Settlement comparison chart.

Table 1 Muskeg settlement observations

Settlement Plate	Chainage and Offset (m)	Approx. Depth of Muskeg (m)	Height of Fill (m)	Actual Settlement (m)	Estimated Settlement (m)
1	10+767, 7m Left	3.5	5.60 8.30	1.70 2.07	1.96 2.06
2	10+757, 8m Right	3.5	5.30 8.30	1.25 1.55	1.92 2.04
3	10+719, 7.7m Left	4.25	5.20 8.67	2.30 2.72	2.34 2.51
4	10+688, 7.5m Right	7.5	5.40 9.90	2.90 3.41	4.12 4.50
5	10+659, 7.5m Left	6.0	6.20 11.00	2.40 2.83	3.36 3.60
6	10+645, 7m Right	6.0	7.00 11.85	2.70 3.14	3.42 3.60

CONCLUSION

This project proved that construction of high fills through muskeg areas is feasible in the most cost effective manner by following simple construction procedures as listed in the paper. However, in order to help dissipate excess pore water pressures, the 2-3 week time intervals followed in the initial construction sequence of this project appears to be inadequate and longer time intervals may be more helpful.

ACKNOWLEDGEMENTS

The author wishes to express his sincere thanks to his senior and junior Colleagues both in the field and the Head Office of Alberta Transportation for their support during the instrumentation monitoring phase and subsequent preparation of this paper. The views expressed in this paper are entirely his own and do not reflect any opinion or policy on the part of Alberta Transportation.

REFERENCES

Millions, K. 1977. Internal correspondence on "muskeg settlement estimates".

Prediction and Performance in Geotechnical Engineering / Calgary / 17-19 June 1987

Compressibility and cementation characteristics of some calcareous sea-floor oozes

Jack Kao
Global Engineering & Testing Ltd., Calgary, Alberta, Canada
J.R.Bell
Oregon State University, Corvallis, USA

ABSTRACT: The effects of cementation and grain crushing on the compressibility of the sea-floor calcareous oozes are studied. The extreme nature of these soils emphasizes these effects and the investigation's results can be used in foundation design and to improve our abilities to predict the performance of terrestrial soils with cemented structures and/or fragile grains. Classification tests, consolidation tests, and microscopic examinations are performed on the calcareous oozes with high percentages of grains consisting of hollow Foraminifera susceptible to mechanical crushing. Three-phase compression theory developed for the calcareous oozes reveals that the consolidation is primarily due to relaxation of cementitious bonds followed by continuous crushing of the calcareous shells and fragments. All the time rates of consolidation curves of the calcareous sediment show that secondary compression governs most of the consolidation and the void ratio-log effective stress curves are concave downward throughout the stress range tested.

1 INTRODUCTION

The calcareous sediments cover 48% of the area of the sea-floors of the world. These minute organisms are composed mainly of nonplastic sand-to-silt size particles which are hollow and crushable. Influences of terrigenous clay content in the calcareous sediments make the soils exhibit different degrees of plasticity. Because of the tendency of the shells to crush, the calcareous sediments exhibit special engineering and compressibility properties.

The purpose of this paper is to investigate the compressibility properties of calcareous oozes containing large percentages of hollow skeletons of Foraminifera. Special attention is given to the effects of the cementitious bond and the crushing of the calcareous shells on the apparent overconsolidation stress, the shape of the consolidation curves, and secondary consolidation.

A series of Atterberg limit tests, consolidation tests, sieve analyses, and microscopic examinations were performed on soil samples obtained from the Panama Basin. Microphotographs show the conditions of individual calcareous shells before and after consolidation. A linear relationship

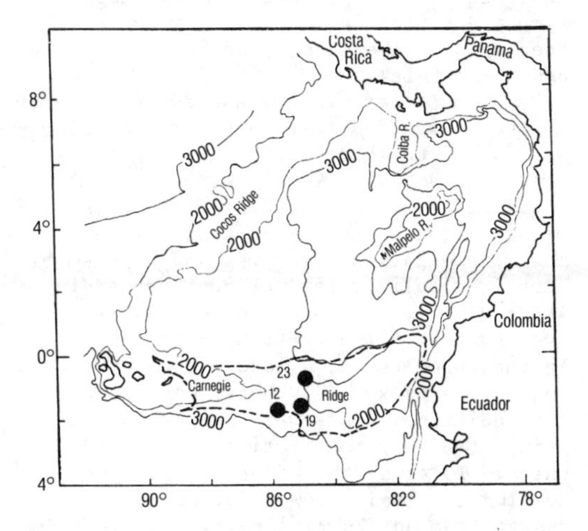

Figure 1. Location map and sampling sites

and three-phase compression theory are developed to relate the compression index to the void ratio and shell crushing. The results indicate that for higher stresses, the crushing of the calcareous shell fragments control the settlement behavior. In addition, a method for estimating the apparent overconsolidation stress by plotting the increment of the rate of secondary compression versus consolidation stress is demonstrated.

2 SAMPLING AND LABORATORY STUDIES

Five samples were selected from calcareous sediment cores collected from three locations at a water depth of approximately 2400 m on the Carnegie Ridge of the Panama Basin. Refer to the site location map in Figure 1. Samples were taken from Core 12 at depths of 1.35, 4.46, and 9.12 m below the sea-floor. One sample each from Core 19 at 8.77 m and Core 23 at 4.05 m were also tested. The color of the samples were tan to light green. The cores were taken with a piston sampler containing a plastic liner of 57 mm inside diameter. After the cores were recovered, the liners were removed, sealed with wax, and refrigerated until time for testing.

Laboratory tests of water content, specific gravity, plastic limit, liquid limit, grain size distribution, and consolidation were performed in accordance with the American Society for Testing and Materials (1964). Due to the high sensitivity and extremely soft nature of the soil, a special trimming device was designed for preparing the samples and lubricated consolidation rings were used to limit sample disturbance. Consolidation specimens were nominally of 50 mm diameter and 25 mm thick.

Because of very low pressures involved in the consolidation test, a special loading system capable of transmitting pressure as low as 0.0125 kg/cm^2 was utilized. A number of thin lead plates of different weights were used for very low consolidation pressures and a pneumatic Conbel consolidometer was used for adding the higher loads, as shown in Figure 2. Four consolidation tests were conducted. In the three tests, a load increment ratio of approximately unity was used. Two additional tests were performed using load increment ratios of two and three. To study the effect of time, the loading interval allowed for compression ranged from one to five days. Initial time readings were taken as soon as possible within the first few seconds of loading. Seawater was used in the test to maintain the same soil salinity as in situ. Consolidation loading began with a low pressure of 0.0125 kg/cm^2 and completed at different pressures for different samples. The lead plates for each sample were selected according to the desired load increment ratio until a consolidation stress 0.8 kg/cm^2 was reached. Then the Conbel

Figure 2. Consolidation equipment

consolidometer was used to add the load with the deformation dial readings to 0.00025 cm. After loading to the maximum desired pressure, the consolidometer was unloaded to 0.8 kg/cm^2, 0.1 kg/cm^2, and 0.0123 kg/cm^2. At the completion of each test, the sample was removed, weighed, and sieved.

Microphotographs were taken of intact specimens and of grains separated by sieving before and after consolidation testing.

3 TEST RESULTS

The results of the index test are summarized in Table 1. Data include unit weight, void ratio, specific gravity, liquid limit, plastic limit, and natural water content.

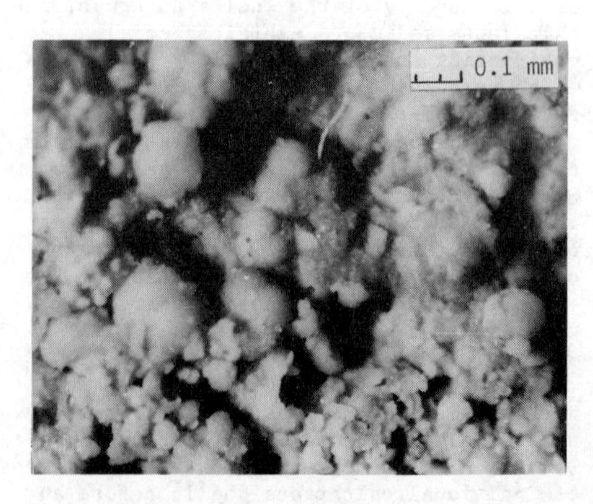

Figure 3. Natural calcareous sediment

Table 1. Index Properties of Calcareous Samples from the Panama Basin

Core Number	Sample Depth (m)	Unit Weight (kg/m)	Initial Void Ratio (e)	Grain Specific Gravity	Liquid Limit (%)	Plastic Limit (%)	Natural Water Content (%)
12	1.35	1400	3.22	2.60	--	--	119
12	9.12	1500	2.38	2.58	70	54.5	91
12	4.46	1400	3.16	2.58	--	--	125
23	4.05	1500	2.82	2.71	62	NP	104
19	8.77	1500	2.71	2.60	62	60	107

Table 2. Sieve Analysis before and after consolidation (in percent)

	Sample 23			Sample 19		
	Retained on #100 Sieve	Retained on #200 Sieve	Passing #200 Sieve	Retained on #100 Sieve	Retained on #200 Sieve	Passing #200 Sieve
Before consolidation	33	38	29	10	56	34
After consolidation	9	10	82	8	10	82
Changes -	-24	-28	+53	-2	-46	+48

(a) (b)
Figure 4. Foram particles retained on #100 sieve (a) before and (b) after consolidation

(a) (b)
Figure 5. Foram particles retained on the #200 sieve (a) before and (b) after consolidation

Table 2 shows the results of the sieve analyses before and after consolidation testings. Representative microphotographs are shown as Figures 3 to 6. These pictures clearly indicate the crushing of individual calcareous shells and the natural calcareous sediments before and after the consolidation tests.

Void ratio-log effective stress curves from four consolidation tests are presented in Figures 7 and 8. Apparent overconsolidation and in situ stresses are indicated on these curves. All the consolidation curves are concave downward without a straight line portion.

To show secondary consolidation characteristics, time rate of consolidation curves are presented in Figures 9 and 10. The results indicate a linear relationship between compression and time (log scale), even though the elapsed times and load increment ratios are different.

Figure 6. Foram particles passing #200 sieve before consolidation

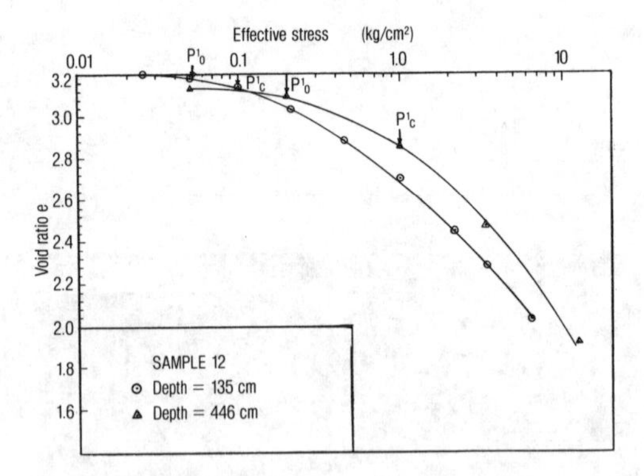

Figure 7. The void ratio vs log effective stress curves of Sample 12 specimens

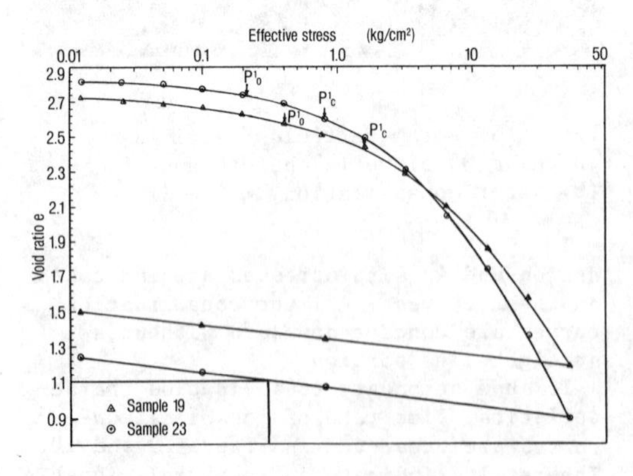

Figure 8. The void ration vs log effective stress curves of Samples 19 and 23

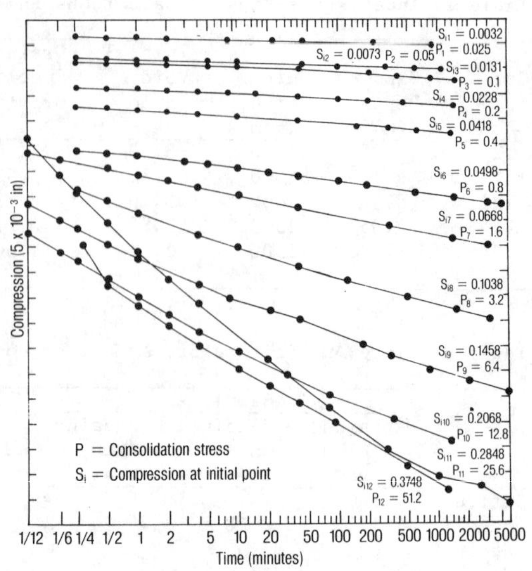

Figure 9. Time rate consolidation curves of Sample 19

4 DISCUSSION OF RESULTS

4.1 General physical properties

Generally the calcareous oozes have low unit weights and specific gravities. The high natural water content, natural void ratio, liquid limit, and low plastic index reflect the sensitive nature of these soils (Miller and Richards, 1969). The low values of both initial void ratio and natural water content of Sample 12 at depth 9.12 m are due to disturbance before testing. Microscopic examinations reveal that the ooze samples consist predominantly of microskeletal shells and fragments of the Foraminiferas with minor clay content (see Figure 3). These minute organisms are composed of nonplastic sand-to-silt size shells which are hollow and crushable. Sample 12 possesses a significant plasticity while Samples 19 and 23 are essentially nonplastic. Because the interest in this study is on calcareous cementitious and crushable soils and because these test results show that Sample 12 contains some amount of clay, Samples 19 and 23 have been chosen for investigating the compressibility properties of these sediments.

4.2 Primary and secondary consolidation

The consolidation test results, sieve analyses, and microscopic examinations all reveal that the continuous crushing of the

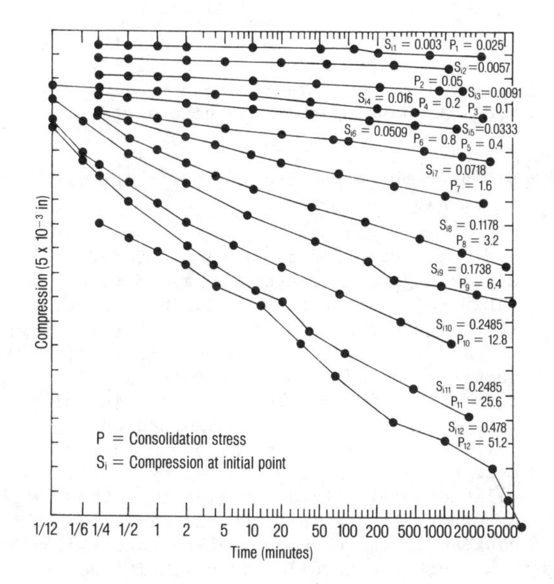

Figure 10. Time rate of consolidation curves of Sample 23

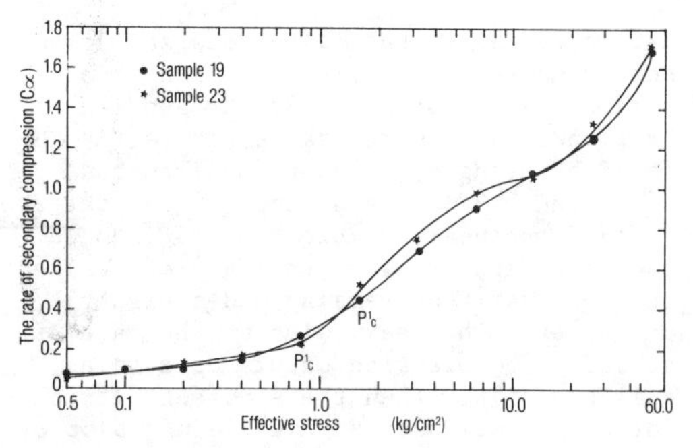

Figure 11. Effective stress vs the rate of secondary compression curves

calcareous particles is important to the consolidation process. These same general processes also take place during the compression of the granular soils, in which crushing and fracturing of granular particles begin when the compressive stress is increased to a high level (Forouki and Winterkorn, 1964, Roberts and deSouza, 1958).

The void ratio (e)-log effective stress (P) curves in Figures 7 and 8 show a common shape which is concave downward throughout the test. No straight line portion occurs, even at the highest consolidation stress of 52 kg/cm^2. This concave curve is typical of compression associated with grain crushing (Valent, Altshaeff and Lee, 1982, Demars and Chaney, 1982). In these tests the particles larger than 0.074 mm in diameter is decreased about 50 percent in both Samples 19 and 23 during the consolidation. The e-log P curves are plotted using the 60 minute compression values. Two series of time-rate of consolidation curves are represented in Figures 9 and 10. Both show a linear relationship between the compression and the elapsed time (log scale). They tend to become steeper with increasing consolidation stress. These time-rates of consolidation curves indicate that secondary compression governs most of the consolidation of these calcareous soils. The linear relationship started at the fifth second and last up to 5000 minutes. There is no evidence that the rate of secondary compression will decrease with longer time.

These trends have been observed by others (Noorany and Gizienski, 1970, Singh and Yang, 1971) for calcareous oozes and have also been reported by Roberts and deSouza (1958) for some sands. Apparently the creep deformation and grain crushing after the first few seconds occur continuously and slow enough that pore water dissipation does not affect the rate of consolidation.

4.3 Apparent overconsolidation stress and cementitious bond

The calcareous oozes tested all show an apparent overconsolidation stress-even Sample 12, taken at the shallow depth of 1.35 m. There is no evidence that this sea-floor sediment has had any overconsolidation history. The sediments at the site should still be in a normally consolidated state due to the slow and gradual sediment deposition. It appears that the apparent overconsolidation stresses exist mainly due to the formation of cementitious bonds between the calcareous particles. This geochemical cementing and bonding of particles has been found and supported by many geologists and geotechnical engineers (Richards, Hamilton, and Edwin, 1967, and Robertson, 1967).

The calcareous sediments have been observed by putting pieces of them into water. Even after a period of two days they still were not disaggregated.

A possible interpretation of the production of cementitious bonds is as follows: Sea-floor dwelling organisms produce CO_2, which then dissolves to form carbonic acid. This carbonic acid reacts with the Ca++ions contained in the calcareous shells. Then the calcium carbonate forms into the calcareous cement and results in a deposit with cementitious bonds among calcareous particles.

As mentioned before, the e-log P consolidation curves of the calcareous samples show concave downward curves without a straight line portion. So the apparent overconsolidation stress is not easily found by using the conventional method. It was noted by Singh and Yang (1971) that the rate of secondary compression (C_α) would increase with pressure to a maximum value and then start decreasing under higher pressures. The peak value of the rate of secondary compression occurs at a value slightly higher than the apparent overconsolidation stress. Figure 11 is a plot of the rate of secondary compression against consolidation stress (log scale) for Samples 19 and 23. The apparent overconsolidation stress is still not distinct enough to be located easily. The increments of the rate of secondary compression (ΔC_α) is plotted against the effective stress in Figure 12. Samples 19 and 23 both have a first peak (lowest point shown on the graphs). The apparent overconsolidation stress is located at the first stress lower than the one of the maximum ΔC_α. These apparent overconsolidation stresses found in Samples 19 and 23 both also correspond to a value slightly lower than the stress of the first inflection point in the curves of the rate of secondary compression vs consolidation stress (Fig. 11).

It is postulated that well below the apparent overconsolidation stress the consolidation is controlled by the gradual relaxation of cementitious bonds accompanied by only minor grain crushing. As the apparent overconsolidation stress is approached, consolidation accelerates dramatically due to complete collapse of the cementation bonds and grain reorientation.

Beyond the apparent overconsolidation stress level, grain crushing dominates

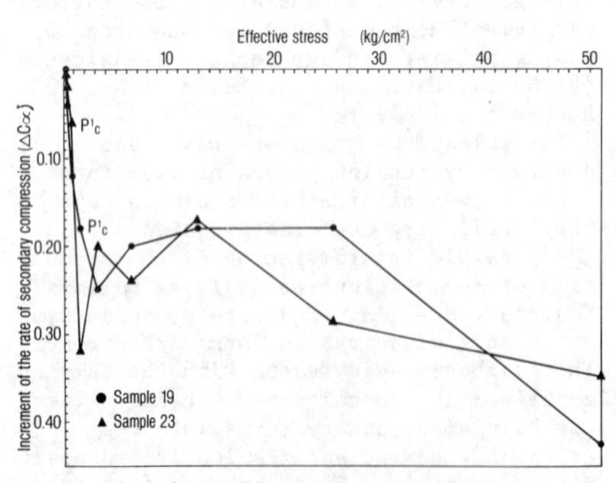

Figure 12. Effective stress - increment of the rate of secondary compression

consolidation and consolidation rates also increase accordingly. A quantitative theory for this three-phase consolidation model is developed in the next section.

5 THREE-PHASE COMPRESSION THEORY OF THE CALCAREOUS OOZES

Since the compression index (Cc), defined as the negative rate of change of void ratio (e) with respect to log of effective stress (-de/dlog p'), is constantly varying; it is desirable to have a theory relating compression index to other factors. The compression index-void ratio relationships for Samples 19 and 23 are plotted on Figure 13. The two sets of data show good agreement and are well represented by a bilinear relationship. It is recalled that the Atterberg limits indicated little clay content in these samples. The compression behavior is controlled by cementitious bonds and the character of the Foram shells.

If a dense packing of mixed sizes of spheres is considered, the void ratio is about 0.28 (Forouki, 1964 and deSouza, 1958). The calcareous oozes consist of approximately spherical hollow shells. The microscopic examination shows that the ratio of outside to inside diameter of the shells averages approximately 1.125. For this geometry the void ratio due to external voids becomes 0.91 and 2.36 due to internal voids. Thus, the total void ratio of a dense packing of unbroken hollow Foram shells is calculated to be approximately 3.27 (Kao, 1973).

The compression of a dense packing of spheres is very low (i.e. Cc = 0.0) unless the spheres are crushed. When the section of the plot on Figure 13 (based upon the

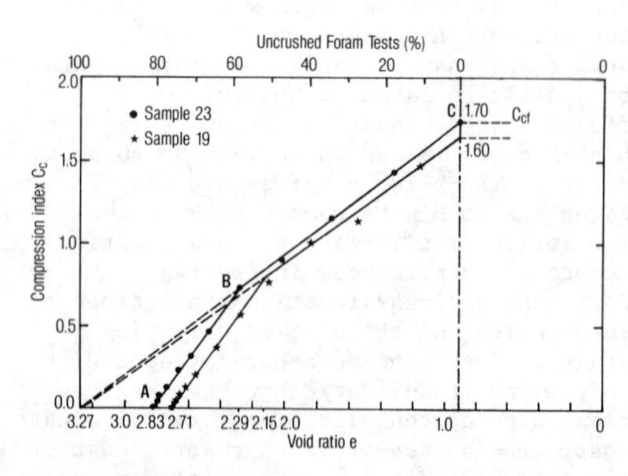

Figure 13. The relationship between the compression index, void ratio (from consolidation results) and percentage of uncrushed Foram shells

consolidation tests) from C to B is projected back to a compression index of zero, the corresponding void ratio is remarkably close to the predicted value of 3.27.

The sieve data and microphotographs show that after consolidation to a void ratio of 0.9 (Point C on Figure 13), Sample 23 has only about 2% unbroken hollow shells. Beyond point C the soil is essentially all shell fragments and compression mainly results from continued crushing of fragments. Without hollow shells, therefore, it is expected that the compression index is nearly constant (Ccf) beyond the point C. The e-log effective stress curves show tendencies to become straight lines at very high stresses.

Equation 1 describes the

$$C_c = \frac{100-F\%}{100} (Ccf) \; .... \; [I]$$

relationship of Figure 13 for the line from the origin to Point C. Ccf is the compression index of the fragments and F% is the percentage of uncrushed Foram shells. This suggests the actual C_c is a function of the compressibility and amount of fragments. Whole shells are essentially floating in a matrix of fragments. Whole shell crushing acts to increase the percentage of fragments.

The actual data of the Sample 19 slightly deviates from the relationship of Equation 1 at point B. This point should correspond to about 60% whole shells. Below this percentage it is probable that there is considerable direct contact between whole Foram shells and the compressibility of the soil structure is primarily controlled by the crushing strength of the whole shells and their cementation bonds, not the fragments.

In summary, there are three ranges of compression which have different factors controlling the compression index. In the first phase compression is dominated by the crushing of whole hollow Foram shells and their cementation bonds. In the second range both the compression and crushing of shell fragments and the continued crushing of the remaining whole shells act together to determine the value of compression index. Finally, at low void ratios, there are only fragments remaining and their continued compression and fracturing gives a nearly constant compression index. Each of these ranges can be represented by simple linear expressions.

6 ENGINEERING APPLICATION

In the design of foundations to be placed on cementitious calcareous soils, the strength of the cementitious bond should be determined. In order to minimize the consolidation settlement, the structure loading should be distributed in such a way that the loading pressure is less than the strength of the cementitious bond of the calcareous soil. Alternatively, preloading may be used to crush the calcareous shells. The actual loading pressure, which is smaller than the preloading pressure, can then be applied. It is suggested that a minimum factor of safety of 2.5 be applied in the preloading design. The preloading method can be applied to the foundation design in the calcareous sediment due to the nature of very low rebound during unloading of the consolidation tests.

7 CONCLUSIONS

1. There are four important factors which significantly influence engineering and compressibility properties of the calcareous-sediments: cementitious bonds, calcareous shells and fragments, clay content, and initial void ratio.
2. The consolidation characteristics of the calcareous oozes tested are independent of the load increment ratio.
3. Secondary consolidation dominates the total compressibility of the calcareous soil.
4. The void ratio -log effective stress consolidation curves maintain a concave downward shape throughout the stress range tested without a straight-line portion.
5. The calcareous sediments tested have an apparent overconsolidation stress higher than the in situ stress. The apparent overconsolidation stress is mainly due to the formation of the cementitious bond. The apparent overconsolidation stress is not easily found by using the conventional method. This apparent overconsolidation stress may, however, be found from the consolidation stress vs the increment of the rate of secondary compression.
6. For sediments which initially consist solely of hollow calcareous shells, the relationship between the compression index and the void ratio has three distinct ranges. The first range is dominated by crushing of the hollow shells and the cementitious bonds. The second range is a transition range controlled both by crushing of whole shells and shell fragments. The third range has relatively constant compression index and is the crushing of fragments after all hollow shells have been broken.
7. The compression index-void ratio relationship in the calcareous oozes can be expressed in simple mathematical

expressions.
8. Studying sea-floor oozes can give insights into the behavior of other cemented and/or crushable soils that will be useful in the foundation design and settlement prediction.

8 ACKNOWLEDGEMENTS

The research reported in this paper was conducted in the Oregon State University, Civil Engineering Laboraties. The project funding was provided by the Office of Naval Research of the United States of America. The research was part of a larger project with the College of Oceanography of the Oregon State University, under the direction of Dr. Tjeerd H. van Andel.

REFERENCES

American Society for Testing and Materials, 1964. Procedures for testing soil. ASTM Committee D-18, Philadelphia.
Demars, K.R. and Chaney, R.C. 1982. Unique Engineering properties and compression behavior of deep-sea calcareous sediments. ASTM STP 777, American Society for Testing and Materials: 97-112.
Forouki, Omar T. and Winterkorn, Hans F. 1964. Mechanical properties of granular systems. Highway Research Record, 52: 10-42.
Kao, Y.T. 1973. Consolidation characteristics of calcareous deep sea-floor sediments from the Panama Basin. M.Sc. Thesis. Oregon State University.
Miller jr., Donald G. and Richards, Adrian F. 1969. Consolidation and sedimentation-compression studies of calcareous core, Exuma Sound, Bahamas. Sedimentology, 12: 301-316.
Noorany, I. and Gizienski, J.F. Sept. 1970. Engineering properties of submarine soils: State-of-Art Review. Proceedings, journal of soil mechanics and foundations division, ASCE: 1735-1762
Richards, Andrian F. and Hamilton, Edwin L. 1967. Investigation of deep-sea sediment coves, Ill. Consolidation Masure Geotechnique.
Roberts, J.E. and deSouza, J.M. 1958. The compressibility of sand. Proceedings of the American Society for Testing Materials, Vol. 58: 1269-1277.
Robertson, Eugene C. 1967. Laboratory consolidation of carbonate sediment, Marine Geotechnique: 118-123.
Singh, A. and Yang, Zan. 1971. Secondary compression characteristics of a deep ocean sediment. Proceedings of the international symposium on the engineering properties of sea-floor soils and their geophysical identification. Sponsored by UNESCO, National Science Foundation, and University of Washington: 121-129.
Valent, P.J. Altschaeff, A.G. and Lee, H.J. 1982. Geotechnical properties of two calcareous oozes. ASTM STP 777, American Society for Testing and Materials: 79-95

Prediction of swell of black cotton soils in Nairobi

F.J.Gichaga, B.K.Sahu & T.G.Visweswaraiya
University of Nairobi, Kenya

SYNOPSIS: The black cotton soils of Kenya are of volcanic origin and have undergone tropical weathering. These soils exhibit swelling pressures of the order of 740 to 1100 kN/m² under optimum conditions of density and water content. The depth of formation of black cotton soils varies from 4m to 20m. Based on the clay fraction of the soils and their plasticity indices black cotton soils are classified as soils of high swellability. In the present investigation, soil samples were collected from four different locations. Swelling pressures and swelling potentials were determined on samples remoulded at various densities and water contents. Emperical correlations were developed for swelling pressure and swelling potential as functions of dry density, liquid limit and initial water content. Predicted values of swelling pressure and swelling potential compare fairly well with the experimentally observed values.

1. INTRODUCTION

The soils of Nairobi are of volcanic origin transformed into black cotton soils and red coffee soils due to tropical weathering. The black cotton soils exhibit swelling because of the presence of mineral montmorillonite, where as the red coffee soils exhibit collapse characteristics because of the mineral halloysite. There are some soils in the intermediary stages of formation which exhibit both swelling and collapse. These are slightly brownish in colour.

Since the experiments to determine the swelling of soils are very time consuming it is desirable to have empirical relationship to establish swelling pressure and swell potential relating density of the soil and the water available to participate in the process of swelling. This paper presents results from experimentation carried out on Nairobi soils. The results were used to derive relationships which can be used to compute swell potential on the basis of dry density and water content of the soil. These relationships have also been used in the field prediction of swell considering the depth of desiccation.

Topography and climate of Nairobi: latitude and longitude

The Nairobi area is included by 36° 24' and 37°4' east longitude: 1°12' and 1°28' south latitude. The city centre of Nairobi is defined by 36° 50' east longitude and 1°17' south latitude.

Topography

The country surrounding Nairobi is a small segment of eastern ramp of Kenya's Rift Valley comprising of plateau sloping upto its edge and its sheetfaulted flank. In general, the inclined plateau of the ramp rises in elevation at an average gradient ranging from 1 percent to 3 percent in lifts varying from 6 to 12m from Athi River.

Climate

Nairobi experiences no real summer or winter due to the fact that it is so near the Equator. For greater part of the year the days are sunny and the nights are cool. The hottest periods are January to March and September/October. The long rains occur from mid-March to June and the short rains from mid-October to mid-December. The average rainfall is about 875mm of which about 450mm falls during the long rains, 125 to 250mm during short rains and the balance in isolated showers at different times of the year. The sky is usually some what overcast during the months of July and August. The temperature rarely rises above 27^{0}C in the middle of the day and is usually about 10^{0}C during the night.

Geology, soil formation and geotechnical problems:

Geology

The city of Nairobi is in an area of successive volcanic activity. The type of geologic formation in this area is called the Basement System because of the volcanic rock formations on the sedimentary rock which form the base. These sedimentary rocks are hard, fine grained and black in colour. They are classified as black trap.

The volcanic rocks formed above are the sedimentary rock base formed in the following order: 1. phonolite 2. trachyte 3. tuff 4. clay stone. The phonolites at the bottom are the hardest, the clay stones at the top are the softest. This sequence of formation stands to reason because the dense lava must have been deposited at the bottom and the lightest at the top. It is reported that following another volcanic activity in Nairobi area some formations of phonolite and trachyte have taken place on the originally formed tuff. This is termed Lake Bed.

Soil formation

Since the geotechnical engineer is concerned with weathered soil and tuff which form the foundation material their genesis and morphology are briefly discussed here. They are mostly based on parent material, topography and the type of weathering. The topmost clay stone layer, due to physical and chemical weathering has given rise to black cotton soils, red coffee soils and secondary laterites depending upon temperature and drainage conditions.

According to Gonzalez de Vallego (1), the parent material has transformed into different minerals in the following sequence: Volcanic Ashes; Montomorillonites (Black Cotton Soils); Halloysites (Red Coffee Soils)and Kaolinites.

During the investigations it was found that the weathering is upto halloysites. In some places halloysites are partially weathered and in the others they are not. This is the reason why some of the red soils exhibit slight swelling reflecting the characteristics of black cotton soils.

2. GEOTECHNICAL PROPERTIES OF BLACK COTTON SOILS

The mecahnism of swelling is attributed to the total osmotic and matrix suction of the soil.

The black cotton soils of the Nairobi area exhibit swelling characteristics because of the constituent clay mineral montmorillonite ranging from 60 to 70%. Further the pH value lies between 5 and 6 as against 6.5 to 8.5 in other countries indicating a different weathering process. These are reflected by the high values of liquid limit ranging from about 70% to 85%. The past and the present stress history is depicted by the density of the soil. Thus the density of the soil and the available water are taken as the necessary variables in the process of swelling.

Sampling and testing of the soil:

Sampling

Disturbed soil samples were collected from four different locations in Nairobi area: 1) Main Campus of the University of Nairobi (in the centre of Nairobi), 2) Langata (Southwest of Nairobi), 3) Buru Buru (East of Nairobi)and 4) Dagoretti (West of Nairobi). These sites were selected on the basis of location of different terrain patterns and drainage conditions.

Testing

Identification tests, like specific gravity, granulometry and Atterberg limits were performed according to B.S.S. 1377 (4). The results of these tests are shown in table 1.

TABLE 1

No.	LOCATION	Sp. Gr	Grain Size [%] Clay	Silt	Sand	Aterberg Limits LL	P.L	P.1	ACTIVITY
1.	MAIN CAMPUS	2.65	62	21	17	86	40	46	0.74
2.	LANGATA	2.48	68	21	11	78	38	40	0.59
3.	BURU BURU	2.64	58	31	17	88	37	51	0.88
4.	DAGORETTI	2.65	68	26	6	80	35	45	0.66

TABLE 2

TESTS RESULTS

No.	Dry Sensity γ_d [kN/m³]	Water Content ω %	Liquid Limit W_{LL} [%]	$(W_{LL} - \omega)$ %	Swell Pressure σ_s [kN/m²]		Deviation %	Swell Potential $h:\frac{h}{ho}$ (%)		Deviation
					Measured	calculated		Measured	Calculated	
	1.MAIN CAMPUS									
1.	-13	2.5	86	83.5	1043	1057	+0.38	26.0	26.42	+1.6
2.	-14	7.0		79.0	1023	1001	-2.15	25.0	25.00	0
3.	-15	12.5		73.5	955	935	-2.09	23.5	23.37	-1.8
4.	16.1	18.0		68.0	912	900	-1.32	22.8	22.50	-1.32
5.	+15	25.0		61.0	631	520	-17.43	15.7	13.00	-17.20
6.	+14	29.0		57.0	331	344	+3.92	8.3	8.60	+3.6
7.	+13	32.5		53.5	224	228	+1.78	5.6	5.70	+1.79
	2.LANGATA									
1.	-13	2.0	78	76.0	697	719	+3.16	17.4	17.98	+3.3
2.	-14	5.0		73.0	758	773	+1.98	19.0	19.30	+1.58
3.	-15	7.5		70.5	813	834	+2.58	20.3	20.85	+2.70
4.	16.5	16.0		62.0	741	712	-3.91	18.5	17.80	-3.78
5.	+15	25.0		53.5	331	338	+2.11	8.27	8.45	+2.18
6.	+14	29.0		49.0	219	210	-2.72	5.48	5.25	-4.20
7.	+13	32.5		45.5	145	134	-7.58	3.63	3.35	-7.71
	3.BURU BURU									
1.	-13	2.5	88	85.5	1150	1057	-8.0	28.75	26.43	-8.0
2.	-14	6.0		82.0	1175	1130	-3.8	29.40	28.25	-3.91
3.	-15	10.0		78.0	1202	1161	-3.4	30.00	30.00	0
4.	16.3	17.0		71.0	1095	1072	-2.1	27.00	26.80	-0.74
5.	+15	24.0		64.0	575	608	+5.74	14.40	17.00	+18.0
6.	+14	28.0		60.0	380	407	+7.10	9.50	10.18	+7.15
7.	+13	31.0		57.0	263	281	+6.80	6.60	7.03	+6.52
	4.DAGORETTI									
1.	-13	2.0	80	78.0	77.6	783	+0.9	19.40	19.50	+0.52
2.	-14	5.0		75.0	831	844	+1.56	20.70	21.00	+1.45
3.	-15	7.5		72.5	891	914	+2.77	22.30	22.85	+2.47
4.	16.1	15.0		65.0	758	776	+2.37	19.00	19.40	+2.10
5.	+15	25.0		55.0	363	370	+1.93	9.00	9.25	+2.78
6.	-14	29.0		51.0	240	239	-0.42	6.00	5.98	-0.3
7.	+13	34.0		46.0	151	139	-7.95	3.80	3.48	-8.42

Table 3.

No	Location	Liquid Limit W_{LL} %	Shrinkage Limit W_{SL} %	Shrinkage Index $I_s = W_{LL} - W_{SL}$ %
1.	Main Campus	86	16	70
2.	Langata	78	12	66
3.	BuruBuru	88	16	72
4.	Dagoretti	80	14	66

Table 4

Swellability	$I_s = W_{LL} - W_{SL}$ %
Small	0 - 20
Moderate	20 - 50
High	30 - 60
Very high	> 60

The swell tests were carried out according to the suggestion made by Horn (2). In this work the swell potential is defined as the percentage swell for 1 kN/m² of surcharge pressure (it is noted that Ranganathan and Satyanarayana defined [3] swell potential on the basis of surcharge pressure of 1 psi while Chen Fu Hua [5] defined it at surcharge pressure of 1000 psf), and the swell pressure as the pressure exerted on the sample when the swelling is prevented. The soils were compacted in a standard Proctor Mould. Samples were taken at optimum, dry and wet of optimum moisture content. The samples were consolidated at different surcharge loads and flooded with water. The maximum swelling due to the surcharge load was needed. These relationships are shown in Fig. (1).

Figure 1 illustrates that for a particular condition of the soil, having a combination of the values of density, γ and moisture content ω the swell potential is determined for different surcharge loads σ_{s_1} σ_{s_2}, σ_{s_3}. The line joining these points are extended to cut the swell potneital and the surcharge load axes. Points A, B, and C give the surcharge loads for zero swell potential and the corresponding values A', B', and C' the swell potential for 1 kN/m² surcharge pressure. The inclinations of these lines α_1, α_2, α_3 are designated as the swell moduli.

The results of these tests on samples taken from the various locations in Nairobi are shown in Table (2).

Results of liquid limit and shrinkage limit tests are shown in Table 3. The value of the shrinkage index (I_s) is defined by Ranganathan B.V. and Satyanarayana B [3] as the difference between Liquid Limit (W_{LL}) and the Shrinkage Limit (W_{SL}). The value of shrinkage index is used to predict the swell pressure.

The grain size curves of black cotton soils tested in the laboratory are demarcated in the area shown in figure (2). They contain 40% to 70% clay. The percentage of silt and sand are comparatively small. The high values of liquid limit ranging between 78% to 88% are a reflection of high percentage of clay.

The shrinkage indices I_s, are classified by Ranganathan and Satyanarayana (3) as shown in Table 4.

On the basis of the criteria in Table 4, it is evident that Nairobi's black cotton soils have very high swellability.

3. CRITERIA TO CLASSIFY THE SWELL_ ABILITY

Figures (3) to (5) show the relationship of the basic soil properties with their swelling behaviour from these relationships several criteria are drawn to define the classification of these soils based on their swellability in terms of (1) moderate (2) high and (3) very high swellability. It is suggested that the use of a single criterion may not be adequate in classifying a soil. The criteria drawn from the test results are shown in Table (5).

TABLE 5

Soil Parameter	Classification		
	Moderate Swellability	High Swellability	Very High Swellability
1. Dry Desnity γ_d [kN/m³]	<15	$15 \le \gamma_d \le 15.75$	>15.75
2. Clay Content <0.002mm [%]	<40	40≤0.002≤55	>55
3. Liquid Limit W_{LL} [%]	<48	48≤LL≤65	>65
4. Plasticity Index PI [%]	<30	30≤P.I≤40	>40
5. Shrinkage Index $I_s = W_{LL} - W_{SL}$	0 to 20 (small) 20 to 30 (moderate)	30 to 60	>60
6. Swell Pressure σ_s [kN/m²]	<120	$120 \le \sigma_s \le 600$	>600
7. Swell Potential $\frac{\Delta h}{h_o} = h'$ [%]	<4.5	h'>4.5 <13	h'>20% ≮13

RELATIONSHIP BETWEEN SURCHARGE LOAD AND SWELL POTENTIAL FOR A GIVEN DENSITY AND MOISTURE CONTENT

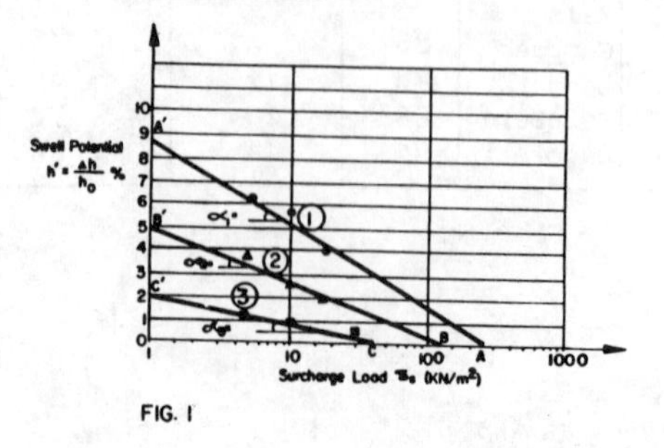

FIG. I

4. DEVELOPMENT OF AN EMPIRICAL RELATIONSHIP TO PREDICT SWELL PRESSURE AND SWELL POTENTIAL

While developing an expression to predict swelling pressure, the density of the soil and the available water to participate in the swelling are considered. The available water is taken as the difference between its initial moisture content and the liquid limit. Here the liquid limit is taken as the ultimate moisture content, because at that stage the moisture absorption is maximum and beyond this limit the soil behaves like a liquid.

The method of regression was adopted in developing this relationship. In figure (6) the relationship between the swell pressure σ_s [kN/m²] and ($W_{LL}-\omega_i$) are drawn for different values of dry density γ_d [kN/m³]. The curves are linear and parallel to one another for the four densities selected making an angle of 73° with the horizontal axis.

The value of tan 73° which is 3.27 is taken as the potence of ($W_{LL}-\omega_i$).

The intercepts 'K' made by the curves on the σ_s axis at ($W_{LL}-\omega_i$) = 10 for different values of γ_d are shown in figure (7). The relationship is linear making an angle of 70° with the γ_d axis. The value of tan 70° which is 2.75 forms the potence of the density γ_d. Thus the equation for swelling pressure is given by:

$$\sigma_s = C \times (\gamma_d)^{2.75} \times (W_{LL}-\omega_i)^{3.27} \text{ [kN/m}^2\text{]}$$

The value of constant 'C' is obtained by equating the above equation with the measured values of swell pressure [σ_s] and taking the average, the value of 'C' was found to be 4.4×10^{-7}. The final form of the equation for swell pressure is thus given by:

$$\sigma_s = 4.4 \times 10^{-7} \times (\gamma_d)^{2.75} \times (W_{LL}-\omega_i)^{3.27} \text{ kN/m}^2$$

As can be seen in figure (8) the relationship between swell pressure σ_s and swell potential is linear having an inclination of 65°. From this the following relationship is deduced:

$$h' = \frac{\sigma_s \times 2.5}{100} \text{ \%}$$

The values of the measured and the ones calculated using the above relationship are shown in Table 2, from where it is seen that the calculated values are reasonably close to the measured values.

Verification of the relationship with other measured values

Similar tests were performed on undisturbed soil samples taken from other locations. The results and their verification are shown in Table 6 from where it is once again evident that the calculated values using the expression derived for swell pressure are close to the measured values.

Table 6

LOCATION	γ_d [kN/m³]	W_{LL} %	ω_i &	Swell Pressure kN/m³		Deviation %
				Measured	Calculated	
Hostel block for diplomacy programme U.O.N. (Center of Nairobi)	13.5	84	30	250	261	-2.44
Pangani (E. Nairobi)	13.4	70	3C	98	96	+2.05

Verification of the results from more samples and through field tests are desirable. However, because of the limitations of laboratory experiments, many authors suggest field values could be one-third of the laboratory values.

5. FIELD PREDICTION OF SWELLING

Using empirical relationship derived earlier the prediction of swelling pressure and swelling potential in the field can be estimated.

Figure 10 shows the variation of the overburden pressure based on the calculated value of the field density (γ_d) and depth (d). Figure (9b) shows the variation of moisture content with depth measured during the dry and the wet seasons. The area between these two curves in Figure (9b) gives the water available to participate in swelling. Figure

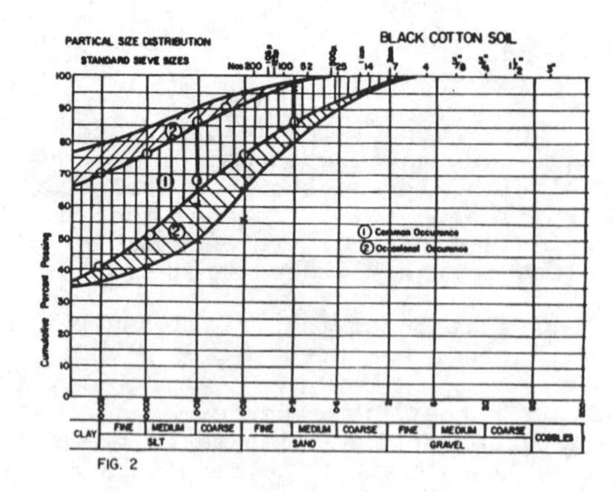

FIG. 2

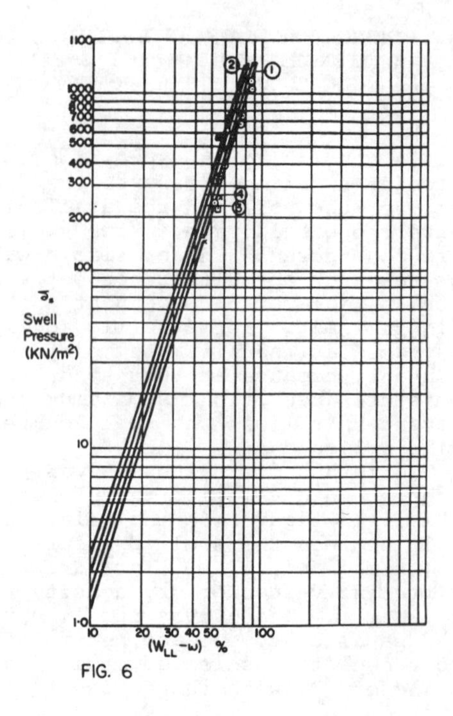

FIG. 6

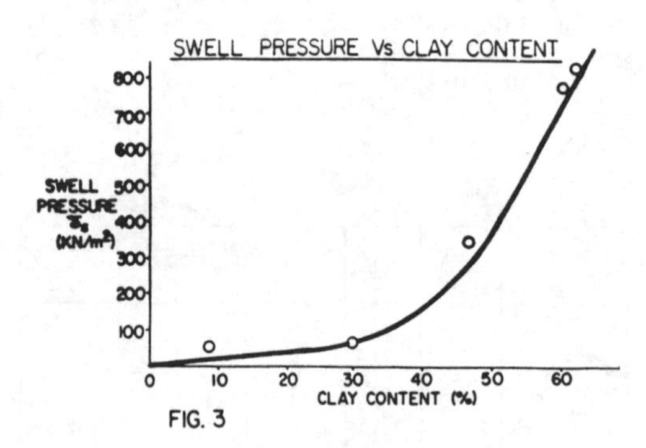

FIG. 3

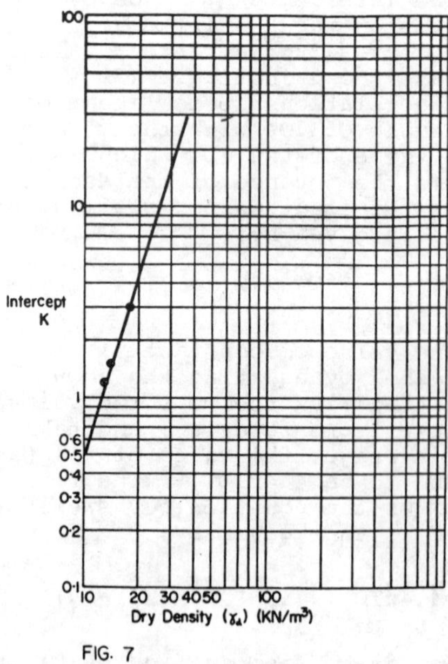

FIG. 7

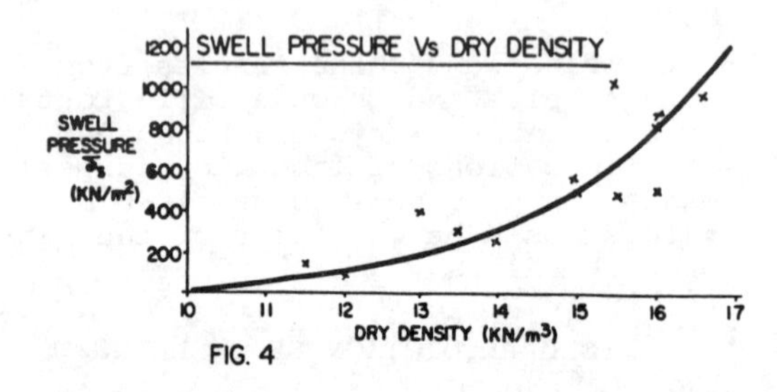

FIG. 4

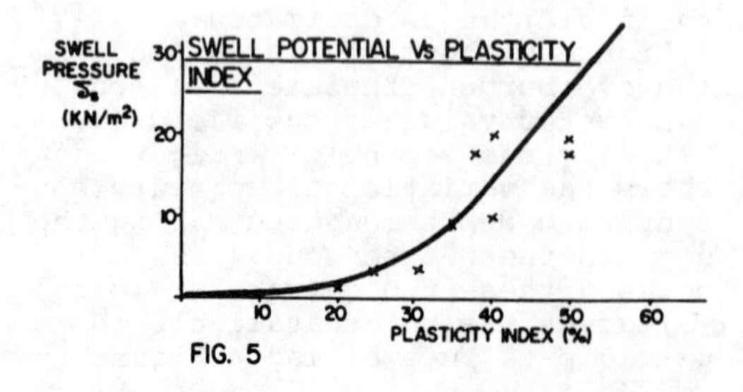

FIG. 5

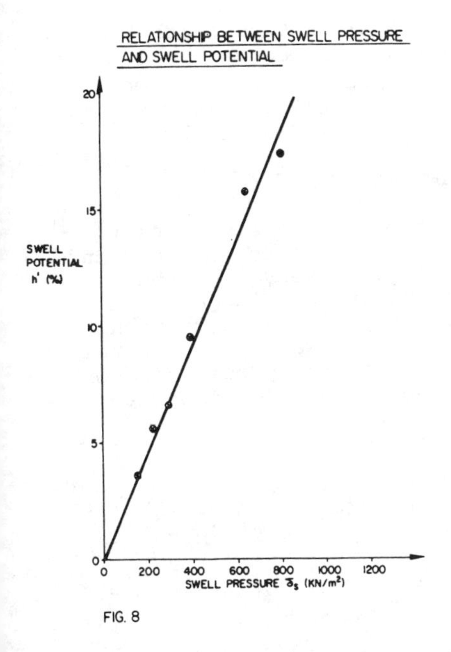

FIG. 8

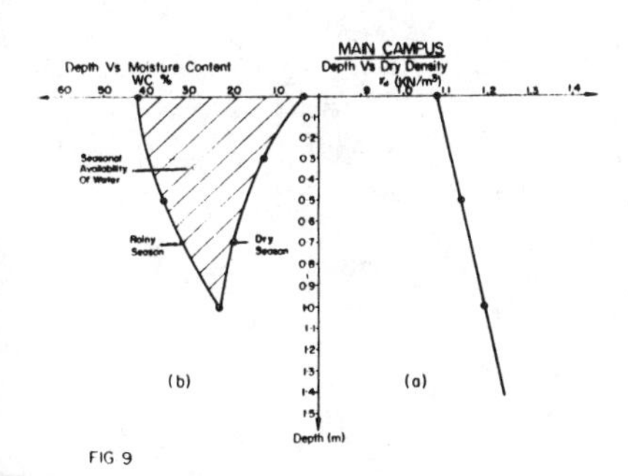

FIG 9

(10) shows the calculated swelling pressure using the empirical relationship derived earlier. The interposed over burden pressure curve gives rise to the following three regions:- (1) where swelling pressure is greater than the overburden pressure $\sigma_s > \sigma$. (2) where swelling pressure is equal to overburden pressure $\sigma_s = \sigma$. (3) where swelling pressure is less than overburden pressure $\sigma_s < \sigma$.

It has been observed from the investigations that the depth of desiccation in Nairobi extends upto 1 to 1.2m. It is reported that in other countries it could extend upto 2.5m or even more. Due to the high swelling pressures exhibited by the Nairobi black cotton soils, the overburden pressure has very little effect on swelling.

Figure (9) shows the variation of dry density and water content at the Nairobi University site. Using this data and the empirical relationship derived earlier swell pressures are calculated, upto the dessication depth giving results which are illustrated in Figure (10). Using the values of swell pressure the swell potential is calculated using the derived expression giving results as shown in Figure (11). It will be observed that there is a difference in swell potential between the calculated value using the total swell pressure in figure (10) and the integrated value from figure (11). This is attributed to assuming the curvilinear portions of the graphs as straight lines in the calculations.

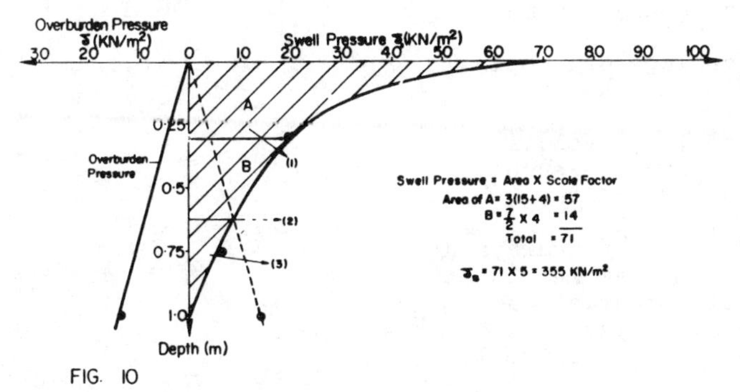

FIG. 10

PREDICTION OF SWELL POTENTIAL IN THE FIELD

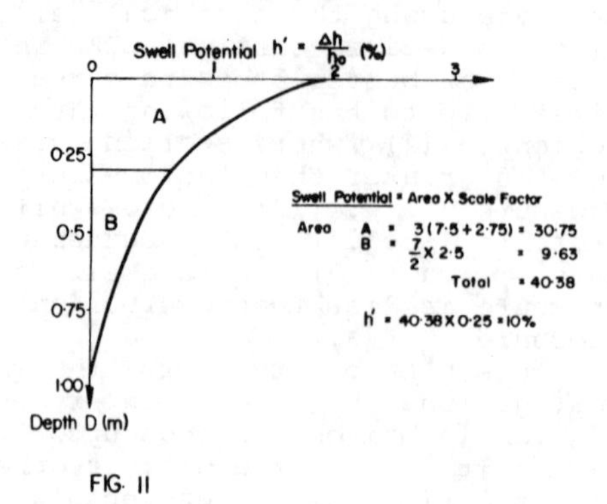

FIG. II

6. CONCLUSIONS

In this work attempts have been made to derive empirical equations which can be used to calculate the swell pressure and swell potential of black cotton soils in Nairobi, using their density and the water content. The latter is calculated as a difference between their liquid limit and the initial water content of the soil. The following expressions were developed for use to calculate the swell pressure and swell potential for Nairobi black cotton soils.

Swelling pressure, $\gamma_s = 4.4 \times 10^7 (\gamma_d)^{2.75} \times (W_{LL} - \omega_i)^{3.27}$ (kN/m²)

Swelling potential, $h' = \frac{\sigma_s \times 2.5}{100}$ (%)

7. REFERENCES

Gonzalez de Vallego L.I., Jiminez Sales J. A. and Leguey Jiminez. (1981).
- Engineering geology of tropical volcanic soils of La Laguna Tenarife.
- Engineering geology, Vol. 17.

Horn A (1976), - Schwellversuche an Expansive Boden. Heft 32 Degebo Berlin.

Ranganathan B. V. and Satyanarayana B (1965). - A rational method of predicting swell potential of compacted expansive clay.
- Proc 6th ICSM & F.E. Montreal, Vol. 1p92

B.S. 1377 (1975). British Standards Institute, London.

Chen, Fu Hua, Foundations on Expansive Soils. Elsevier Scientific Publishing Co. 1975.

8. ACKNOWLEDGEMENT

The authors sincerely thank the National Council for Science and Technology of Kenya for their financial support to carry out this work. They also wish to thank the University of Nairobi for permitting them to carry out this work in their Soil Mechanics Laboratories of the Department of Civil Engineering.

Prediction and Performance in Geotechnical Engineering / Calgary / 17-19 June 1987

An experimental investigation of transient pore pressure behaviour in soils due to gas exsolution

J.C.Sobkowicz
Syncrude Canada Ltd., Alberta, Canada

N.R.Morgenstern
University of Alberta, Edmonton, Canada

ABSTRACT: The transient behaviour of gassy sediments is observed in the laboratory during undrained loading. A macroscopic theory is developed to predict this behaviour, which gives excellent results.

1. INTRODUCTION

When a soil is loaded or unloaded, its volume change and pore pressure response may be affected by the presence of gas in the pore fluid, both in a free and in a dissolved state. If the gas exerts a dominating influence on such behaviour, the soil is classified as a gassy sediment. Sobkowicz and Morgenstern (1984) introduce the concept of gassy sediments, provide case histories, and report on a laboratory investigation which demonstrates typical equilibrium behaviour.

Gas exsolution in soils is a transient phenomenon, however, and it is the transient aspects of the pore pressure response in a gassy soil that are often of concern and of interest in a practical problem. The physical processes responsible for time-dependent behaviour are examined in Sobkowicz and Morgenstern (1987), wherein some theoretical predictions are made regarding bubble pressures, bubble sizes, and the affects of one of the transient processes, gas diffusion.

The objective of this paper is to provide a practical method of predicting transient gassy soil behaviour. Laboratory test techniques for measuring relevant soil parameters are described, a macroscopic theory of gas exsolution is proposed, laboratory results are reported and a comparison is made between measured and predicted results.

The present discussion is purposely limited to soils which are undrained, thereby separating out the effects of gas exsolution from those of consolidation. The reader will find a more complete discussion of a combined gas exsolution/ consolidation theory, and its application to more complex physical problems in Sobkowicz (1982).

2. LABORATORY TESTS

To examine both the transient and equilibrium behaviour of a gassy soil, samples of dense sand with a water/CO_2 pore fluid were tested in a triaxial cell. The water/CO_2 combination was used because it was non-corrosive, easy to handle, easy to control during sample preparation, and CO_2 had a high solubility in water which permitted working at low test pressures. The samples themselves were meant to be representative of dense soils with high gas content generally, and were designed to be relatively easy to fabricate and test, so that attention could be focussed on sample behaviour rather than sample idiosyncrasies. The laboratory testing of actual field samples, or of samples representative of actual field conditions, is a more complicated matter, the discussion of which is beyond the scope of this paper.

The laboratory samples were subjected to various stress paths, including isotropic unloading/loading, anisotropic unloading/ loading, and loading to failure by decreasing mean normal stress (p') or increasing shear stress (q) as shown in Figure 1.

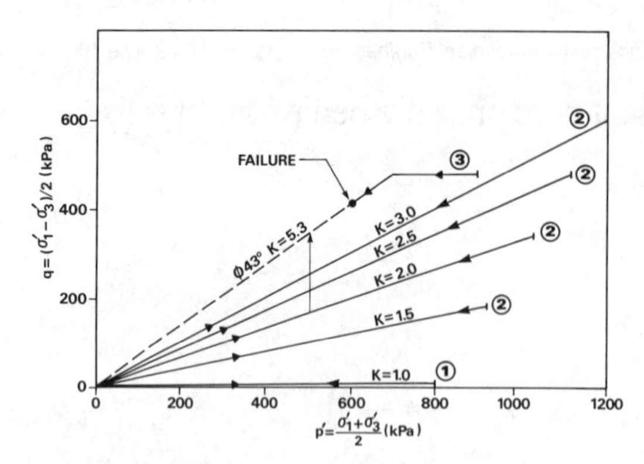

Figure 1. Stress paths for laboratory tests.

Important aspects of equipment design and test procedures are given below, although space constraints do not allow a detailed discussion of these topics, (further information is available in Sobkowicz [1982]). This is followed by a presentation and discussion of the test results.

2.1 Equipment design

2.1.1 Membrane and cell fluid considerations

A typical laboratory sample contained 1100cc of gas (measured at 100 kPaa) dissolved in 250cc of pore liquid, (S = 1 at $P_{l/g}$ = 510 KPaa.) Because of large differences in dissolved gas concentration between the pore liquid and the cell liquid, there was a tendency for gas to diffuse through the membrane surrounding the sample. It was necessary to prevent significant gas losses from the sample for the duration of a test, normally a week to 10 days.

The diffusion process can be limited in one of two ways:

a) by surrounding the sample with a liquid in which the gas has a low solubility. This technique was used by Bishop and Donald (1961) and Fredlund (1973), who surrounded their samples with mercury.

b) by providing a membrane that is nearly impermeable to gas. Dunn (1964) used this approach by enclosing the sample in a double latex rubber membrane, with two sheets of slotted aluminum foil sandwiched between them, and lubricated with silicone grease.

The cell fluid/membrane combination must also meet other design requirements:

a) thin but tough, to allow good contact between the lateral strain indicator and the sample, but prevent membrane puncture.

b) stretchable, to accommodate the large strains that occur at the end of an unloading cycle, (ε_{vol} may reach 10-15%).

c) the sample fluid should be non-conductive, to allow internal instrumentation, and non-corrosive, non-volatile, and non-toxic.

Only one cell fluid was judged capable of meeting the design requirements, glycerol ($C_3H_8O_3$), with H = 0.03 for CO_2 at T = 25°C and an electrical conductivity less than 0.1 micromhos/cm. Several membrane types were examined, including double latex, either by itself, or with a separating layer of aluminum foil or polyethylene; neoprene; and butyl rubber.

Various combinations of sample membrane, cell fluid, and sample fluid were tested. Gas diffusion rates were assessed by monitoring pressure drop (or increase) in the sample with time. It was clear that the double latex membrane with glycerol cell fluid was the only reliable method of preventing gas diffusion which also satisfied the other design constraints.

2.1.2 Measurement of strain or volume change

Various techniques were investigated for measuring vertical strain, and lateral or volumetric strain. It was necessary for the instruments to be capable of measuring very small strains during the early portion of the tests ($\Delta\varepsilon \approx 2 \times 10^{-6}$) but to have a large strain range to accommodate sample volume changes towards the end of the test, (as high as 5% in any one direction).

The final selection of instrumentation is illustrated in Figures 2 through 4. Vertical strains were measured using one or several LVDTs mounted inside the cell, attached to the sample top cap. Lateral strains were measured by a circular, caliper type mechanism, which directly contacted the sample at two points. Displacement of the caliper arms was measured by an LVDT mounted on one arm, with the core attached directly to the other arm. All LVDTs were extremely accurate, allowing reliable measurements of displacements as small as 0.25μm.

2.1.3 Other equipment design considerations

a) Cell and sample pressures were measured using CELESCO strain-gauged

Figure 2. Vertical strain indicator for isotropic test.

Figure 3. Horizontal strain indicator, LVDT mounting.

diaphragm transducers, with an accuracy of ± 0.1 kPa.

b) All tests were stress-controlled. Deviator stresses were applied using a diaphragm-operated air cylinder, and loads were measured external to the triaxial apparatus with an in-line load cell. Corrections were made for ram friction.

c) The test cell was constructed of aluminum and the drain lines of copper tubing to prevent gas loss by diffusion.

d) The complete test apparatus was insulated to protect against minor temperature fluctuations, and cell fluid temperature was monitored.

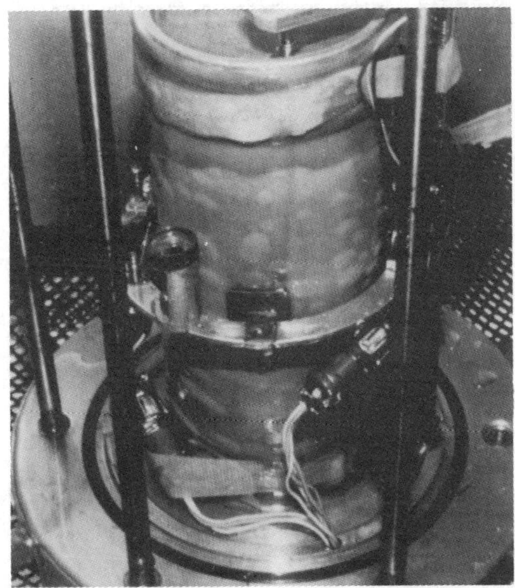

Figure 4. Horizontal strain indicator, sample contact.

e) All electrical wires passing through the cell wall were sealed by means of a specially developed connector using a teflon plug, which sealed around bare wire.

f) The cell was designed so that the sample could be set up on the base and the transducers attached and checked for operation. Figure 5 illustrates an early isotropic test sample just prior to cell assembly, (note the vertical LVDT and hinged contact on the lateral strain indicator, which were later replaced by the configuration shown in Figures 3 and 4).

2.2 Test procedures

The laboratory program consisted of a series of undrained tests on dense samples of Ottawa sand (n_i = 0.32). These samples were initially totally saturated with water (S = 1) which contained dissolved CO_2 gas. All samples were initially unloaded to low effective confining pressures, some experienced stress cycling, some were loaded to failure. Typical stress paths have been shown previously (Figure 1).

All tests were performed by applying several increments of stress, and allowing the sample to equilibrate between stress increments. Transient behaviour due to gas exsolution typically lasted from 1-2 hours. A full test, comprising 10-15

Figure 5. Early isotropic test configuration.

stress changes, plus sample set-up and preparation, took a week to 10 days.

The sample preparation is summarized below:

a) A uniform, medium-fine Ottawa sand (no fines) was compacted to a dry density of 1.8 Mg/m^3 (or n = 0.32). Compaction was performed in a mold placed around the base of the triaxial cell, partly filled with water, and vibrated.

b) The top cap, membranes, and drainage lines were attached and then a small suction (50 kPa) applied to the sample. Sample dimensions were recorded, strain indicators attached, the cell was assembled and the cell filled with glycerol, and a nominal cell pressure applied.

c) CO_2 gas was passed through the sample, displacing all water and entrapped air, then the sample was saturated using de-aired water. Since CO_2 is very soluble in water, complete saturation occurs very quickly and under low back pressures.

d) While the sample saturation was ongoing, the sample was cycled through several sets of isotropic stress changes to cause further compaction and provide information on the operation of the strain indicators. Typical volumetric compressibilities of $\beta_t = 2\text{-}4 \times 10^{-6}$ kPa^{-1} were obtained.

e) Once saturation was complete, the degree of saturation was determined using a B test. Values of S$>$99.98% were usually obtained.

f) The last stage of sample preparation was to replace the de-aired pore liquid by water saturated with CO_2 at some known back-pressure, which had previously been prepared in a bubble chamber.

Once sample preparation was complete, initial sample stresses were set (σ_3 about 1200 - 1400 kPa; u about 600 kPa and above the liquid/gas saturation pressure), drainage lines were closed, and then the first decrease in external stresses was applied. Readings of pore pressure and vertical and lateral strain were monitored with time until a new sample equilibrium had been established. The procedure continued with each stress change until the end of the test.

2.3 Test results

The process of gas exsolution dominated the observed behaviour in all tests and produced a behaviour common to all, that may be understood by examining Figure 6, (Test No. 11). The isotropic unloading portion of this test was conducted in 9 stress decrements (phases A through J).

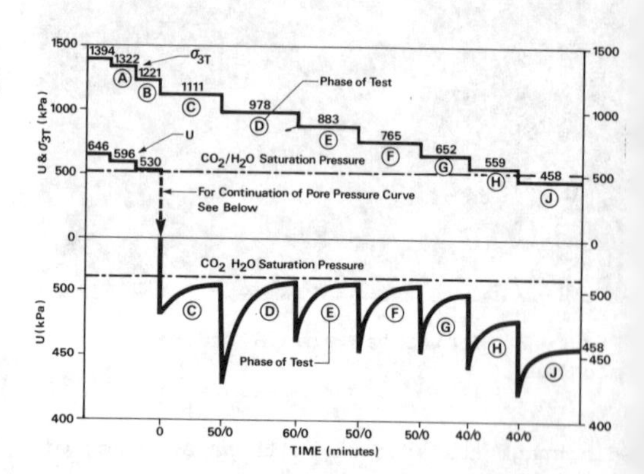

Figure 6. Pore pressure response, Test No.11

Initial stresses were σ_3 = 1394 kPa, u = 646 kPa and a liquid/gas saturation pressure ($u_{l/g}$) of 510 kPa.

During phases A and B of the test, the sample behaved as a "normal" saturated, undrained material, with pore pressures decreasing commensurately with total stress changes, and no transient effects evident. However, in phase C, the pore pressure decreased below $u_{l/g}$, which immediately instigated the gas exsolution process. As gas began to exsolve, the pore pressure increased since no sample drainage was allowed. Small increases in volume occurred as load was transferred from the soil skeleton to the pore fluid. The magnitude of volume change was governed by the soil compressibility and the pore pressure change. Since initially β_t was very small, only a small volume of gas needed to exsolve before u was driven back to a value close to $u_{l/g}$. The total process for phase A was complete, and a new sample equilibrium established after about 1 hour.

A similar behaviour was observed in the ensuing phases of the test, except that after phase F the equilibrium pore pressure started dropping significantly below $u_{l/g}$. This was due to a lower β_t with decreasing σ'_3, which allowed larger amounts of gas to exsolve.

It is interesting to note that for a considerable portion of the test (phases C through F), although the sample was being tested undrained, it was behaving as if it were drained, with an almost constant pore pressure equivalent approximately to the liquid/gas saturation pressure of the pore fluid.

The equilibrium pressure achieved at the end of each phase of the test is dependent on the soil and liquid compressibilities, on the existing stresses, on the porosity and saturation of the soil, and on the gas/liquid solubility relationship. Equations for predicting this value are provided in Sobkowicz & Morgenstern (1984).

It should be noted that at the end of phase J of the test, $\sigma'_3 = 0$. Further decreases in σ_3 force equivalent changes in u, large volumes of gas are released, and the sample effectively behaves in a drained manner. The volume change response during such a "drained" test, (phase H of Test No. 7), is shown on Figure 7. It is clear that this response is very similar to the pore pressure response in the undrained portion of the test, the rate being high at short times and decreasing at longer times as system equilibrium is approached. Such behaviour is typical of processes where the rate of change of the measured behaviour is directly proportional to the distance from some equilibrium condition.

Other samples were unloaded/loaded anisotropically, and also failed by decreasing σ_3. The observed pore

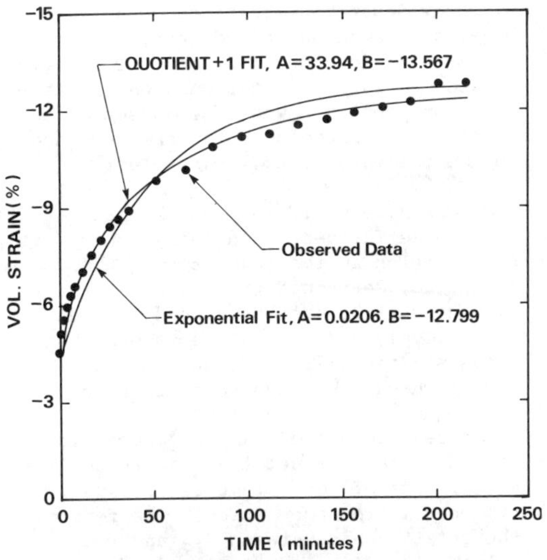

Figure 7. Volumetric strain response, Test No.7H.

pressure responses and volume changes were similar to those described for the isotropic Test No. 11 above, and are discussed in detail in Sobkowicz (1982).

3.0 MACROSCOPIC MODEL OF GAS EXSOLUTION

3.1 Volume - time response

A number of microscopic processes have been identified as contributing to the transient nature of gas exsolution, including gas sorption (mainly influenced by diffusion), bubble nucleation, chemical reaction between solvent and solute, and viscosity (and thus temperature) effects in the pore liquid. The latter two are not sigificant factors with respect to the current testing. The effects of the former two factors are discussed in Sobkowicz & Morgenstern (1987); it was concluded that although they play an important role in gas exsolution, their effects are not entirely quantifiable given the state of existing behavioural theories. It was not considered feasible to develop a general model describing gas exsolution directly from an understanding of the physical processes active at the microscopic level.

The approach taken in developing the macroscopic model was essentially an empirical one, augmented by theoretical considerations. The model could attempt to directly describe either pressure-time behaviour in an undrained test, or volume-time in a drained test. The latter was chosen as it was the only one of the two responses that was independent of soil properties, (although both were dependent on the presence of soil particles).

Unfortunately, it is not a simple task to obtain the fully-drained behaviour of gassy soil samples in the laboratory. The technique, described previously at the end of section 2.3, was "developed" quite fortuitously. "Drained" tests were run in 5 cases, (tests 7H, 11K, 9L, 9M and 9N), one of which was illustrated in Figure 7.

The measurement of sample volumetric strain in the drained tests is actually a measurement of the volume of free gas produced in the pore space. The form of the relationship between V_{fg} and t is illustrated diagrammatically in Figure 8. It was found that the experimental data could best be fit by an equation of the form:

$$V = V_1 + (V_2 - V_1) * t/ (t+A) \quad [1]$$

A non-linear 'least squares' fit using this function is shown for test 7H in Figure 7, labelled "Quotient + 1 Fit". The agreement between observed and predicted values is excellent.

The function suffers one disadvantage, however, in that its differential, DV_{fg}/DT still contains a dependency on t,

$$DV_{fg}/DT = (V_2 - V_{fg}) / (t + A) \quad [2]$$

which made it difficult to use in the development of other theory (not discussed in this paper). A more useful function would be one in which the derivative was independent of time, and indeed one has been hinted at previously, namely:

$$DV_{fg}/DT = E * (V_2 - V_{fg}) \quad [3]$$

or, in other words,

$$V_{fg} = V_2-(V_2-V_1)* \exp [-E*t] \quad [4]$$

A non-linear "least-squares" fit using this function is also shown on Figure 7. Although the fit is not quite as good as equation [1], it is acceptable.

Equations [3] and [4] were then selected to model the drained gas exsolution (volume vs time) response, as they fit the data well, and were suited to other theoretical development.

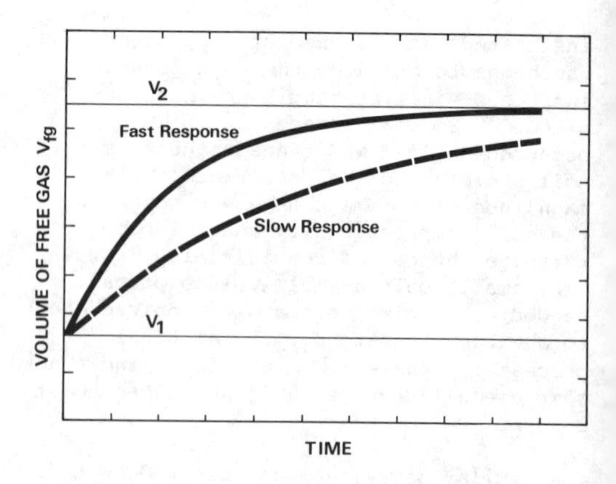

Figure 8. Gas exsolution response in a drained test.

3.2 Pressure-time response

The results of the previous section can be incorporated into a model of the pressure-time response for an undrained test, as described in this section.

For a sample of gassy soil tested in an undrained mode, the total volume change V_T may be calculated for a particular time interval Δt as:

$$\Delta V_T = \Delta V_1 + \Delta V_{fg} \ (\Delta t) \quad [5]$$

assuming the soil grains are themselves incompressible. We know as well that

$$\Delta V_T = -\beta_T \ * V_T * (\Delta \sigma - \Delta u \ [\Delta t]) \quad [6]$$

and that $\Delta V_1 = -\beta_1 * V_1 * \Delta u \ (\Delta t)$ [7]

The ΔV_{fg} in equation [5] must include the influence of both:

a) a compression or expansion of the existing free gas according to Boyle's law:

$$(\Delta V_{fg})_1 = - V_{fg} * \Delta P/(P + \Delta P) \quad [8]$$

b) exsolution. In calculating this volume it is assumed that P = constant over Δt (even though ΔV_{fg} is being used to calculate ΔP), which is reasonable for small Δt, so that from [3]

$$[\Delta V_{fg}]_2 / \Delta t = E * (V_2-V_{fg}) \quad [3b]$$

Defining an additional constant K', equivalent to the total equilibrium volume of dissolved and free gas in the pore space at any pressure p', multiplied by P', we have:

$V_2 = K'/P - H * V_w$ [9]

and hence:

$[\Delta V_{fg}]_2 = E*\Delta t*(K'/P - H * V_w - V_{fg})$ [10]

Combining [8] and [10],

$$\Delta V_{fg} = - V_{fg} * \Delta P/(P + \Delta P) + E*\Delta t*(K'/P - H*V_w - V_{fg}) \quad [11]$$

Substituting equations [7], [6] and [11] into [5], and rearranging, a quadratic equation is found for Δu:

$$A * \Delta u^2 + B * \Delta u + C = 0 \quad [12]$$

where $A = \beta t + n * S * \beta_1$

$D = E * \Delta t * (K'/P/V_T - n * [1 - S + S * H])$

$B = n * (1-S) + P * A - D$

$C = - P * D$

This is the uncoupled solution for Δu over a time interval Δt, given values of n, S, V_T and p at the beginning of the interval. It is approximately correct, with accuracy being better for small Δt.

In a similar manner, a coupled solution to this problem is developed in Sobkowicz (1982) which is about four times more accurate than [12]. The overall quadratic equation and terms A, D and C are the same as above, but the term B is modified as :

$$B = P * A + n*(1-S + S*H * [E * \Delta t/2]) \quad [12b]$$

3.3 Comparison of predicted and observed behaviour

The method of modelling the undrained response given in section 3.2 was coded for computer analysis, the flow chart for which is given in Figure 9.

A comparison of predicted vs observed pore pressure response for three phases of test no. 11 (C, D and E) is given in Figures 10 - 12. The agreement seen here, and in fact in all tests, varies from moderately good to quite good, indicating that the macroscopic model of exsolution behaviour can be applied with a good degree of accuracy.

It should also be noted that the predicted transient response (which is based on equations which are approximately correct) converges on the equilibrium response (exact equations) developed in a

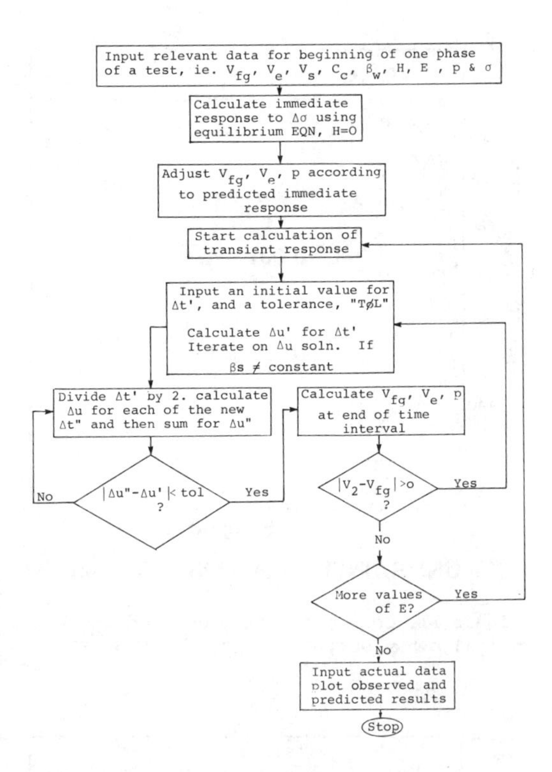

Figure 9. Flow chart for computer prediction of transient pore pressure response.

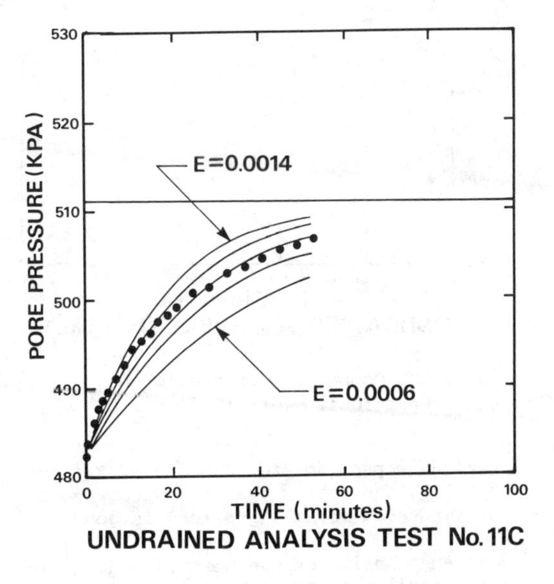

Figure 10. Comparison of predicted and actual behaviour, Test 11C.

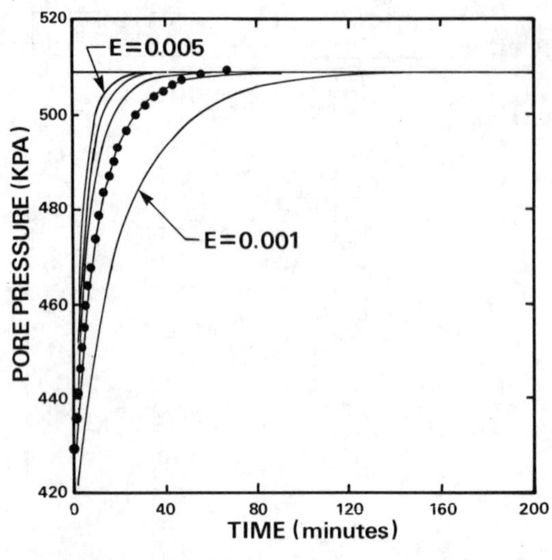

Figure 11. Comparison of predicted and actual behaviour, Test 11D.

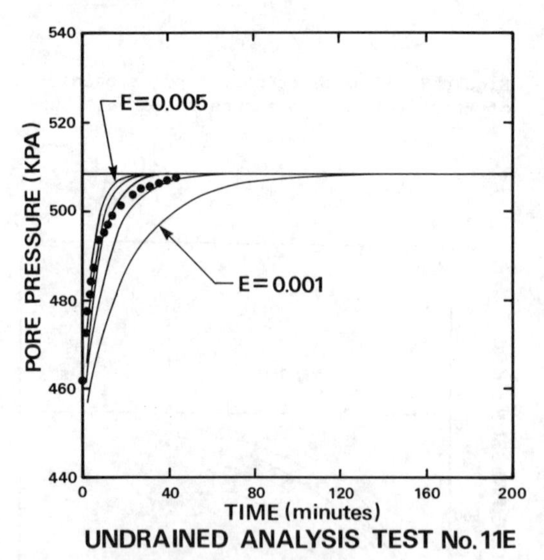

Figure 12. Comparison of predicted and actual behaviour, Test 11E.

previous paper (horizontal line of Figures 10 - 12), indicating that the accuracy of the suggested modelling method is good.

An examination of the best fit curves for all transient tests also indicates that the gas exsolution parameter E is a function of gas saturation. The values obtained in the

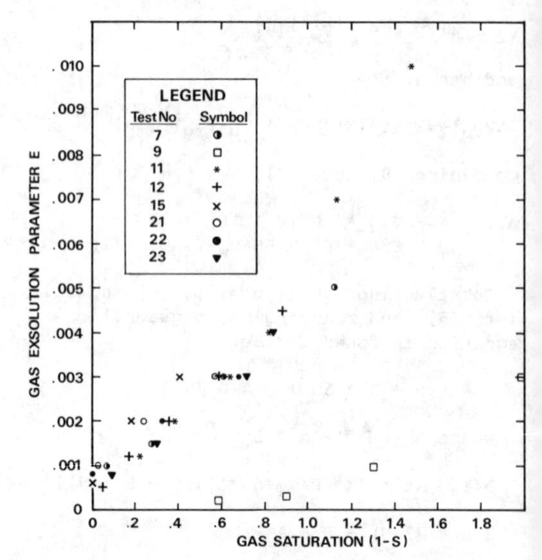

Figure 13. Variation of gas exsolution parameter (E) with gas saturation (1-S; expressed in % units).

laboratory tests are shown in Figure 13. It is hypothesized that the increase of E with S_g is due to an increased number of bubbles in the soil sample, leading to both an increased surface area of liquid/gas contact, and to a decreased length of diffusion path.

4 CONCLUSIONS

A method of predicting rate of gas exsolution in soils has been proposed which agrees well with observed pore pressure behaviour in undrained tests and volume behaviour in "drained" tests. The theory has been empirically developed, but is consistent with underlying physical processes.

SYMBOLS NOT DEFINED IN TEXT.

- H Henry's constant for solubility of gas in liquid (volume of gas/volume of liquid, measured at existing pressure).
- n Porosity
- P Pore fluid pressure (absolute, kPaa)
- S Saturation of liquid in pore space
- t Time
- u Pore Fluid pressure (gauge, kPa)
- V Volume
  Subscripts fg = free gas
  T = total
  l = liquid
- ε Strain
- σ Total stress
- σ' Effective stress

REFERENCES

Bishop, A.W. and Donald 1961. The experimental study of partially saturated soil in the triaxial apparatus. Proc. 5th ICSMFE. 1:13-21.

Dunn 1964. Developments in the design of triaxial equipment for testing compacted soils. Proc. Symp. on the Economic use of soil testing in site investigation. Birmingham. 3:19-25.

Fredlund, D.G. 1973. Volume change behaviour of unsaturated soils. PhD thesis, University of Alberta, Edmonton.

Sobkowicz, J.C. 1982. The mechanics of gassy sediments. PhD thesis, University of Alberta, Edmonton.

Sobkowicz, J.C. and N.R. Morgenstern 1984. The undrained equilibrium behaviour of gassy sediments. Canadian Geotechnical Journal, 21:439-448.

Sobkowicz, J.C. and N.R. Morgenstern 1987. Fundamentals of transient behaviour in gassy sediments. Submitted to Canadian Geotechnical Journal.

Settlement prediction, monitoring and analysis for tower silos in the St.Lawrence Lowlands

R.Gervais
University of Alberta, Edmonton, Canada

J.P.Morin
University of Sherbrooke, Canada

ABSTRACT: Tower silos are very common storage structures on farms in Québec. A research project started in 1975 at Université de Sherbrooke has investigated 108 silos built over Champlain clays in the St-Lawrence Lowlands. Bearing capacities were based on a total of 94 Nilcon vane profiles. Settlement monitoring was done for 28 silos. Readings for 19 of them are available after 10 load cycles. Field sampling and laboratory testing was done at 15 locations. Design of new silos should be based on a FS of 2.5 applied to the bearing capacity derived from a Nilcon vane test. The associated 10 load cycle settlement is estimated at 75 mm.

1 INTRODUCTION

Agricultural tower silos are often present on farms in the Province of Québec in Eastern Canada. It is however frequent to find silos that settle and/or tilt beyond acceptable limits. Insufficient attention is given by the contractors and the farmers to the estimation of bearing capacity and settlement of the soil. A few failures have occurred involving property losses.

A research project was carried out at Université de Sherbrooke on tower silos located in the St-Lawrence Lowlands. Figure 1 shows the region of interest for this paper. The area, approximately 150 km long and 60 km wide, is located East of Montréal in the Southern part of Québec.

Champlain Sea Clays form the major portion of the subsoil for most of this region. These soft and sensitive clays have been extensively described in the geotechnical literature. For the purposes of this paper, it will be sufficient to remember that these soils are weak and compressible. The shear strength can be as low as 10 kPa underneath the dessicated crust.

The economics of silo construction have a detrimental effect on foundation design. A typical silo may cost twenty to forty thousand dollars. While this amount of money is large for the farmer, it does not usually warrant a full scale site investigation and foundation design by an engineering consultant. This often leaves contractors and farmers in a poor situation with respect to allowable bearing capacity determination.

This paper suggests a more adequate foundation design alternative from engineering and economic points of view.

2 PREVIOUS WORK

Silo design and performance aspects were investigated in the Province of Ontario by various authors: Eden and Bozozuk (1962), Bozozuk (1972, 1974, 1976, 1979a, 1979b), Lo and Becker (1979, 1980).

In the Province of Québec, the Ministère de l'agriculture sponsored a research program that started in 1975 at Université de Sherbrooke. Morin (1977), Authier (1980) and Gervais (1980) presented their findings in their respective thesis. A report by Morin and Gervais (1980) was submitted to the Government. A summary of this report was published by Morin and Bozozuk (1983). Gervais (1985a) used probabilistic considerations in silo foundation design. The tenth load cycle settlement survey was documented by Gervais (1985b). Morin and Gervais (1985) compared these measurements with the initial settlement predictions of Morin (1977).

3 GLOBAL SURVEY

As reported by Morin and Gervais (1980), 138 silos were investigated. Site selection criteria are listed in the previous

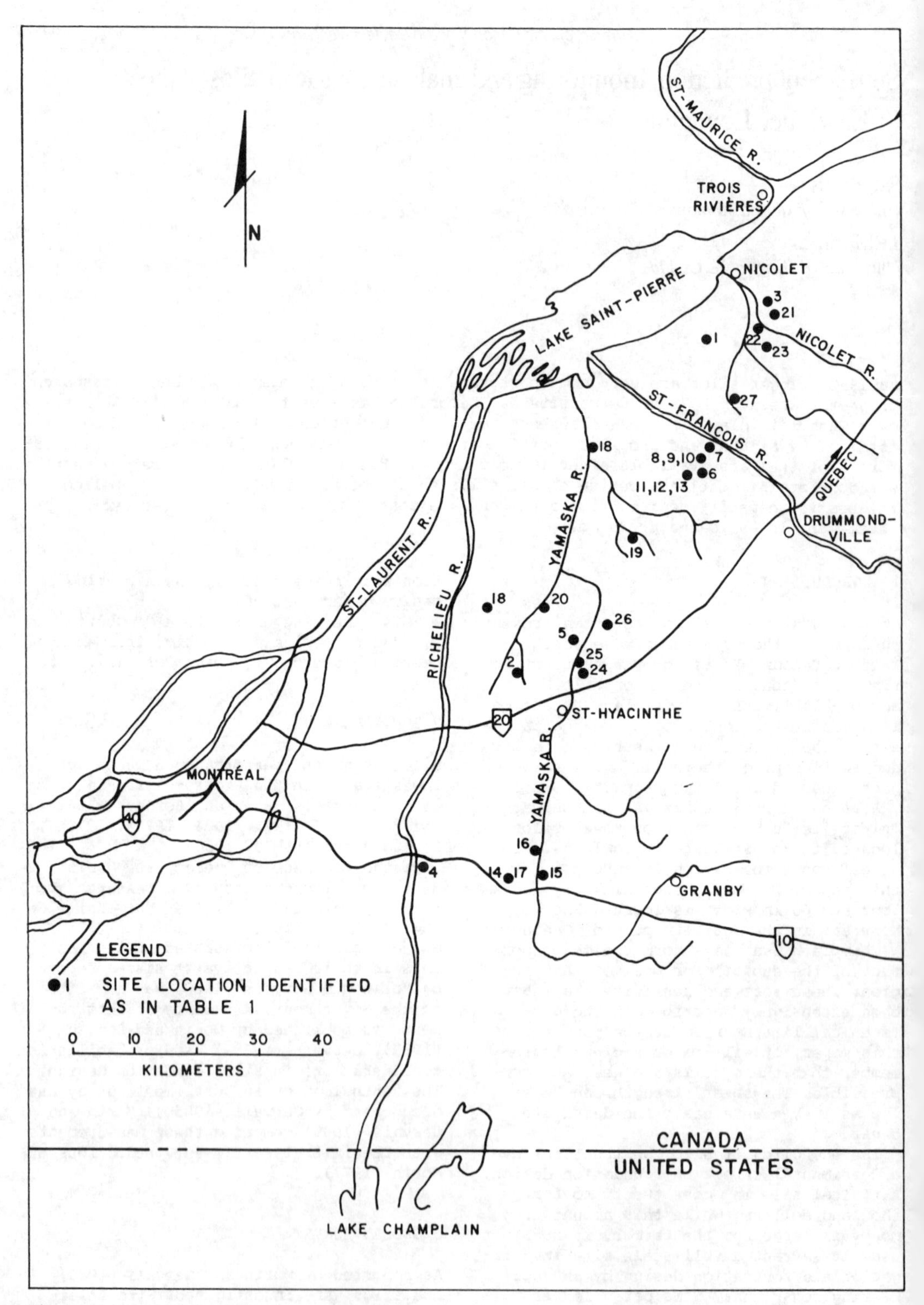

Figure 1. Regional map.

Table 1. Description, observed settlement and tilt of concrete tower silos.

Location no. (see Fig. 1)	Silo dimensions (diam.x ht.) (m)	Silo type (see legend)	Foundation dimensions (th.x o.d.) (m)	Foundation type (see legend)	Total settlement (10 cycles) (mm)	Differential settlement (10 cycles) (mm)	Silo tilt (% of height)	Factor of safety
1	5.5x18.3	CS	0.6x 8.2	ADN	247	na	na	2.33
2[1]	6.1x18.3	CP	0.6x 8.8	BSFDE	na	na	na	2.53
3	6.1x18.3	CS	0.6x 9.4	ADN	49	37	0.60	3.50
4	6.1x18.3	CP	0.6x 7.6	BSFDE	91	21	0.35	2.43
5[2]	6.1x18.3	CP	0.6x10.7	BSFDN	518	21	0.35	1.56
6[3]	5.5x14.6	CP	0.9x 9.1	HSFDN	na	na	na	2.36
7	4.9x17.1	CP	0.5x 7.9	BSFDN	67	9	0.19	3.04
8[4]	5.5x15.8	CP	0.6x 9.4	BSFDE	na	na	na	1.65
9[5]	5.5x18.3	CP	1.1x 7.3	BSCDE	131	15	0.28	2.13
10	5.5x17.1	CP	0.6x 8.5	BSDFE	171	24	0.44	3.20
11	5.5x15.8	CP	0.6x 9.4	BSFDE	183	55	1.00	3.52
12[6]	4.9x17.1	CP	0.6x 8.5	BSFDE	na	na	na	1.58
13[7]	6.1x17.1	CP	0.9x 9.4	BSFDN	na	na	na	1.57
14	6.1x18.3	CP	0.8x 8.5	BSFDE	497	88	1.45	2.29
15	6.1x21.9	CP	0.6x 9.0	BSFDE	110	61	1.00	2.21
16	6.1x16.8	CS	0.6x 7.9	ADN	15	34	0.55	3.92
17[8]	6.1x18.3	CP	0.6x 9.1	BSFDE	na	na	na	2.48
18[9]	6.1x18.3	CP	0.6x12.2	BSFDN	na	na	na	1.61
19	4.9x18.3	CP	0.6x 7.6	BSFDN	268	46	0.94	1.63
20[10]	4.9x19.5	CP	0.8x 6.4	BSFDE	na	na	na	3.13
21[11]	6.1x15.2	CS	0.6x 8.2	BDN	na	na	na	1.72
22	5.5x16.8	CS	0.6x 8.1	ADN	37	15	0.28	5.12
23	6.1x15.2	CS	0.6x 9.1	ADCN	110	9	0.15	3.38
24	5.5x15.2	CS	0.6x 8.5	ASDN	67	na	na	3.18
25	7.3x24.4	CP	0.6x11.0	BSFDN	58	9	0.13	2.70
26[12]	7.3x23.2	CP	0.6x 9.4	BSFDE	274	na	na	1.29
27	5.5x18.3	CS	1.2X 8.5	ADCN	265	85	1.56	1.43
28	4.9x15.2	CS	0.9x 6.7	ADN	21	12	0.25	2.39

NOTES:

1. Instrumentation destroyed October 1977.
2. Foundation reinforcement done approximately four years ago (1983).
3. Last settlement observation May 1979.
4. Last settlement observation May 1979.
5. Silo built one year before most of the present structures, but loaded to full capacity on second year.
6. According to owner, load applied was only dead structural load on first year, increasing to full capacity by the fourth year; factor of safety shown applies to latter case. Last settlement observation May 1979.
7. Last settlement observation May 1979.
8. Structure straightened after third loading. Last settlement observation May 1979. See also note 5.
9. Last settlement observation before third loading; instrumentation destroyed November 1978.
10. Last settlement observation May 1979. Applied load varied over the years; safety factor shown for full capacity.
11. Last settlement observation before fourth loading; instrumentation destroyed November 1978.
12. Structure straightened after second loading.

LEGEND:

CS Precast concrete stave.
CP Cast in-place concrete.
A Annular foundation with small concrete wall at base of silo.
B Plain annular foundation.
H Modification of previous wooden silo foundation.
S Steel reinforcement in footing.
F Concrete floor at bottom of silo.
D Drainage provided for silage juices.
C Cracked foundation by visual observation.
N Natural soil undisturbed; construction may or may not be on a small fill.
E Excavation of top part of natural soil; excavation usually less than thickness of annular foundation.
na Information not available.

reference. The sample size obtained represents about 1.4% of all silos in Québec as per December 1976 estimates. Both old and new structures as of 1975-1976 are included. From the total sample, 108 silos were retained as representative of structures built over clay. This group is called the global survey. It is important to note that in this population, one out of three owners is not satisfied with the performance of his silo.

A total of 94 Nilcon vane shear strength profiles were carried out. The bearing capacity for each site can thus be estimated.

An attempt to regionalize bearing capacities was made in the village of St-Zéphirin. Seven sites located within 4 km from each other have average shear strengths ranging from 14.7 to 29.4 kPa. This variation being so large, it was not considered safe to propose regional values of bearing capacity for design.

A questionnaire was filled for each

Table 2. Soil parameters, predicted and measured settlements.

Location	Baieville	Nicolet	St-Barnabé Sud	St-Bonaventure	St-Césaire	St-Guillaume	St-Jude	Ste-Monique	Ste-Rosalie
No. as per fig. 1	1	3	5	13	14	19	20	21	25
Factor of safety	2.33	3.50	1.56	1.57	2.29	1.63	3.13	1.72	2.70
C_2/C_1	0.35	0.32	0.43	0.88	0.26	0.53	0.47	0.34	0.41
d/B	0.17	0.29	0.12	0.21	0.29	0.14	0.35	0.21	0.34
$\Delta p'/p_c'-p_o'$	1.81	1.11	1.82	2.03	1.15	1.45	1.26	2.33	0.96
Predicted settlement (mm)	486	100	505	495	342	212	97	983	66
Observation date	85 05	85 05	85 05	79 05	85 05	85 05	79 05	78 06	85 05
Elapsed time (days)	3543	3556	3524	1326	3546	3530	1318	1007	3493
Measured settlement (mm)	247	49	518	354	497	268	18	402	58
Ratio of measured to predicted	0.51	0.49	1.03	0.72	1.45	1.26	0.19	0.41	0.88

LEGEND:

C_1 = Average shear resistance in the crust as measured by Nilcon vane.

C_2 = Average shear resistance in the underlying soft layer as measured by Nilcon vane.

C_1/C_2 = Ratio of preceding quantities.

d = Thickness of crust as interpreted from vane test profile.

B = Diameter of footing.

d/B = Ratio of preceding quantities. Indication of relative thickness of crust to size of loaded area.

$\Delta p'$ = Effective surcharge load applied to the soil.

p_c' = Preconsolidation pressure.

p_o' = At rest earth pressure.

silo. Structural dimensions served to estimate the total applied load. Agricultural silos apply a cyclic load on the soil due to the filling and emptying processes. For calculation purposes, the maximum load is used. The first day full load is reached defines time zero for settlement history. Most foundations are annular in shape. The applied load is assumed to be distributed uniformly across the area circumscribed by the outer diameter ot the foundation annulus. Discussion on the limitations of this assumption are presented in Morin and Gervais (1980). Each silo has a safety factor against overturning that can be calculated using the ratio of bearing capacity over applied stress.

Estimates or measurements of settlement and tilt were done on silos built before 1975. The quantity and quality of the information gathered during the survey varies according to the individual farmer. For some of them, the silo was judged only on a fail or pass criteria with reference to their satisfaction with the performance of the structure. In other cases, settlement estimation of farmers can be more accurate due to indirect evidence and good personal observation.

Table 3. Performance criteria for tower silos.

Rating	Performance	Total settlement (mm)	Differential settlement (mm)	Silo tilt (degrees)	Silo tilt (%slope)
A	Excellent	Below 25	Below 20	Below 0.2	Below 0.3
B	Slight problems	25-75		0.2-0.5	0.3-0.8
C	Important Problems	75-150		0.5-1.0	0.8-1.7
D	Serious problems	150-300		1.0-1.5	1.7-2.5
E	Very serious problems	Over 300		Over 1.5	Over 2.5

Note. A differential settlement of 20 mm on a 6.1x18.3 m typical silo is equivalent to 0.2 degree or 0.3% of tilt.

4 GROUP WITH SETTLEMENT MONITORING

At the time the global survey was done in 1975-1976, 28 silos were chosen for settlement monitoring. Figure 1 shows the location of each silo. Table 1 gives details on the structures. Each silo is referenced against one and usually two benchmarks. Measurements at four locations around the silo provided total and differential settlement values for each site. Readings were taken twice a year for the first four years. These results are reported by Morin and Gervais (1980). Figure 2 shows the total settlement after the first load cycle as a function of the safety factor. The individual points follow a trend indicated by the continuous curved line. The observed point scatter is discussed in Morin and Gervais (1980).

A reading from 19 out of the original 28 silos could still be taken after the tenth load cycle. Table 1 lists the recorded values for each site. This survey was documented by Gervais (1985b). Morin and Gervais (1985) discussed some considerations relative to long term readings at these sites.

Figure 3 superposes the settlement vs FS trends for load cycles 1, 2, 3, 4 and 10. The curves evolve as a function of time since consolidation is underway. However the yearly change in the trend line position is slowing down as can be expected.

On figures 2 and 3, the total settlement axis is split in five performance regions, going from A to E. Table 3 reproduces these boundaries plus other requirements on differential settlement as a guide to silo performance qualification. Ratings D and E imply sufficiently poor behaviour that these silos are receiving failing grades even if no failure occured. Ratings A, B and C describe sites that have passed the minimum performance requirements for farm silos.

5 SETTLEMENT ESTIMATION

Field sampling and laboratory testing were done by Morin (1977) and Authier (1980) at 15 locations. A total of 11 sites with complete field and laboratory data is included in the monitored group. Unfortunately long term settlement readings could not be carried out at two of these sites. The anticipated settlement is estimated using the consolidation test results. Table 2 summarizes these predictions for the remaining 9 structures. Six of these have settlement history in the range of 3500 days. On average for these cases, measured settlement reaches 94% of the predicted value after almost ten years of consolidation.

Site investigation, sampling and laboratory testing are not economically acceptable propositions in routine agricultural silo construction even though they provide necessary information for research purposes.

6 PREVIOUSLY BUILT SILO PERFORMANCE

The silos included in the global survey range from well-designed and assembled silos to constructions executed by amateurs. Cases where the silo already had to be

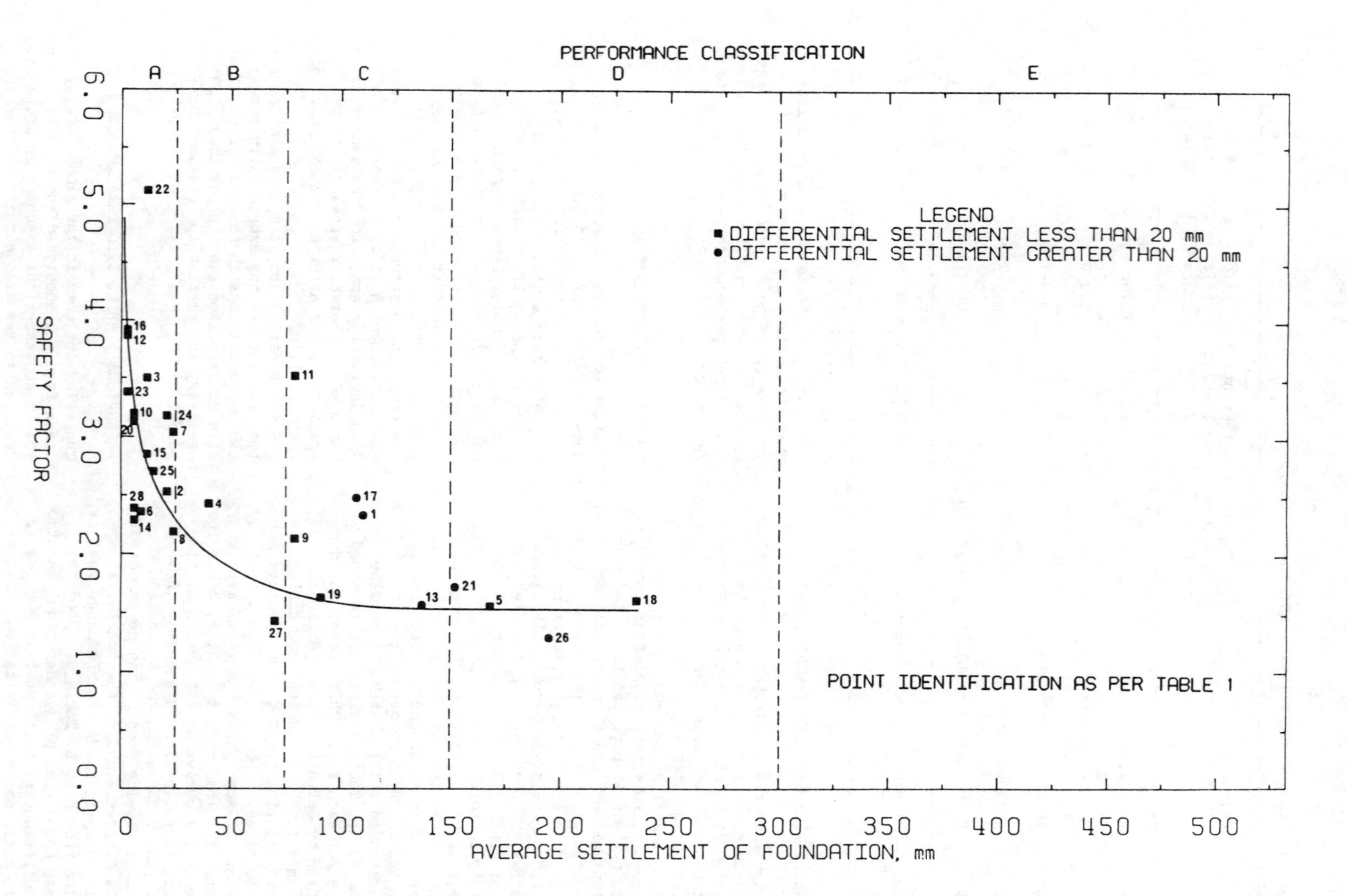

Figure 2. Relation between settlement and safety factor at load cycle no. 1.

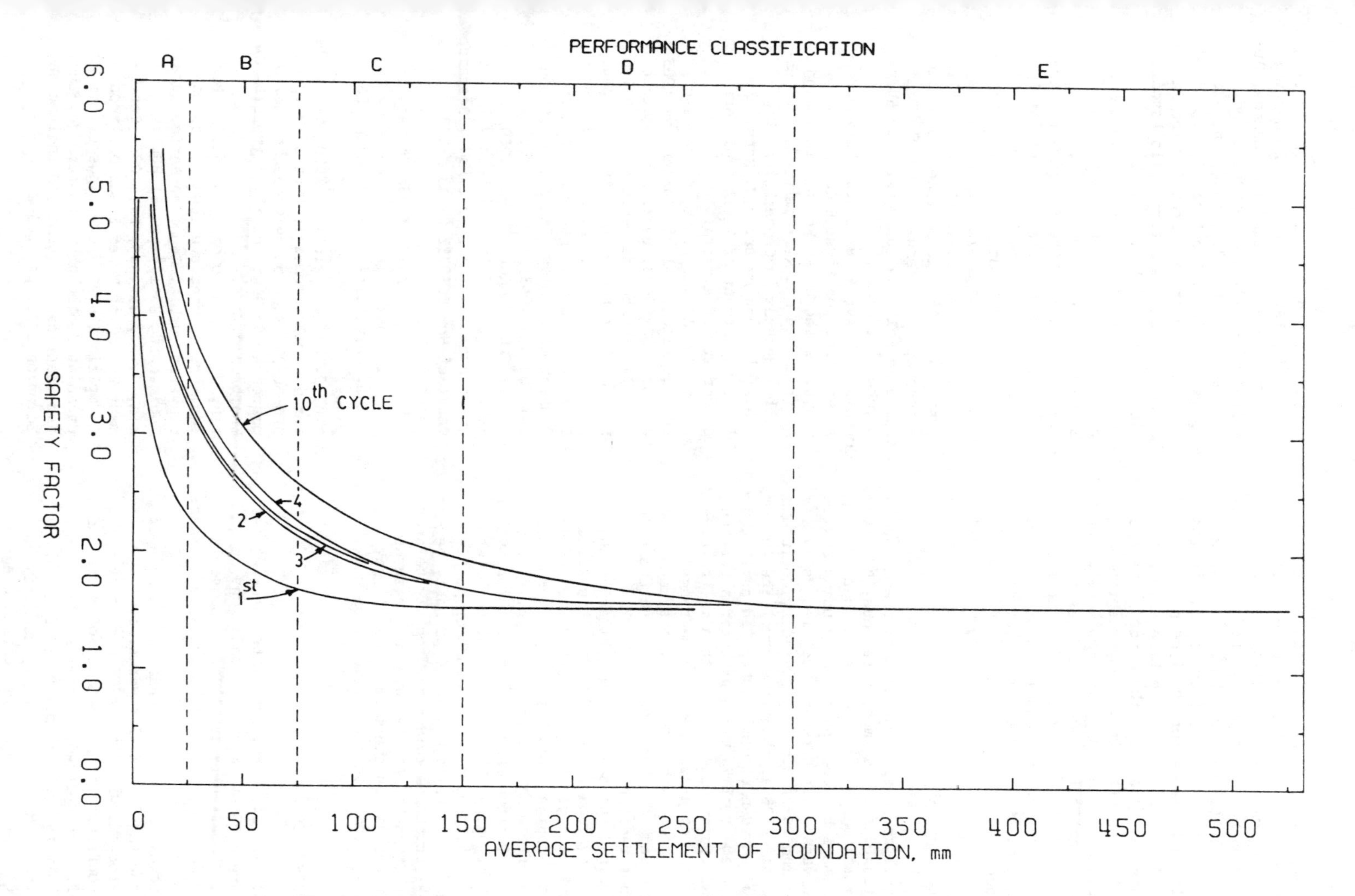

Figure 3. Average relation between settlement and safety factor for load cycles 1 to 10.

straightened, or had an evidently cracked footing, or were built after excavating the dessicated surface crust were excluded from the next step of this study.

Gervais (1980) took the fourth cycle settlement vs FS trend and estimated the silo performance for 49 sites based on as-built FS. At the same time silo performance was rated from the owner's answers to the questionnaire. The perception of the farmers and the performance estimation were identical in 69% of the cases. The safety factor in combination with the settlement vs FS trend lines can thus be accepted as a good estimator of performance.

7 RECOMMENDATIONS

A full scale site investigation and foundation design is impractical due to cost limitations. It is recommended here to estimate the bearing capacity using a Nilcon vane shear strength profile. Using a minimum FS of 2.5 as proposed by Morin and Gervais (1980), the diameter of the footing can then be dimensioned. An estimated settlement of 75 mm is expected after ten load cycles. The silo is then in the middle of acceptable performance range.

The preceding recommendations cannot be extended to other types of structure or to other regions.

Doing vane testing and estimating the bearing capacity would add to the initial cost of a given silo. However the subsequent design would have a very good chance of performing satisfactorily while in the present situation, one out of three owners is not happy. There is a cost associated with failures or maintenance of poorly performing silos. It would probably be less expensive to design properly to start with than to provide remedial measures afterwards. Furthermore, it would make sense from a scientific point of view.

REFERENCES

Authier, J., 1980. Les contraintes et les déplacements sous les fondations de silos construits sur l'argile sensible. Mémoire de maîtrise, Université de Sherbrooke.

Bozozuk, M., 1972. Foundation failure of the Vankleek Hill tower silo. Proceedings of the ASCE Specialty Conference on Performance of Earth and Earth-Supported Structures, Purdue University 1(2), pp. 885-902.

Bozozuk, M., 1974. Bearing capacity of clays for tower silos. Canadian Agricultural Engineering 16(1), pp. 13-17.

Bozozuk, M., 1976. Tower silo foundations. National Research Council of Canada, Division of Building Research, Canadian Building Digest 177, Ottawa, Ont.

Bozozuk, M., 1978. The Hammond silo: preliminary report. Prepared for the Annual Conference of the Ontario Silo Association, Niagara Falls, Ont.

Bozozuk, M., 1979a. Problems with concrete tower silos. Canadian Agricultural Engineering 21(2), pp. 69-77.

Bozozuk, M., 1979b. The instrumented tower silo at Hammond, Ontario. Canadian Agricultural Engineering 21(2), pp. 79-84.

Eden, W. J. and Bozozuk, M., 1962. Foundation failure of a silo on varved clay. Engineering Journal 45(9), pp. 54-57.

Gervais, R., 1980. Comportement des silos construits sur l'argile sensible. Mémoire de maîtrise, Université de Sherbrooke, 654 pages.

Gervais, R., 1985a. Practical Design Considerations for Tower Silos on Soft Clays. Presented at Seminar following short course on 'Reliability and Probability Methods in Geotechnical Engineering' given by Dr. M.E. Harr, May 1985, Department of Civil Engineering, University of Alberta.

Gervais, R., 1985b. Projet de recherche sur les silos, Relevé d'arpentage 1985. Rapport GEO-85-03, Section de Géotechnique, Département de Génie civil, Université de Sherbrooke, août 1985, 75 pages.

Lo, K. Y. and Becker, D. E., 1979. Pore-pressure response beneath a ring foundation on clay. Canadian Geotechnical Journal 16(3), pp. 551-566.

Lo, K. Y. and Becker, D. E., 1980. Settlement analysis of intermittently loaded structures. Agriculture Canada, Research Contract No. DSS07SZ.01843-8-0827, 106 pages.

Morin, J. P., 1977. Etude du tassement des silos sur l'argile. Mémoire de maîtrise, Université de Sherbrooke, 261 pages.

Morin, J. P., Gervais, R., 1980. Les silos sur l'argile, Tomes 1, 2 et 3. Rapport GEO-80-03, Section de Géotechnique, Département de Génie civil, Université de Sherbrooke, avril 1980.

Morin, J. P. and Bozozuk, M., 1983. Performance of concrete tower silos on clays in Quebec. Canadian Agricultural Engineering 25(1), pp. 81-88.

Morin, J. P., Gervais, R., 1985. Settlement Record of Tower Silos built on Champlain Clays, Proceedings of the 38th Canadian Geotechnical Conference, Theory and Practice in Foundation Engineering, Edmonton 1985, pp. 285-294.

Prediction and Performance in Geotechnical Engineering / Calgary / 17-19 June 1987

Embankment on soft fissured clay

Robert Alperstein
Woodward-Clyde Consultants, Chicago, Ill., USA
Larry Holish
Sargent & Lundy, Chicago, Ill., USA

ABSTRACT: A 17-meter-high embankment for storing cooling water for a power station was under construction near Edmonton, Alberta, when a longitudinal crack appeared in the crest at the 12-meter level. The embankment was being constructed over soft to medium fissured clay. Inclinometer data, piezometer data, and data from a previous nearby embankment failure were available. Additional undisturbed soil samples were obtained and direct-simple shear tests (DSS) were conducted. Evaluation of the failure data suggested that C.C. Ladd's SHANSEP approach would be worthwhile in judging the stability of the 12-meter-high embankment and projecting to the 17-meter height. Recommendations were given for improving the stability to meet design criteria, including the use of prefabricated vertical drains to dissipate high (10 meters above embankment grade) excess pore pressures beneath the embankment. This paper describes the data obtained, judgments made in evaluating the data, analyses performed, basis for the recommendations made, and lessons learned (or confirmed).

1 INTRODUCTION

The stability of embankments on soft clays has received much attention in recent years (Bjerrum 1972; MIT 1975). Although most geotechnical engineers understand that shear strength of soils is governed by effective stresses, most embankment stability calculations are conducted using total stress methods. Effective stress analyses require knowledge of pore pressure, its distribution, and horizontal stress. Reliable field equipment for horizontal stress measurement is not yet available. Effective stress analysis can incorrectly estimate path-dependent aspects of strength generation. Also, the bulk of U.S. experience from engineering practice comes from total stress analysis. For the above reasons and because of the simplicity of the method, total stress analysis is preferred by most U.S. practitioners.

The following case history describes an embankment on a soft clay foundation that was on the verge of failure and some of the engineering aspects of selecting appropriate strengths, analyses, and remedial measures required to preserve stability.

The project consists of about 15 kilometers of earth dikes enclosing a perched cooling water pond for a fossil fuel power plant. The project is located near Edmonton, Alberta, Canada. The homogeneous clay dikes vary in height between 2 and 25 meters and were constructed on a glaciolacustrine clay foundation. The clay foundation is 8 to 16 meters thick, generally soft and of high plasticity, and slickensided. The slickensides or fissures are believed to have been caused by past desiccation and rebound. The dikes were to be stage constructed over a period of 2 years. Dikes less than 12 meters high were to be constructed in a single stage with side slopes of 3 horizontal to 1 vertical. A similar central cross section was used for embankments from 12 to 25 meters high; however, to control the rate of construction, stabilizing berms were added and the dissipation of pore pressures was monitored. Construction of the second stage was to have proceeded when 60% of the excess pore pressure from the first stage had dissipated.

The dike alignment also crosses five major zones of peat varying in thickness from 1 to 4.8 meters. The peat covers 48% of the project site or approximately 3.2

km^2. The peat was drained and removed while frozen from beneath water retention dike foundations during the winter months. The peat remained in place beneath baffle dikes and stabilizing berms.

Construction began in June 1982 with the preparation of the foundation at the highest dike and proceeded through September 1982 with construction of a 12-meter-high embankment. Instrumentation consisted of pneumatic piezometers, inclinometers, and settlement plates installed in August 1982 to monitor the embankment/foundation response. Pore pressure response was immediate. The piezometric level in the foundation clay was about equal to the height of the fill. The results of the monitoring showed no dissipation of pore water over the winter through July of 1983 when stabilizing berms were constructed. With the completion of the stabilizing berms and the installation of additional instrumentation, the construction schedule demanded commencement of the second embankment stage in September 1983. No measurable dissipation of pore pressure had occurred, although minor lateral spreading of the foundation clay was noted under the stabilizing berms.

Embankment construction proceeded with foundation pore pressures rising in cadence with the fill height through October 1983. With the crest 3 meters below completion, cracking of the upstream slope was observed along a 200-meter section of the highest dike. The dike was 17 meters high and foundation clay movement was monitored with downstream inclinometers at a rate of 4 millimeters per day, suggesting the entire embankment was involved from the upstream face to the downstream toe. Dike construction was suspended for the winter, more stabilizing berms were added, and the monitoring was continued. At that time, the authors were invited to review the project data and provide recommendations for remedial measures.

2 DEFINITION OF THE PROBLEM

The 1.3-kilometer-long homogeneous embankment consists of approximately 2 million cubic meters of glaciolacustrine clay compacted in 300-millimeter loose lift thicknesses to 95% of standard Proctor density at moisture contents ranging from -1% to +6.5% of optimum moisture content. A simplified stratigraphic section and soil properties are shown in Figure 1.

The critical soft foundation clays contained slickensided zones attributed to the combined action of delayed rebound of the underlying clay till and the differential shrinkage and swelling of the glaciolacustrine high-plasticity clay (May 1978). The designers postulated that embankment construction on a slickensided clay could produce deformations and creep movement that, if large enough, could mobilize the residual shear strength. Residual failure conditions persist in the Edmonton area and are normally encountered in excavated slopes greater than 6 meters high and may be accelerated through exposure to freezing and weathering effects (McRoberts 1974; Thomson 1985).

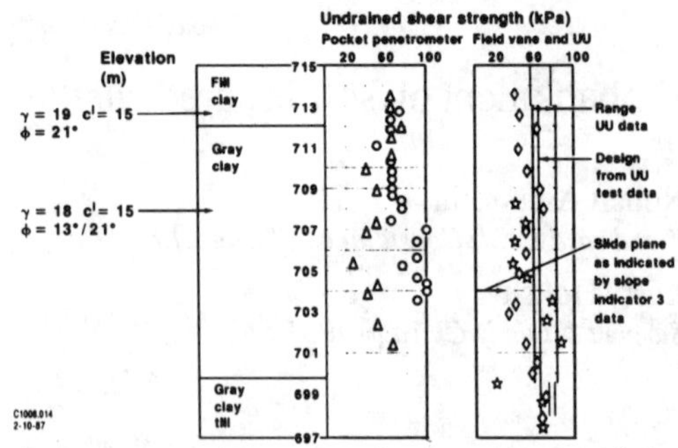

Figure 1. Soil properties profile - high dike section.

Soft near-surface soil was removed and replaced with controlled compacted fill. The removed soils had a pocket penetrometer strength equal to or less than 35 kPa.

The shear strength of the embankment soils was determined from a limited number of triaxial tests on relatively undisturbed thin-wall tube samples of compacted soil. The shear strengths of the foundation soils were determined from in situ vane tests and triaxial tests on selected "undisturbed" samples. The parameters used for the original dike stability assessment are shown in Figure 1. These parameters are based on the original laboratory testing program.

Figure 2 shows the approximate cross section as of November 14, 1983, when a 200-meter-long crack was noted in the upstream side of the embankment, with the crest at elevation 731 meters. The locations of instruments in the vicinity of the distressed area are shown in Figure 3. The crack opening was increasing in width by about 3 millimeters per day. Slope indicator data indicated pronounced movements on a well-defined plane at about elevation 705 meters (refer to Figure 2). The maximum shear strains indicated by the slope indicators were approximately 11%.

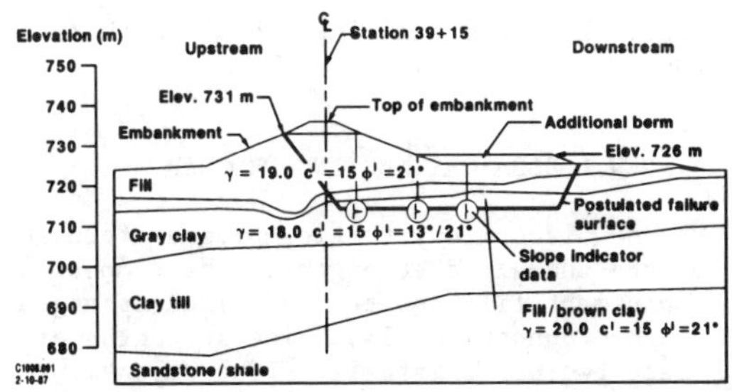
Figure 2. Typical cross section - high dike area (November 14, 1983).

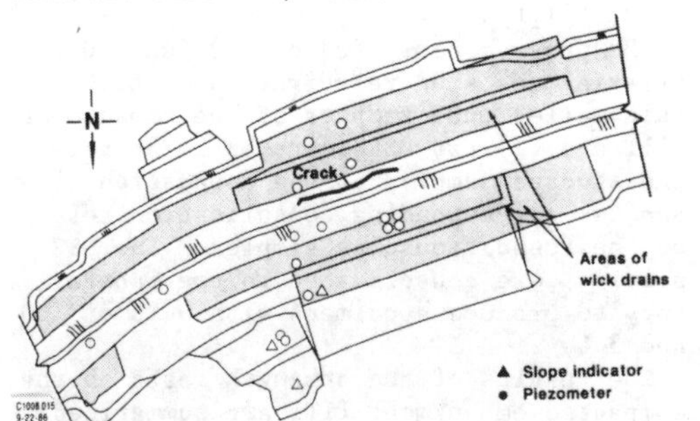
Figure 3. Plan view - high dike.

The downstream berm was raised in response to the increased slope indicator movements. The rate of movement slowed down soon after berm placement and the crack opening effectively ceased. The instruments appeared to respond to construction events about 2 weeks after the event. This was surprising and remains unexplained. Two slope indicators (Nos. 3 and 5) showed continued movements with no signs of reduced rates of movement. Creep appeared to be occurring in these two slope indicators. Typical movements for Nos. 3 and 5 are shown in Figure 4. Typical movements outside the potential slide area were measured and found to be very small.

Piezometers showed little to no pore pressure dissipation since construction began (refer to Figure 5 for typical piezometer measurement). While the dike crest was below about elevation 726 meters (Phase 1), the recorded excess pore pressures were about 50% of the vertical embankment stress. When the dike crest exceeded elevation 724 meters (Phase 2), the changes in pore pressure were about 85% to 100% of the changes in vertical total stress. Piezometers installed in the berm area registered piezometric levels greater than 10 meters above the berm elevation. Piezometer levels in the embankment area were also considerably higher than the embankment grades and remain so at the time of this writing.

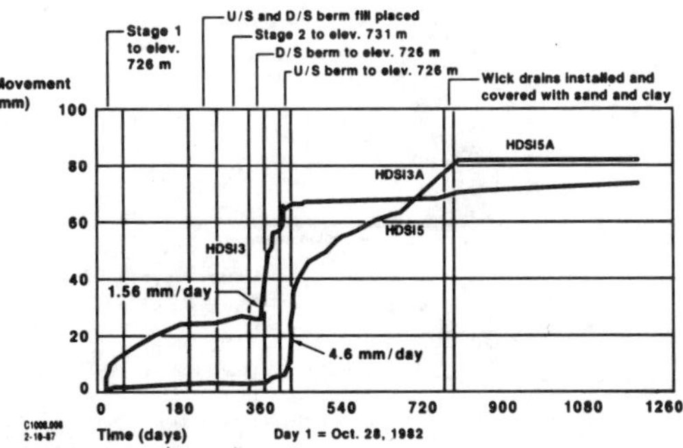
Figure 4. Slope indicator data in area of movement.

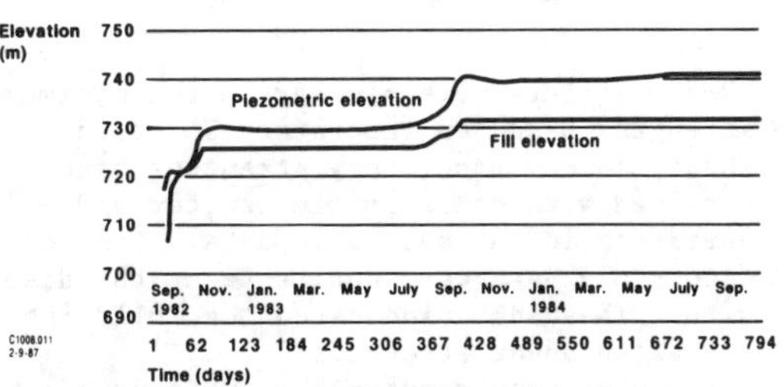
Figure 5. Centerline piezometer data - no dissipation.

The piezometer data indicated that the embankment deformations and cracking were occurring under essentially undrained conditions.

3 EARLIER NEARBY FAILURE

A failure of the nearby SR-770 approach embankment occurred during the initial dike construction. This embankment, located approximately 2 kilometers from the dike site, was approximately 15 meters high at the time of failure. It failed approximately 0.5 meter below its design height.

The writers received a report of the failure (Golder Associates 1982) and compared the apparent soil properties at both sites. Figure 6 shows soil properties based on laboratory tests obtained from Golder Associates (1982). A total stress analysis by Golder Associates (1982), with the assumption that the undrained strength of the clay's crust is approximately 40 kPa, leads to a back-calculated strength of the soft clay of about 20 to 25 kPa. This is in reasonable agreement with the minimum average laboratory unconfined unconsolidated (UU) test results. Figure 1 shows results of laboratory tests that were available at the dike site.

The back-calculated strengths from the

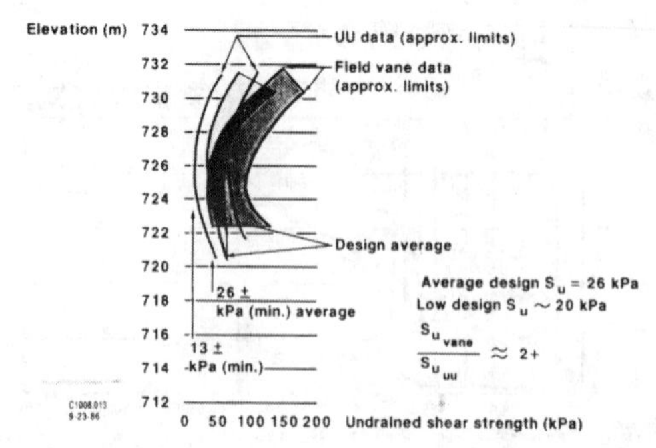

Figure 6. Summary of undrained shear strength - north approach fill failure, SR-770 relocation.

SR-770 failure were similar to the minimum strengths based on laboratory UU tests. These minimum laboratory strengths were obtained with considerable scatter and variation in the available data. Limited laboratory data were available at the dike site. These data indicated the following undrained shear strengths:

- pocket penetrometer: 40 to 60 kPa
- field vane: 35 to 80 kPa
- UU: 55 to 85 kPa

These strengths are similar to those at SR-770, but the minimum values are higher than those at SR-770.

Maximum past pressures estimated from two consolidation tests are shown in Table 1.

Table 1. Maximum past pressures estimated from consolidation tests.

Elevation (meters)	$\bar{\sigma}_{vo}$ (kPa)	$\bar{\sigma}_{vm}$ (kPa)	OCR*	Remarks
712.5	65	350 to 450	5 to 7	Above water table
708.8	115	340 to 530	3 to 5	Below water table

*OCR - Overconsolidation ratio.

After reviewing the failure report for SR-770`and evaluating the limited design soil data, the writers recommended to the owner that additional undisturbed samples be obtained for specialized laboratory testing. Direct-simple shear (DSS) tests were proposed, utilizing the SHANSEP approach developed by Ladd and Foott (1974). The writers were of the opinion that these normalized testing techniques could reduce the scatter and provide a rationale for estimating the strength as a function of depth.

4 NEW LABORATORY TESTING PROGRAM

The new laboratory testing program focused on the undrained strength of the compacted embankment fill and the undrained strength of the foundation clay. The strengths of these two basic materials (the embankment fill and the foundation clay) govern the computed safety factor and the evaluation of stability.

The program consisted of (1) four UU triaxial tests on relatively undisturbed thin-walled tube samples of the embankment fill and (2) two DSS tests on relatively undisturbed samples of the foundation clay and two corresponding consolidation tests on the foundation clay samples. The DSS samples were consolidated in the laboratory to produce specimens with OCRs of 1.0 and 3.0.

The results of the strength tests on the compacted embankment fill are summarized in Figure 7. Also shown on the figure are the probable upper ranges of water content as allowed by specifications and the observed average and standard deviation of the water content of the fill. Based on these data, the writers assumed an "analysis" value of 48 kPa. This value is strongly influenced by the maximum water content allowed by specifications and the average-plus-one-standard-deviation water content. This analysis value is significantly lower than the UU data. Nevertheless, the writers believe that the wide variation in observed water contents and the water contents allowed by the construction compaction specifications suggest a conservative approach.

DSS test results are summarized and compared to similar test data for other clays in Figure 8 (Ladd and Edgers 1972). The ratio of undrained shear

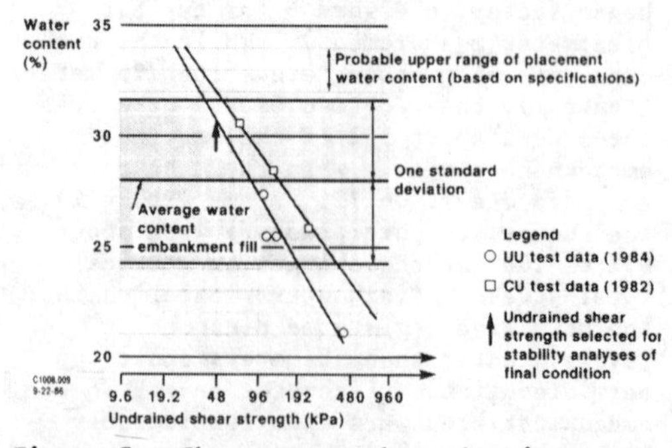

Figure 7. Shear strength evaluation - embankment fill.

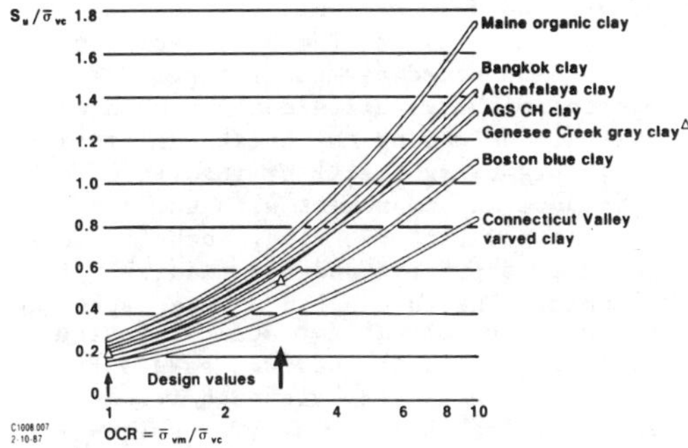

Figure 8. Overconsolidation ratio.

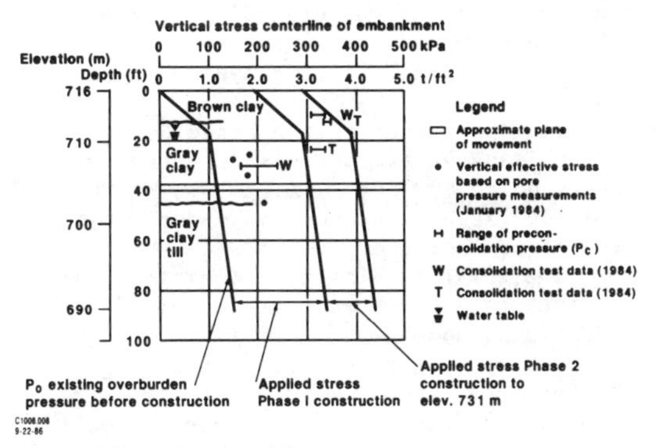

Figure 9. Stress history.

strength to vertical consolidating stress ($S_u/\bar{\sigma}_{vc}$ or c/p) was determined as 0.22 for an artificially induced normally consolidated specimen. A lightly overconsolidated specimen (artificially induced OCR of 3) had a cohesion/pressure (c/p) of 0.56. The effective friction angle for the normally consolidated specimen as indicated by the DSS tests is about 17° to 18°. The shear strains at failure were approximately 12% and 7% for the normally consolidated and overconsolidated specimens, respectively.

The two consolidation tests indicated that the gray clay within the zone of movement appeared to be normally consolidated under the present effective stresses. The upper portions of the gray clay appear to be lightly overconsolidated. The maximum past pressures and present effective stresses are plotted versus depth in Figure 9.

The analysis parameters in Table 2 were selected based primarily on the new laboratory testing program.

Table 2. Analysis parameters.

Material	Unit weight (kN/m^3)	Undrained strengths (kPa)	c/p
Embankment fill	17.7	50	-
Foundation clay	18.6	45*	0.22 (slide plane)

*Based on measured field pore water pressures and c/p from DSS tests with OCR=1. Varied with estimated pore pressures and embankment load.

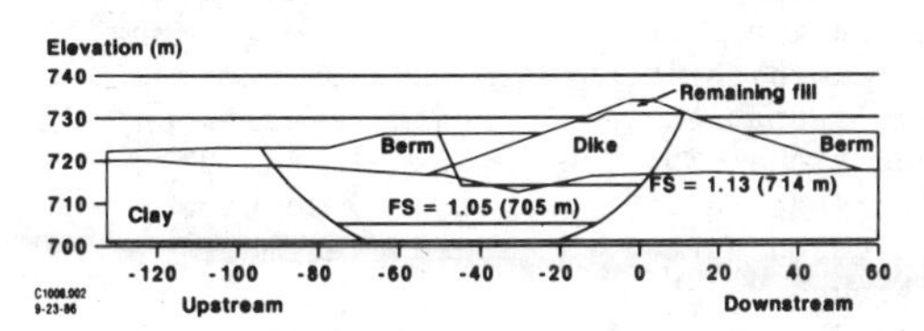

Figure 10. Factor of safety - deep sliding surfaces.

5 ANALYSES

First, analyses were conducted of the conditions at the time the crack developed. Circular analyses using Bishop's (1955) simplified method of slices and Morgenstern and Price (1965) analyses of wedge-type failure on a preferred plane were conducted. The apparent failure geometry and the results of these analyses soon indicated that the wedge-type failure was the more critical of the two. Effective stress analyses with field-measured pore pressures and total stress analyses with undrained strengths were conducted for a "short" and "long" wedge-type failure. The results are summarized in Figure 10 and Table 3.

The writers concluded that the computed

Table 3. Stability analysis results at time of cracking.

Analysis	Failure surface	Computed safety factor
Total stress	short wedge	1.0
Total stress	long wedge	1.1
Effective stress	short wedge	1.0
Effective stress	long wedge	1.3

safety factor of the high dike at the time the crack opened was probably in the range of 1.0 to 1.2. This is based on the following:

• the observance of the crack;

• no massive failure was observed, as occurred at SR-770;

• the large shear strains observed in the slope indicators;

• the shear strains are consistent with laboratory shear strain at failure in the DSS tests; and

• the stability analyses using an "interpreted" embankment shear strength.

Similarly, analyses were made to evaluate the stability of the embankment if it were to be completed to design grade without remedial measures. These analyses produced computed safety factors of less than 1. Therefore, the writers concluded that the likelihood of an embankment failure, similar to the SR-770 failure, would be very high.

The original design criterion for the dikes was a safety factor of 1.4. The writers concurred with this criterion, based on experience, and recommended that the same criterion be maintained to complete construction. Remedial measures would be required to meet the design criterion.

6 EVALUATION OF POTENTIAL REMEDIAL MEASURES

The following primary construction options were available to stabilize the embankment and complete the dike in the next construction season:

• Load the toe (berm) and/or extend berms upstream and downstream using an additional 355,000 m^3 of soil (a 20% increase in dike quantities) to be constructed during the months December through March.

• Provide internal drainage (utilizing 310,000 lineal meters of prefabricated wick drains with surface trenching/drains to lower water levels). The vertical drains, however, were to be predrilled through the existing stabilizing berms containing sandstone boulders and were to remain functional through the Canadian winter.

• Unload the crest.

• Use lightweight fill for the remaining construction.

• Do a combination of the above.

Comprehensive computer-aided analyses indicated that berms would have to be extensive in order to achieve stability. Also, the phased construction program to date had been developed by adding berms in response to field observations of pore pressure and deformation. A crack and the current stability problem had developed in spite of the added berms. The writers had the opinion that a different approach was required. Unloading the crest was impractical. Lightweight fill by itself could not produce an embankment with the required safety factor. Also, such fill was relatively expensive and not readily available. The internal drainage approach appeared to be the most practical (calculations indicated the desired safety factor could be achieved) although it was more expensive than berms. The writers judged the reduction in pore pressures by the drainage option to be significantly more effective than berms in enhancing the stability of the dike. Therefore, the drainage option was recommended to the owner. Berms were selected as a backup contingency plan in the event that the internal drainage could not be achieved before the planned lake filling 9 months later.

The internal drainage option had the following uncertainties:

• What was the appropriate drain spacing to achieve the desired pore pressure reduction within a 9-month time frame? This was a function of drain type, installation procedures, and soil type.

• Will soil properties be reasonably constant during the drainage process, or will the rate of dissipation be nonestimable?

• What effective head should be considered in the analysis of dissipation?

• What type of construction schedule for installing the drains can be expected?

• How would drains perform through the Edmonton winter?

In order to answer these and other construction-related questions, the writers recommended that a wick drain test section be installed.

7 TEST SECTION - WICK DRAINS

Figure 11 shows a plan of the test section in relation to the overall dike. The test section was approximately 14 x 14 meters in plan dimensions with Alidrains spaced on a 2-meter grid. Eight piezometers (pneumatic and vibrating wire-type) were installed to monitor the dissipation of pore pressure. This number of piezometers was deemed necessary as a means of statistically countering the unknown below-grade position of the drains and the piezometers. Three surface settlement observation points were installed. Although settlements were indicated, they were considered nonrepresentative due to the

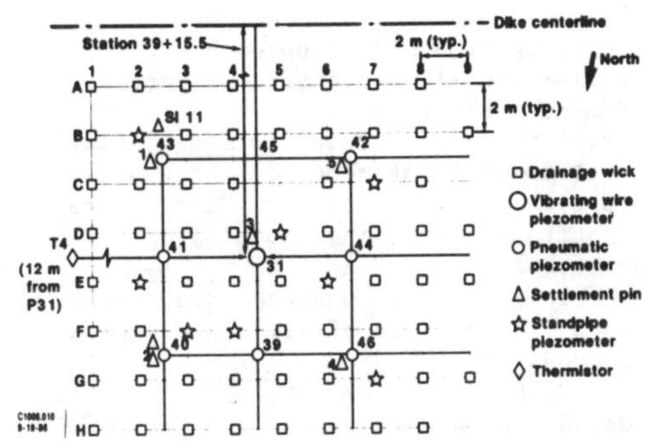

Figure 11. Wick drain test section plan.

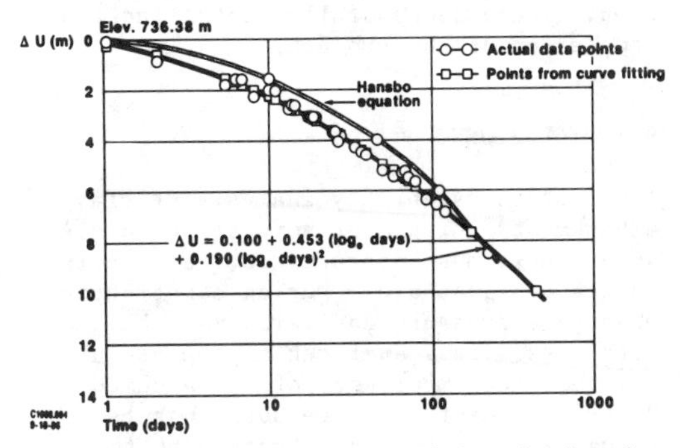

Figure 12. Theory versus field dissipation test section.

effects of ground freezing and construction activities.

The spacing of the drains was selected by using Hansbo's theory (Hansbo 1960; 1981) and the c_y derived from the laboratory consolidation tests. Figure 12 shows the estimated pore pressure dissipation versus time for a 3-month period. Also shown on Figure 12 are the observed pore pressure versus time. The data show that the Hansbo-Lab c_v approach reasonably estimated the pore pressure dissipation at about 3 months. However, at early stages of the dissipation process the rate of dissipation appeared to be significantly faster than the estimated rate. Gradually, the actual rate slowed until the reasonable agreement at 9 months was noted. The gradual slowing of the rate may be caused by consolidation of the clay near the drains, changes in hydrostatic head at the drains, or by clogging of the drains (only clear water was observed at the drainage outlets).

The writers developed an empirical method projecting the rate of consolidation based on the field observations. A polynomial equation was curve fitted to the average pore pressure versus time data over a significant time range (say, 3 months). The equation was used to forecast new data and the forecast was checked against the actual data. The equation was continually adjusted, as necessary, as new data were acquired. This proved to be a reasonable and practical method of forecasting drain performance.

The observed settlement performance was less dramatic than the pore pressure performance and clouded by environmental effects and construction conditions. Nevertheless, the writers concluded that the settlement performance confirmed pore pressure dissipation was occurring.

Based on the test section, wick drain spacings of 2.5 and 2.75 meters were selected to achieve pore pressure dissipation to the berm levels within an 18-month time period. The closer drain spacings were used where the embankment height was greater than 12 meters. The empirical equation-curve fitting was used to monitor performance.

8 PROTOTYPE PERFORMANCE

Typical pore pressure versus time plots compared to estimates based on the test section are shown in Figure 13. In most instances the test section evaluation reasonably forecast the pore pressure dissipation. At some locations, however, slower dissipation than forecast was observed. The reasons for this anomalous behavior are not clearly understood. However, the writers suspect that localized soil conditions along the 1.3 kilometers of dike may be the primary reason for the anomalous behavior.

As of this writing, the embankment has been topped out. No excessive movement nor further cracking have been observed. Criteria for evaluating performance were developed from data and measurements obtained during the failure investiga-

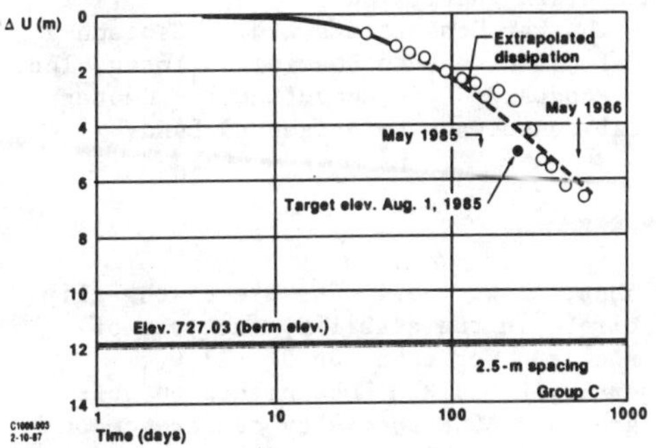

Figure 13. Typical pore pressure dissipation.

tion. These successfully controlled the completion of the project.

9 CONCLUSIONS

Total stress stability analysis of clay embankments built over soft fissured clay subgrade is most appropriate, especially in the design phase. During construction, when pore pressure data are available, effective stress analyses may be useful. However, the importance of the assumed distribution of pore pressure must be recognized. Also, the stress path to failure cannot be considered in an effective stress analysis. The total stress analysis based on an undrained strength automatically accounts for the stress path effects.

The embankment described herein was approaching a failure condition when the crack appeared. Continued filling to design grade likely would have resulted in failure.

The wick drains were effective in stabilizing the embankment. The wicks performed well, but the ability to forecast the rate of pore pressure dissipation during the complete time frame was less than desired. The use of empirical curve-fitting techniques enabled the writers to overcome this shortcoming.

Caution is in order when forecasting long-term pore pressure dissipation on the basis of a relatively short-term test section.

ACKNOWLEDGEMENTS

The writers are indebted to Edmonton Power Co. for permission to publish this paper. The following firms had a role in the project:

- Shawinigan-Integ, Inc. - Engineers for Edmonton Power Co.
- Sargent & Lundy - Consultant to Shawinigan-Integ, Inc.
- Thurber Consultants Ltd. - Geotechnical consultant to Shawinigan-Integ, Inc.
- Woodward-Clyde Consultants - Geotechnical consultant to Sargent & Lundy

REFERENCES

Bishop, A. W. 1955. The use of the slip circle in the stability analysis of slopes. Geotechnique 5:7-17.

Bjerrum, L. 1972. Embankments on soft ground. ASCE specialty conference on performance of earth and earth supported structures, Lafayette, Indiana, 2:1-54.

Golder Associates 1982. Geotechnical investigation north approach fill failure SR-770 relocation, report to Edmonton Power Co.

Hansbo, S. 1960. Consolidation of clay, with special reference to influences of vertical drains. Swedish Geotechnical Institute Proceedings No. 18.

Hansbo, S. 1981. Consolidation by vertical drains. Geotechnique, March 1981, Institution of Civil Engineers, London: 45-66.

Ladd, C. C. and Edgers, L. 1972. Consolidated-undrained direct-simple shear tests on saturated clays. Research Report R72-82, No. 284, Department of Civil Engineering, Massachusetts Institute of Technology, Cambridge.

Ladd, C. C. and Foott, R. 1974. New design procedure for stability of soft clays. Journal Geotechnical Engineering Division ASCE, No. GT 7. 100:763-786.

Massachusetts Institute of Technology 1975. Proceedings of foundation deformation prediction symposium. U.S. Department of Transportation, Report FHWA-RD-75-515, Vol. 1 and 2.

May, R. W. 1978. The geology and geotechnical properties of till and related deposits in the Edmonton, Alberta area. Canadian Geotechnical Journal 15:362-370.

McRoberts, E. C. 1974. The stability of thawing slopes. Canadian Geotechnical Journal 11(4):449-469.

Morgenstern, N. R. and Price, V. E. 1965. The analysis of the stability of general slip surfaces. Geotechnique 15:79-93.

Thomson, S. 1985. A study of delayed failure in a cut slope in stiff clay. Canadian Geotechnical Journal 22:286-295.

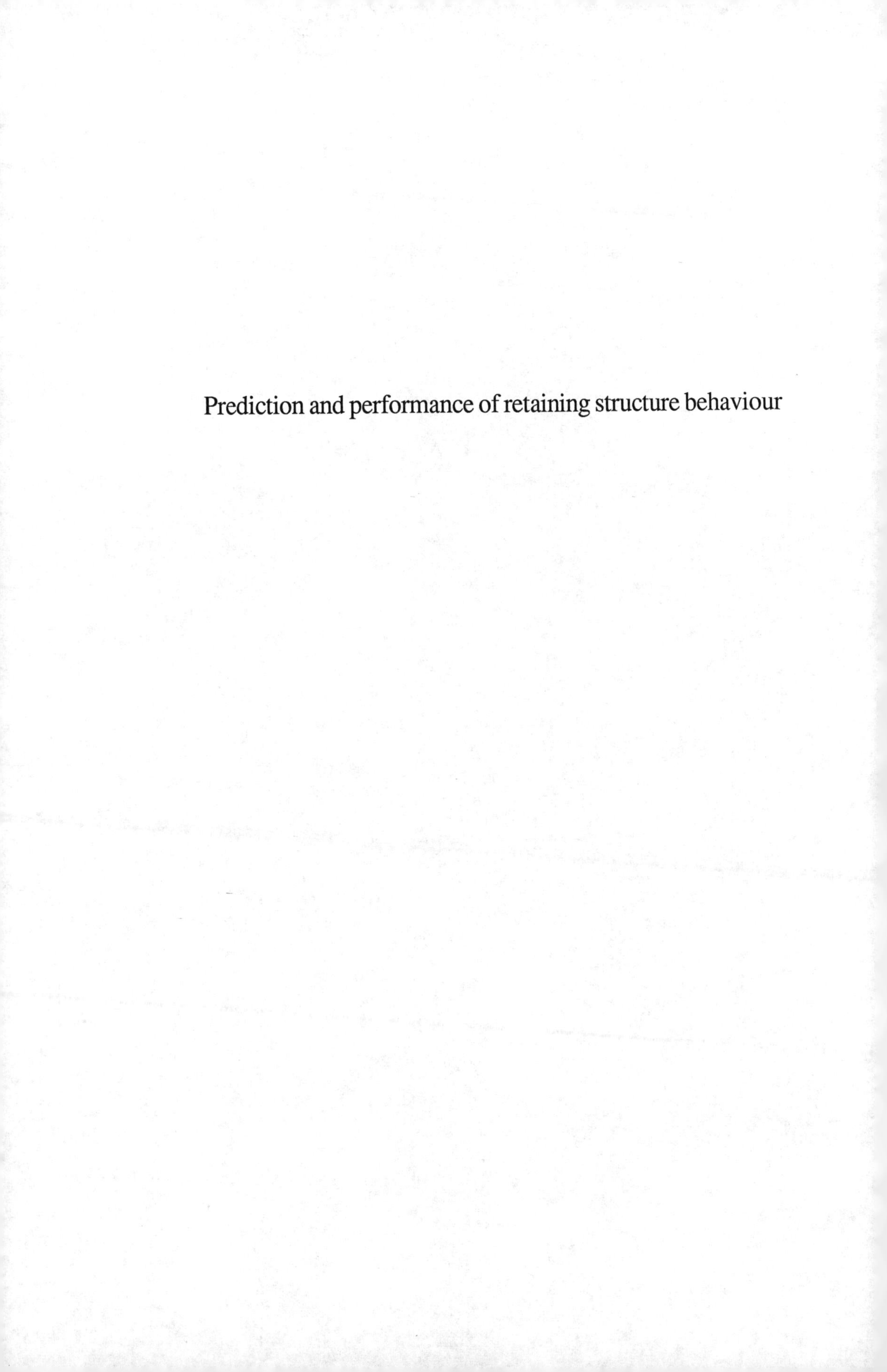

Prediction and performance of retaining structure behaviour

Construction and performance of a permanent earth anchor (tieback) system for the Stanford Linear Collider*

Mark N. Obergfell
Stanford Linear Accelerator Center, Stanford University, Calif., USA

ABSTRACT: The Stanford Linear Collider is the newest addition to the high-energy physics research complex at the Stanford Linear Accelerator Center. One of the many unique features of this project is the large, underground pit, where massive particle detectors will study the collision of subatomic particles. The large, open pit utilizes nearly 600 permanent earth anchors (tiebacks) for the support of the 56 ft (17 m) high walls, and is one of the largest applications of tiebacks for permanent support of a structure. This paper examines the use of tiebacks on this project with emphasis on their installation and performance.

1 INTRODUCTION

The Stanford Linear Collider (SLC) is a new, high-energy physics project located at the Stanford Linear Accelerator Center (SLAC), about thirty miles south of San Francisco on the peninsula between San Francisco Bay and the Pacific Ocean. SLAC is a high-energy physics research center operated by Stanford University for the United States Department of Energy (DOE). SLAC is dedicated to basic, fundamental, particle physics research. The main component of SLAC is the two-mile (3.2 km) long linear accelerator [housed in a concrete tunnel 25 ft (7.6 m) underground], which creates high energy electron and positron beams. The Collider will take the two beams from the accelerator, bend them around two tunnel arcs and collide the beams head on in the experimental hall pit. The two-mile accelerator, along with the pit excavation (at bottom of photo), is shown in Figure 1. In the pit massive, 3300 ton detectors will study the subatomic particles resulting from the collision.

In general, tiebacks have gained widespread acceptance for use in temporary earth-retaining structures. They eliminate the need for large, open excavations with sloped sides, or the need for internally braced sheeting. For large, deep excavations, tiebacks become economically feasible. Relative to conventional cantilever retaining walls, tieback walls do not need large footings and the costly over-excavation and backfill. The use of tiebacks has become popular for tight construction sites in urban areas where space is limited by adjacent structures, and in subway, bridge abutment, and highway retaining wall construction. Their use in temporary structures, and more recently in permanent structures, has been documented by Weatherby (1982) and Anderson (1984).

2 DESIGN

2.1 Preliminary Studies

During preliminary feasibility studies for this project, several schemes for the pit were considered. The pit itself is 233 ft (71 m) long, 65 ft (20 m) wide, and 56 ft (17 m) deep. A pit of this depth with the requirements for clear floor space presented a few problems in the

Figure 1. Aerial Photograph of SLAC site

*Work supported by the Department of Energy, contract DE-AC03-76SF00515.

design of the high retaining walls. The depth of the pit was dictated by primarily two factors: 1) the tunnel had to be at least 40 ft (12 m) below ground at an adjoining park, and 2) the tunnel could have a slope of no more than 10%. Therefore, the pit had to be deep underground to intercept the two tunnel arcs. Conventional cantilever wall construction would have required walls 4 to 6 ft thick (1.2 to 1.8 m) in addition to the over excavation and temporary retaining walls with soldier piles and tiebacks. At that point, it was a logical move to consider permanent tiebacks for this project.

When the tieback system was first proposed, it was met with a fair amount of skepticism and heavy questioning. This was due in part to the relative newness of the technique and limited experience in the United States for permanent wall construction as an integral part of a building. It should be noted that a fair amount of questioning came from the physics and research community and not from the engineering and construction community. Permanency of the tiebacks and their ability to hold the loads without excessive movement were the primary topics of discussion. After numerous meetings, contacts with industry representatives and with consultants in Europe familiar with tiebacks, the system was finally approved for design and construction. In addition to conservative engineering design practices, several other items were implemented to assure the adequacy of the system. The anchors were to be double-corrosion protected, and an extensive testing and monitoring program was to be undertaken. These measures are more typical of tiebacks in cohesive soils, rather than the competent sandstone at this site. Other measures will be discussed later.

2.2 Details of Project

To my knowledge, this is the largest permanent building tieback installation in North America. The pit, as shown in Figure 2 at completion of construction, consists of the central area, intercepted by the two tunnel arcs, where the particle collisions will be observed, and the east and west garage areas where the huge particle detectors will be assembled and serviced. In the central pit area, the large struts between the two walls were an extra measure of protection to insure that the walls would not move inward and pinch the removable concrete shielding walls that would be installed later. Around the pit, there are large rim beams designed to support the weight of the heavy detector parts as they are moved into place and lowered into the pit, and to support the 39 inches (1 m) thick concrete shielding planks. To avoid placing a surcharge on the walls, the rim beams, as well as the main structural steel for the high bay area, are supported on belled caissons which extend to a depth below the pit slab.

Figure 2. Pit at completion of construction

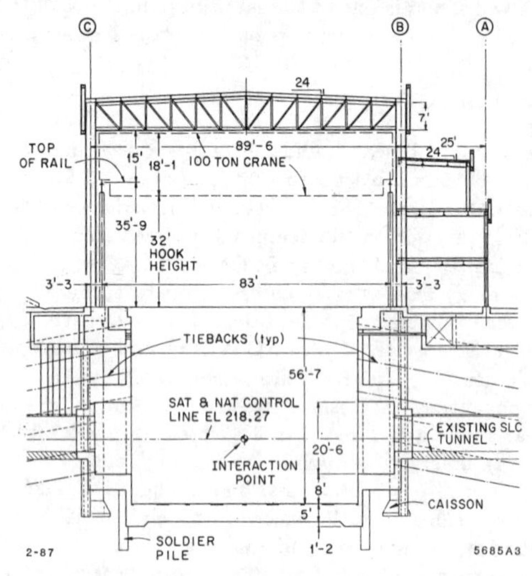

Figure 3. Cross–Section of Pit

There are a total of 584 tiebacks on the four pit walls. They are installed on soldier piles spaced at 5 to 8 ft (1.5 to 2.4 m) on center, with a total of seven tiebacks on each soldier pile. The soldier piles consist of double steel channels embedded in concrete. The walls are conventionally reinforced, cast-in-place concrete, 15 inches (38 cm) thick. A cross-section through the pit (showing the tunnels, tiebacks, and high bay steel structure above) is shown in Figure 3.

The anchors are double-corrosion protected, 1–3/8 inches (3.5 cm) in diameter made of ASTM A722 steel. The corrosion protection system consists of grout encased in a corrugated PVC sheath. This entire assembly was installed as a unit in the drilled hole, then grouted in place. The anchor length of the tiebacks is 25 ft (7.6 m). The predominant soil

Table 1. Engineering properties and geotechnical design parameters for miocene sandstone at the site.

Design parameter	Sandstone type: Uncemented to weakly-cemented	Sandstone type: Moderate to well-cemented
Effective cohesion C'	0 psi	16 psi
Effective friction angle ϕ'	35°	42°
Dynamic effective friction angle ϕ'_E	38°	46°
Density	125 pcf (19.6 kN/m^3)	130 pcf (20.4 kN/m^3)
Coefficient of active earth pressure K_A	.27	.20
Dynamic coefficient of active earth pressure K_{AE}	.24	.16
Coefficient of lateral earth pressure at rest K_o	.43	.33
Compressional wave velocity at low strain level V_p	1928 fps (588 m/s) (top 35 ft) (10.6 m)	3370 fps (1027 m/s) (35 ft–90 ft) (10.6 m–27.4 m)
Shear wave velocity at low strain level V_s	1225 fps (373 m/s) (top 35 ft) (10.6 m)	1952 fps (595 m/s) (35 ft–90 ft) (10.6 m–27.4 m)

at the site is weakly to well cemented miocene sandstone; as such, it was an ideal candidate for permanent tiebacks.

2.3 Testing Program

In order to determine the suitability of tiebacks in the intended location and to determine design parameters, a field testing program was undertaken. A total of five tiebacks were installed at the project site in the initial open cut excavation. The first prototype failed miserably, and this once again led to questioning of the entire system. Upon closer examination, it was determined that the drilling procedure was to blame; water was used for the drilling and this caused a thin layer of mud to form along the hole. Subsequent prototypes gave the desired results after the drilling procedure was modified. Two of the tests, using the final configuration of the tieback, yielded a load of 189 kips (840 kN) before failure. This correlated to a bond stress of approximately 33 psi (228 kPa). After the load testing, two of the tiebacks were locked off at the design load and monitored for creep for a period of four months. When projected out, this showed a loss of less than 10% in the initial load over a period of 25 years.

2.4 Design Parameters

In addition to the precautions taken with the tiebacks themselves, design loads were conservatively determined, and some of the beneficial effects of cemented sands were neglected. The walls and anchors were designed for static and seismic earth pressures. The static pressures are active pressures using a combined rectangular and triangular pressure distribution. The walls were designed for a major earthquake resulting from the San Andreas Fault, which is located approximately 3 miles (4.8 km) to the west of the site. The design earthquake is expected to produce horizontal ground accelerations of six-tenths gravity (0.6 g). The soil pressure diagrams and the design parameters are shown in Figure 4 and Table 1 respectively. The basis for the design criteria was the active failure wedge (treating the material as a soil), which was also used in part to determine the unbonded lengths of the anchors. The earthquake pressures were developed using a pseudostatic force developed from the active wedge, which was then uniformly distributed over the wall (Tudor 1984a). The combined static and seismic loads resulted in a total working design load of 120 kips (534 kN) for a typical tieback.

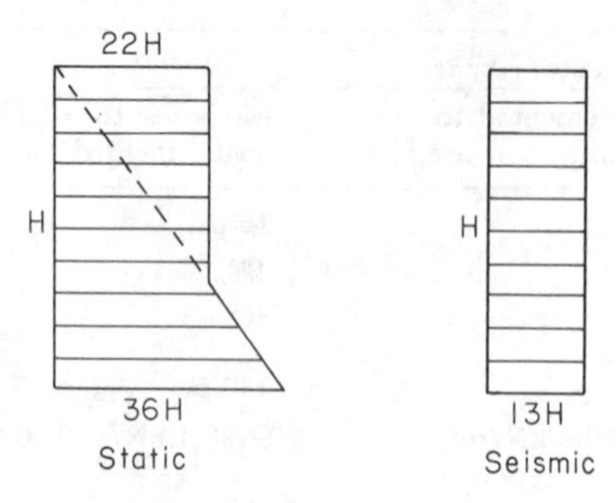

Note: Above distributions assume that groundwater is below bottom of excavation or that material is free-draining
H = vertical height of wall in feet
$\gamma_{total} = 130$ pcf
$\phi' = 42°$ (Static)
$\phi' = 46°$ (Dynamic)

2-87 5685A5

Figure 4. Earth Pressure Diagrams

The anchor capacity was estimated using the following equation:

$$P = \pi DLQ$$

where

P = allowable anchor load ,

D = anchor diameter ,

L = anchor (bond) length ,

Q = bond stress .

The bond stress was conservatively estimated at 5000 psf (239 kPa) for soft, sedimentary rock.

In addition to the above criteria, an overall stability analysis was performed assuming that the tiebacks tie the soil together into a rigid mass. In other words, they prestress the soil so that the failure wedge is prevented from developing Appropriate factors of safety were considered for tieback stress, failure of the active wedge, as well as creep of the anchors. The design also allowed for any one anchor to fail, except the top anchor, without causing overall failure. The walls themselves were designed using working stress, rather than ultimate strength procedures for concrete.

3 CONSTRUCTION

Construction of the two tunnel arcs began in the Fall of 1983, using two road-header tunneling machines purchased from Germany for use on the project. A small portion of the work was done by conventional cut and cover methods in areas where the tunnel was at a shallow depth. At the same time as the tunnel work, a contract was let to begin the initial open cut excavation at the site of the Collider Hall and pit. This involved removing the large amount of overburden down to the future grade level of the building. The actual work for the building and pit began in the Fall of 1984 as the piles and caissons were drilled and placed, and the excavation for the pit started.

Figure 5. Pit Excavation

At shallow depths, excavation was accomplished using a hydraulic excavator with the soil removed via dump trucks and a ramp out of the pit. At greater depths, the soil was removed with a clamshell bucket. Wire mesh was installed between soldier piles to prevent sloughing and to catch loose debris. No wood lagging was used.

The tiebacks were installed in layers as the excavation proceeded. The general procedure was to excavate, install tiebacks and grout, and then proceed down to excavation of the next level when the anchor grout had gained sufficient strength. The holes were drilled using an 8 inch (20 cm) diameter, hollow-stem auger mounted on a tracked rig, with the auger modified to scarify the hole. The excavation and the drilling rig are shown in Figure 5. The tunnel has just been uncovered in the center of the photo, while the tiebacks and soldier piles are visible around the perimeter. Note the drilling rig and the stratification of the sandstone. With this type of set-up used by the contractor, including the use of no drilling water, the anchors developed bond stresses nearly twice that determined from the testing program. A pull-out test done by the contractor in the better sandstone at the pit yielded a bond stress of 65 psi (448 kPa). Once the hole was drilled, the tiebacks were installed in the hole and grouted. Next, the tiebacks were proof-tested to 133% of the design load, and then locked off. A small percentage of the tiebacks were also subjected to performance tests. Concrete wall construction began once the excavation was completed and the base slab was poured.

4 PERFORMANCE AND MONITORING

4.1 System Description

An extensive monitoring program is underway to measure the performance of the tiebacks. Load cells have been installed on 67 tiebacks at various locations on the four walls, and are currently being monitored. During the early monitoring stages, ongoing construction activities limited access to the pit. It was difficult to get access to certain areas due to the large amount of work going on in such cramped quarters. Once construction was complete, the onslaught of the equipment installation further hampered monitoring. It took many reminders to prevent the installation crews from permanently blocking access to the tiebacks with pipes, conduit, and cable trays. Once construction and equipment installation was completed, monitoring could be carried out on a regular basis.

As a part of the construction contract, the contractor was required to perform lift-off tests and any necessary re-tensioning after completion of the excavation, immediately before pouring the concrete walls, and just before completion of the contract (Tudor 1984b). These measures were employed to address the anticipated creep and the adjustment of the soil due to excavation and imposed loads. In addition to the load cells, extensometers were placed on each wall to act as benchmarks for detecting and mapping movements of the walls. These extensometers consist of steel rods anchored into the sandstone 10 ft (3 m) beyond the end of the tiebacks.

4.2 Construction Monitoring

Shortly after the excavation intercepted and passed the previously bored tunnel, cracks began to appear in the fiber-reinforced, shotcrete tunnel lining. There were cracks in both the north and south tunnels, but the ones in the south side were larger and more numerous. Extensometers were installed in the south tunnel to monitor the movement of the cracks. Maximum recorded movements were 0.18 inches (4.6 mm), with the average approximately 0.08 inches (2 mm). Movement stabilized by the time excavation was completed, but it did cause some concern. The cracking and movement was undoubtedly due to the dip of the geological bedding in towards the pit at the south side, and the fact that the tiebacks directly above the tunnel had not been installed or stressed until the excavation was below the level of the tunnel. This was necessitated because the tiebacks had to be installed almost horizontally so they would not intercept the tunnel. The contractors rig had to be at a lower level in order to drill nearly horizontal (the typical slope of a tieback is 20° from the horizontal).

Lift-off tests and load cell monitoring during construction found that loads had increased on the order of 5% since initial installation and lock-off. The largest increases were seen in the east wall and in the easterly half of the south wall. An independent geotechnical consultant's review attributed the increase to the surcharge affect of the 2:1 slope rising to the east and south of the site. The slope rises on an average of 50 ft (15 m) in elevation above the top of the pit, and is setback approximately 30 ft (9 m) from the edge of the pit. The north and west sides of the site are fairly level and little increase was seen in tieback loads in these areas except for the northwest corner of the pit. Contrary to earlier expectations of creep problems, there appeared to be none.

Although there was some concern, it was decided to wait and continue to monitor the load cells. The unfavorable surcharge, combined with the relaxation and readjustment of the earth due to the excavation and overburden removal, were the prime suspects. It was the general consensus that as activities subsided and equilibrium was reached, the load increases would stabilize and the tiebacks would perform as designed.

4.3 Post-Construction Performance

After construction was completed and the facility was turned over to SLAC, a reading was made on all load cells once again. The readings showed a continuing increase in tieback stresses. Some tiebacks were approaching the manufacturer's recommended maximum lock-off load of 165 kips (734 kN). Figures 6–8 show typical load-time histories for the tiebacks. The tiebacks were initially installed and locked-off at 140 kips (623 kN) to allow for some creep and resultant decrease in anchor tension on the assumption that the stress would decrease to the design of 120 kips. Again, the largest increases were seen in the south wall. There was immediate concern, and after several meetings with the consulting engineers and geotechnical engineer, it was decided to embark on a program of reducing all tiebacks on the south, east, and west walls to $\approx$ 125 kips (556 kN). Note from Figure 7 that the east wall could not be reset at the same time due to the large detector blocking access. The north wall however was not seeing such a rapid rate of increase. Loads had held fairly steady at the 140 kip level, and it was decided to leave the north wall alone. Creep was definitely not a problem and the factor of safety in the original design was being encroached upon by the ever increasing loads. Although construction activities had subsided, the tiebacks and earth were still adjusting to the new conditions. Installation activities, such as the rolling-in of the large detector and other heavy equipment use around the pit probably contributed to the changes still being experienced.

Since the program of resetting the tiebacks in the summer of 1986, the load increases have levelled out. Installation activities have subsided as the facility is about ready to begin operation. Readings taken in January, 1987, have shown no significant increases. In the south wall, loads have held steady at or near the 125 kip (556 kN) lock-off load. The north wall has experienced no change whatsoever; readings are identical

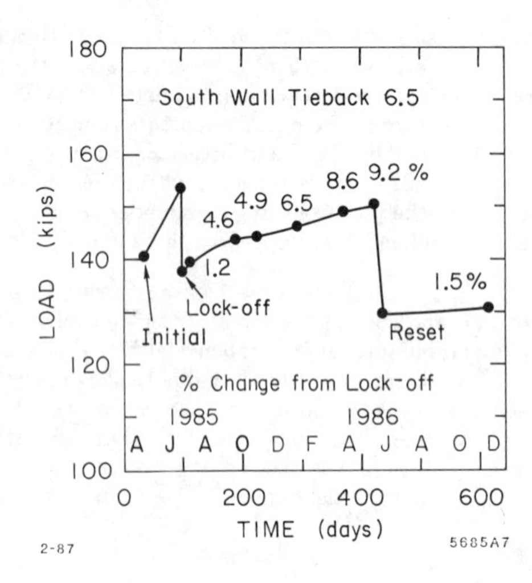

Figure 6. Typical South Wall Tieback

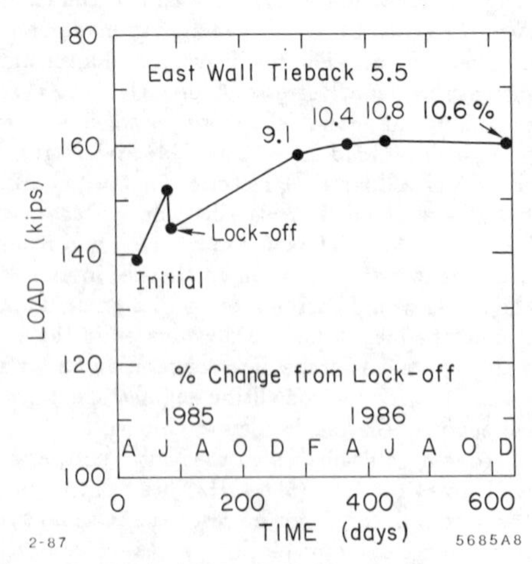

Figure 7. Typical East Wall Tieback

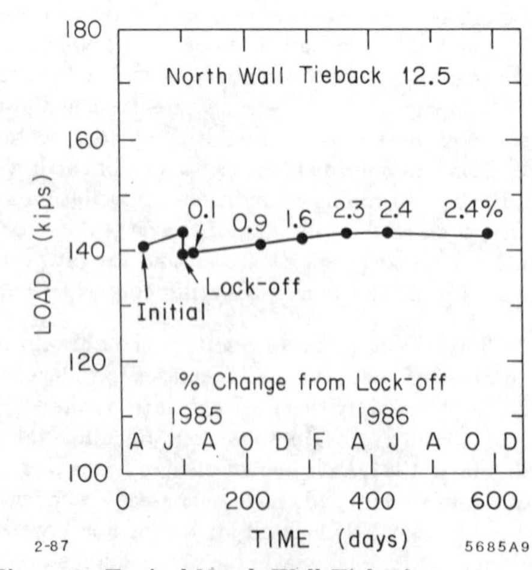

Figure 8. Typical North Wall Tieback

to those taken in the summer of 1986. The east and west walls have stabilized also. To this date, the extensometers in the walls have not been used to map any movements. There have been no indications of movement in the walls and there have been absolutely no cracks or other signs of distress. Implementation of a surveying system to map the relative position of the wall and to detect movements on the order of 1/16 inch (1.6 mm) would be expensive, not to mention the clutter of piping and other equipment that make lines of sight difficult. Since the loads have stabilized, confidence in the system of tiebacks for permanent use has been restored. There have been no moderate or major earthquakes to give the system a true design test, although this will likely happen during the life of the project. Monitoring will be continued on a regular basis; eventually there will be a complete and thorough record of the performance of this unique engineering and construction accomplishment.

5 CONCLUSIONS

• The initial selling of the tieback system for permanent construction to a non-engineering (but highly scientific and educated) community was a difficult task due to the limited amount of information on the system for permanent use.

• Corrosion concerns and long-term performance characteristics were the primary questions that had to be addressed in the initial design.

• Marked differences were seen between field prototypes and the actual installed anchors. Higher bond stresses were attained by the contractor, and creep proved to be no problem.

• The removal of the overburden, the sloping surcharge, and the dip of the geological bedding played significant roles in the changes seen in the tieback stresses. Equilibrium was not reached as soon as expected. Load increases were not expected.

• Recent performance has been favorable. Loads have stabilized and the system is performing as intended.

• The use of tiebacks resulted in a substantial cost savings, estimated to be approximately $1,600,000 (U.S.) by Mueller (1984).

• SLAC is a pioneer in the physics and research field with many important discoveries. It was appropriate therefore, that engineering and construction should play an important role in making this new project possible. It is hoped that the experience and data collected on the tieback system will be helpful to others on future projects.

ACKNOWLEDGEMENTS:

Special thanks to Jon Y. Kaneshiro of Earth Sciences Associates, Palo Alto, California. This paper is based on the presentation given by he and the author entitled: Site Conditions Which Influenced the Design And Construction of Permanent Tieback Walls for the SLAC Linear Collider Project—A Case History, presented at the Association of Engineering Geologists 29th Annual Meeting, San Francisco, October 5-10, 1986

- Consulting Engineer: Tudor Engineering Company, San Francisco, California
- Geotechnical Engineer: Earth Sciences Associates, Palo Alto, California
- General Contractor: Dickman-Nourse, Inc., Mountain View, California
- Tieback Subcontractor: DBM Contractors, Inc., Federal Way, Washington

REFERENCES:

Anderson, T.C. 1982. Earth–Retention Systems—Temporary and Permanent. Minneapolis: Proceedings of 32nd Annual Soil Mechanics and Foundation Engineering Conference.

Mueller, H. 1984. Preliminary Comparative Design Study on Detector Pit Retaining Structures (letter to SLAC under Subcontract 515–S–924.) San Francisco: Tudor Engineering Company.

Tudor Engineering Company 1984. Collider Experimental Hall—Title I Design Report for Stanford Linear Accelerator Center. Prepared under subcontract 515–S–924 (includes geotechnical report prepared by Earth Sciences Associates). San Francisco.

Tudor Engineering Company 1984. Project Manual for the Construction of the SLAC Linear Collider—Collider Experimenal Hall, Subcontract 515–S–1101. San Francisco.

Weatherby, D.E. 1982. Tiebacks. U.S. Department of Transportation, Federal Highway Administration Report No. FHWA/RD–82/047.

Prediction and Performance in Geotechnical Engineering / Calgary / 17-19 June 1987

Performance of twin-tee test wall for Interstate Highway 70 near Glenwood Canyon, Colorado

Tzong H.Wu
University of Colorado, Denver, USA

Nelson N.S.Chou
Colorado Department of Highways, Denver, USA

ABSTRACT: A study was undertaken to investigate the performance of a test wall for Interstate Highway 70 near Glenwood Canyon, Colorado. The test wall is composed of ten panels of post-tensioned face using precast double-tee sections on cast-in-place concrete base. The test site topography consists of a hillside slope of relatively stable talus covered by grass and patches of oak brush. In order to place the base of the counterfort wall, existing slope has to be excavated. The operation is difficult and costly. The primary purpose of the study is to examine the effect of the base length on the deformations and stability of the wall.

The test wall was instrumented with 20 survey points, 12 soil pressure cells along the bottom of the base, 6 manometers and 3 inclinometers. Upon backfilling to the wall height, 18 Jersey barriers were used to surcharge the wall.

Analytical study was carried out by the finite element method. The stress-strain-strength characteristics of the soil was represented by a nonlinear stress-dependent constitutive law. The analyses were performed by an incremental technique to simulate the actual construction sequence including the application of the surcharge loads.

Due to composition of the talus material, it was very difficult to obtain representative soil properties. As a result, field measured data were employed to calibrate soil parameters by the finite element analysis. The parameters were then used to predict the performance of the retaining wall under reduced base length condition. The study provided a reliable basis for redesigning the retaining structure.

1 INTRODUCTION

Counterfort twin-tee wall was used extensively on Interstate Highway 70 in Glenwood Canyon, Colorado. It composed of a precast, twin-tee section panel for the face and a cast-in-place concrete slab for the base, as shown in Figure 1. The wall was constructed by excavation of on-site slope and placement of the base slab, followed with attachment of the face panel and backfilling of the road cut.

A 100 ft long test section of the twin-tee wall consisting of 10 panels was instrumented with soil pressure cells, manometers, inclinometers, and surveying stations for measurement of deflection and settlement of the wall. After the backfill reached the top of the wall panel, additional surcharge was applied to load-test the wall. The measurement data indicated that the wall, even during the load test, experienced very small displacement. The site topography consists of a hillside slope of relatively stable talus covered by grass and patches of oak brush. If the wall can perform satisfactorily with reduced base length, the saving will be very significant, since the operation of excavating on-site slope for placement of the base slab is difficult and costly.

The study described herein aims at predicting the performance of the test wall under reduced base length condition. The analysis was performed by the finite element method. A finite element computer code, SSTIP, developed at the University of California, Berkeley was used for the analysis. The finite element analysis procedures used in this study have been successfully applied to a number of soil engineering problems including calculation of stresses and movements in embankments and excavated

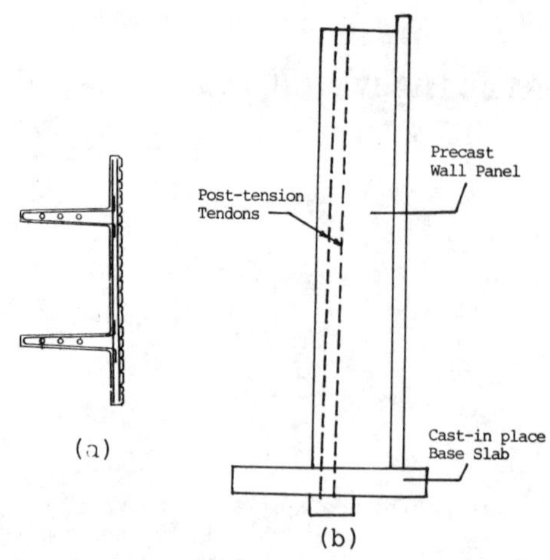

Figure 1. (a) Typical Cross Section of Twin-tee Panel, (b) Elevation View of Typical Twin-tee Wall

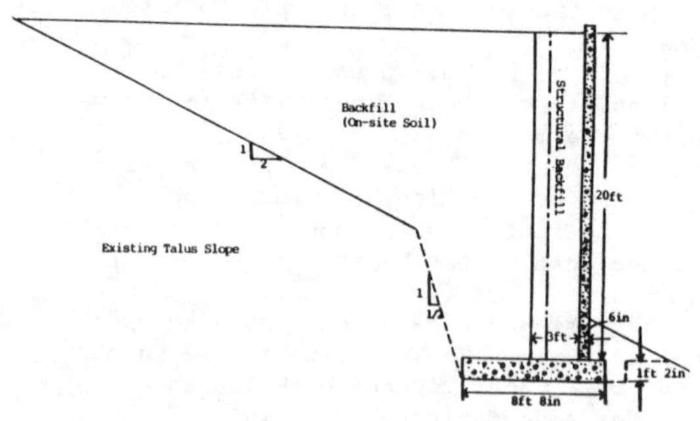

Figure 2. Cross Section of the Test Wall

slops, earth loads on U-frame lock structures, earth pressures on buried structures and anchored sheet pile walls.

Due to the composition of the on-site talus material, it was very difficult to obtain the soil properties from laboratory tests. Consequently, a two-phase study was conducted. The first phase study involved using the field measurement data to calibrate the most representative soil parameters. In the second phase, the soil parameters were employed to assess the stability and displacements of the wall under reduced base length condition.

2 SITE CONDITION AND THE TEST WALL

The test wall is located near the west end of Glenwood Canyon section of Interstate Highway 70, about 150 miles west of Denver, Colorado. The wall is situated between the Colorado River and a series of quartzite step cliffs. The intervening hillside is covered by grass and patches of oak brush. Geologic investigations indicate that subsurface material consists of 10 ft. to 20 ft. of moderately dense to dense talus material. The talus consists primarily of angular fragments of quartzite ranging from gravel to boulder size, intermixed with silt and sand-size particles, and contains areas of empty or very loosely filled interconnected voids up to several inches in diameter. The talus is underlain by denser deposits of river sands and gravels at depths ranging from about 30 ft. to 40 ft. No groundwater was observed during drilling operations.

The test wall is composed of ten panels of twin-tee counterfort wall. Each of the panels is 10 feet wide and 20 feet high. The base slab is 1 ft. 2 in. thick and 8 ft. 8 in. long. Figure 2 illustrate a typical cross-section of the test wall. On-site soils were used as backfill except for the region immediately behind the wall where filter material was used. The walls were instrumented with 20 survey points, 12 soil pressure cells, 6 manometers and 3 inclinometers. The layout of the instrumentations is illustrated in Figure 3.

3 FIELD MEASUREMENTS

Upon close examination of the field measurement data, it was concluded that the data from the soil pressure cells placed along the bottom of the base slab and the vertical displacements of the wall obtained from a number of survey points might be used for the back analysis. These data were synthesized and presented in Figure 4 and Table 1.

Figures 4(a), (b), and (c) show the measured vertical pressure as a function of the fill height at the front edge, at 2.5 feet from the front edge, and at 5 feet from the front edge, respectively. Except for Panel No. 5 (see Figure 3), the pressure in each of the wall panels across the same section did not vary significantly. These plots of pressures at the three sections thus constituted very useful "bounds" for the subsequent back calculations.

The record of the vertical displacements at four survey stations is summarized in Table 1. Also shown in the table are the changes in the vertical displacements after the panel is in-

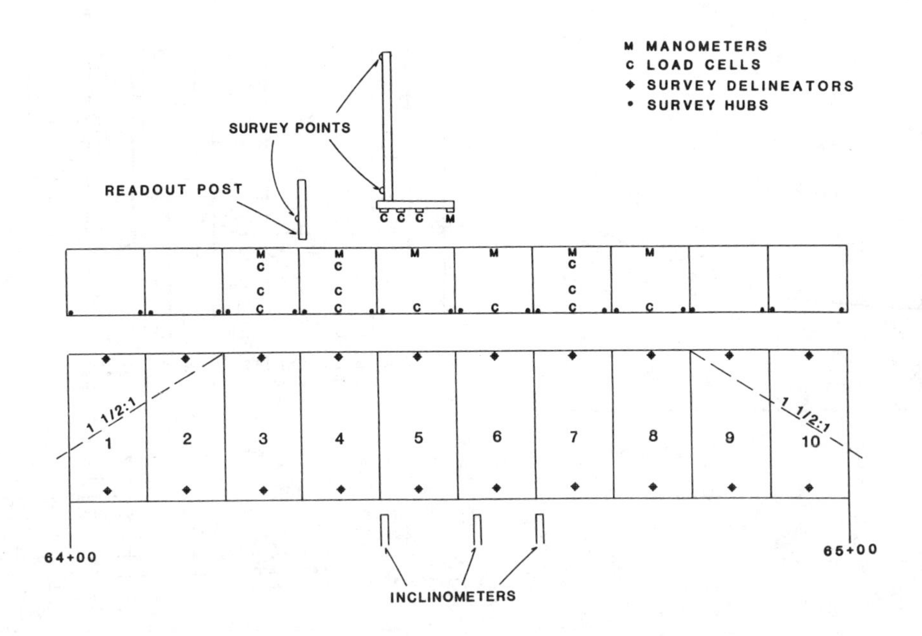

Figure 3. Layout of the Field Instrumentations.

Table 1. Measured Vertical Wall Displacements

Dates (1982)	Fill Height (ft.)	Vertical Displacments (ft.) 64+31*	64+41*	64+51*	64+61*
7.9	1.0	58.09	58.17	58.24	58.25
7-12	1.0	58.10	58.18	58.25	58.26
7-13	1.0	58.10	58.16	58.25	58.26
7-14	1.0	58.10	58.18	58.25	58.26
7-23	1.0 (panel set)	58.11	58.20	58.27	58.27
7-26	1.0	58.10 (0.01)	58.19 (0.01)	58.26 (0.01)	58.27 (0.01)
7-28	1.0	58.10 (0.01)	58.18 (0.02)	58.26 (0.01)	58.27 (0.01)
8-3	3.0	58.11 (0.00)	58.19 (0.01)	58.27 (0.00)	58.28 (0.00)
8-4	3.5	58.08 (0.03)	58.17 (0.03)	58.23 (0.04)	58.25 (0.03)
8-5	7.5	58.09 (0.02)	58.16 (0.04)	58.24 (0.03)	58.25 (0.03)
8-6	10	58.10 (0.01)	58.18 (0.02)	58.25 (0.02)	58.26 (0.02)
8-9	17	58.08 (0.03)	58.16 (0.04)	58.23 (0.04)	58.26 (0.02)
8-10	20	58.07 (0.04)	58.25 (0.05)	58.23 (0.04)	58.23 (0.05)

* Survey station number
** Net change in vertical displacement after panel is in-place

place, δ_y. The values of δ_y up to 20 feet of fill height are in the range of 0 to 0.5 inch, with strong fluctuations as the fill height increases. These data indicated that the displacements were very small. The actual values, however, could not be used directly for the back calculations, since they were within the order of possible survey error.

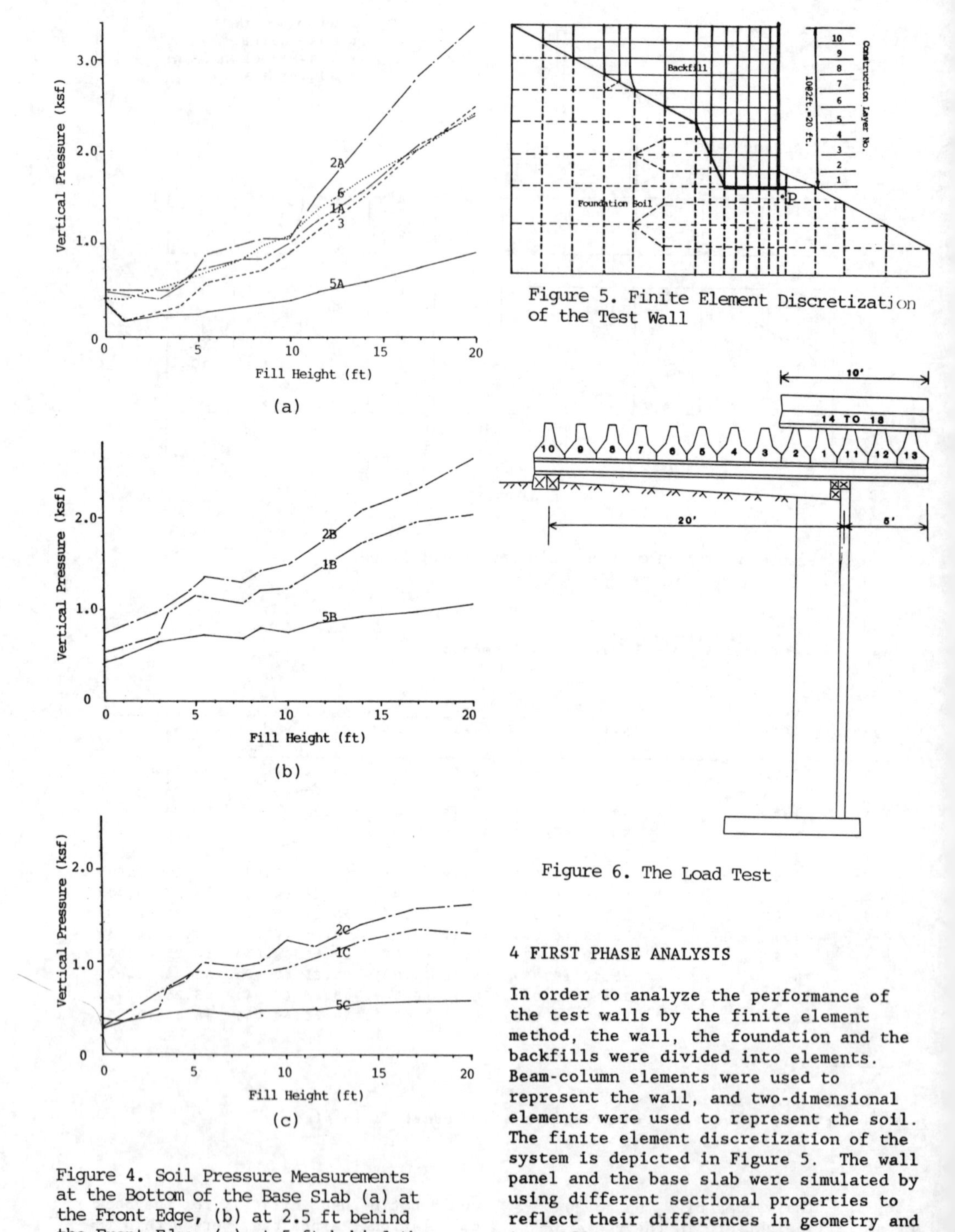

Figure 4. Soil Pressure Measurements at the Bottom of the Base Slab (a) at the Front Edge, (b) at 2.5 ft behind the Front Edge, (c) at 5 ft behind the Front Edge.

Figure 5. Finite Element Discretization of the Test Wall

Figure 6. The Load Test

4 FIRST PHASE ANALYSIS

In order to analyze the performance of the test walls by the finite element method, the wall, the foundation and the backfills were divided into elements. Beam-column elements were used to represent the wall, and two-dimensional elements were used to represent the soil. The finite element discretization of the system is depicted in Figure 5. The wall panel and the base slab were simulated by using different sectional properties to reflect their differences in geometry and in stiffness. Due to the lack of information about the relative stiffness/ strength of the soil, the foundation and backfills were considered the same soil type in the analyses.

The analyses were carried out by an incremental technique to account for the actual construction sequence. The foundation was treated as a pre-existing part. The retaining wall was first placed, succeeded by soil backfilling. The backfilling was performed in 2 foot lifts. It took a total of 10 construction lifts to reach the top of the wall, at 20 feet above the base slab.

The stress-strain behavior of the soil was simulated by using the Duncan-Chang Model (Duncan and Chang, 1970). To simplify the calibration, the Poisson's ratio was taken as a constant, rather than a stress-dependent parameter, in the analyses. The most representative, yet conservative, soil parameters were determined as $\gamma = 120$ pcf, $\mu = 0.3$, $K = 2500$, $n = 0.7$, $R_f = 0.85$, $c = 0.05$ ksf, $\Phi = 42^o$. Using the parameters, the displacements of the wall were on the order of 0 to 1 inch, and the soil pressures along the bottom of the base slab during backfilling agreed well with the upper bound measured values depicted in Figure 4.

5 LOAD TEST

After the fill height reached the top of the wall the backfilling continued for an additional 6 feet. Subsequently, the fill above the top of the wall was removed and Jersey barriers were used as surcharge to load-test the wall. There were a total of 18 barriers used. The sequence of barrier placement is shown in Figure 6, indicated by the numbers on the barriers.

A finite element analysis, using the soil parameters deduced from the first phase analysis, was performed to confirm the result of the calibration study. The analysis was carried out in three steps: step 1 involved placement of barriers no. 1 to 10, step 2 involved barrier no. 11 to 13, while barriers no. 14 to 18 were simulated as step 3. The idealized loading sequence is illustrated in Figure 7.

Figure 8 shows a comparison between the measured changes in the vertical pressure at the center of the front edge during the load test and the results of the finite element analysis. The agreement was very good. This confirmed the fact that the soil parameters determined in the calibration study were good representation of the on-site soil.

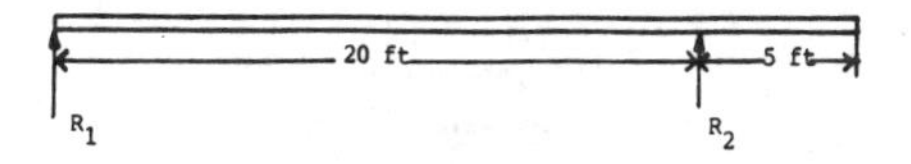

STEP 1 (Barriers 1 to 10)

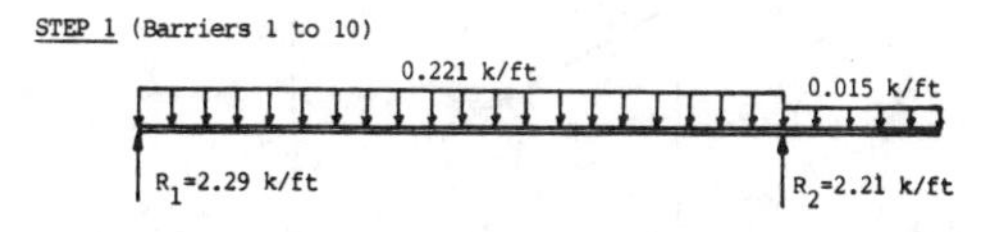

STEP 2 (Barriers 11 to 13)

STEP 3 (Barriers 14 to 18)

Figure 7. Idealized Loading Sequence of the Load Test

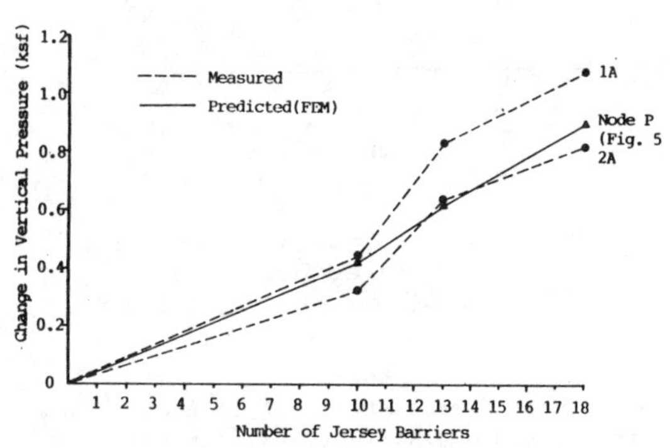

Figure 8. Measured versus Predicted Soil Pressure at the Front Edge of the Wall during the Load Test

6 SECOND PHASE ANALYSIS

In the second phase analysis, the length of the base slab was reduced from 8 ft. 8 in. to 7 ft. 8 in. and to 6 ft. 8 in. The analysis procedures were the same as before. The soil parameters deduced from the first phase analysis were employed.

The predicted displacements of the walls at the end of the load test are shown in Figure 9. The displacements were not significantly different from those of 8 ft. 8 in. base length.

Figure 10 shows the lateral and vertical pressure distributions at the end of the

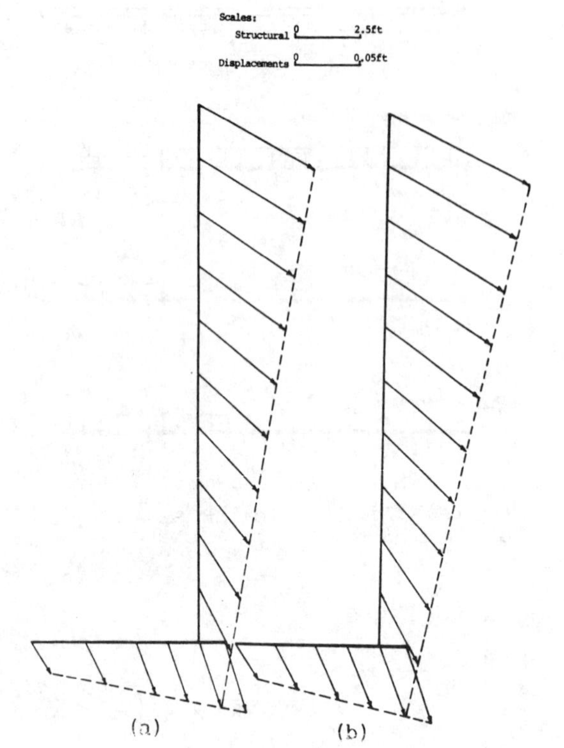

Figure 9. Predicted Displacements at the End of the Load Test for Base Lengths of (a) 7ft 8in, (b) 6ft 8in.

load test for the two base lengths. From the pressure distributions, the resultant forces as well as their points of action can be computed, and the stability of the walls against sliding and overturning can be evaluated. The results of the second phase analysis are summarized in Table 2. In comparison with the generally accepted minimum factor of safety of 1.5, the factors of safety of the walls with reduced base length were well above the criterion.

7 CONCLUSIONS

The results of this study suggest the following conclusions:

1. The field measurement data of wall displacements and soil pressure along the bottom of the base slab during backfilling provide invaluable information for back analysis of the on site talus material properties which are very difficult to determine in laboratory.

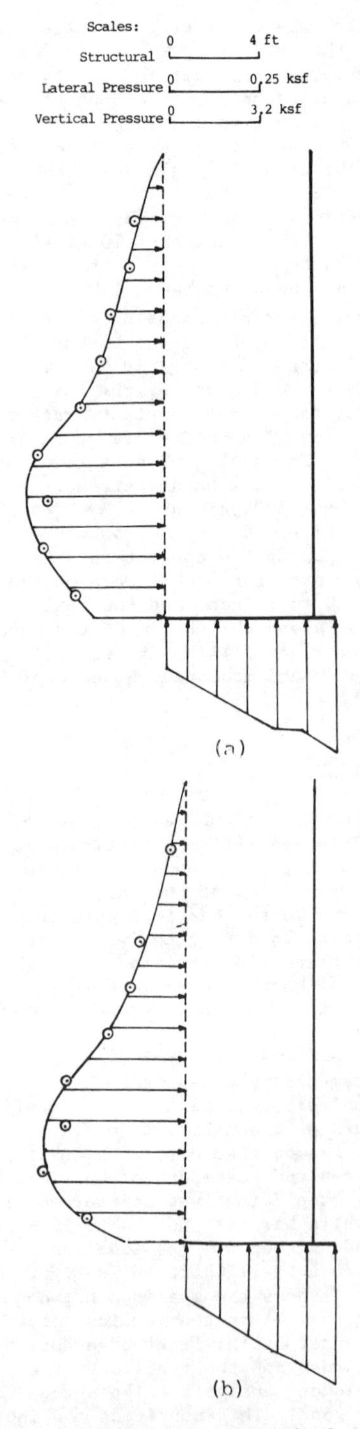

Figure 10. Lateral and Vertical Pressure Distributions at the End of the Load Test for base Lengths of (a) 7 ft 8 in, (b) 6ft 8in.

Table 2. Factors of Safety and Maximum Wall Displacements of the Twin-Tee Test Wall

End of the Load Test		Base Length		
		8'8"	7'8"	6'8"
Disp. At The Top of The Wall	X-Disp (in.)	1.1	1.1	1.3
	Y-Disp (in.)	0.58	0.60	0.63
Fs (Sliding)		3.0~3.2	2.6~2.8	2.4~2.6
Fs (Overturning)		2.7~2.9	2.0~2.2	1.7~1.9

2. A comparison of measured versus predicted soil pressure at the bottom of the base slab during the load test indicate that the soil parameters deduced from finite element back analysis are representative of the on-site material.

3. Using the finite element analysis procedure with the calibrated soil parameters, the factors of safety of the twin-tee wall against sliding and overturning are predicted to be higher than the generally accepted value when the base length is reduced by up to 2 ft. The wall displacements with the reduced base length are only slightly higher than that with the designed base length.

8 REFERENCE

Duncan, J.M., and C.Y. Chang: Non-linear Analysis of Stress and Strain in Soils, J. Soil Mech. Found. Div. ASCE, vol. 96, no. SM5, pp. 1629-1653, September 1970.

9 ACKNOWLEDGEMENT

The study described in this paper was supported by a research contract from the Colorado Department of Highways. The writers gratefully acknowledge the assistance of Robert K. Barrett and John B. Gilmore of CDOH.

Prediction and Performance in Geotechnical Engineering / Calgary / 17-19 June 1987

Prediction of time-dependent movement of tiebacks in cohesive soils

R.Y.K.Liang
University of Akron, Ohio, USA

ABSTRACT: A finite element program which incorporates a time-dependent constitutive model for clays has been used to predict the long-term tieback deformations and load-transfer behaviors measured in a full-scale tieback testing program. The results of comparisons indicate that the numerical program is capable of modeling soil tieback interactions at low load-level. At higher load-level, a contact element is needed in the numerical program to simulate the "slippage" at soil-grout interfaces. The overall numerical predictions of cumulative deformation vs. load, creep rates vs. time, and distributions of strain along the tendon are satisfactory.

1 INTRODUCTION

One particular concern with tiebacks anchored in cohesive soils is their significant time-dependent movements. The prediction of such movement is important as the tiebacks are often designed to permit as little yield as possible to safeguard the vital operation of adjacent structures. Furthermore, continued tieback movement may soften the clay over a long period of time and lead to creep rupture. Presently, the common practice is to design the tieback so that the rate of creep is low enough that the preload capacity will not be reduced below the design load. A performance test which includes a creep test is often conducted to establish the creep characteristics of tiebacks. However, a creep test is extremely difficult and costly to perform due to the fact that a constant load must be maintained and the measured movement is relatively small. On the other hand, a well verified numerical technique can be efficiently used to simulate the time-dependent soil-anchor interaction, thus providing necessary design parameters. This paper presents a finite element technique for predicting the time-dependent movement of a tiebach anchored in cohesive soil.

2 TIME-DEPENDENT TIEBACK MOVEMENT

The time-dependent tieback movements may be caused by (1) creep of the steel tendon, (2) bond degradation at the tendon-grout interface, (3) grout creep, (4) grout cracking, (5) temperature-induced movements, and (6) movements in the soil and at the grout-soil interface (Ludwig, 1984). Among these contributors, the movements in the soil and at the grout-soil interface are the most significant factors. The finite element program should therefore possess the capability to at least model these two aspects of tieback behavior in order to accurately predict the time-dependent tieback movements.

3 CONSTITUTIVE MODEL FOR COHESIVE SOILS

The constitutive model used in the present finite element program is a bounding surface plasticity model with a consideration of both hydrodynamic consolidation and creep effects (Liang, et al, 1987). The model is a modification of the conventional, elasto-plastic formulation originally developed by Borja and Kavazanjian (1985).

The present model divides the total deformations into immediate and delayed components, as suggested by Bjerrum (1967). The immediate deformations refer to those caused by the change in effective stresses; whereas, the delayed deformations refer to those induced by the viscous response of clays. Notice that because of hydrodynamic lag, the change of effective stress and, thus, immediate deformation in a fully saturated clay deposit is a time-dependent process.

The immediate elastic deformations are described by an hyperbolic formulation. The immediate plastic deformations are obtained by the consistency requirement on the bounding surface and the assumed associative flow rule. The shape of the bounding surface adopted in the model is similar to the one proposed by Dafalias and Herrmann (1980).

The delayed deformations are obtained from the Singh-Mitchell creep equation (Singh and Mitchell, 1968) and the assumed associative flow rule. The "aging" or "quasi-preconsolidation" effect under sustained effective stress is accounted for by expanding the bounding surface with time. For further descriptions of the constitutive equations, readers should refer to Borja (1984).

A finite element program that incorporates the described constitutive model has been developed for solving two-dimensional boundary value problems. The program includes libraries of isoparametric elements and different material models. A predictor-corrector type nonlinear solution scheme is used in the program.

The general validity of the constitutive model and the accompanied finite element program has been established through a series of evaluation process, which includes (1) "single" element type comparison with laboratory test results, (2) centrifuge test evaluation, and (3) a case study of an embankment on soft clays (Borja, 1984; Liang, 1985;and Kavazanjian, et al, 1985).

4 TIEBACK TESTING

A full-scale load testing program of tiebacks in cohesive soils has been conducted at Washington state. The reported test results (Zipper, et al, 1985) were used as a basis for evaluating the numerical model's predictability.

The general arrangement of tiebacks is shown in Fig. 1. Basically the tieback holes were drilled by an auger, with a nominal diameter of 12 inches (305 mm). The 1-3/8 (35 mm) diameter Dywidag threadbars were used as the tieback bars. The anchor of the tieback was formed by pea-gravel grouts which were pumped into the hole by a tremie hose. The remaining portion of the holes was backfilled with a bentonite-sand mixture.

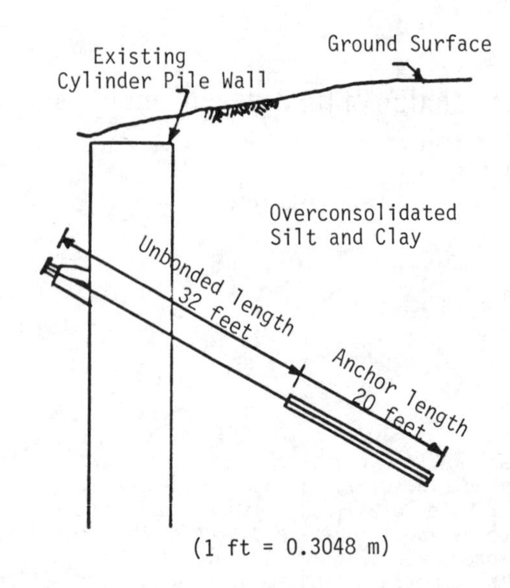

Fig. 1 The Arrangement of Tiebacks

The test site soil deposits, as reported by Zipper, et al (1985), consist of overconsolidated silts and clays. No water table is observed in the test site. The basic soil properties are: water content = 36%, LL = 56, PL = 29, LI = 0.3. The unconsolidated-undrained strength of the soil is in the neighborhood of 3900 psf (186.7 kPa).

Since no detailed information was available on other soil properties, the soil parameters needed for input in the finite element analysis were interpreted from other soil's data which exhibited similar basic index properties (Febres-Cordero and Mesri, 1974). These soil parameters are: recompression index, $\kappa = 0.05$; virgin compression index, $C_c = 0.413$; hyperbolic parameters, $a = 0.005$, $b = 1.8$, $R_f = 0.92$; secondary compression index, $\psi = 0.0026$; Singh-Mitchell creep parameters, $A = 4.4 \times 10^{-5}$, $m = 0.942$, $\bar{\alpha} = 3.2$; the slope of critical state line, $M = 0.8$; permeability, $k_{xx} = k_{yy} = 1.02 \times 10^{-4}$ cm/sec.

5 NUMERICAL PREDICTION

Fig. 2 shows the finite element mesh and boundary conditions used to simulate the tieback. The reaction wall was assumed to be rigid; also, the wall friction was not considered in the analysis.

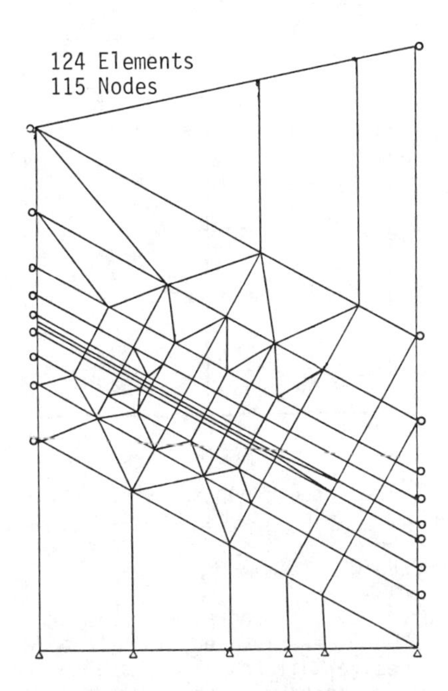

Fig. 2 Finite Element Mesh Used in Numerical Analysis

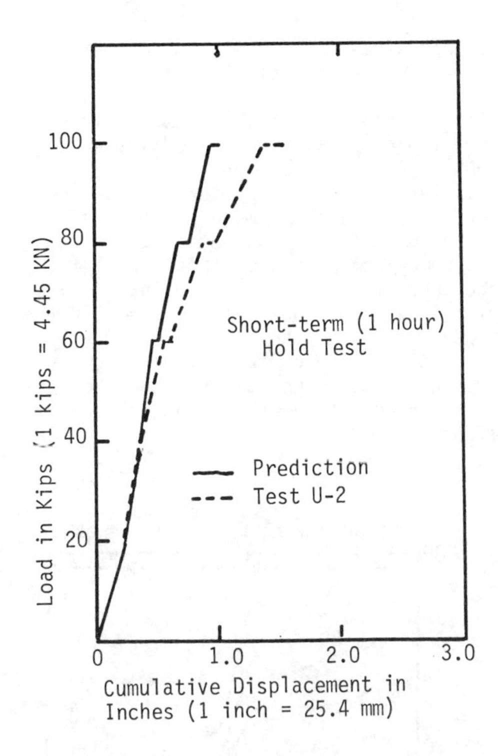

Fig. 3 Comparison of Predicted and Measured Load-Displacement Curve

In the short-term creep test, each anchor was loaded incrementally to pull-out and each load level was held for up to an hour. Fig. 3 shows the comparison between predicted and measured load-displacement curve. It can be seen that in the load level below one-half of the pull-out load, the numerical program predicts quite well both immediate amd delayed tieback movements. At higher stress level, the numerical program underpredicts the deformation, however. This discrepancy may be caused by the fact that the numerical program, in its present form, does not incorporate a contact element to account for the "slippage" between soil and anchor.

Fig. 4 shows the comparison of creep rates during the short-term creep test. Although the numerical model tends to underpredict the creep rate to some extent, the overall prediction is quite satisfactory, considering that the model only considers the soil creep while neglecting other contributors mentioned in section 2.

The predicted distribution of strain in the tendon is shown in Fig. 5. There is a significant amount of decrease in strain along the unbonded length. This decrease may be attributed to two facts: the presence of friction along the unbonded length and the presence of bending in the tendon. No significant time-dependent increase in strain in the unbonded length was predicted by the model. However, a significant amount of time-dependent dissipation of friction was observed by Ludwig (1984) in his full-scale tieback testing.

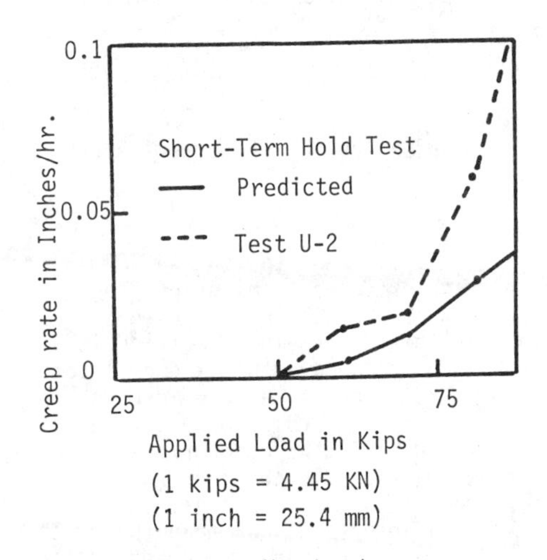

Fig. 4 Creep Rates Vs. Load

Further study on this subject is needed.
The distribution of strain in the anchor length of the tendon is fairly linear, which is contradictory to most field observations (Ludwig, 1984; Ludwig, et al, 1985). Furthermore, the progressive load transfer mechanism described by Ludwig for explaining the time-dependent deformation of tiebacks was not predicted by the numerical program. This inability of the model to predict the strain distribution (therefore the load transfer) in the anchor length of the tendon is largely due to the lack of a contact element in the program. The on-going research is directed at investigating the "contact element" effects.

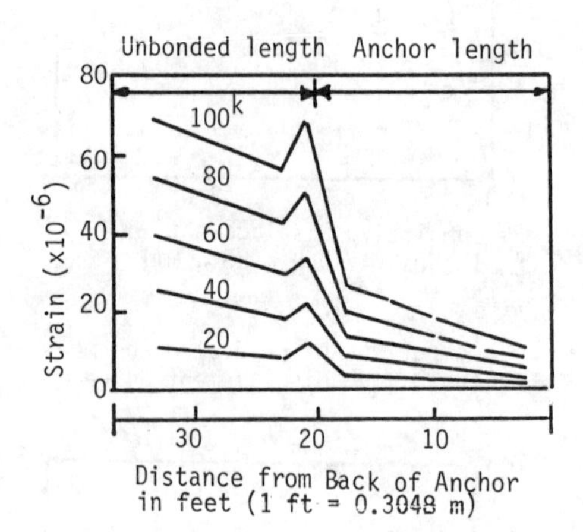

Fig. 5 Distribution of Strains Along the Tendon

The long-term creep monitoring program is still in progress. The measured creep rate in the first 7-months period is in the range of 0.004 inches/log cycle of time. Fig. 6 shows the predicted creep rates up to 10 years. Two observations can be made: (1) the creep rates decrease with time, and (2) the creep rates measured during short-term creep test (Fig.4) are significantly higher than those measured during the long-term creep tests. Thus, the current practice of extrapolating the short-term creep test results to predict the long-term creep deformations seems to be very conservative.

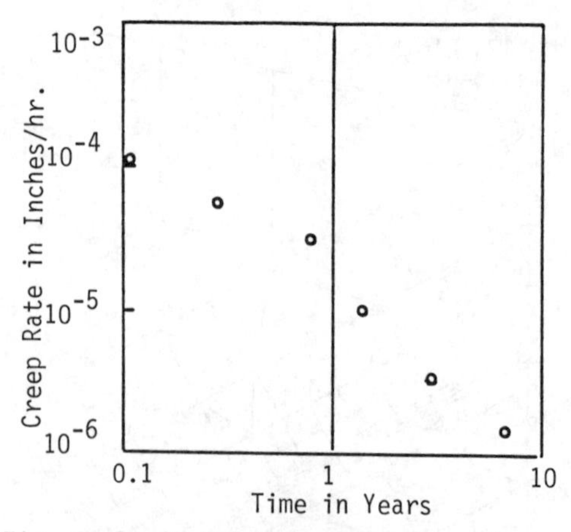

Fig. 6 Predicted Creep Rates Vs. Time (1 inch = 25.4mm)

6 CONCLUSION

A finite element program which incorporates a well verified constitutive model for cohesive soils has been used to predict the long-term tieback deformations and load-transfer behaviors measured in a full-scale tieback testing program. Based on the study presented, the following conclusions can be made.

(1) A contact element is needed in the numerical program in order to accurately predict the load-deformation curve at higher load levels.
(2) A significant amount of friction develops at the unbonded length of the tendon. Interpretations of tieback test results should properly take this friction loss into consideration.
(3) The numerical prediction indicates that creep rates decrease with time.
(4) The creep rates measured in the short-term creep tests (up to 1-hour load hold) are significantly higher than those measured in the long-term creep tests. It seems very conservative to extrapolate short-term creep rates to obtain long-term deformations.

REFERENCES

Bjerrum, L. 1967. Engineering geology of normally consolidated marine clays as related to settlements of buildings. Geotechnique. 17.2:82-118.
Borja, R.I.. 1984. Finite element analysis of the time-dependent behavior of soft clays. Ph.D Thesis, Stanford, University.

Borja, R.I. & Kavazanjian, Jr. E. 1985. A constitutive model for the stress-strain-time behavior of wet clays. Geotechnique 35.3: 283-298

Dafalias, Y.F. & Herrmann, L.R. 1980. A generalized bounding surface constitutive model for clays. Proc. Limit Equilibrium, Plasticity and Generalized Stress-Strain Applications in Geotechnical Eng. 78-95.

Febres-Cordero, E. & Mesri, G. 1974. Influence of testing conditions on creep behavior of clay. Report No. FRA-ORD & D-75-29.

Kavazanjian, Jr. E, Borja, R.I. & Jong, H.L. 1985. Time-dependent deformation in clay soils. 11th ICSMFE. 535-538.

Liang, R.Y.K. 1985. Centrifuge evaluation of a constitutive model for clay. Ph.D Thesis, University of California, Berkeley.

Liang, R.Y.K., Sobhanie, M. & Timmerman,D. 1987. A bounding surface plasticity model for the stress-strain-time behavior of clays. 2nd Int. Conf. Constitutive Laws for Eng. Materials. 699-706.

Ludwig, H. 1984. Short-term and long-term behavior of tiebacks anchored in clay. Ph.D Thesis, McGill University.

Ludwig, H., Weatherby, D.E. & Schnabel,H. 1985. Research on tiebacks anchored in cohesive soils. 11th ICSMFE. 1721-1724.

Singh, A. & Mitchell, J.K. 1968. Generalized stress-strain-time function for soils. J. Soil Mech. Found. Eng.ASCE. 94. SM1:21-46.

Zipper, J.E., Horvitz, G.E. & Fuglevand, P.F. 1985. Ground anchor creep in glacially consolidated clay. 11th ICSMFE. 2145-2148.

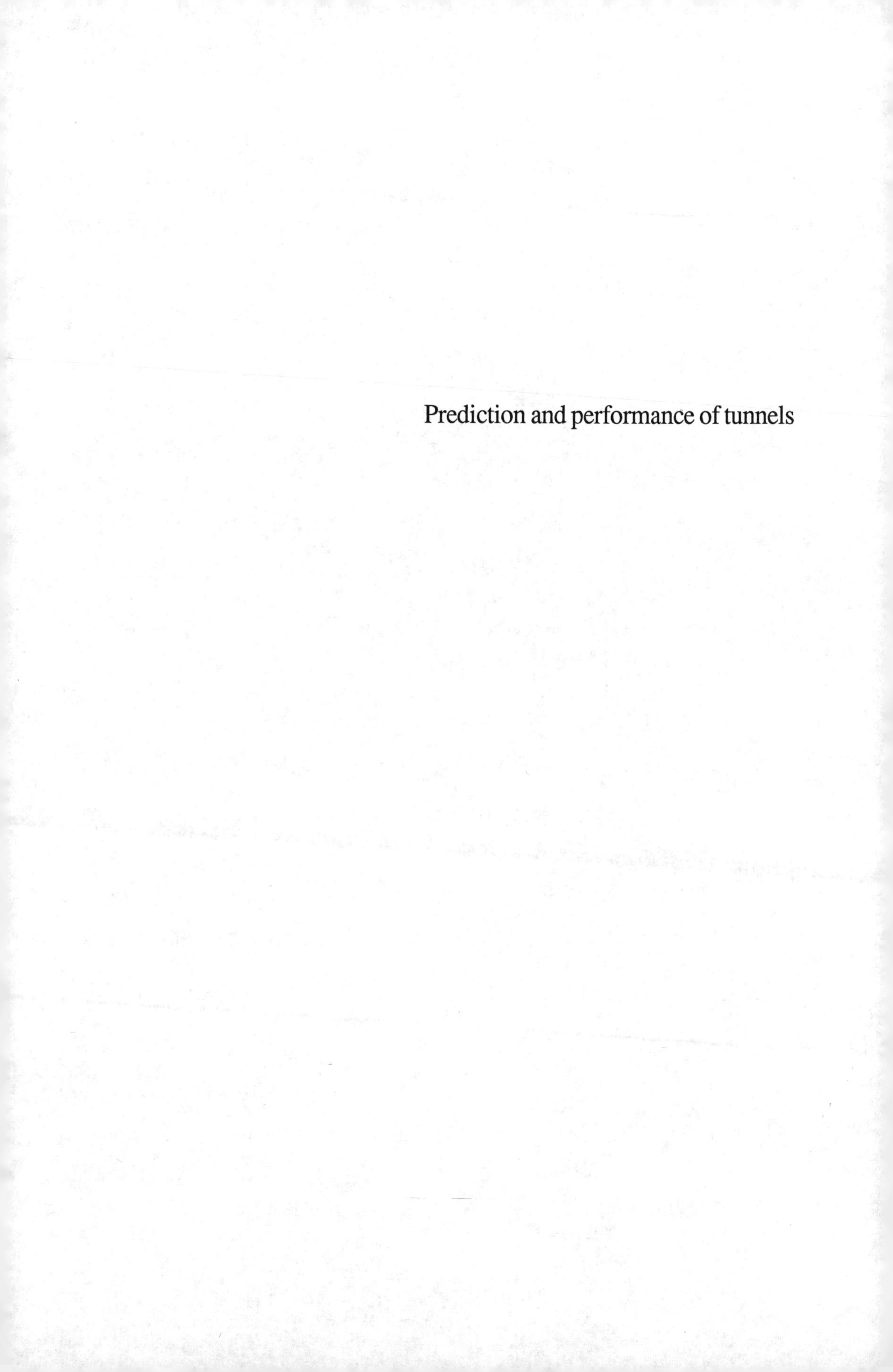

Prediction and performance of tunnels

Prediction and Performance in Geotechnical Engineering / Calgary / 17-19 June 1987

Use of stress change measurements to assess performance of underground excavations

D.R.Korpach
Hardy BBT Ltd., Calgary, Alberta, Canada

P.K.Kaiser
Department of Civil Engineering, University of Alberta, Edmonton, Canada

ABSTRACT: The value of deformation monitoring data from underground excavations can be significantly enhanced if displacement measurements are combined with stress change observations. The concept and methodology of this approach is introduced and illustrated by results from two case histories. The paper deals specifically with aspects of data interpretation from monitoring deep excavations near the face for the purpose of in situ stress determination, rock mass property determination and support evaluation. Results from laboratory tests of the stress change gauges used during field work are presented to document the reliability and expected accuracy of stress change measurements.

1. INTRODUCTION

In recent years there has been an increased awareness of the benefits incurred from monitoring underground excavations. Monitoring provides a method for validating empirical design, evaluating stability and predicting modes of ground behaviour around an underground opening.

The objectives of in situ measurements during the construction of an underground opening may be grouped into three classes:

A. Construction Supervision:
   1. Quality control
   2. Safety evaluation
   3. Performance documentation (for future considerations).

B. Verification of Design Assumptions and Performance Prediction:
   1. Determination of rock mass properties
   2. Assessment of actual deformation mechanism, verification of design model and identification of failure modes.
   3. Comparison of observed with predicted behaviour.

C. Construction Optimization:
   1. Dimensioning of temporary support (e.g., supplemental bolting)
   2. Dimensioning of final support
   3. Selection of optimum support installation point.

The following deals primarily with aspects of Class B. For the purpose of verifying design models, an in situ monitoring program must be laid out in such a manner that all parameters required for a proper performance model are measured. Ideally, the following quantities should be measured or determined from measurements:

1. In situ stress field: σ_o, K_o
2. Rock mass (not intact rock) properties: E, μ, σc, ϕ, ...
3. Support strength and deformation properties: E_s, μ_s, σ_{cs}, ...
4. Pre-support ground deformations: u_o.

The need for these properties can be easily justified on the basis of the principles of the Convergence/Confinement Method illustrated by the schematic diagram in Fig. 1a. This figure demonstrates that the behaviour of an underground opening is an interactive problem and that, at any stage during the excavation process, deformations and stresses are related. Determination of deformations alone, or changes in deformations, does not provide sufficient information to assess the complex interaction between the ground and support. Not even the GCC can be determined unless stress changes and displacement changes are measured simultaneously (see Fig. 1a). Hence, measurement of stress changes in conjunction with displacement measurements are essential and do provide valuable information for determining the in situ stress field, for back calculating rock mass deformation properties and for evaluating the long term behaviour and support requirements.

It is important that instrumentation provide reliable and sufficient data to ensure confidence in interpretation. This paper discusses the importance of stress change measurements and their interpretation for determining the in situ

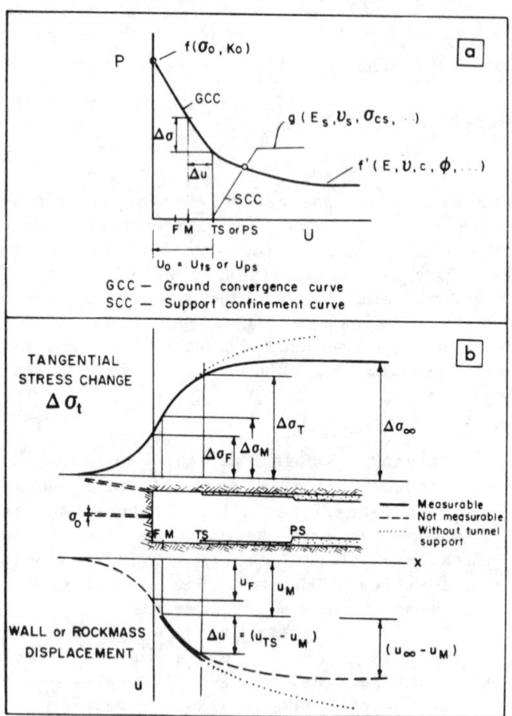

Fig.1. Schematic diagrams of:
(a) Convergence/Confinement Curves; and
(b) New tunnel face stress and displacement patterns

state of stress and for the interpretation of displacement measurements. Two case histories to illustrate proper monitoring procedures are presented.

2.0 STRESS MEASUREMENTS

Knowledge of the in situ stress field is important for many aspects of construction of underground openings including stability, stand up time, deformation and behaviour of ground, support requirements, excavation velocity, wear of tools and amount of explosives.

It is normally not possible to predict the in situ state of stress in rocks. Hence, stresses are measured by one of two techniques; hydraulic fracturing or stress relief methods. Hydraulic fracturing, discussed in detail by Zoback and Haimson, 1982, provides an indication of the in situ stress, but cannot be used to determine stress changes. The basic principles of the second technique, stress relief methods, were adopted to measure stress changes and the in situ stress field for the two projects described later. Instead of overcoring an inclusion, the change in stress was recorded during "undercoring" or rather, during excavation of an opening between stress change gauges. There are several advantages associated with this technique of stress measurement.

First, stress changes can be measured as the excavation face advances. This is of significance because it is difficult to predict the amount of stress change and the associated displacement that occurs ahead of the face due to the existence of a large stress gradient near the excavation face. This has recently been documented by excellent stress change measurements during the sinking of the shaft at the Underground Research Laboratory at Pinawa (Lang and Thompson, 1986). Their measurements show that over 50% of the total measured stress change at a given location occurred within ± 1 radius of the face.

Curtis (1978) and Egger (1980) have shown that the percentage of displacement occurring at the opening face may vary considerably (Fig. 2: between 4 and 71%).

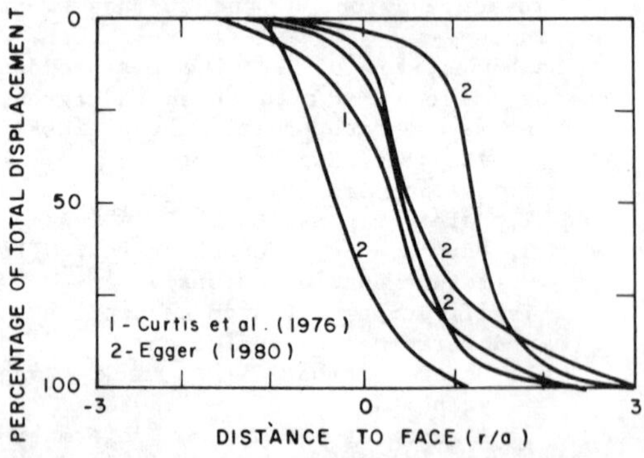

Fig.2. Radial displacements caused by the advancing tunnel face

It is often not possible to measure the full displacement curve, especially in deep tunnels where measurements cannot be obtained ahead of the tunnel face. Therefore, it is difficult to accurately predict the rock mass deformation properties from displacements alone. Only by combining the measured displacements with stress change observations can the accuracy of the back calculated rock mass deformation properties be improved.

This point is further illustrated in Fig. 1b for a hypothetical case. In a deep tunnel deformation measurements, at convergence or multi-extensometer points can only be initiated close to the tunnel face (F) at Location (M) (normally: x_M = 0.25a to 0.75a; where a = tunnel radius). As soon as the temporary support (TS) is installed the stress and displacements are affected by the support action and, hence, no longer provide a pure reflection of the ground response to excavation. Consequently, displacements occurring after Point (TS), and particularly after installation of the permanent support at (PS), are dominated by the support stiffness. In addition, the displacements ahead of the face, u_F, or the displacements before measurements commence, u_M, are not recordable in deep tunnels. For the purpose of back-analysis and rock mass performance evaluation, only the brief window providing $\Delta u = u_{TS} - u_M$ is available for interpretation. A reliable assessment is only possible if the stress change causing these deformations is known. As explained earlier, stress changes or displacements ahead of the face cannot be predicted with great reliability. Hence, stress changes should be measured.

Numerical simulations of near tunnel face conditions show that the shape of the tangential stress change curve recorded near the tunnel wall is similar to the convergence or rock displacement curve (Fig. 1b) for elastic or slightly overstressed rock: Δu is approximately proportional to $\Delta \sigma$. This similarity can be used to extrapolate displacement records to determine the complete displacement curve and to back analyze the rock mass modulus, if the tangential stress change is recorded. Fortunately, it is possible to install stress change gauges ahead of the face at locations where the ground is not yet affected by the advancing tunnel. Even if the absolute amount of stress change cannot be determined because of localized stress fluctuations, the relative change (as a percentage of total change) is sufficient to extrapolate the displacement records.

In summary, stress change measurements in conjunction with limited deformation measurements can be used to reconstruct the displacement history. The rock mass deformation characteristics can then be back-calculated with a much higher degree of accuracy and the value of the other field measurements is enhanced.

Second, gauges can be strategically positioned around the opening to enhance the value of stress change measurements and subsequently, improve the accuracy of in situ stress and ground behaviour predictions. Measured stress changes can be related to the far-field stresses in the manner described by Kaiser and MacKay (1982) and may be used to verify in situ stress measurements by the other techniques.

Third, long term stress change measurements can be obtained when the face is well beyond the monitoring location and when displacement records are impractical and unreliable due to the action of the support. These long term records are of benefit for evaluating the effectiveness of a support system and for extrapolating to predict long term support loads.

Fourth, stress change measurements are necessary to determine or confirm the mode of ground behaviour around an underground opening. Deformation measurements may be misinterpreted and provide misleading information if the actual mode of rock mass behaviour around the opening is not known. The mode of ground behaviour around the Wolverine Tunnel (Section 4.1) could not be properly interpreted from extensometer measurements alone. For example, the extent of a blast damaged zone and the existence of localized shear zones can be detected more readily by stress change measurements.

The reliability and accuracy of the data obtained must be high to allow for confident interpretation. Thus, it is important to place instrumentation at optimal locations around an opening. A minimum redundancy factor of 2 to 3 is required to provide sufficient confidence for stressmeter interpretation.

The stress change measured by a stiff inclusion cell, $\Delta \sigma_c$, is proportional to the change in rock mass stress, $\Delta \sigma_R$, near the instrument. For a vibrating wire stress cell the change in wire stress, σ_w, is proportional to the change in rock stress: (Hawkes and Hooker, 1974).

$$\Delta \sigma_w = \alpha . \Delta \sigma_R \tag{1}$$

where α is the uniaxial gauge sensitivity factor. While the magnitude of total stress change may not be of importance for the ultimate objective of the stress change measurements, confidence must be gained that the loading history does not affect the measurement. For this purpose and for confirmation of the gauge factor, α, a detailed laboratory test program was undertaken to measure reliability, accuracy and repeatability of the gauge measurements.

3. LABORATORY TESTING PROGRAM

The instrument used for measuring stress changes was the Irad Gage Stressmeter which consists of a hollow steel cylinder with a vibrating wire strain gauge. For a linear elastic material, the relationship between the uniaxial rock stress change in the direction of the loading axis of the gauge and the wire tension is given by Eqn. 1. Interpretation of results from gauge measurements has been discussed in detail by Kaiser and Korpach (1986). It is of importance to note that the gauge records an "equivalent uniaxial stress" that is affected by changes in ground stresses both parallel and perpendicular to the gauge axis.

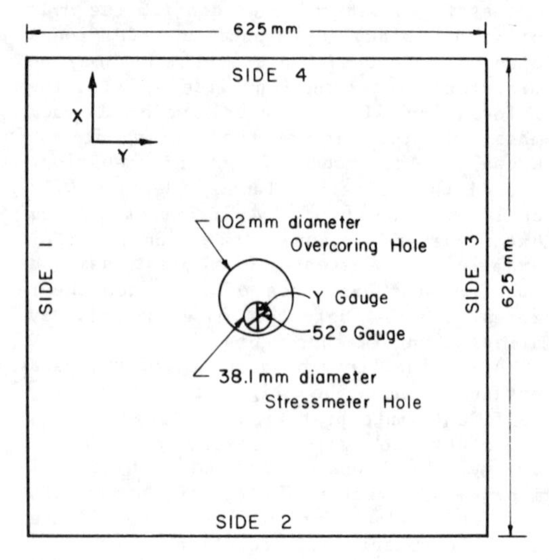

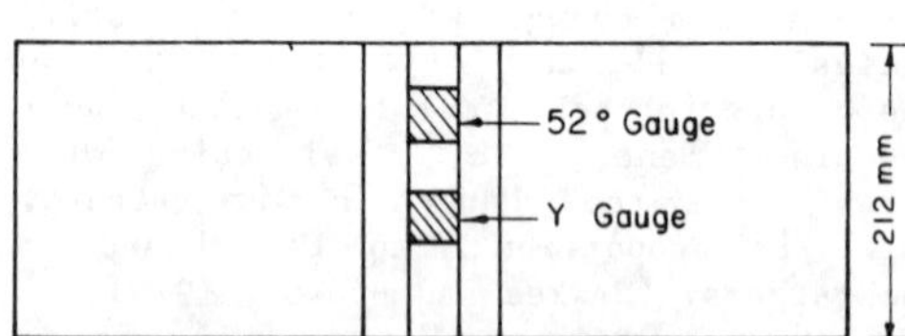

Fig.3. Stressmeter locations in sample without opening

Since completion of this project, other types of stress change measurement cells have been developed and applied successfully (e.g. Lang and Thompson, 1986). For reliable results, it is essential that the best and the most appropriate instrument is selected for a given rock type.

Several preliminary tests were conducted on a large coal sample in a true triaxial apparatus to determine representative material properties. For the purpose of verifying stressmeter data, only the Young's modulus and the Poisson's ratio in the plane of isotropy are of interest. These values were found to be 1.47 $\pm$ 0.32 GPa and 0.225 $\pm$0.064, respectively.

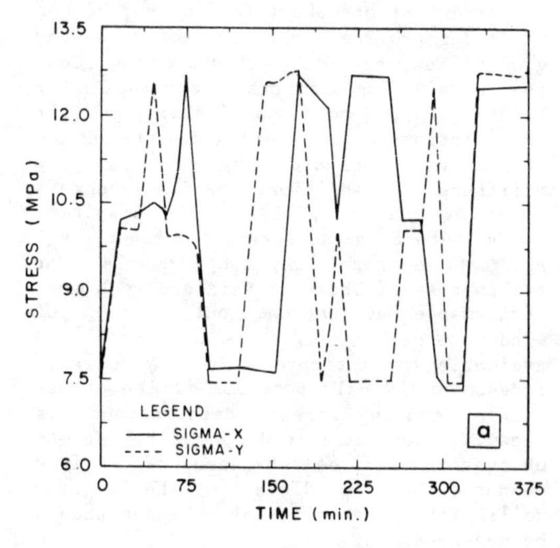

Fig.4a. Loading history for stressmeter calibration test without central opening.

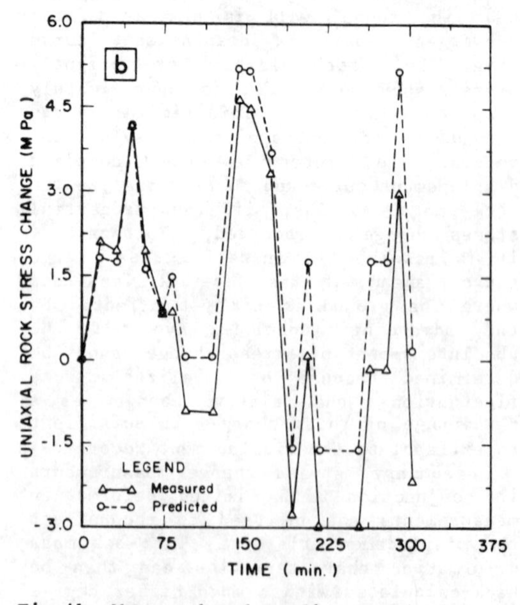

Fig.4b. Measured and predicted uniaxial rock stress change for Y gauge

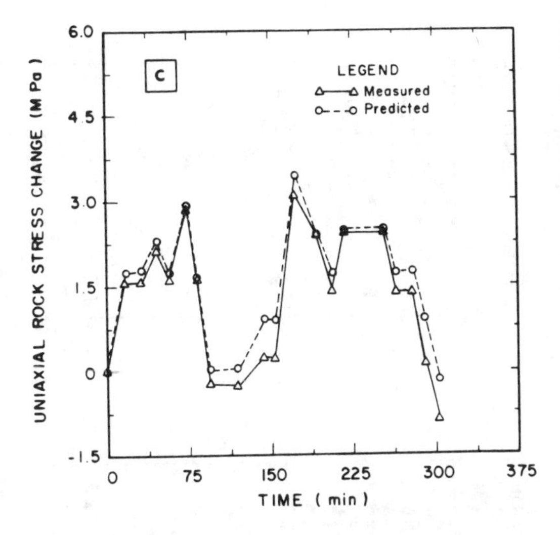

Fig.4c. Measured and predicted uniaxial rock stress change for 52° gauge

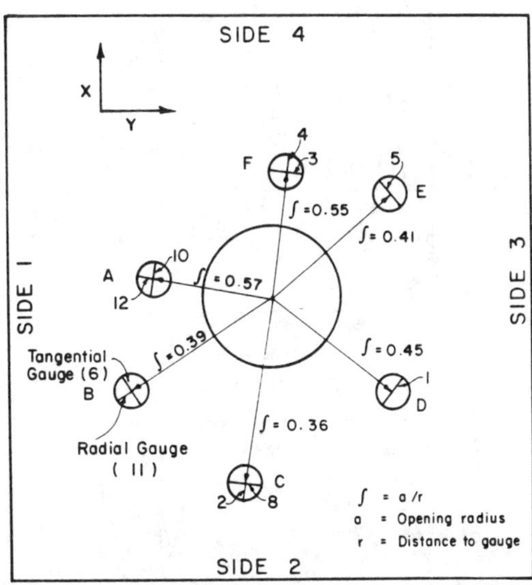

Fig.5. Stressmeter locations in sample with opening

Two sets of gauge calibration tests were performed. First, two stressmeters were installed in a 38.1 mm diameter hole in the center of the sample as shown on Fig. 3. With these gauges in place, the sample was subjected to a complex loading sequence, including rotation of principal stresses, to evaluate the stressmeter response. The sample loading history is presented on Fig. 4a. Plane strain conditions were maintained throughout the test. A comparison of the measured stress changes with those predicted is shown on Figs. 4b and 4c. The measured stress change refers to that recorded by the gauge while the predicted is the expected stress change based on elastic theory.

Close examination of the results shows that there is generally excellent agreement between measured and predicted values for the first one or two loading cycles. A difference between measured and predicted values arises when the sample is unloaded to its initial stress level.

Others (Jaworski et al. 1982) also found discrepancies between measured and predicted stresses after the first loading cycle, but found consistency for future loading cycles.

A second series of calibration tests was conducted with stressmeters placed around a central opening in the sample as shown on Fig. 5. Tangential gauges were placed in all stressmeter holes and radial gauges were placed in Holes A, B, C, and F.

The sample was again subjected to a complex loading sequence, under plane strain conditions, shown by Fig. 6a.

For comparison of the measured and predicted stress changes, sets of typical tangential and radial gauge responses are presented in Figs. 6b and c, and 7a and b.

Fig. 6 (Hole A) shows good agreement between the measured and predicted values during the first two loading cycles. At this point the measured stress changes begin to differ considerably from those predicted by elastic theory.

The tangential gauge measures stress changes much larger than those predicted while the radial gauge measured values lower than predicted. Discrepancies between the measured and predicted values from the radial gauge are not as great as those of the tangential gauge because the maximum stress level experienced by the radial gauge was lower. The discrepancies in observed stresses correspond clearly with non-elastic straining that was recorded in the area of Hole A.

The remaining six gauges were located in areas that appeared to deform elastically. Figs. 7a and b show results from two other gauges. The agreement between the measured and predicted values for these gauges is good for the first two loading cycles. Later in the loading history the magnitude of the measured stress change again differs from the predicted, but the trends are followed closely, i.e., if an increase is predicted the gauge measured an increase. This difference was attributed to a change in the uniaxial gauge sensitivity factor due to local yielding at the gauge contacts.

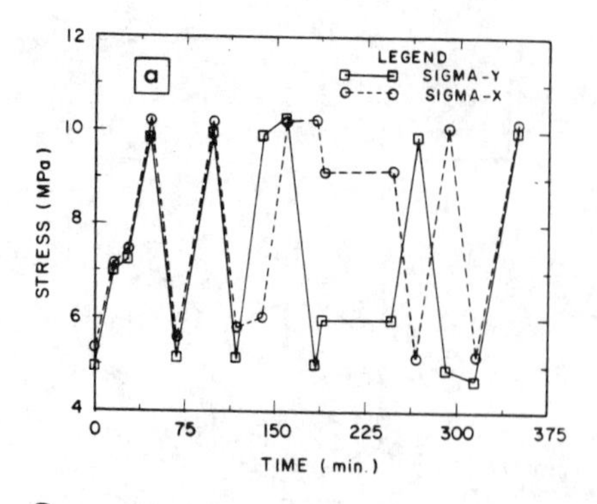

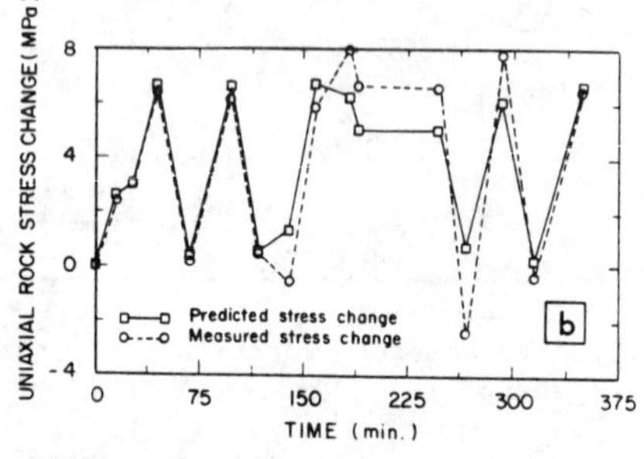

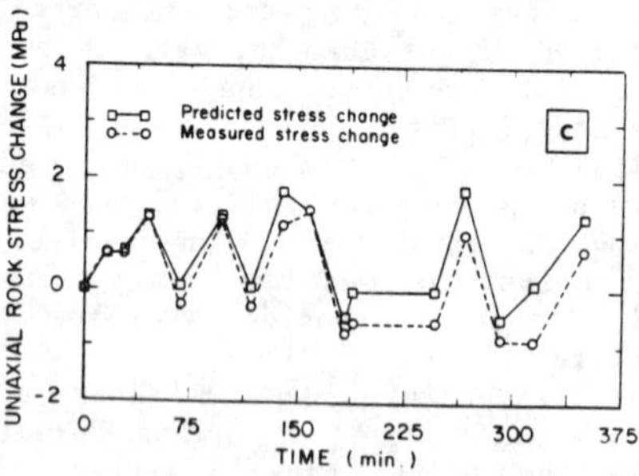

Fig.6a,b,c. (a) Loading history for stressmeter calibration test with central opening. Measured and predicted uniaxial rock stress change for: (b) Gauge 10; and (c) Gauge 12.

In actual field monitoring the gauges are only subjected to one loading or unloading phase. The testing has shown that the gauges provide accurate results over one or two loading cycles. Hence, the findings from the laboratory testing program provides increased confidence that stressmeters can provide accurate and reliable data to enhance other field measurements and to back analyze the in situ state of stress during "underexcavation".

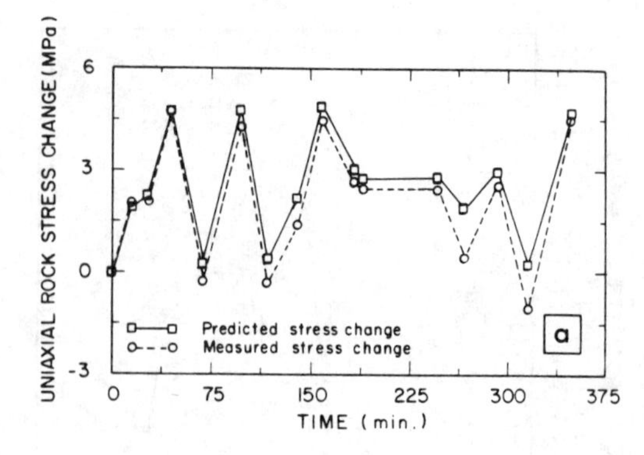

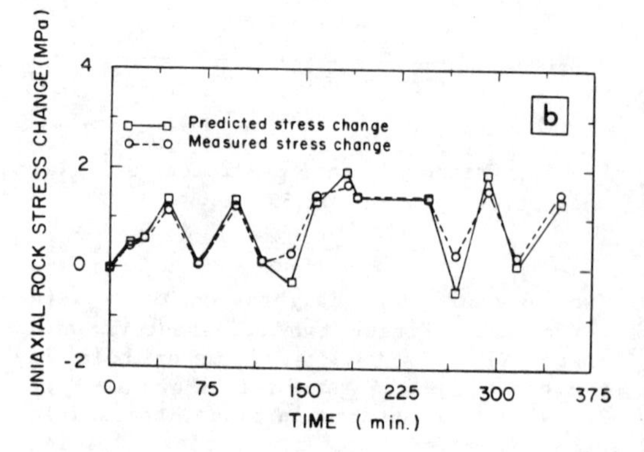

Fig.7a,b. Measured and predicted uniaxial rock stress changes for: (a) Gauge 5; and (b) Gauge 4.

4. CASE HISTORIES

The Irad Gauge Stressmeter was utilized at two projects to test its applicability for measuring stress changes and, in turn, for back calculating the in situ state of stress and the rock mass deformation characteristics. Results from both case histories have been presented previously by Kaiser and MacKay (1982) and Kaiser and Korpach (1986).

4.1 Instrumentation of a Tunnel At Tumbler Ridge, B.C.

In 1983, the deformations and stress changes near the tunnel face were measured during construction of the Wolverine Tunnel (British Columbia, Canada). The tunnel was horseshoe-shaped; 5.4 m wide and 8.4 m high. It penetrated primarily sedimentary rocks, triassic limestones, dolomites, calcareous sandstones and siltstones, and quartzites. The rock quality was generally classified as good to excellent (Kaiser et al 1986). The tunnel was driven simultaneously from two headings using conventional full-face drilling and blasting techniques. The western heading was completed in April 1983, approximately one month before breakthrough. This difference in completion time provided an excellent opportunity to install instruments ahead of the West Face for the purpose of monitoring stress changes during breakthrough.

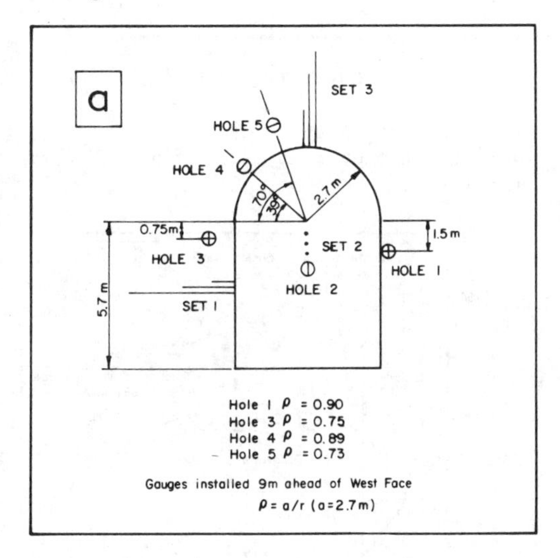

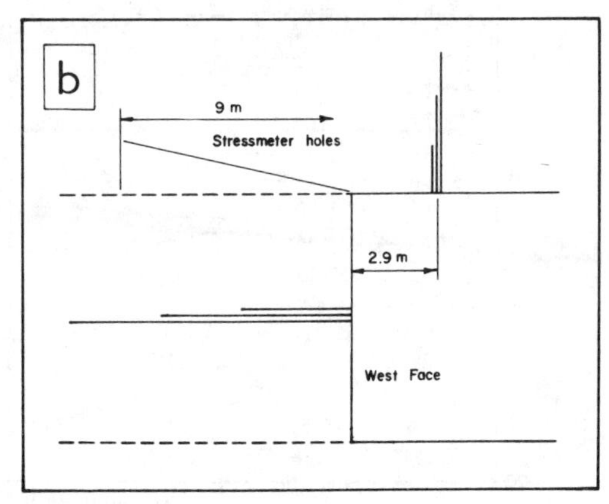

Fig.8. Stressmeter and extensometer installations at Wolverine Tunnel.

Seven stress change gauges were installed 9 m ahead of the face as shown on Fig. 8; radial and tangential gauges in Boreholes 1 and 3, and tangential gauges only in Boreholes 4 and 5. Hole 2 was drilled directly ahead of the West Face and one vertically orientated gauge was positioned. Several borehole extensometers were installed as shown in the same figures. Monitoring commenced when the East Heading was almost 50 m from the plane of measurement and continued until breakthrough when all stressmeters were destroyed.

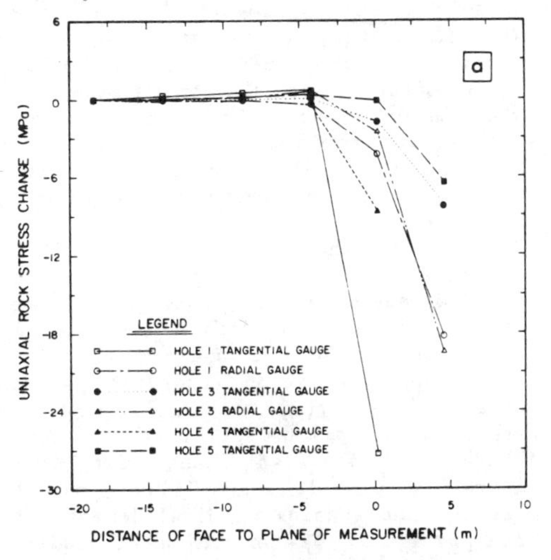

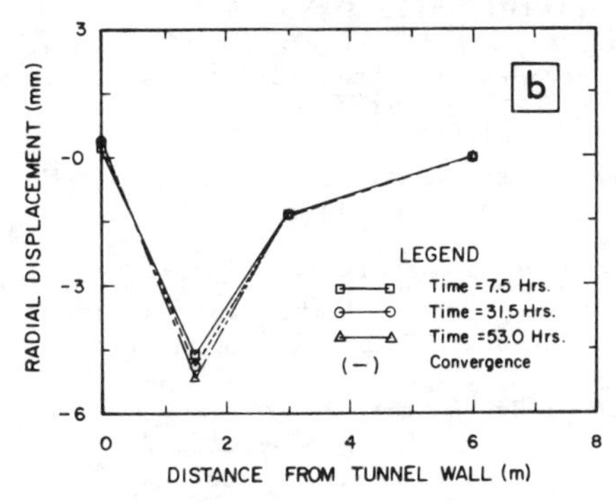

Fig.9a,b. (a) Uniaxial stress changes near tunnel face during breakthrough; and (b) radial displacement relative to deepest anchor point for springline (north wall).

Uniaxial rock stress changes and radial displacements measured are shown on Figs. 9a and b. It can be seen that all stress change gauges, both radial and tangential, recorded a reduction in stress as the excavation reached and moved past the gauge. This rather unexpected result can only be explained by an extensively yielded or softened zone created by blast damage ahead of and adjacent to the tunnel wall. Finite element analyses were conducted to estimate the in situ stress field at the site based on the stress change measurements. The in situ stress field could not be predicted accurately since all gauges were located within the blast damaged zone. For a more accurate determination of the in situ state of stress, gauges should be placed both within and outside the estimated blast damaged zone. However, results were extremely useful for assessing the mode of ground behaviour around the opening. The stress change data in combination with extensometer measurements showed that softened or yielded rock extended for about 2 m beyond the opening at the crown and for about 1 m beyond the opening at the springlines.

The results show clearly that a back-analysis of rock mass properties from these extensometer measurements is not possible at this location because the in situ stress field and the stress change during the deformation recording periods were unknown. While these findings are negative, they demonstrate that the stress change may differ significantly from that theoretically predicted and hence, must be measured for conclusive displacement data interpretation. The extensometer readings would be completely useless (except for safety monitoring), if assessed independent of the actual stress state and stress change.

4.2 Stress Measurements in a Shaft Near Lethbridge, Alberta

In 1980, a 4.32 m finished diameter concrete lined shaft was sunk, using conventional shaft sinking methods, to a depth of 235 m at a coal mine site near Lethbridge, Alberta (Kaiser and Mackay, 1982). The lining generally followed one to two shaft diameters behind the shaft bottom. The rock was seldom unsupported for more than 16 to 20 hours.

The shaft penetrated about 60 m of glacial clay till underlain by about 6 m of saturated basal sands and gravels, and the Upper Cretaceous marine Bearpaw Formation composed of clay shales, siltstones and shales with frequent coal seams in the uppermost member.

Seven IRAD stress change gauges were placed at the 152 m level at 1.37 m (West) and 1.52 and 2.68 m (South) from the shaft wall (Kaiser and Mackay, 1982). The stress change gauges were installed 10 m ahead of the shaft bottom in steeply inclined boreholes drilled into the shaft wall. The stress change gauges were oriented radially and tangentially with respect to the shaft perimeter. Several radial multipoint extensometers were installed as close as possible to the shaft face at this same depth level.

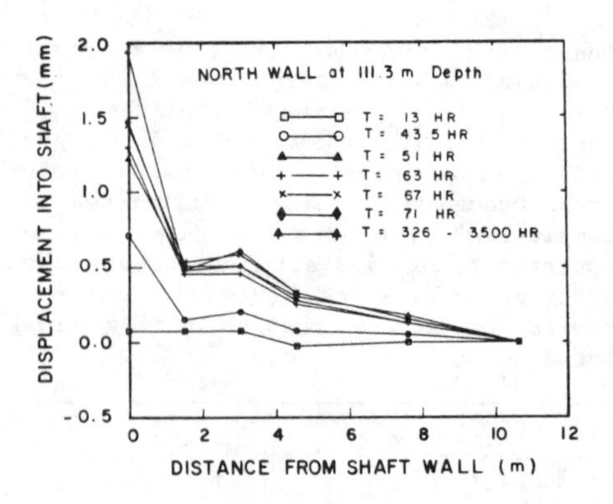

Fig.10. Depth-Displacement plot for north wall extensometer at 111 m depth (Mackay, 1982)

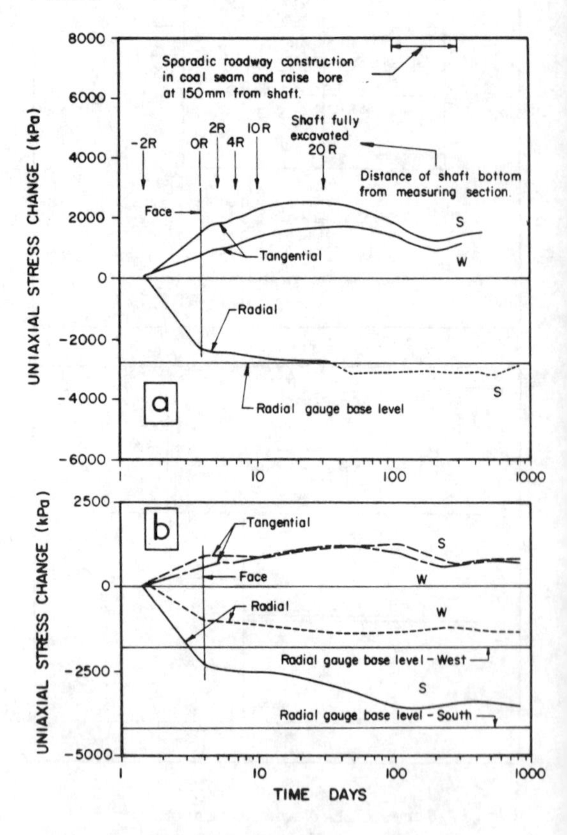

Fig.11. Long term uniaxial stress change for: (a) Stressmeter at = 0.65; and (b) Stressmeter at ρ= 0.51 (S) and ρ= 0.67 (W); S = south, W = west wall

Results from stress change measurements were analyzed in a manner similar to those from the Wolverine Tunnel. The magnitude of the major principal stress was predicted with confidence to within ±15% and the orientation to within ±10°. Fig. 10 presents a typical set of multi-point extensometer readings from one extensometer. Again, it was possible to establish, based on stressmeter and extensometer data, the approximate extent of blast damage around the opening.

Fig. 11 shows the measured uniaxial stress change at several gauge locations around the shaft. The most important observation is that more than half of the total measured stress change occurred before the face reached the gauge locations. Thus, the majority of displacements would also have occurred before this time and prior to the commencement of measurement.

On average, 54% (42 to 72%) of the tangential stress change occurred ahead of the face and only about 10% (on average) of the total stress change caused the deformations shown in Fig. 10. Consequently, the total time-independent movement outside the blast damaged zone at 1.5 m from the wall would be about 5 mm. If the rock mass modulus were back-calculated based on an assumed rather than measured stress change during deformation measurements, the modulus would be underestimated by a factor of about three to four.

5. CONCLUSIONS

It is proposed that deformation measurements in underground openings should be combined with stress change measurements if the purpose of monitoring is to determine the rock mass properties or to evaluate the support design.

The combined monitoring technique described in this paper has been shown to produce much more useful results. While they may not be sufficient to permit a conclusive back analysis, the combined measurements provide valuable information on the in situ state of stress and for determining the mode of behavior around the opening. More importantly, the stressmeter provides a practical tool for measuring stress changes ahead of an underground opening face, thus, providing a basis for data extrapolation. Although there may be some error in the absolute magnitude of the measured stress change, the percentage of total change recorded is very reliable. This aspect is of great importance for interpreting deformation records obtained from instruments installed behind the advancing face.

Stress change data in conjunction with deformation measurements constitutes a reliable method for back calculating rock mass deformation characteristics. This technique is often useful for determining tunnel support requirements. It is currently limited to elastic or slightly overstressed rock. For most useful data, stress change gauges must be placed both inside and outside the zone of overstressed, damaged or weakened rock.

6. ACKNOWLEDGEMENTS

The execution of the laboratory testing program was a success due to the dedication of Mr. G. Cyre. The research projects were funded by Petro Canada Exploration Inc., Calgary, the B.C. Railway Company and supplemented by funds from the National Reseach Council (Institute for Research in Construction) and the National Sciences and Engineering Research Council of Canada.

7. REFERENCES

Curtis, D.J., L.M. Lake, W.T. Lawton and D.E. Crook, 1976. In situ ground and lining studies for the Channel Tunnel Project. Tunnelling '76, London, pp. 231-242.

Egger, P., 1980. Deformation at the face of the heading and determination of the cohesion of the rock mass. Underground space, Vol. 4, No. 5, pp. 313-318.

Hawkes, I. and V.E. Hooker, 1974. The vibrating wire stressmeter. 3rd Congress of the International Society of Rock Mechanics, Denver, Vol. 2B, pp. 439-444.

Jaworski, G.W., B.C. Dorwart, W.F. White and W.R. Beloff, 1982. Behaviour of a rigid inclusion stressmeter in an anisotropic stress field. 23rd U.S. Symposium on Rock Mechanics, pp. 211-218.

Kaiser, P.K. and C. Mackay, 1982. Development of rock mass and liner stresses during sinking of a shaft in clay shale. 1st International Conference on "Stability in Underground Mining", Vancouver, Ch. 36, pp. 790-809.

Kaiser, P.K. and D. Korpach, 1986. Stress Changes Near the Face of Underground Excavations. International Symposium on Rock Stress and Rock Stress Measurement (ISRM) Stockholm, pp. 635-645.

Kaiser, P.K., C. Mackay and A.D. Gale, 1986. Evaluation of rock classification at B.C. Rail Tumbler Ridge Tunnels. Journal of Rock Mechanics and Engineering, Vol. 19, pp. 205-234.

Lang, P.A. and P.M. Thompson, 1986. Geomechanics experiments during excavation of the URL Shaft. Technical Record, TR 375, Vol. 1, pp. 125-145.

Zoback, M.D. and B.C. Haimson, 1982. Status of the hydraulic fracturing method for in situ stress measurements. 23rd U.S. Symposium on Rock Mechanics, Berkeley, Ch. 15, pp. 143-156.

Prediction and Performance in Geotechnical Engineering / Calgary / 17-19 June 1987

Prediction of ground movements above shallow tunnels

R.C.K.Wong
Esso Resources Ltd., Calgary, Alberta, Canada

P.K.Kaiser
Department of Civil Engineering, University of Alberta, Edmonton, Canada

ABSTRACT: Conventional design methods, based on semi-empirical correlation, usually treat the two most important design aspects, (1) prediction of the induced displacement field and (2) evaluation of the lining pressure distribution, separately. However, they are interrelated in reality. A conceptual model has been developed and is proposed to determine the dependence between the vertical settlements and the tunnel support pressure. This model is verified by comparisons with results obtained from field measurements and numerical simulations, and is used to evaluate conventional design methods. This evaluation revealed that the relationship between the settlement and the support pressure is a continuously changing function reflecting the ground/support interaction for the entire spectrum of tunnel behaviour. Separation of the two design aspects, settlement and support pressure, is therefore not justified.

1. INTRODUCTION

For the design of a tunnel there are two major concerns: (1) the induced displacement field and (2) the lining pressure distribution. Although these two aspects are interrelated, the conventional design methods which are often based on empirical correlations and simplistic approximations, treat these two design criteria separately. The induced settlement pattern above a tunnel depends on the tunnel wall displacement which in turn depends on the support pressure provided during construction. The interdependency of tunnel wall displacement and support pressure has long been recognized by many, e.g., Ferner (1939), Pacher (1964), and Rabcewicz and Golser (1973). Eisenstein and Negro (1985) have expanded this ground reaction concept to relate the far field or ground surface settlements and the support pressures. Wong and Kaiser (1986) have then extended this approach to interpret different behavioural modes of shallow tunnels. In this paper, an attempt is made to expand this concept to relate other important parameters such as the surface/crown settlement ratio, (S_s/S_c), and the location, i, of the maximum settlement gradient.

2. REVIEW OF CONVENTIONAL DESIGN METHODS

2.1 Displacement Prediction

2.1.1 Surface Settlement

The shape of the ground surface settlement trough is normally approximated by an error function:

$$S = S_s \exp\left(\frac{-x^2}{2i^2}\right) \qquad (1)$$

where: S = surface settlement at a transverse distance x from the tunnel centre line; S_s = maximum settlement (at x = 0); and i = location of maximum settlement gradient or point of inflexion.

Eqn. 1 was first put forward by Martos (1958), based on a statistical analysis of field observations of subsidences above mine openings. Others (e.g., Schmidt (1969) and Peck (1969)) have proven its adequacy for approximating the surface settlement profile above shallow tunnels. The two unknown parameters, i and S_s, must be estimated. As outlined below, a significant amount of research involving field observations and model tests was devoted to this subject.

(a) Estimation of i:

The location of the maximum gradient, i, is normally correlated with the overburden depth, H, to the tunnel axis, and the tunnel radius, a (Fig. 1). For example:

- Schmidt (1969) and Peck (1969) found from field observations:

$$i/a = (H/2a)^n \tag{2}$$

where: n = 0.8 to 1.0

- Atkinson and Potts (1977) determined from model tests and field measurements:

$$i = 0.25\ (H + a) \tag{3.a}$$
$$i = 0.25\ (1.5H + 0.5a) \tag{3.b}$$

Eqn. 3.a is intended for loose sand with surface surcharge and Eqn. 3.b for dense sand and overconsolidated clay without surface surcharge.

- O'Reilly and New (1982) found from field observations:

$$i = 0.43\ H + 1.1 \tag{4.a}$$
$$i = 0.28\ H - 0.1 \tag{4.b}$$

where: H is to be given in metres. Eqns. 4.a and b are for cohesive and granular soils, respectively.

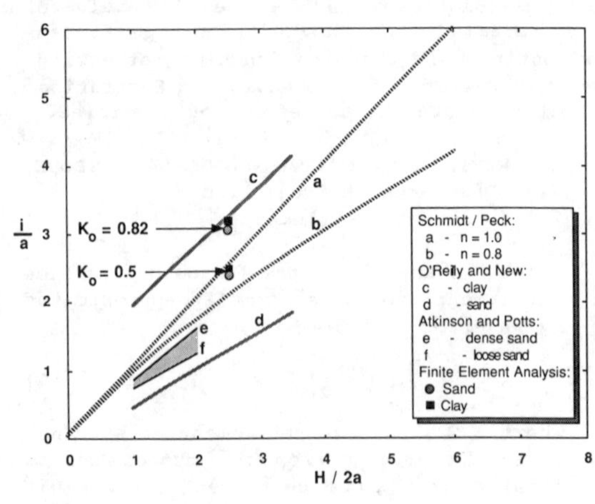

Fig. 1. Settlement trough width parameter, i, versus normalized tunnel depth H/2a.

(b) Estimation of S_s:

The maximum settlement S_s can be expressed in terms of i and V_s from integration of Eqn. 1 with respect to the transverse distance x.

$$S_s = \frac{V_s}{2.5i} \tag{5}$$

where: V_s = volume of settlement trough per unit length along tunnel axis.

Several researchers attempted to correlate V_s to the volume of ground loss during excavation, V_{exc}. With a knowledge of V_{exc}, which depends on the construction method, an estimate of V_s and thus S_{max} can be obtained. Some notable attempts are listed as below.

- Peck (1969):

$$V_{exc} = f\ \left(\frac{p_o - p_i}{c_u}\right) \tag{6}$$

where: p_o = overburden pressure (above tunnel axis); p_i = internal or support pressure in excess of hydrostatic pressure; and, c_u = undrained shear strength (for clay).

- Atkinson and Potts (1977):

$$\frac{V_s}{V_{exc}} = 2\sqrt{\frac{2}{\pi}}\ \left(\frac{i}{2a}\right)\ \left(\frac{S_s}{S_c}\right) \tag{7}$$

where: S_c = the vertical settlement at the crown of the tunnel.

- O'Reilly and New (1982):

$$\frac{V_s}{V_{exc}} = m \tag{8}$$

where: m = 0.5-3% (stiff fissured clay; shield or none); m = 2-2.5% (glacial deposits; shield in free air); m = 1-1.25% (glacial deposits; shield in compressed air); m = 30-45% (silty clay deposits; c_u = 10 to 40 kPa; shield in free air); and, m = 2-20% (silty clay deposits; c_u = 10 to 40 kPa; shield in compressed air).

- Attewell and Farmer (1975), based on some field observations, suggested to correlate S_s with H/a.

For H/a ≥ 8: $S_s/2a$ = 0.2-0.35% (9.a)

For H/a ≤ 4: $S_s/2a \geq 1.0\%$ (9.b)

2.1.2 Subsurface Settlement

Atkinson and Potts (1977), based on results from model tests and field observations, proposed a relationship between the settlements at the ground surface and at the crown for shallow tunnels of limited depth ratio, H/a (Fig. 2):

$$\frac{S_c}{S_s} = 1.0 - \alpha\ \left(\frac{H - a}{2a}\right) \tag{10}$$

where: α = 0.57 (dense sand at low stresses; large dilation); α = 0.40 (loose sand and dense sand at high stresses; small dilation); and α = 0.13 (overconsolidated clay; no or small volume change).

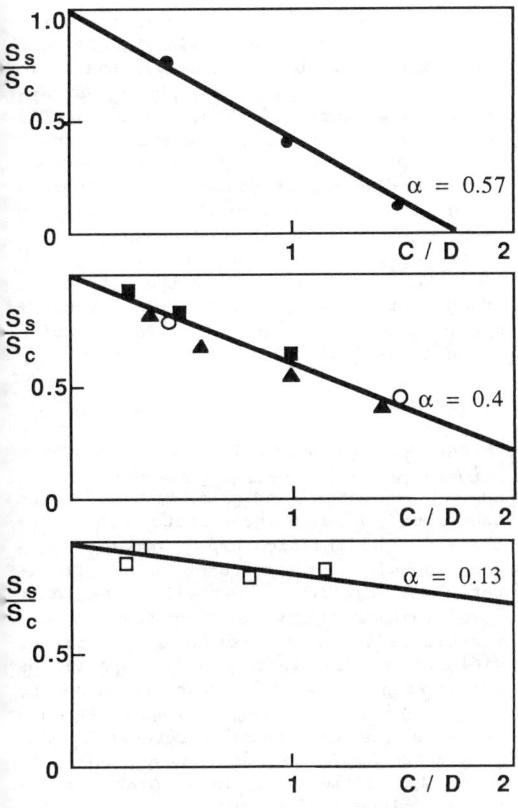

Fig. 2 Surface/crown settlement ratio versus normalized tunnel depth (after Atkinson and Potts (1977)).

2.2 Pressure Prediction

Terzaghi (1943) investigated arching of soil above a tunnel and adopted the 'trap door' approach to calculate the vertical pressure on a tunnel.

Duddeck and Erdman (1982) recommended based on a review of common design approaches that the tunnel lining should be designed against the following pressures: i) For $H/a \leqslant 6$: full overburden pressure at the crown; ii) For $4 \leqslant H/a \leqslant 10$: full primary stresses may be applied to the lining; and, iii) For $H/a \geqslant 6$: appropriate reduction of ground pressure permitted (50% or more is debatable).

3. CONCEPTUAL MODEL FOR PREDICTION OF GROUND SETTLEMENT (with respect to Support Pressure)

The surface settlement profile above a tunnel is reasonably well defined by the maximum surface settlement, S_s, and the infection point at i. This surface settlement profile depends on the crown settlement, S_c, or the support pressure, p_i. A generally applicable model should describe this interrelationship.

3.1 Settlement/Support Relationship for Cohesionless Soils

3.1.1 p_i - S_c and p_i - S_s Relationships

Wong and Kaiser (1986) derived a relationship between S_c and p_i for a circular tunnel in cohesionless soil, as schematically shown in Fig. 3.a. The support pressure, p_i, originating at the in situ stress, p_o, first decreases due to elastic ground response. At Point 1,

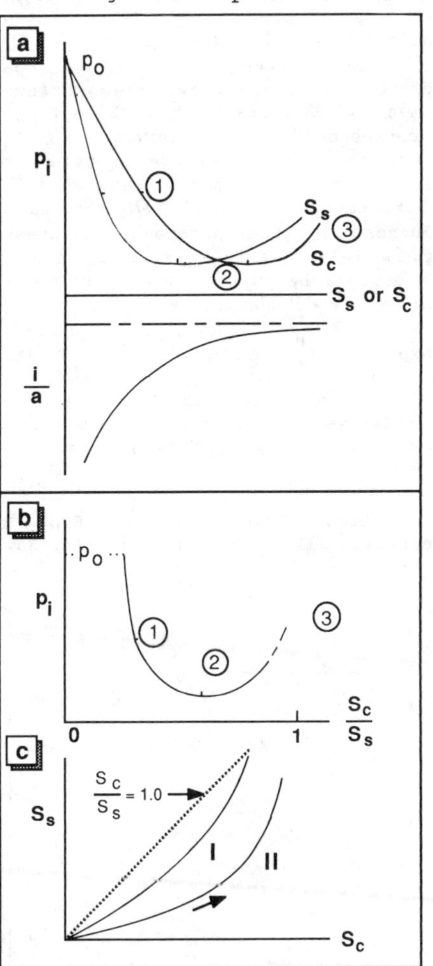

Fig. 3 Conceptual model for settlement/pressure relationship.

yielding occurs and plastic deformations take place. Point 2 denotes the minimum support pressure at which collapse at the crown may be initiated. Further displacement of the crown will cause a collapse mechanism to propagate toward the ground surface (in shallow tunnels), and a higher support pressure (Point 3) is required to maintain equilibrium because the effect of arching above the crown has been disturbed. The crown causes settlements and the surface settlement, S_s, are related (Fig. 3.a).

3.1.2 S_s - i/a Relationship

Since the maximum settlement, S_s, is dependent on p_i, it is to be expected that the location of the point of inflection, i, also is a function of p_i or S_c. The surface settlement profile, shown schematically in Fig. 4.a, is defined by S_s and i for a given p_i. If more crown settlement is permitted, the surface settlement will increase from S_s to S_s' and, correspondingly, i moves to i'. Assuming that the surface settlement profile outside a distance w (between 2.5i and 3i; Cording et al., 1976) remains uninfluenced by the progressive stress relief, a relationship between S_s and i can be derived by putting $S = S_o$ at $x = w$ (where: $w = H \cot \beta + a$):

$$\frac{S_s}{S_o} = \exp\{\frac{1}{2}\ [(1+(\frac{H}{a})\ \cot\beta)/(\frac{i}{a})]^2\} \qquad (12)$$

where: β = angle defining the trough width relative to the tunnel depth which depends on the mode of yielding and K_o (Wong, 1986).

The relationship resulting from Eqn. 12 is schematically shown in Fig. 3a. Results of numerical examples (H/a = 3 to 10, ϕ = 20° to 40° and β = 70° to 80°) are plotted in Fig. 4b. In general, the ratio i/a decreases with increasing S_c or S_s. For shallow tunnels, i/a decreases rapidly to a limiting value at relatively small ground or crown displacements because little stress relief causes the yield zone to propagate quickly to the surface. For deep tunnels, i/a also decreases as S_s increases but at a slower rate. Fig. 4.b indicates that the shape of the settlement trough for cohesionless ground is dependent on H/a and p_i (or S_c). The friction angle ϕ has less influence.

3.1.3 Development of Settlements (S_s/S_c)

Fig. 3.a shows the strong dependency between p_i and S_c or S_s. The ratio S_s/S_c is a parameter indicating the ground behaviour of a unique state above the tunnel. The relationship p_i-(S_s/S_c) is a continuously changing function as illustrated by Fig. 3.b. Initially, the ratio S_s/S_c remains fairly constant for a linear elastic media but it increases steadily as yielding occurs, and rapidly approaches unity when collapse propagates after p_i has reached the minimum value. Fig. 3.c presents the relationship between S_s and S_c and shows that S_s will become equal to S_c at the ultimate collapse state, i.e., during 'plug-like' failure.

The ground behaviour near a soft ground tunnel (Wong and Kaiser, 1986) may be characterized by two distinct modes of

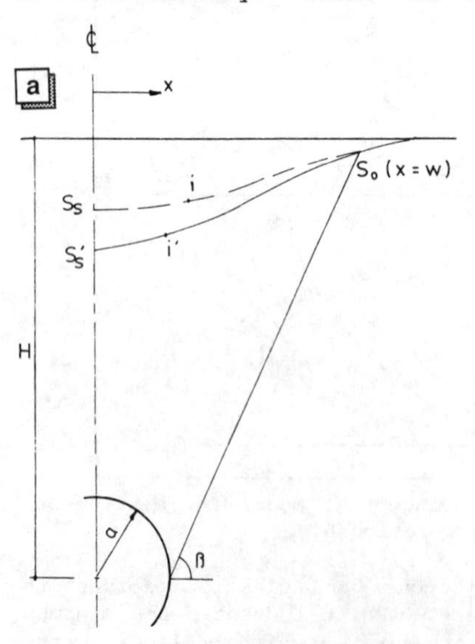

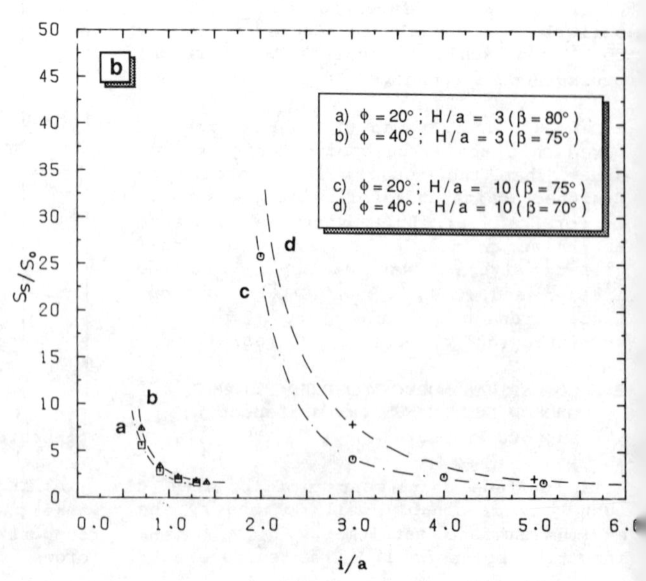

Fig. 4 Surface settlement profile

yielding (Modes I and II), separated by a critical K_o-value (K_{cr}). For Mode I ($K_o < K_{cr}$), yielding induced by stress relief is initiated at the shoulders of a tunnel, and localized yield zones propagate to the surface with further stress relief (Fig. 5.a). For Mode II ($K_o > K_{cr}$), a continuous yield zone surrounds the tunnel opening and no localized shearing takes place (Fig. 5.b). It is intuitively expected that Modes I and II will display distinctly different features in their settlement profiles because they represent different subsurface displacement patterns. For the subsurface settlement profiles presented schematically in Fig. 5 equal crown settlements were assumed for Mode I and II. If S_c is small, the extent of the yield zone is also small and the shape of the vertical settlement profiles is very similar, but the magnitude of settlement is larger in Mode I than in Mode II because of differences in arching above the opening. For large or excessive crown settlements two localized shear planes develop in Mode I and the soil block above the tunnel displaces as a rigid body toward the opening. It does not yield and the differential displacement between the crown and the ground surface is small. Hence, the ratio of surface to crown displacement (S_s/S_c) approaches unity (Fig. 3.c). Near collapse S_s/S_c tends toward unity for both modes, but at a faster rate for Mode I.

With respect to surface settlements, Wong (1986) has shown that Mode I exhibits a narrower trough width than Mode II. Furthermore, if the same displacements are allowed during tunnel construction, larger surface settlements will be induced in Mode I because ground with low K_o-values is more difficult to control than ground with a K_o close to unity.

3.3 Settlement/Support Relationship for Cohesive Soils

Similar reasoning, as used for cohesionless soils can be applied to cohesive soils. For simplicity, only one case with $K_o = 1$ and constant cohesive strength, independent of depth, will be considered in the following assessment.

4. COMPARISON WITH NUMERICAL SIMULATIONS

4.1 Introduction

Numerical examples were generated by the finite element method to investigate the validity of the conceptual model described

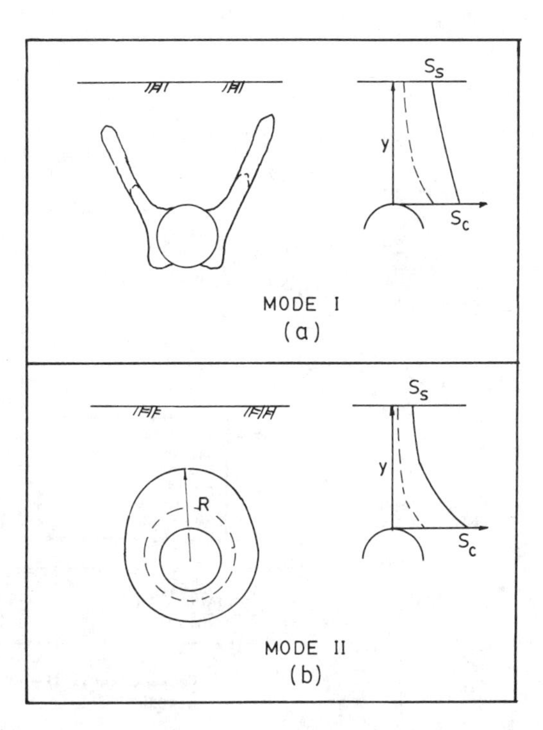

Fig. 5 Subsurface settlement profiles above tunnels for Modes I and II.

in the previous section. The finite element program SAFE, developed by Chan (1985), was employed for all analyses. Details of the finite element mesh, the imposed boundary conditions, and the tunnel excavation simulation procedure were described by Wong and Kaiser (1987).

Six analyses were performed for tunnels in purely cohesionless and cohesive soils. Excavation was simulated by incremental stress-relief steps. The input data for each analysis are listed in Table 1. For the cohesionless soil, one set of typical soil parameters and two K_0-values 0.5 and 0.82, were chosen to observe the effect of yield zone localization. The assumed

Table 1. Input Data for Finite Element Analyses

No. of Analysis	Soil Model	Soil Properties					K_o	H/a	Stress relief simulation
		ϕ (°)	q_u (kPa)	E (MPa)	γ (kN/m^3)	ν			
ST1	Elastic, perfectly plastic, associated flow rule, Mohr-Coulomb	30	0	40	20	0.4	0.5	5	P
		(constant with depth)							
ST2	as above	30	0	40	20	0.4	0.82	5	P
DT1	as above	30	0	40	20	0.4	0.5	18	P
DT2	as above	30	0	40	20	0.4	0.82	18	P
STC1	Elastic, perfectly plastic, Von Mises	0	30	40	20	0.495	1.0	5	P
STC2	as above	0	60	40	20	0.495	1.0	5	P

P - excavation simulated by unloading proportional to initial in situ stresses

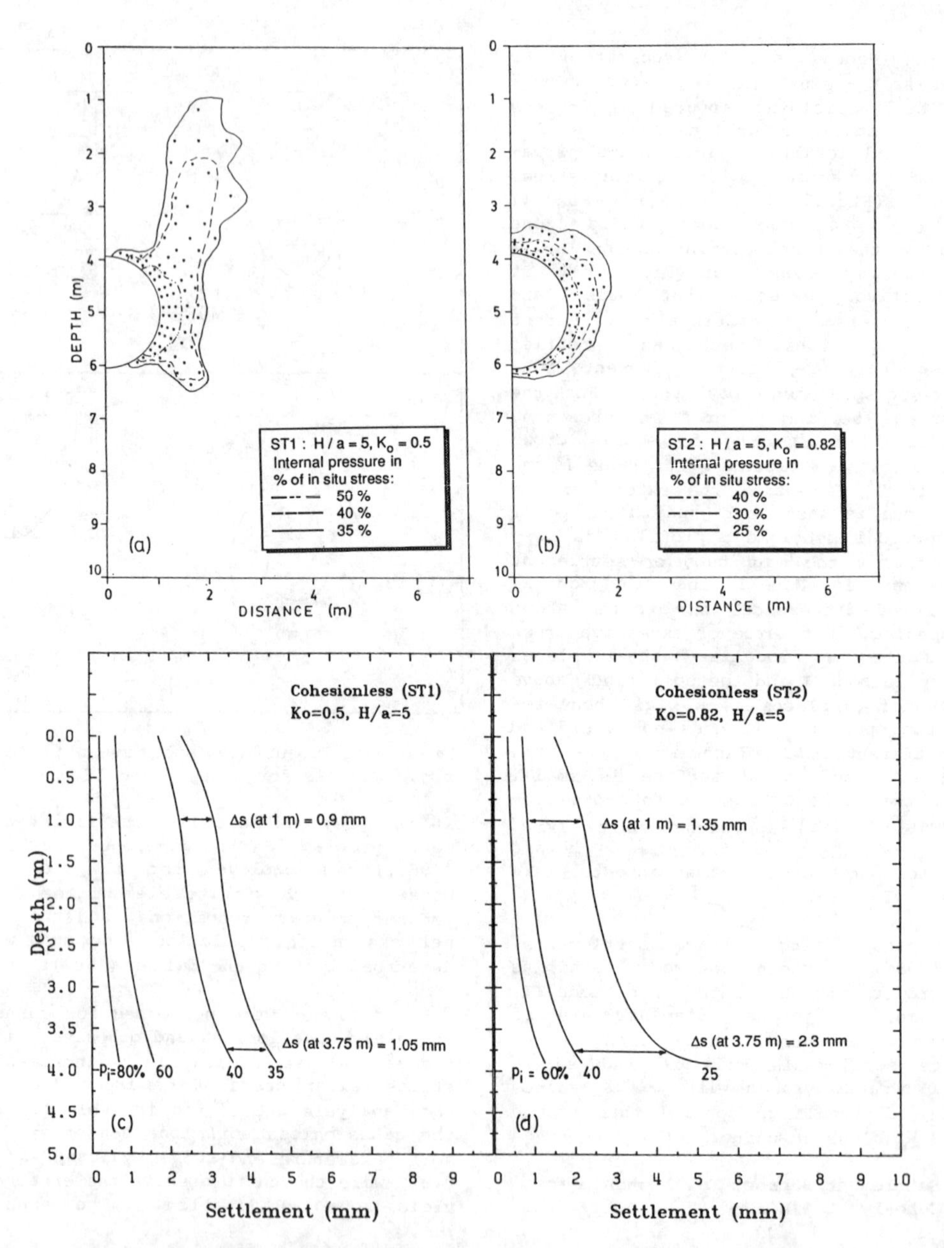

Fig. 6. Subsurface settlement profiles above a shallow tunnel in cohesionless soil (ST1, ST2)

associated flow rule for cohesionless soils will overpredict dilation, and larger than actually expected displacements are calculated. For the cohesive soil, the unconfined compression strength of the ground was assumed to be constant with depth and a K_o-value of unity was used.

4.2 Interpretation of Results

The important aspects of yield zone propagation and stress redistribution were already covered by Wong and Kaiser (1986) and a detailed interpretation of the entire displacement field was presented by Wong (1986). The following discussion will concentrate on displacements of the ground above the tunnel and of the surface settlement.

4.2.1 Subsurface Settlement Profile above the Tunnel

It was mentioned earlier that localization of shearing may occur at $K_o < K_{cr}$ in cohesionless soils. This is illustrated for Cases ST1 in Fig. 6.a. At $K_o > K_{cr}$ a continuous yield zone is observed as shown in Fig. 6.b.

The corresponding subsurface settlement profiles above the crown for various stress relief levels are plotted in Fig. 6.c and d, and those of four other cases in Figs. 7 and 8. They are to be compared with the schematic diagrams in Fig. 5.

For shallow tunnels in cohesionless soil (Cases ST1 and ST2), two distinctly different profiles can be observed. For Mode I (ST1), a localized yield zone forms, the soil above the roof does not yield and moves downward as a coherent block. The differential settlement with depth is small and only due to the elastic deformations. Thus, the vertical displacement increases only slightly with depth (Fig. 6.a). After the yield zone propagates toward the ground surface (at p_i < 45% of p_o) the settlement increment, Δs, is almost independent of depth.

In Mode II where the opening is surrounded by a continuous yield zone, the settlement increases gently in the elastic zone and accelerates rapidly within the plastic zone (Fig. 6.b). An abrupt change in settlement gradient can be observed at the elastic/plastic boundary (for p_i = 25% at about 3.5 m depth). The settlement increment near the tunnel is much larger than near the ground surface.

For deep tunnels (Cases DT1 and DT2; Fig. 7) localization does not progagate to the surface. The settlement profiles are similar because the extent of the yield zones are relatively small. However, at equal stress relief (e.g. p_i = 40% of p_o) larger settlements are experienced in Case DT1. A rapid increase in subsurface settlement is observed near the tunnel.

Differences between subsurface settlement profiles are less distinguishable for shallow tunnels in cohesive soil (Fig. 8; K_o = 1) because zero volume change was assumed during plastic flow. However, it is interesting to observe (Fig. 8.a) that decrease in support pressure from 50% to

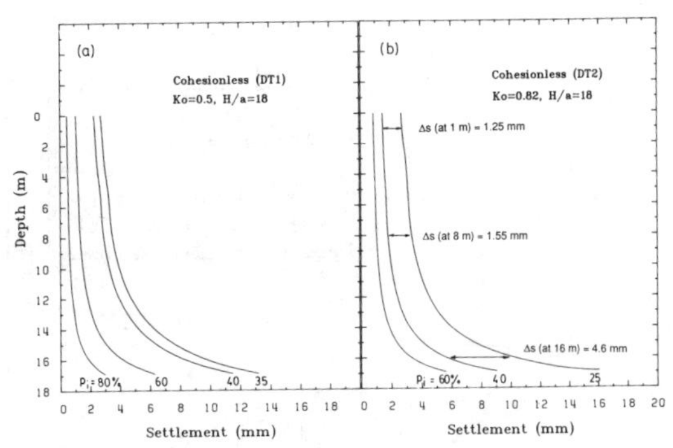

Fig. 7. Subsurface settlement profile above a deep tunnel in cohesionless soil (DT1, DT2).

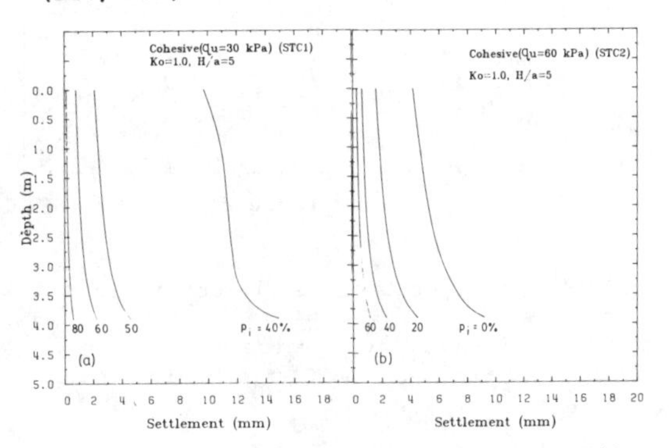

Fig. 8. Subsurface settlement profile above a tunnel in cohesive soil (STC1, STC2).

40% induces a sudden, almost constant, increase in vertical displacement indicating initiation of a "plug-like" collapse mechanism. This mode of failure does not occur in the stronger soil (Fig. 8.b).

The development of the settlement ratio, S_s/S_c, during progressive support pressure reduction is presented for all six cases in Fig. 9. It confirms the earlier finding that the settlement ratio for Mode I is greater than for Mode II. As schematically shown in Fig. 3.b, the ratio is initially almost constant and increases generally at lower support pressures. Since the ultimate behaviour (near collapse) could not be simulated by the FE analysis, the results should not be directly compared with those from conditions near collapse. Values calculated from the relationship proposed by Atkinson and Potts (1977) are also plotted on Fig. 9. Their findings, deduced from model tests at or after collapse, correspond to the ultimate state (Point 2 on Fig. 3.b). An α-value of 0.13 seems to predict the

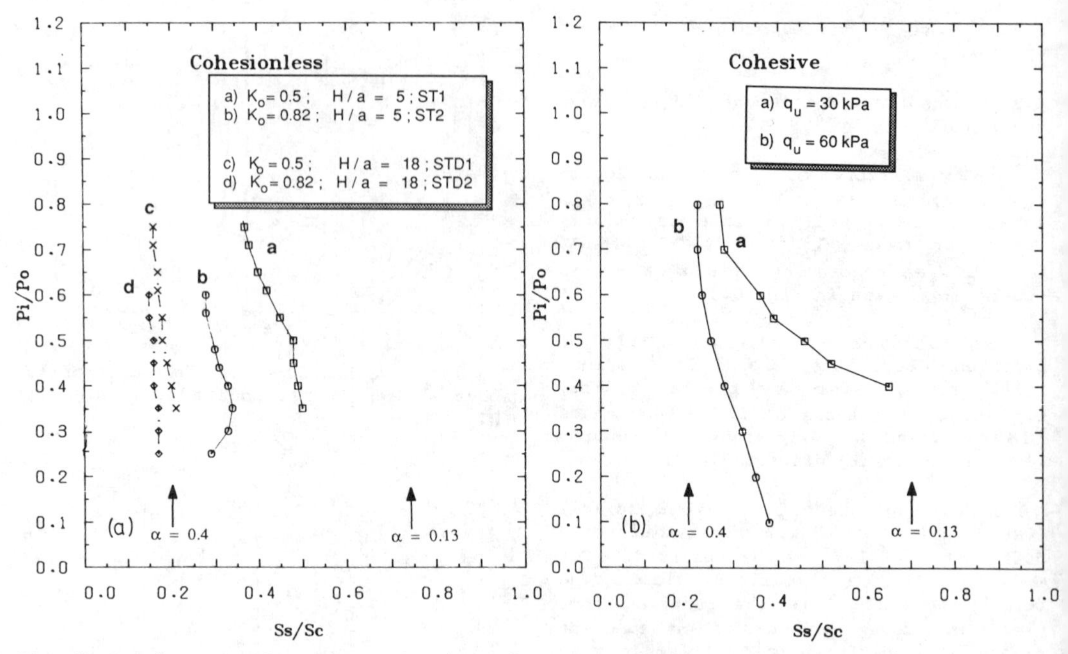

Fig. 9 Development of settlement ratio S_s/S_c.

ultimate behaviour which was only approached in Case STC1 (cohesive soil with q_u = 30 kPa).

4.2.2 Surface Settlement Profile

In tunnel design one is also concerned with the damage causing differential settlement, i.e., the gradient at the inflection point and the trough width. Fig. 10 presents the surface settlements profiles calculated for all six cases. Case ST1 is the most critical in terms of differential settlements. The large maximum settlement and narrow trough is due to the relatively low K_o which reduces the effect of stress arching to resist downward motion. The narrow trough can also be attributed to the formation of two nearly vertical, localized yield zones with downward movement of the soil above the tunnel. As K_o increases, arching above the roof increases, localized yield zones are suppressed, the trough is widened and the settlements are reduced. For deep tunnels, the trough is wider (Fig. 10.b).

As was shown earlier, a minimum support pressure is required to maintain stability in Case STC1 whereas in Case STC2 the tunnel may stand open without support. At p_i = 40% (Fig. 10.c), the maximum settlement for STC1 is more than ten times that for STC2 and more than twice the settlement of STC2 at zero support pressure. The most distinct feature is the difference in settlement gradient.

The (i/a)-ratios calculated at the final incremental excavation step (ST1, ST2, STC1 and STC2) are plotted in Fig. 1 for comparison with the conventional design methods. These i/a values fall in a range bounded by the relations given by Schmidt (1969), Peck (1969) and O'Reilly and New (1982) for clay. Since the soil above the tunnels has not yet reached the collapse mode in the finite element analyses, this implies that predictions based on the work of these authors may lead to unconservative predictions if the initiation of a collapse mechanism is not be prevented in the field. The (i/a)-ratios suggested by O'Reilly and New (1982) are lower than those suggested by Atkinson and Potts (1977) whose findings were based on model tests on tunnels near collapse. It may be inferred that the ground control in tunnels reported by O'Reilly and New (1982) may not have been adequate to prevent the development of yield zones and the initiation of collapse mechanisms.

It must be concluded that a satisfactory prediction of settlements can only be made if the behavioural modes of a tunnel can be predicted, i.e., the extent of yielding and the proximity of collapse has been assessed for the actually applied support pressure.

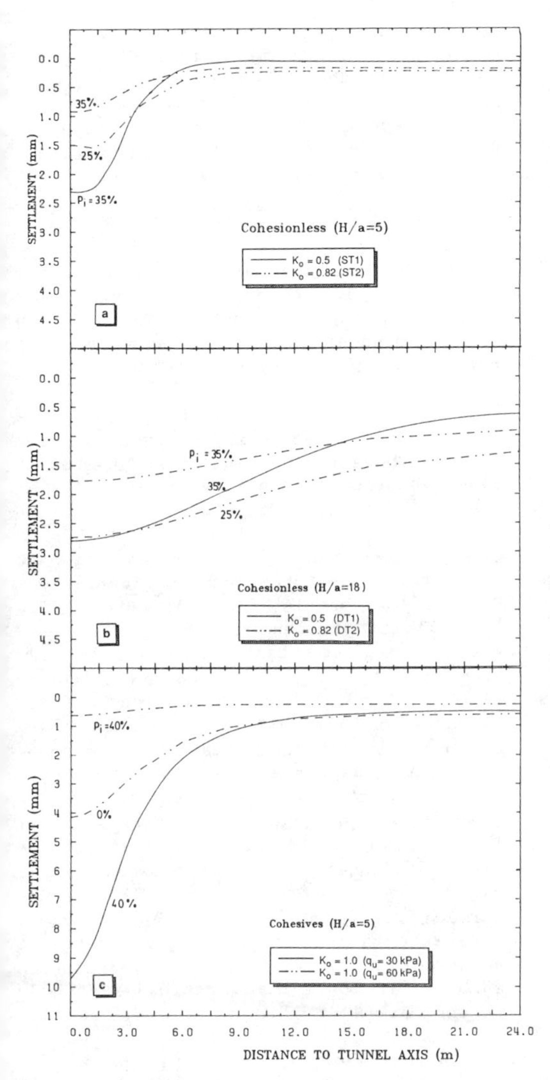

Fig. 10. Surface settlement profiles.

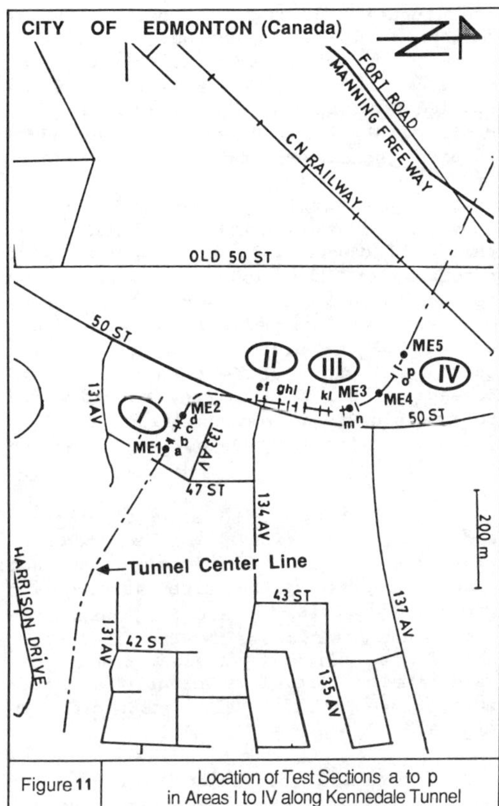

Figure 11 Location of Test Sections a to p in Areas I to IV along Kennedale Tunnel

5. COMPARISON WITH CASE HISTORIES

5.1 Description of Kennedale Tunnel

This tunnel was constructed to convey stormwater (Fig. 11). The excavated diameter of the tunnel was 3.2 m over a total length of 1,670 m, with an average overburden cover to the tunnel axis of 13.7 m. The tunnel was bored by Lovat Tunnel Boring Machine (TBM) and the advance rate varied from 6 m to 25 m per day depending on the ground conditions encountered. The primary lining consisted of segmented steel ribs and wood lagging spaced at 0.9 to 1.2 m. A plain, cast-in-place concrete liner was installed after tunnelling was completed. Details of the construction sequence and the soil conditions are reported by Corbett (1984) and Wong and Kaiser (1986).

An extensive field instrumentation program was devised to monitor the relationship between the maximum vertical settlements and the actual support pressures induced by tunnelling. Twenty-six centreline surface settlement points and five multi-point extensometers were installed to detect the vertical settlement profiles along the tunnel axis. In addition, the crown pressure was measured by the calibrated segmental laggings in 16 sections designated by 'a' to 'p' in Fig. 11 (Corbett, 1984). Some of these measurements are used in the following to illustrate the dependency between support pressure and settlement as described in the previous section.

5.2 Interpretation of Measurements at Kennedale Tunnel

Five multi-point extensometers (ME1 to 5; Fig. 11) were installed to detect the subsurface settlement profile above the tunnel in regions of distinctly differing ground behaviour. Fig. 12 presents the final subsurface displacement profiles together with the maximum surface settlements, S_s, measured at about 2 m depth. The records of ME3, 4 and 5 require some explanation. At ME3, excessive ground loss created a large void, approximately 2 m high and 3 m wide above the crown. Eventually collapse of this void caused a large sudden downward movement of the lower probes. The decreasing vertical movement with depth observed in ME4 is believed to be a result of placing the probes in sloughing fine sand. ME5 was lost four days after the passage of the tunnel face, probably due to the signifcant amount of ground loss (note: different scales are used on Fig. 12).

The settlements monitored at the surface and near the crown are plotted in Fig. 13.a. These measurements do not exactly represent the true surface and crown settlements, S_s and S_c, because of the anchor positions. The measured S_s/S_c should be slightly smaller than the calculated. Fig. 13.a clearly supports the conceptual model described in Fig. 3.c.

The measured settlement ratios are combined with the observed support pressures (normalized to the overburden pressure) and plotted in Fig. 13.b. This figure confirms the strong dependency between p_i and S_s/S_c as shown in Fig. 3.b. The settlement ratio is initially constant, then increases rapidly as yielding propagates and eventually approaches unity when collapse occurs.

5.3 Other Case Histories

Semi-empirical approaches for predicting the surface settlement profile (e.g., Peck, 1969) are based largely on field measurements from various case histories. In these approaches the location of the point of inflection, i, is assumed to be dependent on the depth ratio, H/a, only. Because many important factors governing the tunnel behaviour such as strength and deformation properties of ground as well as stress relief and boundary conditions are not considered, a wide scatter in the observed data is to be expected. Results from histories reported by Attewell (1978), Branco (1981), Atkinson and Potts (1977), O'Reilly and New (1982), El-Nahhas (1980), and Cording et al. (1976) were grouped in cohesionless and cohesive soils, and will be used to further illustrate the dependence of i/a and S_s/a (or (S_c/a or p_i/p_o) and to support the surface settlement/support pressure relationship (i/a)-(S_s/S_c)-(p_i/p_o) introduced earlier in Section 2 (Fig. 3.a and 4.6).

5.3.1 Cohesionless Soils

Results from tunnels in cohesionless soils, grouped in four depth ranges, are plotted in Fig. 14.a and b. The 'shallow tunnel' boundary H/a = 5 is shown and the following aspects may be observed:

(1) Fig. 14.a, a plot of H/a versus i/a with the expected range given by Peck (1969), shows that a significant number of case histories (especially deeper tunnels; H/a > 5) lies outside this range.

(2) By comparing Fig. 14.b and Fig. 4.b similar features can be identified. For shallow tunnels, i/a decreases rapidly to a limiting value of between 0.7 and 2 with a small increase in S_s, while, for deep tunnels, i/a decreases at a slower rate. The results from model tests (squares on Fig. 14) tend to the lower bound of (i/a)-values because the measurements were taken near collapse. For depth (H/a < 5), the ratio i/a approaches its limiting value when S_s/a exceeds 1.0%. This can abe explained by the fact that a small displacement at the crown causes a yield zone to propagate to the ground surface (e.g., at S_c/a = 0.4%, the yield zone reached the ground surface in the FE analysis at Case ST1). For deep tunnels (H/a >> 5), the minimum ratio i/a is only reached at (S_s/a)-values in excess of 5%.

The influence of K_o could not be assessed because K_o was seldom recorded.

5.3.2 Cohesive Soils

Cohesive soils generally exhibit a wider variation in strength and deformation properties than cohesionless soils. Hence, the case histories were subdivided into two categories: (1) c_u < 100 kPa and (2) c_u > 100 kPa. The results are plotted in Figs.15 and 16. Notable aspects of practical significance are:

(1) For c_u < 100 kPa (Fig. 15), similar conclusions as for cohesionless soils can be drawn. For shallow tunnels (H/a < 5), i/a becomes independent of the magnitude of surface settlement when S_s/a exceeds about 1% but for deep tunnels i/a

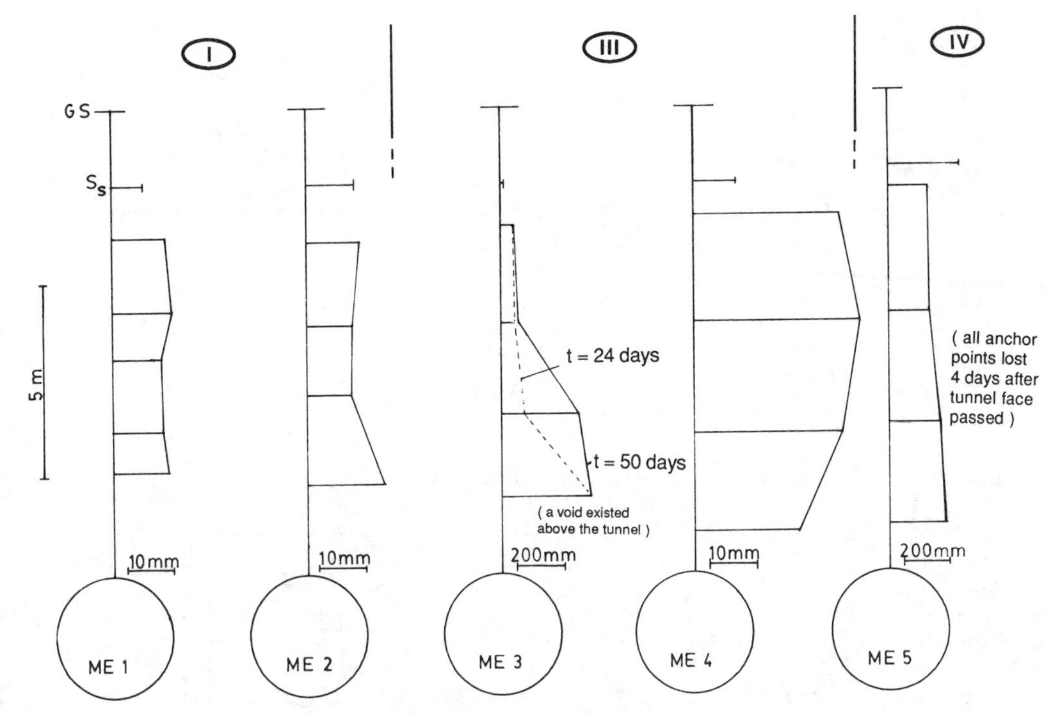

Fig. 12 Subsurface displacement profiles measured by multi-point extensometers (Kennedale Tunnel).

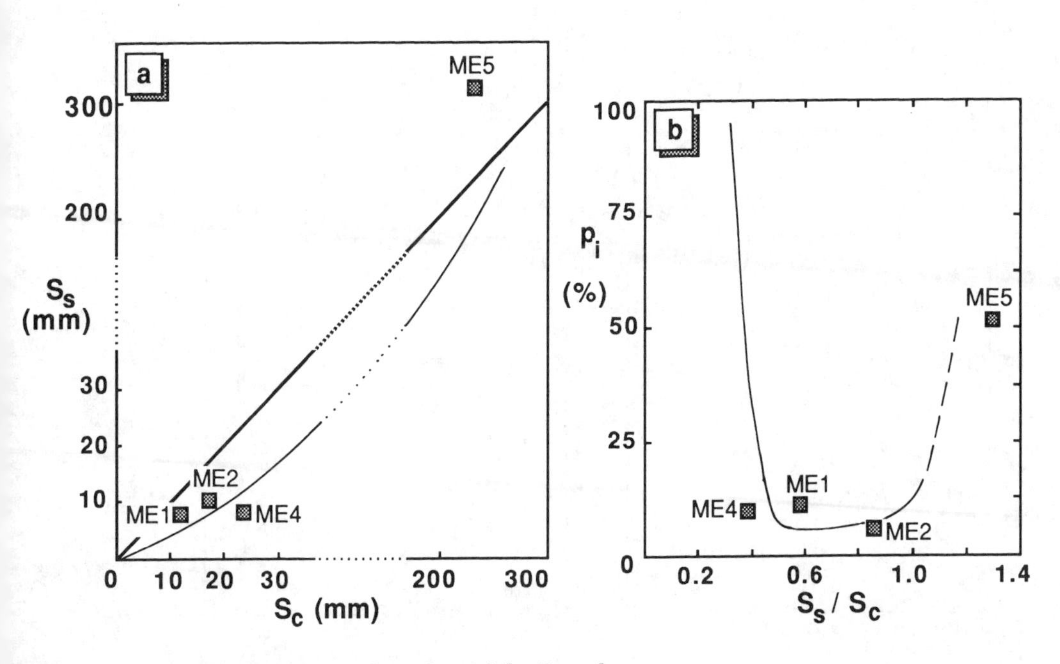

Fig. 13. Field measurements from Kennedale Tunnel.

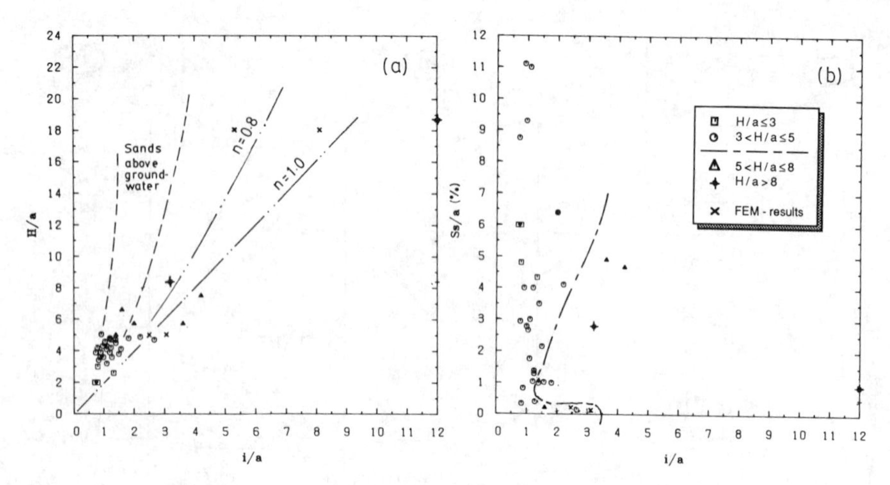

Fig. 14. Field measurements from tunnels in cohesionless soils.

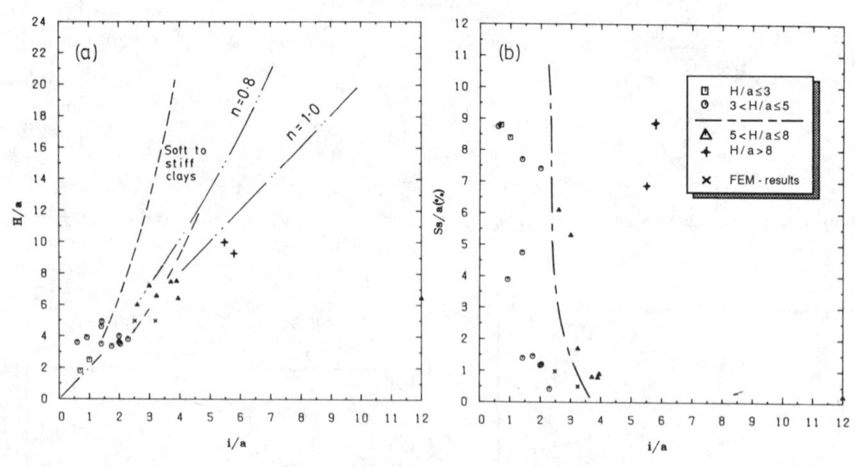

Fig. 15. Field measurments from tunnels in cohesive soils (c_u < 100 kPa).

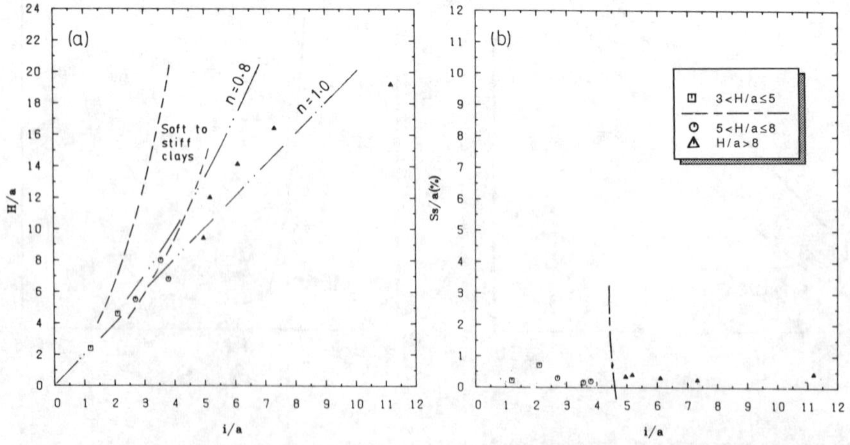

Fig. 16. Field measurements from tunnels in cohesive soils (c_u > 100 kPa).

changes more gradually to a limiting value at $S_s/a > 5\%$. The relationship (H/a)-(i/a) is almost linear but (S_s/a) varies widely.

(2) For $c_u > 100$ kPa (Fig. 16), the surface settlement is relatively small (< 1.0%) due to high soil strength (or stiffness). The relationship between H/a and i/a is still linear but little settlement is experienced.

The data presented in Figs 14 to 16 confirms the equation proposed by Schmidt (1969):

$$\frac{i}{a} = \left(\frac{H}{2a}\right)^n$$

Examples for n = 0.8 and 1.0 are shown for comparison. The fact that this relationship is independent of p_i does not contradict earlier findings of support pressure dependency. It demonstrates that the support pressure relief caused during construction (especially for shallow tunnels) is always sufficient to reach the ultimate i/a-value. Hence, in practice only the magnitude of settlement depends on the support pressure.

6. CONCLUSIONS AND PRACTICAL IMPLICATIONS

1. A conceptual model, verified by numerical simulations and field measurements, is proposed to relate the surface and subsurface settlements to the tunnel support pressure. The p_i-(S_s/S_c)-(i/a) relationship is a continuously changing function reflecting the entire spectrum of tunnel behaviour from the initial elastic response to the ultimate collapse state. The shape of this function depends on the initial in situ stresses, the tunnel depth, the soil strength and deformation parameters, and the construction sequence. Any design method, which does not consider these factors could lead to unsatisfactory settlement predictions.

2. It has been demonstrated that the conventional design methods for settlement predictions based on only one or two parameters could lead to a conservative prediction because the most important factor, the actual support pressure, is ignored. The applicability of the correlations can be much improved if the behavioural mode of a tunnel or the extent and type of yielding is identified.

3. Since the actually mobilized support pressure at the crown and the settlement are interrelated, the design loads suggested, for example by Terzaghi (1943) or Duddeck and Erdman (1980), should not be used to predict the expected displacement field around a tunnel. These recommended loads correspond to one specified situation and are intended to predict the extreme stresses in a support but not the expected ground movements.

4. While the width of the settlement trough was shown to be dependent on the support pressure, field measurements indicate that during most tunnel construction procedures (particularly in shallow tunnels) sufficient pressure relief is created to reach the ultimate, minimum (i/a)-ratio. Hence, i/a can be predicted by the relationship proposed by Schmidt (1969) and only the maximum settlement, S_s, must be established as a function of the support pressure, p_i, for a reliable prediction of the surface settlement trough.

5. The subsurface settlement profile above the tunnel is an excellent indicator to identify the actual mode of tunnel behaviour. It should be employed to verify the model adopted for performance prediction.

6. The in situ stress ratio not only governs the mode of yielding but also influences the displacements of soil around the tunnel. A critical K_o value, separating Mode I ($K_o < K_{cr}$) of localized yielding and Mode II ($K_o > K_{cr}$) of general yielding, exists. The surface settlement in Mode II is much smaller than that in Mode I at an equal support pressure. In addition, Mode I has a narrower settlement trough than Mode II. This implies that the potential for damage due to surface settlement becomes more critical for Mode I than for Mode II. For normally consolidated soils where $K_o = 1 - \sin\phi$, the expected mode of yielding is generally Mode I. Hence, great care should be taken under these conditions to minimize crown settlements by applying relatively high support pressures near the crown (e.g. by expanding segments or by immediate pressure grouting).

7. REFERENCES

Atkinson, J.H. and Potts, D.M., 1977. Subsidence above shallow tunnels in soft ground. Journal of Geotechnical Engineering, ASCE, GT4, pp. 307-325.

Attewell, P.B. and Farmer, I.W. 1975. Ground settlement above shield driven tunnels in clay. Tunnels and Tunnelling, Vol. 7, No. 1, pp. 58-62.

Attewell, P.B., 1978. Ground movements caused by tunnelling in soil. Conference on 'Large Ground Movements and Structures', Cardiff, Pentech Press, pp. 812-948.

Branco, P., 1981. The Behaviour of Shallow Tunnels in Edmonton Till. M.Sc. Thesis, Department of Civil Engineering, Univ. of Alberta, Edmonton, Alberta, 331p.

Chan, D., 1985. Finite Element Analysis of Strain-Softening Materials. Ph.D. Thesis, Department of Civil Engineering, Univ. of Alberta, Edmonton, Alberta, 355p.

Corbett, I., 1984. Load and Displacement Variation along a Tunnel. M.Sc. Thesis. Department of Civil Engineering, Univ. of Alberta, Edmonton, Alberta, 246p.

Cording, E.J., Hansmire, W.H., MacPherson, H.H., Lenzini, P.H. and Vonderohe, A.P., 1976. Displacement around Tunnels in Soils. Report prepared for Dept. of Transportation, Washington, Civil Engineering Department, University of Illinois.

Duddeck, H. and Erdmann, J., 1982. Structural design models for tunnel. Tunneling '82, pp. 83-91.

Eisenstein, Z. and Negro, A., 1985. Comprehensive Design Method for Shallow Tunnels. ITA/AITES, International Conference on Underground Structures in Urban Areas, Prague, pp. 3-20.

El-Nahhas, F., 1980. The behaviour of tunnels in stiff soils. Ph.D. Thesis, Department of Civil Engineering, Univ. of Alberta, Edmonton, Alberta, 305p.

Fenner, R., 1939. Untersuchungen zur Erkenntnis des Gebirgsdruckes. Glueckauf, Ann. 74:32,33.

Martos, F., 1958. Concerning an approximate equation of the subsidence trough and its time factors. International Strata Control Congress, Leipzig. (Berlin: Deutsche Akademie der Wissenschaften zu Berlin, Sektion für Bergbau, pp. 191-205.

O'Reilly, M.P. and New, B.M., 1982. Settlements above Tunnels in the U.K. - their magnitude and prediction. Tunnelling '82, pp. 173-181.

Pacher, F., 1964. Measurements of deformations in a test gallery as a means of investigating the behaviour of the rock mass and specifying lining requirements. Rock Mechanics and Engineering Geology, Suppl. I, pp. 149-161.

Peck, R.B., 1969. Deep excavations and tunnelling in soft ground. 7th International Conference on Soil Mechanics and Foundation Engineering, State-of-the-Art Volume, Mexico City, pp. 225-290.

Rabcewicz, L.V. and Golser, J., 1973. Principles of dimensioning the supporting system for the New Austrian Tunnelling Method. Water and Power, Vol. 25, pp. 88-93.

Schmidt, B., 1969. Settlements and Ground Movements associated with Tunnelling in Soil. Ph.D, thesis, University of Illinois.

Terzahgi, K., 1943. Theoretical Soil Mechanics. John Wiley and Sons, 502p.

Wong, R.C.K., 1985. Design and Evaluation of Tunnels and Shafts. Ph.D. thesis, Department of Civil Engineering, University of Alberta, Edmonton, Alberta, 315p.

Wong, R.C.K. and Kaiser, P.K., 1986. Ground behaviour near soft ground tunnels. International Congress on Large Underground Openings, Florence, Vol. 1, pp. 942-951.

Wong, R.C.K. and Kaiser, P.K., 1987. Effect of support pressure distribution on surface settlement. Tunnels and Tunnelling (in press).

APPENDIX A
NOTATIONS

a = radius of tunnel
D = diameter of tunnel
H = depth above tunnel centre
x = transverse distance from tunnel axis
C = cover above tunnel roof (= H - a)
γ = unit weight
ν = Poisson's ratio
E = Young's modulus
ϕ = angle of internal friction
q_u = unconfined compression strength of cohesive soil
c_u = shear strength of cohesive soil

p_i = internal support pressure (radial stress) at tunnel wall
p_o = vertical initial in situ stress
K_o = stress ratio at rest
K_a = stress ratio in active state (= $\tan^2 (45° - \phi/2)$)
K_{cr} = critical stress ratio to distinguish Mode I from Mode II type yielding
K = coefficient of tangential arching above the roof of a tunnel (introduced by Terzaghi)
S_o = surface settlement parameter
S_s = maximum ground surface settlement
S_c = vertical displacement at the crown of tunnel
i = point of inflection on surface settlement trough
V_s = volume of settlement trough per unit length along tunnel
V_{exc} = volume of ground loss per unit length of advance

Environmental geotechnology

Prediction and Performance in Geotechnical Engineering / Calgary / 17-19 June 1987

Effect of caustic soda contamination on volume change characteristics of bentonite

A.Sridharan, Sudhakar M.Rao & V.S.Gajarajan
Indian Institute of Science, Bangalore

ABSTRACT : The present work investigates the effect of caustic soda contamination on the plasticity and volume change characteristics of bentonite clay. Lack of sufficient information on the action of aggressive acidic and alkaline contaminants on soil mechanical properties necessitated the study. Interaction with the strong alkali increases the net particle negative charge indicated by the enhanced exchangeable sodium content of the treated specimen. Such an increase results in an overall expansion of the diffuse double layer thickness responsible for the higher liquid limit, compressibility, swelling magnitude and swelling pressure of the treated clay specimen. Also the increased negative charge facilitates a more oriented particle arrangement containing pores of narrower radius which cause greater capillary stresses and lower shrinkage limit.

1 INTRODUCTION

Modification of soil mechanical properties on chemical contamination from disposal of hazardous wastes and accidental spillage or leakage of highly aggressive industrial effluents form an important aspect of environmental problems relating to geotechnical engineering. Case histories of foundation failures, structural damage to light industrial buildings on contamination of subsoils by acids and alkalies (Lukas et al 1972, Sridharan et al 1981, Kumapley and Ishola 1985) emphasize the importance which needs to be attached to the consideration of the above problem.

Environmental geotechnics is realized to be more of a planning and decision tool to be used for forecasting geotechnical problems that may ultimately arise rather than solving, after the event has occurred (Sembenelli and Ueshita 1981). Such a requirement calls for a fore-knowledge of the effect of industrial contaminants on the geotechnical properties of soil materials; the contaminants include inorganic and organic solutes in aqueous solutions and organic fluids. Although, the role of inorganic solutes and organic fluids in influencing the geotechnical properties of clayey soils has been the consideration of several workers (Michaels and Lin 1954, Bolt 1956, Mesri and Olson 1971 a and b, Sridharan and Venkatappa Rao 1973, Fernandez and Quigley 1985, Sridharan et al 1986), insufficient information exists on the action of aggresive contaminants, such as acids and alkalies, on the soil mechanical properties.

The present paper discusses the effect of a strong alkali, caustic soda, on the volume change characteristics of bentonite clay. The clay was chosen for study as it is frequently encountered in foundation soils and is also used as a pollution control liner in waste impoundment sites.

2 EXPERIMENTAL PROCEDURE

2.1 Treatment with caustic soda

Bentonite clay from Kolar district., Karnataka State was used in the studies. Batches of 500 g of the clay specimens were equilibrated with N/2 sodium hydroxide solutions (solid: solution ratio, 1:5) with constant agitation for a period of 7 days. The pH of bentonite-caustic soda suspension was measured to be 13.1 at the end of the equilibration. On completion of treatment, the excess alkali solution was discarded and the treated clay repeatedly washed with distilled water to remove adhering salts. It is realized that removal of excess alkali from the clay-pore water system following equilibration, is not representative of field condition. The presence of a high salt concentration in the clay-pore water system is however known to significantly influence the plasticity and volume change characteristics of montmorillonite clay (Bolt 1956, Olson and Mesri 1970, Mesri and Olson 1971). Hence, to avoid interference from pore salt concentration effect, removal of excess alkali and washing of the treated clay specimen was resorted to.

The washed alkali treated clay specimen was air-dried, pulverized to pass a 36 mesh British Standard sieve and stored in polythene bags. The amount of alkali interacted with bentonite clay was estimated from the differences in hydroxyl ion concentrations prior to and after the equilibration by titrating the hydroxyl ions against standardized hydrochloric acid solution.

2.2 Exchangeable cation analysis

The exchangeable cation analysis of the untreated and treated clay specimens were performed by displacing the exchangeable cations with ammonium ions on successive washings with neutral 1 N ammonium acetate solution and estimating the displaced calcium, magnesium, sodium and potassium ions by standard methods (Hesse 1972). The cation exchange capacity, CEC, was computed from the sum of individual exchangeable ions present.

2.3 Index properties

The specific gravity of the untreated and alkali treated bentonite specimens were determined by pycnometer method and the reported values are an average of three tests; individual determination differed from the mean by < 0.01. Liquid limits of the clay specimens were obtained by Casagrande's procedure. The liquid limit test was repeated twice for each specimen and the average of the two values considered. The variation between two trials was 1%. The plastic and shrinkage limits of the clay specimens were determined by standard procedures. The shrinkage limits reported are an average of three determinations; the variations between individual determinations were < 2% (of the reported value).

2.4 Oedometer tests

The clay specimens were tested in standard fixed ring consolidometers using brass rings, 7.6 cm in diameter and 2.5 cm height. The inside of the ring was coated with silicone grease to minimize side friction between the ring and the clay specimen. The oedometer tests were performed in a room maintained at a uniform temperature of 20°C.

In the one dimensional consolidation test procedure, untreated and alkali treated bentonite specimens were mixed into soft easily remouldable states at moisture contents corresponding to 55% of the individual liquid limits, placed under vaccum and agitated periodically to remove entrapped air. Following de-aeration, the moist clay specimens were hand remoulded into the consolidation rings to a thickness of 1.8 cm, taking care to prevent any air entrapment. Using a load increment ratio of unity, consolidation tests were performed in the usual manner. Each specimen was loaded to a maximum of 400 kPa and unloaded to the seating pressure of 6.25 kPa. Loads were left in place till primary compression was complete and the rate of secondary compression became very small. Time-deformation curves obtained at each load increment were used to determine the coefficient of consolidation by the rectangular hyperbola method (Sridharan and Sreepada Rao 1981). The coefficient of permeability was computed from the C_v values. Repetition of oedometer tests with identically prepared clay specimens gave consistent data.

To evaluate the magnitude of maximum swelling (in divisions) and the swelling pressure of the untreated and caustic soda treated clay specimens, the 'free-swell oedometer test' procedure (Fredlund 1969) was adopted. The untreated and treated clay specimens were packed in their loosest possible condition in standard oedometer rings corresponding to initial thickness of 0.56 cm and dry densities of 1.16 g/cm^3 (initial void ratios were approximately 1.38) and allowed to swell on addition of water to equilibrium positions under a token load of 6.25 kPa. Subsequent to the attainment of maximum swelling, loads were added in conventional increments (load increment ratio = 1) and the specimens consolidated at each load. The load required to bring the clay specimen back to its original void ratio (~1.38) is defined as the 'swelling pressure'.

3 RESULTS AND DISCUSSIONS

3.1 Index properties

Table 1 presents the index properties of the untreated and caustic soda treated bentonite specimens. The specific gravity value of the alkali treated specimen is slightly lower than that of the untreated specimen. Treatment with caustic soda increases the liquid limit of bentonite clay from 495% to 570%. In comparison the plastic limits are slightly affected. The shrinkage limit of bentonite decreases from 13.1% to 11.1% on alkali treatment.

The liquid limit of bentonite is known to be controlled by the thickness of the diffuse double layer surrounding the clay particles in an aqueous environment (Sridharan and Venkatappa Rao 1975, Mitchell 1976). Earlier Sridharan and co-workers (1986) had investigated the relations between liquid limit of montmorillonite soils and soil composi-

Table 1. Index properties of bentonite specimens.

Clay type	G	w_l %	w_p %	w_s %
Bentonite	2.81	495	49.0	13.1
Bentonite treated with caustic soda	2.72	570	44.7	11.1

tional factors that effect the diffuse double layer thickness. Results showed that the liquid limit has a strong positive relationship with the amount of exchangeable sodium ions present. The diffuse double layer thickness and hence the liquid limit of montmorillonite is a function of the exchangeable sodium content owing to the ion's greater ease of dissociation from the surface; the other exchangeable cations namely calcium, magnesium and potassium are strongly adsorbed by the clay surface and do not contribute significantly to the diffuse double layer thickness.

Treatment with caustic soda increases the exchangable sodium content of bentonite clay from 58.6 meq/100g to 83.0 meq/100g(Table 2). The reason for such an increase is attributed to the high pH (13.1) of the bentonite-alkali suspension. The high pH encourages dissociation of hydroxyl groups at the particle edges and increases the negative charge (Schofield and Samson 1954) as -

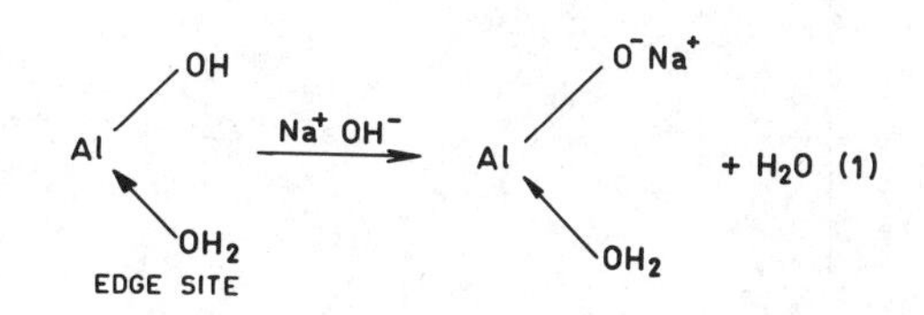

The negative charge developed at the particle edges is compensated by adsorption of additional sodium ions from the ambient solution. Support for this hypothesis is found from the near 1:1 correspondence between the amount of hydroxyl ions consumed (26.5 meq/100 g) and the increase in exchangeable sodium content (24.4 meq/100 g) on caustic soda interaction with bentonite clay. The increase in exchangeable sodium content would result in an overall increase in the diffuse double layer thickness surrounding the clay particles and the liquid limit of the treated clay specimen.

The decrease in shrinkage limit of bentonite clay on equilibration with caustic soda is explained as follows. Treatment with alkali was observed to increase the exchangeable sodium content resulting in an overall expansion of the diffuse double layer thickness. Such a situation lends to the development of greater inter-particle repulsion and a more oriented particle arrangement (Yong and Warkentin, 1966). The latter arrangement contains pores with narrower radius allowing for greater development of capillary stresses during shrinkage and also provides for relatively easier movement of particles and particle groups, both conditions

Table 2. Physico-chemical properties of bentonite specimens.

Clay type	Amount of hydroxyl ion consumed meq/1ØØg	Cation exchange capacity meq/1ØØg				
		Calcium	Magnesium	Sodium	Potassium	Total
Bentonite	--	39.Ø	15.2	58.6	Ø.7	113.5
Bentonite treated with caustic soda	26.5	36.8	14.6	83.Ø	1.Ø	135.4

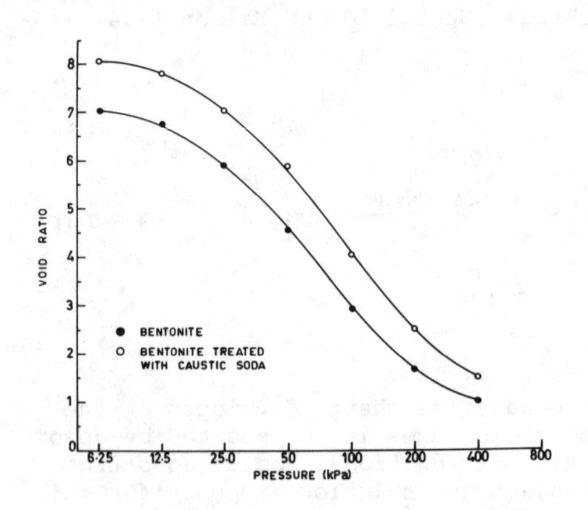

Fig. 1 Void ratio - pressure relations for bentonite and caustic soda treated bentonite specimens.

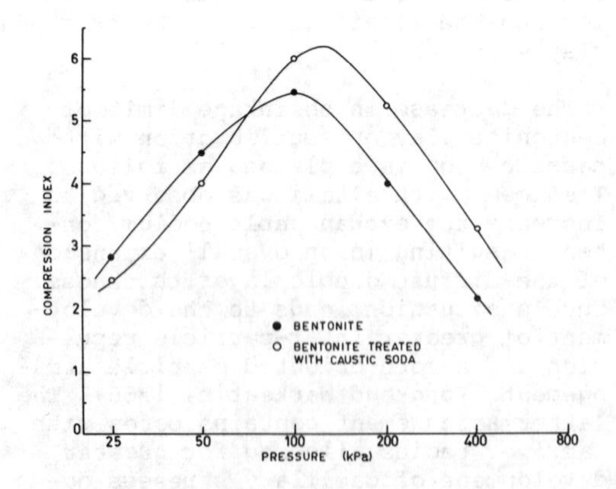

Fig. 2 Compression index versus pressure plots.

condusive to low shrinkage limits (Sridharan and Venkatappa Rao 1971, Mitchell 1976).

3.2 Oedometer test results

Figure 1 presents the void ratio-pressure plots for the untreated and alkali treated clay specimens. The consolidation curve of the alkali treated specimen is positioned above that of its untreated counterpart at any applied pressure. The equilibrium void ratio sustained by montmorillonite at a given pressure is governed by the physico-chemical repulsion force arising from interaction of adjacent diffuse double layers (Bolt 1956, Olson and Mesri 197Ø, Sridharan and Venkatappa Rao 1973). Presumably the increase in exchangeable sodium content on caustic soda treatment which mobilizes relatively expanded diffuse double layers and greater inter-particle repulsion is responsible for the higher void ratio sustained by the alkali treated specimen.

Figure 2 presents compression index as a function of applied pressure for untreated and treated bentonite specimens. Upto pressures of approximately 7Ø kPa, the alkali treated specimen exhibit lesser compressibility than the untreated specimen; at higher pressures the trend is reversed.

Figure 3 plots the coefficient of permeability as a function of void ratio for the untreated and treated clay specimens. At a given void ratio, the alkali treated specimen exhibits a lower permeability coefficient than the untreated clay. Caustic soda treatment increases the net negative charge of bentonite clay. Such an increase enhances particle dispersion (Yong and

Warkentin 1966), resulting in narrower pores between particles and a more tortuous flow path (Quigley and Thomson 1966). Further the available channels would be relatively more constricted by the expanded diffuse double layers (in comparison to the untreated specimen) reducing the effective void space for water flow. Apparently, the increased tortuosity factor and greater constriction of flow channels cause lower k values for the alkali treated specimen.

Figure 4 plots the time-swelling curves obtained for the untreated and caustic soda treated bentonite specimens. The plots readily depict the markedly higher maximum swelling attained by the alkali treated specimen at equilibrium. The void ratio-pressure curves obtained on subsequent loading of the clay specimens (following attainment of maximum swelling) by the conventional procedure are given in Figure 5. The loads required to bring the untreated and alkali treated specimens back to the original void ratio (1.38) are 570 kPa and 730 kPa respectively. The swelling behaviour of montmorillonite clay is recognised to be controlled by diffuse double layer repulsion

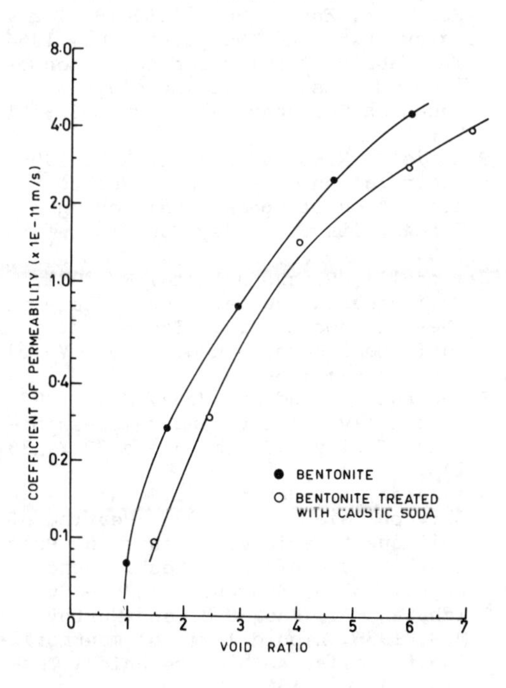

Fig.3 Coefficient of permeability versus void ratio plots.

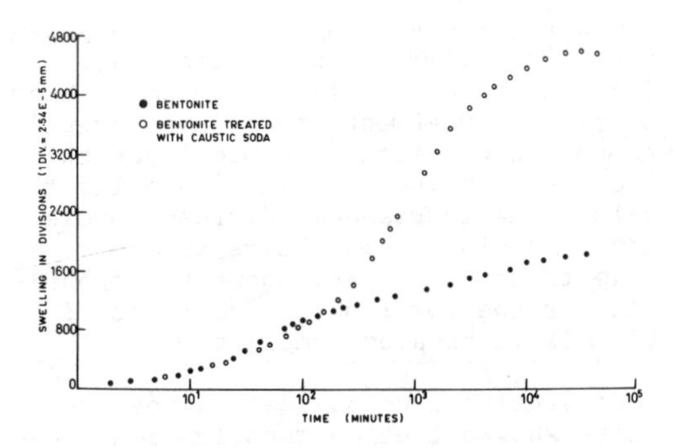

Fig. 4 Swelling versus time curves.

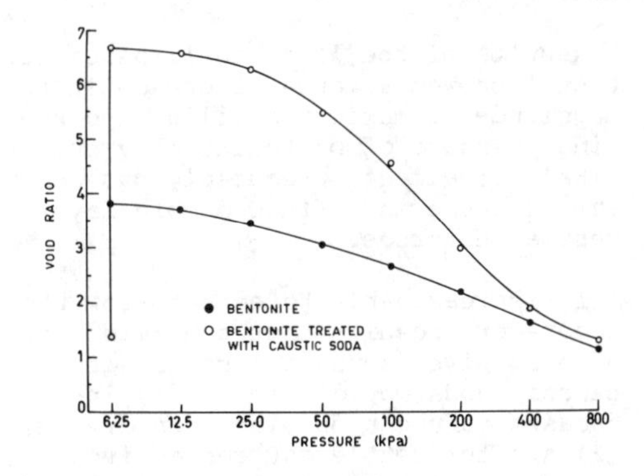

Fig. 5 Oedometer free - swell test results

forces (Bolt 1956, Warkentin and Schofield 1962, van Olphen 1963, Sridharan and Jayadeva 1982). Since treatment with caustic soda increases the exchangeable sodium content and the diffuse double layer repulsion force, the higher magnitude of maximum swelling and swelling pressure exhibited by the alkali treated specimen is understandable.

4 CONCLUSIONS

Interaction of caustic soda with bentonite influences the plasticity and volume change characteristics of bentonite. The high pH of the clay-alkali suspension favours dissociation of the edge hydroxyl groups increasing the net negative charge, reflected in the higher exchangeable sodium content of the alkali treated specimen.

Treatment with caustic soda increases the liquid limit of bentonite clay. The increase in exchangeable sodium content on alkali treatment leads to an overall expansion of diffuse double layer thickness and an increase in liquid limit value. The increase in diffuse double layer repulsion also leads to a more oriented particle arrangement responsible for the lower shrinkage limit of the alkali treated specimen.

Conventional consolidation test results showed that on caustic soda treatment, the bentonite clay sustains the external load at higher equilibrium void ratios. The enhanced diffuse double layer thickness is also responsible for the higher C_c values of the alkali treated specimens beyond stresses of 70 kPa.

Results of the 'free-swell oedometer tests' showed a marked increase in the magnitude of maximum swelling and swelling pressure of bentonite clay on alkali treatment, presumably caused by the increase in diffuse double layer repulsion forces.

It is reasonable to expect from the laboratory results that contamination of expansive foundation soils with caustic soda would lead to (1) increased volume reduction upon drying, (2) greater settlement for a given stress level as a result of reduction in equilibrium void ratio, particularly in case of more heavily loaded foundations and (3) increased heave on inundation with water, for a given stress level; all the three situations detrimental to the stability of structures founded on the contaminated expansive soil. The decrease in permeability coefficient on caustic soda treatment would however have a beneficial influence on the effectivensss of clay liner used to mitigate contaminant migration in waste impoundment sites.

REFERENCES

Bolt, G.H. 1956. Physico-chemical analysis of the compressibility of pure clays. Geotechnique 6:86-93.

Fernandez, F. and Quigley, R.M. 1985, Hydraulic conductivity of natural clays permeated with simple liquid hydro-carbons. Can. Geotech. J. 22:205-214.

Fredlund, D.G. 1969. Consolidometer test procedural factors affecting swell properties. Proc. Second International Conference on Expansive clays, Texas:435-456.

Hesse, P.R. 1972. A text-book of soil chemical analysis. New York : Chemical Publishing Co.

Kumapley, N.K. and Ishola, A. 1985. The effect of chemical contamination on soil strength. Proc. Eleventh Int. Conf. Soil Mech. Found. Engg., Vol.3, San Francisco : 1199-1201.

Lukas, R.G. and R.J. Gnaedinger Jr. 1972. Settlement due to chemical attack of soils. Proc. ASCE Speciality Conf. on the Performance of Earth and Earth-supported structures, Vol.1, Purdue Univ., Lafayette, Indiana : 1087-1104.

Mesri, G. and Olson, R.E. 1971. Consolidation characteristics of montmorillonite. Geotechnique 21 : 341-352.

Michaels, A.S. and Lin, C.S. 1954. Permeability of kaolinite. J. Industrial and Engineering Chemistry 46 : 1239-1246.

Mitchell, J.K. 1976. Fundamentals of soil behaviour. New York : Wiley.

Olson, R.E. and Mesri, G. 1970. Mechanisms controlling compressibility of clays. J. Soil Mech. Fdns. Div. Am. Soc. Civ. Engrs. 96 : 1863-1878.

Quigley, R.M. and Thompson, C.D., 1966. The fabric of anisotropically consolidated sensitive marine clay. Canadian Geotechnical Journal 3 : 61-73.

Schofield, R.K. and Samson, R.H. 1954. Flocculation of kaolinite due to attraction of oppositively charged forces. Disc. Faraday Soc. 18 : 135-145.

Sembenelli, P. and Ueshita, K. 1981. Environmental geotechnics : State-of-the-art report. Proc. Tenth Int. Conf. Soil Mech. Found. Engg., Vol.4, Stockholm : 335-394.

Sridharan, A. and Jayadeva, M.S. 1982. Double layer theory and compressibility of clays. Geotechnique 32 : 133-144.

Sridharan, A., Nagaraj, T.S. and Shivapullaiah, P.V. 1981. Heaving of soil due to acid contamination. Proc. Tenth Int. Conf. Soil Mech. Found. Engg., Vol.6, Stockholm : 383-386.

Sridharan, A., Rao, S.M. and Murthy, N.S. 1986. Liquid limit of montmorillonite soils. ASTM Geotechnical Testing. J. 9 : 156-159.

Sridharan, A., Rao, S.M. and Murthy, N.S. 1986. Compressibility behaviour of homo-ionised bentonites. Geotechnique 36 : 551-564.
Sridharan, A. and Sreepada Rao, A. 1981. Rectangular hyperbola fitting method for one dimensional consolidation.ASTM Geotech. Test. J. 4 : 161-168.
Sridharan, A. and Venkatappa Rao, G. 1971. Effective stress theory of shrinkage phenomena. Can. Geotech. J. 8 : 503-512.
Sridharan, A. and Venkatappa Rao, G. 1973. Mechanisms controlling volume change of saturated clays and the role of effective stress concept. Geotechnique 23 : 359 - 382.
Sridharan, A. and Venkatappa Rao, G. 1975. Mechanisms controlling liquid limit of clays. Proc. Istanbul Conf. on Soil Mech. Found. Engg., Vol.1, Istanbul : 75-84.
van Olphen, H. 1963. An introduction to clay colloid chemistry. New York : John Wiley.
Warkentin, B.P. and Schofield, R.K. 1962. Swelling pressure of Na-montmorillonite in NaCl solutions. J.Soil. Sci. 13 : 98-105.
Yong, R.N. and Warkentin, B.P. 1966. Introduction to Soil Behaviour. New York : Macmillan.

Treatment of high water-content sludge using powdery wastes

Motoharu Tamai
Kinki University, Higashi-Osaka, Japan

ABSTRACT: A huge amount of sludge (bottom deposit) is piled up in the rivers, lakes, and bays connected with Japanese cities; the treatment of such sediment with cement or lime will need a great expense. On the other hand, the discharge of powdery wastes has recently increased markedly, the treatment or disposal of which becoming a source of anxiety among the persons concerned.

This report will discuss the treating method of high water-content sludge making use of the properties of powdery wastes having high-temperature hysteresis.

1 INTRODUCTION

The industrial structure of Japan has recently been converting to that of resouce-saving and intelligence-intensive industries. Therefore, the amount of general industrial wastes seems to remain on the same level or tends to decrease depending upon the type of the industry, but the amount of powdery wastes such as dust, residual ashes, and so on is increasing.
The following causes are considered for the above tendency:
1) Recovery of fine particular dust in order to prevent air pollution.
2) Incineration of the sewage sludge resulting from the promotion of water pollution control.
3) Expanded use of coal as a substitution of oil.
4) Recovery of the heat of combustion obtained from combustible wastes.
5) Incineration of various dewatered sludges and municipal refuse, resulting in the reduction of their quantities which will be desirable because of the insufficient land for the final disposal.
This study indicated the chemical and physical properties of the powdery wastes. The some methods for the treatment of high water-content sludge using the incineration ashes of the sewage sludge which had a comparatively stable property among the above-mentioned powdery wastes were also presented together with the indoor and outdoor experiment data.

2 PROPERTIES AND PRESENT STATE OF POWDERY WASTES

2.1 Dust

The dust discharged from various blast furnaces of iron and steel industry contains metals of high value and is hence mostly re-melted as a raw material of the metals: the collected dust in the ceramic industry is completely recycled as a raw material of cement.
Collected dust of the pulverized-coal firing boiler in the power plants is sieved and is mostly utilized as fly ashes for use in fertilizers, soil-improving agents, cement additives, etc.
The EP dust coming from the incineration furnace of the municipal refuse is mostly disposed by earth filling: the treating and the utilizing methods of the EP dust are thought to be most delayed, or out of date, in development.

2.2 Coal ashes

Coal ashes are the greatest in quantity among the waste ashes discharged from industries and are supposed to grow to a huge amount in future when used as a substitution of oil. The specific gravity of coal ashes is 2.2 - 2.4 and the major chemical components are SiO_2 and Al_2O_3.

2.3 Municipal refuse incineration ashes

The refuse coming from each home is mainly composed of combustible waste and garbage which are large in volume or have putrefactive smell; hence the incineration treatment is most promising in order to reduce the volume of waste and protect

Table 1. The properties of EP dust, municipal refuse incineration ashes and etc.

Sample	Specific gravity	Unit weight of volume (t/m³)	Mean diameter of particles (mm)	Chemical composition (%)					
				SiO_2	Fe_2O_3	Al_2O_3	CaO	Na_2O	Others
EP dust	2.80	0.54	0.04	60.4	4.5	12.0	14.7	2.6	5.8
MRIA*	2.20	0.52	0.65	39.5	11.7	17.4	13.5	3.9	14.0
Sewage ashes	2.95	0.67	1.25	24.8	18.0	11.4	32.5	0.6	12.7

* Municipal refuse incineration ashes

the environment. The properties of these incineration ashes are in general as shown in Table 1, being different depending upon the discharged place and season.

2.4 Sludge incineration ashes

The sludges obtained from the sewage or the waste-water treatment of pulp and paper industry, etc. contain a large amount of organic materials; hence the composting is also proceeding. However, many obstacles in operation such as the transportation cost to the agricultural zone and the seasonal variance in the demand have made the composting impossible: hence in such districts the incineration is mostly carried out or planned for the purpose of volume-reduction (1/8 - 1/10) and the environmental protection.

In order to dewater the sludge into the cake form at the sewage treating plant in general, the coagulant of 0.5 - 5% to the suspended solid (S.S.) in the sludge and the coagulant aid of 20 - 30% as $Ca(OH)_2$ to the S.S. are added and mixed; then the sludge is filtered by vacuum or pressure to give cake containing 50 - 80% of water.

Incineration of the cake in the multi-stage furnace operating at 900°C or in the rotary kiln leaves only the inorganic substances, as the free water evaporates and the organic substances burn. Especially, the combind water of the clayish mineral comes out at 400 - 500°C, accompanied by its conversion to SiO_2 and Al_2O_3 which can be easily activated. In case $Ca(OH)_2$ is used for coagulant aid, it decomposes to CaO and H_2O at 580°C and is converted to a silicate or aluminate compound containing CaO in the temperature range above 800°C, thus showing a weak hydraulic property.

In general, specific properties of the incineration ashes are:

1) Being particular and low bulk specific density of 0.5 - 0.8 t/m³.

2) High water absorption or water retention, absorbing water in the range of 60 - 90% by weight.

3) Showing pozzolanic reactivity because of the incineration temperature higher than 800°C.

4) Miscibility of additives such as lime and cement, depending upon the usage.

3 MIXING AND STIRRING METHOD

3.1 Indoor experiment

Two kinds of river bottom deposits (hereinafter described as sludge) as shown in Table 2 were treated with the sewage incineration ashes (hereinafter described as ash) shown in Table 1, and tests concerning transportation properties, compressive strength, etc. of the treated mixture were carried out.

Table 2 Physical properties of sludge

Sample name	Sludge A	Sludge B
Gravel fraction 2000μm %	0	2
Sand fraction 74-2000μm %	32	51
Silt fraction 5-74μm %	39	24
Clay fraction 5μm under %	29	23
Max. particle diameter mm	2.0	4.0
Liquid limit wl %	79.8	68.4
Plastic limit wp %	39.1	33.7
Plasticity index Ip	40.7	34.7
Japan soil classification	CH	SC
Soil name	Clay	Clay sand
Specific gravity	2.643	2.516
Water content %	106.0	103.4
Color	Black	Black
Smell	H_2S	H_2S

1) As to the transportation properties, the flow test of JIS R 5201 was carried out and the mixtures with a flow value less than 105 were recognized as transportable.

The tests revealed that the addition of ashes more than 25% by weight to sludge A or the addition of the ashes, mixed with 5% of CaO, more than 18% by weight to sludge A enabled the mixture to be transported by dump truck.

In the case of sludge B, the same required amounts of the added ashes were

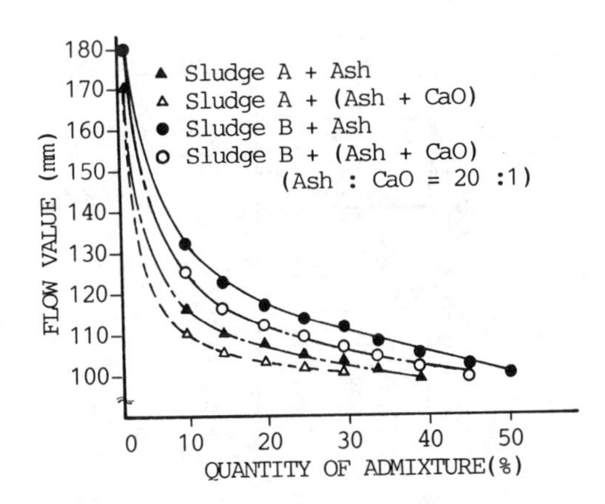

Fig.1 Flow value VS. quantity of admixture

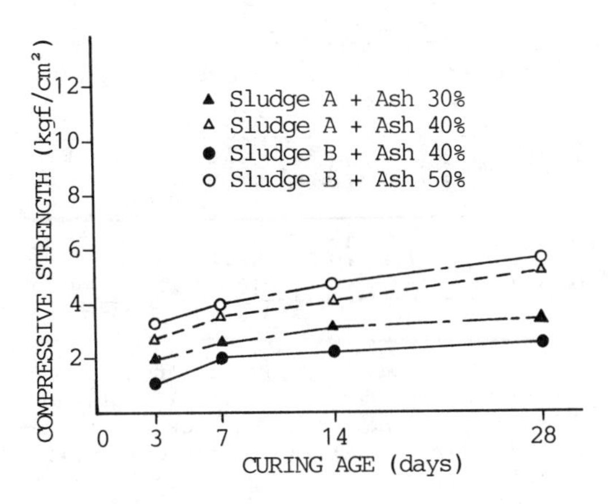

Fig.2 Compressive strength of the mixture

38% and 30%, respectively. (Fig. 1)
2) The compressive strength test was carried out according to the testing method of JIS A 1216 preparing test-pieces of ø5 x 10cmH.
The amount of ashes to be mixed with sludges were decided to meet the formulation in which immediate transportation became possible.
When sludge A is mixed with 30% and 40% of ash or sludge B is mixed with 40% and 50% of ash the strength after wet curing become 2 - 5 kgf/cm^2 for sludge A and 1 - 5 kgf/cm^2 for sludge B as shown in Fig. 2.

When dry curing is carried out, the strength increases sharply with the curing age and the strength for the curing age of 28 days becomes three to four times greater than that for the wet curing.
The use of the ashes mixed with 5 - 10% of cement or lime improves the strength of the treated mixture: this increase of strength is thought to be due to the formation of CSH.
3) The CBR test was carried out in accordance with JIS A 1211. The same formulation as that in the compressive strength test was used for each sample, and kept standing for more than 7 days.
The water content of each sample is adjusted to that of the optimum water-content percentage obtained from the CBR compaction method: test pieces were prepared by 92 times compaction of three layers, and then immediately submitted to the penetration test.
The value of $CBR_{2.5}$ becomes 5 - 9% when sludge A is treated with 30% or 40% of ashes and 12 - 16% when sludge B is treated with 40 - 50% of ashes.

3.2 Outdoor experiment

Sludge A treating experiments in field scale were carried out using the treating system shown in Fig. 3.

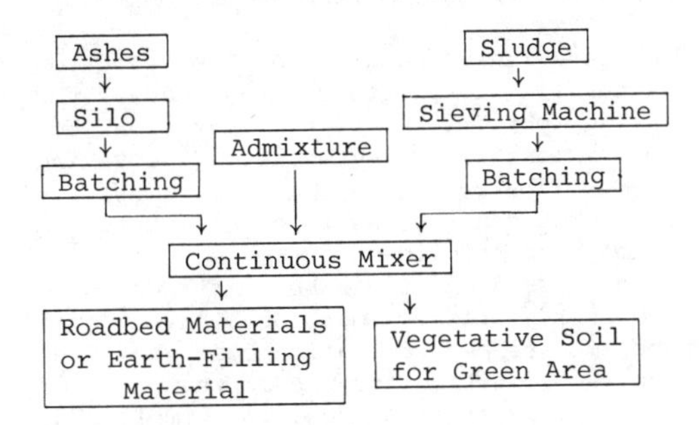

Fig.3 Flow chart of treating system of sludge use of the ashes

The experimental field consisted of five blocks; each block, to which the treated mixture is charged, was 4x4x0.5m in size. Table-3 shows the results of the cone penetration test and the CBR test of the treated mixtures together with their formulation for each block. These test results were almost the same as those of indoor experiments; the reproducibility of the field test was confirmed.
In general, it has been thought difficult to transport by truck, etc. the treated mixture of the fluid sludge with cement, lime, etc. immediately after the treatment: however this method using ashes of high water-absorption power enabled the above-mentioned immediate transportation. The mixing of sludge with powder often be-

Table 3. Mix proportions of each block and properties of the treated mixture (curing age : 7 days)

Block Mix	A Ash30%	B Ash40%	C Ash30% +CaO	E Ash30% +Cement	E Ash40% +CaO
$CBR_{2.5}$ (%)	5.0	15.0	43.6	14.3	42.0
$CBR_{5.0}$ (%)	7.1	20.0	48.1	18.5	48.6
Qc kgf/cm²	25	30	10	27	8

comes impossible because of the tunnel phenomenon caused by the possible plastic property of the mixture: however this method uses a special continuous mixer (treating capacity: 40 m³/hr) resulting in a good mixing.

In case the treated mixture is used for vegetative soil, it is recommended to reduce the ash mixing ratio to 20 - 25 % for to softening the strength and to add a small amount of ferrous sulfate, flower of sulfur, etc. for the adjustment of pH.

4 SURFACE LAYER STABILIZING METHOD

The strewing of natural earth or sand on the surface of the fluid sludge causes the heaving phenomenon and the operation becomes impossible. Therefore, covering the surface of the sludge with cloth or net, as done in the rope net method, FM method, etc., followed by strewing earth or sand on it has been introduced.

However, these methods have such problems as high operation cost and the required long time for the improvement of the under layer of sludge. A simple and low cost method of stabilizing the surface layer is by strewing light powdery substances having some hygroscopic properties on very weak ground.

4.1 Indoor experiment

Kaolinite muddy water having water content of 150% was prepared artificially, and the powdery ashes was strewed on the muddy water in the experimental apparatus shown in Fig. 4 to carry out punching rupture tests.

A study on the equation of the supporting force along the strewed surface gives the following equation as most appropriate:

$$P = A \cdot \tau \cos\theta + A' \cdot \tau' + B$$

where,

P: load on the surface layer
A: shearing area of the strewed ashes
τ: shear strength of the strewed ashes
θ: shear angle

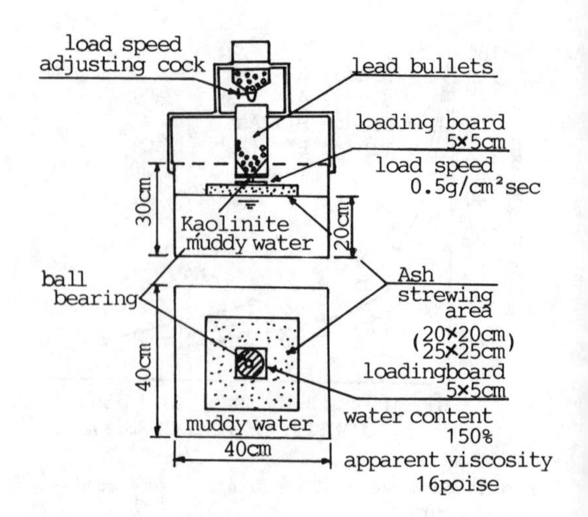

Fig.4 Experiment apparatus

A': shearing area of the sludge layer
τ': shear strength of the sludge layer
B: floating force due to deformation to strewed layer

Next, the influence of the specific gravity and the water content percentage of the strewed ashes on the compressive strength is shown in Fig. 6, indicating that the highest strength is obtained when the water content percentage is about 50% and the specific gravity is comparatively low.

The shear strength of the strewed surface may be regarded as half of the compressive strength.

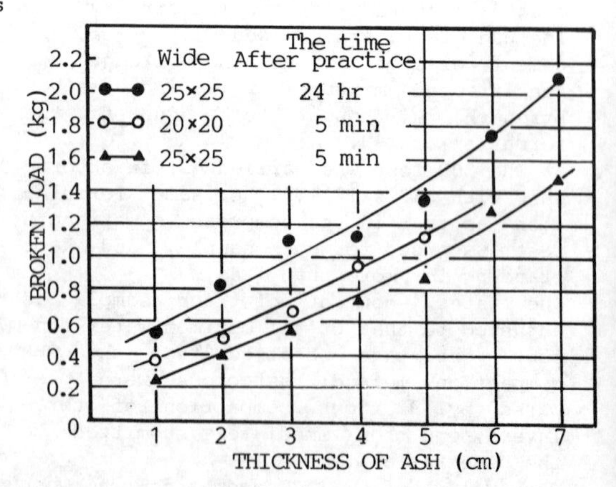

Fig.5 Strewing thickness of ashes vs. broken load

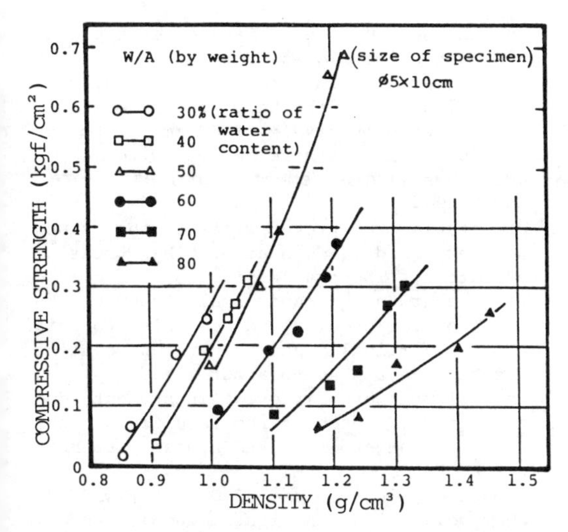

Fig.6 Influence of the density and the water content percentage of the strewed ashes on the compressive strength

4.2 Outdoor experiment

There was a weak-ground block of 2,000 m² in some dumping site: 600 m² of which was treated by this method for the stabilization of the surface layer.

1) State before improvement and treating plan (make reference to Fig. 7)

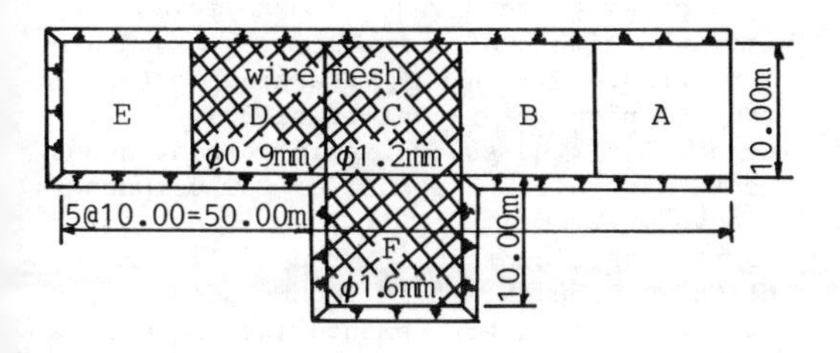

Fig.7 Ground plane of treating plan

Table-4 shows the water content percentages for the ground of A - F blocks and the shear strength measured by the vane testing apparatus. The particle size analysis of the earth of the ground indicates that it has 84% of particles smaller than 0.074 mm dia. and is thought to be a sandyish clay. The thickness of the strewed ashes layer was set to average 80 cm, and the wire-mesh looking like tortoise shell was installed 20 cm below the surface. A marshy-type bulldozer, D20P type, (bearing pressure : 0.21 kgf/cm²) was used as a strewing machine.

Table 4. Water content and shear strength for the ground

Block		A	B	G	D	E	F
Water content (%)	10cm*	74	78	105	96	167	159
	30cm	83	72	97	87	54	120
	50cm	99	95	87	73	63	107
Shear Strength (kgf/cm²)	10cm	25	25	10	15	--	10
	50cm	17	30	15	15	--	10
	70cm	19	37	29	20	--	10

* depth from the surface

2) Results of operation

These operations were carried out very successfully like the indoor experiment. Even if there is a water layer of about 10 cm in thickness on the ground, the treating operation is possible.

The water content percentage for the ground of each block decreased by 20% 80% by the water absorption capacity of the strewed ashes seven days after the operation, indicating the proceeding improvement of the ground.

The strewed ashes, absorbing water in the weak ground, sometimes lower the water content percentage to 90%, but cause no decrease of the strength in the long run. Table 5 shows the results of the plate bearing test for the strewed surface of each block. The results indicate that the ashes increase in strength and show weak hydraulic property with an increase of the curing age.

Table 5. Results of the plate bearing test K_{30} (kgf/cm³)

Block		C	D	E	F
Age (day)	7	8.5	4.0	10.2	4.2
	14	11.2	4.7	11.1	5.3
	28	14.3	9.0	12.8	11.9

In order to evaluate the installed effect of the wire mesh, wire strain gauges were fitted up to measure the stress distribution; the effective width of the bearing load could be prolonged by about 2m from the center of the gravity towards the circumference, thus indicating that this method could further increase the pontoon effect.

5 DRY PILING METHOD

5.1 Indoor experiment

In order to examine the water-absorbing effect of the ashes, four casings made of polyvinylchloride (10 cm dia. x 30 cmL) were put in two kinds of containers (30

Table 6. Measurement of water-absorbing performance

	Initial water content (%)	Distance of column (cm)	Passed age (days) and water content (%)			
			1	3	5	7
Sand	140	20	124	121	120	119
		15	102	96	94	93
	120	20	108	106	104	103
		15	90	83	81	81
	100	20	89	87	85	84
		15	75	73	71	69
Ash	140	20	120	115	113	112
		15	96	92	86	84
	120	20	100	99	97	95
		15	81	72	68	66
	100	20	86	84	73	80
		15	66	62	56	55

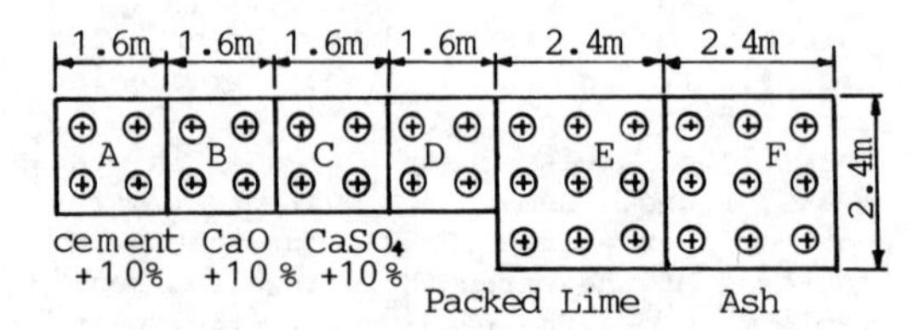

Fig. 8 Ground plane of dry piling

cm x 30 cm and 40 cm x 40 cm, both 30 cm high) with intervals of 15 cm and 20 cm respectively: the ashes were packed in the casings and the kaolinite muddy water of 100, 120, and 140% of the water content percentages were poured into the outside of the casings: after the casings were pulled out, the water content ratios of the muddy water in the center of the containers were measured with an elapsed period of time.
For comparison, sand having 50% of voids was also used as an experimental material instead of the ashes.
Table 6 shows that the ashes have better water-absorbing performance than that of sand: the reason may be that the column of the ashes has fine voids continuously formed and thus the capillary action is still maintained.

5.2 Field experiment

As shown in Fig. 8, two blocks of 2.4m x 2.4 m and four blocks of 1.6m x 1.6m were set for the experiment: steel sheet piles of 5m length were driven into surround the blocks and after digging the inside of the blocks, the fluid sludge of 200% water content was fed up the depth of 3 m. After the ashes were strewed up to 50 cm in thickness on the surface of each block, the ash columns of 37 cm dia. and 2.50 m length were driven into the under layer of the sludge by the help of steel pipe casings with intervals of 80 cm: as to the column ashes, those mixed with additives such as cement, lime, etc. were also used.
The stable or improved state of the ground was examined 14 days after the treating operation. The examination results clarified that the ground of all the blocks was so improved as to be able to dig almost vertically and the water content was also decreased to about 60%.
The columns were different in strength due to the difference of the additives: the list in the order of the higher strength was A, B, C, F, D, E.
Especially in the case of D and E, CaO of 20cm dia. was, in its columnar shape, driven into the center of the ash column, with the result that CaO slaked due to the excess water and some part was observed changing to creamy state due to partial saturation of water. This suggests that the use of CaO simple substance is undesirable for the super high water content sludges.

6 CONCLUSIONS

1) The powdery wastes of high-temperature histeresis are very fine in size with low bulk density and have pozzolanic reactivity, high water retentivity, and are appropriate for use as treating material for high water-content sludge.
Especially the sewage incineration ashes containing a large amount of CaO compounds are very effective.
If the ashes contain no or a small amount of CaO compounds, it is recommended to add 10 - 20% of CaO or cement-like materials before use.
2) In case the powdery wastes contain some toxic substances which may leach out, the addition of cement-group materials which will form a large amount of Ettringite is desirable.
3) Although the treatment of fluid sludge is thought to be difficult, the utilization of the various properties of the powdery wastes enabled to develop a new method suggesting the reasonable treatment of sludge at markedly-low cost will be possible.
4) If the sludge is regarded as muddy waste water, this treating method suggests the possibility of a complex treatment of wastes, and will give a guide for environmental protection.

Prediction and Performance in Geotechnical Engineering / Calgary / 17-19 June 1987

The concept of the master profile for tailings dam beaches

G.E.Blight
University of the Witwatersrand, Johannesburg, South Africa

ABSTRACT: The profile of an hydraulic fill beach may be non-dimensionalized to give a master profile which is the same for all beaches of a specific material deposited at a specific solids concentration. Hydraulic particle sorting along a beach may also be predicted from a simple expression and hence the variation of tailings permeability may be found. From this, the position of the phreatic surface in a dam may be predicted.

INTRODUCTION

Studies in Russia (eg Melent'ev et al 1973) which are not generally known or available in the West, have shown that the profile of an hydraulic fill beach for a particular fill material, solids concentration at placing and (probably) rate of placing, can be represented by a single dimensionless "master profile". The master profile applies regardless of the length of the beach or (within limits) the difference in elevation between the point of deposition and the pool. Reference to this phenomenon was discovered by the author and colleagues in an English translation of Melent'ev's paper. As the concept appeared to be a most useful one, research into the subject was started.

This paper is a summary of the present state of knowledge of master profiles (Blight and Bentel (1983), Blight, Thomson and Vorster (1985), Abadjiev (1985)).

There are two other phenomena which are closely related to the master profile concept:

(i) The profile appears to be generated by gravitational sorting of particle sizes as the tailings slurry moves down the beach.

(ii) The resultant gradient of particle sizes down the beach produces a gradient of permeability, the permeability of the tailings deposit decreasing from the point of deposition towards the pool.

The latter phenomenon, in particular, has important consequences for the stability of a tailings deposit.

THE MASTER PROFILE

Figure 1 shows beach profiles measured at six different locations on a large complex of platinum tailings dams. It will be noted that each profile has a different length H and a difference in elevation Y between the point of deposition and the pool. In Figure 2 the profiles have been non-dimensionalized by normalizing the elevation and distance down the beach of each point on the profile. The result is a single master beach profile with the equation

$$h/_{y} = (1-H/_{X})^{n} \quad (1)$$

Each different tailings product can be characterized by a different exponent n, as illustrated in Figure 3. The exponent is also affected by the solids concentration of the slurry and the fineness of the material.

The concept of a master profile enables the designer to predict the position of the pool, to assess the storm-water capacity of the top of the dam and more accurately to assess the tailings storage capacity.

What makes the concept practically useful is the fact that the master profile appears to apply almost regardless of the length of the beach. As a result the master profile for a new product can be established by means of beaching tests in a small laboratory flume. This enables the designer to establish the master profile on samples of tailings produced during pilot plant operations. What is more, the influence of such variables as solids concentration at deposition, fineness of grind and the presence

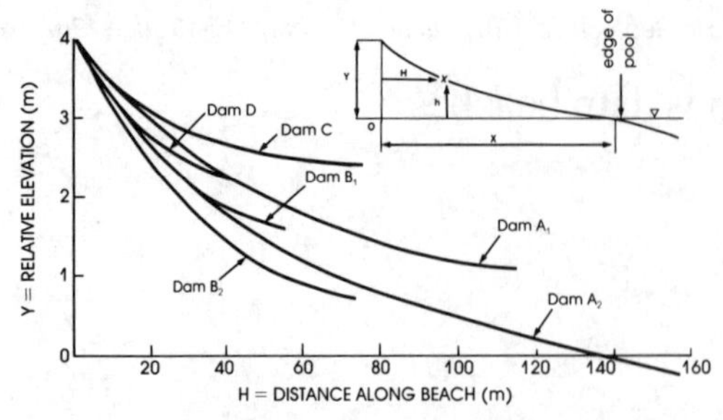

Figure 1: Measured beach profiles on 6 platinum tailings dams

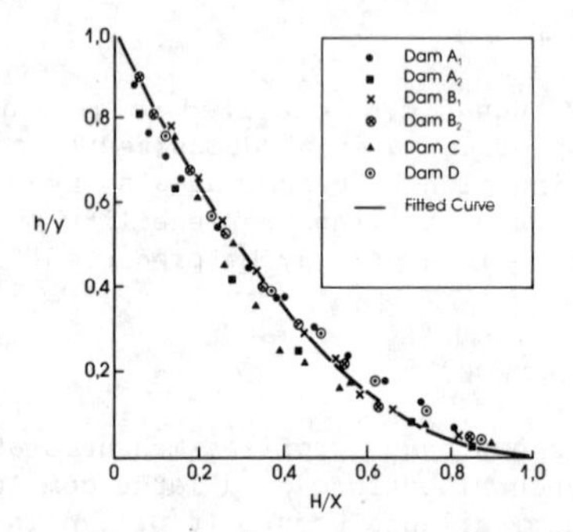

Figure 2: Dimensionless beach profiles for 6 platinum tailings dams

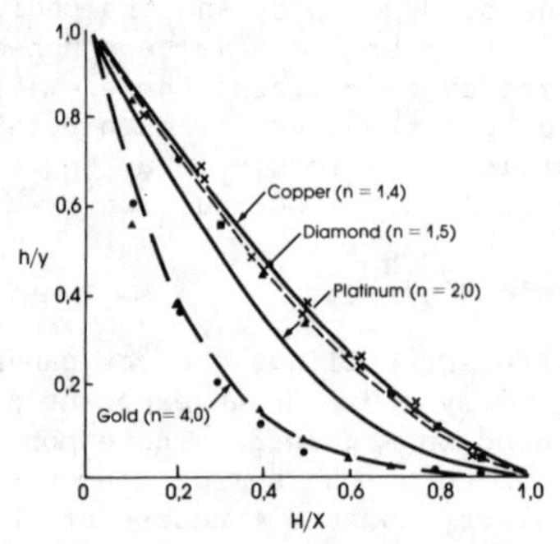

Figure 3: Dimensionless beach profiles for dams of various types of tailings

of flocculation can easily be studied.

Figure 4 shows a set of observations that illustrate the above statement. The field profile in these figures was established on a gold tailings dam where the beach length was 125m and the difference in elevation between the point of deposition and the pool was 0.7m. The solids concentration at deposition was 50 per cent. The other profiles in Figure 4 were established in a laboratory flume only 1.82m in overall length with an actual beach length of 1.5m and a difference in elevation of 100mm.

Figure 4a shows that the model-scale beach profile almost exactly matches the

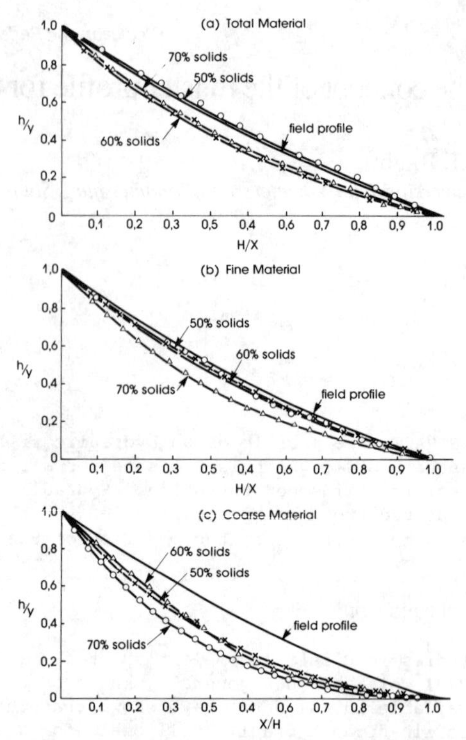

Figure 4: Comparison of field and model beach profiles for gold tailings

field profile, if the same material and solids concentration are used. As the solids concentration is increased, the exponent n in the equation to the profile (see Figure 3) increases. Varying the solids concentration produces the same effect if material either finer or coarser than the total tailings of the prototype beach is deposited, (Figures 4b and 4c). A beach of finer material has, in this example, almost the same profile as a beach of total tailings, whereas a beach of coarser material has a greater exponent n.

MECHANICS OF PROFILE FORMATION

As the tailings slurry flows down the beach there is a tendency for particles to gravitate to the bottom of the slurry stream and to deposit out. According to the laws of gravitational settling, larger particles will settle out higher up the beach while finer particles will travel further towards the pool. At any point along the beach, the beach slope i will be given by the equation for the stability of an infinite slope

$$i = \frac{1}{2}\sin^{-1}\frac{\tau_o}{\gamma\delta} \qquad (2)$$

in which τ_o is the shear strength of the

just-settled tailings, γ is its unit weight and δ the thickness of material having shear strength τ_o. Because the coarser material will tend to drain more easily, it will have a higher strength as it deposits and the slope i of the beach will decrease continuously from the point of deposition to the pool.

Figure 5 shows the variation of shear strength down a model beach of gold tailings just after deposition. Unfortunately, the corresponding beach slopes cannot be calculated via equation (2) as the thickness δ is not known with any precision. It is known, however, that the average slope of a beach, expressed by Y/X in Figure 1, increases with the solids concentration. Also, as the solids concentration increases, there is an increasing degree of interference between particles in the slurry which inhibits the settling process described

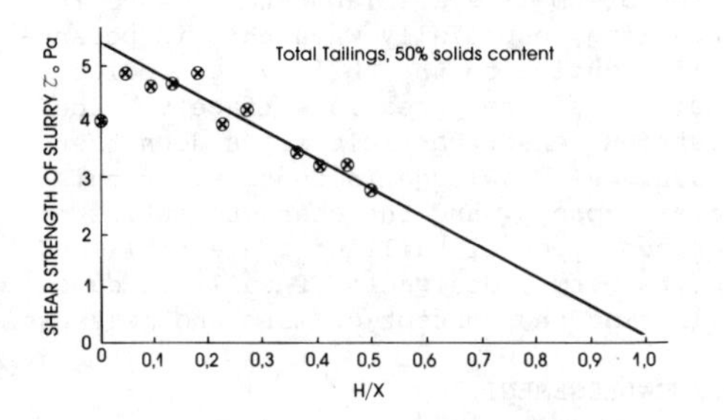

Figure 5: Variation of shear strength down a model beach of gold tailings

above. Ultimately, beyond a certain solids concentration, the flow regime changes and instead of progressive particle sorting occurring, with the average particle size decreasing towards the pool, the slurry starts to flow as a homogeneous material. This change of regime is illustrated in Figure 6 which shows the variation of Y/X of model beaches as the solids concentration of a fly ash slurry is progressively increased. The pronounced maximum slope that occurs at a solids concentration of 40 per cent marks the change from a "particle settling" to a "mud-flow" regime. Note the sudden decrease in Y/X as the "mud-flow" condition is established between solids concentrations of 40 and 50 per cent. As the solids concentration is increased further, it is surmised that Y/X will again increase progressively. To the left of the maximum slope in Figure 6, the particle settling regime, under which most conventional tailings dams operate, applies. To the right, the mud-flow regime, under which thickened discharge type dams (eg Robinsky (1978)) operate, applies.

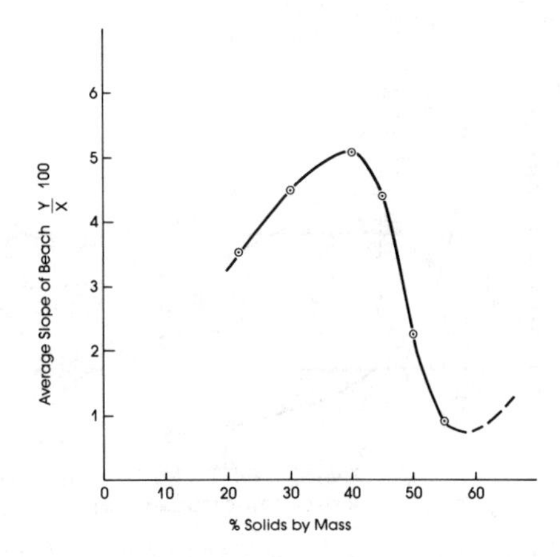

Figure 6: Variation of average beach slope with solids concentration for model beaches of fly ash slurry

It appears likely that as the rate of deposition on a beach increases, as measured in terms of tonnage discharged per unit of beach area, so the discontinuity between regimes will occur at progressively lower solids concentrations. This would follow because the velocity of flow down the beach would increase, and probably also the depth of flow. There would therefore be less opportunity for a "particle settling" regime to become established.

PARTICLE SORTING ALONG A BEACH AND ITS CONSEQUENCE

Figure 7 illustrates the particle size sorting that has taken place along a 280m long beach of diamond tailings. The inset shows similar information for the same beach taken 12 months later and expressed in dimensionless form. It will be noted that, although there is a lot of scatter, the profile of relative particle sizes is very similar in shape to that of a beach profile.

A study of particle size profiles such as those shown in Figure 7 has indicated that the size of particle at a distance H along an hydraulic fill beach from the point of deposition can be roughly predicted from the relationship

$$A = e^{-B.H/X} \tag{3}$$

in which A is given by

$$A = \frac{\text{D50 (at H down the beach)}}{\text{D50 (of total product)}}$$

B is a characteristic of the tailings (and probably also of the rate of deposition)

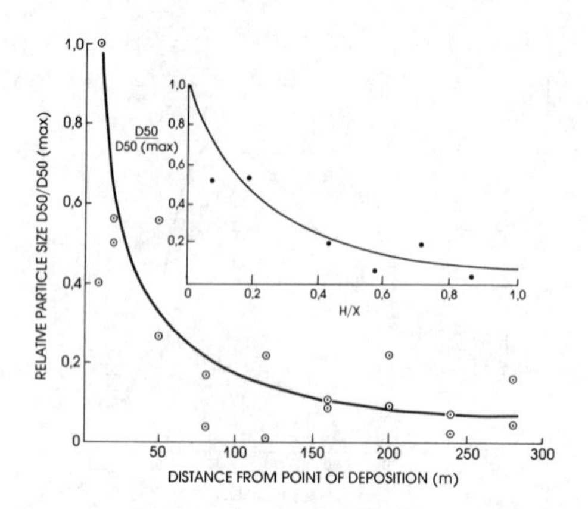

Figure 7: Particle size sorting that occurs on a hydraulic beach

X is the length of the beach.

For the distribution shown in Figure 7, B has a value of 2.5

The most important consequence of the particle size sorting is that the permeability of the tailings mass will decrease continuously from the point of deposition to the pool. The variation of permeability will follow a relationship such as

$$k = ae^{-bH} \tag{4}$$

in which a and b are characteristics of the beach and tailings and H is the distance down the beach from the point of deposition

At present, the variation of k with H can be calculated from the distribution of particle size down the beach via the well-known Hazen expression or by the expression given more recently by Sherard et al (1984):

$$k = 0.35(D_{15})^2 \tag{5}$$

(D_{15} in mm and k in cms^{-1})

As a result of this variation in permeability, the phreatic surface of the dam will be depressed as shown in Figure 8. The explanation of the depression is quite simple: Continuity of flow requires that the flow rate of water from the pool be the same through the lower permeability material near the pool and the higher permeability material near the point of deposition. It follows from D'Arcy's law that the flow gradient will progressively decrease from the pool outwards, ie the slope of the phreatic surface must progressively flatten, thus producing the depression shown in Figure 8.

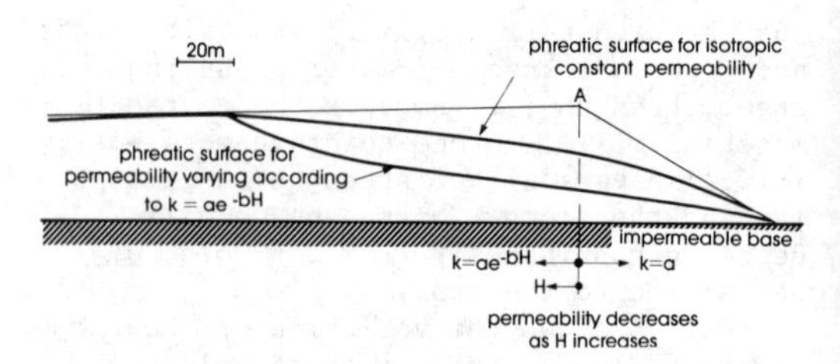

Figure 8: Depressed phreatic surface resulting from hydraulic particle sorting on a tailings beach

CONCLUDING REMARKS

This paper has served to introduce the concept of the master profile for tailings dam beaches. The idea that the profile of an hydraulic fill beach can be predicted by means of small-scale laboratory tests is appealing, especially when this is potentially coupled to the ability to predict the position of the phreatic surface via the distribution of particle sizes down the beach. As knowledge in the area of this paper expands, and the characteristics of various types of tailings are established in its terms, designers of tailings dams will find the concept of more and more use.

ACKNOWLEDGEMENT

The data shown in Figure 6 was established by personnel of ESCOM working in the laboratory of the Civil Engineering Department at the University of the Witwatersrand.

REFERENCES

Abadjiev, C.B.(1985). Estimation of the physical characteristics of deposited tailings in the tailings dam of non ferrous metallurgy, Proceedings, 11th International Conference on Soil Mechanics and Foundation Engineering, San Francisco, vol.3, p.1231-1234.

Blight, G.E. & G.M. Bentel (1983). The behaviour of mine tailings during hydraulic deposition, Journal of the South African Institute for Mining and Metallurgy, vol.83, no.4, p.73-86.

Blight, G.E., R.R. Thomson & K. Vorster (1985). Profiles of hydraulic fill tailings beaches and seepage through hydraulically sorted tailings, Journal of the South African Institute for Mining and Metallurgy, vol.85, no.5, p.157-161.

Melent'ev, V.A., N.P. Kolpashnikov & B.A. Volnin (1973). Hydraulic fill structures, Energy, Moscow. (English translation of

original Russian).
Robinsky, E.I. (1978). Tailings disposal by the thickened discharge method for improved economy and environmental control, Proceedings, 2nd International Conference on Tailings Disposal, Denver, p.75-92.
Sherard, J.L., L.P. Dunnigan & J.R. Talbot (1984). Basic properties of sand and gravel filters, Journal of Geotechnical Engineering, vol.110, no.6, p.684-700.

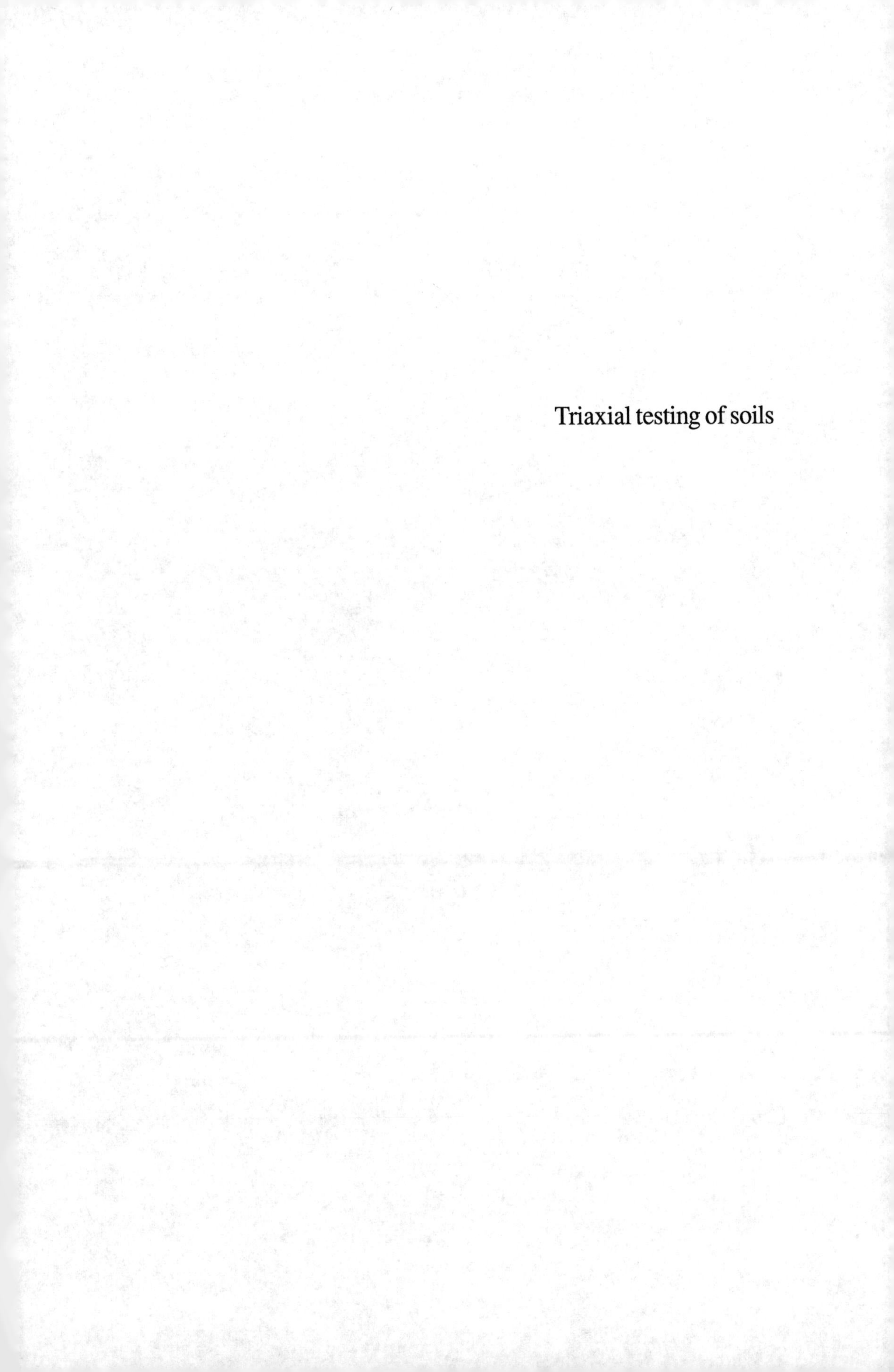

Triaxial testing of soils

Prediction and Performance in Geotechnical Engineering / Calgary / 17-19 June 1987

Prediction of plane-strain angles of shear resistance from triaxial test results

A.M.Hanna, N.Massoud & H.Youssef
Concordia University, Montreal, Quebec, Canada

ABSTRACT: Semi-empirical approaches were developed for predicting the angle of shearing resistance for plane-strain condition from the conventional triaxial test results. In order to verify the theories developed, tests were conducted on three types of sand using triaxial and plane-strain apparatus. The predicted plane-strain angles of shearing resistance agreed well with the measured values.

INTRODUCTION

Design of structures such as long earth dams and retaining walls, requires knowledge of the shear strength parameters of the soil in plane-strain shear failure conditions. Determination of the plane-strain angle of shearing resistance experimentally, usually requires special equipment which could be available or may be complicated and time consuming. Thus, prediction of these angles, using indirect methods by utilizing conventional testing technique and equipment, could be of interest. The purpose of this paper is to present two approaches to determine the angles of shearing resistance for plane-strain conditions using triaxial test data.

To verify the developed approaches, experimental studies using plane-strain and triaxial compression tests were performed on three types of sands under the same conditions of confining pressures and strain rates. The plane strain results were compared with the predicted values.

SOIL TESTED AND TESTING PROCEDURES

The soil used for this investigation is an air dry uniform silica sands of three types: the physical properties are presented in Table [1]. The triaxial tests were performed on 76 mm in length by 38 mm in diameter on cylindrical samples: the method utilized for testing is essentially the same as described by Bishop and Henkel [1978] and Bowles [1978]. Typical results are presented in Fig. [1].

The plane-strain tests were carried out on prismatic samples with the following nominal dimensions 92 mm, 38 mm, 75 mm in width, length and height respectively. The plane-strain apparatus is a very specialized device utilized mainly in research; a detailed description is given by Massoud [1981]. Figure [2] presents a schematic drawing of the horizontal projection of the apparatus, as well as an overview of the experimental set-up. As can be seen from Fig. [2.a], the two parallel ends of the samples are fixed to provide zero lateral strain [$\varepsilon_2 = 0$]; the intermediate principal stress [σ_2] was measured through two pressure transducers. Figure [3] presents typical results obtained from plane-strain compression tests.

THEORETICAL DEVELOPMENT

In order to develop the theoretical model, the dilatancy theory reported by Rowe [1962, 1969], Cole [1967], Proctor [1967] and Wightman [1967] was utilized as follows:

$$R = K \cdot D \qquad [1]$$

where R is the stress ratio at failure for both triaxial and plane-strain loading conditions, furthermore:

$$R = \frac{\sigma_1}{\sigma_3} \qquad [2]$$

and D is the dilatancy rate, where:

$$D = 1 - \frac{d\varepsilon_v}{d\varepsilon_1} \qquad [3]$$

$d\varepsilon_v$ and $d\varepsilon_1$ are the volumetric and axial strains increments and:

$$K = \tan^2\left[45° + \frac{\phi_f}{2}\right] \qquad [4]$$

where [ϕ_f] is defined as Rowe's frictional angle.

The state of strain for triaxial and plane-strain loading conditions is described in Eqns. [5 and 6]. In plane-strain compression tests, the state of strains can be described as:

$$d\varepsilon_v = d\varepsilon_1 + d\varepsilon_3 \quad [5]$$

where $d\varepsilon_3$ is the minor principal strain increment and $d\varepsilon_2 = 0$ is the intermediate principal strain. In triaxial compression conditions:

$$d\varepsilon_v = d\varepsilon_1 + 2d\varepsilon_3 \quad [6]$$

Utilizing Cole's [1967] definition for the angle of dilatancy [α], the following relationship is obtained for triaxial stress conditions:

$$\cos 2\alpha = \frac{d\varepsilon_v}{d\varepsilon_1 - d\varepsilon_3}, \text{ and} \quad [7]$$

$$\sin 2\alpha = \sqrt{1 - \cos^2 2\alpha} \quad [8]$$

Taking into consideration Mohr-Coulomb's criteria, combined with the definition of dilatancy, the following relationships are obtained:

$$\frac{\sigma_1 - \sigma_3}{2} = \frac{\tau}{\sin 2\alpha}, \text{ and} \quad [9]$$

$$\frac{\sigma_1 + \sigma_3}{2} = \sigma - \tau \cot 2\alpha, \text{ hence} \quad [10]$$

$$\frac{\sigma_1 + \sigma_3}{\sigma_1 - \sigma_3} = \frac{\sigma}{\tau}\sin 2\alpha - \cos 2\alpha \quad [11]$$

$$\frac{R+1}{R-1} = \frac{\sigma}{\tau}\sin 2\alpha - \cos 2\alpha, \text{ or} \quad [12]$$

Substituting equation [5] and [3] into equation [7] we have:

$$\cos 2\alpha = \frac{d\varepsilon_v}{2d\varepsilon_1 - d\varepsilon_3}$$

$$\cos 2\alpha = \frac{1-D}{1+D}, \text{ or} \quad [13]$$

and hence equation [8] can be written in the form:

$$\sin 2\alpha = \sqrt{1 - \frac{[1-D]^2}{[1+D]^2}} \quad [14]$$

Substituting equations [13] and [14] into equation [12]:

$$\frac{R+1}{R-1} = \frac{\sigma}{\tau}\sqrt{1 - \frac{[1-D]^2}{[1+D]^2}} - \frac{[1-D]}{[1+D]} \quad [15]$$

Equation [15] can be written in the form:

$$\frac{\tau}{\sigma} = \sqrt{\frac{1}{D}} \cdot \frac{[R-1]}{\left[\frac{R}{D} + 1\right]} \quad [16]$$

Since [D = R/K] from Rowe's [1962] dilatancy theory, hence:

$$\frac{\tau}{\sigma} = \sqrt{\frac{K}{R}}\left[\frac{R-1}{K+1}\right] \quad [17]$$

For plane-strain compression $\phi_f = \phi_{cv}$ then:

$$K = \tan^2\left[45° + \frac{\phi_{cv}}{2}\right] \quad [18]$$

where ϕ_{cv} is the angle of friction at constant volume and describes the critical state [$\phi_{cv} = \phi_{crit}$], Rowe [1969] and Bolton [1986]. Accordingly, equation [17] can be written in the form [Rowe, 1969]:

$$\left[\frac{\tau}{\sigma}\right]_{p.s.} = \tan \phi_{p.s.} \cdot \cos \phi_{cv} \quad [19]$$

For triaxial compression tests, assuming that the Mohr-Coulomb's criteria for plane-strain and triaxial strain compression are identical, we have:

$$\cos 2\alpha = \frac{d\varepsilon_1 + 2\,d\varepsilon_3}{d\varepsilon_1 - d\varepsilon_3} = \frac{[1-D]}{[1+0.50\,D]}, \text{ and} \quad [20]$$

$$\sin 2\alpha = \sqrt{1 - \frac{[1-D]^2}{[1+0.50\,D]^2}} = \frac{\sqrt{12\,D - 3D^2}}{[2+D]} \quad [21]$$

Substituting equations [20] and [21] in Eqn. [12], we have:

$$\frac{R+1}{R-1} = \frac{\sigma}{\tau}\left[\frac{\sqrt{12D - 3D^2}}{[2+D]}\right] - \frac{[1-D]}{[1+0.50\,D]} \quad [22]$$

which can be written in the form:

$$\left[\frac{\tau}{\sigma}\right]_T = \frac{(R-1)\sqrt{12D - 3D^2}}{4R - RD + 3D} \quad [23]$$

Since R = K . D [Rowe's Dilatancy Theory], equation [23] can be written in the following form:

$$\left[\frac{\tau}{\sigma}\right]_T = \frac{[KD-1]\sqrt{12D - 3D^2}}{4\,KD - KD^2 + 3D} \quad [24]$$

According to Rowe [1969], the stress-dilatancy theory [R = K . D] is applicable for both triaxial and plane-strain conditions, therefore:

$$\left[\frac{\tau}{\sigma}\right]_{p.s.} = \left[\frac{\tau}{\sigma}\right]_T \quad [25]$$

Therefore, by equating equations [19] and [24], we have:

$$\tan \phi_{p.s.} \cdot \text{co}\,\phi_{cv} = \frac{[KD-1]\sqrt{12D - 3D^2}}{4KD - KD^2 + 3D} \quad [26]$$

It was found that Eqn. [26] slightly underestimates the experimental correlations. Following Bolton [1986], an empirical coefficient was found equal to 1.035 which does correlate the experimental data with good agreement :

$$\tan \phi_{p.s.} \cdot \cos \phi_{cv} = 1.035 \frac{[R - 1]\sqrt{12D - 3D^2}}{4R - RD + 3D} \qquad [27]$$

Based on the developed theoretical model; Eqn [27], two approaches are proposed for prediction of $[\phi_{p.s.}]$ from the triaxial results parameters, namely, ϕ_{cv}, R and D as defined before. The results for both triaxial and plane-strain compression are reported in Tables [2] and [3].

APPROACH I

1. The triaxial tests are to be performed on the field as well as on predetermined arbitrary sand prosities. From the results of these tests, plot the relationship between ϕ_T and $[d\varepsilon_v/d\varepsilon_1]$ at failure, thus determine the value of ϕ_{cv} at $[d\varepsilon_v/d\varepsilon_1 = 0]$.
2. Calculate the parameter $R = \sigma_1/\sigma_3$.
3. Knowing the values of the relative prosity of the sand tested, determine the value of the dilatancy rate D from Fig. [4].
4. Substitute the values of D, R and ϕ_{cv} in Eqn. [27] to determine the values of $\phi_{p.s.}$.Numerical values for this approach utilizing the present results are presented in Table [4].

APPROACH II

1. Three triaxial tests on sands at the field relative prosity, as well as at the loosest and densest prosities are to be performed; for determination of the angle $[\phi]$ for the three backing conditions; and for the determination of $[\phi_{cv}]$ as explained above. Knowing $[\phi_{cv}]$ determine $[\phi_{\mu}]$ from the theoretical relationship, Horne [1965] as shown in Fig. [5].
2. For the two limits of the sand packing, namely, the loosest [D = 1] and the densest [D = 2] states, and substituting the values of $[R = \sigma_1/\sigma_3]$, and values of $[\phi_{cv}]$ for [D = 1, 2] determine $[\phi_{p.s.}]$:
3. Knowing the values of $[\phi_{p.s.}]$ and $[\phi_{\mu}]$ determine the interlocking angle $[\phi_I]$ for [D = 1, 2]:

$$[\phi_I]_{p.s.} = \phi_{p.s.} - \phi_{\mu} \qquad [28.1]$$

$$[\phi_I]_T = \phi_T - \phi_{\mu} \qquad [28.2]$$

4. Draw the relationship

$$\frac{[\phi_I]_{p.s.}}{[\phi_I]_T}$$

For the present results as well as for data recalled from the literature from the [Mersery River Quartz Sands (Rowe, 1962)] versus the relative prosity as given in Fig. [6].
5. Determine from Fig. [6] the ratio:

$$\frac{[\phi_I]_{p.s.}}{[\phi_I]_T}$$

corresponding to the field prosity.
6. Knowing $[\phi_I]_T$ corresponding to the field prosity and the ratio $[[\phi_I]_{p.s.}/[\phi_I]_T]$ the value of $[\phi_I]_T$ can be calculated.
7. The angle of shearing resistance at the field prosity for plane-strain conditions $[\phi_{p.s.}]$ can be calculated from Eqn. [28.1]. Numerical values for this approach utilizing the present results are given in Table [5].

It is of interest to notice from Fig. [6] that the relationships between $[[\phi_I]_{p.s.}/[\phi_I]_T]$do yield straightline relationship for each type of sand . When these relationships were utilized following the above mentioned method [Approach II] for each relative prosity $[n_r]$, it was possible to calculate $[[\phi_I]_{p.s.}/[\phi_I]_T]$; and by knowing $[\phi_{\mu}]$ for each type of sand, it was possible to estimate the predicted values of $[\phi_{p.s.}]$ from knowledge of the triaxial compression test parameters, namely $[\phi_I]_T$ and ϕ_{μ}. Figure [8] shows the relationship between measured and predicted values of $[\phi_{p.s.}]$ utilizing the second approach; excellent agreements between the measured and predicted values are obtained.

EXPERIMENTAL VERIFICATIONS

The above two approaches have been utilized for prediction of $[\phi_{p.s.}]$ from the present triaxial results, as well as data recalled from the literature. The comparision between the measured and predicted values are presented in Tables [4] and [5] for the above mentioned two approaches and presented graphically in Figs. [7] and [8]. As it can be seen from the figure, good agreements are achieved between the measured and predicted values utilizing the developed theoretical model.

CONCLUSIONS

This paper presents two approaches for determining the angles of shearing resistance for plane-strain conditions from the experimental results deduced from the triaxial testing on the same material. The theories were verified with the experimental data available in the literature, where good agreement was found.

ACKNOWLEDGEMENTS

The financial support of the Natural Science and Engineering Research Council of Canada is gratefully acknowledged.

REFERENCES

Bishop A.W. and Henkel, D. [1978]. The Measurements of Soil Properties in the Triaxial Test. Fourth Edition, Edward Arnold Ltd., London.

Bowles, J.E. [1978]. Engineering Properties of Soils and their Measurements. McGraw-Hill Book Co., N.Y., USA.

Cole, E.R.L. [1967]. Soils in the S.S.A., Ph.D. Thesis, Cambridge University, England.

Hanna, A.M. and Massoud, N. [1981]. Interlocking of Granular Materials in Two and Three-Dimensional Shear Failure. Proc. of the Eighth Canadian Congress of Applied Mechanics, CANCAM'81, Moncton University, N.B., Canada.

Hanna, A.M. and Youssef, H. [1987]. Evaluation of Dilatancy Theory of Granular Materials. Proc. the Int. Symp. of Prediction and Performance in Geotechnical Engineering. The Univ. of Calgary, Alberta, Canada.

Hanna, A.M., Massoud, N., and Youssef, H. [1987]. Prediction of Plane-Strain Angles of Shearing Resistance from Direct Shear Test Results. Proc. the 11th Canadian Congress of Applied Mechanics, CANCAM'87, The University of Alberta, Edmonton, Canada.

Massoud, N. [1981]. Shear Strength Characteristics of Sands. M.Sc. Thesis, Concordia University, Montreal, Canada.

Rowe, P.W. [1962]. The Stress-Dilatancy Relation for Static Equilibrium of an Assembly of Particles in Contact. Proc. Royal Soc. A. 269, pp. 500-527.

Rowe, P.W. [1969]. The Relation Between Strength of Sands in Triaxial Compression, PLane-Strain and Direct Shear, Géotechnique 19, No. 1, pp. 75-76.

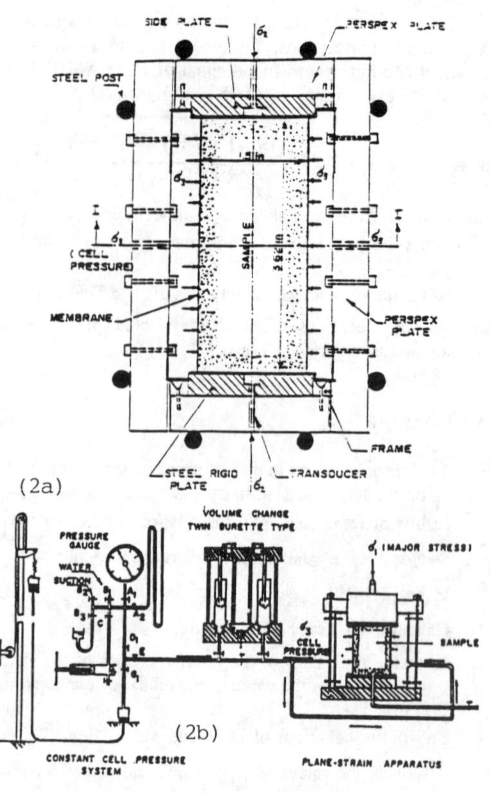

FIG. 2 Schematic drawing of the Plane-strain apparatus

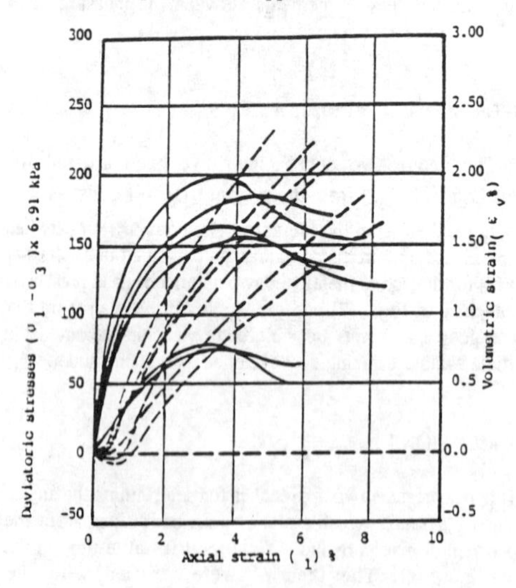

FIG. 1 Typical triaxial compression results

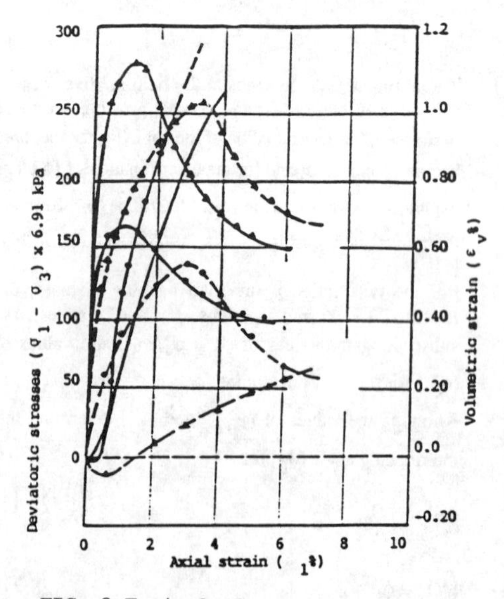

FIG. 3 Typical plane-strain results

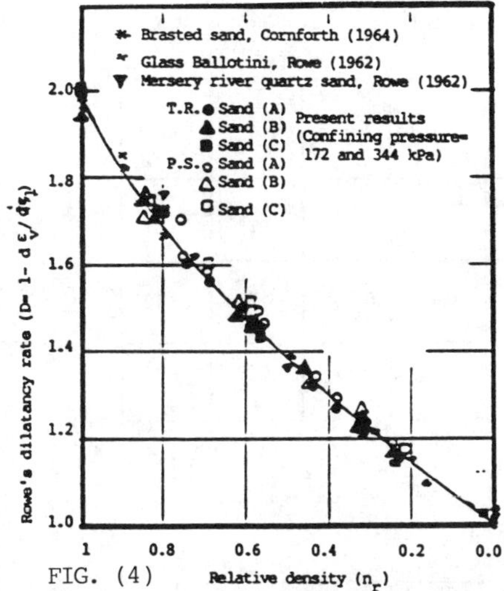

FIG. (4)
Variation of the dilatancy rate (D) with the relative prosity (n_r) for triaxial and plane-strain compression results
(Note T.R. and P.S. are triaxial and plane-strain results)

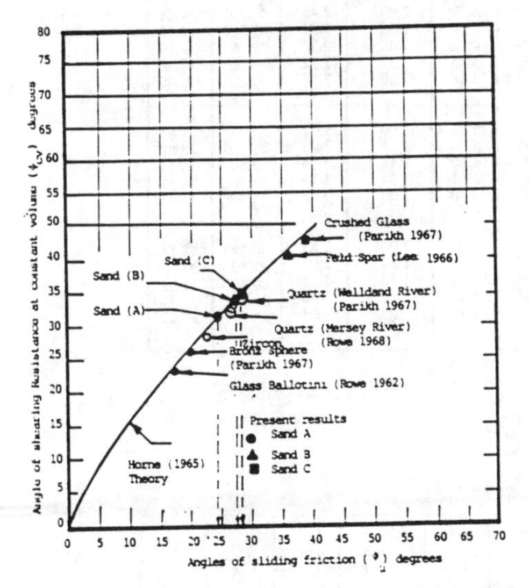

FIG. 5 Horne theory and the experimental results for angle of sliding friction

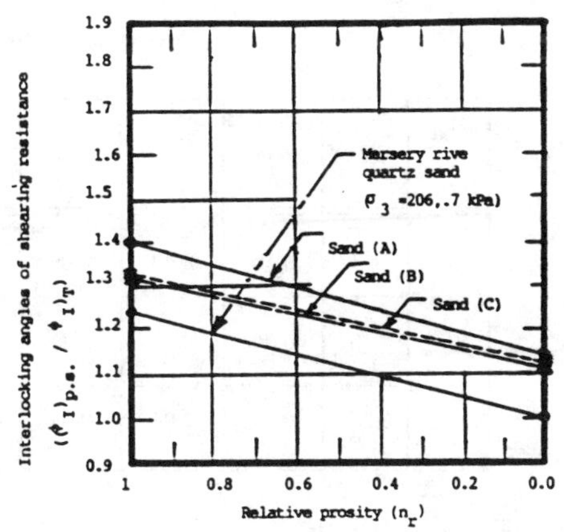

FIG. 6 Relationship between $((\phi_I)_{p.s.}/(\phi_I)_T)$ vs. (n_r)
Present results of sands A,B and C for confining pressure of 172 and 344 kPa.

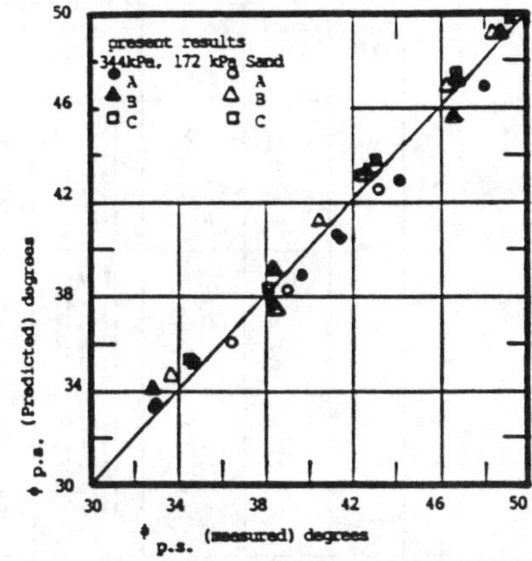

FIG. 7 Comparison between the measured and predicted plane-strain angles utilizing the developed theoretical model (Approach I)

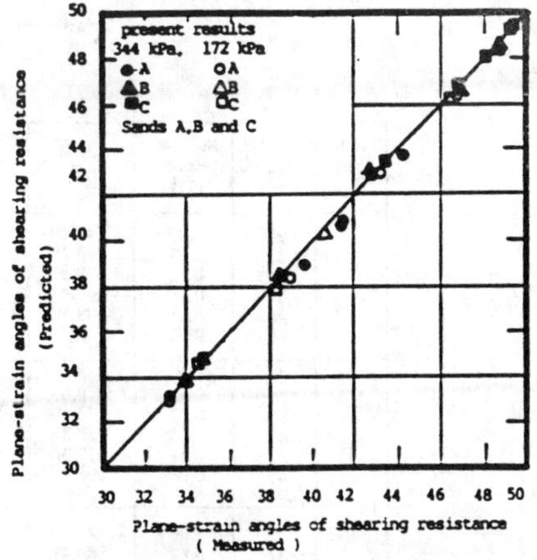

FIG. 8 Comparison between the measured and predicted Plane-Strain angles of shearing resistance utilizing the developed theoretical model (Approach II)

TABLE [1]
Physical Properties of the Tested Sands [Cohesionless Soil]

Type of Sands	D_{60}	D_{10}	C_u	G_s	e_{max}	e_{min}	n_{max}	n_{min}	γ_{max}	γ_{min}
A	0.24	0.10	2.40	2.65	0.80	0.40	0.44	0.29	18.75	15.81
B	0.70	0.30	2.33	2.63	0.90	0.50	0.47	0.33	17.77	15.49
C	0.70	0.35	2.00	2.64	0.95	0.40	0.49	0.29	17.28	14.34

TABLE [2.a]
Triaxial Compression Results and Analysis
[σ_3= 344 KPa]

SAND TYPE	Prosity [n]	Relative Prosity [n_r]	σ_1 KPa	$d\varepsilon_1$ %	$d\varepsilon_v$ %	R = σ_1/σ_3	D = $[1 - \frac{d\varepsilon_v}{d\varepsilon_1}]$	K = R/D	$[\phi^\circ_f]_{max}$	ϕ°_f
A	28	1.00	1682.16	2.70	-2.70	4.89	2.00	2.45	41.33°	24.85°
	32	0.75	1490.89	3.20	-1.92	4.34	1.60	2.71	38.70°	27.45°
	35	0.56	1384.48	3.85	-1.73	4.03	1.45	2.78	37.00°	28.10°
	37	0.44	1325.72	4.00	-1.28	3.85	1.32	2.92	36.00°	29.30°
	44	0.00	1121.44	4.20	0.00	3.26	1.00	3.26	32.00°	32.00°
B	33	1.00	1867.92	2.55	-2.55	5.43	2.00	2.72	43.55°	27.54°
	35	0.85	1734.55	3.50	-2.65	5.04	1.75	2.88	42.00°	28.98°
	38	0.62	1546.18	3.95	-1.90	4.50	1.48	3.04	39.50°	30.33°
	42	0.31	1354.62	4.80	-1.15	3.95	1.24	3.19	36.50°	31.51°
	46	0.00	1190.24	5.50	0.00	3.46	1.00	3.46	33.50°	33.48°
C	29	1.00	1943.60	3.50	-3.50	5.65	2.00	2.83	44.37°	28.54°
	32	0.82	1772.46	4.38	-3.07	5.15	1.70	3.03	42.50°	30.25°
	36	0.59	1610.55	5.50	-2.48	4.68	1.45	3.23	40.40°	31.82°
	42	0.24	1342.28	9.00	-1.44	3.90	1.16	3.36	36.30°	32.77°
	46	0.00	1217.76	9.30	0.00	3.54	1.00	3.54	34.00°	34.02°

TABLE [2.b]
Triaxial Compression Results and Analysis
[σ_3= 172 KPa]

SAND TYPE	Prosity [n] %	Relative Prosity [n_r]	σ_1 KPa	ε_1 %	ε_v %	R = σ_1/σ_3	D = $[1 - \frac{d\varepsilon_v}{d\varepsilon_1}]$	K = R/D	$[\phi^\circ_f]_{max}$	ϕ°_f
A	28	1.00	841.10	2.55	-2.55	4.89	2.00	2.45	41.33°	24.85°
	33	0.69	740.25	3.40	-1.90	4.30	1.56	2.76	38.50°	27.91°
	35	0.56	690.51	3.55	-1.53	4.01	1.43	2.80	37.00°	28.27°
	38	0.38	648.88	3.85	-1.04	3.77	1.27	2.97	35.50°	29.75°
	40	0.25	609.50	3.95	-0.60	3.54	1.15	3.08	34.00°	30.65°
	44	0.00	560.72	4.00	0.00	3.26	1.00	3.26	32.00°	32.00°
B	33	1.00	934.00	2.50	-2.50	5.43	2.00	2.72	43.55°	27.54°
	35	0.85	870.00	2.55	-1.91	5.06	1.75	2.89	42.00°	29.07°
	38	0.62	773.10	2.67	-1.25	4.50	1.47	3.05	39.50°	30.41°
	40	0.46	723.83	2.80	-0.99	4.21	1.35	3.12	38.00°	30.97°
	42	0.31	677.38	2.95	-0.68	3.94	1.23	3.20	36.50°	31.59°
	46	0.00	595.12	3.06	0.00	3.46	1.00	3.45	33.50°	33.48°
C	29	1.00	971.90	2.00	-2.00	5.65	2.00	2.83	44.37°	28.54°
	32	0.82	888.41	4.06	-2.92	5.17	1.72	3.01	42.50°	30.08°
	35	0.59	806.30	5.38	-2.45	4.68	1.46	3.21	40.40°	31.66°
	42	0.24	671.14	7.20	-0.95	3.90	1.13	3.45	36.30°	33.41°
	46	0.00	608.88	8.00	0.00	3.54	1.00	3.54	34.00°	34.02°

TABLE [3.a]
Plane-Strain Compression Results and Analysis
[σ_3 = 344 KPa]

SAND TYPE	Prosity [n]	Relative Prosity [n_r]	σ_1 KPa	σ_2 KPa	ε_1 %	ε_v %	R = σ_1/σ_3	D = $[1 - \frac{d\varepsilon_v}{d\varepsilon_1}]$	K = R/D	$[\phi^\circ_{p.s.}]^*_{max}$	ϕ°_f
A	28	1.00	2332.32	600.00	0.50	-0.45	6.78	1.90	3.57	48.00°	34.22°
	32	0.75	1918.61	583.40	0.65	-0.45	5.58	1.70	3.28	44.10°	32.20°
	35	0.56	1679.53	571.04	0.70	-0.35	4.88	1.50	3.25	41.30°	32.00°
	37	0.44	1551.32	563.19	0.81	-0.28	4.51	1.34	3.37	39.60°	32.85°
	44	0.00	1166.16	531.28	1.84	-0.00	3.39	1.00	3.39	33.00°	33.00°
B	33	1.00	2414.88	602.21	0.50	-0.48	7.02	1.95	3.60	48.65°	34.42°
	35	0.85	2204.15	595.12	0.60	-0.42	6.41	1.70	3.77	46.90°	35.50°
	38	0.62	1806.00	577.92	0.61	-0.32	5.25	1.52	3.45	42.90°	33.41°
	42	0.31	1466.53	557.28	0.90	-0.23	4.26	1.25	3.41	38.30°	33.13°
	46	0.00	1207.44	535.45	2.60	-0.00	3.51	1.00	3.51	33.80°	33.80°
C	29	1.00	2518.08	605.94	1.25	-1.38	7.32	2.08	3.52	49.40°	33.89°
	32	0.82	2204.15	595.12	1.44	-0.32	6.41	1.80	3.56	46.80°	34.20°
	36	0.59	1854.08	580.33	1.75	-0.39	5.39	1.54	3.55	43.40°	34.09°
	42	0.24	1471.30	557.62	2.90	-0.46	4.28	1.24	3.69	38.40°	35.00°
	46	0.00	1259.04	540.36	5.94	-0.00	3.66	1.00	3.66	34.80°	34.81°

* Calculated using Eqns. [16 and 17].

TABLE [3.b]
Plane-Strain Compression Results and Analysis
[σ_3 = 172 KPa]

SAND TYPE	Prosity [n]	Relative Prosity [n_r]	σ_1 KPa	σ_2 KPa	ε_1 %	ε_v %	R = σ_1/σ_3	D = $[1 - \frac{d\varepsilon_v}{d\varepsilon_1}]$	K = R/D	$[\phi^\circ_{p.s}]^*_{max}$	ϕ°_f
A	28	1.00	1166.16	299.78	0.48	-0.50	6.78	2.04	3.32	48.00°	32.48
	33	0.69	920.20	289.83	0.60	-0.22	5.35	1.62	3.39	43.20°	33.00
	35	0.56	842.75	285.69	0.65	-0.36	4.90	1.55	3.16	41.40°	31.30
	38	0.38	755.22	280.19	0.85	-0.25	4.39	1.29	3.40	39.00°	33.10
	40	0.25	679.49	274.51	1.20	-0.04	3.95	1.19	3.32	36.60°	32.48
	44	0.00	583.08	265.64	1.80	-0.05	3.39	1.03	3.29	33.00°	32.26
B	33	1.00	1207.74	301.12	0.45	-0.47	7.02	2.04	3.44	48.65°	33.34
	35	0.85	1069.88	295.36	0.55	-0.45	6.22	1.81	3.44	46.30°	33.34
	38	0.62	899.65	288.79	0.58	-0.35	5.23	1.60	3.23	42.80°	32.10
	40	0.46	813.67	283.97	0.85	-0.37	4.73	1.44	3.56	40.60°	34.15
	42	0.31	733.26	278.64	1.25	-0.36	4.25	1.29	3.55	38.30°	34.10
	46	0.00	603.72	267.72	2.50	-0.05	3.51	1.02	3.44	33.80°	33.34
C	29	1.00	1259.04	302.65	1.22	-1.28	7.32	2.05	3.57	49.40°	34.22
	32	0.82	1100.18	297.49	1.40	-1.09	6.40	1.78	3.60	46.50°	34.39
	35	0.59	920.06	289.82	1.65	-0.99	5.35	1.59	3.34	43.20°	32.63
	42	0.24	728.52	278.30	2.35	-0.42	4.24	1.21	3.59	38.20°	34.35
	46	0.00	629.52	270.04	5.10	-0.10	3.66	1.02	3.59	34.80°	34.35

* Calculated using Eqns. [16 and 17].

Table [4]
COMPARISON OF THE CALCULATED AND MEASURED PLANE-STRAIN ANGLES OF SHEARING RESISTANCE [APPROACH I]

SAND	Relative Prosity n_r	Plane-Strain Angles of Shearing Resistance				Remarks
		$[\phi_{p.s}]$, σ_3 = 172 KPa		$[\phi_{p.s}]$, σ_3 = 344 KPa		
		Measured	Theoretical	Measured	Theoretical	
A	1.00	48.00°	46.80°	48.00°	46.80°	
	0.75	–	–	44.10°	42.76°	ϕ_{cv} = 32°
	0.69	43.20°	42.44°	–	–	
	0.56	41.40°	40.39°	41.30°	40.32°	ϕ_{μ} = 24.8°
	0.44	–	–	39.50°	38.80°	
	0.38	39.00°	38.10°	–	–	
	0.25	36.60°	36.02°	–	–	
	0.00	33.00°	33.22°	33.00°	33.17°	
B	1.00	48.65°	49.19°	48.65°	49.19°	
	[illegible]	[illegible]	[illegible]	[illegible]	[illegible]	
	0.85	46.30°	46.78°	46.90°	45.56°	ϕ_{cv} = 33.5°
	0.62	42.60°	43.10°	42.90°	43.11°	
	0.46	40.60°	41.10°	–	–	ϕ_{μ} = 27.5°
	0.31	38.30°	37.40°	38.30°	39.14°	
	0.00	33.80°	34.63°	33.80°	34.10°	
C	1.00	49.40°	50.10°	49.40°	50.10°	ϕ_{cv} = 34°
	0.82	46.50°	47.20°	46.80°	47.02°	
	0.59	43.20°	43.87°	43.40°	43.80°	ϕ_{μ} = 28.5°
	0.24	38.20°	38.20°	38.40°	37.51°	
	0.00	34.80°	35.30°	34.80°	35.29°	

Table [5.a]
COMPARISON OF THE CALCULATED AND MEASURED PLANE-STRAIN ANGLES OF SHEARING RESISTANCE [APPROACH II - σ_3 = 172 KPa]

SAND	n_r	ϕ°_μ	$[\phi^\circ_1]_T$	$[\phi^\circ_1]_{p.s.}$	$\frac{[\phi^\circ_1]_{p.s.}}{[\phi^\circ_1]_T}$ Measured	$\frac{[\phi^\circ_1]_{p.s.}}{[\phi^\circ_1]_T}$ Calculated	$\phi^\circ_{p.s.}$ Measured	$[\phi^\circ_{p.s.}]$ Calculated
A	1.00	24.80°	16.53°	23.20°	1.40	1.40	48.00°	48.00°
	0.69		13.70°	18.40°	1.34	1.32	43.20°	42.88°
	0.56		12.20°	16.60°	1.36	1.28	41.40°	40.42°
	0.38		10.70°	14.20°	1.33	1.25	39.00°	38.20°
	0.25		09.20°	11.80°	1.28	1.20	36.60°	35.84°
	0.00		07.20°	08.20°	1.14	1.14	33.00°	33.00°
B	1.00	27.50°	16.05°	21.15°	1.32	1.32	48.65°	48.65°
	0.85		14.50°	18.80°	1.30	1.29	46.30°	46.21°
	0.62		12.00°	15.30°	1.28	1.28	42.80°	42.86°
	0.46		10.50°	13.10°	1.25	1.21	40.60°	40.21°
	0.31		09.00°	10.80°	1.20	1.20	38.30°	38.30°
	0.00		06.00°	06.30°	1.05	1.05	33.80°	33.80°
C	1.00	28.50°	15.87°	20.90°	1.32	1.32	49.40°	49.40°
	0.82		14.00°	18.00°	1.29	1.28	46.50°	46.42°
	0.59		11.90°	14.70°	1.24	1.28	43.20°	43.67°
	0.24		07.80°	09.70°	1.24	1.17	38.20°	37.60°
	0.00		05.50°	06.30°	1.15	1.15	34.80°	34.80°

Table [5.b]
COMPARISON OF THE CALCULATED AND MEASURED PLANE-STRAIN ANGLES OF SHEARING RESISTANCE [APPROACH II - σ_3 = 344 KPa]

SAND	n_r	ϕ°_μ	$[\phi^\circ_1]_T$	$[\phi^\circ_1]_{p.s.}$	$\frac{[\phi^\circ_1]_{p.s.}}{[\phi^\circ_1]_T}$ Measured	$\frac{[\phi^\circ_1]_{p.s.}}{[\phi^\circ_1]_T}$ Calculated	$\phi^\circ_{p.s.}$ Measured	$[\phi^\circ_{p.s.}]$ Calculated
A	1.00	24.80°	16.53°	23.20°	1.40	1.40	48.00°	48.00°
	0.75		13.90°	19.30°	1.39	1.35	44.10°	43.60°
	0.56		12.20°	16.50°	1.35	1.28	41.30°	40.42°
	0.44		11.20°	14.80°	1.32	1.25	39.60°	38.80°
	0.00		07.20°	08.20°	1.14	1.14	33.00°	33.00°
B	1.00	27.50°	16.05°	21.15°	1.32	1.32	48.65°	48.65°
	0.85		14.50°	19.40°	1.34	1.31	46.90°	46.50°
	0.62		12.00°	15.40°	1.28	1.27	42.90°	42.74°
	0.31		09.00°	10.80°	1.20	1.19	38.30°	38.21°
	0.00		06.00°	06.30°	1.05	1.05	33.80°	33.80°
C	1.00	28.50°	15.87°	20.90°	1.32	1.32	49.40°	49.40°
	0.82		14.00°	18.00°	1.31	1.28	46.80°	46.42°
	0.59		11.90°	14.90°	1.25	1.24	43.40°	43.30°
	0.24		07.80°	09.90°	1.27	1.21	38.40°	37.94°
	0.00		05.50°	06.30°	1.15	1.15	34.80°	34.80°

Multiaxial testing of compacted shales

S.M.Sargand & G.A.Hazen
Ohio University, Athens, USA

K.M.Miller
Ohio Department of Transportation, Columbus, USA

ABSTRACT: In this investigation a high pressure multiaxial device has been used for testing of compacted shales. Tests were conducted by following a number of significant straight stress paths. The study focused on shear strength and stress-strain relations of compacted shales. Anisotropic behavior of the shale specimens, induced by compaction process, has been investigated; and the results are discussed.

1 INTRODUCTION

Since construction of highways requires the utilization of the most economical material available near the site and shale happens to be very abundant in the Midwest, highway embankments and subgrades have been constructed of shale materials. However, the lack of understanding about the properties of compacted shales has led to serious problems, such as embankment failures and excessive settlement. Therefore, a comprehensive study of the load response properties of compacted shales is necessary.

A number of investigators (Abeyesekera et. al. 1978, Strohm, Jr., W.E., 1978) have examined the strength properties of compacted shales using triaxial test devices, and direct shear devices that are not capable of characterizing the influence of three-dimensional stress field on the material parameters.

In this investigation a high pressure cubical system was modified to enable testing of compacted shale.

All shale specimens were compacted with identical amounts of energy and were tested at the same moisture content. Tests were conducted by following conventional triaxial compression (CTC), conventional triaxial extension (CTE), simple shear (SS), and hydrostatic compression (HC) stress paths. During the tests no drainage was allowed.

2 MULTIAXIAL DEVICE

Data that are obtained using conventional laboratory equipment to determine the constitutive relations are of questionable value since it cannot represent true, homogeneous boundary conditions, and true material parameters. Thus in this investigation a multiaxial cubical device has been used.

In a multiaxial device a cubical specimen of 102 mm (4") is loaded on six orthogonal faces by six hydraulically pressurized flexible membranes. Care has been taken not to introduce shear stress on the specimen faces during loading. Thus the three applied stresses in the direction of the cell axes are considered to be principal stresses. This loading will insure that a properly prepared sample will float between unconstrained deformations (Atkinson, R.H., 1972).

Sample deformations are measured by a total of six linear variable differential transformer (LVDT) probes, each being located on one face inside the pressurized membranes.

3 SAMPLE PREPARATION

Samples for this study were collected from a slope at the construction site of the Ohio River Bridge interchange in Chesapeake, Ohio. They can be simply described as specimens collected from a

weathered red shale. For these samples the following Atterberg limits were found:
Liquid limit (LL) = 28.80%
Plastic limit (PL) = 21.25%

The following procedure was utilized in the preparation of compacted specimens.

The sample rocks were crushed with a carpenter's hammer and seived through the U.S. Standard No. 4 (4.75 mm opening) sieve. The passed portion was then oven-dried to constant weight at about 110°C. The standard compaction test specified in ASTM D698 was performed on these dry samples, and a compaction curve was determined. The maximum dry density was (19.17 kN/m³). In this study the specimens used had a dry density of 98% of the maximum. Two moisture contents corresponding to this dry density were found to be 8.0% and 13.2%. In this investigation, 13.2% water in weight was always added to the dry samples.

To prepare the specimens for testing a cubic mold was built, and each sample was compacted inside this special mold, maintaining the same nominal compaction energy (586 kJ/m³).

4 TESTING PROGRAM AND TEST RESULTS

With a multiaxial testing device, more variety of loading/unloading conditions can be produced in the specimen for more realistic analysis of soils. A total of thirteen samples were used to study the three dimensional stress-strain-strength behavior of the compacted red shale. In each test, the initial density and initial moisture content of the sample were determined. Table 1 presents a summary of this testing program. An ultimate strength level plateau was reached in CTC, CTE and SS tests.

4.1 Hydrostatic compression

Figure 1 shows HC test results. Almost identical responses were found in the x- and y-directions, but the z-direction (compacted direction) exhibited the largest compressibility under isotropic compression. The anisotropic behavior of the shales due to the compaction process was noticeable.

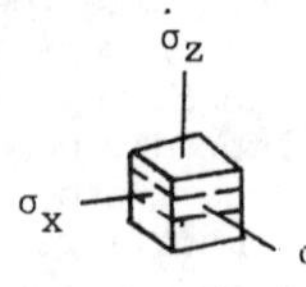

Table of Summary: Multiaxial Testings of Compacted Red Shale

Type of Stress Path	γ_o(kN/m³)	ω(%)	σ_o(kPa)	Failure Condition σ_1KPa		σ_2KPa		σ_3kPa	
HC	21.93	13.319	34.47						
CTC-(I)	21.37	12.95	68.95	413.69				68.95	
	22.33	12.27	137.90	517.11	σ_z			137.90	σ_x, σ_y
	21.93	12.37	206.84	586.05				206.84	
CTC-(II)	22.23	11.76	68.95	344.74				68.95	
	22.12	12.25	137.90	413.69	σ_y			137.90	σ_x, σ_z
	22.05	12.72	206.84	517.11				206.84	
CTE	22.06	13.089	68.95	344.74				68.95	
	22.13	12.954	137.90	448.16	σ_x, σ_y			137.90	σ_z
	22.11	12.39	206.84	517.11				206.84	
SS	22.22	12.873	206.84	344.74		206.84		68.95	
	21.96	12.246	275.79	448.16	σ_y	275.79	σ_x	103.42	σ_z
	12.27	12.629	344.74	517.11		344.74		172.37	

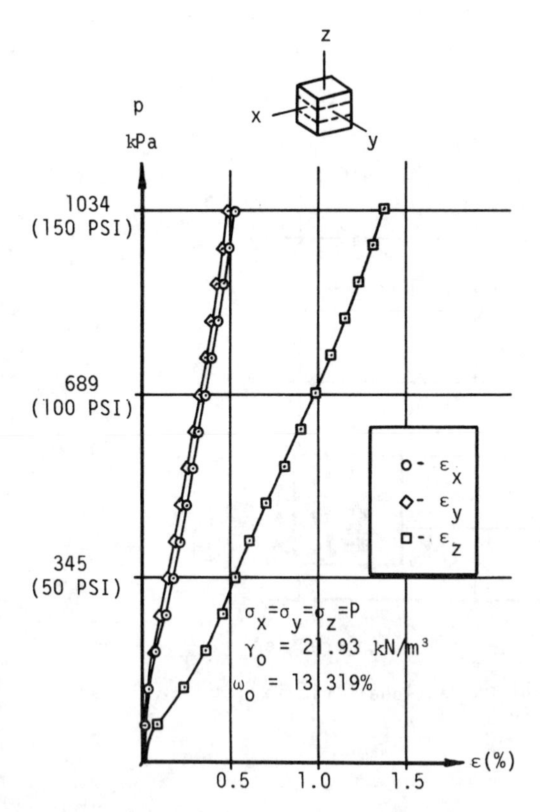

Fig. 1 Stress-Strain Response Curves for Hydrostatic Compression Test on Compacted Red Shale

4.2 Conventional triaxial compression

In conventional triaxial stress path, the specimen is loaded hydrostatically to the desired continuing pressure level and then the stress of two directions are held constant while in the third direction stress is increased. Figures 2 to 5 show results from CTC tests, including unloading and reloading loops. In the first series of tests the stress in the direction of compaction increased to the failure, however, in the second series the load was applied in the direction perpendicular to the compaction. In the Figures the nonlinear behavior of the samples are well observed where the hysteresis loops tend to become wider at higher stress levels. The samples reveal high ultimate strength and high flexibility along the compaction direction.

4.3 Conventional triaxial extension (CTE)

In a CTE test, $\sigma_3(=\sigma_z)$ is held constant while $\sigma_1(=\sigma_x)$ and $\sigma_2(=\sigma_y)$ are increased equally. A typical result of the CTE test is shown in Fig. 6. From results of the CTE test it can be determined that the stiffness of the sample increases with increasing minor principal stress. In the x- and y-directions isotropy is observed. The compacted shale exhibits low ultimate strength along the CTE stress path. In Fig. 6 the initial confining pressure is 138 kPa and the ultimate strength is (448 kPa).

4.4 Simple shear (SS)

In a SS test, $\sigma_2(=\sigma_x)$ is held constant while $\sigma_1(=\sigma_y)$ and $\sigma_3(=\sigma_z)$ are increased and decreased respectively by the same amount. Figure 7 shows results from SS test. In this series of tests the influence of the intermediate principal stress on behavior of compacted shales is investigated. It is interesting to note that the intermediate strains moved from the extension side to the compression side as the intermediate stress increased.

In Figure (8), a p-q diagram is defined corresponding to all the failure states observed in the CTC and CTE tests. And a K_f-line is drawn by applying the linear regression technique to these points. Two strength parameters d and α_f are obtained directly from the diagram.

Corresponding friction angle α_f and cohesion intercept c are equal to 8.6° and 110.4 kPa respectively.

CONCLUSIONS

The results presented in this paper indicated that red shales compacted to 98% of maximum dry density may develop considerable shear strength. If the strength is described by the empirical Mohr-Coulomb law, then the parameter is 8.6° and c is 116.9 kN/m² (16.95 psi). In this study, it is demonstrated that the strength and stress-strain relations of compacted red shales are remarkably influenced by

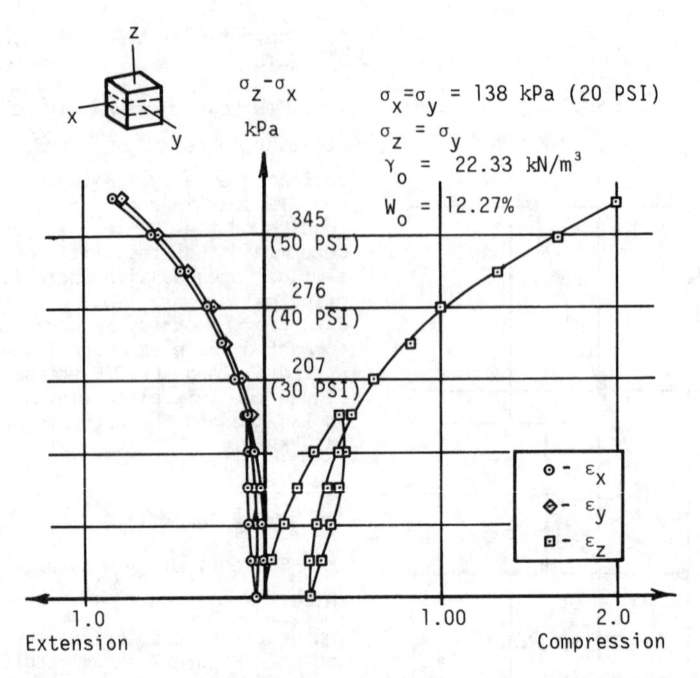

Stress-Strain Response Curves for Conventional Triaxial Compression Test on Compacted Red Shale

Fig. (2)

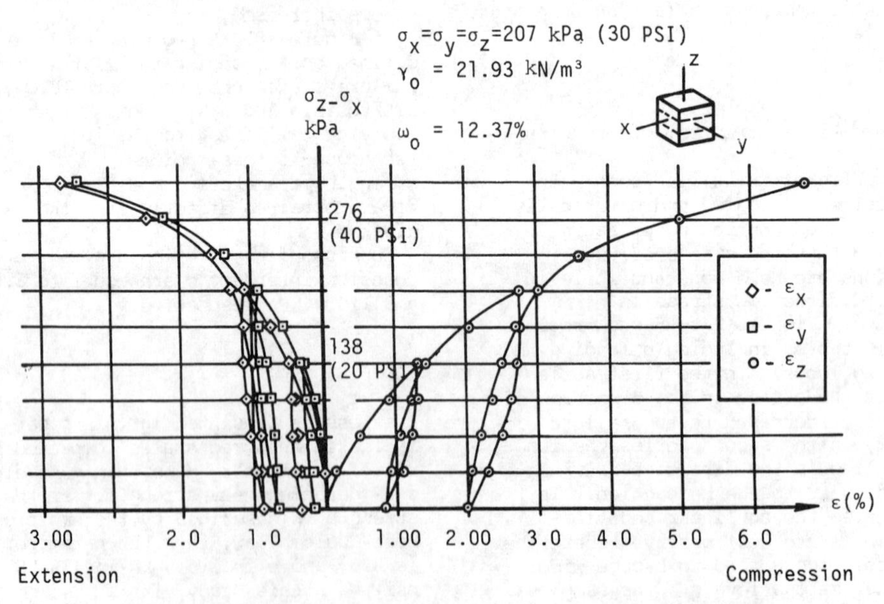

Stress-Strain Response Curves for Conventional Triaxial Compression Test on Compacted Red Shale

Fig. (3)

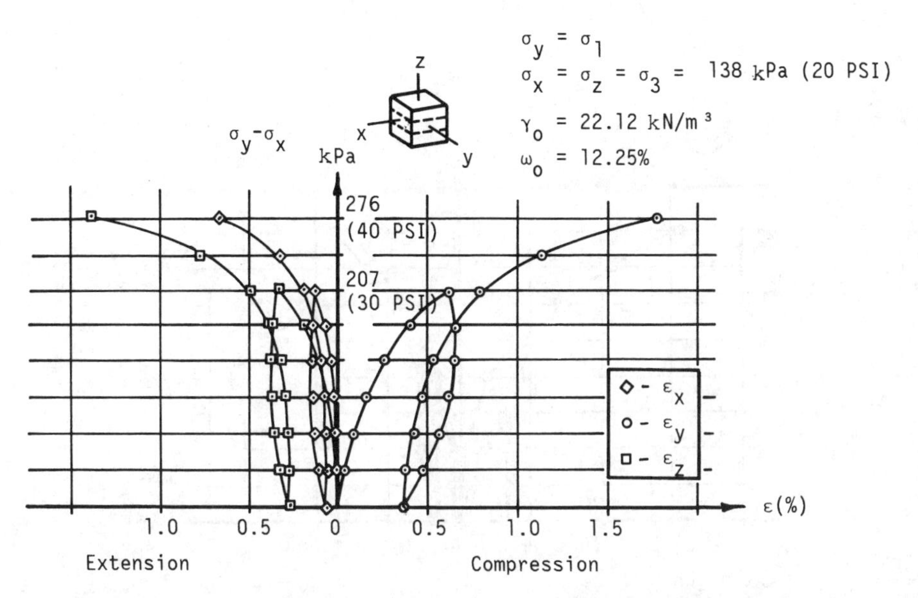

Stress-Strain Response Curves for Conventional Triaxial Compression Test on Compacted Red Shale (Loading Parallel to the Layers)

Fig. (4)

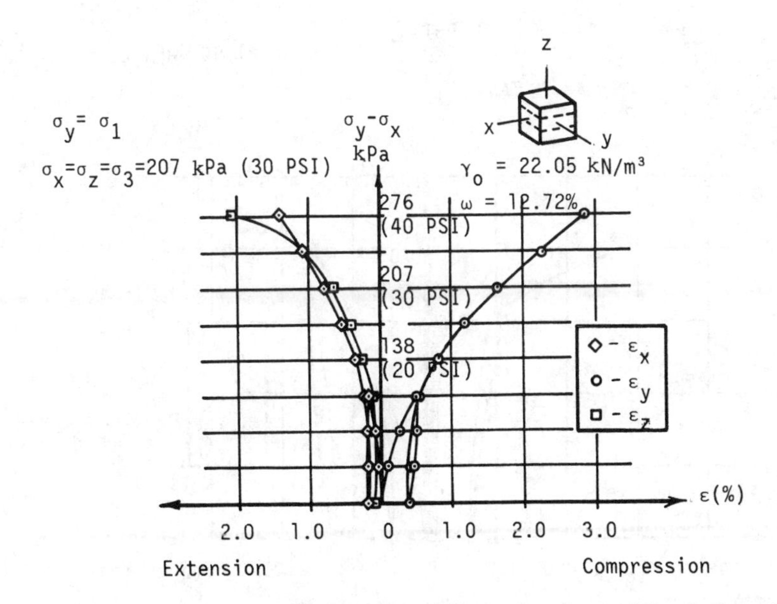

Stress-Strain Response Curves for Conventional Triaxial Compression Test on Compacted Red Shale (Loading Parallel to the Layers)

Fig. (5)

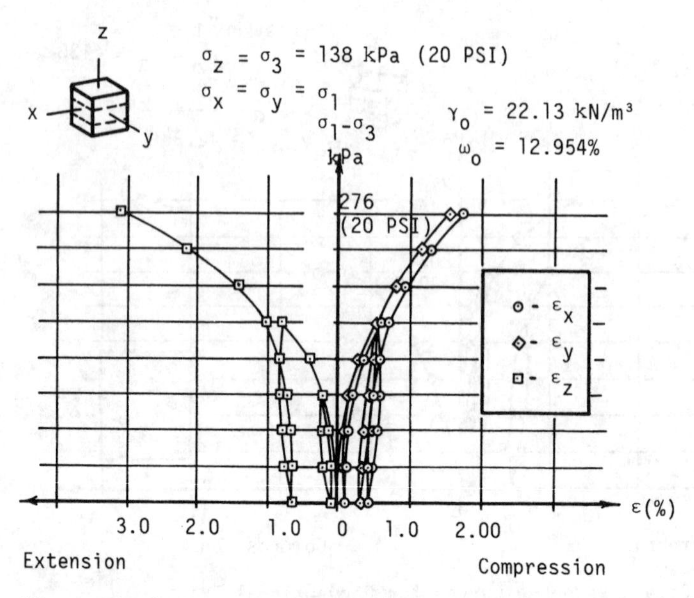

Stress-Strain Response Curves for Conventional Triaxial Extension Test on Compacted Red Shale

Fig. (6)

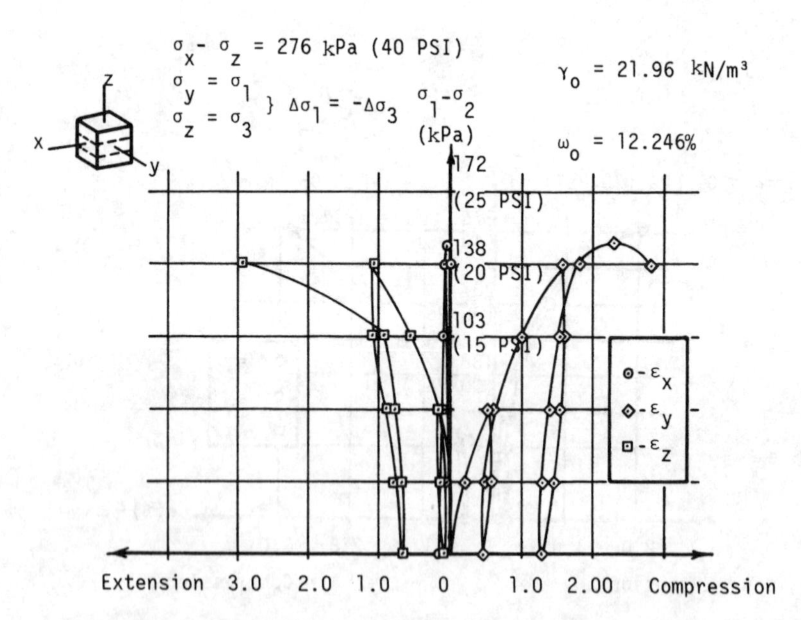

Stress-Strain Response Curves for Simple Shear Test on Compacted Red Shale

Fig. (7)

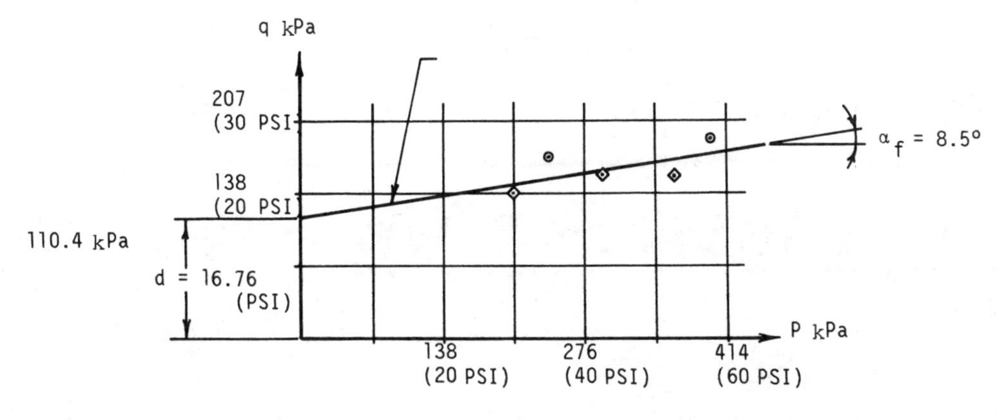

p-q diagram

Fig. (8)

the stress path. For instance, the SS test results exhibit lower shear strength than the CTC test results.

Based on multiaxial test results, it is concluded that the red shales show an anisotropic behavior due to the compaction and display higher strength and flexibility in the direction of compaction.

ACKNOWLEDGEMENTS

The investigations reported herein were performed under a research grant No. 14375(0) by the Ohio Department of Transportation and the Federal Highway Administration.

REFERENCES

Abeyesekera, R.A., Lowell, C.W. and Wood, L.E. 1978. Stress-deformation and strength characteristics of a compacted shale, in clay fills. Institute of C.E., London: 1-14.

Atkinson, R.H. 1972. A cubical test cell for multiaxial testings of material. Ph.D. dissertation, University of Colorado.

Strohm, Jr., W.E. 1978. Design and construction of compacted shale embankments, Field and Laboratory Investigations, Vol. 4, Rep. No. FHWA-RD-78-140, U.S. Army Engineer Waterways Experiment Station, Vicksburg, M.S.

Centrifuge model testing

Prediction and Performance in Geotechnical Engineering / Calgary / 17-19 June 1987

Crown pillar subsidence prediction from centrifuge models

L.Wizniak & R.J.Mitchell
Queen's University, Kingston, Ontario, Canada

Where high extraction is to be achieved in shallow, gently dipping ore bodies of significant areal extent, the overburden stresses must be supported by backfill in order to prevent caving of the crown pillar. Ground subsidence is to be expected as the overburden stresses are transferred from stiff rock pillars to relatively soft backfill materials. In order to minimize the surface effects due to subsidence it is necessary to place fill in continuous contact with the crown pillar, allowing the fill to take up some overburden stress as bending progresses in the crown pillar. An ideal situation would be one in which the crown pillar and backfill share the support function -- the crown pillar in elastic bending and the backfill in confined compression.

Three series of centrifuge model studies were carried out using plain concrete to model rock pillars and various depths of crown pillars. A conventional cemented tailings backfill was used in the models and lead shot was used to simulate the overburden mass. Crown pillar beam depths, backfill depths and backfill cement contents were varied in an attempt to model a variety of situations. The results of these model tests show that tight filling with conventional backfill can delay beam caving and reduce subsidence. Similitude between the model and prototype conditions is discussed and it is shown that direct similitude models can be realistically achieved.

The Hetenyi theory of deformations in beams on an elastic foundation has been adopted to apply to the mine subsidence problem and a computer analog was written to calculate crown pillar subsidence for the realistic situation of a non-linear backfill subgrade modulus. Predictions from this numerical model are shown to compare favourably with the results of the centrifuge physical models. Two design graphs produced by the computer analog are included in the paper.

Introduction

Considerable research has been carried out on ground subsidence due to seam mining in weathered rock (Brauner, 1973) but there is a lack of published information on subsidence due to bulk metaliferrous mining operations. In most cases, bulk mining is carried out at great depths in relatively narrow ore bodies and there is no surface subsidence except in the extreme case of hanging wall failure. In some cases, however, where high extraction rates are to be achieved in shallow ore bodies of relatively large areal extent, there is a risk of subsidence extending to the ground surface and a major concern must be to limit the subsidence of crown pillars in the mine. What effect mine backfill has in reducing potential subsidence has been largely a matter of conjecture. Since the stiffness modulus of typical backfills is several orders less than that of a rock mass, engineers have generally assumed that the backfill effect is marginal. From considering an adaption of the Hetenyi (1958) beam on an elastic foundation theory, Mitchell (1983) suggests that tight backfill (placed in contact with the back or crown pillar) may result in the backfill and crown pillar sharing the overburden stresses and, hence, reduce subsidence to a minimum. The centrifuge model studies presented in this paper were carried out to evaluate this approach to subsidence control.

Crown Pillar Subsidence Analysis

A crown pillar in a shallow mine, when directly underlain by backfill, can be considered analagous to a beam supported by a subgrade as shown on Figure 1. All overburden materials are considered flexible and the backfill support is represented by a subgrade modulus, K. Applying the theoretical developments due to Hetanyi (1958), the moments and deflections in the beam are given by:

End moments (fixed ends) $$M_e = \frac{-\ell\sigma}{2\lambda^2}\left(\frac{d}{c}\right) \quad (1)$$

Central moment, $$M_c = \frac{\ell\sigma}{\lambda^2}\left(\frac{b-a}{c}\right) \quad (2)$$

Central deflection, $$Y_c = \frac{\sigma}{K}\left(1-\frac{2(a+b)}{c}\right) \quad (3)$$

where

σ = overburden stress

$\lambda = (\ell K/4EI)^{1/4}$ is a characteristic

$I = \ell d^3/12$ is the moment of inertia

ℓ = unit beam width

$a = \sinh(\lambda L/2)\cos(\lambda L/2)$

$b = \cosh(\lambda L/2)\sin(\lambda L/2)$

$c = \sinh\lambda L + \sin\lambda L$

L = beam length

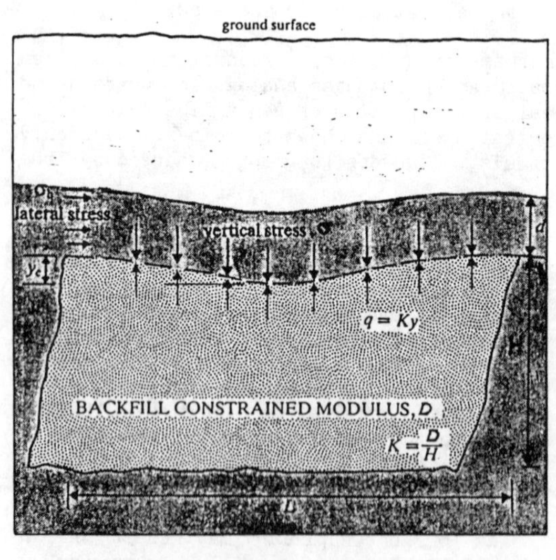

Figure 1 - Crown pillar subsidence

The modulus of subgrade reaction, K, is given as D/H where D is the constrained modulus of a backfill of height H. Data plotted on Figure 2 show that the constrained fill modulus increases with increased strain. Thus, when the beam length, L, is relatively large with respect to the fill height, H, the backfill may begin to carry significant load during elastic bending of the beam. Lateral stresses , σ_h, in the crown pillar would act to maintain a plastic beam moment, $Mp = d^2\sigma_h/6$ after large deflections such that the beam and the backfill would continue to share the stresses (i.e. $q < \sigma$). Comparison of equation (3) with the more complex equation for a fixed end beam with an external prestress (Hetenyi, 1958) indicates that horizontal prestress has little effect on the subsidence prediction.

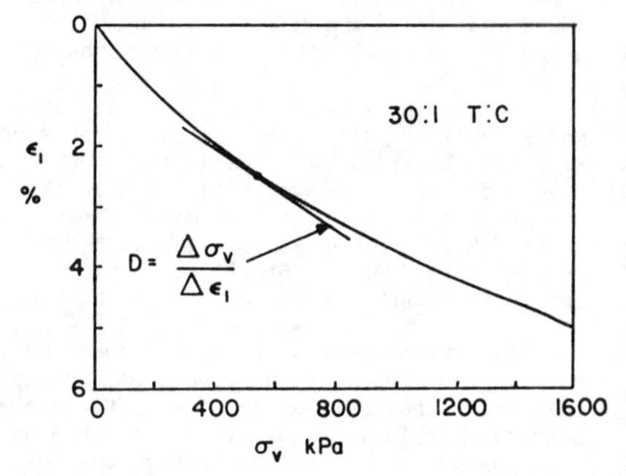

Figure 2 - Backfill constrained moduli

A computer analog called MINE B/F was written for the IBM compatible Zenith PC to calculate crown pillar deflections before and following the attainment of the fully plastic moment in the crown pillar. This numerical model incorporates a user selected step increase in overburden stress so that variations in subgrade modulus can be accommodated. The programs are written in microsoft FORTRAN 77 v.3.20 and use a graphics interface developed in Queen's Civil Engineering Department. The source code and subroutine schematics are contained in a report by Wizniak (1986).

Centrifuge Models

The Queen's University geotechnical centrifuge is a 3 m radius, 30 g-tonne machine powered by a 45 kW hydraulic motor and it is equipped with strobe video monitoring as well as slip ring access for transducer instrumentation. It is described in detail by Mitchell (1986). Figure 3 shows one of the centrifuge subsidence models used in this study. The crown pillar was modelled using fixed-end precast concrete sections having a stiffness modulus of E = 26 GPa (measured in compression), a compressive strength of 30 MPa and a tensile strength of σ_t = 3300 KPa (measured in direct tension tests and beam bending tests reported by Wizniak, 1986). Actual mine tailings, at tailings to cement (T:C) ratios of 40:1, 30:1, 20:1 and 10:1 and cured for 28 days, were used for the model backfill. The models were set in strong boxes on a thin bedding layer of quick-setting plaster of paris and a 1 mm space was left at each end of the beam to accommodate beam elongation during bending. Lead shot of about 1.5 mm diameter were placed to a height of about 0.1 m above the beam to represent the flexible overburden.

Figure 3 - Typical subsidence model

More realistic mine crown pillar models might be created using fractured laminated materials but it would be necessary to incorporate a mechanism to impose a lateral prestress, σ_h, on the end of the beam. The concrete beam offers the advantage of a known tensile strength, σ_t, which allows the beam to develop bending stresses in the absence of a lateral prestress. Thus, the tensile strength of the concrete beam effectively models the lateral prestress in the prototype up to the point of cracking of the beam. Unlike the prototype, however, the concrete beam does not maintain a plastic moment and, following beam failure, the beam subsidence can be estimated as

$$Y_c = \sigma/K \qquad (4)$$

The main purposes of the centrifuge models were, as mentioned earlier, to examine the application of the 'beam on an elastic foundation' approach to subsidence prediction and to determine whether ordinary backfills would provide any significant resistance to crown pillar subsidence. The concrete models are considered suitable for these purposes. Models were tested by increasing the centrifuge speed in increments and monitoring the central beam deflection using a direct current differential transformer (DCDT). Model beams of 0.2 m x 0.2 m top surface (0.14 m clear span) and of 0.28 m x 0.1 m top surface (0.23 m clear span) were used.

Centrifuge Test Results

Three series of centrifuge model studies were carried out using plain concrete to model rock pillars and various depths of crown pillars. Conventional tailings backfills were used in the models. A mass of 22.7 kg of lead shot was used to produce a theoretical stress

Table 1 - Overburden Stress Measurements

Description of calibration test	Test result and conclusion
4 kg mass placed on 500 kg load cell resting on bottom of strongbox	Theoretical forces reproduced exactly, including minor correction for mass of load cell
22.7 kg lead shot placed to 0.1 m height over 500 kg load cells resting on bottom of strongboxes and surrounded by wooden blocking. Three cells, one on C and two close to sides of box; flush mounted and covered by thin rubber membrane; active cell area approx. 6 cm^2	Measured stresses higher than theoretical by 30% at C and 40% near box boundaries. Cell more rigid than surrounding blocking hence local arching above load cells resulted in enoneously high stress readings
22.7 kg lead shot placed to 0.1 m height over plate into which 1000 kPa pressure transducers were threaded, covered by thin rubber membrane. Three transducers, one on C and two near sides of box. Active strain guaged faces approx. 0.6 cm^2	With flush mounted active face, measured stresses averaged 25% lower than theoretical, with central stress slightly higher than those near box boundaries; stresses increased to equal theoretical if transducer face mounted slightly above plate. Transducer mounting and compliance critical to achieving correct measurement
22.7 kg lead shot placed to 1.0 m height above plate resting on three springs having a total stiffness of k = 1.2 kN/mm	Measured force of 1.2 kN per mm of deflection equal to theoretical force of 0.22 a/g kN at all accelerations up to 10 mm of deflection. Wall friction and arching across box found to be negligable.

level of 5.57 (a/g) kPa on a 0.2 m by 0.2 m model surface (a = centrifugal acceleration, g = gravitational constant) or 7.96 (a/g) kPa on a 0.28 m by 0.1 m model surface. Three methods were used to check that these stress levels were achieved during model testing and Table 1 contains the details of this calibration testing. The results of the calibrations indicated that, although stress level was difficult to measure using ordinary load cells or transducers, the stress levels attained during centrifuge testing were close to the theoretical levels.

The first set of centrifuge tests carried out on 0.2 m by 0.2 m model crown pillars having a clear span of L = 0.14 m, depths of 15 mm and 30 mm and using 30:1 T:C backfill were, as reported by Wizniak (1986), not completely successful due to several technical problems. Foremost amongst these problems was horizontal stresses induced into the beams by elongation due to bending because these beams were tightly fitted into the strong boxes. This and the other technical problems were corrected as a result of the first set of tests.

The second set of tests were carried out on models having 0.14 m clear span model crown pillars of 20 mm depths with backfill heights from 60 mm to 150 mm. Tight fill placement was used in four models and four models were created with a 4mm space between the beam and the backfill in order to simulate standard backfill practice. Two extra model beams were accelerated to collapse without fill support. Centrifuge testing was carried out by conditioning the models for about 5 minutes at 60 rpm and then increasing the speed in increments corresponding to acceleration increases of about 10 gravities. This represents overburden stress increments of about 56 kPa. During these incremental increases in stress, the central subsidence developed quite rapidly and deflections attained apparent equilibrium in about five minutes (before beam cracking) to about 10 minutes (following beam cracking).

Figure 4 shows typical models after testing. The thin black lines on the beams were penned onto a white painted surface and allowed a gross physical measurement of the subsidence after testing. The white lines spray painted onto the backfill were to aid video observation in flight. Central subsidence data from the second set of model tests are plotted on Figure 5. The results shown on Figure 5 indicate that unsupported beams consistently collapsed under overburden stresses of 170 to 190 KPa. Using the normal fixed end beam formulae, collapse is predicted at

$$\sigma = 2\sigma_t \left(\frac{d}{L}\right)^2 \qquad (5)$$

for d/L = 1.143 and σ_t = 3300 kPa, the theoretical stress at abutment cracking is 135 kPa. Abutment cracking was noted in some models at about 150 kPa, prior to central cracking and collapse at about 180 kPa. The theoretical central deflection at the point of cracking is given as

$$Y_c = \frac{1}{384} \frac{\sigma L^4}{EI} \qquad (6)$$

Using σ = 180 kPa, the value of E required to obtain correlation with the 0.25 mm of observed subsidence is found to be 10^8 kPa. This value is roughly four times the measured compressive modulus of 26 GPa but this value is used in the bending analyses because it is possible that the tensile modulus is higher than the compression modulus. The difficulties in measuring a modulus this high in concrete samples are substantial.

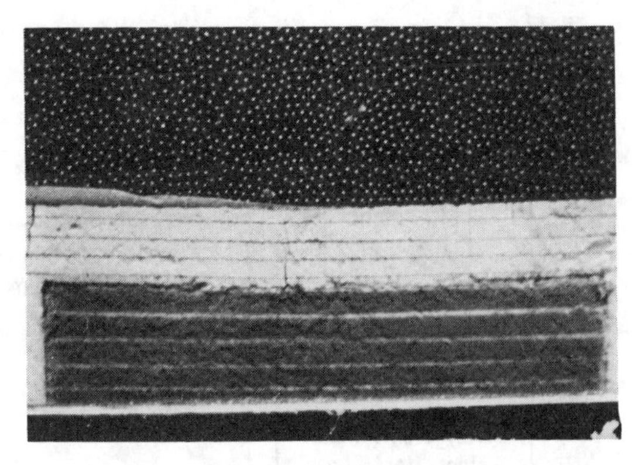

(a) cracking and subsidence in 0.14 m span

(b) cracking and subsidence in 0.23 m span

Figure 4 - Subsidence models after testing

It is clear from the data on Figure 5 that very high subsidence is associated with caving of the beam when the fill is not placed tightly. Even tightly placed fills that are relatively deep (where H ≥ L) show significant subsidence.

Following cracking of the beam the subsidence predictions plotted on Figure 5 follow the equation

$$Y_c = \frac{\sigma}{K} + \rho\left(\frac{a}{g}\right) H^2/2D \qquad (7)$$

where the second term accounts for backfill self-weight compression and

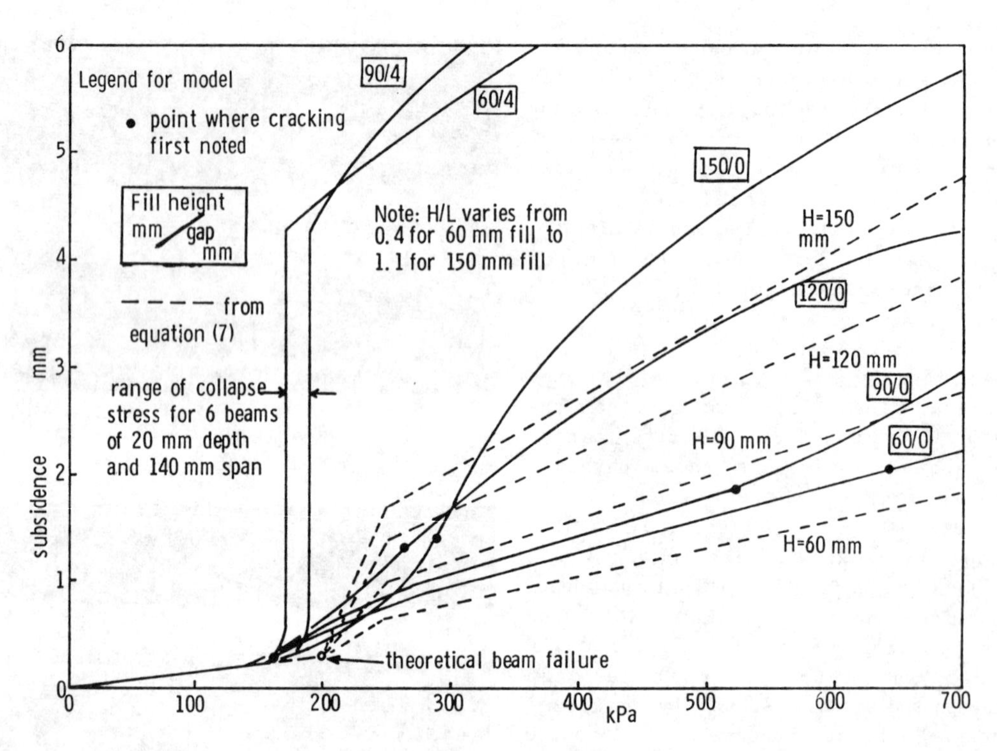

Figure 5 - Subsidence results from 0.14 m model spans

varies, at σ = 700 kPa, from 0.14 mm for the 60 mm fill height to 0.85 mm for the 150 mm fill height. A backfill constrained modulus of $D = 2.4 \times 10^4$ kPa, the average value of the 30:1 T:C fill in the range $\sigma \leq 700$ kPa, was used to calculate the predicted subsidence. The predictions assume that stresses carried by beam action are transferred to the backfill, following beam failure, during the stress increment from 200 to 250 kPa. The predicted and observed subsidence agree reasonably well although it is noted that the beam cracking was not observed until the subsidence reached 1.2 to 2.0 mm and the transfer of stress took place over a fairly large stress range. Some minor differences between the theoretical and observed subsidence may also result because rotation of the broken beam segments induces a non-uniform stress distribution in the backfill.

The third set of centrifuge tests were carried out on models that were 0.28 m long by 0.1 m wide having a beam depth of 30 mm and a tight backfill depth of 80 mm. Four different backfill T:C ratios were used and the constrained moduli are listed in Table 2. The free span of the beams was 0.23 m which gives, from equation (5), a vertical stress of 110 kPa at first cracking. The actual collapse of unsupported beams occurred at σ values of 120 and 140 kPa. Using equation (6) with $E = 10^8$ kPa gives a central subsidence of 0.42 mm at σ = 130 kPa. The subsidence results are plotted on Figure 6 and it is seen that for this geometry, the backfill

Table 2 - Constrained moduli of backfills

	Average Modulus, D, kPa for	
T:C mix	σv < 800 kPa	σv > 800 kPa
40:1	2.7×10^4	3.8×10^4
30:1	2.8×10^4	4.0×10^4
20:1	3.3×10^4	4.5×10^4
10:1	5.0×10^4	7.0×10^4

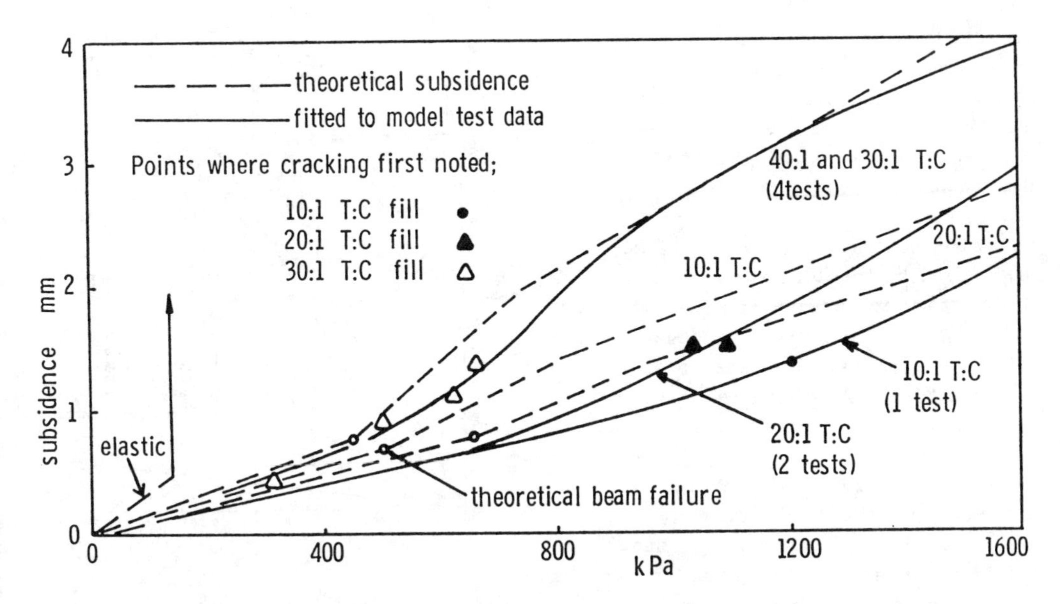

Figure 6 - Subsidence results from 0.23 m model spans

takes up load immediately and restricts the early subsidence to values less than the elastic deflections of an unsupported beam. Subsidence predictions are plotted on Figure 6 for the 40:1 T:C, 20:1 T:C and 10:1 T:C backfills. The 30:1 T:C backfill modulus is very similar to the 40:1 T:C backfill modulus at all stress levels. Predictions prior to yielding of the beam (for $M_c \leq \sigma_h d^2/6$) are given by equation (3) and following beam cracking, by equation (7). It is assumed that stress transfer from the broken beam takes place over a 300 kPa stress range following yielding. The observed subsidences on Figure 6 agree reasonably well with the predicted values.

From the results shown on Figures 5 and 6 it is concluded that the 'beam on an elastic foundation' theory provides an excellent framework in which to interpret the data from centrifuge subsidence models. It may then be useful for prototype similitude analyses and for the prediction of subsidence at the field scale.

Dimensional Analysis and Design Charts

In order to evaluate prototype subsidence, dimensionally similar models can be constructed in the centrifuge. It can be shown (Wizniak, 1986) that the ratio of subsidence to beam length, Y_c/L, is a function of d/L, σ/KL and λL. The model value of λ must increase according to the linear scale and this is achieved because the moment of inertia, I, decreases according to the fourth power of the linear scale if d/L is maintained constant. The value of ℓK remains the same in model and prototype if the same backfill is used. Therefore the prototype crown pillar stiffness modulus should be used in designing the model beam although variations in fill height can be used to compensate for differences in either modulii. The horizontal prestress in the model should be equal to the estimated field horizontal prestress and the model should be accelerated such that the overburden vertical stress is simulated. Allowance should be made to accommodate the self-weight compression in the model backfill as the model is accelerated -- this could be done by allowing the abutments of the beam to rest on the fill and monitoring the differential settlement between the centre of the beam and the abutments. This differential settlement would be considered to represent the central subsidence of the model and, when multiplied by the linear scale would give the expected prototype subsidence. Research with improved models having externally applied horizontal prestress is required to verify whether correlation can be achieved for direct similitude models.

The program MINE B/F was used to generate two design aids, Figure 7 and Figure 8: Figure 7 shows the safe beam depths and expected subsidence for various spans; Figure 8 is a dimensionless plot of subsidence parameters. These two design aids assume a constant value for the backfill constrained modulus, D.

Conclusions

From the results of analyses using the theory of a beam on an elastic foundation combined with centrifuge model studies, three conclusions are reached:

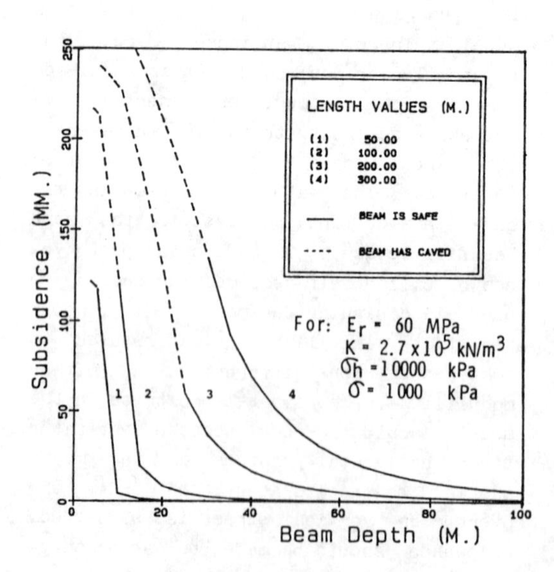

Figure 7 - Beam depths and subsidence

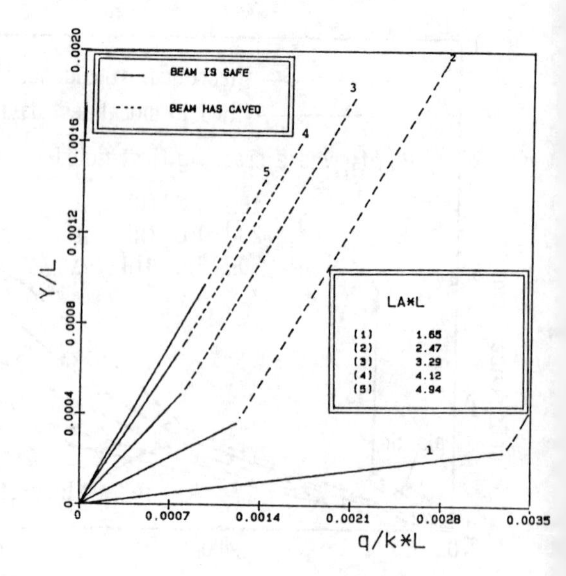

Figure 8 - Subsidence chart

(1) Centrifuge models indicate that tight backfilling with conventional tailings backfills can delay the development of plastic moments in crown pillar beams and reduce crown pillar subsidence.

(2) The beam on an elastic foundation approach has been applied successfully to analyze subsidence in centrifuge models and may be useful in estimating prototype subsidence.

(3) Dimensionally similiar centrifuge models can be developed to model a variety of prototype mine subsidence situations. Such models should be tested and analyzed using available analytical models or compared with field subsidence data, if available.

A computer program called MINE B/F has been developed for assisting in subsidence calculations and was used to generate two design guides included in this paper.

Acknowledgments

Financial support to construct the centrifuge was provided by the Natural Sciences and Engineering Research Council of Canada (NSERC), Westmin Resources Ltd., and Falconbridge Ltd. on a Shared Facilities grant. Mr. Jim Roettger helped to construct the centrifuge and to cast and test the models. Continued financial support from NSERC for operation of the centrifuge facility is also acknowledged.

References

Brauner, G. (1973) Subsidence due to Underground Mining. Rep. 8571, US Bureau of Mines, Washington, D.C.

Hetenyi, M. (1958) Beams on Elastic Foundations. University of Michigan Press, Chicago.

Mitchell, R.J. (1983) Earth Structures Engineering. Allen and Unwin, Boston.

Mitchell, R.J. (1986) Centrifuge Model Tests on Backfill Stability. Canadian Geotech. J. 23:3:341-345.

Wizniak, L.M. (1986) Experimental and Numerical Modelling of a Crown Pillar Beam in a Mine. M.Sc. project report, Civil Engineering Department, Queen's University.

Centrifuge modeling of bearing capacity of sand under concentric and eccentric loading

Abbas Abghari, Bruce L.Kutter & J.A.Cheney
University of California, Davis, USA

ABSTRACT: The results of centrifuge model tests on the bearing capacity of concentrically and eccentrically loaded circular footings in dense sand (Monterey 0/30 sand) are presented and compared with the conventional bearing capacity equations. The Meyerhof procedure for analysis of eccentric loading is shown to be adequate. The stress dependency of the bearing capacity factor N_γ and the friction angle ϕ is demonstrated. This fact necessitates a rational choice of friction angle for bearing capacity prediction which accounts for the stress dependency. An alternative method for bearing capacity prediction in sand utilizing an "apparent" cohesion, C and friction angle ϕ is presented. It is found that either method of predicting bearing capacity is adequate, but the latter method is simpler to apply. The use of C for cohesionless sand appears inappropriate, but a better fit to the triaxial test data is obtained if a "cohesion" is included. Modeling of models show good agreement, thereby verifying the scaling laws and test procedure.

1 INTRODUCTION

Terzaghi (1943) developed the following simple equation for the ultimate bearing capacity of a shallow strip foundation or footing:

$$q_u = P_u/BL = CN_c + \gamma D_f N_q + .5 \gamma B N_\gamma \qquad (1)$$

where P_u = ultimate load, B = width of the foundation, L = length of the foundation, C = cohesion of soil, γ = effective unit weight of soil, D_f = depth of foundation, and N_c, N_q and N_γ are bearing capacity factors for cohesion, surcharge, and self weight which depend primarily on the friction angle of the soil. In this equation the shearing resistance of the soil above the bottom of the foundation is neglected.

Since then many researchers (Meyerhof 1948, 1951, 1953, Hansen 1970, Muhs and Weiss, 1973) have investigated the bearing capacity problem and have modified the above equation for different shapes of footing and different loading conditions. The general bearing capacity equation according to Hansen (1970) for the case of horizontal ground surface is

$$q_u = CN_c\, s_c\, d_c\, i_c + \gamma D_f N_q\, s_q\, d_q\, i_q + 0.5\, \gamma B N_\gamma\, s_\gamma d_\gamma i_\gamma \qquad (2)$$

where s = shape factors to account for the shape of the foundation in developing a failure surface, d = depth factors to account for embedment depth and the additional shearing resistance due to the soil above the foundation level, and i = inclination factors to allow for both horizontal and vertical foundation loads. Empirical relationships have been established for shape factor s, depth factor d, and inclination factor i by several researchers (Hansen 1970, De Beer 1970, Vesic 1973).

For the case of eccentric loading the concept of effective area was introduced by Meyerhof (1953). He proposed that only the portion of the footing area which is symmetrical on either side of the load be considered effective and the rest of the area simply neglected.

The majority of the experimental data that has been used to verify the bearing capacity equations was obtained on small scale models with relatively low stress levels compared to the size and stress levels of real prototype scale footings. In this report the results obtained from centrifuge tests are used to check the validity of bearing capacity equations at stress levels comparable to the real prototype.

2 TESTING PROGRAM

Tests were carried out with circular footings of diameters 1.5 and 2 inches placed on the surface of Monterey 0/30 dense sand. Tests covered concentric, eccentric, and moment loading. All tests were performed on the U.C. Davis one meter radius Schaevitz centrifuge at centrifugal accelerations of 18.75, 25 and 50 g.

2.1 Sample preparation

The sample container was 13 in. by 10 in. by 10 in. deep. To obtain a homogeneous sample, the samples were prepared by pluviating the sand from an elevated storage hopper into the container. The densities of the samples were 105.5 to 106 pcf with a void ratio of .56 to .55. The depth of the samples was approximately 6.5 in.

2.2 Soil properties

The sand used in all samples was a poorly graded Monterey 0/30 sand (Figure 1) with a coefficient of uniformity of $C_u = 1.48$, specific gravity $G_S = 2.64$, and a maximum and a minimum void ratio of 0.892 and 0.535 respectively. The samples were prepared with void ratios in the range of .56 to .55 corresponding to relative densities D_r of 93 to 96 percent.

The angle of internal friction of a sand not only varies with void ratio, but also for a given void ratio it varies with the mean normal pressure. Terzaghi (1951) found that in his tests the coefficient of internal friction decreased with increasing pressure. Since then many researchers (Bishop 1957, De Beer 1967, James 1984), have investigated the effect of stress level on friction angle.

Knowing that the internal friction angle is stress dependent, triaxial tests with varying confining pressures were performed to obtain the friction angle as a function of stress level on the test material.

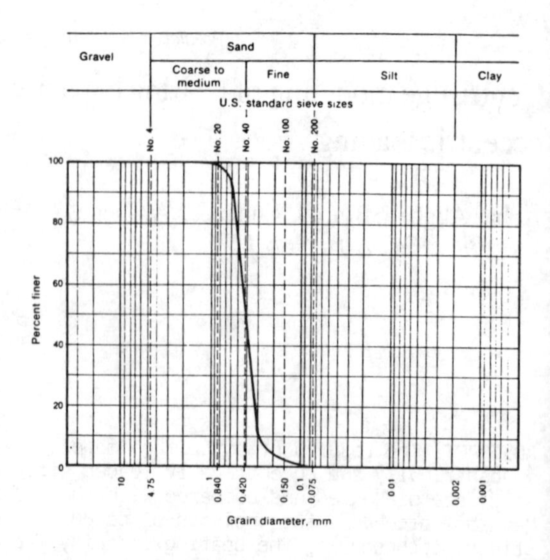

Fig. 1 Grain size distribution of Monterey 0/30 sand

The range of confining pressures were from 0.2 Ksc to 6.9 Ksc. Samples were prepared by pluviation and had void ratios in the range of .55 to .56 corresponding to relative densities D_r of 93 to 96 percent, identical to that obtained in the centrifuge samples.

2.3 Triaxial Tests Results

Table 1 gives a summary of the triaxial test results. ϕ is the maximum friction angle corresponding to the maximum deviatoric stress, i.e., peak strength. Figure 2 gives the Mohr's Circles for each of the test results. It can be seen that the Mohr-Coulomb failure envelope is not a straight line, but curves, turning toward the σ axis.

Figure 3 shows a plot of the variation of friction angle ϕ versus the mean normal stress $p = (\sigma_1 + 2\sigma_3)/3$. Also shown are the results obtained by Bagge and Christensen (1977) on Danish normal sand. Test results show that in the range of confinement from 0.21 Ksc to 6.9 Ksc, the maximum friction angle changes from 52.4 to 41.9 degrees.

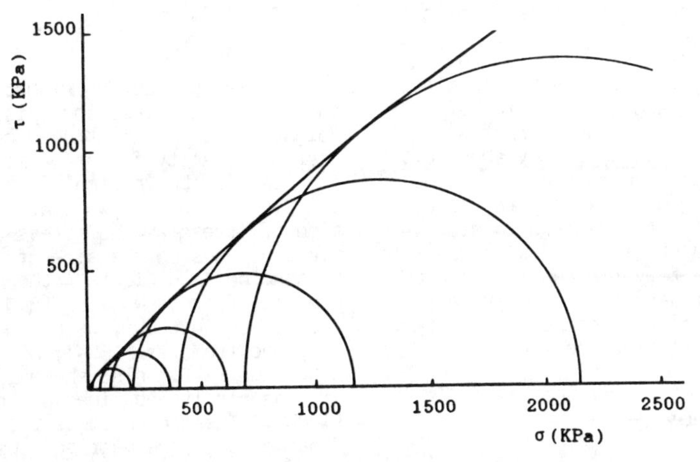

Fig. 2 Mohr failure envelope for Monterey 0/30 dense sand (dry)

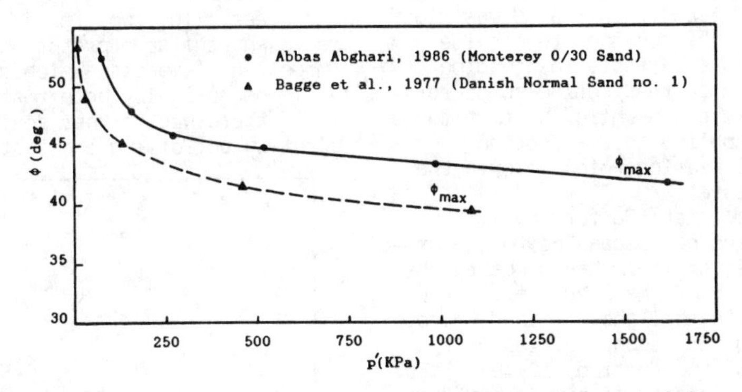

Fig. 3 Variation of peak friction angle with mean normal stress at failure for two sands. The Monterery sand tested for this study was dry and dense ($D_r \approx 95\%$).

Table 1. Triaxial Compression Test Results on Monterey 0/30 Dry Dense Sand

Test	σ_3(Kpa)	σ_1(KPa)	ϕ_{max}(deg)
1	21	184.5	52.4
2	53	361	48.07
3	100	610	45.92
4	200	1162	44.9
5	400	2155	43.38
6	690	3460	41.87

2.4 Membrane Correction

With very low confining pressure σ_3 = 0.21 Ksc in the triaxial test the membrane stretch may increase the effective confinement pressure which could be interpreted as yielding a higher friction angle. Thus a drained test on a saturated sample with the same density was run at the same confining pressure (0.21 Ksc). This allows volumetric strain to be measured and then the circumferential stretch of the membrane could be determined. Also a test was done on the membrane to determine the modulus of elasticity of the membrane. Then the pressure due to the membrane corresponding to the strain at maximum deviatoric stress was calculated using elastic theory. This calculation yielded an increased confining pressure of 0.004 Ksc. This additional pressure reduces the calculated friction angle by 0.3 degrees. It should be noted that this measurement is approximate because the deformation of the membrane is not uniform. To check this hypothesis a test was carried out with two membranes at the same confining pressure. If membrane had a larger effect, this test would result in a higher friction angle. In point of fact, this test gave a lower friction angle than the previous test by about 0.5 degrees. Apparently errors owing to the samples not being made exactly the same, and also other errors involved in measuring stresses and strains at low stress contributes more to variation in measured ϕ than the membrane effect. Nevertheless a small correction for the membrane effect was made in determining the ϕ.

2.5 Test Procedure

The loading apparatus had a capacity of 1000 lbs. Concentric or eccentric loading can be applied to the footings. The footings were made from structural steel with diameters of 1.5 and 2 in. Sand was glued to the base of the footings to provide a rough base. In some tests the loading shaft was fixed to the footing so no rotation could take place while in other tests the load was applied to the footing through a ball bearing which allowed the footing to rotate.

Tests were done at 50g for the 1.5 inch diameter footing corresponding to the prototype diameter of 75 inches. Due to the limited capacity of the load cell, it was not possible to fail the 2 inch diameter footing at 50g for eccentricities of r/6 (r is radius of the footing) or less. So the concentric loading of the 2 inch diameter footing was done at 18.75g corresponding to a 37.5 inches diameter prototype, and the eccentric loading of eccentricity r/6 was done at 25g which corresponds to a 50 inches diameter prototype.

Moment and shear loading were also carried out while keeping the normal stress constant. For this case the apparatus was modified so that a lateral force could be applied at a height h above the footing. The range of bearing pressure in these tests was 18 to 114 psi. All moment loading tests were performed at 50g. Thus the model footings correspond to prototype diameters of 75 and 100 inches.

3 INTERPRETATION

For the case of concentric loading of circular footings on cohesionless soil in the centrifuge the bearing capacity equation becomes:

$$q_u = 0.3\ (n\gamma)\ DN_\gamma \tag{3}$$

where D is the footing diameter, n is the number of gravities, and γ is the unit weight at 1g. As mentioned before the bearing capacity factor N_γ is mainly a function of the internal friction angle, ϕ. Since it was shown that ϕ is a function of stress, N_γ is stress dependent, i.e., depends on the size of the foundation. On Figure 4 the bearing capacity factor, $N_\gamma = q_u/(0.3\ n\gamma D)$ is plotted against $Dn\gamma$, using the results obtained from tests BS-1, BS-2, BS-3, BS-4, and BS-5. Also shown in Figure 4 are the results obtained by James et al. (1984). The figure shows the stress dependence of the bearing capacity factor, N_γ, which reduced by a factor of three by going from a D of 2.0 inches, to a D of 75.0 inches. This corresponds to a reduction of internal friction angle ϕ of about 5 degrees. The two tests BS-1 and BS-3 represent the same prototype diameter (modeling of models) which give very close N_γ. Theoretically both must give the same N_γ if there was neither scale effect nor error in performing the tests.

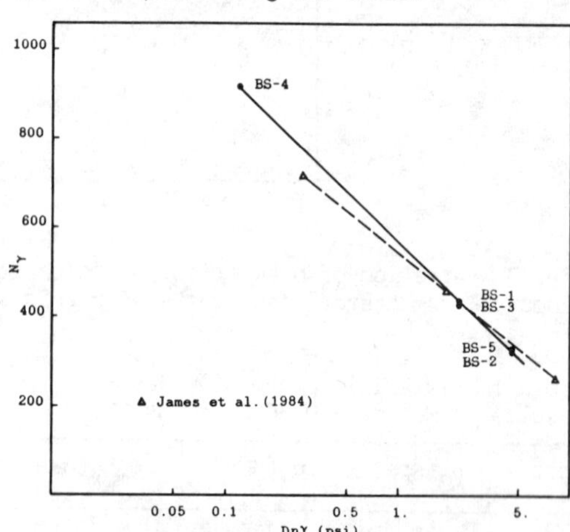

Fig. 4 Variation of bearing capacity factor N_γ as a function of $Dn\gamma$ (size effect)

Nondimensional load versus displacement ($F/n\gamma D^3$ vs. δ/D) are shown in Figure 5 for tests BS-1, BS-2, BS-3, BS-4, and BS-5. Good agreement is observed between the results from Tests BS-1 and BS-3 which represent the same prototype, modeling of models.

Figure 4 shows how a small change (about 5 degrees) in friction angle drastically changes the bearing capacity

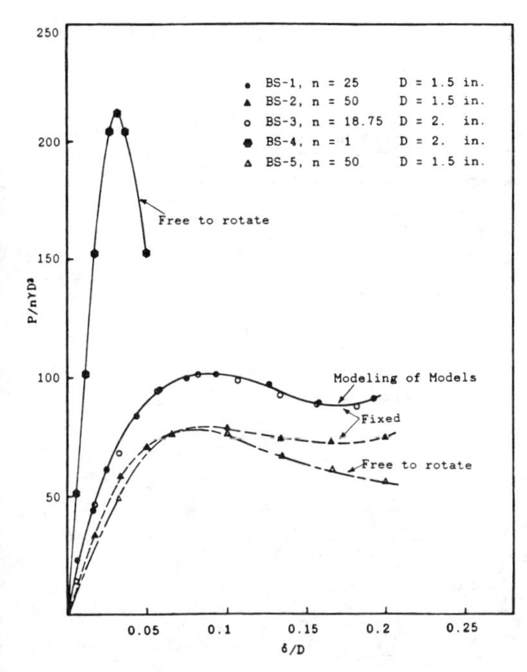

Fig. 5 Dimensionless load versus displacement for circular footing on Monterey 0/30 dense sand

factor, N_γ (by one third), and consequently the ultimate bearing capacity. So it is very important to choose the appropriate friction angle.

Stress varies with depth under the foundation. Along the failure surface the stress changes from point to point, and so the friction angle is not a constant value along the failure surface. Since in the derivation of the ultimate bearing capacity equation the variation of friction angle has not been taken into account, but considered a constant value, the question is; what friction angle corresponding to what stress level should be chosen? Meyerhof (1950) assumed that the mean value of normal stress along failure surface was 1/10 of the ultimate bearing load q_u and chose the friction angle corresponding to the secant connecting the origin to this point (Figure 6, point M) on the Mohr-Coulomb failure envelope. From this De Beer (1967) derived a semi-empirical formula for the mean value of the mean normal stress, p_m, along the slip surface:

$$p_m = (q_u + 3\gamma D_f)(1 - \sin\phi)/4 \qquad (4)$$

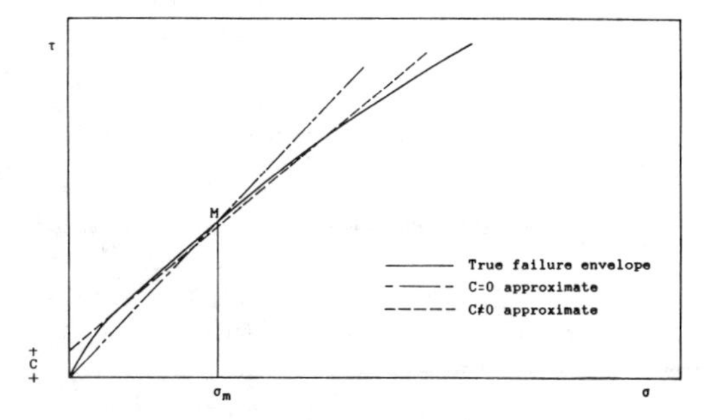

Fig. 6 Schematic failure envelope for sands with exaggerated curvature, and illustrations of C = 0 and C ≠ 0 approximations to the failure envelope

The Meyerhof and De Beer methods both require an iterative procedure to determine the bearing capacity.

As an alternative to a curved failure envelope (varying friction angle) for the sand tested a least squares fit straight line has been drawn through the origin and triaxial test results which are in the range of stresses along the failure surface of the bearing capacity tests (Figure 6). This resulted in a friction angle, ϕ = 43.1 degrees and an apparent cohesion, C = 13.01 Kpa(1.89 psi). Although the use of an apparent cohesion for cohesionless sand appears inappropriate, a better fit of the Mohr-Coulomb criterion in the range of interest is obtained if a cohesion is used. Using these values and the bearing capacity factors suggested by Hansen (1970) in the general bearing capacity equation (Eq. 2), the bearing capacities were calculated for the centrifuge tests. Table 2 compares the results of the centrifuge tests with the predictions using the above methods. Also shown in Table 2 are the results from Bagge et al. (1977) and predictions using these methods.

Table 2 shows that agreement between the test results and all three methods of prediction is adequate for the centrifuge tests. The 1-g test could not be predicted by either Meyerhof or De Beer's method because triaxial data was not available at the very low stresses corresponding to these tests (triaxial tests were not carried out at lower stresses because as previously explained the errors could be significant at low stresses). Stresses along the failure surface are very low for the bench test, so C and ϕ

Table 2

RESULTS FROM CENTRIFUGE TESTS AND PREDICTIONS

Test	Centrifuge Acceleration n (g)	Prototype Foundation Diameter (m)	Measured Bearing Capacity (MN/m²)	Predicted Bearing Capacity (MN/m²)			
				Meyerhof method**	De Beer method**	Proposed method	
BS-1	25	0.96	2.06	1.50	2.11	2.32	c = 13.01 KPa, φ = 43.1
BS-2	50	1.91	3.05	2.14	2.90	2.99	
BS-3	18.75	0.96	2.01	1.50	2.11	2.32	
BS-4	1	0.051	0.23	*	*	*	
BS-5	50	1.91	3.09	2.14	2.90	2.99	
		Maximum error		31%	6%	15%	
F + B***		0.25	0.404	0.35	0.54	0.48	c = 1.49 KPa, φ = 44.9
G + C***		0.50	0.776	0.59	0.87	0.724	
P + L + H***		1.0	1.519	1.01	1.21	1.17	c = 9.69 KPa, φ = 39.35
		Maximum error		34%	34%	23%	

* Triaxial test data were not available for these low stresses.

** Using Meyerhof's procedure, φ is 47.3 for tests BS-1 and 3 and is 45.6 for BS-2 and 5. For De Beer's method, φ is 48.9 for tests BS-1 and 3 and 47.1 for BS-2 and 5.

*** Bagge, G. and Christensen (1977).

Note: 1 MN/m² = 145 psi.

obtained from data points including high stress cannot be used for this test. De Beer's method gives the best results, but the disadvantage of the the method is the iterative nature of the procedure.

Figure 7 shows the experimental failure envelope for combined axial and moment loading. P_u is the bearing capacity at zero eccentricity, D is the footing diameter, and P and M are the vertical load and the moment at failure obtained from the moment loading and eccentric loading tests. It shows that at about $P/P_u = 0.5$ the failure mode changes from tilting (lift off) to vertical bearing failure. The envelope also indicates that at vertical loads close to the bearing failure the moment capacity is very low.

In the tests where the footing was restricted from rotating, the load capacity was increasing after failure. This fact can be explained by the observation that after the footing was sunk into the sand, the bearing capacity increased due to surcharge, i.e., the second term in Equation 1. In the case where the footing was allowed to rotate, the footing rotated at failure and so the load dropped. This fact can be due to the low moment capacity at the state close to bearing failure. In either case, the load-displacement curves up to the first peak stress were indistinguishable, meaning that the bearing capacity is unaffected by foundation rotational fixity.

For the case of eccentric loading of strip footings on sand, the bearing capacity equation based on Meyerhof's effective area concept (1953) can be written as:

$$q = (1 - 2e/D)^2 \, (0.5) \, \gamma D N_\gamma \qquad (5)$$

$(1 - 2e/D)^2$ is the reduction factor due to the load eccentricity. If q_u is the bearing capacity for the case of zero eccentricity (e = 0), then:

$$q/q_u = (1 - 2e/D)^2 \qquad (6)$$

On Figure 8 q/q_u is plotted against e/D (D is the footing diameter). The figure shows that Equation 5 is the lower estimate for the case of eccentric loading, but for the case of moment loading, test results fall below Equation 6, which is due to the fact that the foundation is

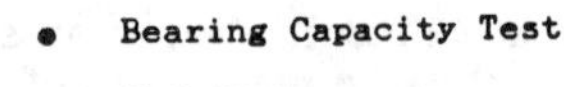

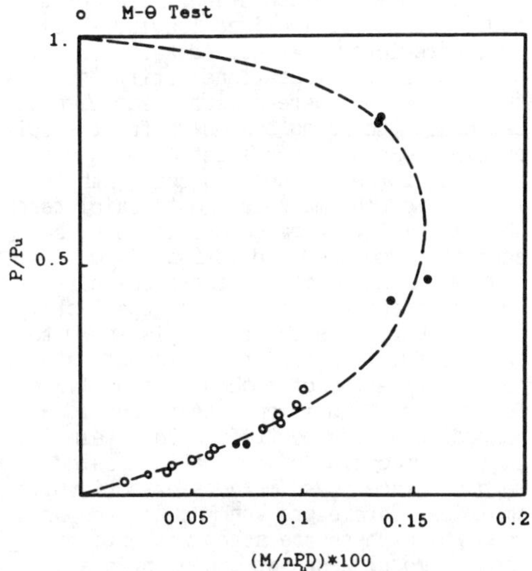

Fig. 7 Failure envelope of moment loading for Monterey 0/30 sand (dense)

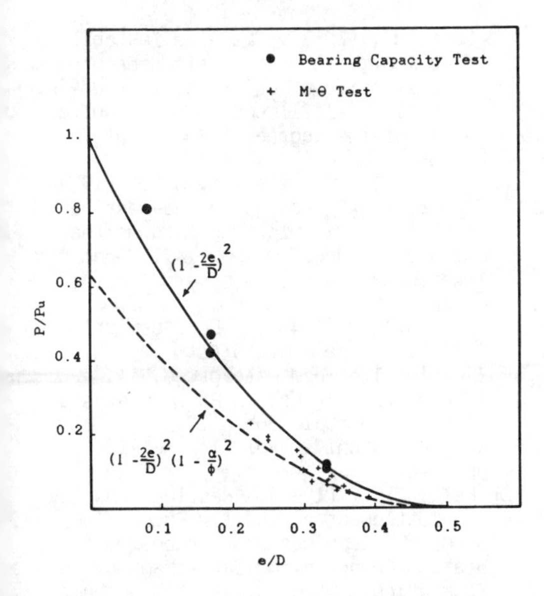

Fig. 8 Bearing capacity of eccentrically loaded footing

subjected to shear stress (inclined loading) in which case Equation 6 must be modified as the following (Meyerhof 1956):

$$q/q_u = (1 - 2e/D)^2 \, (1 - \alpha/\phi)^2 \qquad (7)$$

where α is the inclination of the load, $\alpha = \tan^{-1} (P_h/P_v)$ and ϕ is the friction angle (P_h and P_v are the horizontal and vertical components of the applied load). On Figure 8 the dashed line represents Equation 7 where α = 9 degrees is the maximum inclination for the moment loading tests. The angle of inclination in moment loading tests varied between 6 to 9 degrees. For $\alpha < 9$ as expected the data falls between the dashed line and the solid line.

Borowicka's Equation (1943) for rotational stiffness, $K_\theta = dM/d\theta$ of a semi-infinite elastic isotropic material subjected to a circular loading area is:

$$K_\theta = Ed^3/(6(1 - \nu^2)) \qquad (8)$$

Using the initial slope of the moment-rotation relationships obtained in the moment loading tests as the rotational stiffness, the modulus of elasticity, E was calculated using Equation 8 (Poisson's ratio, ν was assumed to be 0.3). The results are plotted in Figure 9. The mean normal stress, σ_m, was determined from

$$\sigma_m = \frac{(1 + 2K)\sigma_v}{3} \qquad (9)$$

with σ_v being the normal stress at the surface of the soil and K was assumed to be 0.4. The initial tangent modulus of elasticity, E was also determined from the triaxial test results as the slope of the deviatoric stress, q versus axial strain. Figure 9 shows good agreement between these two methods of calculating E. A straight line has been drawn through the data points which shows that the modulus of elasticity, E, is proportional to $\sigma_m^{0.65}$. Also, the equation proposed by Seed and Idris (1970):

$$G = 1000K_2 \, (\sigma_m)^{1/2} \qquad (10)$$

was used to calculate the shear modulus, G for shear strains of 0.1% and .5%. The modulus of elasticity, E was then calculated using a Poisson's ratio of 0.3. The two dashed lines on Figure 9 are plots of

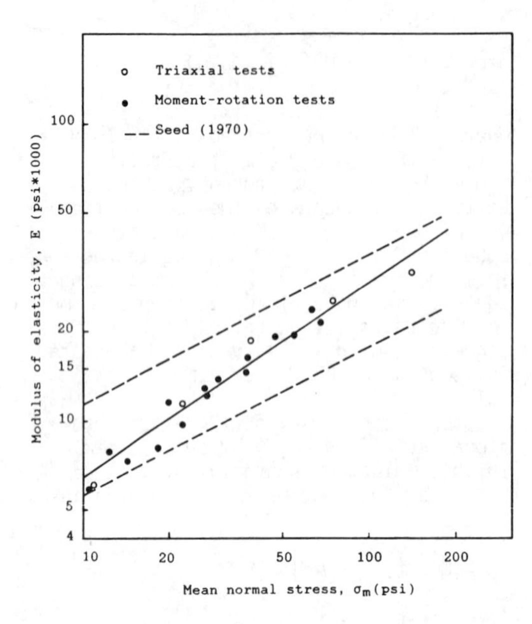

Fig. 9 Logarithm of modulus of elasticity, E versus Logarithm of mean normal stress, σ_m for Monterey 0/30 dense sand

the modulus of elasticity, E for strains of .1% and .5%. Good agreement between the modulus of elasticity, E obtained from triaxial tests and moment loading using Equation 9 is observed.

4 CONCLUSIONS

The centrifuge has been used to obtain data to verify commonly used theories regarding the behavior of foundations on dense sand.

1. The method of De Beer for determining the appropriate friction angle for prediction of bearing capacity is shown to be accurate, but the procedure requires an iterative calculation since the bearing capacity will influence the confining stresses and the confining stresses influence the friction angle.

2. A proposed alternate procedure utilizing an apparent cohesion for dry sand eliminates much of the stress dependency of the strength parameters. This means that iteration is not necessary, and the same strength parameters can be used to analyze footings with a range of sizes and bearing capacities.

3. Meyerhof's effective area concept for moment and eccentric loading was tested and it appears to work satisfactorily. In tests where the moment was produced by application of a horizontal force at some height above the foundation, Meyerhof's modification for inclined loading successfully predicted the test data.

4. The initial rotational stiffnesses of the foundations were measured and used to calculate the Young's modulus for the soil by use of Borowicka's equation. It was found that these results agreed with the Young's modulus measured in triaxial tests and fell within a range predicted by an equation from Seed and Idriss (1970).

5. Modeling of models tests demonstrated the consistency of the centrifuge tests.

In summary, the centrifuge is shown to be a useful tool for production of data for verification of methods for analyzing foundations. Additional tests should be conducted on the centrifuge to obtain further verification of foundation design methods since these methods often involve empirical parameters and much of the data upon which these are based was obtained from 1g model tests at inappropriate stress levels.

5 REFERENCES

Abghari, A., (1987). "Leaning Instability of Tall Structures," thesis presented to the University of California, at Davis, in partial fulfillment of the requirements for the degree of Doctor of Philosophy.

Bagge, G., and Christensen, S.N., (1977). "Centri fugal Testing on the Bearing Capacity of Circular Footings on the Surface of Sand," DIALOG 1-77, Danmarks Ingeniorakademi, Bygningsafdelingen, Lyngby.

Borowicka, H., (1943). "Uber Ausmitting Belaste Starre Platten Auf Elastich-Isotropm Untergrund," ("Eccentrically Loaded Rigid Plates on Elastic Isotropic Subsoil"), Igenieur-Archive, Vol. 14, No. 1, pp. 1-8.

De Beer, E.E., (1967). "Bearing Capacity and Settlement of Shallow Foundations on Sand," Proceedings of a Symposium on Bearing Capacity and Settlement of Foundations, Duke University, Durham, North Dakota.

Hansen, J.B., (1970). "A Revised and Extended Formula for Bearing Capacity," Bulletin, No. 28, The Danish Geotechnical Institute, pp. 5-11.

James, R.G., and Tanaka, H., (1984). "An Investigation of the Bearing Capacity of Footings Under Eccentric and Inclined Loading on Sand in a Geotechnical Centrifuge," Proceedings of the Symposium on Recent Advances in Geotechnical Centrifuge Modeling, University of California, Davis, pp. 88-115.

Meyerhof, G.G., (1948). "An Investigation of the Bearing Capacity of Shallow Footings on Dry Sand," Proceedings of the Second International Conference on Soil Mechanics, Vol. 1, p. 237.

Meyerhof, G.G., (1953). "Some Recent Research on the Bearing Capacity of Foundations," Canadian Geotechnical Journal, Vol. 1, No. 1, pp. 16-25.

Muhs, H., and Weiss, K., (1973). "Inclined Load Tests on Shallow Strip Footings," Proceedings of the Eight International Conference on Soil Mechanics and Foundation Engineering, Vol. 1, Part 3, pp. 173-179.

Ovesen, N.K., (1975). "Centrifuge Testing Applied to Bearing Capacity Problems of Footings on Sand," Geotechnique, Vol. 25, No. 2.

Seed, H.B., and I.M. Idriss, (1970). "Soil Moduli and Damping Factors for Dynamic Response Analysis," Earthquake Engineering Research Center, Report No. EERC 70-10, University of California, Berkeley, December.

Terzaghi, K., (1951) Theoretical Soil Mechanics, John Wiley and Sons, Inc., New York, 510 pp.

Vesic, A.S., (1973). "Analysis of Ultimate Loads on Shallow Foundations," Journal of the Soil Mechanics and Foundations Division, ASCE, Vol. 99, No. SM1, pp. 45-73.

Prediction and Performance in Geotechnical Engineering / Calgary / 17-19 June 1987

Centrifugal test on pullout resistance of buried anchor in sand

K.Tagaya
Mitsubishi Heavy Industries Ltd., Hiroshima, Japan

R.F.Scott
California Institute of Technology, Pasadena, USA

H.Aboshi
Hiroshima University, Japan

ABSTRACT : This paper describes, using sandy soil as an object, 1) the scale effect in a buried anchor pullout test by the centrifugal technique, 2) the theoretical formula of pullout resistance of a buried anchor, and 3) systematic centrifuge model tests on pullout resistance of buried anchor and a comparison between the theoretical values and the results of an elastoplastic finite element analysis. This study has made clear : 1) the centrifugal technique generally satisfies the law of similarity of dry sand without causing scale errors at the magnitude of acceleration greater than a certain value, and 2) the designing method of a buried anchor for sand.

1. INTRODUCTION

Centrifugal technique can be used in soil mechanics model tests such as anchor pullout resistance, bearing capacity of foundation, earth pressure, stability of slope, etc. for the frictional material. It is said that the use of the same material as the actual soil in this technique will ensure the law of similarity, eliminating the scale error due to difference in the failure condition of the prototype and the model. However, sufficient care should be exercised in using the centrifugal technique, which will also cause a problem of the law of similarity with respect to the grain size of sand and acceleration of the test field. In this paper, the scale effect and the law of similarity on the pullout resistance of anchor buried in dry sand will first be described in detail.

Next, the anchor pullout resistance will be discussed based on the theoretical analysis by plasticity, centrifugel model tests and elastoplastic finite element analyses.

The study on the vertical pullout of horizontal shallow anchor was initiated by Balla (1961), who obtained the formula of the pullout resistance by solving Kötter's equation on the assumption of the arciform failure surface. Matsuo (1969) extended Balla's method by assuming the combined failure surface with a logarithmic spiral and a straight line. Vesić (1971 and 1972) performed the analysis of vertical pullout of shallow anchor by the theory of plasticity with the theory of cavity expansion.

On the other hand, Meyerhof and Adams (1968), and Meyerhof (1973) offered a general solution of shallow and deep anchors, by the theory of plasticity, indicating a well agreement between the theoretical value and the experimental value with respect to the shallow anchor. In the case of deep anchors, however, the theoretical value becomes larger than the experimental value. For the purpose of advancing a proposal on the design formula of the pullout resistance of anchor buried in sand in an arbitrary angle and depth, the theoretical investigations, centrifuge model tests and elastoplastic finite element analyses were performed by the authors of this paper.

2. SCALE EFFECT IN ANCHOR PULLOUT TEST BY CENTRIFUGAL TECHNIQUE

2.1 Law of similarity on Pullout Resistance of Buried Anchor

The law of similarity on geometry, material and stress in ground of a $S = 1/\lambda$ scaled model test performed in the conventional 1 g field and that of a $S = 1/\lambda$ scaled model test in a centrifugal λ g field using a centrifugal technique are obtained by Buckingham's π-theorem as shown in Table 1. In this case, the model test is to be performed on the assumption that

Table 1 Law of similarity for anchor pullout test

	Prototype Scale: S = 1/1 Acceleration of test field: 1 g	Conventional model test Scale: S = 1/λ Acceleration of test field: 1 g		Centrifugal model test Scale: S = 1/λ Acceleration of test field: λg	
Similarity on geometry	$\frac{Df}{B}$	$\frac{Df}{B}$	similar	$\frac{Df}{B}$	similar
	α	α	similar	α	similar
Similarity on material	e	e	similar	e	similar
	$\frac{d}{B}$	$\frac{d}{\frac{B}{\lambda}}=\frac{d}{B}\lambda$	not similar	$\frac{d}{\frac{B}{\lambda}}=\frac{d}{B}\lambda$	not similar
	γ	γ	similar	γ	similar
	ϕ	ϕ	similar	ϕ	similar
	$\frac{c}{\rho g B}$	$\frac{c}{\rho g\frac{B}{\lambda}}=\frac{c}{\rho g B}\lambda$	not similar	$\frac{c}{\rho\lambda g\frac{B}{\lambda}}=\frac{c}{\rho g B}$	similar
	$\frac{E}{\rho g B}$	$\frac{E}{\rho g\frac{B}{\lambda}}=\frac{E}{\rho g B}\lambda$	not similar	$\frac{E}{\rho\lambda g\frac{B}{\lambda}}=\frac{E}{\rho g B}$	similar
	ν	ν	similar	ν	similar
	$\frac{E_g}{\rho g B}$	$\frac{E_g}{\rho g\frac{B}{\lambda}}=\frac{E_g}{\rho g B}\lambda$	not similar	$\frac{E_g}{\rho\lambda g\frac{B}{\lambda}}=\frac{E_g}{\rho g B}$	similar
	ν_g	ν_g	similar	ν_g	similar
	$\frac{\sigma_g}{\rho g B}$	$\frac{\sigma_g}{\rho g\frac{B}{\lambda}}=\frac{\sigma_g}{\rho g B}\lambda$	not similar	$\frac{\sigma_g}{\rho\lambda g\frac{B}{\lambda}}=\frac{\sigma_g}{\rho g B}$	similar
Similarity on initial ground stress	$\frac{\sigma_B}{\rho g B}$	$\frac{\sigma_B}{\rho g\frac{B}{\lambda}}=\frac{\sigma_B}{\rho g B}\lambda$	not similar	$\frac{\sigma_B}{\rho\lambda g\frac{B}{\lambda}}=\frac{\sigma_B}{\rho g B}$	similar

Model soil: Same material and same condition as for prototype.

using a $1/\lambda$ scaled model, the same soil as that of the prototype is used under the same compacting condition in the centrifugal field of the acceleration λ g.

In Table 1, the law of similarity on material cannot be satisfied in the conventional model test performed in the acceleration field of 1 g. That is, the conventional model test requires to multiply the grain size, cohesion of soil, Young's modulus of soil, Young's modulus of soil particle, breaking strength of soil particle, etc., by $1/\lambda$ of those of the prototype and would be impossible to realize it as far as the same material as for the prototype is used. Accordingly, the law of similarity on initial stress in soil, as shown in Table 1, and response (anchor pullout resistance, stress in soil due to pullout of anchor and anchor displacement) would neither be satisfied. In this case, the law of similarity could be satisfied to make similar to each other both stress in soil and response due to weight of soil if only such a material as having the same void ratio, effective unit weight of soil, angle of shear resistance, Poisson's ratio of soil and Poisson's ratio of soil particle, and the grain size of soil particle, cohesion of soil, Young's modulus of soil, Young's modulus of soil particle and breaking strength of soil particle multiplied by $1/\lambda$ would be used. In reality, there would be no such a material existing. Even if it could exist and would be used in a model test at a scale of S = 1/10 - 1/100, very poor testing accuracy would result because of extremely small response and small stress in soil.

On the contrary, in the centrifugal model test, all physical quantities will be satisfied, except the grain size of soil if the gravity of λ g is applied. Hence, if the phenomena are not affected by the non-similarity of the grain size, the centrifugal model test of anchor pullout resistance will become highly significant.

In the next section, the effect of grain size of soil on test result, the soil mechanics failure criteria, and scale error due to difference in failure condition of soil between the prototype and the model will be discussed.

2.2 Discussion on the Law of Similarity of Pullout Resistance of Buried Anchor

1) Soil mechanics failure criteria

In general, Mohr-Coulomb's failure criteria expressed by the following equation is used for the soil failure criteria:

$$\tau_f = c + \sigma' \tan \phi \quad (1)$$

As discussed in the previous section, if a model test using a scaled model of S = $1/\lambda$ and the same material under the same compacting condition is performed in the field of λg acceleration, $c_m = c_p$, $\phi_m = \phi_p$ and $\sigma'_m = \sigma'_p$, and therefore, $\tau_{fm} = \tau_{fp}$, satisfying the failure criteria of Mohr-Coulomb(suffix p : quantity for prototype, suffix m : quantity for model). On the other hand, since $\phi_m = \phi_p$ and $\sigma'_m = 1/\lambda \cdot \sigma_p$ in the field of 1 g acceleration, $c_m = c_p/\lambda$ and $\tau_{fm} = 1/\lambda \cdot \tau_{fp}$ must be met. However, it would be impossible to satisfy this relation as far as the same material is used under the same compacting condition and nearly impossible to select the material of $c_m = c_p/\lambda$ and $\tau_{fm} = 1/\lambda \cdot \tau_{fp}$. In the test in 1 g field, moreover, the failure mechanism is different between the prototype and the model, as described later.

2) Grain size

If the grain size is $d_m = d_p/\lambda$ meeting the scale in length, it will be about λ = 20-200 in a test by the existing centrifugal testing machine. Even when sand is used for the prototype, the sand will become silt or clay for the model, which is a cohesive material with the characteristics of sand lost. That is, the change of a factor that governs the phenomena will make the failure mechanism different between the prototype and the model.

The failure mechanism of sand, a granular material, is governed by the relative magnitude of the grain size of sand to the width of the foundation or anchor rather than by the absolute grain size. The average grain size of soil used in centrifugal model tests is 0.1 - 0.3mm, while the width of the foundation model or anchor model is 10-100 mm, which can be considered to be sufficiently large to the grain size of soil.

For the purpose of investigating the effect of grain size, Ovesen (1981) performed anchor pullout tests in the vertical and 45° directions in the centrifugal field and in the site as changing λ and B so that the product λB of the acceleration of the test field and the width of the anchor would become constant, and demonstrated that there were no scale errors in case of the ratio of the anchor width to the average grain size being greater than 25. In this reference, he also added that the conventional model test might provide overestimated values for the prototype.

Yamaguchi, Kimura and Fujii (1977) performed a bearing capacity test of a shallow foundation with 20 - 40 mm width in the 10 - 40 g acceleration field using glass ballotini having two different grain sizes, demonstrating that there was no difference in bearing capacity. The ratio of the foundation width to the average grain size in this case was 36-286.

From the above, it is considered that anchors having a width or diameter of about 15 - 48 mm will cause no scale errors on grain size (average grain size being 0.1 - 0.2 mm), as described later.

3) Scale effect in soil mechanics model test

Scale effect must be sufficiently examined in any model test. Especially, bearing capacity, earth pressure, anchor pullout resistance, etc., in soil mechanics, should show the different status of soil failure between the prototype and the model due to the initial stress condition of soil and the compressibility of soil corresponding thereto.

Ovesen (1981) performed vertical pullout tests of a circular anchor of 29.1 mm in diameter in the 1 g field and in the 50 g centrifugal field, and gave Fig.1. In this figure, his original graph is modified by taking γGB as abscissa, which is connected by a similar curve to that shown in the test results by the authors of this paper. The ordinate shows the dimensionless ultimate pullout resistance $Q_u/A\gamma GD_f$. This figure shows the conclusions that there are scale errors to exist in the conventional test of a scaled model in the 1 g field and that the prediction

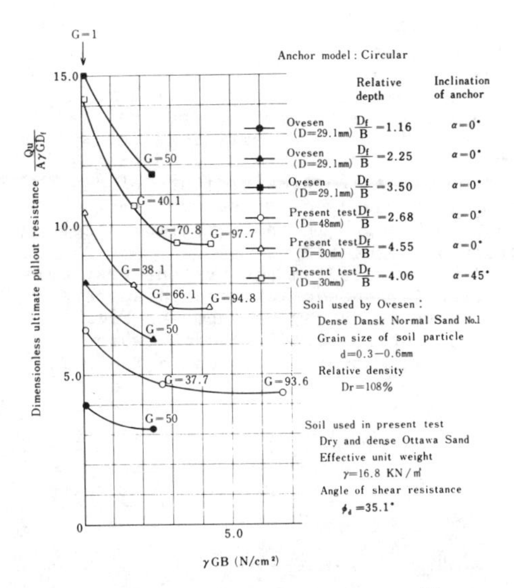

Fig.1 Scale effect in anchor pullout test by Ovesen (1981) and present test

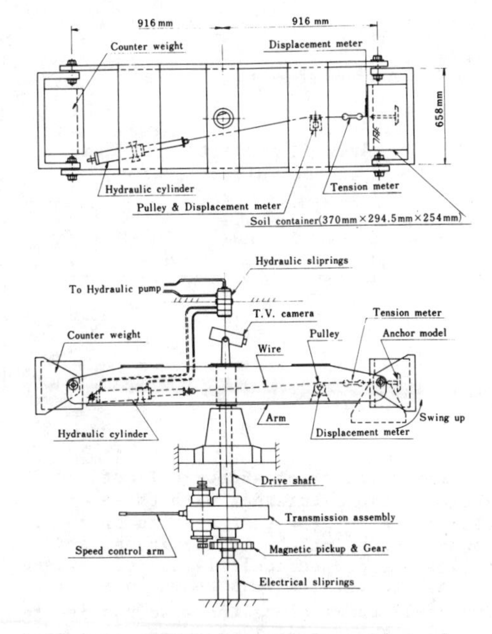

Fig.2 Centrifugal testing machine of California Institute of Technology

for the prototype from the results obtained in the conventional test is in overestimation.

The authors of this paper obtained the similar results in the conventional test in 1 g field and the centrifuge tests. In these tests, the centrifugal testing machine of California Institute of Technology was used, whose outline is shown in Fig.2. It has a maximum acceleration of about 175 g.

The models used in the present test were the circular anchors of which diameters were 30 mm and 48 mm as shown in Fig.3. Dry dense Ottawa Sand was used, whose average properties are shown in Table 2.

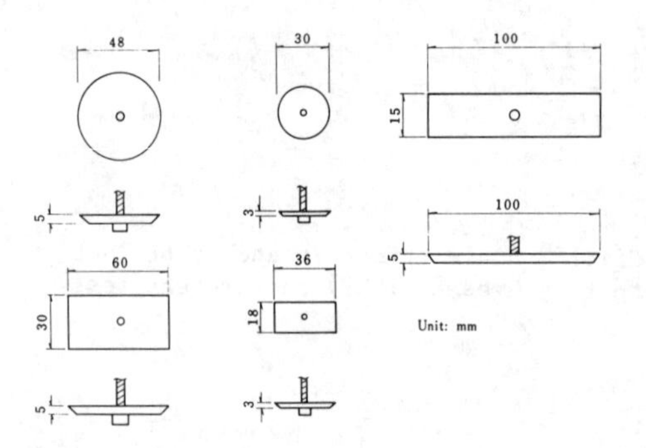

Fig.3 Anchor models

Table 2 Average properties of model ground for anchor pullout test

Test sand	50 percent grain size D_{50}(mm)	Coefficient of uniformity Uc	Specific gravity Gs	Effective unit weight of soil γ(kN/m³)	Angle of shear resistance ϕ_d (degree)
Ottawa Sand	0.47	2.16	2.69	16.2	31.6 (loose)
				16.8	35.1 (dense)
Toyoura Sand	0.17	1.49	2.64	15.5	42.0 (dense)

The test results of anchor pullout are shown in Fig.1 together with those of Ovesen. Also, the test results of an anchor buried at 45° to horizontal and pulled out perpendicularly to its surface (at 45° to vertical) is shown in Fig.1. The small range along the abscissa is the range of conventional test and the greater range of λGB is the range of the prototype or the centrifugal test. From Fig.1 the conventional model test involves scale errors, which become greater as the buried depth of anchor becomes greater, and provides greater test results. It is obvious from Fig.1 that a danger will be caused by the overestimation if a prototype is designed using the results obtained in the conventional test.

Yamaguchi, Kimura and Fujii (1977) performed bearing capacity tests of shallow foundations using model soils of Toyoura Sand and glass ballotini and model footings of 20 mm, 30 mm and 40 mm in width and at accelerations of 10 g, 20 g and 40 g in the centrifugal field and in the 1 g gravitational field and with the depths of embedment at 0, 0.5B and 1.0B (B : width of foundation) and obtained the same result as above. They indicated that the coefficient of bearing capacity will become greater as the acceleration of test field becomes smaller, causing scale errors, that this tendency will be increased as the depth of embedment becomes greater, and that Toyoura Snad, used in their test, eliminates the scale errors for the foundation width greater than about 900 mm in 1 g field.

From Fig.1 and the above conclusions by Yamaguchi, Kimura and Fujii, scale errors caused in the range of small acceleration, even if the centrifugal testing machine is used, will affect the test result.

As described above, such scale errors are inherent to the conventional small model tests including anchor pullout test, bearing capacity test, earth pressure test, etc., performed in the 1 g field and the test results will provide design values in the dangerous side.

4) Summary of law of similarity on pullout resistance of buried anchor

As discussed above, the use of centrifugal technique with the same material and under the same compacting condition as for the prototype for dry sand completely satisfies the law of similarity on pullout

Table 3 Model-prototype conversion table for anchor pullout test

	Physical quantity	Symbol	Dimension	Conversion ratio from model to prototype
Geometry	Length	L_r	L	λ
	Area	A	L^2	λ^2
	Inclination of anchor	α	O	1
External condition	Pullout force	P	F	λ^2
	Pullout velocity	V	LT^{-1}	1
	Time	t	T	λ
	Acceleration of field (Acceleration level)	G	LT^{-2}	λ^{-1}
Response	Ultimate pullout resistance	Qu	F	λ^2
	Stress	σ	FL^{-2}	1
	Strain	ϵ	O	1
	Displacement	δ	L	λ

Scale: $S = \frac{1}{\lambda}$

resistance of a buried anchor without scale errors and a conversion table from model to prototype can be obtained, as shown in Table 3.

In this table, the scale of models and the acceleration in the centrifugal field must be $S = 1/\lambda$ and λg, respectively

However, to obtain a practical solution of pullout resistance of buried anchor in sand, the test should be performed within a range that provides greater λGB and constant values of the dimensionless ultimate pullout resistance $Q_u/A\gamma GD_f$. In case of Ottawa Sand, $\gamma GB \cong 3.0 N/cm^2$ with respect to the anchor pullout resistance.

3. PULLOUT RESISTANCE OF BURIED ANCHOR IN SAND

3.1 General Solution of Anchor Pullout Resistance

Buried anchors are divided into shallow and deep anchors by the difference in their failure mechanism, as shown by Vesić (1971), Meyerhof (1973), and others. The soil failure due to pullout of the former occurs in a range of small dimensionless buried depth (relative depth) D_f/B and extends to the ground surface. The soil failure due to pullout of the latter occurs only within the soil when the dimensionless buried depth D_f/B is great. The ultimate pullout resistance of the shallow and deep anchors by the theory of plasticity are shown, as follows.

1) Shallow Anchor

The most sophisticated solution of the shallow anchor by the theory of plasticity has been reported by Meyerhof (1973). Since the solution best explains the test values described later and is generally applicable, it is adopted in this paper. The ultimate pullout resistance on a two dimensional (strip) anchor buried in sand at any angle is expressed by the following equation:

$$Q_{us} = \frac{1}{2}\gamma\frac{D_0^{\ 2}}{B}K_b A + (W_s + W_A)\cos\alpha \qquad (2)$$

The shallow anchor uplift coefficient K_b was shown by Meyerhof.

2) Deep Anchor

The failure mechanism of a deep anchor is very similar to that of the tip bearing capacity of a pile, as shown by Vesić (1975). He analyzed the point bearing of pile by the theory of plasticity into which the concept of cavity expansion was introduced. This point bearing mechanism is made upside down for a deep anchor to obtain its pullout resistance by the two-dimensional failure mechanism in Fig.4.

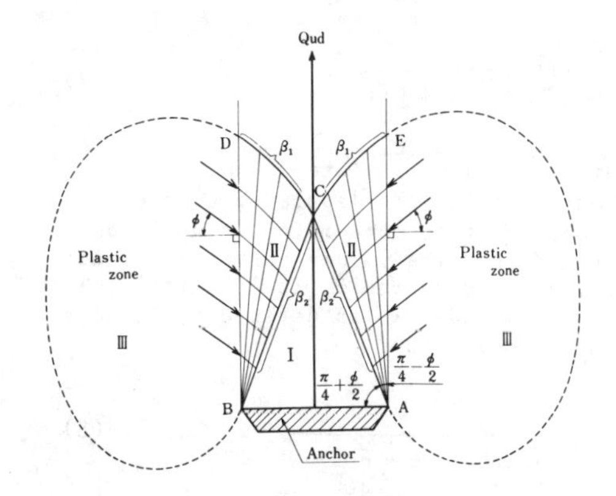

Fig.4 Physical plane of deep anchor

In Fig.4, the advancement of anchor into soil is made possible by lateral expansion of the soil along the long rectangular plates AE and BD by introducing the concept of cylindrical cavity expansion. The ultimate pullout resistance can be shown as follows:

$$q_0 = \bar{\sigma}\tan^2\left(\frac{\pi}{4}+\frac{\phi}{2}\right)exp.\left(\frac{\pi}{2}-\phi\right)\tan\phi \qquad (3)$$

With respect to the pile tip bearing capacity, Vesić assumed $\bar{\sigma}$ as the product of the mean effective ground stress of the plastic zone III and the cavity expansion factor. In this paper, his assumption is followed, as

$$\bar{\sigma} = \bar{q}\overline{F_q} \qquad (4)$$

$\bar{q}$ is expressed two-dimensionally by the coefficient of earth pressure at rest K and the initial vertical ground stress σ_v, as

$$\bar{q} = \frac{1+K_0}{2}\sigma_v \qquad (5)$$

For the cavity expansion factor that is also to be considered two-dimensionally, the cylindrical cavity expansion factor is used and expressed by

$$\overline{F_q} = (I'rr\sec\phi)^{\frac{\sin\phi}{1+\sin\phi}} \qquad (6)$$

I'_{rr} is the reduced rigidity index for cylindrical cavity defined by Vesić, which is expressed by the rigidity index of soil I_r, volumetric change parameter Δ of plastic zone III and the angle of shear resistance ϕ, as

$$I'_{rr} = \frac{I_r}{1 + I_r \Delta \sec\phi} \qquad (7)$$

Since the anchor plate makes a volume behind it for the displaced volume in front of it to move into, it would be argued that the volume change parameter Δ, as used by Vesić, is 1 (unity). The rigidity index I_r is defined by Vesić as shown in Eq. (8).

$$I_r = \frac{E}{2(1+\nu)(c + \bar{q}\tan\phi)} \qquad (8)$$

According to Vesić (1965), the rigidity index I_r is 100 to 500 for loose to dense sand. Since the deep anchor is considered in this study, the rigidity index I_r has the greater number (300 - 500) for the greater $\bar{q}$ and medium to dense sand. Therefore, for the entire practical range of I_r' the value of $\bar{F}_q$ becomes 1, and Eqs. (7) and (6) become, as follows, respectively:

$$I'_{rr} = \frac{1}{\sec\phi} \qquad (9)$$

$$\bar{F}_q = 1 \qquad (10)$$

From Eqs. (4), (5) and (10), Eq. (3) becomes, as follows:

$$q_0 = \frac{1+K_0}{2}\sigma_V \tan^2\left(\frac{\pi}{4} + \frac{\phi}{2}\right) \cdot \exp.\left(\frac{\pi}{2} - \phi\right)\tan\phi \qquad (11)$$

The above analysis relates to the vertical pullout of a deep anchor buried horizontally. Pullout of such an anchor at an arbitrary angle is considered, as follows: On the basis of the initial horizontal confining stress σ_H in the vertical pullout, an anchor being pulled out at an angle of α to vertical will be modified by the ratio of the initial confining stress σ_n acting in perpendicular to its pullout direction to σ_H. From this assumption, Eq. (11) can be expressed, as

$$q_0 = \frac{1+K_0}{2}\sigma_V \frac{\sigma_n}{\sigma_H} \tan^2\left(\frac{\pi}{4} + \frac{\phi}{2}\right) \cdot \exp.\left(\frac{\pi}{2} - \phi\right)\tan\phi \qquad (12)$$

Generally, σ_n can be expressed by

$$\sigma_n = \frac{1}{2}(\sigma_1 + \sigma_3) - \frac{1}{2}(\sigma_1 - \sigma_3)\cos 2\alpha \qquad (13)$$

Since an anchor buried in soil having a horizontal ground surface is considered in this study, $\sigma_1 = \sigma_V$ and σ_3 is expressed by Jaky, as

$$\sigma_3 = \sigma_H = K_0\sigma_V = (1 - \sin\phi)\sigma_V \qquad (14)$$

From Eqs. (12)-(14), and since $\sigma_V = \gamma D_f$ and $Q_{ud} = q_0 \times A$, Eqs. (15) and (16) can be finally obtained.

$$Q_{ud} = N_q \gamma D_f A \qquad (15)$$

Where

$$N_q = \frac{(2 - \sin\phi)\{2 - (1 + \cos 2\alpha)\sin\phi\}}{4(1 - \sin\phi)} \cdot \tan^2\left(\frac{\pi}{4} + \frac{\phi}{2}\right)\exp.\left(\frac{\pi}{2} - \phi\right)\tan\phi \qquad (16)$$

N_q is called herein the deep anchor uplift coefficient.

From Eq. (15), it is clear that the dimensionless quantity $Q_{ud}/A\gamma D_f$ of the ultimate pullout resistance will become constant when the angle of shear resistance of soil ϕ and the pullout direction α have been determined.

The deep anchor uplift coefficient N_q obtained with respect to various values of ϕ and α is shown in Fig.5. In an actual design, the effective anchor weight in the pullout direction can be added.

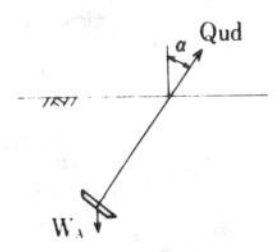

$$Qud = Nq\,\gamma\,D_f A S_h + W_A \cos\alpha$$

$$Nq = \frac{(2-\sin\phi)\{2-(1+\cos 2\alpha)\sin\phi\}}{4(1-\sin\phi)} \tan^2\left(\frac{\pi}{4}+\frac{\phi}{2}\right) \exp.\left(\frac{\pi}{2}-\phi\right)\tan\phi$$

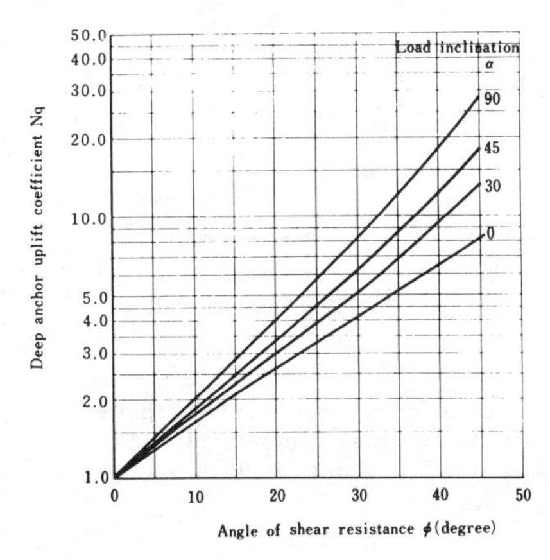

Fig.5 Deep anchor uplift coefficient

3) Shape Factor

This section will discuss the three-dimensional effect of an ordinary buried anchor having finite length by introducing the shape factor. Meyerhof and Adams (1968) considered that the shape effect for the vertical pullout of horizontal rectangular anchor was similar in B/2 on both ends to that of a circular anchor and in the middle of (L_A - B) to that of a strip anchor. Meyerhof (1973) also proposed based on experiment that the uplift coefficient of a rectangular anchor could be obtained by interpolating those values of a strip and a square anchor by the width to length ratio B/L_A and that a circular anchor could be analyzed as a square anchor having the same area.

As above, the shape effect of a buried anchor has not yet been definitely theorized, but only possibly by experiments. From the practical point of view, the shape effect of a buried anchor may be defined, as follows:

According to Meyerhof and Adams' concept, when the shape factor of a strip anchor is taken as 1, a square anchor will have the number of vertical planes passing through the anchor edges twice that of the former giving the latter's shape factor as 2. Considering that the shape factor of a rectangular anchor is obtained by interpolation between the shape factor of a square anchor and that of a strip anchor, the general shape factor S_h can be expressed by:

$$S_h = 1.0 + \frac{B}{L_A} \tag{17}$$

For a square and a circular anchor, S_h = 2.0.

Although no sufficiently reviewed data available, the variation in the shape effect by the anchor pullout angle α is little observed. A linear increase to the relative depth D_f/B of the shape factor, as presented by Meyerhof (1973), will not be considered in this paper because such a result has not been obtained in the laboratory test of a buried anchor in a centrifugal field, as described later.

Since the above idea can relatively well explain the test results, it is applied to a shallow and a deep anchor to obtain the general equation for pullout resistance of a buried anchor, as follows:

For a shallow anchor:

$$Q_{us} = \frac{1}{2}K_b\gamma\frac{D_0^2}{B}AS_h + (W_S + W_A)\cos\alpha \tag{18}$$

For a deep anchor:

$$Q_{ud} = N_q\gamma D_f AS_h + W_A\cos\alpha \tag{19}$$

3.2 Centrifuge Model Test on Pullout Resistance of Buried Anchor in Sand

The results of the centrifuge model tests that were performed to clarify the pullout resistance of an anchor buried in sand were compared with the solutions of the theory of plasticity and the results of the elastoplastic finite element analyses based on the Lade's constitutive law (1972, 1975 and 1976). Meanwhile, the data without scale error in the centrifuge model tests were obtained based on the previous chapter in which the scale effect was discussed in detail.

1) Models

Circular and rectangular anchor models of steel as shown in Fig.3 were used in the test. The scale was assumed to be about 1/90 - 1/100. Their side surfaces were cut at 45° to eliminate the effect of side friction.

2) Sand Used

Ottawa Sand and Toyoura Sand were used as test sand. Their soil characteristics are shown in Table 2. These Sands were used in the furnace-dried condition.

3) Centrifugal Testing Machine and Instrumentation

In these tests, two centrifugal testing machines of California Institute of Technology, whose outline is shown in Fig.2, and Tokyo institute of Technology were used. The former is provided with two soil containers of 370 mm x 294.5 mm x 254 mm (parallelepiped) and 152 mmϕ x 617 mm (cylindrical) having a radius of rotation 916 mm and rotating at a maximum acceleration of 175 g, and the latter, with one soil container of 500 mm x 100 mm x 300 mm having a radius of rotation 1180 mm and rotating at a maximum acceleration of 300 g. The instrumentation systems is also shown in Fig.2.

4) Test Procedure

Outline of the test procedure is as follows:

(1) The test container of the centrifugal testing machine is filled with the specified amount of soil. The anchor model is set at the specified position. A counterweight is placed on the opposite side of the test container. The instrumentation systems are provided for pullout of anchor and measuring forces and displacement.

(2) The centrifugal testing machine is rotated at the specified number of revolutions. The pullout force and the displacement of the anchor are measured as the anchor is being pulled out by remote control.

5) Test Cases

The tests were performed by changing the buried depth of anchor (shallow and deep), the inclination of anchor (horizontal and 45°), the density of soil (dense and loose), the type of anchor (rectangular and circular). The anchor was always pulled out perpendicular to the anchor plane through its center.

6) Test Results and Discussion

(1) Method of data analysis
In this paper, the pullout resistance of anchor only will be discussed by performing a data analysis with the shaft resistance and the weight of anchor and shaft excluded based on the preliminary test.

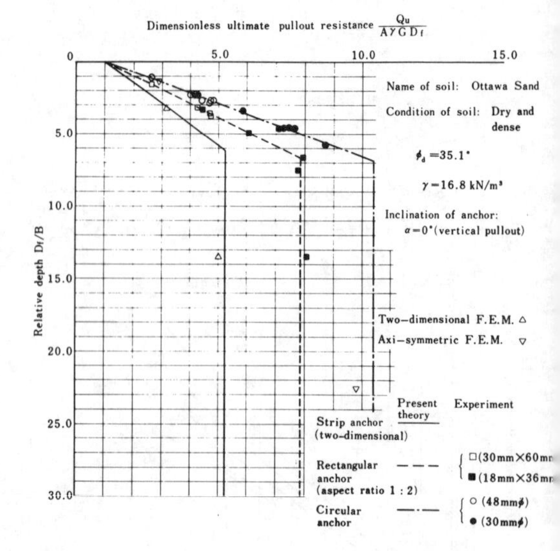

Fig.6 Relation between dimensionless ultimate pullout resistance and relative depth (Vertical pullout, dry dense Ottawa Sand, ϕ_d =35.1)

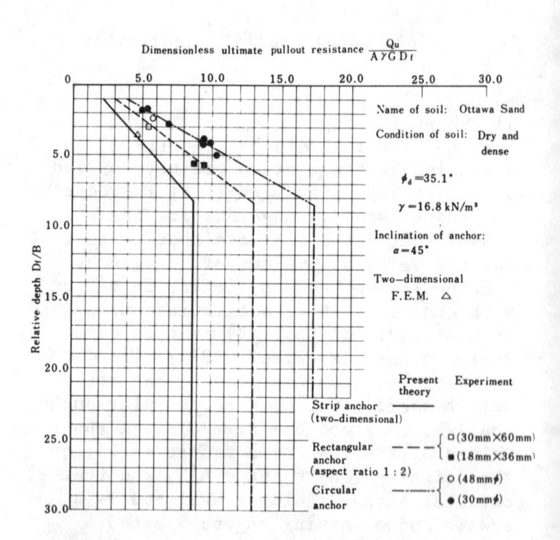

Fig.7 Relation between dimensionless ultimate pullout resistance and relative depth (45° -pullout, dry dense Ottawa Sand, ϕ_d =35.1)

(2) Relation between ultimate pullout resistance and buried depth
Figs. 6 and 7 show the relation between the dimensionless ultimate pullout resistance $Q_u/A\gamma GD_f$ and the relative buried depth D_f/B of the anchor. These figures involve the theoretical lines by the

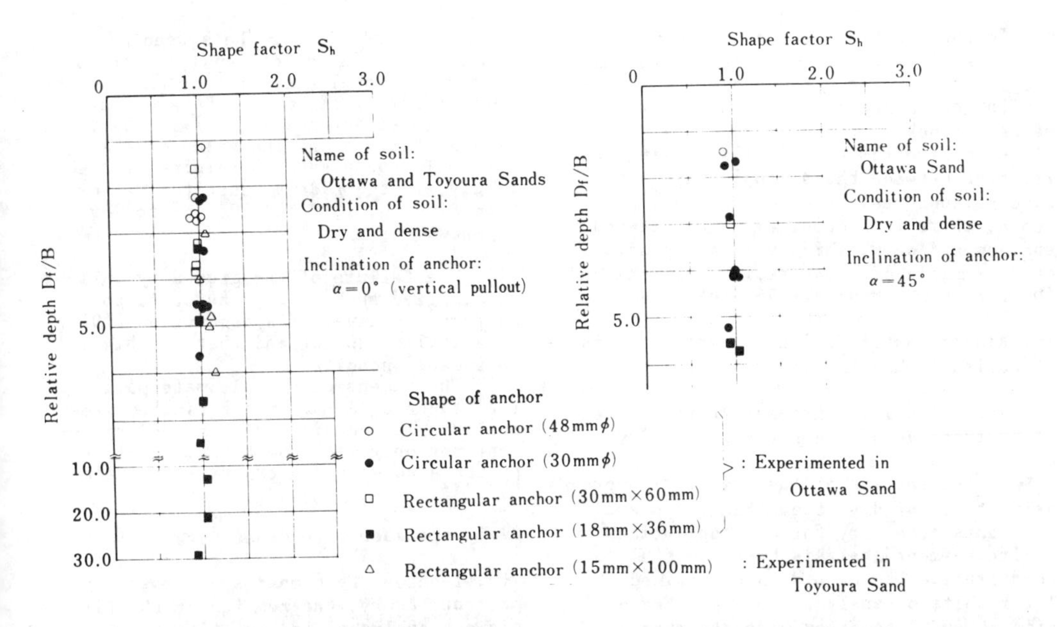

Fig.8 Three-dimensional effect in anchor pullout (shape factor) in depth direction (Dry Ottawa Sand and Dry Toyoura Sand)

theory of plasticity and the analyzed values by the elastoplastic finite element method. These theoretical values also include the three-dimensional shape effect, as described before. The angle of shear resistance to obtain these theoretical lines has been obtained from the results of the triaxial compression test.

From the Ovesen's proposeal (1981) that a circular anchor is equal in pullout resistance to a square anchor having the equal area to the former, the side length of the square anchor having the same area as the circular anchor was taken as a representative length in the data analysis ($B = \sqrt{\pi/4} = 0.886D$). Fig.8 shows all the data converted into the shape factor of the two-dimensional anchor (to be 1.0) using the equation of shape factor given by Eq. (17). This figure indicates the three-dimendional effect remains almost unchanged in depth.

From the above, the following can be said:

a) The three-dimensional effect (shape factor) of the buried anchor in dense Ottawa Sand will not be changed in depth.

b) The shape factor S_h taken as $S_h = 1.0 + B/L_A$ from the above discussion can well explain the experimental values and be sufficiently applicable to design.

Also, the equivalent side length of a circular anchor B can be expressed by 0.886D.

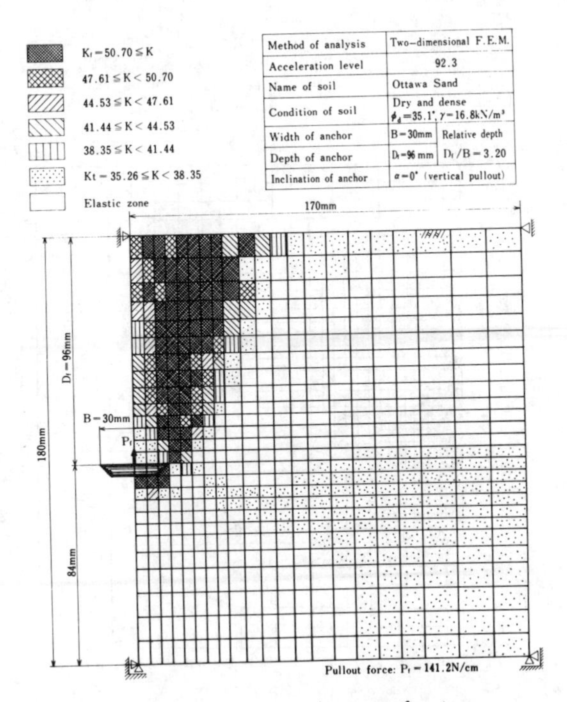

Fig.9 Spread of plastic zone by two-dimensional finite element analysis (Vertical pullout of shallow horizontal anchor)

c) In the relation between the dimensionless ultimate pullout resistance and the relative depth, the theoretical values are in good agreement with the experimental values.

d) As clear from the figures showing the relation between the dimensionless ultimate pullout resistance and the relative depth, there is a critical depth between shallow and deep, which varies depending on the angle of shear resistance achieved $(D_f/B)cr = 6 - 8$ at $\phi = 35^\circ - 42^\circ$.

7) Elastoplastic Finite Element Analysis of Buried Anchor in Ottawa Sand

Two-dimensional and axi-symmetric finite element analysis were performed on dry Ottawa Sand.

From the results of the triaxial compression tests of dry Ottawa Sand, the soil constants necessary for elastoplastic finite element analysis besed on the constitutive law by Lade are obtained. The results of analysis were plotted on Figs. 6 and 7 together with the experimental values.

Figs. 9 and 10 show the spread of plastic zone in typical analysis cases (two-dimensional analysis for shallow and deep anchors.) K, K_f and K_t are the stress level defined as I_1^3/I_3, the value of K at failure and the value of K at threshold for plastic straining, respectively, in the Lade's constitutive law. From Figs. 6, 7, 9 and 10, the following is shown:

(1) The failure of the ground by pullout of a buried anchor are considered to support the Meyerhof's assumption (1973) on a shallow anchor and that of this study on a deep anchor.

(2) The dimensionless ultimate pullout resistance $Q_u/A\gamma GD_f$ by the finite element analysis is comparatively in good agreement not only with the theory of plasticity but also with the experimental values.

8) Comparison with Other Data

Fig.11 shows the comparison between the present theory, the results of the finite element analyses, and experimental data reported by other researchers. These

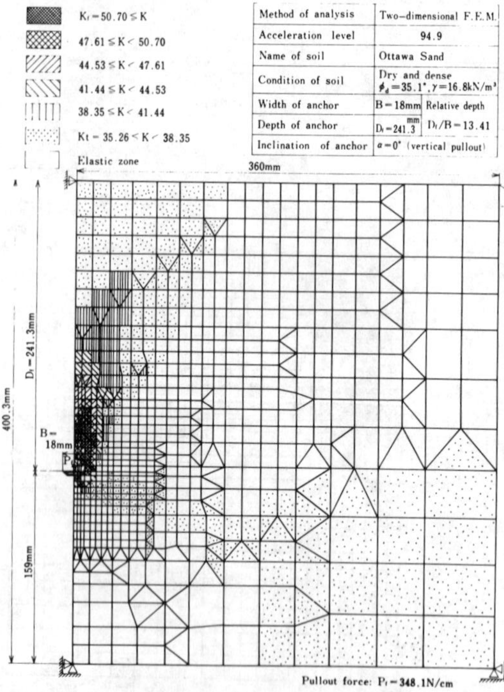

Fig.10 Spread of plastic zone by two-dimensional finite element analysis (Vertical pullout of deep horizontal anchor)

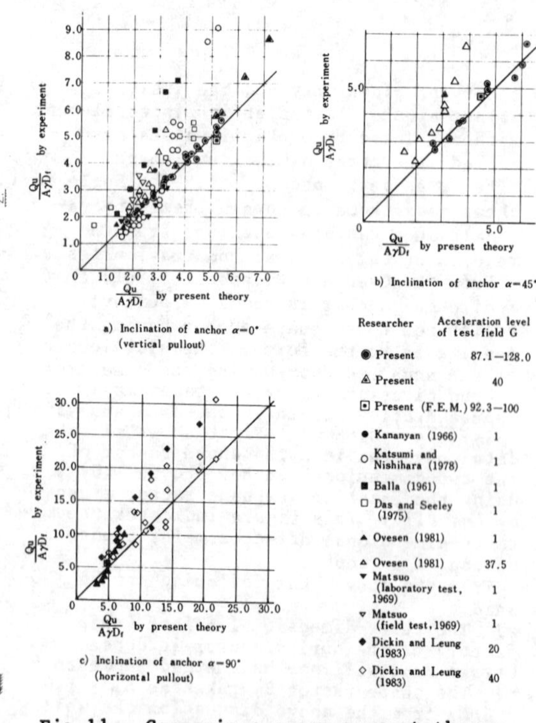

Fig.11 Comparison among present theory, present test data, present F.E.M. analysis and various data by other researchers (The data converted into those of the two dimensional anchor)

experimental data by other researchers include those which have been obtained by a conventional model test in 1 g field, by a field test using a relatively small model ($D \leqq 1100$ mm) and by a centrifugal technique at a low acceleration level. These data, most of which are concerned with a three-dimensional anchor, have been plotted for comparison purposes after converting them by the shape factor into those concerned with a two-dimensional anchor (strip anchor). Fig.11 shows that most of the data by other researchers, giving dangerous values in design, are greater than the values by the present theory, present data and the results of the finite element analyses, probably because of the evaluation of angle of shear resistance ϕ , the effect of the compaction of back-fill, etc.,whereas some are in agreement with those by the present theory.

4. CONCLUSIONS

In this paper, the scale effect in anchor pullout test for frictional material by centrifugal technique and the pullout resistance of buried anchor in sand were discussed. The main conlclusions are as follows:

1) Scale Effect in Anchor Pullout Test by Centrifugal Technique

(1) The actual anchor pullout resistance estimated from the results of the conventional model test, that is subject to scale errors, will stand on the dangerous side.
(2) The centrifugal technique used at an acceleration greater than a certain value will eliminate the scale errors on dry sand including grain size of soil particle.

2) Pullout Resistance of Buried Anchor in Sand

(1) Two estimation formulas of anchor pullout resistance have been proposed for the two failure mechanisms (shallow and deep). And the shape factor of an anchor of finite length showing the three-dimensional effect has also been proposed.
(2) These proposed formulas have proved good agreement with the results of elasto-plastic finite element analysis and the experimental values by the centrifugal technique.

NOTATION

A = area of anchor
B = width of anchor or effective width of circular anchor
c = cohesion of cohesive soil
D = diameter of circular anchor
D_f = depth of anchor center
(D_f/B) = critical relative depth of anchor
D_0 = maximum depth of anchor
D_{50} = 50 percent grain size
d = grain size of soil
E = Young's modulus of soil as continuum
E_g = Young's modulus of soil particle
$\overline{F}_q$ = cavity expansion factor
G = acceleration level in centrifugal field
G_s = specific gravity of soil particle
g = acceleration of field
I_1 = first stress invariant
I_3 = third stress invariant
I_r = rigidity index of soil
I'_{rr} = reduced rigidity index of soil
K = stress level defined as I_1^3/I_3 in Lade's constitutive equation
K_b = shallow anchor uplift coefficient
K_f = value of K at failure in Lade's constitutive equation
K_0 = coefficient of earth pressure at rest
K_t = value of K at threshold for plastic straining in Lade's constitutive equation
N_q = deep anchor uplift coefficient
P_f = pullout force
Q_u = ultimate pullout resistance of anchor
Q_{ud} = ultimate pullout resistance of deep anchor
Q_{us} = ultimate pullout resistance of shallow anchor
q_0 = ultimate pullout resistance per unit area of anchor
$\overline{q}$ = mean effective ground stress
S = scale
S_h = shape factor of anchor
S_r = degree of saturation
U_c = coefficient of uniformity
V = pullout velocity
W_A = effective weight of anchor
W_S = effective weight of soil above anchor
I = highly compressed zone in front of anchor plate in physical plane of anchor
II = radial shear zone in physical plane of anchor
III = plastic zone in physical plane of anchor
α = inclination of anchor or load inclination from vertical (degrees)
β_1 = family of first slip lines
β_2 = family of second slip line
γ = effective unit weight of soil at 1 g field
Δ = volumetric change parameter
δ = displacement of anchor or deformation of soil

ε = strain of soil
λ = magnification factor of prototype to model in length or acceleration in centrifugal field
ν = Poisson's ratio of soil as continuum
ν_g = Poisson's ratio of soil particle
ρ = mass density of soil
σ = stress in soil
σ' = effective normal stress acting on slip surface
$\bar{\sigma}$ = ultimate pressure needed to expand cavity
σ_B = stress in ground due to weight of soil
σ_g = breaking strength of soil particle
σ_H = initial horizontal confining stress or initial horizontal ground stress
σ_n = initial normal stress
σ_v = initial vertical ground stress
$\sigma_1, \sigma_2, \sigma_3$ = principal stresses
τ_f = shear stress at time of failure
ϕ = angle of shear resistance
ϕ_d = angle of shear resistance by consolidated-drained test

REFERENCES

Balla, A. (1961): "The resistance to breaking-out of mushroom foundations for pylons," Proceedings of the Fifth International Conference on Soil Mechanics and Foundation Engineering, Paris, France, Vol. I, pp. 569-576.

Lade, P.V. (1972): The Stress-Strain and Strength Characteristics of Cohesionless Soils, Thesis presented to the University of California at Berkeley, California, for the Degree of Doctor of Philosophy.

Lade, P.V. and Duncan, J.M. (1975): "Elastoplastic stress-strain theory for cohesionless soil," Journal of the Geotechnical Engineering Division, American Society of Civil Engineers, Vol. 101, No. GT10, October, pp. 1037-1053.

Lade, P.V. and Duncan, J.M. (1976): "Stress-path dependent behavior of cohesionless soil," Journal of the Geotechnical Engineering Division, American Society of Civil Engineers, Vol. 102, No. GT1, January, pp. 51-68.

Matsuo, M. (1969): Study on Foundation Subjected to pullout Force and Bearing Capacity of Composite Foundation, Thesis presented to Kyoto University for the Degree of Doctor of Engineering (in Japanese).

Meyerhof, G.G. and Adams, J.I. (1968): "The ultimate uplift capacity of foundations," Canadian Geotechnical Journal, Vol. V, No. 4, pp. 225-244.

Meyerhof, G.G. (1973): "Uplift resistance of inclined anchors and piles," Proceedings of the Eighth International Conference on Soil Mechanics and Foundation Engineering, Moscow, Union of Soviet Socialist Republics, pp. 167-172.

Ovesen, N.K. (1981): "Centrifuge tests of the uplift capacity of anchors", Proceedings of the Tenth International Conference of Soil Mechanics and Foundation Engineering, Stockholm, Sweden, Vol. 1, pp. 717-722.

Vesić, A.S. (1965): "Cratering by explosives as an earth pressure problem", Proceedings of the Sixth International Conference on Soil Mechanics and Foundation Engineering, Montreal, Canada, pp. 427-431.

Vesić, A.S. (1971): "Breakout resistance of objects embedded in ocean bottom," Journal of Soil Mechanics and Foundation Division, Proceedings of American Society of Civil Engineers, Vol. 97, No. SM9, pp. 1183-1205.

Vesić, A.S. (1972): "Expansion of cavities in infinite soil mass," Journal of Soil Mechanics and Foundation Division, Proceedings of American Society of Civil Engineers, Vol. 98, No. SM3, pp. 265-290.

Vesić, A.S. (1975): Principles of pile Foundation Design, Soil Mechanics Series No. 38, School of Engineering, Duke University.

Yamaguchi, H., Kimura, T. and Fujii, N. (1977): "On the scale effect of footings in dense sand," Proceedings of the Ninth International Conference on Soil Mechanics and Foundation Engineering, Tokyo, Japan, Vol. 1, pp. 795-798.

Prediction and Performance in Geotechnical Engineering / Calgary / 17-19 June 1987

Seismic response of pile foundations in a centrifuge

W.B.Gohl & W.D.Liam Finn
University of British Columbia, Vancouver, Canada

ABSTRACT: Simulated earthquake loading tests were conducted on model pile foundations in sand in the centrifuge at the California Institute of Technology. Details of experimental test procedures are described. For the first time, the in-situ moduli of the foundation sands were measured during operation of the centrifuge. Response of single piles and pile groups were measured and the dynamic interaction between piles was investigated.

1 INTRODUCTION

Centrifuge model tests on pile foundations in dry sand were conducted at the California Institute of Technology (Caltech) to provide a data base for checking existing methods of seismic response analysis. (Finn and Gohl, 1986). Single piles were subjected to both sinusoidal and random earthquake motion. The excitation levels ensured a range in response from linear elastic to nonlinear. The excitation levels in tests on pile groups were kept low enough to ensure elastic response since current methods of dynamic response analysis of pile groups usually employ elastic interaction factors. Further tests are planned to explore the response of pile groups in the nonlinear range.

Analyses of free field ground motions and pile response require a knowledge of the shear stiffnesses of the foundation soils. The distribution of in-situ shear moduli was determined while the centrifuge was in flight. Moduli were deduced from in-situ shear wave velocities which were measured using piezoceramic bender elements. This procedure is an important improvement in centrifuge testing as it allows a more accurate measure of the moduli which were previously estimated from correlations with relative density.

This paper is chiefly concerned with describing the instrumentation and test procedures. However, selected data on the response of both single piles and pile groups are presented to illustrate fundamental aspects of pile response and demonstrate the kind of information that can be obtained from centrifuge model tests.

2 PRINCIPLES OF CENTRIFUGE MODELLING

The centrifuge test has become the most convenient and effective procedure for evaluating methods for the dynamic analysis of soil structures and soil-structure interaction systems. The models can be fully instrumented to measure response parameters of interest. Full scale structures are rarely affected by seismic motions and usually are not adequately instrumented to document seismic response fully.

In a centrifuged model, stresses at the same levels that exist in a full scale structure can be produced at corresponding points by spinning the centrifuge arm at an angular speed such that the centrifugal acceleration at the location of the model on the arm is ng where g is the acceleration due to gravity and n is the model scale. This ability to create prototype stresses in the model is important since soil properties are dependent on effective stresses. Since the static stress levels in both model and prototype are similar at corresponding points, each soil element in the centrifuged model may be expected to undergo the same response history as corresponding elements in the prototype for a given excitation.

The scaling relationships used in centrifugal modelling studies are summarized in Table 1.

Table 1. Scaling Relations

Quantity	Prototype	Model at n g's
Linear Dimension	1	1/n
Stress	1	1
Strain	1	1
Force	1	$1/n^2$
Acceleration	1	n
Velocity	1	1
Time	1	1/n
Frequency (Dynamic Problems)	1	n

3 DESCRIPTION OF CALTECH CENTRIFUGE

The Caltech centrifuge, described in detail by Allard (1983), has a 2.03 m (80 in.) diameter aluminum alloy arm which rotates in the horizontal plane and is rated at 44.5 g-kN (10,000 g-pounds) payload capacity. A magnesium swing basket which carries the soil container is located at the end of the arm (Figure 1). During centrifuge spin-up the soil container swings into an approximately horizontal position so that the vertical axis of the model lies along the axis of the centrifuge arm. Electrical and hydraulic power, air pressure and signals to and from the rotating arm or basket are conducted through electrical slip-rings and rotating unions.

In the test program described below the rotational speed of the centrifuge arm was selected so that the acceleration at the centre of the model was 60 g.

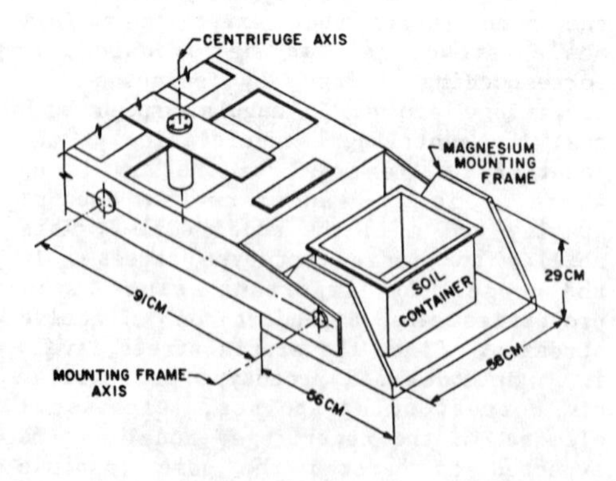

Figure 1. Schematic drawing of centrifuge arm (after Allard, 1983).

Since each model is of finite size, different parts of the model are at different radii from the rotational axis of the centrifuge. Therefore, at any speed different parts of the model will be subjected to different gravitational intensities. Calculations indicate that the acceleration varied from 55 g at the surface of the model to 68 g at the base in the present study. An average centrifuge scale factor, n, equal to 60, was used in converting model test quantities to prototype scale.

Prescribed input motions are delivered to the base of the model by a hyraulic actuator controlled by a servo-valve. The valve is regulated by a feedback system which compares the prescribed motion with the actual motion. Earthquake motions are input to the servo-valve controller by computer. Sinusoidal input was supplied to the controller by a function generator.

4 CENTRIFUGE TEST PROCEDURES

A level sand foundation was prepared in the centrifuge soil container (see Figure 1) prior to pile insertion. Two 12.5 mm (0.5 in) thick styrofoam pads were placed at each end of the container to reduce wave reflection from the sides of the box perpendicular to the direction of base motion. Two glass plates were fitted on the other two sides to minimize side friction during the tests. The plan area of the soil container after the above inserts were installed was 529 mm × 172 mm (20.8 x 6.8 in) with a total depth of 254 mm (10 in). During sand placement a vertical array of piezoceramic bender elements was installed in the sand to measure shear wave velocities, as described later. An accelerometer was placed in the surface of the sand a distance of 190 mm (20 pile diameters) from the centre of the single pile (or pile group) to measure the free field surface accelerations. Another accelerometer was placed on the base of the soil container to measure input accelerations.

A scaled drawing of the pile used in the single pile tests is shown in Figure 2. The model pile was made of stainless steel tubing having a 0.25 mm (0.010 in) wall thickness and an outside diameter of 9.52 mm (0.375 in). Eight pairs of foil type strain gauges were mounted on the outside of the pile and wired in a half bridge configuration to measure bending strains. Lead wires from each strain gauge pair were threaded through holes drilled in the tube and brought up inside

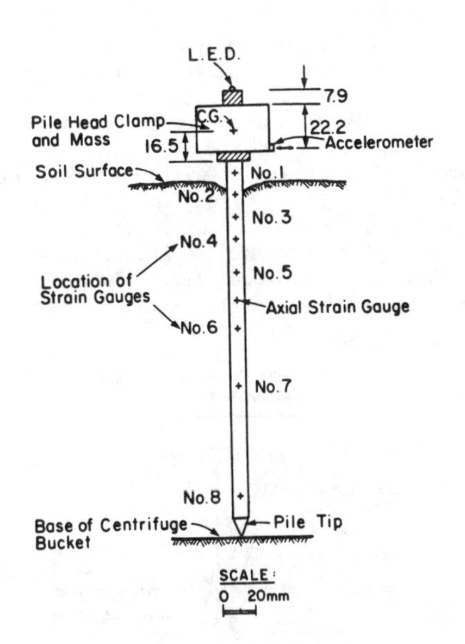

Figure 2. Single pile showing instrumentation layout.

the tube to the pile head. The wires were passed through a slot in the pile head mass and connected to bridge completion and amplifier circuitry mounted on the centrifuge arm. The strain gauge signals were calibrated using transverse dead loading of the test pile when it was clamped at one end and free at the other.

A mass was screwed to a clamp attached to the head of the pile to simulate the influence of superstructure inertia forces acting on the pile during excitation. The pile head mass was instrumented using a non-contact photovoltaic displacement transducer manufactured by United Detector Technology Inc. The displacement transducer tracks the location of a light emitting diode placed on top of the pile head mass (Figure 2) to an accuracy of ±0.0025 mm (±0.0001 in.). Pile head displacements were measured with respect to the moving base of the soil container. An Entran miniature accelerometer was also placed on the pile head mass.

Pile tests were carried out in both "loose" and "dense" Nevada sand to examine the influence of significant density differences on pile response. The Nevada sand has a D_{50} of 0.13 mm. Loose sands with an initial average void ratio of 0.83 were prepared by spreading constant weights of dry sand uniformly over a perforated screen placed on the base of the container or on top of the previous sand layer. The newly placed sand was levelled by hand and then bulked by pulling the screen up through the sand. Dense sands were prepared using high frequency low level vibration so that an average void ratio of 0.57 was achieved. Instrumented piles were pushed into the soil by hand. In the case of dense sands a low level vibration of the sand foundation was used to assist penetration.

Tests on pile groups, consisting of two piles each at various spacings, were conducted to evaluate pile interaction effects. A typical pile group configuration is shown in Figure 3. Both piles were instrumented to measure bending strains as described earlier. In addition, one pile was instrumented to record axial strains caused by rocking of the pile foundation during shaking. The axial strain gauges were calibrated by applying axial dead load to the model pile.

The piles in the group were rigidly attached to a pile cap whose design allowed the pile spacing to be varied. An additional mass was then bolted to the pile cap to simulate the effects of a superstructure as in the case of a single pile. The pile cap mass was instrumented with an accelerometer and a displacement L.E.D. as described earlier for the single pile. The locations of the accelerometer and L.E.D. are shown in Figure 3.

After pile insertion and connection of all associated instrumentation, four

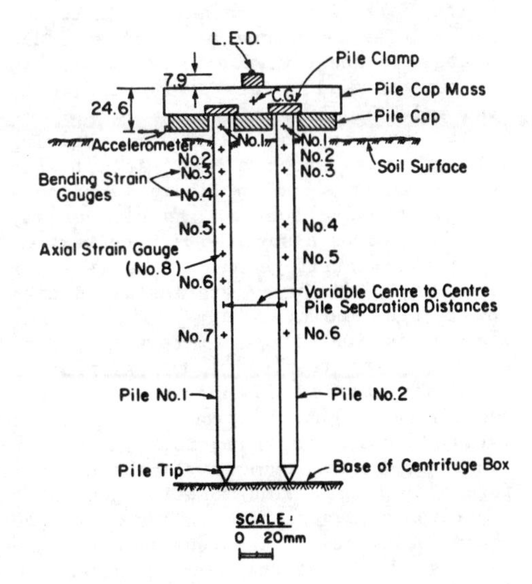

Figure 3. Pile group showing instrumentation layout.

lightweight settlement plates were placed a minimum of eight pile diameters from the center of any pile in order to measure surface settlement. The settlement results from two causes; settlement due to the increase in self-weight of the soil during spin-up of the centrifuge and that due to the cyclic shear strains generated by the base motion. Prior to starting the centrifuge, the balance of the centrifuge arm was assessed and a final check run on all instrumentation. The centrifuge was brought up to the test speed corresponding to a centrifugal acceleration of 60 g at the centre of the model. The centrifuge was then brought to rest. The average void ratio of the foundation layer was found to be 0.78, a reduction of 0.05 from the void ratio in the 1 g environment. Under similar circumstances, void ratio changes in the dense sand were negligible. The stick-up of the pile cap above the soil surface was also measured on spin-down.

The centrifuge was now brought up to test speed again, and shear wave velocity measurements made in the sand using the bender elements, both before and after excitation of the model pile foundations.

5 MEASUREMENTS OF SHEAR WAVE VELOCITIES

Shirley and Hampton (1978) used piezoceramic bender elements for generating and receiving shear waves in laboratory sediments. Detailed studies of the performance of bender elements in triaxial test equipment have been reported by the Norwegian Geotechnical Institute (1984). The technique has been extended to measure shear wave velocities, V_s, in centrifuged soil models during flight by Finn and Gohl (1987).

By applying an alternating driving voltage to a bender element it vibrates in the manner shown in Figure 4 and acts as a shear wave source. An element may also be used as a shear wave receiver.

In the centrifuge test, the receiver elements in the array are installed with their tips pointed up. The source element is installed tip down with its bearing plate resting on the surface of the sand layer (Figure 5). During centrifuge flight, the source element is excited using a function generator connected to the element via the centrifuge slip rings. The function generator supplies a square wave signal to the source element. The small amplitude bending of the source element under the input generates a shear wave which propagates downward into the foundation soil and is picked up by each of the receiver bender elements numbered R-1 to R-3 (Figure 6). The minute soil vibration caused by the shear wave propagation mechanically excites the receivers, which then respond with an electrical signal. The resultant source and amplified receiver waveforms are stored by a digital storage oscilloscope.

Signals recorded from the receivers due to excitation by a shear wave are

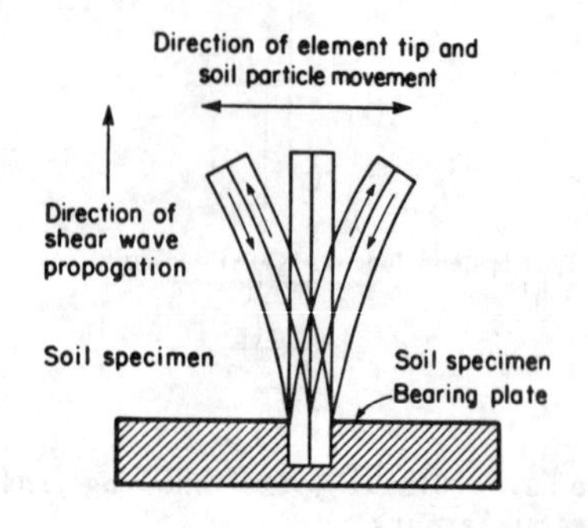

Figure 4. Piezoceramic bender element (after NGI, 1984).

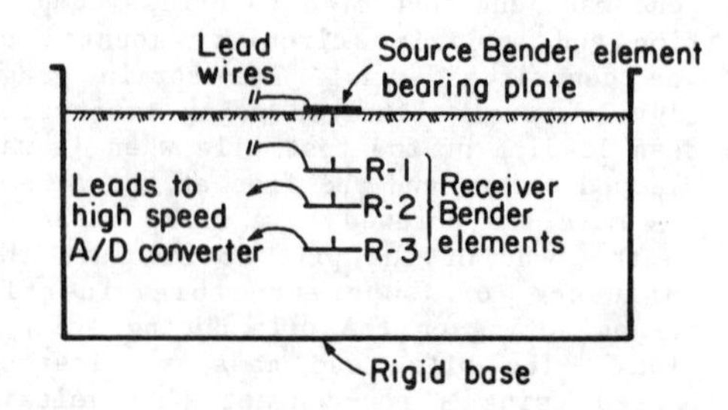

Figure 5. Schematic drawing showing bender element layout.

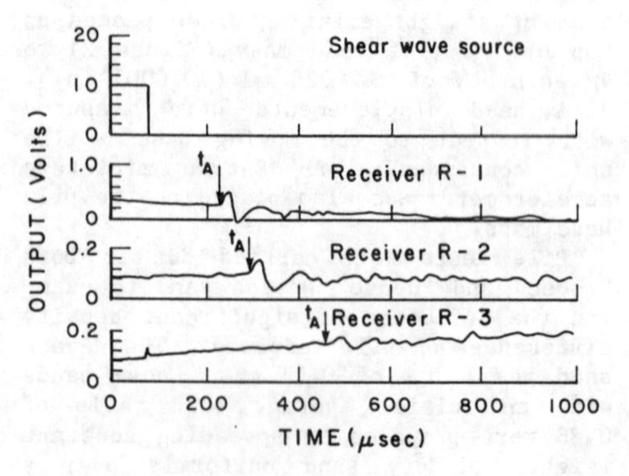

Figure 6. Source and receiver bender element voltage outputs.

as shown in Figure 6. The arrival time t_A is clearly evident from the recorded voltage signal. The distance between the tips of two adjacent bender elements divided by the transit time of the shear wave between them gives the average shear wave velocity over the depth interval in question. The distribution of shear wave velocities determined in this way is shown in Figure 7 for loose sand. It is evident from measurements made before and after each shaking test that the very small void ratio changes which occur

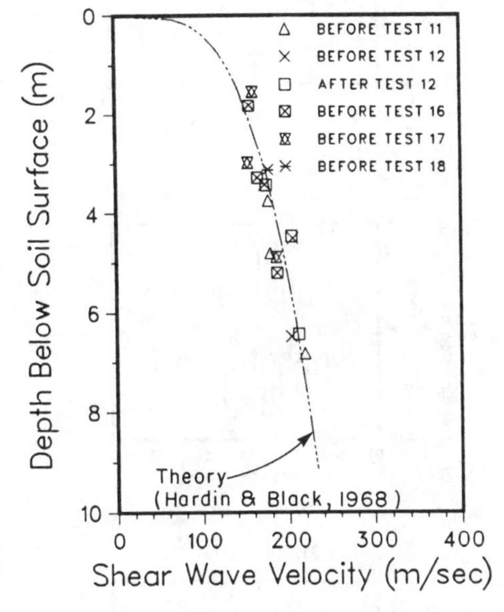

Figure 7. Shear wave velocities during centrifuge flight for loose sand.

during shaking have little effect on the measured shear wave velocity.

The distribution of shear wave velocity for the sand has been estimated on the basis of the void ratio and mean effective stress conditions using the Hardin & Black (1968) equation for G_{max} and converting G_{max} to shear wave velocity using the equation $G_{max} = \rho V_s^2$ where ρ = mass density of the soil and G_{max} = maximum low strain shear modulus. As may be seen in Figure 7 the estimated and computed shear wave velocities are in good agreement.

With the stiffness properties of the foundation soil determined, the centrifuge was again taken up to test speed. When equilibrium conditions had been established, a series of input motions were applied to the base of the model. Output from the instrumentation was amplified as required and then transmitted via slip rings to a multi-channel high speed analogue to digital (A to D) converter. The converter interfaced with a microcomputer and the data were stored on disc.

6 DATA REDUCTION AND ANALYSIS

6.1 Single Pile Response

Data on the response of a single pile in loose sand (test 12) will be presented to illustrate typical single pile response to earthquake excitation. All data are presented at prototype scale. The centre of gravity of the pile head mass was 1.95 m above the soil surface and the pile was subjected to moderate earthquake shaking (peak base acceleration 0.15 g).

The acceleration input at the base of the model and accelerations recorded at the surface in the free field and at the pile head are shown in Figures 8a,b,c. Pile head displacements are shown in

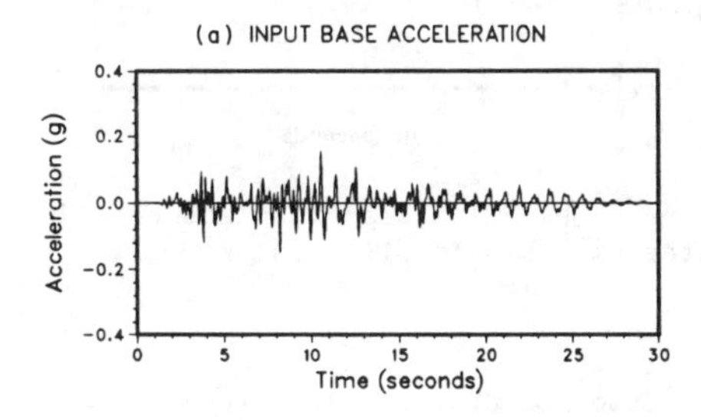

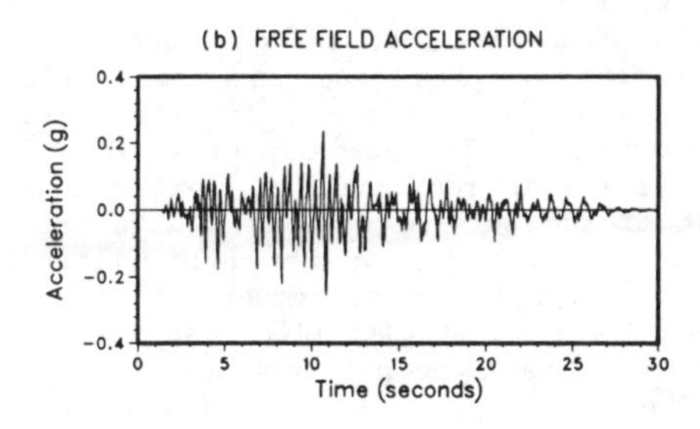

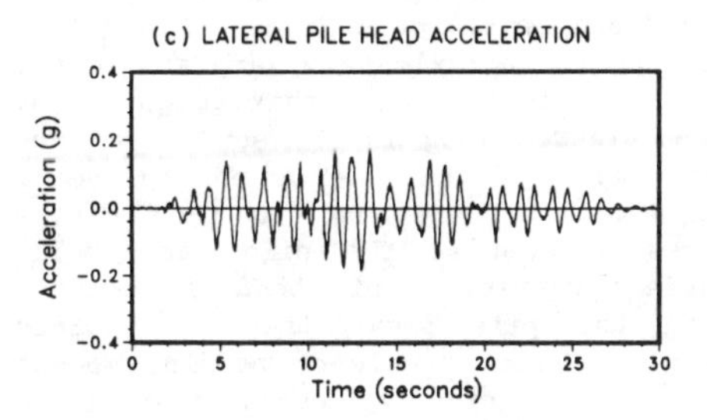

Figure 8a,b,c. Acceleration time histories - single pile test no. 12.

Figures 9a,b and time histories of pile bending moment at various points along the pile are shown in Figures 10a,b,c. The bending moment distribution along the pile at a time when maximum pile head

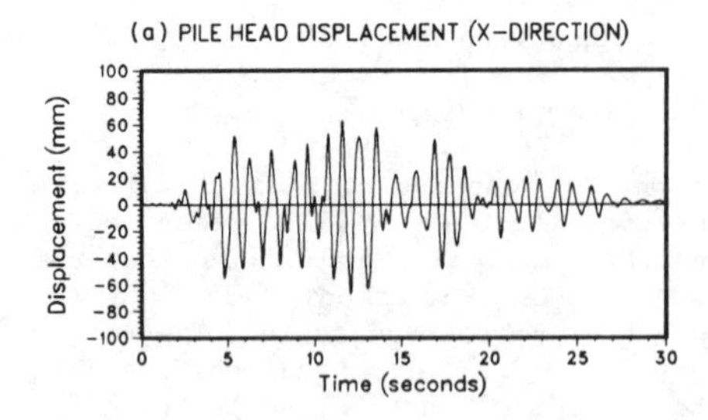

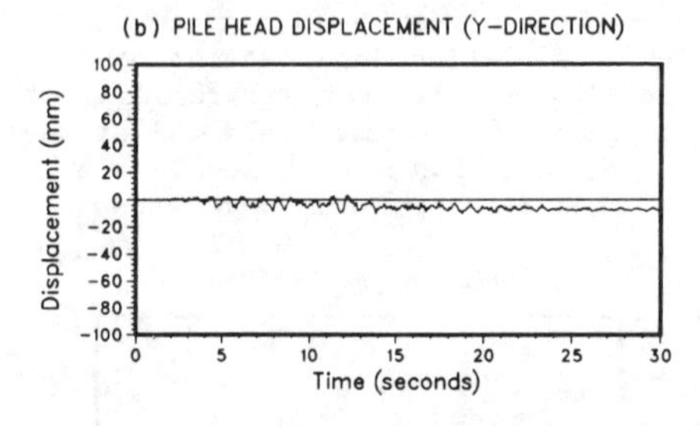

Figure 9a,b. Pile head displacement time histories - single pile test no. 12.

deflection occurs (t = 12.0 sec) is shown in Figure 11.

The data show that the peak accelerations of the pile head and free field were magnified relative to the base acceleration. The predominant period of the pile head response was greater than that of the surface motions in the free field, suggesting that the pile filters out the high frequency components of the ground motion and that pile head inertia forces dominate the pile response.

The bending moments (Figure 11) increase to a maximum near strain gauge 4 at a depth of 4.4 pile diameters and then decrease to approximately zero at greater depths. This indicates that the pile may be considered long in the sense that the lower parts of the pile do not influence the pile head response to the inertia forces applied at the pile head. The spatial variation of bending moments along the pile shows that all points along the pile experience bending moment of the same sign at every instant in time, suggesting that the free headed pile is vibrating in its fundamental mode.

6.2 Pile group response

The pile groups consisting of two piles at various spacings connected by a rigid pile cap were tested at low levels of excitation using a sinusoidal base

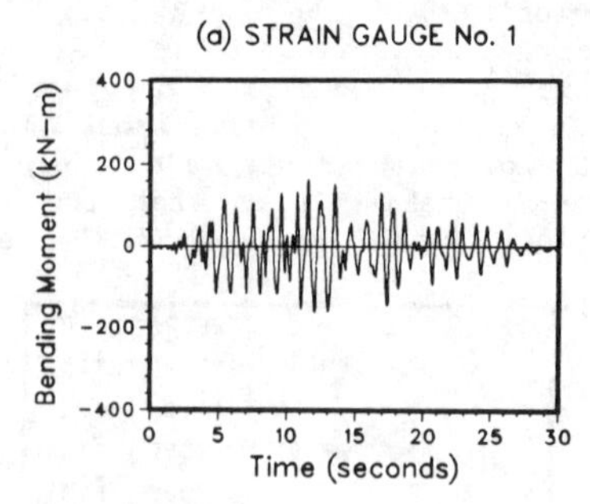

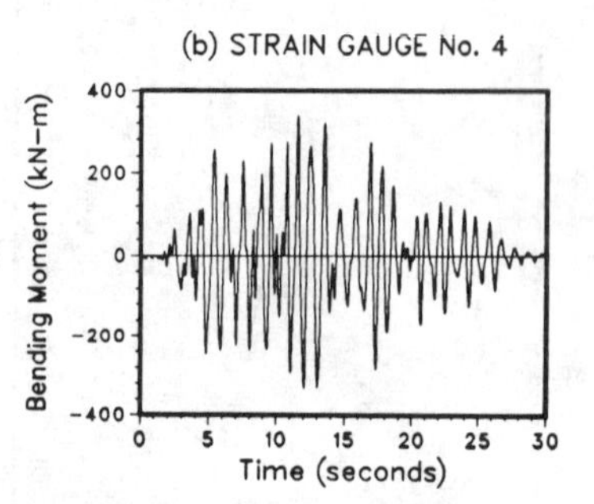

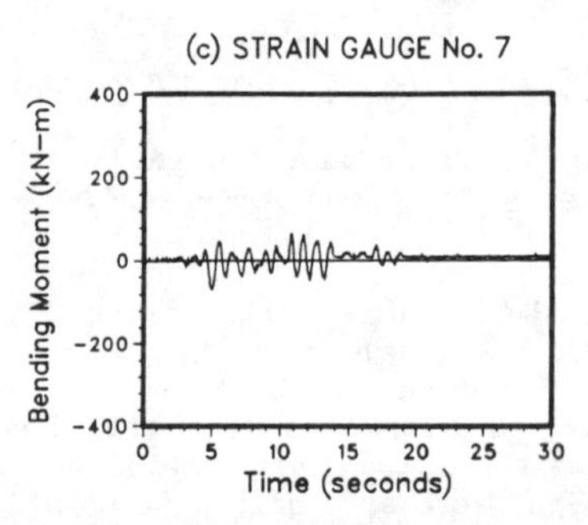

Figure 10a,b,c. Bending moment time histories - single pile test no. 12.

motion. To obtain estimates of the amount of interaction between piles in the group and its dependence on orientation of the piles with respect to the direction of shaking, tests were carried out with the inertia forces inline with the pile group or offline at right angles to the group axis.

Bending moment distributions in the two piles in the group were obtained for both inline and offline loading. For the offline tests, bending moment distribu-

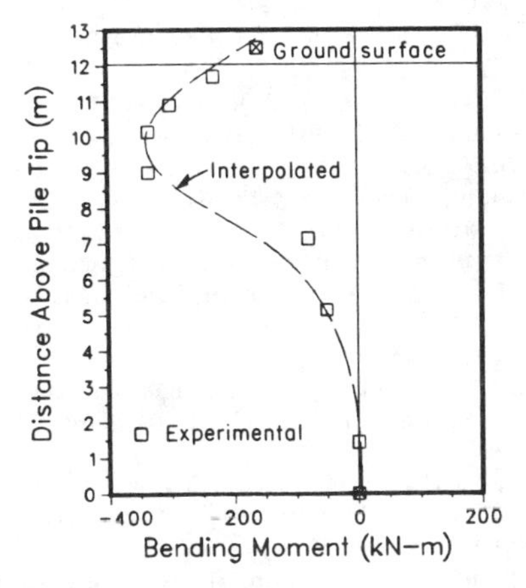

Figure 11. Bending moment distribution - single pile test no. 12.

tion was of the same sign along the pile and is similar to that for a single free headed pile. This suggests that the pile cap provides no restraint during offline loading. The data also show that both piles in the group have approximately the same bending moment distribution indicating that the piles are loaded equally.

During inline loading, the bending moment changes sign indicating restraint by the pile cap against rotation (Figure 12). The moment distributions in the two piles are sufficiently different to suggest non-uniform load sharing.

The peak pile cap displacements are plotted against the pile spacing ratio, s/b, in Figure 13 for both offline and inline shaking. The displacements have been scaled in each case to a common peak acceleration; 0.04 g for inline and 0.06g for offline shaking. This was done to account for variations in the pile cap acceleration due to differences in peak input motion. Such scaling is considered appropriate because of the very low levels of excitation.

The variability in load sharing between piles in the group and in the peak pile cap displacements for pile spacings less than six pile diameters indicates strong interaction between the piles for inline loading. At a spacing of 6 pile diameters the loads and moments are approximately equal in each pile. The deflection of an isolated single pile under these same loading conditions was computed using a Winkler model with a lateral stiffness proportional to the square root of depth (Franklin and Scott, 1979). The proportionality constant was determined from test data on single piles. The isolated single pile deflection for each loading case is shown in Figure 13. The comparisons between single pile and group deflections confirm that there is very strong interaction between piles in the group for spacings less than six pile diameters. On the other hand, there appears to be little interaction between piles during offline loading except possibly at very close spacings.

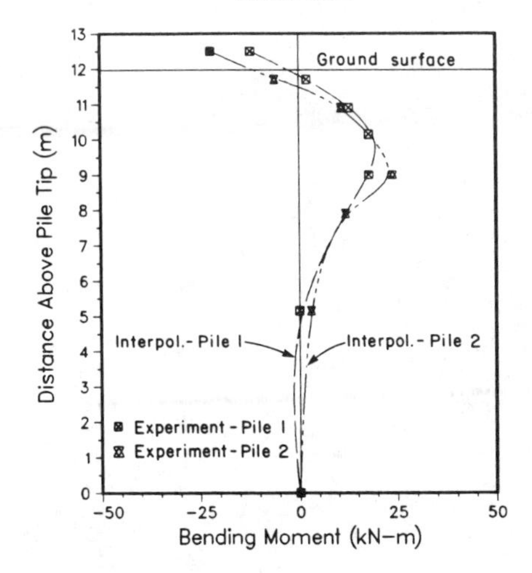

Figure 12. Bending moment vs depth; inline loading - group test no. 21.

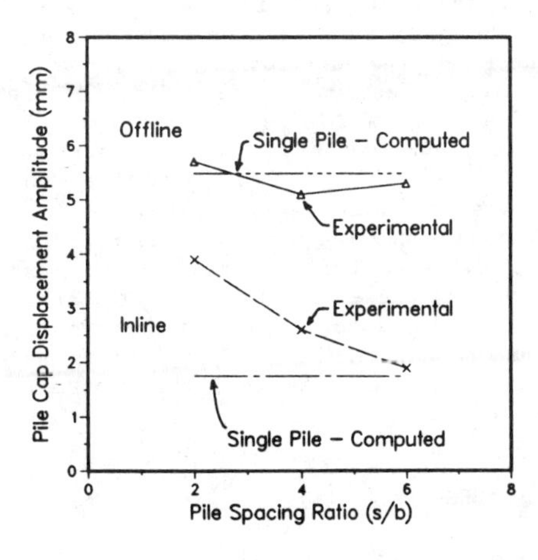

Figure 13. Influence of pile interaction on pile cap displacement.

7 CONCLUSION

Simulated earthquake loading tests were conducted on single piles and pile groups in dry sand to provide data for checking current methods for the analysis of the seismic response of pile foundations. Centrifuge test procedures and equipment have been described. A unique feature of these tests is that, for the first time, the in-situ distribution of shear moduli in the soil was measured during flight using piezoceramic bender elements.

Displacement data from tests on groups of two piles at various spacings showed strong interaction at close spacings for inline loading. Interaction effects gradually reduced with distance and were negligible at spacings over six pile diameters. Bending moment diagrams show reversal of the sign of the bending moment indicating the effective restraint of the pile cap against pile rotation for inline excitation.

Data from offline loading indicated little interaction. Bending moments retained the same sign with depth along the pile showing that the piles behaved similarly to a single free headed pile.

The experimental procedures adopted in the program and the instrumentation worked very well throughout. Centrifuge modelling represents a flexible and cost effective means of simulating the effects of earthquake loading on pile foundations.

8 ACKNOWLEGEMENTS

The geotechnical centrifuge and earthquake simulator used in the experiments were designed by R.F. Scott and his co-workers at the California Institute of Technology, Pasadena. The success of the testing program is due in major part to the help and co-operation of Scott and John Lee. Their help, advice and encouragement are deeply appreciated.

The financial support of Imperial Oil Ltd., Calgary, the National Science and Engineering Research Council of Canada and Franki Canada Ltd. is gratefully acknowledged. The test descriptions and figures are extracted from the report to Imperial Oil.

REFERENCES

Allard, A. 1983. Caltech Centrifuge Manual, Soil Mechanics Laboratory, California Institute of Technology, Pasadena, California.

Finn, W.D. Liam and W.B. Gohl, 1986. Seismic Response of Piles in Centrifuge Model Tests, Report to Imperial Oil Ltd.

Finn, W.D. Liam and W.B. Gohl. 1987. Centrifuge Model Studies of Piles Under Simulated Earthquake Lateral Loading, Proc. of the Symp. on Dynamic Response of Pile Foundations: Experiment, Observation and Analysis, ASCE National Convention, Atlantic City, NJ (preprint).

Franklin, J.N. and R.F. Scott 1979 . Beam Equation with Variable Foundation Coefficient, Proc. ASCE, 105, EM5, pp. 811-827.

Hardin, B.O. and W.L. Black. 1968 . Vibration Modulus of Normally Consolidated Clay, ASCE, J. Soil Mechanics and Foundations Division, Vol. 94, pp. 353-369.

Norwegian Geotechnical Institute 1984 . The Use of Piezoceramic Bender Elements to Measure G_{max} in the Laboratory, Internal Report No. 40014-3 by R. Dyvik.

Shirley, D.J. and L.D. Hampton. 1978 . Shear Wave Measurements in Laboratory Sediments, J. Acoustical Society of America, Vol. 63, No. 2, pp. 607-613.

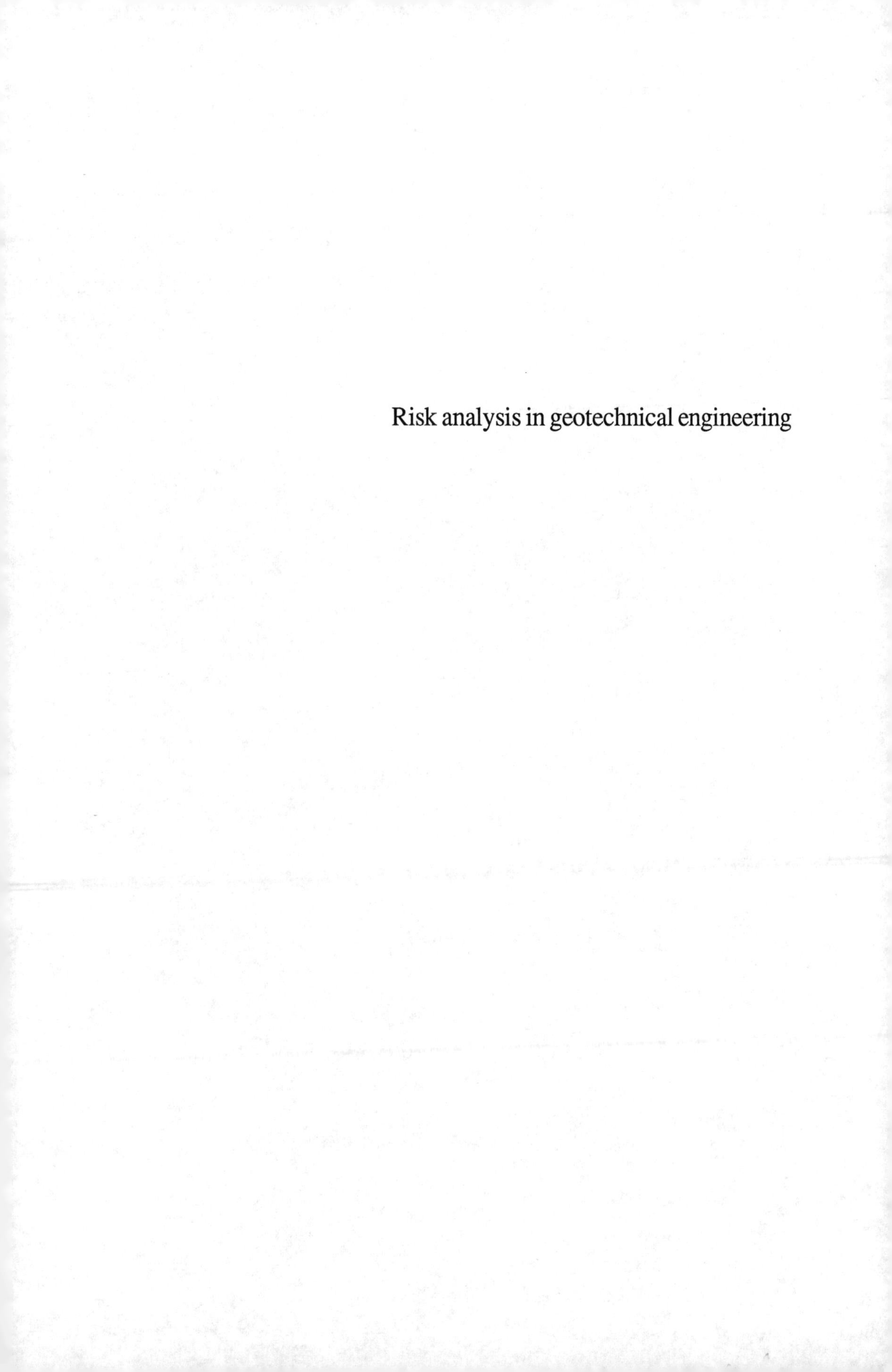

Risk analysis in geotechnical engineering

Prediction and Performance in Geotechnical Engineering / Calgary / 17-19 June 1987

Spatial variation estimation of soil properties in one dimension for regularly spaced data

P.H.S.W.Kulatilake & R.K.Southworth
Department of Mining and Geological Engineering, University of Arizona, Tucson, USA

ABSTRACT: A procedure is suggested to estimate spatial variability of soil properties in one dimension for regularly spaced data. It is developed for statistically homogeneous layers. In applying this technique to a stratified soil deposit, first it is necessary to separate the deposit into statistically homogeneous layers. Then the technique can be applied to each layer separately. Soil property data were considered to consist of a non-stationary portion and a stationary portion. Methods are given to investigatethe stationarity. Polynomial regression analysis was used to model the non-stationary portion. Box-Jenkins time series analysis was used to model the stationary portion. Overall mean estimations were calculated by combining the mean estimations from the non-stationary portion with estimations from the stationary portion. First order, second-moment, Taylor series method was used to estimate the variances associated with the mean estimations. Estimation of total random error is also given. An example is given to illustrate the use of the suggested method.

1 INTRODUCTION

In deterministic analysis of geotechnical engineering problems, it is common to model the soil profile at a site in terms of homogeneous layers with constant soil properties. However, even within apparently homogeneous soil layers, engineering soil properties may show considerable variation from point to point. Risk analysis can incorporate the spatial variability of soil properties and is important in quantifying the reliability of the performance prediction of a geotechnical structure. Quantification of spatial variability of soil properties is also helpful in the design of soil exploration programs and in the evaluation of their effectiveness.

Uncertainties in soil properties arise from three major sources. The first source of uncertainty is due to the natural heterogeneity of the soil deposit. A second source of uncertainty can be attributed to the limited availability of data of the soil property under consideration. This leads to statistical uncertainty. This can be decreased at the expense of additional testing. A final source of uncertainty is due to the difference between the measured values, either in the laboratory or in the field, and the actual values. Sample disturbance, test imperfections, instrument errors and human errors are the main causes of this uncertainty. This third type of uncertainty is also possible when empirical correlations are used to estimate engineering properties from index properties.

The primary focus of this paper is on the first type of uncertainty. The study is limited to one dimensional (1D) modeling of the spatial variability of regularly spaced soil property data. Several methods of soil investigation generate regularly spaced soil property data; two examples are the Dutch cone test and the standard penetration test.

Random field theory in conjunction with regression analysis has been used to model spatial variability of soil properties in statistically homogeneous deposits. (Alonso and Krizek 1975; Lumb 1975; Vanmarcke 1977; Tabba and Yong 1981). The models have considered the soil property which is of concern to consist of a non-stationary trend and a stationary stochastic component. In these models, either the stationarity has been assumed or very little attention has been paid in checking into the stationarity. The procedure developed in this paper provides several methods to check the stationarity. Linear unweighted regression analysis with polynomial functions has been used to model the non-stationarity trend. In performing the unweighted regression analysis, it is assumed that the variance of the residuals is a constant. However, in some soil deposits, properties show non-stationarity due to non-constant variance (Lumb 1975). This issue also has been taken into account in the model suggested in this paper.

Previous investigators have limited the applicable functions to first or second order autoregressive functions in modeling the stationary stochastic portion. However, these simple functions may not be sufficient to model the autocorrelation structure in all soil deposits. Therefore, in order to broaden the applicable functions, in this paper all types of autogregressive moving average (ARMA) models (Box and Jenkins 1976) are considered as possible functions to model the autocorrelation structure.

In addition to modeling the non-stationary and stationary portions of the spatial variation, this paper provides a technique to estimate the mean soil properties and their estimation variances.

2 OUTLINE OF TECHNIQUE

Soil property values with depth can be considered as a time series data set (Lumb 1975). In general, the measured values of a soil property may be divided into three components as given by the following equation.

DATA = TREND + SIGNAL + NOISE (1)

The TREND can be considered as the non-stationary component of the data. It contains the global systematic change of the spatial variation along with a possible bias in the test method. The bias may be evaluated by comparing the property measured by the test method, with that determined by more reliable methods. The SIGNAL plus the NOISE represent the stationary stochastic portion of the data. The SIGNAL can be considered as the stationary correlated portion and provides the local systematic change of the spatial variation. Global and local trends are relative terms whose interpretation depends on site dimensions and availability of data with the spatial dimension. The NOISE can be considered as the random uncorrelated portion. Rapid variations in the soil property over distances shorter than the sampling interval and random testing error are probable causes of NOISE. It is not necessary for soil property values to contain all three components given in equation (1). This is very clearly shown through the flow chart of the suggested technique given in Figure 1.

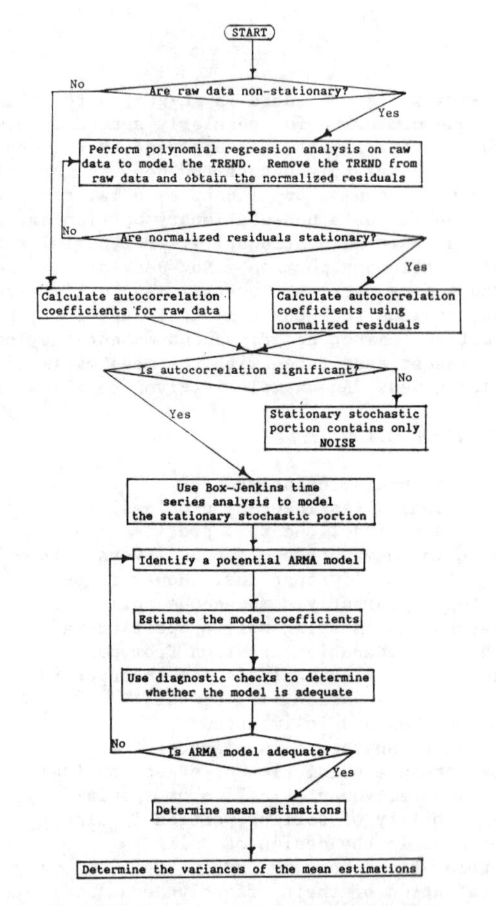

Fig. 1 Flow chart of the suggested technique.

Related to Figure 1, the following should be noted: (1) The non-stationary portion of data can be checked through the following three methods: (a) plotting raw data versus depth, (b) plotting the variogram for raw data, (c) plotting the autocorrelation function for raw data. (2) If the normalized residuals are found to be non-stationary, then it is necessary to go back to the previous step and try higher order polynomials to model the trend of raw data until stationary residuals are obtained. (3) In obtaining mean estimations and the variances of the mean estimations it is necessary to consider the contributions from both TREND and SIGNAL portions.

3 DETECTION OF NON-STATIONARITIES IN SPATIAL DATA

Presence of a trend in a data set can usually be detected by visual inspection of the plotted data. For example, Figure 2 shows measured values of the cone resistance from a Dutch cone test plotted against the depth for a soil deposit (Schmertmann 1969). This graph clearly indicates that the mean of the soil property changes with depth. Therefore it violates the stationary requirements. In addition to visual inspection, plotting of variogram function or autocorrelation function for raw data can be used to detect non-stationarities in raw data.

Let x be the depth and soil property data are available at regularly spaced interval of Δx. Let t be the integer variable which satisfy $x = t.\Delta x$ where t = 1, 2, ...n. The variogram is a graph between the semivariance function, $\gamma(k)$, and the lag k. The $\gamma(k)$ for raw

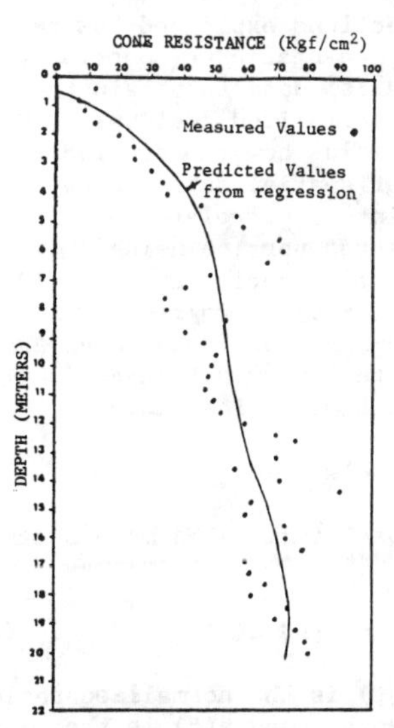

Fig. 2 Dutch cone resistance measured values and predicted values from fourth degree unweighted polynomial regression analysis.

data can be estimated by (Journel and Huijbregts 1978)

$$\hat{\gamma}(k) = \frac{1}{2(n-k)} \sum_{t=1}^{n-k} [y(t+k)-y(t)]^2 \qquad (2)$$

in which k=1, 2....K where K is the maximum lag number, y(t) is the soil property value at integer depth t, y(t + k) is the soil property value at integer depth (t + k) and $\wedge$ denotes the estimated value. Since the reliability of the estimates decreases with increasing K, the value of K should be restricted to about 25 percent of n (David 1977). If the raw data is stationary then the $\gamma(k)$ levels off as k increases (Journel and Huijbregts 1978). If the variogram does not level off for high k values, it implies existence of non-stationarities in raw data.

The autocorrelation function, $\rho(k)$, for raw data can be estimated by

$$\hat{\rho}(k) = \frac{\frac{1}{n-k}\sum_{t=1}^{n-k}[y(t+k) - \bar{y}]\,[y(t) - \bar{y}]}{\frac{1}{n}\sum_{t=1}^{n}[y(t) - \bar{y}]^2} \qquad (3)$$

in which k=1,2,....K. In this case too, value of K should be restricted to about 25 percent of n (Box and Jenkins 1976). For lags greater than 25 percent of n, the estimates are not reliable. Vandaele (1983) suggests that the stationarity of a data set should be carefully examined if, for lags greater than 5, absolute value of $\rho(k)$ is greater than 0.7.

4 MODELING THE NON-STATIONARITY IN SPATIAL DATA

Polynomial regression analysis can be used to model the non-stationary trend. A package program such as P5R of BMDP (1985) can be used to accomplish this task. Two types of polynomial regression analyses are possible with P5R: (a) unweighted (b) weighted. First, unweighted polynomial regression may be performed according to the model given below.

$$y_i = \beta_o+\beta_1x_i+\beta_2x_i^2+..+\beta_px_i^p+e_i \qquad (4)$$

in which y_i is the i th soil property value, x_i is the i th depth value, β's are the regression coefficients, e_i is the i th residual value, and p is the degree of the polynomial. It is assumed that e_i is a random error with a mean zero and a variance σ^2. It is also assumed that e_i's and e_j's are uncorrelated. Polynomial degrees one through a certain maximum degree may be tried in modeling the trend. Statisticians usually use a minimum of seven to ten data points for each coefficient estimated. This rule may be used to select the maximum degree which should be tried. Multiple R-square and Residual mean square (RMS) parameters (BMDP 1985) may be used in determining the best polynomial to model the trend. Draper and Smith (1981) provide details related to using the afore-mentioned parameters in finding the best fit polynomial. Brook and Arnold (1985) suggest that the multiple R-square should be at least 0.5 for a significant regression fit.

The residuals coming out of the best fit polynomial should be checked for the constant variance assumption of the residuals. This can be done by plotting the magnitude of residuals versus depth. If the variance of the residuals is constant, a graph of the magnitudes of residuals plotted against the depth should fall in a horizontal band. For further details related to this aspect, the reader is referred to Draper and Smith (1981). If the constant variance assumption is violated, then it is necessary to perform weighted polynomial regression analysis (BMDP 1985) on raw data to model the trend. The basic idea here is to assign a weight to each observation which is inversely proportional to the variance of the corresponding residual. The variance of each residual is the square of the standard deviation for each residual. The standard deviations of residuals may be estimated by the best fit polynomial between the magnitude of residuals versus depth.

The residuals are normalized in one of the two ways depending on which method of regression analysis is used to remove the trend from raw data. When the unweighted regression method is used, the residuals are normalized by dividing the residual values by the standard deviation of the residuals. The estimated value of the standard deviation of the residuals can be obtained from the P5R output. If the weighted regression method is used, then each residual coming out of weighted regression analysis is divided by its respective standard deviation to obtain the normalized residual.

5 CHECKING FOR STATIONARITY OF NORMALIZED RESIDUALS

Let z(t) be the normalized residual at integer depth t. Using z(t) instead of y(t) in equations (2) and (3), the $\gamma(k)$ and $\rho(k)$ for normalized residuals can be determined. These calculated values can be used to check the stationarity of normalized residuals as explained in Section 3.

For stationary data, it can be shown that $\gamma(k)$ relates to $\rho(k)$ through the following expression (Agterberg 1970).

$$\sigma^2(z) = \frac{\gamma(k)}{1 - \rho(k)} \qquad (5)$$

where $\sigma^2(z)$ is the constant variance of the random variable z. According to this, if z is stationary, then $\hat{\gamma}(k)/[1 - \hat{\rho}(k)]$ values should be approximately constant. This fact also can be used in checking for stationarity of normalized residuals.

6 MODELING THE STATIONARY PORTION OF SPATIAL DATA

Previous sections explained how to remove the non-stationary portion from the raw spatial data to obtain the stationary normalized residuals. This section explains how the stationary component of spatial data can be separated into a correlated portion and an uncorrelated portion using Box-Jenkins time series analysis (Box and Jenkins 1976). Program P2T in BMDP (1985) allows a user to build an ARMA model for stationary data according to Box-Jenkins time series analysis.

6.1 ARMA Models

In the compact form, ARMA models can be given by

$$\phi[B]\, z(t) = \theta[B]\, a(t) \qquad (6a)$$

in which z(t) is the normalized residual value at depth t and a(t) is the uncorrelated portion of the stationary data at depth t, which is assumed to be a normally and independently distributed random variable with a mean of zero and a constant variance. The polynomials in B can be written in multiplicative form as

$$\phi[B] = (1 - \phi_1 B)(1 - \phi_2 B^2) \ldots \quad (6b)$$

$$\theta[B] = (1 - \theta_1 B)(1 - \theta_2 B^2) \ldots \quad (6c)$$

in which ϕ's are the autoregressive (AR) parameters, θ's are the moving average (MA) parameters and B represents the backshift operator and is defined as

$$B^k z(t) = z(t - k) \quad (6d)$$

$$B^k a(t) = a(t - k) \quad (6e)$$

6.2 Identification of ARMA Models

The identification of a potential ARMA model is the first step in modeling the stationary portion of the data. Two tools used in the identification of ARMA models are the estimated autocorrelation function (acf) and the estimated partial autocorrelation function (pacf). Estimation of the acf using z(t) values was discussed under Section 5. For the expression to estimate pacf we refer the reader to Box and Jenkins (1976: p.64). The program P2T in BMDP (1985) can be used to obtain plots of acf and pacf for z(t). In order to use acf and pacf to identify potential ARMA models, it is essential to know which acf and pacf coefficients are significantly different from zero. The standard errors of acf and pacf estimations provide a way of determining this. The expressions to calculate these standard errors are given in Box and Jenkins (1976). If any acf or pacf coefficient is less than two times its standard error, the coefficient can be considered as zero at the 5% significance level (Pankratz 1983). Therefore, an estimated acf or pacf coefficient can be considered to be statistically different from zero if it is greater than two times its standard error. The program P2T provides the two times standard error values in acf and pacf plots.

Every ARMA model has a theoretical acf and pacf associated with it. These theoretical functions contain distinguishing characteristics which enable them to be used to identify ARMA models. At the identification stage we compare the estimated acf and pacf with various theoretical acfs and pacfs. We then choose the model whose theoretical acf and pacf most closely resemble the estimated acf and pacf of the data series. The writers found the following two guidelines to be very useful in identifying potential ARMA models. (i) A significant acf coefficient can be considered to indicate the presence of the corresponding MA coefficient for that lag. (ii) A significant pacf coefficient can be considered to indicate the presence of the corresponding AR coefficient for that lag.

6.3 Estimating the Model parameters

Once a potential model has been identified, the next step is to estimate its parameters. The program P2T uses two methods to estimate the model parameters: one is by conditional least squares; the other by unconditional least squares, also known as the backcasting method (Box and Jenkins, 1976). Once the parameters of the model have been estimated, various diagnostic checks should be performed to determine whether the model adequately describes the stationary data.

6.4 Testing for Adequacy of Fit

A good model should satisfy the following requirements (Pankratz, 1983): (1) When $\phi[B]$ is expressed in the multiplicative form as given in equation (6b), to satisfy the stationary requirement, the magnitude of ϕ_i for all i should be less than 1. (2) When $\theta[B]$ is expressed in the multiplicative form as given in equation (6c), to satisfy the invertibility condition, the magnitude of θ_i for all i should be less than 1. (3) AR and MA parameters should have absolute t-values greater than 2 for them to be significantly different from zero at the 5% significance level. (4) The estimated parameters should have absolute correlations less than 0.9 for them to stable. (5) The residuals arising from the estimated ARMA model should be uncorrelated. If all the coefficients in acf and pacf functions for the residuals are less than two times their standard errors, then statistically it can be concluded that the residuals are uncorrelated. Uncorrelation of the residuals also can be accepted when Ljung-Box Q Statistic (Vandaele 1983) is less than the table chi-square value at the chosen significance level.

If the model is found to violate any of the conditions 1 through 5 given above, then a new model is chosen and the whole procedure of estimation and diagnostic checking is done for the new model. This procedure is repeated until an appropriate model is found. If several models are found to satisfy the conditions 1 through 5 given above, the model having the lowest residual mean square should be used.

7 SPATIAL VARIATION ESTIMATION WITH DEPTH

This section explains how mean estimations and the amount of uncertainty associated with the mean estimations are determined for the soil property with depth.

7.1 Mean Estimations

The mean estimations for the soil property are composed of two components: the mean estimations from the non-stationary component; and the estimated value from the stationary component. When the variance of the residuals is constant, the following formula provides the mean estimation for the soil property.

$$\hat{P}(t) = \hat{y}(t) + [\hat{z}(t)][R] \quad (7)$$

in which $\hat{P}(t)$ is the mean estimation for the soil property at index depth t, $\hat{y}(t)$ is the predicted value at index depth t from the fitted regression function for the global trend, $\hat{z}(t)$ is the estimated value at index depth t from ARMA model and R is the standard deviation of the residuals obtained from the regression analysis on raw data.

When the variance of the residuals is non-constant, the following formula is used to obtain the mean estimation for the soil property.

$$\hat{P}(t) = \hat{y}(t) + [\hat{z}(t)]\,[\hat{R}(t)] \quad (8)$$

where $\hat{R}(t)$ is the estimated standard deviation of the residual at index depth t for the non-constant variance residual case.

7.2 Variances of the Mean Estimations

When the variance of the residuals is constant, under the assumption of negligible correlation between $\hat{y}(t)$ and $\hat{z}(t)$, the variance of the mean estimation, Var $[\hat{P}(t)]$, can be given by

$$\mathrm{Var}[\hat{P}(t)] = \mathrm{Var}[\hat{y}(t)] + R^2\mathrm{Var}[\hat{z}(t)] \quad (9)$$

According to the propagation of error technique (Ang and Tang, 1975), the first order approximation of Var $[\hat{y}(t)]$ can be written as

$$\mathrm{Var}[\hat{y}(t)] = \sum_{i=0}^{p}\sum_{j=0}^{p} [x(t)]^{i}[x(t)]^{j} (\rho_{ij})_{\beta}[\mathrm{Var}(\hat{\beta}_i)]^{1/2}[\mathrm{Var}(\hat{\beta}_j)]^{1/2} \quad (10)$$

in which $(\rho_{ij})_\beta$ is the correlation coefficient between $\hat{\beta}_i$ and $\hat{\beta}_j$. Output of P5R provides the values for $(\rho_{ij})_\beta$ and $[\mathrm{Var}(\hat{B}_i)]^{1/2}$. Similarly the first order approximation of Var $[\hat{z}(t)]$ can be written as

$$\mathrm{Var}[\hat{z}(t)] = \sum_{i=1}^{N}\sum_{j=1}^{N} c_i c_j (\rho_{ij})_\alpha [\mathrm{Var}(\hat{\alpha}_i)]^{1/2} [\mathrm{Var}(\hat{\alpha}_j)]^{1/2} \quad (11a)$$

$$c_i = \frac{\partial[\hat{z}(t)]}{\partial\hat{\alpha}_i} \quad (11b)$$

$$c_j = \frac{\partial[\hat{z}(t)]}{\partial\hat{\alpha}_j} \quad (11c)$$

in which α's are the coefficients of the ARMA model, N is the total number of coefficients and $(\rho_{ij})_\alpha$ is the correlation coefficient between $\hat{\alpha}_i$ and $\hat{\alpha}_j$.

Output of P2T provides the values for $(\rho_{ij})_\alpha$ and $[\mathrm{Var}(\hat{\alpha}_i)]^{1/2}$. The c_i's can be calculated according to equations (11b) and (11c).

When the variance of the residuals is non-constant, under the assumption of negligible correlation between the three variables $\hat{y}(t)$, $\hat{z}(t)$ and $\hat{R}(t)$, the variance of the mean estimation can be given by

$$Var[\hat{P}(t)]=Var[\hat{y}(t)]+\{Var[\hat{z}(t)]\}\{Var[\hat{R}(t)]\}$$
$$+Var[\hat{z}(t)]\hat{R}^2(t)+Var[\hat{R}(t)]\hat{z}^2(t) \qquad (12)$$

The calculation for Var $[\hat{R}(t)]$ is similar to the calculation for Var $[\hat{y}(t)]$.

8 APPLICATION

The purpose of the application is to illustrate the use of the developed estimation model. The soil property data considered in this section come from Dutch cone penetration tests performed in sand at Eglin Air Force Base, Florida (Schmertmann, 1969). The general conclusion from the site investigation was that the entire site area can be considered as a statistically homogeneous, clean, coarse-grained sand deposit for the depth considered. Figure 2 shows the penetration results plotted with depth for one of the sounding logs. Cone resistances at a regular interval of spacing of 0.4 meters were used to show an application of the developed model. In this data, measured values contain a component due to random testing error in addition to the spatial variation. However, it is reasonable to assume absence of a bias component in these measured values.

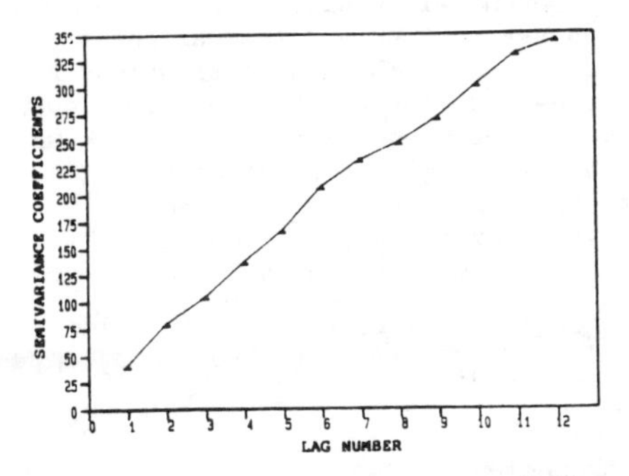

Fig. 3 Variogram for cone resistance data.

Figure 2 clearly shows that the data contain a global trend. The presence of this non-stationarity is also depicted by the variogram for the raw data (Fig. 3). Altogether this profile has about fifty data points. Unweighted polynomial regression analysis was performed between cone resistance and depth using degrees one through five for this data. This allowed a minimum of ten data points for each coefficient estimated. The results of the regression analysis (Table 1) indicate that the fourth degree polynomial represents the data best. This polynomial has the lowest RMS value and has a significant (0.5) multiple R-square value. Figure 2 shows a graph of the values predicted by the fourth degree polynomial against the depth. The plot between the magnitude of the residuals versus the depth for this polynomial showed that the variance of the residuals is approximately constant with depth. Therefore, it is not required to perform a weighted regression analysis for this set of data to model the global trend. The RMS value of 183.2 found in the regression analysis (Table 1) provides an estimate for σ^2.

Table 1. Results of the regression analysis on cone resistance data.

Degree of Polynomials	RMS	Multiple R-Square
1	230.2	.442
2	233.5	.445
3	215.2	.499
4	183.2	.582
5	187.0	.583

As the next step, the normalized residuals were obtained for this data by dividing each residual value by $\hat{\sigma}$. Variogram function for normalized residuals (Figure 4) shows that the variogram levels off and indicates the stationarity of the normalized residuals. The second column of Table 2 shows the autocorrelation coefficients for the normalized residuals. None of these autocorrelation coefficients is larger than 0.7 after the fifth lag. This also indicates the stationarity. Lastly, the relationship between the autocorrelation function and the semivariance function was examined using equation (5). This relationship is shown in the last column of Table 2. The values are approximately one. Since this is a constant variance case, the variance of the normalized residuals should be one. Therefore, the values satisfy equation (5) quite well and confirm the stationarity of the normalized residuals.

The program P2T was then used on normalized residuals to model the stationary component of the cone resistance data. The estimated acf and pacf for the normalized residuals are shown in Figures 5a and b respectively.

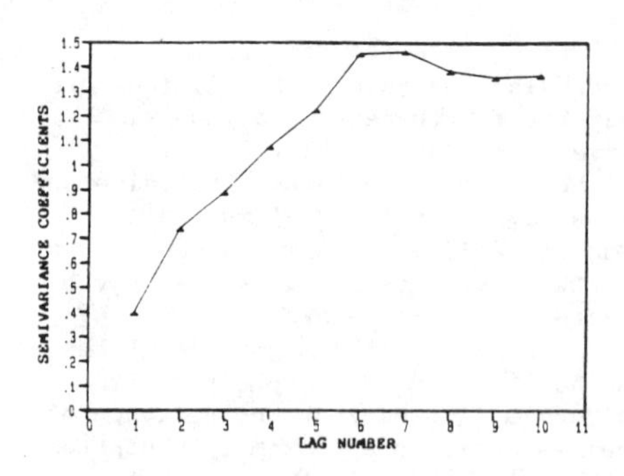

Fig. 4 Variogram for normalized residuals.

Table 2. Autocorrelation coefficients and the relationship between the autocorrelation coefficients and the semivariance coefficients for normalized residuals.

Lag	Autocorrelation Coefficients	Semivariance/ (1-Auto. Coeff.)
1	.5725	.9231
2	.3279	.9737
3	.1251	.9924
4	-.0440	1.0076
5	-.2273	1.0517
6	-.3683	1.0210
7	-.3639	1.0288
8	-.4547	1.0354
9	-.3250	1.0439
10	-.1691	1.0543
11	-.0555	1.0833
12	-.1913	.9839
13	-.1796	.9890

The acf plot has a significant coefficient at lag 1. The pacf plot has significant coefficients at lags 1 and 6. According to the guidelines given for identification, it seems that the possible models are AR(1)(6), MA(1)AR(6) and MA(1)AR(1). For each of these models, the model parameters were estimated and checks 1 through 5 listed under Section 6 were performed. Summary of the results is given in Table 3. For the considered normalized residuals MA(1)AR(6) with $\hat{\theta}_1 = -0.7868$, $\hat{\phi}_6 = -0.3458$ and Var[a(t)] = 0.60 was found

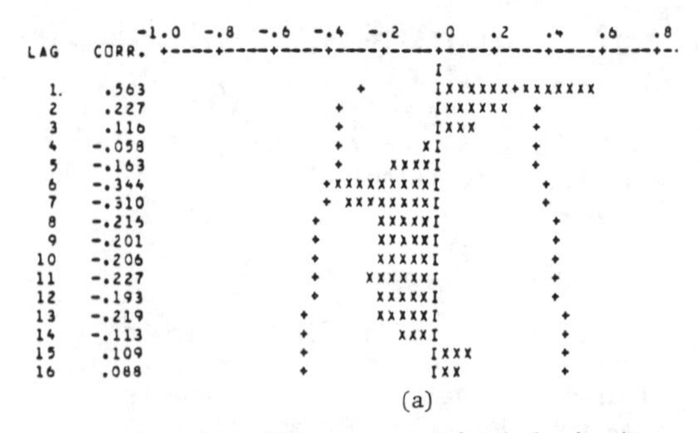

(a)

note: + represents two standard deviation values

```
             -1.0 -.8  -.6  -.4  -.2  .0   .2   .4   .6   .8
LAG   CORR.  +----+----+----+----+----+----+----+----+----+--
                                      I
  1    .563                     +      IXXXXXX+XXXXXXX
  2   -.132                     +   XXXI      +
  3    .067                     +      IXX    +
  4   -.196                     + XXXXXI      +
  5   -.054                     +     XI      +
  6   -.319                    X+XXXXXXI      +
  7    .099                     +      IXX    +
  8   -.104                     +   XXXI      +
  9   -.031                     +     XI      +
 10   -.206                     + XXXXXI      +
 11   -.128                     +   XXXI      +
 12   -.192                     + XXXXXI      +
 13   -.229                     +XXXXXXI      +
 14    .020                     +      I      +
 15    .094                     +      IXX    +
 16   -.255                     +XXXXXXI      +
```

(b)

Fig. 5 Estimated acf and pacf for normalized residuals (a) acf (b) pacf.

to be the only suitable model out of the tried potential models.

Mean estimations and the variances associated with the mean estimations for the cone resistance data are shown in Fig. 6. Measured values are also shown in the same figure for comparison purposes. An estimate for standard deviation of NOISE can be obtained by finding the square root of the product between $\hat{\sigma}^2$ and Var[a(t)]. This value turned out be 10.5 Kgf/cm^2 for this example.

9 CONCLUSIONS

From the study presented here, it can be concluded that Box-Jenkins time series

Table 3 Summary of the results obtained for different ARMA Model fits

ARMA Model	Checks for adequacy of fit				
	Condition 1	Condition 2	Condition 3	Condition 4	Condition 5
AR(1) (6)	satisfied	not applicable	satisfied	satisfied	did not satisfy
MA(1) AR(6)	satisfied	satisfied	satisfied	satisfied	satisfied
MA(1) AR(1)	satisfied	satisfied	satisfied	satisfied	did not satisfy

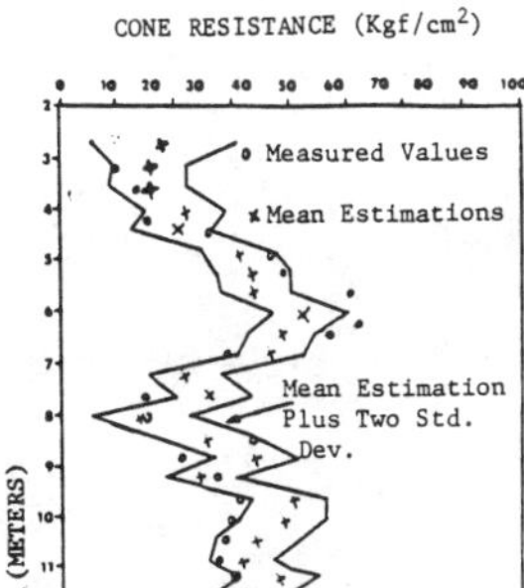

Fig. 6 Estimated values of spatial variation for cone resistance.

analysis in conjunction with regression analysis provide a suitable technique to estimate the spatial variability of soil properties in one dimension for regularly spaced data. In applying this technique to a stratified soil deposit, first it is necessary to separate the deposit into statistically homogeneous layers. Then the technique can be applied to each layer separately.

If one is interested in finding the lowest degree polynomial suitable to remove the non-stationary portion of the data, the following technique can be utilized. Start with the first degree polynomial to model the global trend and check if the residuals produced by it are stationary. If not, increase the polynomial degree by one and check its residuals for stationarity. This should be repeated until stationary residuals are found. The last polynomial degree tried can be taken as the lowest degree polynomial suitable to remove the trend.

ACKNOWLEDGMENT

The financial support of the U.S. Army Engineer Waterways Experiment Station is gratefully acknowledged.

REFERENCES

Agterberg, F.P. 1970. Autocorrelation funtions in geology. D.F. Merriam (ed.). Geostatistics, p. 113-141. New York: Plenum Press

Alonso, E.E., and Krizek, R.J. 1975. Stochastic formulation of soil properties. Proc. Second Int. Conf. on Appl. of Stat. and Prob. in Soil and Structural Engr., 9-32.

Ang, A. and Tang, W. 1975. Probability concepts in engineering planning and design, Vol. 1. New York: John Wiley and Sons.

BMDP Statistical Software Manual. 1985. Berkeley: University of California Press.

Box, G.P. and Jenkins, G.M. 1976. Time series analysis. San Francisco: Holden-Day, Inc.

Brook, R.J. and Arnold, G.C. 1985. Applied regression analysis and experimental design. New York: Marcel Dekker, Inc.

David, M. 1977. Geostatistical ore reserve estimation. Elsevier Scientific Publ. Co.

Draper, N.R., and Smith, H.Jr. 1981. Applied regression analysis, 2nd ed. New York: John Wiley and Sons, Inc.

Journel, A.G. and Huijbregts, C. 1978. Mining Geostatistics. Academic Press.

Lumb, P. 1975. Spatial Variability of soil properties. Proc. 2nd Int. Conf. on Appl. of Stat. and Prob. in Soil and Struct. Eng., Aachen, Germany, 397-421.

Pankratz, A. 1983. Forecasting with univariate Box-Jenkins models. New York: John Wiley and Sons, Inc.

Schmertmann, J.H. 1969. Dutch friction-cone penetrometer exploration of research area at field 5, Eglin AFB, Florida. Contract Report S-69-4, U.S. Army Engineer Waterways Experiment Station, Vicksburg, Miss.

Tabba, M.M. and Yong, R.M. 1981. Mapping and predicting soil properties: Theory. J. Engrg. Mech. Div., ASCE, 107: 773-791.

Vandaele, W. 1983. Applied time series and Box-Jenkins models. Orlando: Academic Press, Inc.

Vanmarcke, E.H. 1977. Probabilistic modeling of soil profiles. J. of Geotech. Engrg. Div., ASCE, 103: 1227-1246.

Prediction and Performance in Geotechnical Engineering / Calgary / 17-19 June 1987

Risk updating for rainfall-triggered spoil failures

R.N.Chowdhury & V.U.Nguyen
University of Wollongong, N.S.W., Australia

ABSTRACT: The importance of rainfall-triggered slope failures and especially of failures of spoil piles associated with strip mining and the difficulties of prediction are emphasized. Methods for prediction or for estimation of factor of safety or of risk of failure are reviewed. A proposed method for updating risk on the basis of updating the estimated service life of spoil piles is discussed. An illustrative example is presented to show how the estimated service life may be updated everytime some failures occur.

INTRODUCTION

Prediction of the behaviour of earth masses is often difficult and it is, therefore, important to supplement analytical and laboratory studies with assessments of observed field behaviour. Both natural and man-made slopes are influenced significantly by intense or prolonged rainfall. Therefore, the behaviour of slopes during rainfall and the occurrence of failures should receive special attention. Rainfall-triggered failures of natural slopes have often been reported in geotechnical and other landslide literature and such failures are often very significant in tropical areas with high rainfall. Relatively little attention has been given to rainfall-triggered failures of mining spoil piles. There is a perception that in the Bowen Basin of Queensland (Australia) spoil failures of spoil piles from strip coal mines are associated with intense rainfall. Direct correlations between rainfall intensity and the frequency and magnitude of spoil pile failures have not been established. However, there is a great deal of evidence concerning the role of water and, therefore, attention must be given to the risk of failures triggered by rainfall.

Conventionally, for slopes, one may calculate safety factors for a given set of conditions. On the basis of influencing factors and slope performance over time, these factors can be updated. A recent trend is the assessment of risk, based on calculated failure probabilities. If a risk approach is adopted an engineer should consider different ways of reassessing and updating calculated risk and the use of information theory and Bayes' theorem in this regard has been indicated by Nguyen (1985) and is further discussed in this paper.

Spoil failures may also occur due to causes other than rainfall. For example, failures may be caused as a consequence of blast-induced fractures or during the building-up of spoil piles. These failures are outside the scope of this paper. In general, if a spoil pile survives the construction period, failures would occur as a consequence of rainfall.

This paper is primarily concerned with a review of methods for factor of safety and risk evaluation and also continued development of an approach proposed by Nguyen (1985) the aim of which was to use information theory and Bayes' theorem for updating the estimated service life and hence the risk of failurc of spoil piles.

SLOPE FAILURES IN SPOIL PILES AND THE ROLE OF WATER

A large proportion of coal extraction in Australia and other parts of the world is carried out by the open pit strip mining method. Overburden strata are first explosively fragmented and then cast over the pit by a dragline to form spoil piles. The coal seam is exposed and coal is transported out of the pit by trucks, shovels, conveyor belts, and other related equipment. Coal extraction proceeds down the dip of strata as the above sequence of operations is repeated.

One of the major problems that could threaten strip coal mining operations is spoil pile instability. Spoil pile failures may cause disruptions in mining and involve extra costs in pit clearing operations. Spoil pile and, to a large extent, highwall failures have been a major concern of most surface mines in Australia. With the continual increase in depth as mining is proceeded down dip, spoil pile stability inevitably becomes more important.

It is well known from previous investigations that spoil failures associated with strip coal mines in the Bowen Basin (Queensland) are water-induced and that moisture-softening of the spoil-floor interface is often the key to the occurrence of failure. This failure mode is a consequence of rainfall, a natural event which is beyond human control. Probabilities of failure associated with all other failure modes are usually insignificant and, therefore, attention is focussed on rainfall-induced failures comprising a two-wedge failure mechanism.

There are, in fact, two important aspects of water-induced failures. One is associated with high piezometric levels along the spoil floor, and the other with moisture softening characteristics of spoil and floor materials. Hydrological investigations have concluded that groundwater aquifers of the mound type do not form readily in spoil piles even after a very heavy rain.

Moisture infiltration may extend only to a depth of few metres. Water or rain induced failures, therefore, must assume a progressive mode in which floor material or spoil material in the regions of spoil toe or crest is wetted by moisture ingress, resulting in a strength reduction.

It should be emphasized, however, that groundwater effects in mining spoil piles are often complex. Seepage may occur from sub-floor aquifers, through joints and fractures in the pit floor. Also rain water percolates down to the base of a spoil pile through highly permeable and localised drill zones. Perched water tables and local ponding in depressions have also been noted (see Nguyen, 1985 for literature review).

UPDATING THE SAFETY FACTOR

Recent analytical research at the University of Wollongong has been concerned with computer simulation of spoil pile stability using deterministic and probabilistic methods. Attention has been focussed on deep-seated failures considering the two-wedge mode of failure which has been interpreted from field observation as the dominant mode.

Progressive failure is facilitated by rainfall-infiltration, moisture-softening of spoil and the strain-softening characteristics of spoil materials. The analysis of slope stability which includes a consideration of progressive failure is discussed by Chowdhury et al (1986). Essentially the basic wedge model has been developed further so that the factor of safety can be updated on the basis of the extent of the softened zone which, in turn, may be determined from stress-deformation studies. Alternatively, observed spoil behaviour may allow interpretation of the extent of the softened zone. This is considered in terms of a parameter which is well known in soil mechanics as the 'residual factor'. As failure progresses along a slip surface, the overall factor of safety will reduce. Diagrams showing the reduction of the factor of safety with the increase in the value of the residual factor (from R = 0 to R = 1) can be plotted for specific modes of progression. For example, failure may progress from the toe of a slope or from the crest of a slope and the curves for these modes will, in general, not coincide for a frictional material. In principle, a similar approach can be developed for failure probability or risk provided the shear strength data is analysed to obtain relevant statistical parameters (e.g. the mean and standard deviation of peak shear strength parameters and residual shear strength parameters). In this model the shear strength parameters along the rear of the wedge were recognised to be different from those along the basal plane of the wedge. Thus there would be at least eight parameters in a probabilistic model i.e. $c_p, r_r, \varphi_p, \varphi_r$ for the rear plane and for the basal plane. In addition one would have to consider correlations between the parameters. Therefore, this approach would require considerable amount of investigation and data analysis if a risk model were to be developed. Moreover, in such a model the occurrence of rainfall events is not considered directly and, therefore, it will not be considered further in this paper.

STATISTICAL OR CORRELATION MODELS

It is useful to mention attempts to develop methods of slope failure prediction based on interrelationships among various influencing factors. Okamoto (1977) presented a prediction method for risk rate of slope failure caused by heavy rainfall. Data relevant to slope failure in a particular region were obtained from topographical maps and aerial photographs and other existing information. Except for the risk rate of slope failure and rainfall amount most of the data were categorical and two-factor interactions among items were checked five ways with rainfall amount as the one covariate. The interaction between them was not found to be statistically significant. A prediction of the risk rate of slope failure for any other given rainfall amount may be calculated by the use of scores determined on the basis of site-specific data. The risk rate of slope failure was defined as the number of slope failures per hectare. The actual rate of slope failure was investigated after a severe disaster due to a heavy rainfall. The hourly rainfall distribution of the highest intensity for that rainfall event was taken as one option and the hourly rainfall distribution one hour before was taken as the second option. Very weak interactions may exist between factors such as weathering and ground cover or between weathering and valley density. Apart from rainfall, only the contribution of valley direction to the risk rate was found to be high; those of weathering and ground cover being lower and of valley slope quite small, which is somewhat surprising.

Thus, some aspects of this type of statistical approach are useful while other aspects do not accord with what would have been expected from geotechnical considerations. The shift of rainfall distribution in the study of Okamoto (1977) shows that the area of high risk rate extends across the study region as the centre of rainfall moves towards the highest intensity.

It is doubtful if an approach similar to that used by Okamota (1977) would have any application to spoil piles. However, it is useful to refer to another statistical approach for a prediction model used by Neuland (1985). Stable and unstable sloping sites in tropical soil material throughout Colombia were investigated using discriminant analysis by 36 variables. The variables considered included morphometric, hydrometeorological, ecological, soil-physical, soil-mechanical, stratification and other characteristics. The application of multivariate statistical techniques for the determination of slope stability was the primary aim of this study of 163 slopes. The variables with the highest prediction power were found to be:- depth of weathering, the shear strength of underlying beds, the catchment or drainage area for the site and the density (frequency?) of inferior (low strength?) layers. The reliability and the efficiency of the model was then tested on fifty sites and the conditions (stable or unstable) were predicted accurately for forty five of these.

The merits and shortcomings of this approach were discussed by Chowdhury (1985) and, in particular, he pointed out the difficulty of incorporating the role of individual geotechnical parameters in their proper context. For instance, one may ask: "How does one account for pore water pressure due to water table and due to rainfall?" Due to use of terminology in the paper which is quite different from that used in conventional geomechanics, there is also some confusion about some of the variables considered in the multivariate approach.

A COMBINED SOIL-WATER MODEL

A probabilistic model which includes one-dimensional water infiltration has been presented by Anderson and Howe (1985) and applied to 'infinite slope' type of geotechnical analysis i.e. long slopes in which the expected depth of slip surface is relatively small. Such long, shallow failures have, according to the paper, been observed in the mid levels area of Hong Kong. The influence of soil suction and other parameters and their variability is considered in the analysis. All geotechnical and hydrological parameters are treated as stochastic variables. Three different critical rainstorms, 150mm, 257mm and 371mm, with respective durations of 2,8 and 24 hours and a 10 year return period were used in the study. The model was applied to the Chater Ridge site in Hong Kong. Typically, the coefficient of variation was 45.5% for the effective cohesion c', only 4.3% for the friction angle and 10% for the soil conductivity (permeability). Monte-carlo simulation technique with 50 iterations per analysis was used to obtain the probability distribution of the factor of safety of the slope. This approach is the type which could be incorporated in the updating-strategy proposed below.

RISK-UPDATING APPROACH

Geotechnical engineering decisions concerning rainfall-triggered slope failures require that (a) the risk or probability of failure be assessed and (b) the intensities and return periods of rainstroms which may lead to instability be estimated. These two tasks are inextricably linked. The risk of failure depends on the rainfall intensity and hazard rate is inversely proportional to the return period of a critical rainstorm.

The procedure for risk-updating may be presented as a series of steps as follows:

(I) Calculate the probability of failure p_f which is a probability of the factor of safety F reducing to unity in the event of the rainfall intensity I exceeding a certain critical intensity I_b. (It is also the probability of moisture softened length S of the slip surface exceeding a critical length S_o in the event of rainfall occurrence). One may write:

$$p_f = \sum_{i=1}^{N} P\left(F \le 1 \,\middle|\, I_i \ge I_b\right)\lambda_i \qquad (1)$$

in which

$$P\left(F\le 1 \,\middle|\, I_i \ge I_b\right) = \int_{I_b}^{I_a} P(F\le 1) \,\middle|\, I_i > I) f_R(I)\,dI \qquad (2)$$

where:-

(a) $f_R(I)$ is the probability distribution of critical rainfall intensity which may be modelled as a truncated exponential distribution with I_a and I_b as the lower and upper bounds of critical rainfall intensities (Nguyen, 1985).

(b) λ_i = the mean Poisson occurrence rate of rainfall i.

(c) N is the number of rainfalls exceeding a certain critical intensity I_b during the service life of a spoil pile.

Assuming the mean occurrence rate of all the N critical rainfalls to be the same (i.e. a constant) will lead to a simplification of Eq.(2). Thus assuming:

$$\lambda_i = \lambda_c = \text{a constant, where } i = 1, N \quad .. \quad (3)$$

Eq.2 becomes:

$$p_f = \lambda_c\, P\left(F \le 1 \,\middle|\, I \ge I_b\right) \qquad (4)$$

The calculation of the probability term in Eqs.(2) or (4) may be carried out using either Monte-Carlo simulation or other more economical numerical methods e.g. Taylor series approach or Rosenblueth's point estimate method (Nguyen and Chowdhury, 1984,1985). Consideration would have to be given to adopting moisture-softened length of slip surface as a stochastic variable. More importantly, the influence of rainfall intensity must be provided for.

(II) By definition of critical rainfall intensity, the occurrence rate of critical rainfalls λ_c is related to the expected service life of spoil piles as follows:

$$\lambda_c = \frac{1}{L_e} \qquad (5)$$

Assuming an exponential distribution of critical rainfall occurrence with a mean of λ_c, the principle of maximum entropy within the context of Information Theory and Bayes' theorem may be used to continually update λ_c with new evidence or information and, as shown earlier (Nguyen, 1985), the expression for L_e is:-

$$L_e = \left(\frac{T}{\omega}\right)^{1/2} \frac{K_{n-2}(2(\omega T)^{1/2})}{K_{n-1}(2(\omega T)^{1/2})} \qquad (6)$$

where:- $K_{n-2}(\cdot)$ and $K_{n-1}(\cdot)$ are the modified Bessel functions of the second kind with order (n-2) and (n-1) respectively,

n - is the number of spoil failures

ω - is the previous estimate of $\lambda_c = \frac{1}{L_e}$

T - is the cumulative time of all spoil piles between successive occurences of critical rainfalls.

III. Calculate the probability of failure p_f again as in step I but using the new updated value of λ_c based on Eq.(6).

The updating of λ_c is now illustrated using an example and related issues are then discussed.

Illustrative Example

A surface mine has 9 strips in operations. Recent experience of spoil pile failures associated with rainfall occurrence have been recorded as follows at times t after construction:-

At t = 20 days, number of spoil failures = 2

At t = 55 days, the number of additional spoil failures = 3

At t = 72 days, the number of additional spoil failures = 1

Update the expected service life of spoil piles assuming the following alternatives for initial estimate or assumption:-

(i) L_e = 30 days

(ii) L_e = 60 days

Solution

(I) Given L_e = 30 days; hence $\omega = \frac{1}{30}$

(a) Based on the fact that failures were first observed at t=20 days we may write

$$T = 9 \times 20 = 180$$

Hence $\omega T=6$ and $(\frac{T}{\omega})^{1/2} = (5400)^{1/2} = 73.48$

From Eq.(6), when n = 2

$$L_e = 73.48 \frac{K_0(4.90)}{K_1(4.90)} = \frac{73.48x.00419}{.00452} = 66.96 \text{ days}$$

(b) Based on the fact that failures were next observed after a lapse of 35 more days at t = 55 days.

$$T = 9 \times 35 = 225$$

$$\text{New value of } \omega = \frac{1}{L_e} = \frac{1}{66.96}$$

$$\omega T = 4.71 \text{ and } (\frac{T}{\omega})^{1/2} = 145.23$$

From Eq.6 when n = 3

$$L_e = 145.23 \frac{K_1(4.336)}{K_2(4.336)} = 106.68 \text{ days}$$

(c) Based on the fact that failures were next observed after a lapse of (72-55)=17 days it may similarly be shown from Eq.6 when n = 1 that

$$L_e = 127.57 \frac{K_1(2.391)}{K_0(2.391)} = 152.22 \text{ days}$$

It may be noted that the modified Bessel function has the following two important properties:

$$\text{(i)} \quad K_{n+1}(z) = \frac{2n}{z} K_n(z) + K_{n-1}(z)$$

$$\text{(ii)} \quad K_\nu(z) = K_{-\nu}(z)$$

(II) Using now an initial estimate of L_e = 60 days

$$\omega = \frac{1}{60}$$

(a) T=9 x 20=180, $\omega T=3$, $(\frac{T}{\omega})^{1/2} = 103.92$

when n=2, Eq.6 gives $L_e = \frac{103.92x0.02054}{0.2334}$

= 91.45 days

(b) $\omega = \frac{1}{91.45}$, and at t=55 days, T=35x6

$\omega T = 3.444$ and $(\frac{T}{\omega})^{1/2} = 169.71$

when n=3, $L_e = 169.71 \frac{K_1(3.71)}{K_2(3.71)} = 119.2$ days

(c) similarly at t = 72 days and when

n=1, $L_e = 127.27 \frac{K_1(2.135)}{K_0(2.135)} = 156.31$ days

Relevant comments

It may be noted that after updating three times the estimated service life approaches a similar value (152 days and 156 days) for two quite different initial assumed values (30 days and 60 days). Whether this convergence will hold for other initial assumed values and for other data of observed failures must be further investigated. If it is not simply fortuitous such convergence is very significant and may be a consequence of the Bayesian basis for the formulation.

One difficulty with regard to this approach is that only part of a spoil pile will fail and this is not considered in the formulation. In other words, the length of strips of the surface mine does not form part of the relevant equations. The importance of this aspect may be considered as follows.

If each failure is considered as the failure of a whole spoil strip then the total number of spoil piles to be considered in the subsequent iteration will be the initial number of strips minus the number of failures. In the example above this would mean 9 strips before t=20 days, 7 strips before t=55 days and only 4 strips before t=72 days. Thus the magnitude of T would vary at each failure event unless, of course, the failed strips of spoil were rectified in the meantime or new strips were dumped equal in number to recent failures. However, this is an unlikely set of circumstances.

On the other hand, the length of each of the failures may be quite small in relation to the length of a strip of spoil. This is a far more likely situation. Whether the failures have been rectified or not, one can assume confidently that the number of strips is the same at each step of calculation. However the value of T should include the length of strips in some way.

A further development of the present model may, therefore, be made on the basis of 'length units' of spoil pile where a 'length unit' has the dimensions of the most likely length of the failure zone based on past experience or estimated on the basis of an appropriate goemechanics model including probabilistic consideration. Thus the value of T would then be the product of total strip length expressed in appropriate 'length units'. For example 9 strips each of 200m length with average likely failure length of 50m would be (9x200)/50 = 36 units. In the above example the value of T would be T = 36 x 20 at the first iteration.

CONCLUDING REMARKS

The methods used for predicting slope failures triggered by rainfall have been reviewed. The role of water in spoil pile instability and especially of moisture-softening has been emphasized. Further development of a model for updating risk based on an updating of the estimated service life (inverse of critical rainfall frequency) has been considered. An illustrative example for the latter has been presented and a suggestion made for overcoming a remaining limitation.

ACKNOWLEDGEMENTS

The authors would like to acknowledge The University of Wollongong and the Department of Civil and Mining Engineering for support of research activities.

REFERENCES

Anderson, M.G. and Howes, S. 1985 . Development and application of a combined soil-water slope stability model, Q.J.Eng.Geology, London, Vol.18, No.3, 225-236.

Chowdhury, R.N., Nguyen, V.U. and Nemcik, J.A. 1986. Spoil stability considering progressive failure, Mining Science and Technology, Vol.3, 127-139.

Chowdhury, R.N. 1985 . Erosion and slope stability in tropical lateritic and saprolitic soils, General Report to Sessions 3.1 and 3.2, TropicaLS'85, Proc. of First Int.Conf. on Geomechanics in Tropical, Lateritic and Saprolitic Soils, Brasilia, Brazil, Vol.3, 341-362.

Neuland, H. 1985 . A prediction model of landslides in tropical areas by means of discriminant analysis, TropicaLS'85, First Int.Conf. on Geomechanics in Tropical Lateritic and Saprolitic Soils, Brasilia, Brazil, Vol.2, 29-40.

Nguyen, V.U. 1985 . Risk assessment of spoil piles based on information theory Australian Geomechanics News, Number 9, 25-34.

Nguyen, V.U. and Chowdhury, R.N. 1984 . Probabilistic analysis of mining spoil piles - two techniques compared. Int. J. Rock Mech. and Mining Science, Vol. 21, No.6, 303-312.

Nguyen, V.U. and Chowdhury, R.N. 1985 . Risk analysis with correlated variables, Geotechnique, London, Vol.35, No.1, 47-59.

Okamoto, M.B. 1979 . Prediction method for risk rate of slope failure caused by heavy rainfall, Natural Disaster Science, Vol.1, No.2, 25-40.

Probabilistic analysis of shallow foundation settlements

M.A.Usmen
West Virginia University, Morgantown, USA

C.Wang
I.T. Corp., Pittsburgh, Pa., USA

S.C.Cheng
Lafayette College, Easton, Pa., USA

ABSTRACT: Probabilistic analyses are performed for shallow foundation settlements using computer models. Both elastic and consolidation settlements are considered. Analyses indicate that the uncertainties around the input parameters involving soil, site and loading conditions introduce a measurable uncertainty on the prediction of settlement. If probabilistic analysis techniques are to be adopted, it will be necessary to develop probability-based tolerable settlement criteria.

1. INTRODUCTION

Design of shallow foundations for civil engineering structures such as buildings, bridges, silos, etc. involves primarily three steps: subsurface exploration, bearing capacity analysis, and settlement analysis. Subsurface explorations provide information on soil/rock stratification and properties and groundwater conditions. Bearing capacity relates to the shear strength of soil supporting the foundation, while settlements occur with elastic deformation and/or time-dependent consolidation of soils under structural loads imposed on the foundation. Although bearing capacity failures would appear to be most critical relative to the safety of the structure, foundation design is most frequently controlled by the tolerable settlement criteria. Settlement analysis is therefore a pivotal component of shallow foundation design.

In current practice, settlement analysis is based on deterministic procedures, employing single-valued input parameters for soil and site characteristics and loading conditions. The values of these parameters, however, are not known with certainty. Soil and site conditions are often quite variable, and it may be difficult to predict the nature and magnitude of all of the loads acting on the foundation. While these uncertainties are always present, they are not systematically accounted for and incorporated in the deterministic analyses employed in estimating settlements. Also, as a consequence of the uncertainties associated with the soil, site, and loading parameters, an uncertainty will arise on the value of the predicted settlement. This uncertainty is currently dealt with by adopting conservative but economical designs. According to Wahls (1983), for example, the shallow foundation alternative for bridges have sometimes been adandoned in favor of a more costly deep foundation system, partly because questions have been raised on the reliability of the settlement predictions.

The use of probability and statistics principles to account for uncertainty in geotechnical engineering analysis and design has become popular in recent years (Harr, 1984). It is possible to employ the probabilistic analysis techniques in predicting settlements while considering the uncertainties of input parameters. One approach that can be undertaken is to treat the input variables as random variates represented by a mean value, a standard deviation (or coefficient of variation), and a probability density function. All or any desired number of variables contained in the deterministic settlement equations (models) can be treated as random variables. Settlement, being a function of these random variables, also becomes a random variable with a mean, standard deviation, and probability density function, and can be further analyzed probabilistically. This methodology quantifies the uncertainty, and provides a valuable tool for the geotechnical engineer in the sharpening of his/her decision making ability through better understanding of the relationships between the design, risk and economy.

The following section of this paper presents a brief review of the well-known deterministic models (equations) used in predicting settlements in clays and sands as applicable to the study. Next, fundamentals of probabilistic analysis are reviewed as applicable to the study, and the essential features of the probabilistic models developed for the analysis of settlements are described. Two case studies are presented to demonstrate the applications and usefulness of these models. The summary and conclusions of the study appear in the final section of the paper.

2. DETERMINISTIC SETTLEMENT ANALYSIS

Foundation settlements can be categorized as elastic settlements, or consolidation settlements. Elastic settlements occur immediately after construction, while consolidation settlements occur in a time-dependent manner as a result of the dissipation of pore water pressures in soil voids. Elastic settlements are generally significant for granular soils, whereas consolidation settlements carry importance for saturated cohesive soils.

The predictive equations most frequently used for consolidation settlements are based on Terzaghi's classical one-dimensional consolidation theory. As cited by Das (1984), the equation for normally consolidated (NC) clays is:

$$S_c = \frac{C_c H_c}{1+e_o} \log \frac{P_o+\Delta P}{P_o} \quad (1)$$

and, the equation for overconsolidated (OC) clays are:

$$S_c = \frac{C_s H_c}{1+e_o} \log \frac{P_o+\Delta P}{P_o} \quad (2)$$

if $(P_o+\Delta P) < P_c$, and

$$S_c = \frac{C_s H_c}{1+e_o} \log \frac{P_o+\Delta p}{P_o} + \frac{C_c H_c}{1+e} \log \frac{P_o \Delta P}{P_c} \quad (3)$$

if $P_o < P_c < (P_o+\Delta P)$. In these equations;

S_c = Consolidation settlent
H_c = Thickness of compressible layer
C_c = Compression index
C_s = Swell (recompression) index
e_o = Initial void ratio
P_o = Average effective overburden pressure
P_c = Preconsolidation pressure, and
Δp = Average stress increase due to surcharge.

The thickness H_c in this equation is obtained from subsurface exploration. The soil parameters such as C_c, C_s, e_o, and P_c are determined from laboratory tests. P_o is determined by considering soil unit weights, groundwater table and soil stratification. The Δp values are obtained by using the Boussinesq elastic theory. Single point estimates of each of these parameters are used in equations (1), (2), and (3).

Although the Boussinesq theory has also been used in predicting elastic settlements for many years, the elastic settlement formulations recently developed by Schmertmann and his co-workers (1978) based on strain influence factors have gained wide attention. This method is applicable to granular soils but can account for time-dependent creep effects. As cited by Das (1984), the elastic settlement by Schmertmann's method can be calculated by the equation

$$S_e = C_1 C_2 (\bar{q}-q) \sum_{0}^{2B} \frac{I_z}{E_s} \Delta z \quad (4)$$

where,
S_e = Elastic settlement
C_1 = $1-0.5[q/(\bar{q}-q)]$; a correction factor for the depth of foundation embedment
C_2 = $1+0.2 \log$ (time in yrs/0.1); a correction factor to account for creep in soil
$\bar{q}$ = Contact stress of the foundation
q = γD_f; stress due to soil surcharge above the foundation level (γ=Unit wt; D_f= Fnd. depth).
B = Width of the foundation
I_z = Strain influence factor at depth increment Δz
E_s = Young's modulus.

In order to use equation (4), one needs to establish the variation of I_z with depth (Figure 1a), and evaluate the approximate variation of E_s with depth (Figure 1b). The E_s values can be estimated from the results of field tests such as standard penetration or cone penetration. Table 1 shows I_z values at different depths for different L/B ratios for the foundation. These values can be used to construct the appropriate strain influence factor diagrams, and interpolations are needed for intermediate cases. Note that the summation is performed over a depth of 2B in equation (4) representing a square or circular foundation (L/B=1). For strip footings with L/B≥10, the same summation has to be performed over a depth of 4B.

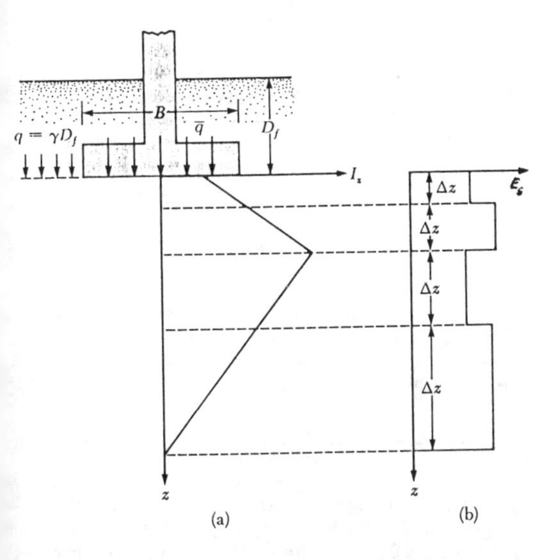

Fig 1. Elastic settlement calculation by using strain influence factor (After Das 1984).

Table 1. Strain influence factors.

Depth, z	I_Z	L/B
0 0.5B 2B	0.1 0.5 0	1
0 B 4B	0.2 0.5 0	≥10

B = Width of the foundation
L = Length of the foundation

3. PROBABILISTIC FUNDAMENTALS

A fundamental principle in the probabilistic analysis is that a random variable (independent, or dependent) can be described by a collection of statistical parameters, rather than a single value. These parameters include the mean, standard deviation, coefficient of variation, coefficient of skewness and coefficient of kurtosis. When two or more random variables are involved in the analysis, correlation coefficients can also be calculated. Associated with each random variable is also a probability density function (pdf) indicating the probability distribution (e.g. normal, lognormal, Beta, etc.). The definitions and standard formulas for these parameters are covered in standard probability/ statistics texts (see for example, Benjamin and Cornell 1970).

When one or more of the independent variables in a model, (or equation), are random variates that are represented by a mean, a standard deviation and a pdf, the values for the same statistical parameters can be established for the dependent variables by methods like Taylor series expansion, or Monte Carlo Simulation; or, the point estimate method (PEM) which is recently developed by Rosenbleuth (1975), can be employed. To find the pdf, one can use the Pearson's system, as described by Elderton and Johnson (1969). Brief descriptions of the PEM technique and the Pearson's system are presented herein.

3.1 The Point Estimate Method (PEM)

Consider y = f(x) to be the probability distribution of the random variable x. Refer to Figure 2, where the continuous distribution is replaced by two discrete point estimates, P_+ and P_-, acting at x_+ and x_-. Then we have:

$$P_+ + P_- = 1 \quad (5)$$
$$P_+x_+ + P_-x_- = E[P(x)] = \bar{x} \quad (6)$$
$$P_+[x_+ - \bar{x}]^2 + P_-[x_- - \bar{x}]^2 = s_{\bar{x}}^2 \quad (7)$$
$$P_+[x_+ - \bar{x}]^3 + P_-[x_- - \bar{x}]^3 = \beta_1 s_{\bar{x}}^3 \quad (8)$$

where: $\bar{x}$ = mean value for variable x
$s_{\bar{x}}^2$ = variance for variable x
β_1 = coefficient of skewness for variable x.

Further,

$$P_+ = 1/2\left[1 \pm \sqrt{1 - 1/(1+(\beta_1/2))^2}\right] \quad (9a)$$
$$P_- = 1 - P_+ \quad (9b)$$
$$x_+ = \bar{x} + s_x \sqrt{P_-/P_+} \quad (9c)$$
$$x_- = \bar{x} - s_x \sqrt{P_+/P_-} \quad (9d)$$

If f(x) is symmetrical, β_1 =0 and equation (9) can be reduced to:

$$P_+ = P_- = 1/2 \quad (10a)$$
$$x_+ = \bar{x} + s_x \quad (10b)$$
$$x_- = \bar{x} - s_x \quad (10c)$$

With the distribution f(x) approximated by the point estimates P_- and P_+, the moment of y= f(x) are

$$E[y] = \bar{y} = P_-y_- + P_+y_+ \quad (11a)$$
$$E[y^2] = P_-y_-^2 + P_+y_+^2 \quad (11b)$$

and in general $E[y^n] = P_- y_-^n + P_+ y_+^n$. (11c)

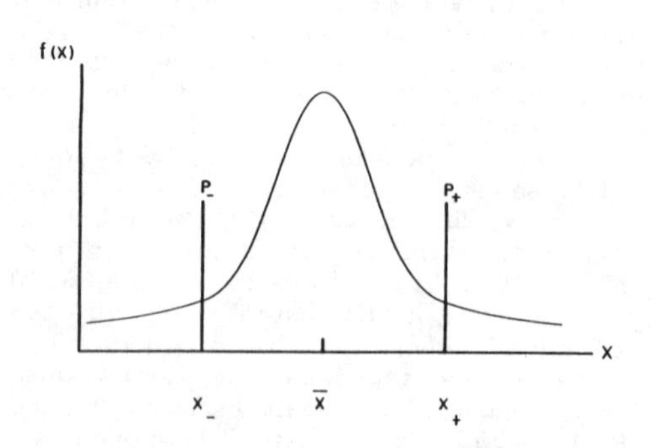

Fig. 2. Point estimate approximation.

For a function of four correlated random variables, say $y = f[x_1, x_2, x_3, x_4]$ or settlement = f[contact stress, unit wt, compression index, and void ratio],

$$E[y^n]=P_{++++}\ y_{++++}^n+P_{+++-}\ y_{++-}^n+P_{++--}\ y_{++--}^n+P_{---+}\ Y_{---+}^n+P_{----}\ Y_{----}^n \quad (12)$$

where, $y(\pm\pm\pm\pm)=f[\bar{x}_1 \pm s_{x1}, \bar{x}_2 \pm s_{x2}, \bar{x}_3 \pm s_{x3}, \bar{x}_4 \pm s_{x4}]$ and, (13)

$$P_{++++}=P_{----}=\frac{1}{2^4}[1+\rho_{12}+\rho_{13}+\rho_{14}+\rho_{23}+\rho_{24}+\rho_{34}]$$

$$P_{+++-}=P_{---+}=\frac{1}{2^4}[1+\rho_{12}+\rho_{13}-\rho_{14}+\rho_{23}-\rho_{24}-\rho_{34}]$$

$$P_{++--}=P_{--++}=\frac{1}{2^4}[1+\rho_{12}-\rho_{13}-\rho_{14}-\rho_{23}-\rho_{24}+\rho_{34}]$$

$$P_{+---}=P_{-+++}=\frac{1}{2^4}[1-\rho_{12}-\rho_{13}-\rho_{14}+\rho_{23}+\rho_{24}+\rho_{34}]$$

$$P_{+--+}=P_{-++-}=\frac{1}{2^4}[1-\rho_{12}-\rho_{13}+\rho_{14}+\rho_{23}-\rho_{24}-\rho_{34}]$$

$$P_{+-+-}=P_{-+-+}=\frac{1}{2^4}[1-\rho_{12}+\rho_{13}-\rho_{14}-\rho_{23}+\rho_{24}-\rho_{34}]$$

$$P_{+-++}=P_{-+--}=\frac{1}{2^4}[1-\rho_{12}+\rho_{13}-\rho_{14}+\rho_{23}+\rho_{24}+\rho_{34}]$$

$$P_{++-+}=P_{--+-}=\frac{1}{2^4}[1+\rho_{12}-\rho_{13}+\rho_{14}-\rho_{23}+\rho_{24}-\rho_{34}]$$

(14)

Here, s_{xi} is the estimated standard deviation of x_i; ρ_{ij} is the coefficient of correlation between x_i and x_j. The sign of ρ_{ij} is determined by the multiplication rule of ij; i.e. i=(-), j=(+) yield ij=(-)(+)=(-). Of course, when all the parameters are uncorrelated then all the P terms become equal to $1/16 = 1/2^n$.

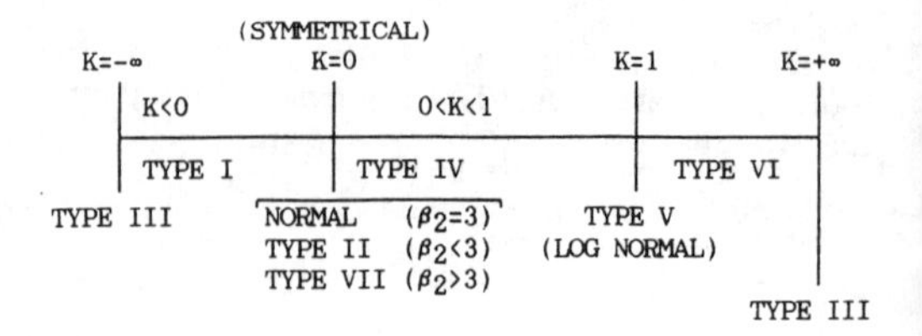

Fig. 3. The K-criterion (after Elderton and Johnson 1969).

3.2 Pearson's System

Pearson's system is based on the finding that a majority of the continuous probability distributions f(x) can be generated from the differential equation

$$\frac{df(x)}{dx} = \frac{(a_0+x)f(x)}{b_0+b_1x+b_2x^2} \quad (15)$$

by the proper selection of the four constants a_0, b_0, b_1, and b_2 (Elderton and Johnson, 1969). The pdf is then decided upon by the K-criterion, expressed as

$$K = \frac{\beta_1(\beta_2+3)^2}{4(2\beta_2-3\beta_1-6)(4\beta_2-3\beta_1)} \quad (16)$$

where, β_1 and β_2 are the coefficients of skewness and kurtosis. This criterion is illustrated in Figure 3.

Armed with the mean, standard deviation, and the pdf of the dependent variable (i.e. consolidation settlement or elastic settlement), one can analyze the uncertainty by an appropriate probabilistic parameter. For example, the proability of the settlement being larger than a given (allowable) value, $P[S>S_{allow}]$ can be considered instead of a deterministic comparison of S and S_{allow}.

4. PROBABILISTIC MODELS

Two probabilistic models have been developed as part of this study in the form of computer codes. The first model (PSCLAY) is based on equations (1), (2), and (3), and is suitable for the probabilistic analysis of consolidation settlements in saturated cohesive soils. The second model (PSSAND) is based on equation (4), and performs probabilistic settlement analysis for granular soils.

The foundation dimensions, soil stratification, and the position of the static groundwater table are considered to be deterministic in both models. All of the remaining variables, which are associated with soil properties, are taken as random variates. The loads acting on the foundation are assumed to be axial but random. Therefore, the contact stress at the base of the foundation and the average stress increase in the compressible layer resulting from these loads are also random variables. All independent variables in the model are further assumed to be normally distributed. The information input to the computer model on each random variable includes the mean and standard deviation (or coefficient of variation). Correlation coefficients between pairs of random variables can also be input to the model. The foundation dimensions, soil layer and groundwater elevations are input as deterministic quantities as noted.

Essentially, the mean values and standard deviations of consolidation or elastic settlements are computed by the models using the PEM technique. Since the skewness and kurtosis of settlements are also computed, it is possible to establish the pdf for settlement using the Pearson's system, which is built into the models. Further probabilistic analyses dealing with the calculation of probabilities of settlement being larger than an allowable value can be performed manually using the appropriate equations, statistical tables, or charts.

5. CASE STUDIES

Two parametric studies are presented herein to demonstrate the application and merit of the computer models PSCLAY and PSSAND described in the previous section. The first case study involves the probabilistic analysis of consolidation settlements in the clay layers shown in Figure 4, and the second case study involves the probabilistic settlement analysis of the sand layers shown in Figure 5. The site stratigraphies, ground water conditions, soil properties and loads (contact stresses) are presented in these figures as deterministic parameters. However, the values given the soil properties γ, C_c, C_s, e_o, P_c, and E_s, and for the contact stress $\bar{q}$ are entered into the models as the mean values of these parameters.

Presented in Tables 2 and 3 are the other inputs to the models. Three levels of uncertainty have been considered, as

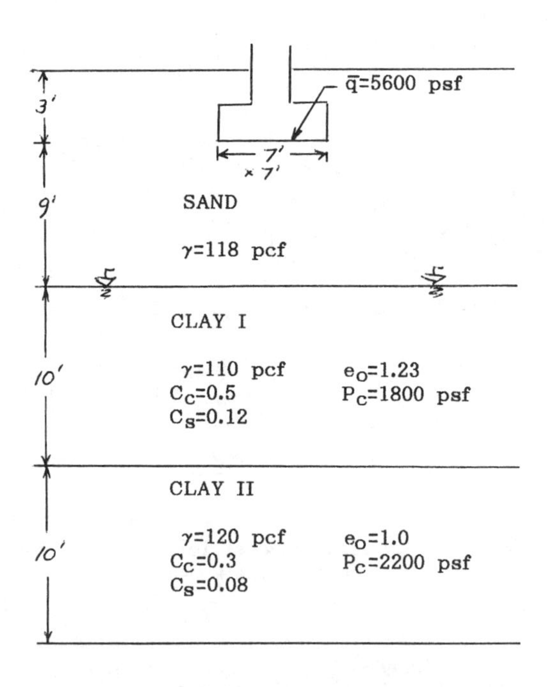

Fig. 4. Foundation, soil and site parameters input to PSCLAY.

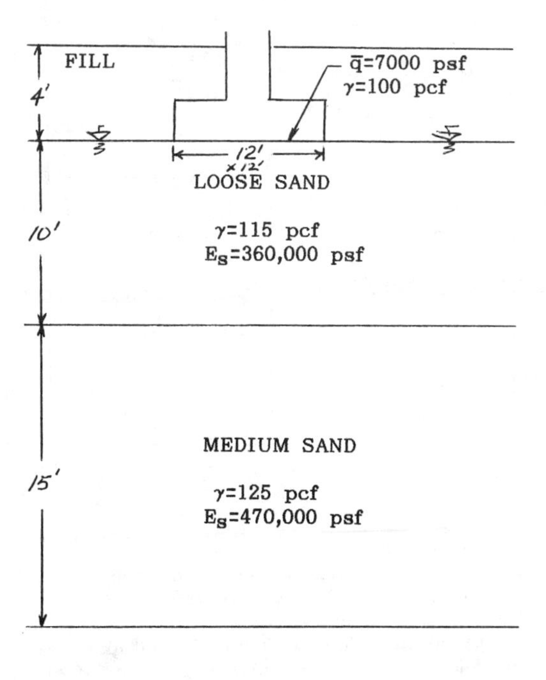

Fig. 5. Foundation, soil and site parameters input to PSSAND.

Table 2. Uncertainty levels used as input to the models.

Random Variable	Low CV, %	Med. CV, %	High CV, %
$\bar{q}$	10	30	50
γ	5	10	15
C_c	25	50	75
C_s	25	35	45
e_o	15	25	35
P_c	20	40	60
E_s	25	50	75

Table 3. Correlation coefficients used as input to the models.

Random Variables	Correlation Coefficient
C_c and C_s	0.9
C_c and e_o	0.84
C_c and P_c	0.2
C_c and P_o	-0.7
C_s and e_o	0.7
C_s and P_c	0.1
C_s and P_o	-0.6
e_o and P_c	-0.18
e_o and P_o	-0.01
P_o and P_c	-0.62
$\bar{q}$ and E_s	0.5

indicated by low, medium and high coefficient of variation (CV) for the input parameters. The CV value for each parameter at each uncertainty level is provided in Table 2. Table 3 contains the correlation coefficients between appropriate pairs of random variables. All other pairs have been assumed to be uncorrelated. The values and ranges

Table 4. Probabilistic analysis of consolidation settlements in clays using PSCLAY.

Parameter	Low CV	Med. CV	High CV
$\bar{S}_c$, in.	2.16	2.01	1.88
σ_{S_c}, in.	1.10	1.45	1.97
CV_{S_c}, %	50.9	72.1	104.8
$P[S_c>1"]$, %	84.4	78.9	77.3
$P[S_c>2"]$, %	54.4	54.2	53.5
$P[S_c>3"]$, %	25.0	31.8	40.9
$P[S_c>4"]$, %	6.0	10.9	31.0
Deterministic S_c, in.	2.73		

Table 5. Probabilistic analysis of elastic settlements in sands using PSSAND

Parameter	Low CV	Med. CV	High CV
$\bar{S}_e$, in.	1.83	2.13	3.15
σ_{S_e}, in.	0.11	0.28	0.79
CV_{S_e}, %	6.0	13.1	25.1
$P[S_e>1"]$, %	99.99	99.9	99.9
$P[S_e>2"]$, %	5.0	75.0	95.0
$P[S_e>3"]$, %	0.1	1.0	60.0
$P[S_e>4"]$, %	0.01	0.1	15.0
Deterministic S_e, in.	1.74		

given in Tables 2 and 3 have been selected based on a previous study conducted at West Virginia University by Frost (1983). Similar information has also been covered by Lee, et. al (1983).

The results of the probabilistic

settlement analyses performed on the clays and sands based on the input data described above are summarized in Tables 4 and 5. The mean values of settlement ($\bar{S}$), standard deviations (σ_s) and coefficients of variation (CV_s) are obtained from the computer output directly. The same is also true for the deterministic settlements (where all CV's are zero for inputs). The probabilities of settlement being larger than an allowable value, i.e. $P[S>S_{allow}]$, have been calculated from charts developed by Harr (1977) for Beta distribution. Four possible levels of allowable settlements, 1 inch, 2 inches, 3 inches and 4 four inches, have been considered in the analyses. Each column in Table 4 or Table 5 designated as low, medium, or high CV refers to the uncertainty levels established in Table 2 with the appropriate numbers used as input.

Some observations can be readily made from the data presented in Tables 4 and 5. It is quite clear that the uncertainty around the independent variables (as quantified by CV) directly reflect upon the uncertainty of the dependent variable, settlement. Since more uncertainties are present around the inputs in predicting consolidation settlements than in predicting elastic settlements, the degree of uncertainty around the consolidation settlement is more than that of the elastic settlement. In both cases, however, as the level of uncertainty of the input variables increases the level of uncertainty of the settlements will increase.

In the probabilistic analysis, the decisions relative to whether or not a tolerable settlement limit have been exceeded by the predicted settlement is based on a "probability of occurance" for that event rather than a deterministic comparison of the predicted and allowable settlement values. As larger allowable settlements are specified, the probabilities for the non-satisfaction of the tolerable limit criteria will get lower. When the probabilities are very high (e.g. 99.9 percent) or very low (e.g. 0.01 percent) the practical implications are quire clear. For example, there is little question in the second case study (Table 5) that the settlement will exceed 1 inch but will be less than 4 inches. It may be more difficult, however, to understand the implications when the probabilities are in the intermediate range (e.g. 30 to 70 percent). Presently, probability-based criteria is not available for tolerable settlements to facilitate the quantitative interpretations of these probabilities.

6. SUMMARY AND CONCLUSIONS

Using the Terzaghi's one-dimensional consolidation equation for clays and Schmertmann's elastic settlement equation for sands, along with probabilistic tolls such as the point estimate method and Pearson's system, probabilistic analyses have been performed on shallow foundations. Parametric studies using computer models indicate that the uncertainties around the input variables like contact stress, γ, C_c, C_s, e_o, P_c and E_s used in the settlement equations directly impact upon the uncertainty of the predicted value of settlement. The higher the uncertainty around the input parameters, the higher the uncertainty around the settlement value.

It is possible with the models developed in this study to compute the probability that the predicted settlement will exceed an allowable settlement value. This probability decreases as the tolerable settlement limit is raised. The quantitative interpretation and practical implication of this parameter, however, is not clearly established. Probability-based tolerable settlement criteria will be needed to make the probabilistic analysis techniques practical and useful design tools. Until such criteria are developed, probabilistic methods can serve as supplements or complements to the existing deterministic settlement analysis techniques.

REFERENCES

Benjamin, J.R. and C.A. Cornell 1970. Probability, statistics, and decision for civil engineers. McGraw Hill, New York.

Das, B.M. 1984. Principles of foundation engineering. Brooks/Cole Engineering Division.

Elderton, W.P. and N.L. Johnson 1969. Systems for frequency curves. Cambridge University Press.

Frost, D.D. 1983. Variability and correlation of soil properties employed in geotechnical engineering applications. Problem report submitted to West Virginia University, College of Engineering in partial fulfillment of the requirements for M.S.C.E. degree. Morgantown, West Virginia.

Harr, M.E. 1977. Mechanics of particulate media - a probabilistic approach. McGraw Hill, New York.

Harr, M.E. 1984. Reliability based design in civil engineering. Twentieth Henry M. Shaw lecture in civil engineering. School of Engineering, N.C. State

University, Raleigh, N.C.
Lee, I.K. et. al. 1983. Geotechnical engineering. Pitman.
Rosenblueth, E. 1975. Point estimates for probability moments. Proceedings, National Academy of Science 72(10): 3812-3814.
Wahls, H.E. 1983. Shallow foundations for highway structures, NCHRP, Synthesis of Highway Practice 107, Transportation Research Board, Washington, D.C.
Schmertmann, J.H., Harman, J.P., and P.R. Brown 1978. Improved strain influence factor diagrams. Technical notes, Journal of the Geotechnical Engineering Division, ASCE, No. GT8, pp. 1131-1134.

ACKNOWLEDGEMENT

The study reported in this paper was performed under a research grant provided by the West Virginia University Senate and Office of Sponsored Programs. This support is gratefully acknowledged.